ACCEPTANCE SAMPLING
IN
QUALITY CONTROL

STATISTICS: Textbooks and Monographs

A SERIES EDITED BY

D. B. OWEN, Coordinating Editor
Department of Statistics
Southern Methodist University
Dallas, Texas

OTHER VOLUMES IN PREPARATION

ACCEPTANCE SAMPLING
IN
QUALITY CONTROL

Center for Quality and Applied Statistics
Rochester Institute of Technology
Rochester, New York

MARCEL DEKKER, INC.   New York and Basel

Library of Congress Cataloging in Publication Data

Schilling, Edward G., [date]
    Acceptance sampling in quality control.

    (Statistics, textbooks and monographs; v. 42)
    Includes bibliographical references and indexes.
    1. Acceptance sampling.   I. Title. II. Series.
TS156.4.S34   658.5'62   81-19606
ISBN 0-8247-1347-8   AACR2

MARCEL DEKKER, INC.
270 Madison Avenue, New York, New York 10016

Current printing (last digit):
10 9 8 7 6 5 4 3

PRINTED IN THE UNITED STATES OF AMERICA

To Jean

O Fortune,
variable
as the moon,
always dost thou
wax and wane.
Detestable life,
first dost thou mistreat us
and then, whimsically,
thou heedest our desires.
As the sun melts the ice,
so dost thou dissolve
both poverty and power.

-Carl Orff, *Carmina Burana*

NOTE FROM THE SERIES EDITOR

The use of acceptance sampling has grown tremendously since the Dodge and Romig *Sampling Inspection Tables* were first widely distributed in 1944. Throughout this period many people have contributed methods and insight to the subject. One of these contributors is the author of this book, which might better be identified as a compendium of acceptance sampling methods. The American Society for Quality Control has recognized Dr. Schilling's contributions by awarding him the Brumbaugh Award four times, first in 1973 and again in 1978, 1979, and 1981. This award is given each year to the author of that paper published in either the *Journal of Quality Technology* or *Quality Progress* which an American Society for Quality Control committee judges has made the largest single contribution to the development of industrial applications of quality control.

Dr. Schilling has been employed both as an educator and as an industrial statistician. This broad experience qualifies him to write this treatise as few others are qualified. The beginner will find much of interest in this work, while the experienced person will also find many interesting items because of its encyclopedic coverage.

I am very pleased with the completeness and clarity exhibited in this book, and it is with great pleasure that I recommend it to others for their use.

D. B. Owen
Department of Statistics
Southern Methodist University
Dallas, Texas

v

FOREWORD

As the field of quality control enters the 1980s, it is having new respon-
sibilities thrust upon it. The public is demanding products free from
defects, and often making these demands in costly court cases. Management
is demanding that all departments contribute to technical innovation and
cost reduction while still continuing to justify its own costs. The quality control specialist is caught like others in this squeeze between perfect
performance and minimum cost. He or she needs all the help that fellow
professionals can give, and Edward Schilling's book is a worthy contribution. Written by one of the foremost professionals in the field, it is
comprehensive and lucid. It will take its place as a valuable reference
source in the quality control specialist's library.

My own first contact with a draft of the book came when I was teaching
a quality control course to industrial engineers. Over the semester I found
myself turning to this new source for examples, for better explanations of
standard concepts, and for the many charts, graphs, and tables, which are
often difficult to track down from references. Acceptance sampling is not
the whole of statistical quality control, much less the whole of quality
control. But Dr. Schilling has stuck to his title and produced a text of
second-level depth in this one area, resisting the temptation to include
the other parts of quality control to make a "self-contained work." The
added depth in this approach makes the text a pleasure for a teacher to
own and will make it a pleasure for students to use. This is one book
that any student should take into the world where knowledge is applied to
the solution of problems.

Colin G. Drury
Department of Industrial Engineering
SUNY at Buffalo
Buffalo, New York

The methods of statistical acceptance sampling in business and industry
are many and varied. They range from the simple to the profound, from the
practical to the infeasible and the naive. This book is intended to present
some of the techniques of acceptance quality control that are best known and
most practical — in a style that provides sufficient detail for the novice,
while including enough theoretical background and reference material to
satisfy the more discriminating and knowledgeable reader. The demands of
such a goal have made it necessary to omit many worthwhile approaches; however, it is hoped the student of acceptance sampling will find sufficient
material herein to form a basis for further explorations of the literature
and methods of the field.

While the prime goal is the straightforward presentation of methods for
practical application in industry, sufficient theoretical material is included to allow the book to be used as a college level text for courses in
acceptance sampling at a junior, senior, or graduate level. Proofs of the
material presented for classroom use will be found in the references cited.
It is assumed, however, that the reader has some familiarity with statistical quality control procedures at least at the level of Irving W. Burr's
Statistical Quality Control Methods (Marcel Dekker, Inc., New York, 1976).
Thus, an acceptance sampling course is a natural sequel to a survey course
at the level suggested.

The text begins with a fundamental discussion of the probability theory
necessary for an understanding of the procedures of acceptance sampling.
Individual sampling plans are then presented in increasing complexity for
use in the inspection of single lots. There follows a discussion of
schemes which may be applied to the more common situation of a stream of
lots from a steady supplier. Finally, specific applications are treated in

the areas of compliance sampling and reliability. The last chapter is concerned with the administration of acceptance control and, as such, is intended as a guide to the user of what sampling plan to use (and when).
Readers having some familiarity with acceptance sampling may wish to read
the last chapter first, to put into context the methods presented.

This book views acceptance quality control as an integral and necessary
part of a total quality control system. As such, it stands with statistical
process quality control as a bulwark against poor quality product, whose
foundations are rooted deep in mathematics but whose ramparts are held only
by the integrity and competence of its champions in the heat of confrontation.

It is fitting that this book on acceptance sampling should begin with
the name of Harold F. Dodge. His contributions have been chronicled and are
represented in the Dodge Memorial Issue of the *Journal of Quality Technology*
(Vol. 9, No. 3, July 1977). Professor Dodge, as a member of that small band
of quality control pioneers at the Bell Telephone Laboratories of the
Western Electric Company, is considered by some to be the father of acceptance sampling as a statistical science. Certainly, he nurtured it, lived
with it, and followed its development from infancy, through adolescence, and
on into maturity. In no small way he did the same for the author's interest
in the field, as his professor and his friend.

Edward G. Schilling

ACKNOWLEDGMENTS

Books are not made — they grow. It is impossible to acknowledge all the help and support which has come from friends and associates in the development and construction of the present volume. A few may be singled out not only for their individual contributions, but also as a sample of those yet unnamed. In particular, I wish to thank Carl Mentch for suggesting the possibility of such an undertaking in September of 1965 and for his unflagging encouragement and help since that time. My thanks also go to Mrs. Lucille I. Johnson, whose technical and editorial assistance helped to bring concept into reality. I must also mention Dr. Lloyd S. Nelson for his continued interest and suggestions, and Dan J. Sommers and Professor Emil Jebe for their constructive comments and theoretical insight. Certainly, my appreciation goes to Dr. Donald P. Petarra, Dr. James R. Donnalley and Dr. Pieter J. von Herrmann of the General Electric Lighting Research and Technical Services Operation for their encouragement and support throughout.

I am indebted to the American Society for Quality Control, the American Society for Testing and Materials, the American Statistical Association, The Institute of Mathematical Statistics, the Philips Research Laboratories, the Royal Statistical Society, and Bell Laboratories for permission to reprint a variety of material taken from their publications. I am also indebted to Addison-Wesley Publishing, Inc., for permission to reprint material from D. B. Owen, *Handbook of Statistical Tables*; to Cambridge University Press for permission to reprint material from E. S. Pearson, *Tables of the Incomplete Beta-Function*; to McGraw-Hill Book Company for permission to reprint material from A. H. Bowker and H. P. Goode, *Sampling Inspection by Variables*, I. W. Burr, *Engineering Statistics and Quality Control*, and J. M. Juran, *Quality Control Handbook*; to John Wiley and Sons, Inc., and Bell Laboratories for permission to reprint material from H. F. Dodge and H. G. Romig,

Sampling Inspection Tables; to Prentice-Hall, Inc., for permission to re-
print material from A. H. Bowker and G. J. Lieberman, *Engineering Statis-
tics*; to Stanford University Press for permission to reprint material from
G. J. Lieberman and D. B. Owen, *Tables of the Hypergeometric Probability
Distribution*, and G. J. Resnikoff and G. J. Lieberman, *Tables of the Non-
Central t-Distribution*, to the European American Music Distributors Corpora-
tion for permission to use the English translation of "O Fortuna" from Carl
Orff's scenic cantata *Carmina Burana*; and to my associates K. S. Stephens,
H. A. Lasater, L. D. Romboski, R. L. Perry, and J. R. Troxel for material
from their Ph.D. dissertations taken at Rutgers University under Professor
Harold F. Dodge, in a unique intellectual environment which was created and
sustained at the Statistics Center under the inspired direction of Dr. Ellis
R. Ott and with the dedicated administrative support of Dr. Mason E.
Wescott.

Finally, these debts of gratitude are in terms of time and talent.
How much more the debt to my wife, Jean, and to my daughters, Elizabeth and
Kathryn, who are as much a part of this book as the author himself.

CONTENTS

Contents   xv

1

INTRODUCTION

Harold F. Dodge (1969b, p. 156) has indicated that in the early days of the development of military standards during World War II, a distinction became apparent between acceptance sampling plans, on the one hand, and acceptance quality control, on the other. The former are merely specific sampling plans, which, when instituted, prescribe conditions for acceptance or rejection of the immediate lot inspected. The later may be compared to process quality control, which utilizes various indicators (such as control charts) and strategies (such as process capability studies) to maintain and improve existing levels of quality in a production process. In like manner, acceptance quality control exploits various acceptance sampling plans as tactical elements in overall strategies designed to achieve desired ends. Such strategies utilize elements of systems engineering, industrial psychology, and, of course, statistics and probability theory, together with other diverse disciplines, to bring pressures to bear to maintain and improve the quality levels of submitted product. For example, in the development of the Army Ordnance Sampling Tables in 1942, Dodge (1969b, p. 156) points out that

> ... basically the "acceptance quality control" system
> that was developed encompassed the concept of protect-
> ing the consumer from getting unacceptably defective
> material, and encouraging the producer in the use of
> process quality control by: varying the quantity and
> severity of acceptance inspections in direct relation
> to the importance of the characteristics inspected,
> and in *inverse* relation to the goodness of the quality
> level as indicated by those inspections.

The resulting tables utilize not just one sampling plan, but many in a
scheme for quality improvement.

This book stresses acceptance quality control in recognition of the
importance of such systems as a vital element in the control of quality.
There is little control of quality in the act of lot acceptance or rejection. While the utilization of sampling plans in assessing lot quality is
an important aspect of acceptance sampling, it is essentially short run in
effect. The long-run consequences of a well-designed system for lot acceptance can be more effective where applicable. Thus, an individual sampling
plan has much the effect of a lone sniper, while the sampling scheme can
provide a fusillade in the battle for quality improvement.

ACCEPTANCE QUALITY CONTROL

Individual sampling plans are used to protect against irregular degradation
of levels of quality in submitted lots below that considered permissible by
the consumer. A good sampling plan will also protect the producer in the
sense that lots produced at permissible levels of quality will have a good
chance to be accepted by the plan. In no sense, however, is it possible to
"inspect quality into the product." In fact, it can be shown (Mood, 1943)
that if a producer continues to submit to the consumer product from a
process with a constant proportion defective, lot after lot, simple acceptance or rejection of the lots submitted will not change the proportion defective the consumer will eventually receive. The consumer will receive
the same proportion defective as was in the original process.

This idea may be simply illustrated as follows. Suppose you are in the
business of repackaging playing cards. You have an abundance of face cards
(kings, queens, and jacks) and so submit an order to the printer for 5000
cards having an equal selection of non-face cards. Any face cards, then,
can be considered as defectives if they are found in the shipment. The
cards are supposed to come to you in packages of 50 resembling standard
52-card decks. Unknown to you, the printer has mixed up your order and is
simply sending standard decks. Your sampling plan is to accept the deck if
a sample of one card is acceptable. The lot size is actually, of course, 52.

What will be the consequences? Twelve of the 52 cards in a standard
deck are face cards; so the probability of finding a face card on one draw
is 12/52 = 0.23, or 23 percent. This means that in 100 decks examined there
should be roughly 23 rejections. Suppose these rejected decks are thrown
into the fire, what will be the proportion of face cards in the accepted
material? Why 23 percent, of course, since all the decks were the same.
Thus, the sampling plan had no effect on the quality of the material accepted

while the process proportion defective remained constant. The proportion
defective accepted is the same as if no inspection had ever been performed.

Suppose, instead, the printer had become even more mixed up. The
printer fills half the order with ordinary playing cards and the other half
with cards from pinochle decks. Pinochle decks are composed of 48 cards,
one-half of which (or 24) are face cards. The printer ships 50 ordinary
decks (2600 cards) and 50 pinochle decks (2400 cards). Inspection of the
50 ordinary decks by the same plan will reject about 23 percent, or about
12 of them. The remaining 38 will pass and be put into stock. Of the 50
pinochle decks, however, half will be rejected and so 25 will go into stock.

Some calculation will show that, with no sampling (i.e., 100 percent
lot acceptance) the stock would consist of

$$(12 \times 50) + (24 \times 50) = 1800$$

face cards out of a total stock of 5000 cards, or

$$\frac{1800}{5000} \times 100 = 36.0\%$$

face cards.

Using the sampling plan, simple and ineffective as it was, stock would
consist of

$$(12 \times 38) + (24 \times 25) = 1056$$

face cards out of a total stock of

$$(52 \times 38) + (48 \times 25) = 3176$$

or

$$\frac{1056}{3176} \times 100 = 33.2\%$$

face cards.

Thus, quality can be improved by the imposition of a sampling plan in
the face of fluctuation in proportion defective since it will tend to selec-
tively screen out the highly defective material relative to the better lots.
Clearly, a larger improvement could have been made in the above example if
a more discriminating sampling plan had been used.

Now, consider the imposition of some rules beyond the single-sampling
plan itself. Suppose the rejected decks are 100 percent inspected with any
face cards found being replaced with non-face cards. Then, in the last part
of the example, the number of face cards in stock would be 1056 out of 3176

as before, since they came from the accepted lots. But, since the 36 rejected lots would have been replaced with perfect product the stock would
be increased by 50 X 36 = 1800 cards to a level of 4976 cards. The stock
would now consist of

$$\frac{1056}{4976} \times 100 = 21.2\%$$

face cards. Here we have a substantial improvement in the level of quality
even when using an extremely loose plan in the context of a sampling strategy - in this case what is called a rectification scheme.

Finally, suppose complete 100 percent inspection were instituted. It
is generally conceded that no screening operation is 100 percent effective
and, in the real world 100 percent inspection of a large number of units
may be only about 80 percent effective. If this is the case, about 20 percent of the defective cards will be missed and the final stock will contain

1800 X 0.20 = 360

defectives, or a percent defective of

$$\frac{360}{5000} \times 100 = 7.2\%$$

at a cost of examining all 5000 cards. Even 100 percent inspection will not
necessarily eliminate defective items once they are produced.

Thus it is that sampling strategies (or schemes) can be developed to
attain far more protection than the imposition of a simple sampling plan
alone. What is required, of course, is a continuing supply of lots from the
same producer to allow the strategy to be effective. It is for this reason
that there are two basic approaches to acceptance quality control, depending
upon the nature of the lots to be inspected. A continuing supply of lots
from the same producer is most effectively treated by a sampling scheme. A
single lot, unique in itself, is treated by sampling plans designed for use
with an "isolated lot." This distinction is fundamental to acceptance sampling, and even the basic probability distributions used in the two cases
are not the same. We speak of "Type A" sampling plans and probabilities
when they are to be used with a single lot and "Type B" when used in the
context of a continuing series of lots produced from the same supplier's
process. Effective acceptance quality control will utilize the schemes and
plans of acceptance sampling to advantage in either case.

ACCEPTANCE CONTROL AND PROCESS CONTROL

Acceptance sampling procedures are necessarily defensive measures, instituted as protective devices against the threat of deterioration in quality.
As such, they should be set up with the aim of discontinuance in favor of
process control procedures as soon as possible.

Process quality control is that aspect of the quality system concerned
with monitoring and improving the production process by analysis of trends
and signals of quality problems or opportunities for the enhancement of
quality. Its methods include various types of control charts, experiment
designs, response surface methodologies, evolutionary operations, and other
procedures including, on occasion, those of acceptance sampling. These
methods are an essential adjunct for effective acceptance control since

1. Quality levels for selecting an appropriate sampling procedure
   should be determined from control chart analysis to ascertain
   what minimum levels the producer can reasonably and economically
   guarantee and what maximum levels can be tolerated by the consumer's process or will fulfill the consumer's wants and needs.
2. Acceptance sampling procedures should be set up to "self-destruct"
   after a reasonable period in favor of process controls on the
   quality characteristic in question. Simultaneous use of acceptance
   quality control and process quality control should eventually lead
   to improvement in quality levels to the point that regular application of acceptance sampling is no longer needed.

Thus, at the beginning and at the end of an acceptance sampling procedure,
process quality control plays an important part.

PROCESS QUALITY CONTROL

With the invention of the control chart by Shewhart in 1924, process quality
control had gained its most valuable tool as well as its genesis. When
samples are taken periodically on a process, the averages of the samples
will tend to cluster about some overall average, or process level, as long
as the process is not affected or changed to some new average, or level.
Such changes in process level may be intentional or completely inadvertent
and unexpected. The control chart is essentially a means for determining
and signaling when the process level has actually shifted to a new level in
the face of chance variations in sample results. Observations are collected

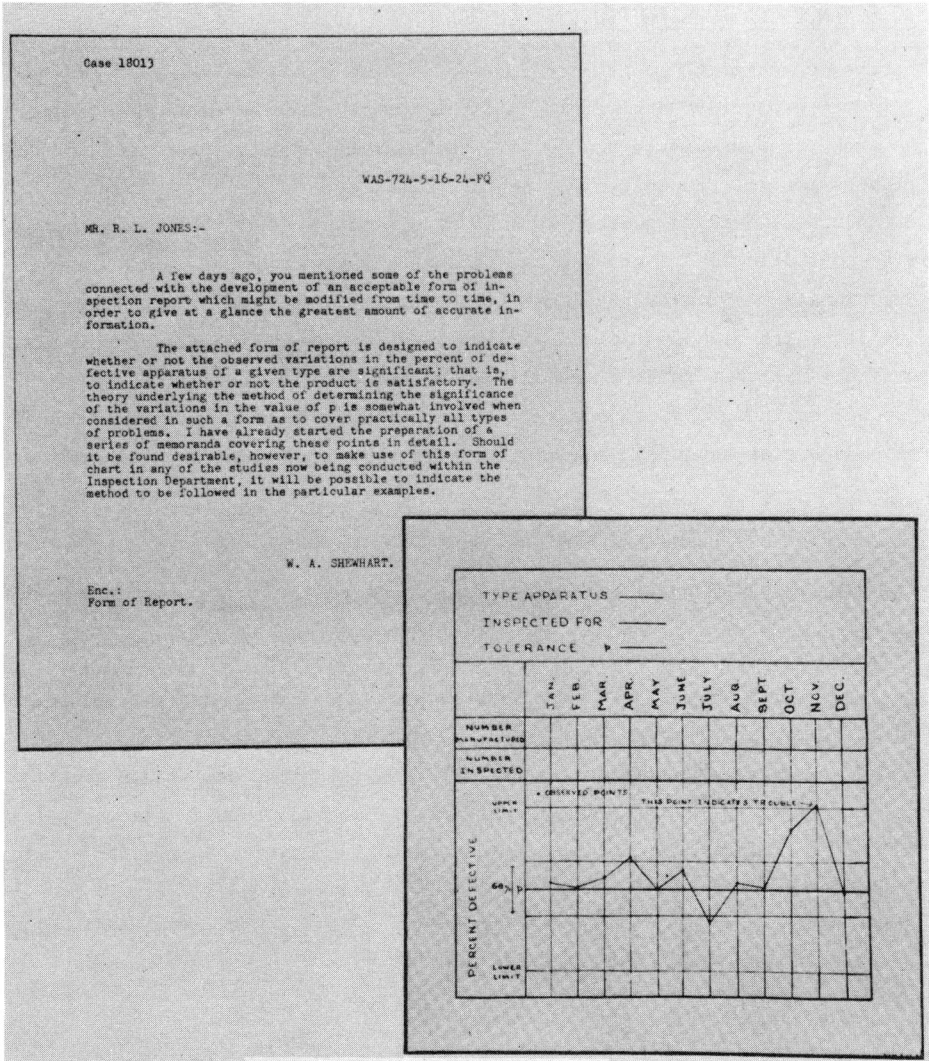

FIG. 1-1 The first Shewhart control chart. (From Olmstead, 1967, p. 72.
Reprinted by permission.)

in what are called *rational subgroups*. These are taken so as to maximize
the opportunity to show the source of a change in the process.

The Shewhart chart consists simply of three parallel lines: two outside
lines, called *upper* and *lower control limits*, and a center line. An example
of such a chart is given in Fig. 1-1. It shows Shewhart's first control
chart and the memorandum which accompanied it (Olmstead, 1967, p. 72). In
practice, sample results are plotted on the chart in sequence. The center
line reflects the average of the data, while the control limits are calcu-

lated to have a high probability (usually 333:1 odds) of the sample data being contained between them if the process is stable. It is, then, very unlikely (3 chances in 1000) that a point will plot outside the limits when the process is running well. In this event it can safely be left alone. If the process level shifts however, points will plot outside the limits, indicating the need for corrective action on the process. In some cases the points may plot outside the limits in a favorable direction. This is an indication of the possibility for process improvement when the source of the process change is detected. The control chart, then, provides control of a process in the face of measurement and other sources of variation in the sense that it shows when the process has significantly degraded or improved. This provides a timely opportunity for assessment of the reasons for the change and hence for positive action on the process.

A control chart, in control for twenty or thirty samples, that is with all the points plotted within the limits, is usually considered evidence of a stable process. The center line of such a chart may be taken as a measure of the process average and used as an input to an acceptance sampling plan. Charts out of control, that is with points outside the limits, are an indication of lack of stability. Such charts can be interpreted to mean that the overall average will not give a true representation of the data plotted on the chart.

Process control engineers and inspectors have at their disposal many auxiliary methods for the analysis of control charts which are used for early detection of an out of control condition before a point plots outside the limits. They are also used to isolate evidence of the nature of the "assignable cause" of an out of control point. These methods, covered in the literature of statistical quality control, should be utilized by qualified individuals to determine the fundamental causes of process shifts before evidence from control charts which are out of control is used in setting up an acceptance sampling plan.

Instructions for constructing a control chart will be found in any basic text on quality control [see, for example, Burr (1979)]. Excellent discussions will be found in Wescott (1959) and Knowler (1946), while factors for determining limits are given in ANSI/ASQC Standard Al(1978). A few of the control chart factors are given in Appendix Table Tl-1 for the convenience of the reader who is familiar with control chart construction and interpretation.

Still another procedure in process quality control is the process optimization study. As defined by Mentch (1980) such studies include

1. Process performance check. A "grab sample" to estimate the characteristics of the process at a given time.

2. Process performance evaluation. An analysis of past data by control chart methods which is used to *estimate* process capability, the limits of process performance if the process were to remain in a constant state of control.

3. Process capability study. An ongoing, real-time, study of the process including correction of assignable causes to bring the process into a state of control so that estimates of process capability can actually be realized or surpassed.

4. Process improvement study. Basic modification of the process through designed experiments and other means when existing process capability is not sufficient to meet required specifications.

Process capability has been defined by Ekvall and Juran (1974, pp. 9-16) as follows: "Process capability is the measured, inherent reproducibility of the product turned out by a process." It is of utmost importance in acceptance sampling since, in no event, should the requirements of a sampling procedure exceed the producer's process capability. When this happens, either a new supplier should be selected or the specifications should be changed. In like manner, it is sometimes the case that a consumer's process can tolerate variation in raw material beyond design requirements imposed by engineering. This provides the opportunity for widening the specifications, with associated economic advantages. It is important to determine what variables must be controlled either through acceptance control or process control to achieve a desirable steady state process level in the consumer's process. Process optimization studies (on the real process as installed) can do this, frequently with large cost savings in the process.

Two important aspects of process quality control, control charts, and process capability studies have been discussed. It should be recognized that successful application of the principles of acceptance control requires intimate knowledge of process control. The reader is well advised to consult basic texts on the subject to take full advantage of the synergism that can be achieved by simultaneous application of both forms of quality control.

BACKGROUND OF ACCEPTANCE QUALITY CONTROL

Development of the statistical science of acceptance sampling can be traced back to the formation of the Inspection Engineering Department of Western Electric's Bell Telephone Laboratories in 1924. The department was compris-

ed of H. F. Dodge, R. B. Miller, E. G. D. Paterson, D. A. Quarles, and W. A.
Shewhart. Later, H. G. Romig, P. S. Olmstead, and M. N. Torrey became
members of the group. It was directed initially by R. L. Jones with G. D.
Edwards becoming its second and long-term director. Applications to shop
operations at the Western Electric Hawthorne plant in Chicago were later
formalized in 1926 by the formation of the Western Electric Committee on
Rating of Quality of Telephone Products and a Special Committee on Inspection Statistics and Economy. Early members of these committees included
J. A. Davidson, A. B. Hazard, M. E. Berry, E. D. Hall, J. M. Juran, C. A.
Melsheimer, S. M. Osborne, C. W. Robbins, W. L. Robertson, and W. Bartkey
(consultant).

Out of this group and its lineage came the following early developments

1924.	The first control chart
1925-26.	Terminology of acceptance sampling (consumer's risk, producer's risk, probability of acceptance, OC curves, lot tolerance percent defective, average total inspection, double sampling, Type A and Type B risks); LTPD sampling tables
1927.	AOQL sampling tables; multiple sampling
1928.	Demerit rating system

among others.

The 1930s saw applications of acceptance sampling within Western
Electric and elsewhere. A Joint Committee for the Development of Statistical Applications in Development and Manufacturing was formed in 1930 by
The American Society of Mechanical Engineers, American Society for Testing
and Materials, American Institute of Electrical Engineers, American Statistical Association and the American Mathematical Society with W. A. Shewhart
as chairman. By the mid-1930s Egon Pearson (1935) had developed British
Standards Institution Standard Number 600, "Application of Statistical
Methods to Industrial Standardization and Quality Control," which helped to
incite interest in England. Also, in England W. J. Jennett and B. L. Welch
(1939) published their paper on variables sampling plans entitled, "The
Control of Proportion Defective as Judged by a Single Quality Characteris--
tic Varying on a Continuous Scale." Meanwhile in the same year in the
United States H. G. Romig (1939) submitted his doctoral dissertation to
Columbia University on "Allowable Averages in Sampling Inspection," presenting variables sampling plans along the lines of the Dodge-Romig tables
which had been in use in Western Electric for some time.

The early 1940s saw publication of the Dodge-Romig (1941) "Sampling Inspection Tables," which provided plans based on fixed consumer risk (LTPD protection) and also plans for rectification (AOQL protection) which guaranteed stated protection after 100 percent inspection of rejected lots.

With the war, quality control, and particularly acceptance sampling, came of age. This included the development by the Army's Office of the Chief of Ordnance (1942) of "Standard Inspection Procedures" of which the Ordnance sampling tables, using a sampling system based on a designated acceptable quality level, were a part. The development of the system was largely due to G. D. Edwards, H. F. Dodge, and G. R. Gause, with the assistance of H. G. Romig and M. N. Torrey. This work later developed into the Army Service Forces (ASF) tables of 1944.

In this period H. F. Dodge (1943) developed an acceptance sampling plan which would perform rectification inspection on a continuous sequence of product guaranteeing the consumer protection in terms of the maximum average quality the consumer would receive (AOQL protection). Also, A. Wald (1943) put forward his new theory of sequential sampling as a member of the Statistical Research Group, Columbia University (1945), which later published applications of Wald's work. This group was responsible for some outstanding contributions during the War. Its senior scientific staff consisted of K. J. Arnold, R. F. Bennett, J. H. Bigelow, A. H. Bowker, C. Eisenhart, H. A. Freeman, M. Friedman, M. A. Girshick, M. W. Hastay, H. Hotelling, E. Paulson, L. J. Savage, H. Solomon, G. J. Stigler, A. Wald, W. A. Wallis, and J. Wolfowitz. Their output consisted of advancements in variables and attributes sampling in addition to sequential analysis. Some of these are documented in the Statistical Research Group (1947) "Techniques of Statistical Analysis." They were active in theoretical developments in process quality control, design of experiments, and other areas of industrial and applied statistics as well. Out of the work of the Statistical Research Group came a manual on sampling inspection prepared for the U. S. Navy, Office of Procurement and Material. Like the Army Ordnance Tables, it was a sampling system also based on specification of an acceptable quality level (AQL) and was later published by the Statistical Research Group (1948) under the title "Sampling Inspection." In 1949 the manual became the basis for the Defense Department's nonmandatory Joint Army-Navy Standard JAN-105; however, a committee of military quality control specialists had to be formed to reach a compromise between JAN-105 and the ASF tables. This resulted in MIL-STD-105A issued in 1950, and subsequently revised as 105B, 105C and 105D. The Statistical Research Group had considered development of a set of variables plans to match the AQL attributes system it had set forth in the Navy manual. However, the group was disbanded on September 30, 1945 before it was possible to construct such tables. Fortunately, the Office of Naval Research supported preparation of such a work at Stanford University resulting in the book by A. H. Bowker and H. P. Goode (1952) which was a milestone in the development of variables sampling plans. The work of the Statistical Research Group has been documented by Wallis (1980).

Work in the area of acceptance sampling did not end with World War II. Many, if not most, of the procedures presented in this volume were developed later. This brief history, however, has been presented to place the rest of the book in context so that each method discussed can, in some sense, be traced to its natural origins. More detailed accounts of the history and development of acceptance sampling will be found in Dodge (1969a,b,c;1970a) and in a series of papers published by the American Statistical Association (1950) under the title "Acceptance Sampling."

PROBLEMS

1. Distinguish between acceptance sampling and acceptance control.
2. Explain why installation of a sampling plan is futile if the level of quality is poor but stable and cannot be improved.
3. Distinguish between Type A and Type B sampling plans.
4. Distinguish between process quality control and acceptance quality control. How is process quality control used in acceptance sampling?
5. What are the odds of an incorrect signal of a process change on a conventional Shewhart chart?
6. What are the four constituents of a process optimization study?
7. Define process capability.
8. What was one of G. D. Edwards principal contributions to quality control?
9. Which came first, the control chart or AOQL sampling plans? Where were they developed?
10. Who invented continuous sampling plans? When?

REFERENCES

American Society for Quality Control (1978), *American National Standard: Definitions, Symbols, Formulas and Tables for Control Charts*, ANSI/ ASQC A1(1978), American Society for Quality Control, Milwaukee, Wisconsin.

American Statistical Association (1950), *Acceptance sampling — A Symposium*, American Statistical Association, Washington, D.C.

Bowker, A. H., and H. P. Goode (1951), *Sampling Inspection by Variables*, McGraw-Hill, New York.

Burr, I. W. (1979), *Elementary Statistical Quality Control*, Marcel Dekker, New York.

Dodge, H. F. (1943), A Sampling Plan for Continuous Production, *Annals of Mathematical Statistics*, 14: 264-279.

Dodge, H. F. (1969a,b,c; 1970a), Notes on the Evolution of Acceptance Sampling Plans, *Journal of Quality Technology*, Part I, 1(2): 77-88; Part II, 1(3): 155-162; Part III, 1(4): 225-232; Part IV, 2(1): 1-8.

Dodge, H. F., and H. G. Romig (1941), Single Sampling and Double Sampling Inspection Tables, *The Bell System Technical Journal*, 20: 1-61.

Jennett, W. J., and B. L. Welch (1939), The Control of Proportion Defective as Judged by a Single Quality Characteristic Varying on a Continuous Scale, *Supplement to the Journal of the Royal Statistical Society*, 6: 80-88.

Juran, J. M., Ed. (1974), *Quality Control Handbook*, 3rd ed., McGraw-Hill, New York.

Knowler, L. A. (1946), Fundamentals of Quality Control, *Industrial Quality Control*, 3(1): 7-18.

Mentch, C. C. (1980), Manufacturing Process Quality Optimization Studies, *Journal of Quality Technology*, 12(3): 119-129.

Mood, A. M. (1943), On the Dependence of Sampling Inspection Plans Upon Population Distributions, *Annals of Mathematical Statistics*, 14: 415-425.

Olmstead, P. S. (1967), Our Debt to Walter Shewhart, *Industrial Quality Control*, 24(2): 72-73.

Pearson, E. S. (1935), *The Application of Statistical Methods to Industrial Standardization and Quality Control*, British Standard No. 600-1935, British Standards Institution, London.

Romig, H. G. (1939), *Allowable Average in Sampling Inspection*, Ph.D. Dissertation, Columbia University, New York.

Statistical Research Group, Columbia University (1945), *Sequential Analysis of Statistical Data: Applications*, Columbia University Press, New York.

Statistical Research Group, Columbia University (1947), *Techniques of Statistical Analysis*, McGraw-Hill, New York.

Statistical Research Group, Columbia University (1948), *Sampling Inspection*, McGraw-Hill, New York.

United States Department of the Army (1944), *Standard Inspection Procedures, Quality Control*, Army Service Forces, Office of the Chief of Ordnance, Washington, D.C.

Wald, A. (1943), *Sequential Analysis of Statistical Data: Theory*, report submitted by the Statistical Research Group, Columbia University, to the Applied Mathematics Panel, National Defense Research Committee.

Wallis, W. A. (1980), The Statistical Research Group, 1942-1945, *Journal of the American Statistical Association*, 75(370): 320-330.

Wescott, M. E. (1959), Fundamental Control Techniques, *Rubber World*, Part I: 252-262; Part IIa: 717-722; Part IIb: 869-872.

PROBABILITY AND THE OPERATING CHARACTERISTIC CURVE

Undoubtedly the most important single working tool in acceptance quality control is probability theory itself. This does not mean that good quality engineers have to be accomplished probabilists or erudite mathematical statisticians. They must be aware, however, of the practical aspects of probability and how to apply its principles to the problem at hand. This is because most information in quality control is generated in the form of samples from larger, sometimes essentially infinite, populations. It is vital then that the quality engineer have some background in probability theory. Only the most basic elements are presented here.

PROBABILITY

It is important to note that the term *probability* has come to mean different things to different people. In fact, these differences are recognized in defining probability, for there is not just one but at least three important definitions of the term. Each of them gives insight into the nature of probability itself. Two of them are objectivistic in the sense that they are subject to verification, while the third is personalistic and refers to the degree of belief of an individual.

Classical Definition

"If there be a number of events of which one must happen and all are equally likely, and if any one of a (smaller) number of these events will produce a certain result which cannot otherwise happen, the probability of this result is expressed by the ratio of this smaller number to the whole number of events" (Whitworth, 1959, Rule IV, p. 130). Here *probability* is defined

14

as the ratio of favorable to total possible equally likely and mutually exclusive cases.

EXAMPLE. There are 52 cards in a deck of which 4 are aces. If the cards are shuffled so that they are equally likely to be drawn, the probability of obtaining an ace is 4/52 = 1/13.

This is the definition of probability which is familiar from high school mathematics.

Empirical Definition

"The limiting value of the relative frequency of a given attribute, assumed to be independent of any place selection, will be called 'the probability of that attribute ...'" (von Mises, 1957, p. 29). Thus, probability is regarded as the ratio of successes to total number of trials in the long run.

EXAMPLE. In determining if a penny was in fact a true coin, it was flipped 2000 times resulting in 1010 heads. An estimate of the probability of heads for this coin is 0.505. It would be expected that this probability would approach 1/2 as the sequence of tosses was lengthened if the coin were true.

This is the sort of probability that is involved in saying that Casey has a .333 batting average. It implies that the probability of a hit in the next time at bat is approximately 1/3.

Subjective Definition

"Probability measures the confidence that a particular individual has in the truth of a particular proposition, for example, the proposition that it will rain tomorrow" (Savage, 1954, p. 3). Thus probability may be thought of as degree of belief on the part of an individual, not necessarily the same from one person to another.

EXAMPLE. There is a high probability of intelligent life elsewhere in the universe.

Here we have not counted the occurrences and nonoccurrences of life in a number of universes. Nor have we sampled universes to build up a ratio of trials. This statement implies a degree of belief on the part of an individual which may differ considerably from one individual to another.

These definitions have immediate applications in acceptance quality control. Classical probability calculations are involved in the determination of the probability of acceptance of a lot of finite size, where all the possibilities can be enumerated and samples taken therefrom. Empirical probabilities are used when sampling from a process running in a state of statistical control. Here, the process could conceivably produce an uncountable number of units so that the only way to get at the probability of a defective unit is in the empirical sense. Subjective probabilities have been used in the evaluation of sampling plans, particularly under cost constraints. They reflect the judgment of an individual or a group as to the probabilities involved. While sampling plans have been derived which incorporate subjective probabilities, they appear to be difficult to apply in an adversary relationship unless the producer and the consumer can be expected to agree on the specific subjective elements involved.

There are many sources for information on probability and its definition. Some interesting references of historic value are Whitworth (1950) on classical probability, von Mises (1957) on empirical probability, and the Savage (1954) work on subjective probability. Since the classical and empirical definitions of probability are objectivistic and can be shown to agree in the long run, and since the empirical definition is more general, the empirical definition of probability will be used here unless otherwise stated or implied. When subjective probabilities are employed their nature will be specifically pointed out.

RANDOM SAMPLES AND RANDOM NUMBERS

Random samples are those in which every item in the lot or population sampled has an equal chance to be drawn. Such samples may be taken with or without replacement. That is, items may be returned to the population once drawn, or they may be withheld. If they are withheld, the probability of drawing a particular item from a finite population changes from trial to trial. Whereas, if the items are replaced or if the population is uncountably large, the probability of drawing a particular item will not change from trial to trial. In any event, every item should have an equal opportunity for selection on a given trial, whether the probabilities change from trial to trial or not.

This may be illustrated with a deck of cards. There are 52 cards, one of which is the ace of spades. Sampling without replacement, the probability of drawing the ace of spades on the first draw is 1 out of 52, while on the second draw it is 1 out of the 51 cards that remain, assuming it was not drawn on the first trial. If the cards were replaced as drawn, the probability would be 1 out of 52 on any draw since there would always be 52 cards in the deck.

Note that if the population is very large, the change in probability when samples are not replaced will be very small and will remain essentially the same from trial to trial. In a raffle of 100,000 tickets the chances of being drawn on the first trial is 1 in 100,000 and on the second trial 1 in 99,999. Essentially .00001 in each case. Few raffles are conducted in which a winning ticket is replaced for subsequent draws.

At the core of random sampling is the concept of equal opportunity for each item in the population sampled to be drawn on any trial. Sometimes special sampling structures are used such as stratified sampling in which the population is segmented and samples are taken from the segments. Formulas exist for the estimation of population characteristics from such samples. In any event, equal opportunity should be provided within a segment for items to be selected.

To guarantee randomness of selection, tables of random numbers have been prepared. These numbers have been set up to mimic the output of a truly random process. They are intended to occur with equal frequency but in a random order. Appendix Table T2-1 is one such table. To use the random number tables:

1. Number the items in the population.
2. Specify a fixed pattern for the selection of the random numbers (e.g., right to left, bottom to top, every third on a diagonal).
3. Choose an arbitrary starting place and select as many random numbers as needed for the sample.
4. Choose as a sample those items with numbers corresponding to the random numbers selected.

The resulting sample will be truly representative in the sense that every item in the population will have had an essentially equal chance to be selected.

Sometimes it is impractical or impossible to number all the items in a population. In such cases the sample should be taken with the principle of random sampling in mind to obtain as good a sample as possible. Avoid bias. Avoid examining the samples before they are selected. Avoid sampling only from the most convenient location (the top of the container, the spigot at the bottom, etc.). In one sampling situation, an inspector was sent to the producer's plant to sample the product as a boxcar was being loaded, since it was impossible to obtain a random sample thereafter. Such strategies as these can help to provide randomness as much as the random sampling tables themselves.

COUNTING POSSIBILITIES

Evaluation of the probability of an event under the classical definition involves counting the number of possibilities favorable to the event and forming the ratio of that number to the total of equally likely possibilities. The possibilities must be such that they cannot occur together on a single draw; that is, they must be mutually exclusive. There are three important aids in making counts of this type: permutations, combinations, and tree diagrams.

Suppose a lot of three items, each identified by a serial number is received, two of which are good. The sampling plan to be employed is to sample two items and accept the lot if no defectives are obtained. Reject if one or more are found. Thus, the sampling plan is n = 2 and c = 0, where n is the sample size and c represents the acceptance number or maximum number of defectives allowed in the sample for acceptance of the lot.

If the items are removed from the shipping container one at a time, we may ask in how many different orders (permutations) can the three items be removed from the box. Suppose the serial numbers are the same except for the last digit which is 5, 7, 8, respectively. Enumerating the orders we have

      578   875
      758   857
      785   587

The formula for the number of permutations of n things taken n at a time is

$$P_n^n = n! = n(n - 1)(n - 2) \cdots 1$$

where n!, or n *factorial*, is the symbol for multiplications of the number n by all the successively smaller integers down to one. Thus

      1! = 1
      2! = 2(1) = 2
      3! = 3(2)(1) = 6
      4! = 4(3)(2)(1) = 24

and so on. It is important to note that we define

   $0! = 1$

In the example, we want the number of permutations of 3 things taken 3 at a time, or

   $P^3_3 = 3! = 3(2)(1) = 6$

which agrees with the enumeration.

In how many orders can we select the two items for our sample? Enumerating again:

   57   87

   75   85

   78   58

The formula for the number of permutations of n things taken r at a time is

   $$P^n_r = \frac{n!}{(n-r)!}$$

Clearly the previous formula for P^n_n is a special case of this formula. To determine the number of permutations of three objects taken two at a time we have

   $$P^3_2 = \frac{3!}{(3-2)!} = \frac{3!}{1!} = \frac{3(2)(1)}{1} = 6$$

This makes sense and agrees with the previous result since the last item drawn is completely determined by the previous two items drawn and so does not contribute to the number of possible orders (permutations).

Now, let us ask how many possible orders are there if some of the items are indistinguishable one from the other. For example, disregarding the serial numbers, we have one defective item and two good ones. The good items are indistinguishable from each other and we may ask in how many orders can we draw one defective and two good items. The answer may be found in the formula for the number of permutations of n things, r of which are alike (good) and n - r are alike (bad)

   $$P^n_{r,(n-r)} = \frac{n!}{r!(n-r)!}$$

and for the example, the answer is

   $$P^3_{2,(2-1)} = P^3_{2,1} = \frac{3!}{2!\,1!} = \frac{3(2)(1)}{2(1)(1)} = 3$$

Enumerating them we have

   B   G   G

   G   B   G

   G   G   B

The reader may notice the similarity of the formula

$$P^n_{r,(n-r)} = \frac{n!}{r!(n-r)!}$$

and the classic formula for the number of combinations (groups) which can be made from n things taken r at a time. That formula is

$$C^n_r = \frac{n!}{r!(n-r)!}$$

and shows how many distinct groups of size r can be formed from n *distinguishable* objects. If we phrase the question, "in how many ways can we select two objects (to be the good ones) out of three," we have

	Good	Bad
Group 1	57	8
	75	8
Group 2	78	5
	87	5
Group 3	85	7
	58	7

or

$$C^3_2 = \frac{3!}{2!(3-2)} = \frac{3!}{2!\ 1!} = 3$$

Thus we see

$$P^n_{r,(n-r)} = C^n_r$$

In general, the combinatorial formula may be used to determine the number of groupings of various kinds. For example, the number of ways (groups) to select 4 cards from a deck of 52 to form hands of 4 cards (where order is not important) is

$$C^{52}_4 = \frac{52!}{4!\ 48!} = \frac{52 \cdot 51 \cdot 50 \cdot 49 \cdot 48!}{4 \cdot 3 \cdot 2 \cdot 1 \cdot 48!} = 270,725$$

Using the classical definition of probability, then, the probability of getting a hand containing all four aces is

$$P(4\text{ aces}) = \frac{\text{number of four ace hands}}{\text{number of four card hands}} = \frac{1}{270,725}$$

Here we have counted groups where order in the group is not important.

In the same way, probabilities can be calculated for use in evaluating acceptance sampling plans. The plan given in the earlier example was sample size 2, accept when there are no defectives in the sample. That is n = 2 and c = 0. To evaluate the probability of acceptance when there is one defective in the lot of N = 3, we would proceed as follows:

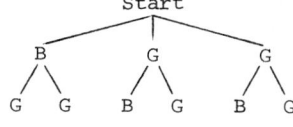

First draw

Second draw

Acceptance decision

FIG. 2-1 Tree diagram.

1. To obtain probability of acceptance, we must count the number of samples in which we would obtain 0 defectives in a sample of two.
2. The probability is the quantity obtained in step 1 divided by the total number of samples of 2 that could possibly be obtained.

Then

1. To obtain samples of 2 having no defectives, we would have to select both items from the two items which are good. The number of such samples is $c_2^2 = 1$.

2. There are $c_2^3 = 3$ different unordered samples. So the probability of accepting P_a with this sampling plan is

$$P_a = \frac{c_2^2}{c_2^3} = \frac{1}{3}$$

The third tool in counting possibilities in simple cases such as this is the tree diagram. Fig. 2-1 shows such a diagram for this example, for the acceptance (A) and rejection (R) of samples of good (G) and bad (B) pieces. Each branch of the tree going downward shows a given sample permutation. We see that 1/3 of these permutations lead to lot acceptance. Counting the permutations we have

$$P_2^3 = \frac{3!}{1!} = 6$$

possible samples, of which

$$P_2^2 = \frac{2!}{0!} = 2$$

lead to acceptance. The probability of acceptance then is

$$P_a = \frac{P_2^2}{P_2^3} = \frac{2}{6} = \frac{1}{3}$$

which shows that the probability of acceptance can be obtained by using
either permutations or combinations.

PROBABILITY CALCULUS

There are certain rules for manipulating probabilities which suffice for
many of the elementary calculations needed in acceptance control theory.
These are based on recognition of two kinds of events.

> Mutually exclusive events. Two events are mutually exclusive if, on
> a single trial, the occurrence of one of the events precludes
> the occurrence of the other.

> Independent events. Two events are stochastically independent if the
> occurrence of one of them on a trial does not change the
> *probability* of occurrence of the other on that trial.

Thus the events head and tail are mutually exclusive in a single trial
of flipping a coin. They are also not independent events since the occurrence of either on a trial drives the probability of occurrence of the
other on that trial to zero.

In contrast the events ace and heart are not mutually exclusive in
drawing cards since they can occur together in the ace of hearts. Further,
they are also independent since the probability of drawing an ace is
$4/52 = 1/13$. If you know that a heart was drawn, the probability of the
card being also an ace is still $1/13$. Note that the events face card and
queen are not independent. The probability of drawing a queen is $4/52 =$
$1/13$; however, if you know a face card was drawn, the probability of that
card being a queen is now $4/12 = 1/3$.

Trials are sometimes spoken of as being independent. This means the
sampling situation is such that the probabilities of the events being
investigated do not change from trial to trial. Flips of a coin are such
as this in that the odds remain 50:50 from trial to trial. If cards are
drawn from a deck and not replaced the trials are dependent, however.
Thus, the probability of a queen of hearts is $1/52$ on the first draw from
a deck, but it increases to $1/51$ on the second draw assuming it was not
drawn on the first.

A. N. Kolmogorov (1950) has developed the entire calculus of probabilities from a few simple axioms. Crudely stated and somewhat condensed,
they are as follows:

1. The probability of an event, E, is always positive or zero, never negative: $P(E) \geq 0$.

2. The sum of the probabilities of events in the universe U, or population to which E belongs, is one: $P(U) = 1$.

3. If events A and B are mutually exclusive, the probability of A or B occurring is

$$P(A \text{ or } B) = P(A) + P(B)$$

From the axioms, the following consequences can be obtained

4. The probability of an event must be less than or equal to one, never greater than one: $P(E) \leq 1$.

5. The probability of the null set (no event occuring) is zero: $P(\text{no event}) = 0$.

6. The probability of an event not occurring is the complement of the probability of the event: $P(\text{not } E) = 1 - P(E)$.

The most useful rules in dealing with probabilities are the so-called

General rule of addition. Shows the probability of A or B occurring on a single trial.

$$P(A \text{ or } B) = P(A) + P(B) - P(A \text{ and } B)$$

Clearly, if A and B are mutually exclusive, the term $P(A \text{ and } B) = 0$ and we have the so-called special rule of addition.

$$P(A \text{ or } B) = P(A) + P(B)$$

for A and B mutually exclusive.

General rule of multiplication. Shows the probability of A and B both occurring on a single trial where $P(B|A)$ is the conditional probability of B given A is known to have occurred

$$P(A \text{ and } B) = P(A) \, P(B|A)$$
$$= P(B) \, P(A|B)$$

Clearly, if A and B are independent, the factor $P(B|A) = P(B)$ since the probability of B is unchanged even if we know A has occurred (similarly for $P(A|B)$). We then have the so-called special rule of multiplication

$$P(A \text{ and } B) = P(A) \, P(B) \quad , \quad \text{A and B independent}$$

This is sometimes used as a test for the independence of A and B since if the relationship holds, A and B are independent.

These rules can be generalized to any number of events. The special rules become

P(A or B or C or D) = P(A) + P(B) + P(C) + P(D)   , A, B, C, D
                                                     mutually exclusive

P(A and B and C and D) = P(A) P(B) P(C) P(D)   , A, B, C, D independent

and so on. These are especially useful since they can be employed to calculate probabilities over several independent trials. The general rule for addition is

$$P(A \text{ or } B \text{ or } C \text{ or } D) = P(A) + P(B) + P(C) + P(D)$$
$$- P(AB) - P(AC) - P(AD) - P(BC) - P(BD) - P(CD)$$
$$+ P(ABC) + P(ABD) + P(ACD) + P(BCD) - P(ABCD)$$

alternating additions and subtractions of each higher level of joint probability, while that for multiplication becomes

$$P(A \text{ and } B \text{ and } C \text{ and } D) = P(A) \ P(B|A) \ P(C|AB) \ P(D|ABC)$$

when there are four events. Each probability multiplied is conditional on those which went before.

These rules may be illustrated using the example given earlier involving the computation of the probability of acceptance P_a of a lot consisting of 3 units when one of them is defective and the sampling plan is n = 2, c = 0. Acceptance will occur only when both items in the sample are good. If we assume random samples are drawn without replacement, the events will be dependent from trial to trial. We need the probability of a good item on the first draw *and* a good item on the second draw.

Let A = {event good on first draw} and B = {event good on second draw} then

$$P(A) = \frac{2}{3} \qquad P(B|A) = \frac{1}{2}$$

since there are only two pieces left on the second draw. Applying the general rule of multiplication:

$$P_a = P(A) \ P(B|A) = \frac{2}{3}\left(\frac{1}{2}\right) = \frac{1}{3}$$

which agrees with the result of the previous section.

Now, what if the items were put back into the lot after inspection and the next sample drawn? This is a highly unusual procedure in practice, but serves as a model for some of the probability distributions developed later. It simulates an infinite lot 1/3 defective since, using this method of

inspection, the lot would never be depleted. Under these conditions the
special rule of multiplication could be employed since the events would be
independent of each other from trial to trial. We obtain

$$P_a = P(A) \; P(B) = \frac{2}{3} \left(\frac{2}{3} \right) = \frac{4}{9}$$

This makes sense since the previous method depleted the lot and made it
more likely to obtain the defective unit on the second draw.

Further, suppose two such lots are inspected using the procedure of
sampling without replacement. What is the probability that at least one
will be accepted? That is, what is the probability that one *or* the other
will be passed? Here, let C = {event first lot is passed} and D = {event
second lot is passed}, then the probability both lots are passed is

$$P(C \text{ and } D) = \frac{1}{3} \left(\frac{1}{3} \right) = \frac{1}{9}$$

using the special rule of multiplication since they are inspected independently. Then the probability of at least one passing is

$$P(C \text{ or } D) = P(C) + P(D) - P(C \text{ and } D)$$
$$= \frac{1}{3} + \frac{1}{3} - \frac{1}{9} = \frac{5}{9}$$

The probability of *not* having at least one lot pass is

$$P(\text{both fail}) = 1 - P(C \text{ or } D) = 1 - \frac{5}{9} = \frac{4}{9}$$

which could have been calculated using the special rule of multiplication
as

$$P(\text{both fail}) = [1 - P(C)] \; [1 - P(D)]$$
$$= \frac{2}{3} \left(\frac{2}{3} \right)$$
$$= \frac{4}{9}$$

Finally, suppose there are five inspectors: V, W, X, Y, Z, each with the
same probability of selection. The lot is to be inspected. What is the
probability that the inspector chosen is X, Y, or Z? Since in this case
the use of the inspectors is mutually exclusive, the special rule of
addition may be used

$$P(X \text{ or } Y \text{ or } Z) = P(X) + P(Y) + P(Z)$$
$$= \frac{1}{5} + \frac{1}{5} + \frac{1}{5}$$
$$= \frac{3}{5}$$

These are a few of the tools of probability theory. Fortunately, they have been put to use by theorists in the design of the methods of acceptance quality control to develop procedures which do not require extensive knowledge of the subject for application. These methods are presented here in subsequent chapters. Nevertheless, to gain a true appreciation for the subtleties of acceptance sampling, a sound background in probability theory is invaluable.

THE OPERATING CHARACTERISTIC CURVE

A fundamental use of probability with regard to acceptance sampling comes in describing the chances of a lot passing sampling inspection if it is composed of a given proportion defective. The very simplest sampling plan is, of course, as follows:

1. Sample one piece from the lot.
2. If the sampled piece is good, accept the lot.
3. If the sampled piece is defective, reject the lot.

This plan is said to have a sample size n of one and an acceptance number c of zero since the sample must contain zero defectives for lot acceptance to occur; otherwise, the lot will be rejected. That is, n = 1, c = 0. Now, if the lot were perfect, it would have no chance of rejection since the sample would never contain a defective piece. Similarly, if the lot were completely bad there would be no acceptances since the sample piece would always be defective. But what if the lot were mixed defective and good? This is where probability enters in. Suppose one-half of the lot was defective, then the chance of drawing out a defective piece from the lot would be 50:50 and we would have 50 percent probability of acceptance. But it might be one-quarter defective leading to a 75 percent chance for acceptance, since there are three-quarters good pieces in the lot. Or again, the lot might be three-quarters defective leading to a 25 percent chance of finding a good piece. Since the lot might be any of a multitude of possible proportions defective from 0 to 1, how can we describe the behavior of this simple sampling plan? The answer lies in the operating characteristic (OC) curve which plots the probability of acceptance against possible values of proportion defective. The curve for this particular plan is shown in Fig. 2-2.

We see that for any proportion defective p, the probability of acceptance P_a is just the complement of p; that is

$$P_a = 1 - p$$

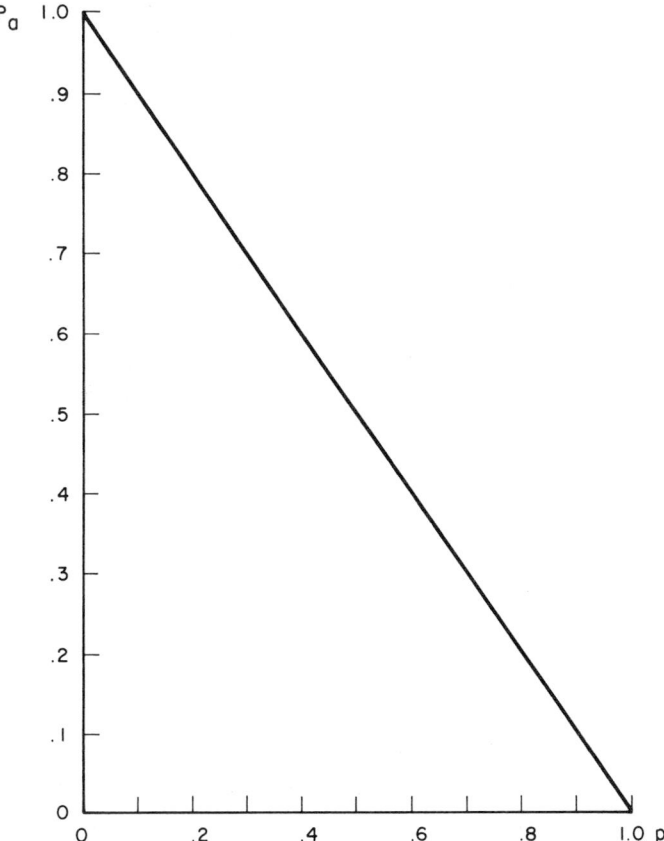

FIG. 2-2 Operating characteristic curve, n = 1, c = 0.

This is only true of the plan n = 1, c = 0. Thus the operating characteristic curve stands as a unique representation of the performance of the plan
against possible *alternative* proportions defective. A given lot can have
only one proportion defective associated with it. But we see from the curve
that lots which have a proportion defective greater than 0.75 have less than
a 25 percent chance to be accepted and those lots with less than 0.25 defective pieces will have greater than a 75 percent chance to pass. The OC curve
gives at a glance a characterization of the potential performance of the
plan, telling how the plan will perform for any submitted fraction defective.

Now consider the plan n = 5, c = 0. The operating characteristic curve
can be easily constructed using the rules for manipulation of probabilities
given above. First, however, let us assume we are sampling from a very
large lot or better yet from the producer's process so the probabilities

will remain essentially independent from trial to trial. Note that the
probability of acceptance P_a for any proportion defective p can be computed
as

$$P_a = (1 - p)(1 - p)(1 - p)(1 - p)(1 - p)$$
$$= (1 - p)^5$$

since all the pieces must be good in the sample of 5 for lot acceptance.
To plot the OC curve we compute P_a for various values of p

p	(1 - p)	P_a
.005	.995	.975
.01	.99	.951
.05	.95	.774
.10	.90	.590
.20	.80	.328
.30	.70	.168
.40	.60	.078
.50	.50	.031

and graph the result as in Fig. 2-3.

We see from Fig. 2-3 that if the producer can maintain a fraction defective less than .01 the product will be accepted 95 percent of the time
or more by the plan. If product is submitted which is 13 percent defective,
it will have a 50:50 chance of acceptance, while product 37 percent defective has only a 10 percent chance of acceptance by this plan. It is conventional to designate proportions defective having a given probability of
acceptance as probability points. Thus a fraction defective having probability of acceptance γ is shown as p_γ. Particular probability points may
be designated as follows:

P_a	Term	Abbreviation	Probability point
.95	Acceptable quality level	AQL	$p_{.95}$
.50	Indifference quality	IQ	$p_{.50}$
.10	Lot tolerance percent defective (10% limiting quality)	LTPD [LQ(.10)]	$p_{.10}$

Designation of these points gives a quick summary of plan performance. The
term *AQL* is commonly used as the 95 percent point of probability of acceptance, although most definitions do not tie the term to a specific point on

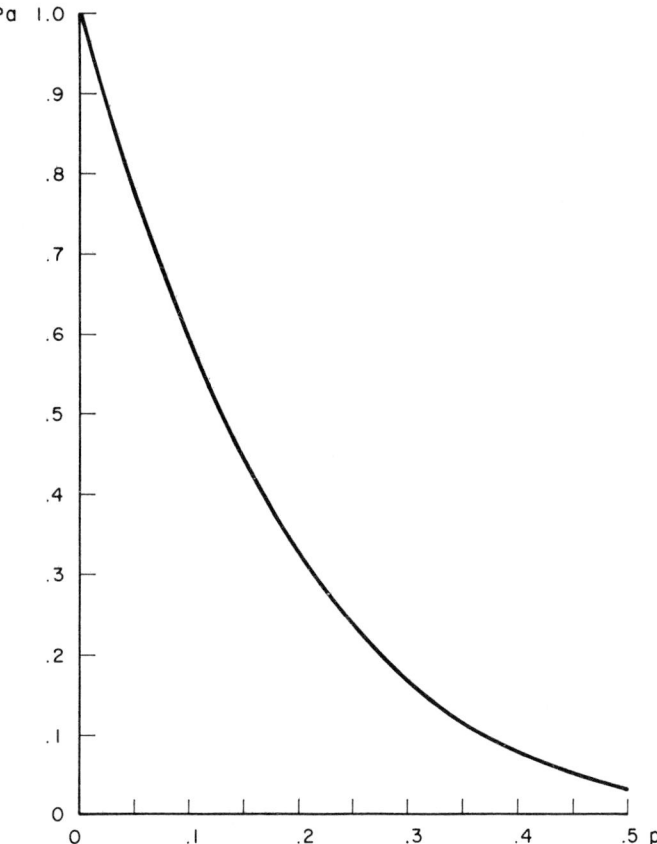

FIG. 2-3   Operating characteristic curve, n = 5, c = 0.

the OC curve and simply associate it with a "high" probability of acceptance. The term is used here as it was used by the Columbia Statistical Research Group in preparing the Navy (1946) input to the JAN-STD-105 standard.   LTPD refers to the 10 percent probability point of the OC curve and is generally associated with percent defective.   The advent of plans controlling other parameters of the distribution led to the term *limiting quality*, usually preceded by the percentage point controlled.   Thus, "10 percent limiting quality" is the LTPD.

The operating characteristic curve is often viewed in the sense of an adversary relationship between the producer and the consumer.   The producer is primarily interested in insuring that good lots are accepted while the consumer wants to be reasonably sure that bad lots will be rejected.   In this sense we may think of a producer's quality level (PQL) and associated

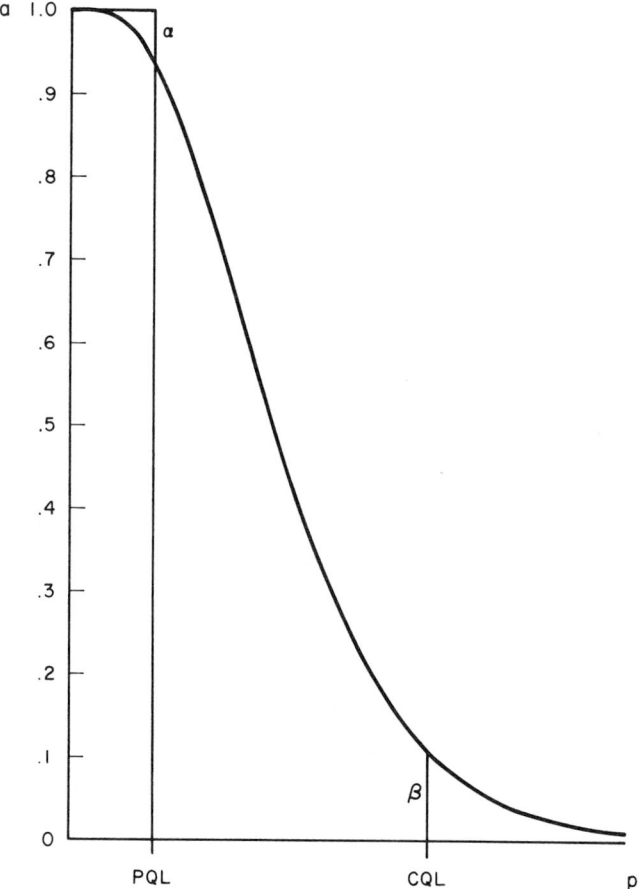

FIG. 2-4 Producer's and consumer's quality levels.

producer's risk α and a consumer's quality level (CQL) with associated consumer's risk β. Viewed against the OC curve the producer's quality level
and consumer's quality level appear as in Fig. 2-4.

Plans are often designated and constructed in terms of these two points
and the associated risks. As indicated above, the risks are often taken as
α = .05 for the producer's risk and β = .10 for the consumer's risk.

The operating characteristic curve sketches the performance of a plan
for various possible proportions defective. It is plotted using appropriate
probability functions for the sampling situation involved. The probability
functions are simply formulas for the direct calculation of probabilities
which have been developed using the appropriate probability theory.

PROBLEMS

1. A lot of 50 items contains one defective unit. If one unit is drawn
    at random from the lot, what is the probability that the lot will be
    accepted?

2. A bottle of 500 aspirin tablets is to be randomly sampled. The tablets
    are allowed to drop out one at a time to form a string, those coming
    out first at one end, those last at the other. A random number from 1
    to 1000 is selected and divided by two, rounding up. The tablet in
    the corresponding numerical position is selected. Is this procedure
    truly random?

3. Two out of six machines producing bottles are bad. The bottles feed
    in successive order into groups of six which are scrambled during
    further processing and packed in six-packs. In how many different
    orders can the two defective bottles appear among the six?

4. Six castings await inspection. Two of them have not been properly
    finished. The inspector will pick two and look at them. How many
    groups of two can be formed from the six castings? How many groups of
    two can be formed from the two defective castings? What is the proba-
    bility that the inspector will find both castings looked at are bad?

5. Form a probability tree to obtain the probability that the inspector
    will find both castings bad in Prob. 4.

6. Use the probability calculus to find the probability that the inspec-
    tor will find two bad castings in selecting two. Why is it not $2/6 \times 2/6 = 4/36 = 1/9$? What is the probability that they are both good?
    What is the probability that they are both the same? What type events
    allow these probabilities to be added?

7. At a given quality level the probability of acceptance under a certain
    sampling plan is .95. If the lot is rejected the sampling plan is
    applied again, "just to be sure," and a final decision is made. What
    is the probability of acceptance under this procedure?

8. Draw the OC curve for the plan n = 3, c = 0. What are the approximate
    AQL, IQ, and LTPD values for this plan?

9. In a mixed acceptance sampling procedure two types of plans are used.
    The first plan is used only to accept. If the lot is not accepted,
    the second plan is used. If both type plans have PQL = .03, CQL = .09
    with α = .05 and β = .10, what is the probability of acceptance of the
    mixed procedure when the fraction defective is .09?

10. At the indifference quality level the probability of acceptance is 0.5.

In five successive independent lots, what is the probability that all
fail when quality is at the indifference quality level? What is the
probability that all pass? What is the probability of at least one
failure?

REFERENCES

Kolmogorov, A. N. (1950), *Foundations of the Theory of Probability*, Chelsea,
     New York.

Savage, L. J. (1954), *Foundations of Statistics*, John Wiley and Sons, New
     York.

United States Department of the Navy (1945), *General Specifications for
     Inspection of Material*, (Superintendent of Documents, Washington, D.C.,
     1946), Appendix X, April 1, 1946; see also U.S. Navy Material Inspec-
     tion Service, Standard Sampling Inspection Procedures, Administration
     Manual, Part D, Chap. 4.

von Mises, R. (1957), *Probability, Statistics and Truth*, 2nd ed., Macmillan,
     New York.

Whitworth, W. A. (1959), *Choice and Chance*, Hafner, New York.

3

PROBABILITY FUNCTIONS

Many sampling situations can be generalized to the extent that specific
functions have proved useful in computing the probabilities associated with
the operating characteristic curve and other sampling characteristics.
These are functions of a random variable X which takes on specific values x
at random with a probability evaluated by the function. Such functions are
of two types:

> Frequency function. Gives the relative frequency (or density) for a
>> specific value of the random variable X. It is represented by
>> the function f(x).

> Distribution function. Gives the cumulative probability of the random
>> variable X up to and including a specific value of the random
>> variable. It can be used to obtain probability over a specified
>> range by appropriate manipulation. It is represented by F(x).

In the case of a discrete, go no-go, random variable,

$$f(x) = P(X = x)$$

and the distribution function is simply the sum of the values of the
frequency function up to and including x

$$F(x) = \sum_{i=0}^{x} f(x) \qquad X \text{ discrete}$$

When X is continuous, i.e., a measurement variable, it is the integral
from the lowest possible value of X, taken here to be $-\infty$, up to x:

$$F(x) = \int_{-\infty}^{x} f(t) \, dt \qquad X \text{ continuous}$$

where the notation

$$\int_{a}^{b} f(t) \, dt$$

may be thought of as representing the cumulative probability of f(t) from
a lower limit of a to an upper limit of b. In either case, these functions
provide a tool for assessment of sampling plans and usually have been suf-
ficently well tabulated to avoid extensive mathematical calculation.

The probability functions can be simply illustrated by a single toss
of a six-sided die. Here the random variable X is discrete and represents
the number of spots showing on the upward face of the die. It takes on the
values 1, 2, 3, 4, 5, 6. This is called the *sample space*. Since the probability of any of these values is constant, namely 1/6, the frequency
function is

$$f(x) = \frac{1}{6} \qquad x = 1, \ 2, \ 3, \ 4, \ 5, \ 6$$

and the distribution function is

$$F(x) = \sum_{i=1}^{x} \frac{1}{6} = \frac{x}{6}$$

With these it is possible to determine the probability of rolling a 1

$$f(1) = \frac{1}{6}$$

or of getting a result of 3 or less

$$F(3) = \frac{3}{6}$$

Values of the random variable over a range may be found by subtraction.
Thus, the probability of throwing a 4 or a 5 is

$$P(4 \text{ or } 5) = P(X \leq 5) - P(X \leq 3)$$

$$= F(5) - F(3) = \frac{5}{6} - \frac{3}{6} = \frac{2}{6}$$

PROBABILITY DISTRIBUTIONS

Using the frequency function, it is possible to find the distribution of
probabilities over all possible values of the random variable X. The
frequency function and the distribution function may then be displayed in
tabular form as follows:

x	f(x)	F(x)
1	1/6	1/6
2	1/6	2/6
3	1/6	3/6
4	1/6	4/6
5	1/6	5/6
6	1/6	6/6

When plotted, the probability distribution is shown in terms of its frequency function in Fig. 3-1 and in terms of its distribution function in Fig. 3-2.

Now consider a continuous distribution. An example might be the position of the second hand of watches when they stop from need of winding. The distribution of these values could be assumed to be rectangular in the interval from 0 to 60. The frequency function of such a distribution is

$$f(x) = \frac{1}{60} \qquad 0 \le x < 60$$

and its distribution function is

$$F(x) = \frac{x}{60} \qquad 0 \le x < 60$$

If measured closely enough, there is an infinity of possible positions at which the second hand might stop. For example, 47.2186327... seconds. The probability of stopping exactly at any given position, specified to an infinity of possible decimal places, is infinitesimally small. This is why

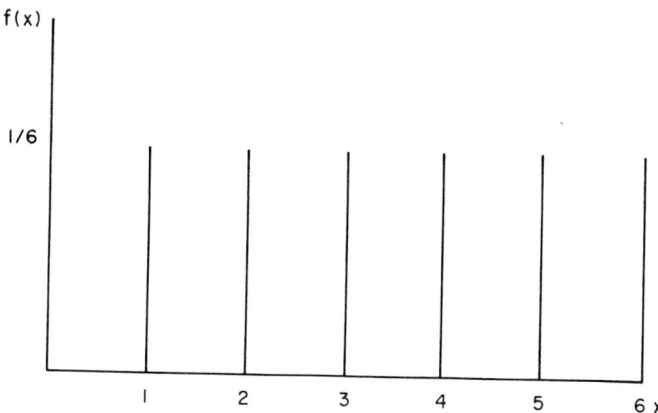

FIG. 3-1 Frequency function for die.

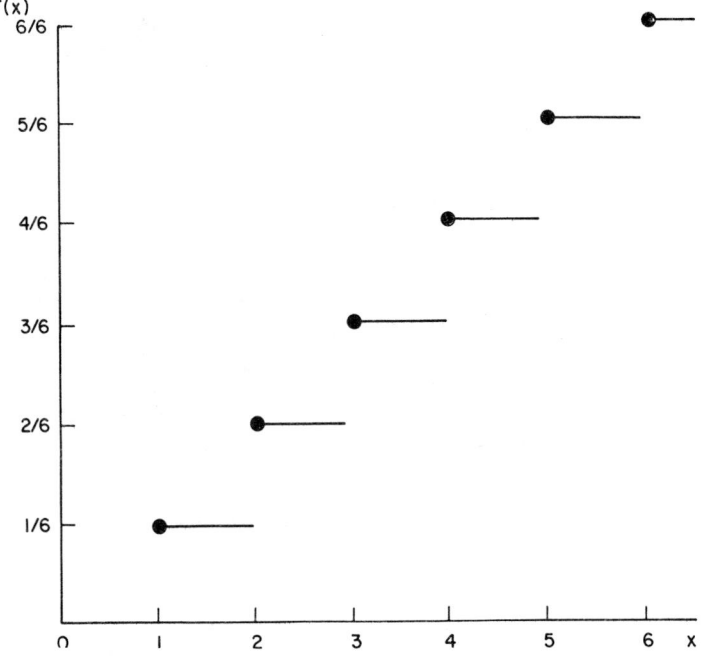

FIG. 3-2 Distribution function for die.

the frequency function is often referred to as a *probability density
function* in the continuous case. It shows density, not probability. This
is true for all continuous distributions. The distribution function cannot
then be obtained by summing the values of the frequency function in the
same sense as with discrete data but requires use of the calculus. Thus,

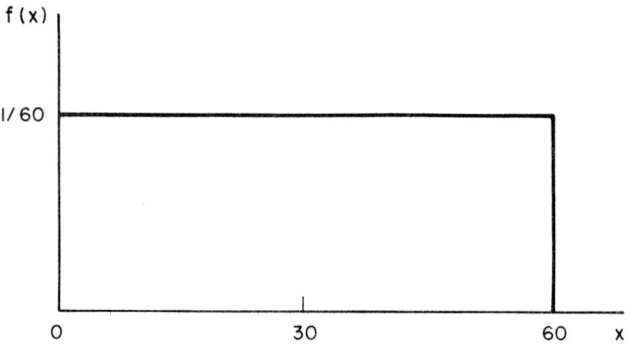

FIG. 3-3 Frequency function for watch stoppage.

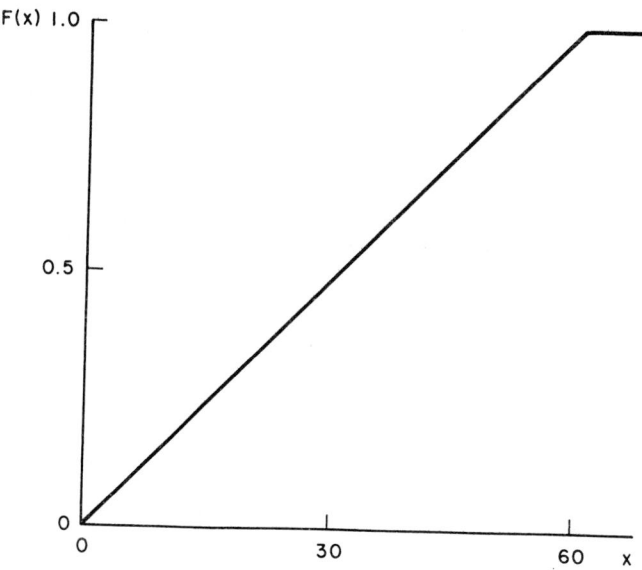

FIG. 3-4 Distribution function for watch stoppage.

$$F(x) = \int_0^x f(t)\ dt$$

$$F(x) = \int_0^x \frac{1}{60}\ dt$$

$$F(x) = \frac{x}{60} \qquad 0 \le x < 60$$

A plot of the probability distribution appears in Fig. 3-3 and a graph of the distribution function is given in Fig. 3-4. Such graphs are useful in visualizing the shape, nature, and properties of distribution functions.

MEASURES OF DISTRIBUTION FUNCTIONS

There are several important measures of distribution functions which show location and spread of the distribution. The most important measure of location, or central tendency, is the first moment of the distribution, or its mean (center of gravity). It gives the arithmetic average of the values in the distribution and is its expected value. The mean of a distribution is calculated by the formula

$$\mu = \sum_{\text{all } x} x f(x) \qquad \text{discrete distribution}$$

or

$$\mu = \int_{-\infty}^{\infty} xf(x) \, dx \qquad \text{continuous distribution}$$

where the limits for the continuous distribution are taken at the extreme values of X.

For the discrete distribution of the results of a toss of the die, we have

$$\mu = 1\left(\frac{1}{6}\right) + 2\left(\frac{1}{6}\right) + 3\left(\frac{1}{6}\right) + 4\left(\frac{1}{6}\right) + 5\left(\frac{1}{6}\right) + 6\left(\frac{1}{6}\right) = 3.5$$

while for the distribution of the second hand

$$\mu = \int_{0}^{60} x \, \frac{1}{60} \, dx = \frac{x^2}{2(60)} \bigg|_{0}^{60} = \frac{3600}{120} = 30$$

Note, that for a finite population of size N,

$$\mu = \frac{\Sigma \, x}{N}$$

Other measures of location of a distribution are the median (middle value) and the mode (most frequently occurring value).

The standard deviation stands as the primary measure of the spread of a distribution. It is the square root of the second central moment about the mean (moment of inertia). For discrete data it is calculated as

$$\sigma = \sqrt{\Sigma (x - \mu)^2 f(x)}$$

For continuous data, the variance, or square of the standard deviation, is calculated as

$$\sigma^2 = \int_{-\infty}^{\infty} (x - \mu)^2 f(x) \, dx$$

so

$$\sigma = \sqrt{\sigma^2}$$

The standard deviation is the root-mean-square average deviation of an observation from the mean. In this sense it can be considered to measure the average distance of an observation from the mean.

For the results of the die we have

$$\sigma = \sqrt{\begin{array}{l} (1 - 3.5)^2\left(\frac{1}{6}\right) + (2 - 3.5)^2\left(\frac{1}{6}\right) + (3 - 3.5)^2\left(\frac{1}{6}\right) \\[2mm] \qquad + (4 - 3.5)^2\left(\frac{1}{6}\right) + (5 - 3.5)^2\left(\frac{1}{6}\right) + (6 - 3.5)^2\left(\frac{1}{6}\right) \end{array}}$$

$$= \sqrt{\frac{6.25 + 2.25 + .25 + .25 + 2.25 + 6.25}{6}}$$

$$= \sqrt{\frac{17.5}{6}} = 1.71$$

Note that for a finite population of size N,

$$\sigma = \sqrt{\frac{\Sigma(x - \mu)^2}{N}}$$

The standard deviation of the continuous distribution of stopping times of the second hand is

$$\sigma^2 = \int_0^{60} (x - 30)^2 \, \frac{1}{60} \ dx$$

$$= \int_0^{60} (x^2 - 60x + 30^2) \, \frac{1}{60} \ dx$$

$$= \frac{x^3}{180} - \frac{60x^2}{120} + \frac{30^2 x}{60} \ \Big|_0^{60}$$

$$= \frac{60^3}{180} - \frac{60^3}{120} + \frac{900}{60} \ 60$$

$$= 1200 - 1800 + 900 = 300$$

so

$$\sigma = \sqrt{300} = 17.3$$

The other principal measure of spread used in acceptance sampling is the *range* (the difference between the highest and lowest observed values). This is not usually applied to populations, but rather to measure the spread in sample data. The primary measures of *sample* location and spread are the sample mean $\overline{X}$ and the sample standard deviation s. The appropriate formulas for a sample of size n are

$$\overline{X} = \frac{\Sigma\ x}{n}$$

and

$$s = \sqrt{\frac{\Sigma(x - \overline{X})^2}{n - 1}}$$

The n - 1 denominator in the standard deviation formula can be shown to be necessary to make the expected (mean) value of s^2 equal to σ^2 when the sample variances are averaged over all possible samples of size n from the population. Often sample estimates are denoted with a carat over the symbol for the parameter. Thus, for example, we have $\hat{\sigma}$ = s with measurements data and use the symbol $\hat{p}$ to represent an estimate of p from attributes data.

HYPERGEOMETRIC DISTRIBUTION

The hypergeometric distribution is fundamental to much of acceptance sampling. It is applicable when sampling an attributes characteristic from a finite lot without replacement. Here

N = lot size, N > 0

p = proportion defective in the lot, p = 0, 1/N, 2/N, ..., 1

q = proportion effective in the lot, q = 1 - p

n = sample size, n = 1, 2, ..., N

x = number of occurrences, x = 0, 1, 2, ..., n

Its frequency function is

$$f(x) = \frac{C_x^{Np} \, C_{n-x}^{Nq}}{C_n^N}$$

where, because of discreteness in the lot, the proportion defective is restricted to one of the values p = 0, 1/N, 2/N, 3/N, ..., 1. A recursion formula to obtain successive values of the hypergeometric distribution is

$$f(x + 1) = \frac{(n - x)(Np - x)}{(x + 1)(Nq + x - n + 1)} \, f(x)$$

The hypergeometric was, in fact, the distribution used in Chap. 2 to obtain the probability of a 4-card hand of aces; there

$$N = 52 \qquad p = \frac{4}{52} \qquad q = \frac{48}{52} \qquad n = 4$$

Now

$$Np = 52 \, \frac{4}{52} = 4$$

and

$$Nq = 52 \, \frac{48}{52} = 48$$

so that Np is just the number of defective units in the lot and Nq is the
number of effective units in the lot. The formula gives

$$f(4) = \frac{C_4^4 \, C_0^{48}}{C_4^{52}}$$

$$f(4) = \frac{\dfrac{4!}{4! \; 0!} \dfrac{48!}{0! \; 48!}}{\dfrac{52!}{4! \; 48!}}$$

$$= \frac{4! \; 48!}{52!}$$

$$= \frac{4 \cdot 3 \cdot 2 \cdot 1}{52 \cdot 51 \cdot 50 \cdot 49}$$

$$= \frac{1}{270725}$$

and we see the usefulness of a ready-made probability function in solving
a problem.

Again, in the acceptance sampling problem of Chap. 2, the plan $n = 2$,
$c = 0$ was applied to a lot of size 3 containing one defective. Since sampling was without replacement the hypergeometric distribution is applicable.
Here

$$N = 3 \qquad Np = 1 \qquad Nq = 2 \qquad n = 2$$

and

$$f(0) = \frac{C_0^1 \, C_2^2}{C_2^3}$$

$$= \frac{1!}{0! \; 1!} \cdot \frac{2!}{2! \; 0!} \Big/ \frac{3!}{2! \; 1!}$$

$$= \frac{2!}{3!} = \frac{2 \cdot 1}{3 \cdot 2 \cdot 1} = \frac{1}{3}$$

as before.

In this simple problem, it may be well to completely specify the
distribution for 0 or 1 defectives in a sample of 2. Using the recursion
formula we can list out the distribution as

x	p(x)
0	1/3
1	2/3

since $f(1)$ may be obtained from $f(0)$ by

$$f(1) = f(0 + 1) = \frac{(2 - 0)(1 - 0)}{(0 + 1)(2 + 0 - 2 + 1)} \frac{1}{3}$$

$$= 2\left(\frac{1}{3}\right) = \frac{2}{3}$$

For the hypergeometric distribution, the mean is

$$\mu = np = 2\left(\frac{1}{3}\right) = \frac{2}{3}$$

Thus, we would expect to get an average of two defective units in every three draws.

The standard deviation is

$$\sigma = \sqrt{npq}\sqrt{\frac{N - n}{N - 1}}$$

$$= \sqrt{3\left(\frac{1}{3}\right)\left(\frac{2}{3}\right)}\sqrt{\frac{3 - 2}{3 - 1}}$$

$$= \sqrt{\frac{1}{3}} = .577$$

which represents the average distance of an observation from the mean using the root-mean-square average. Since the mean is 2/3, we see that one-third of the observations (which are zeros) deviate from the mean by 2/3 and two-thirds of the observations, the ones, deviate from the mean by 1/3. Taking the arithmetic average we obtain the mean deviation

$$MD = \frac{(1/3)(2/3) + (2/3)(1/3)}{1/3 + 2/3} = \frac{4}{9} = .444$$

Also if we had used the mode as an average, the modal deviation is

$$M_o D = \frac{1}{3} = .333$$

The median deviation is also .333. And so the standard deviation can be seen to be just one method of computing the average distance of an observation from the mean.

BINOMIAL DISTRIBUTION

Undoubtedly the most used distribution in acceptance sampling is the binomial. It complements the hypergeometric in the sense that it is employed when sampling an attributes characteristic from an infinite lot (or process) or from a finite lot when sampling with replacement. Here n = sample size, n > 0

p = proportion defective, $0 \le p \le 1$

q = proportion effective, q = 1 - p

x = number of occurrences, x = 0, 1, 2, ..., n

Its frequency function is

$$f(x) = C_x^n p^x (1 - p)^{n-x}$$

$$= C_x^n p^x q^{n-x}$$

The mean of the binomial distribution is

$$\mu = np$$

and its standard deviation is

$$\sigma = \sqrt{npq}$$

Values of the frequency function can be calculated recursively using the formula

$$f(x + 1) = \frac{n - x}{x + 1} \frac{p}{q} f(x)$$

As an illustration of the binomial distribution, consider the sampling plan n = 5, c = 0 presented in Chap. 2. In setting up such a plan it may be desirable to obtain the distribution of the number of defectives in a sample of 5 when p is at the producer's process average proportion defective, say p = .01. We have

$$f(0) = C_0^5 (.01)^0 (1 - .01)^5$$

$$= \frac{5!}{0! \ 5!}(1)(.99)^5 = .951$$

Similarly,

$$f(1) = C_1^5 (.01)^1 (1 - .01)^{5-1}$$

$$= \frac{5!}{1! \ 4!}(.01)(.99)^4$$

$$= 5(.01)(.961) = .048$$

and using the recursion formula for f(2)

$$f(2) = f(1 + 1) = \frac{5 - 1}{1 + 1} \frac{.01}{.99} (.048) = .001$$

so that the distribution of the number defective is

x	f(x)
0	.951
1	.048
2	.001
3	.000
4	.000
5	.000

It is apparent that it is highly unlikely to obtain 3, 4, or 5 defectives in application of this plan. Such a result would be a clear indication that the process proportion defective is higher than .01.

For this distribution, the mean is

$$\mu = 5(.01) = .05$$

and the standard deviation is

$$\sigma = \sqrt{5(.01)(.99)} = \sqrt{.0495} = .22$$

It should be noted that in using tables for this distribution, it is possible to use $1 - p$ as an argument instead of p and vice versa. This is done using the relationship

$$B(x|n,p) = 1 - B(n - x - 1|n, 1 - p)$$

where $B(x|n,p)$ is read as the binomial distribution function evaluated at x given parameters n and p. Thus, for example, $B(1|5,.01) = .999$ which may be obtained as

$$B(1|5,.01) = 1 - B(3|5,.99)$$

$$= 1 - \sum_{x=0}^{3} C_x^5 (.99)^x (.01)^{n-x}$$

$$= 1 - \left[C_0^5 (.99)^0 (.01)^5 + C_1^5 (.99)^1 (.01)^4 \right.$$

$$\left. + C_2^5 (.99)^2 (.01)^3 + C_3^5 (.99)^3 (.01)^2 \right]$$

$$= .999$$

The Larson (1966) nomograph for the binomial distribution, given in Fig. 3-5, is extremely useful in acceptance sampling applications. The probability of c or fewer successes in a sample of n for a specific proportion defective p is characterized by a single line on the chart. The point representing p is set on the left scale, the pair of values n and c determine

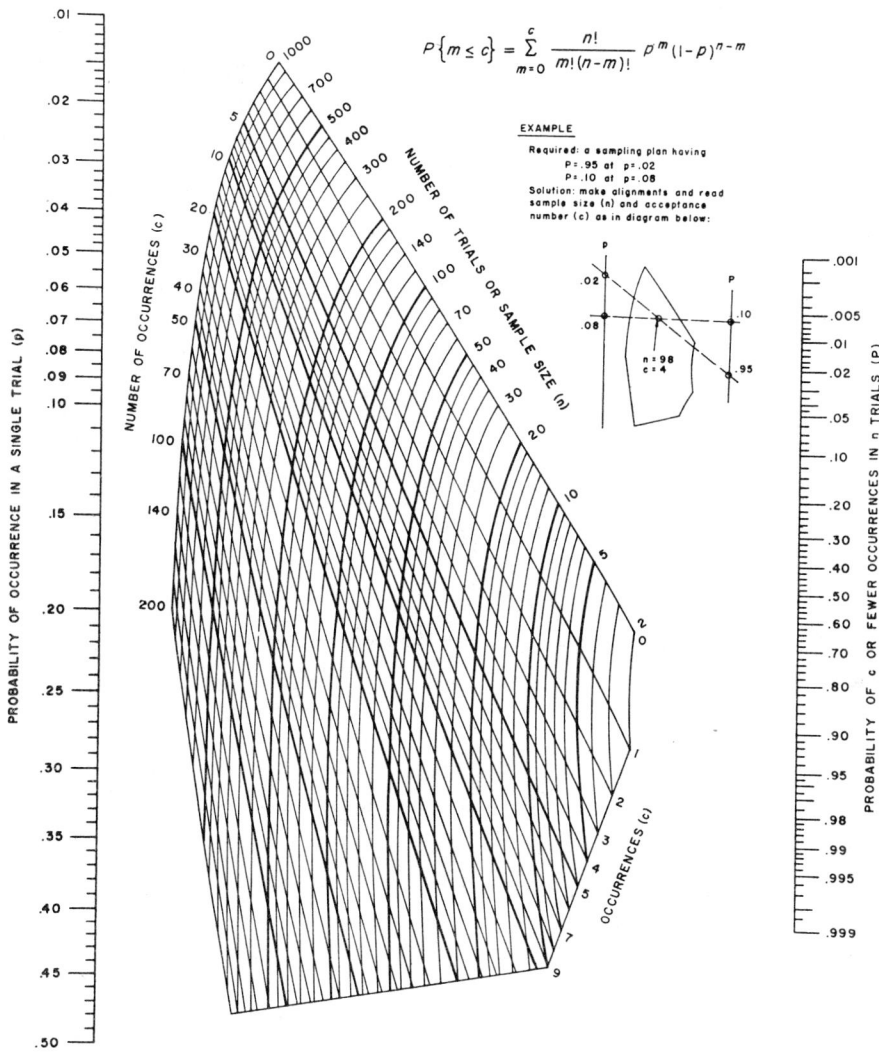

FIG. 3-5 Larson binomial nomograph. (From Larson, 1966, p. 273.
Reprinted by permission.)

a point in the grid, and the cumulative probability P(x ≤ c) is read from
the right scale. Thus, when p = .25, the plan n = 20, c = 6 has P(x ≤ 6)
= .79, which is, of course, the probability of acceptance. A straight
line connecting any two of the points representing p, (n,c) or P(X ≤ c)
will give the third. Thus, the Larson nomograph is a very versatile tool
for use in evaluating acceptance sampling plans.

POISSON DISTRIBUTION

The Poisson distribution is used in calculating the characteristics of sampling plans which specify a given number of defects per unit such as the number of defective rivets in an aircraft wing or the number of stones allowed in a piece of glass of a given size. The parameter in the Poisson distribution is simply μ. Here

μ = mean number of defectives, $\mu > 0$

x = number of occurrences, x = 0, 1, 2, ...

and its frequency function

$$f(x) = \frac{\mu^x e^{-\mu}}{x!}$$

where e = 2.71828$\cdots$. Values of e^{-x} will be found in Appendix Table T3-1. The mean and standard deviation are simply

$$\mu = \mu \qquad \sigma = \sqrt{\mu}$$

Successive values of the Poisson distribution can be calculated using the recursion formula

$$f(x + 1) = \frac{\mu}{x + 1} f(x)$$

Suppose an importer of glassware wishes to insure that the process average of his supplier is no more than a specified 2 bubbles per piece. The number of bubbles would be expected to vary from piece to piece. The Poisson distribution can be used to determine how the number of bubbles per piece would vary if the producer maintained the agreed upon average. Evaluating the Poisson distribution in this case we obtain

$$f(0) = \frac{2^0 e^{-2}}{0!} = e^{-2} = .1353$$

$$f(1) = \frac{2^1 e^{-2}}{1!} = 2e^{-2} = .2707$$

$$f(2) = \frac{2^2 e^{-2}}{2!} = .2707$$

and so on. Using the recursion relationship, subsequent values can be obtained. For example,

$$f(3) = f(2 + 1) = \frac{2}{3} (.2707) = .1805$$

$$f(4) = f(3 + 1) = \frac{2}{4} (.1805) = .0902$$

$$f(5) = f(4 + 1) = \frac{2}{5} (.0902) = .0361$$

$$f(6) = f(5 + 1) = \frac{2}{6} (.0361) = .0120$$

$$P(X > 6) = 1 - .1353 - .2707 - .2707 - .1805 - .0902 - .0361 - .0120$$
$$= .0045$$

Note that there is no upper limit on the number of bubbles that could be obtained, so that the probability distribution is

x	f(x)
0	.1353
1	.2707
2	.2707
3	.1805
4	.0902
5	.0361
6	.0120
>6	.0045

We see that pieces with more than 6 bubbles would be very rare, occurring less than half a percent of the time. On the average we would expect 2 bubbles per piece with a standard deviation of

$$\sigma = \sqrt{2} = 1.41$$

A very useful tool in determining Poisson probabilities is the Thorndyke chart shown in Fig. 3-6. This chart shows probability of x or less on the vertical axis and gives values of μ on the horizontal axis. To use the chart a vertical line is drawn at μ for the Poisson distribution to be evaluated. Its intersection with the curves for x = 0, 1, 2, etc., determines the cumulative probability of x or less defects when read on the probability axis horizontally from the intersection. This chart was developed by Thorndyke (1926) and was subsequently modified by Dodge and Romig (1941).

NEGATIVE BINOMIAL DISTRIBUTION

It is sometimes necessary to determine the number of random trials required to obtain a given number of defectives. The negative binomial is the probability distribution which is used to obtain the probability of given numbers

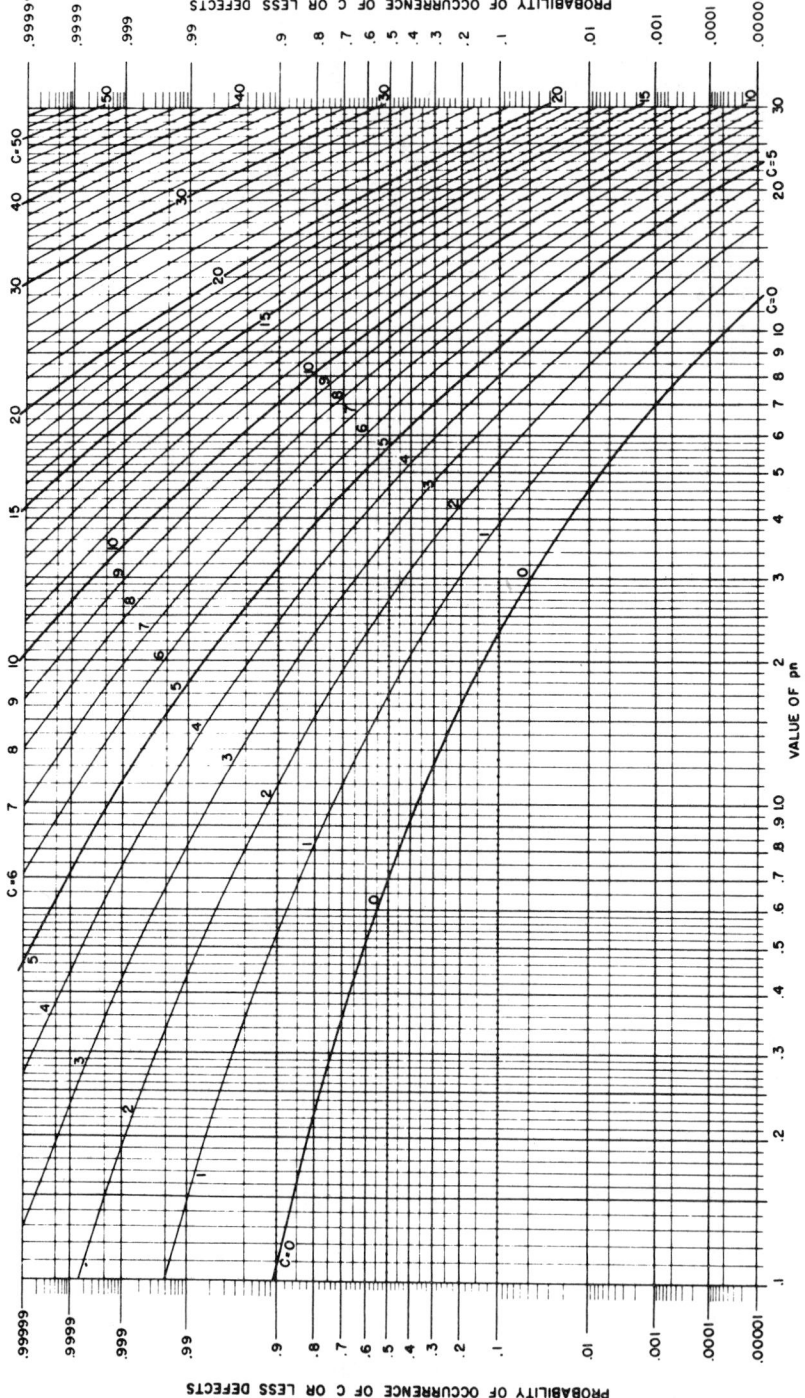

FIG. 3-6 Thorndyke chart. (From H. F. Dodge and H. G. Romig, *Sampling Inspection Tables*, 2nd ed., 1959, p. 35. Copyright©1944 Bell Telephone Laboratories. Reprinted by permission of John Wiley and Sons, Inc.)

of trials up to and including the xth defective. The parameters of the
negative binomial distribution are similar to those of the binomial itself.
Here

> n = number of trials to and including the xth defective, $n \geq x$
> p = proportion defective, $0 \leq p \leq 1$
> q = proportion effective, $q = 1 - p$
> x = number of occurrences, $x = 1, 2, \ldots$

Its frequency function may be represented as $b^{-1}(n|x,p)$ so that

$$f(n) = b^{-1}(n|x,p) = C_{x-1}^{n-1} p^x q^{n-x}$$

with a mean of

$$\mu = \frac{xq}{p}$$

and a standard deviation of

$$\sigma = \frac{\sqrt{xq}}{p}$$

Successive values of the negative binomial distribution may be calculated using the recursion relation

$$f(n + 1) = q \frac{n}{n - x + 1} f(n)$$

The negative binomial gives the number of trials to a fixed number of
successes, rather than the number of successes in a fixed number of trials
as does the binomial. The term *negative binomial* relates to the fact that
the successive values of the frequency function can be determined from an
expansion of

$$\left(\frac{1}{p} - \frac{q}{p}\right)^{-x}$$

which, of course, is of the binomial form.

Sometimes called the *Pascal* or *Pólya distribution*, the cumulative
negative binomial is related to the cumulative binomial by the relation

$$\sum_{i=x}^{n} C_{x-1}^{i-1} p^x q^{i-x} = 1 - \sum_{i=0}^{x-1} C_i^n p^i q^{n-i}$$

which shows that the negative binomial distribution function for up to n
trials to obtain x successes is equal to the complement of the binomial

distribution function for x - 1 successes in n trials. Using $B^{-1}(n|x,p)$
for the negative binomial distribution function and $B(x|n,p)$ for the bino-
mial distribution function, we have

$$B^{-1}(n|x,p) = 1 - B(x - 1|n,p)$$

and individual terms are simply

$$b^{-1}(n|x,p) = \frac{x}{n} b(x|n,p)$$

Consider the sampling plan which was used earlier to illustrate the
binomial distribution when p = .01. Namely, n = 5, c = 0. Suppose we
wish to calculate the probability of 1, 2, 3, 4, 5, or more trials before
finding a defective in random samples from a large lot. Using the negative binomial distribution, we have

$$f(1) = C_{1-1}^{1-1}(.01)^1(.99)^{1-1}$$

$$= C_0^0(.01)(1)$$

$$= 1(.01)(1) = .01$$

also

$$f(2) = C_{1-1}^{2-1}(.01)^1(.99)^{2-1}$$

$$= C_0^1(.01)(.99)$$

$$= 1(.01)(.99) = .0099$$

and tabulating the probability of 3 and 4 trials before finding a defective
using the recursion formula

$$f(3) = (.99)\frac{2}{2 - 1 + 1}(.0099) = .0098$$

$$f(4) = (.99)\frac{3}{3 - 1 + 1}(.0098) = .0097$$

and finally

$$f(5) = C_{1-1}^{5-1}(.01)^1(.99)^4$$

$$= 1(.01)^1(.99)^4 = .0096$$

so that

$$P(>5) = 1 - .01 - .0099 - .0098 - .0097 - .0096 = .951$$

and the distribution is

x	f(x)
1	.01
2	.0099
3	.0098
4	.0097
5	.0096
>5	.951

The expected number of trials to a defective is

$$\mu = \frac{1(.99)}{.01} = 99$$

and the standard deviation of the number of trials on which the first defective occurs is

$$\sigma = \frac{\sqrt{1(.99)}}{.01} = 99.5$$

which indicates a large spread in the number of trials to a defective.

As a check on the probability distribution, we may observe

$$B^{-1}(n|x,p) = 1 - B(x - 1|n,p)$$
$$B^{-1}(5|1,.01) = 1 - B(0|5,.01)$$
$$.049 = 1 - .951$$
$$= .049$$

where the binomial probability was calculated previously in the discussion of binomial probabilities.

EXPONENTIAL AND CONTINUOUS DISTRIBUTIONS

Continuous distributions are extremely useful in acceptance sampling, although somewhat more complicated and restrictive in their use. Sampling plans based on attributes data are typically nonparametric by nature; that is, it is not necessary to know the shape or parameters of the distribution of any measurements involved to use an attributes plan. This is not generally true for the variables sampling plans which are based on measurements (variables) data. These distributions are usually continuous and require specification of shape and parameters, such as measures of location and spread. We will consider two such distributions, the exponential and the normal. We shall also consider the Weibull family of distributions which includes the exponential as a special case and which can be used to approximate the normal.

The exponential distribution is used extensively in evaluating acceptance plans for reliability and life testing. It is distinguished by a constant failure rate. That is, the probability of future failure is constant regardless of how long a unit has been in operation. The parameter of this function is simply μ. Here

μ = mean of the distribution, $\mu > 0$

x = measurement distributed, $x \geq 0$

and its frequency (density) function is

$$f(x) = \frac{1}{\mu} e^{-x/\mu}$$

The density is not as useful as the frequency function for discrete distributions. As a matter of fact, evaluation of this function will not lead to the probability associated with any given point in the continuum, since the probability of a point is zero. Consequently, the density function must be integrated over a range of possible values of the argument to obtain a probability. This may be expressed in terms of the distribution function. The distribution function for the exponential distribution is

$$F(x) = \int_0^x \frac{1}{\mu} e^{-t/\mu} \, dt = \frac{-\mu}{\mu} e^{-t/\mu} \Big|_0^x = -e^{-x/\mu} + 1$$

and

$$F(x) = 1 - e^{-x/\mu}$$

The mean of the exponential distribution is, of course, $\mu = \mu$, while, simply enough, its standard deviation is $\sigma = \mu$ also. Many problems involving the exponential distribution are couched in terms of its (constant) failure rate λ which is simply

$$\lambda = \frac{1}{\mu}$$

Consider a requirement that the mean life of a power transistor must be greater than 5000 hrs. That is, its failure rate (per hr.) must be less than

$$\lambda = \frac{1}{5000} = .0002$$

Suppose a random unit is tested for 500 hrs. What is the probability that

it will fail in that period if it comes from a process with mean life of
exactly 5000 hrs?

$$F(500) = 1 - e^{-500/5000}$$

$$= 1 - e^{-.1}$$

$$= 1 - .905 = .095$$

where the value of $e^{-.1}$ was obtained from Appendix Table T3-1. The mean
and standard deviation of this distribution are both 5000 hrs.

Values of the distribution function of the exponential distribution
may be found by using the Thorndyke chart given in Fig. 3-6 for the Poisson
distribution. Enter with a value of μ on the horizontal axis equal to the
exponent in the exponential distribution and read the cumulative probability
associated with $x = 0$. This is $e^{-\mu}$, which when subtracted from one gives
the exponential distribution function. A check of that chart will show that
in entering the x axis with a value of 0.1 and reading the y value for the
curve $c = 0$, a value of roughly .905 is obtained, subtracting from 1 gives
.095. This agrees with the previous calculation.

WEIBULL DISTRIBUTION

The Weibull distribution may be thought of as a generalization of the exponential distribution incorporating parameters for location, spread, and
shape. The distribution is defined for positive values of x, starting at
zero. The location parameter γ adjusts the distribution to start at a
value γ, other than zero. The scale parameter η or characteristic life,
is the x value for which $F(x - \gamma) = .6321$ for any Weibull shape. The
shape parameter β gives the distribution flexibility in shape so that it
can be used to fit a variety of empirical and theoretical failure distributions. Here

γ = location (minimum life) parameter, $\gamma > 0$

η = scale parameter, $\eta > 0$

β = shape parameter, $\beta > 0$

x = measurement distributed, $x \geq \gamma$

The frequency (density) function is

$$f(x) = \frac{\beta}{\eta}\left(\frac{x - \gamma}{\eta}\right)^{\beta-1} e^{-[(x-\gamma)/\eta]^{\beta}}$$

with distribution function

$$F(x) = 1 - e^{-[(x-\gamma)/\eta]^{\beta}}$$

The mean of the distribution is

$$\mu = \gamma + \eta\Gamma\left(1 + \frac{1}{\beta}\right)$$

and its standard deviation

$$\sigma = \eta\sqrt{\Gamma\left(1 + \frac{2}{\beta}\right) - \Gamma^2\left(1 + \frac{1}{\beta}\right)}$$

where $\Gamma(x)$ is the gamma function such that

$$\Gamma(x) = \int_0^{\infty} e^{-t}t^{x-1} \, dt = (x - 1)!$$

the factorial function for $x - 1$, if x is a positive integer.

The exponential distribution is a special case of the Weibull distribution when $\beta = 1$. It will be seen that when this is the case and the location parameter $\gamma = 0$, $\mu = \eta$ thus the mean occurs at the point at which $F(x) = .6321$. Also $\sigma = \eta$.

The shape parameter β allows the distribution to take on a variety of shapes as shown in Fig. 3-7. Specifically, we have

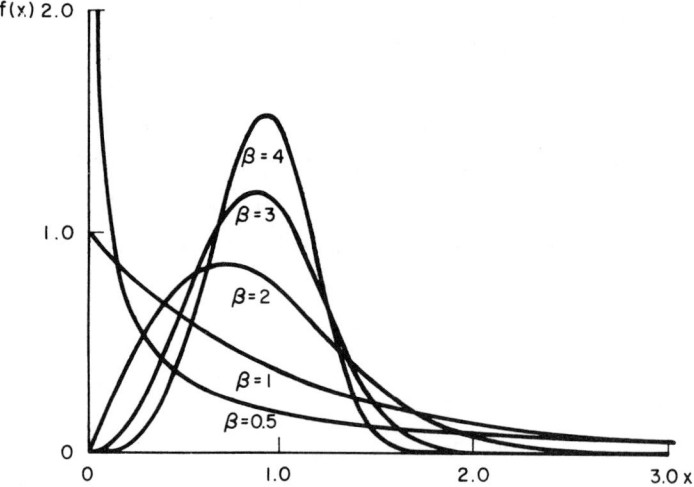

FIG. 3-7 The Weibull frequency (density) function for various values of shape parameters, β ($\gamma = 0$, $\eta = 1$).

Exponential distribution: $\beta = 1$

Rayleigh distribution: $\beta = 2$

Approximate normal distribution: $\beta \simeq 3.44$

The value $\beta = 3.44$ is given as the value of the shape parameter approximating a normal distribution in the sense that when $\beta = 3.44$ the median and the mean of the distribution are equal to each other. When $\gamma = 0$ and $\eta = 1$ this distribution has a mean $\mu = .899$ and standard deviation $\sigma = .289$. Thus, normal data x' with mean μ' and standard deviation σ' should plot approximately as a straight line on Weibull probability paper with $\beta \simeq 3.44$ when transformed using

$$x = .289 \frac{x' - \mu'}{\sigma'} + .889$$

with inverse

$$x' = \sigma'\left(\frac{x - .899}{.289}\right) + \mu'$$

The failure rate of the exponential distribution is constant. The failure rate for the Weibull distribution is decreasing for values of the shape parameter $\beta < 1$ and increasing for $\beta > 1$. Of course, when $\beta = 1$, the failure rate is constant. Since the failure rate changes over the possible values of x (life), it is quoted in terms of the instantaneous failure rate at any chosen value of x. This is called the *hazard rate* h(x), where

$$h(x) = \frac{\beta}{\eta}\left(\frac{x - \gamma}{\eta}\right)^{\beta-1}$$

Reliability specifications for use in acceptance sampling are sometimes written in terms of the hazard rate.

Suppose the example given for the exponential distribution is regarded as a special case of the Weibull distribution. The specified mean in that case was 5000 hrs. and we have

$$\gamma = 0 \qquad \eta = 5000 \qquad \beta = 1$$

so that the probability of a failure before 500 hrs. is

$$F(500) = 1 - e^{-[(500-0)/5000]^1} = .095$$

as before.

NORMAL DISTRIBUTION

No area of statistics seems to have escaped the impact of the normal distribution. This is certainly true of acceptance sampling where it forms
the basis of a large number of "variables" acceptance sampling plans. It
has pervaded other areas of acceptance sampling as well.

The normal distribution is completely specified by two parameters μ
and σ. Here

μ = the mean, $-\infty < \mu < \infty$

σ = the standard deviation, $\sigma > 0$

x = measurement distributed, $-\infty < x < \infty$

Its frequency function is

$$f(x) = \frac{1}{\sigma \sqrt{2\pi}} \, e^{-\frac{1}{2}[(x-\mu)/\sigma]^2}$$

Unlike the exponential and the Weibull distributions, no closed form formula can be obtained for the distribution function. Expressed as an integral,
it is

$$F(x) = \frac{1}{\sigma \sqrt{2\pi}} \int_{-\infty}^{x} e^{-\frac{1}{2}[(t-\mu)/\sigma]^2} \, dt$$

and is shown cumulated over the standard normal frequency function in Fig.
3-8.

Fortunately, values of the distribution function may be obtained from
tables of the standard normal distribution such as is given in Appendix
Table T3-2. The table is for the specific standard normal distribution
with $\mu = 0$, $\sigma = 1$. It is tabulated in terms of standard normal deviates,

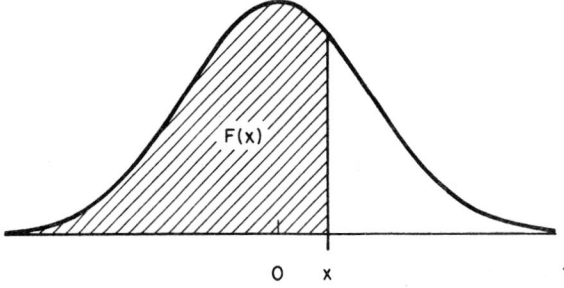

FIG. 3-8 Standard normal frequency (density)
function ($\mu = 0$, $\sigma = 1$).

z, which are simply the x values for the standard normal distribution. To
use the table to obtain probabilities at specific values of x for other
normal distributions (that is, with different means and standard deviations),
it is necessary to transform the x values into the z values given by the
table by using the formula

$$z = \frac{x - \mu}{\sigma}$$

Similarly, if an x value is desired which has a given probability, the
probability may be found in the body of the table in terms of z and the x
value obtained using the transformation

$$x = \mu + z\sigma$$

For example, suppose bolts are manufactured by a process having a mean
of 50 mm and a standard deviation of .1 mm. The distribution of bolt
lengths conforms to the normal distribution, that is

$$\mu = 50 \text{ mm} \qquad \sigma = 0.1 \text{ mm}$$

If it is desired to determine what proportion of the bolts have lengths
less than 49.8 mm (which is, of course, the probability of obtaining such
a bolt in a random sample), we have

$$z = \frac{49.8 - 50}{0.1} = -2$$

and using Appendix Table T3-2

$$P(Z \leq -2) = .0228$$

so

$$P(X \leq 49.8) = .0228$$

Similarly, to determine what length is exceeded by 10 percent of the bolts,
we have

$$P(Z \leq 1.282) = .90$$

and this would give 10 percent above the value z = 1.282. Accordingly

$$x = 50 + 1.282(.1) = 50.1282$$

The probability of obtaining a result between any two specified values
may be found by subtracting cumulative probabilities. For example, 80 percent of the bolts (symmetric about the mean) lie between

$$z = 1.282 \qquad \text{(90 percent below)}$$

and

z = -1.282   (10 percent below)

or between 49.8718 and 50.1282 mm. We also find that the proportion of
bolts between 49.9 mm (z = -1) and 50.3 mm (z = +3) is

z	Cumulative probability
3	.9987
-1	.1587
	.84

or 84 percent of the bolts.

Certainly, a large share of the importance of the normal distribution
in statistics lies in the central limit theorem which can be stated as
follows:

Central Limit Theorem. Let f(x) be any frequency (density) function of a
population with finite mean μ and standard deviation σ. Let $\overline{X}$ be the mean
of a random sample of n from the population. Then the frequency function
of $\overline{X}$ approaches the normal distribution with mean μ and standard deviation
$\sigma/\sqrt{n}$ as n increases without bound.

The theorem is proved in most basic mathematical statistics texts, such as
Mood and Graybill (1963).

It is important to realize that the population distribution is unspec-
ified — the theorem holds for any underlying population having a finite
mean and standard deviation. Thus, for any population we can say that the
distribution of sample means will be approximately normal with

$$\mu_{\overline{X}} = \mu$$

$$\sigma_{\overline{X}} = \frac{\sigma}{\sqrt{n}}$$

as the sample size n becomes large. How closely the distribution of sample
means is said to approach normality depends, of course, on the shape of the
underlying distribution and the magnitude of n. Shewhart (1931) has demonstrated empirically and Schilling and Nelson (1976) have shown mathemati-
cally that in many applications sample sizes of 5 are adequate, 9 are good,
and 25 excellent, in assuring a normal distribution of sample means from a
variety of fairly well behaved underlying distributions. Naturally, when
the underlying distribution is normal, the normality of the distribution

of sample means is assured.

Suppose samples of size n = 25 are taken from the population of bolts
(μ = 50 mm, σ = 0.1) mentioned previously.   Then

$$\mu_{\overline{X}} = 50$$

$$\sigma_{\overline{X}} = \frac{0.1}{\sqrt{25}} = 0.02$$

and, for instance, we can state that 95 percent of the possible sample
averages from this population will be below

$$x = 50 + 1.65(0.02) = 50.03 \text{ mm}$$

SUMMARY OF DISTRIBUTIONS

A summary of the probability distributions presented in this chapter is
given in Table 3-1. For quick reference, the table shows the frequency
(density) function, the distribution function, the mean, standard deviation,
restrictions on the parameters, domain, and use of each distribution.

TABLES OF DISTRIBUTIONS

Many useful tables have been generated for the evaluation of the probability distributions shown here. A convenient notation for the range of the
argument of the tables is x(y)z which indicates the values move from x in
increments of y up to z. A few of the tables applicable in acceptance
sampling are the following.

Hypergeometric Tables

Lieberman and Owen (1961) give tables of the hypergeometric frequency and
distribution functions tabulated in the following notation:

N = lot size

n = sample size

k = number defectives in lot (Np here)

x = argument

P(x) = value of distribution function [F(x) here]

p(x) = value of frequency function [f(x) here]

The tables are complete for N $\leq$ 50 and require interpolation thereafter up
to a maximum lot size of N = 2000. Sufficient values are tabulated through
N = 50 to use the relation

TABLE 3-1   Distributions Useful in Acceptance Sampling

DISTRIBUTION	FREQUENCY FUNCTION	DISTRIBUTION FUNCTION	MEAN	STANDARD DEVIATION	RESTRICTIONS	USE
HYPERGEOMETRIC	$f(x) = \dfrac{C_x^{Np}\, C_{n-x}^{Nq}}{C_n^N}$	$F(x) = \sum_{i=0}^{x} f(i)$	np	$\sqrt{npq}\,\sqrt{\dfrac{N-n}{N-1}}$	$N > 0$ $n = 1,2,\dots,N$ $p = 0,1/N,2/N,\dots,1$ $q = 1-p$ $x = 0,1,2,\dots,n$	Sampling from finite lot without replacement.
BINOMIAL	$f(x) = C_x^n\, p^x\, q^{n-x}$	$F(x) = \sum_{i=0}^{x} f(i)$	np	$\sqrt{npq}$	$n > 0$ $0 \le p \le 1$ $q = 1-p$ $x = 0,1,2,\dots,n$	Sampling from infinite lot or process. Sampling from finite lot with replacement.
POISSON	$f(x) = \dfrac{\mu^x\, e^{-\mu}}{x!}$	$F(x) = \sum_{i=0}^{x} f(i)$	μ	$\sqrt{\mu}$	$\mu > 0$ $x = 0,1,2,\dots$	Sampling defects from area of infinite opportunity for occurrence, with mean occurrence μ.
NEGATIVE BINOMIAL	$f(n) = C_{x-1}^{n-1}\, p^x\, q^{n-x}$	$F(n) = \sum_{i=1}^{n} f(i)$	$\dfrac{xq}{p}$	$\dfrac{\sqrt{xq}}{p}$	$n \ge x$ $0 \le p \le 1$ $q = 1-p$ $x = 0,1,2,\dots$	For number units, n, sampled up to and including the xth success.
EXPONENTIAL	$f(x) = \dfrac{1}{\mu}\, e^{-\frac{x}{\mu}}$	$F(x) = 1 - e^{-\frac{x}{\mu}}$	μ	μ	$\mu > 0$ $x \ge 0$	Life distribution for units with constant failure rate $\lambda = \dfrac{1}{\mu}$.
WEIBULL	$f(x) = \dfrac{\beta}{\eta}\left(\dfrac{x-\gamma}{\eta}\right)^{\beta-1} e^{-\left(\frac{x-\gamma}{\eta}\right)^{\beta}}$	$F(x) = 1 - e^{-\left(\frac{x-\gamma}{\eta}\right)^{\beta}}$	$\gamma + \eta\Gamma\left(1+\dfrac{1}{\beta}\right)$	$\eta\sqrt{\Gamma\left(1+\dfrac{2}{\beta}\right) - \Gamma^2\left(1+\dfrac{1}{\beta}\right)}$	$\gamma > 0$ $\eta > 0$ $\beta > 0$ $x \ge \gamma$	Family of life distributions with decreasing ($\beta < 1$), constant ($\beta = 1$), or increasing ($\beta > 1$) hazard rate $h(x) = \dfrac{\beta}{\eta}\left(\dfrac{x-\gamma}{\eta}\right)^{\beta-1}$
NORMAL	$f(x) = \dfrac{1}{\sigma\sqrt{2\pi}}\, e^{-\frac{1}{2}\left(\frac{x-\mu}{\sigma}\right)^2}$	$F(x) = \int_{-\infty}^{x} f(t)\, dt$	μ	σ	$-\infty < \mu < \infty$ $\sigma > 0$ $-\infty < x < \infty$	Common underlying measurement distribution. Distribution of sample means. Useful as approximation.

$$F(N,n,k,x) = F(N,k,n,x)$$

that is

$$\sum_{i=0}^{x} \frac{C_i^k C_{n-i}^{N-k}}{C_n^N} = \sum_{i=0}^{x} \frac{C_i^n C_{k-i}^{N-n}}{C_k^N}$$

to allow reversing the roles of k and n in obtaining values from the tables. Other symmetries which may be utilized are

$$F(N,n,k,x) = F(N, N - n, N - k, N - n - k + x)$$
$$= 1 - F(N, n, N - k, n - x - 1)$$
$$= 1 - F(N, N - n, k, k - x - 1)$$

All these relationships apply to the frequency functions also.

Binomial Tables

Three tables which cover the binomial distribution are

1. U.S. Department of Commerce Applied Mathematics Series No. 6 (1950) gives values of the binomial frequency function as

$$f(r) = C_r^n p^r q^{n-r}$$

and the cumulative probabilities from r to n, that is

$$1 - F(r - 1) = \sum_{s=r}^{n} C_s^n p^s q^{n-s}$$

These are given for the following values:

$$p = .01(.01).50 \quad q = 1 - p$$
$$n = 2(1)49 \quad r = 0(1)(n - 1)$$

2. Romig (1947) gives values of f(x) and F(x) for $p = .01(.01).50$, $q = 1 - p$, $n = 50(5)100$, x up to $F(x) = .99999$.

3. Harvard Computation Laboratory (1955) also presents tables of the cumulative binomial distribution from r to n, that is,

$$1 - F(r - 1) = E(r,n,p) = \sum_{x=r}^{n} C_x^n p^x (1 - p)^{n-x}$$

for ranges of p between .01 and .50 with $r = 0(1)n$ and $n = 1(1)50(2)100(10)200(20)500(50)1000$.

Procedures and examples useful in applying binomial tables have been given by Nelson (1974).

Poisson Tables

Molina (1942) has tabulated the Poisson distribution in terms of its frequency function

$$f(x) = \frac{a^x e^{-a}}{x!}$$

and cumulative distribution from c to ∞,

$$1 - F(c - 1) = \sum_{x=c}^{\infty} \frac{a^x e^{-a}}{x!}$$

for

$$\mu = a = .001(.001).01(.01).30(.10)15(1)100$$

Another useful set of tables has been prepared by the Defense Systems Department, General Electric Co. (1962). They tabulate

$$f(x) = \frac{U^x e^{-U}}{x!}$$

and the cumulative distribution

$$F(x) = \sum_{r=0}^{x} \frac{U^r e^{-U}}{r!}$$

over an extensive range from U = .0000001 to U = 205.

Negative Binomial Tables

Williamson and Bretherton (1963) present values of the negative binomial in terms of

k = successes (x here)

n = nonsuccesses (n - x here)

p = probability of success

for a total of n* = n + k trials to reach the kth success where n* is the value of n as presented in the formula for the negative binomial distribution given here. They give the frequency function

$$f(n* - k) = P(n) = C_{k-1}^{n+k-1} p^i q^n$$

and distribution function

$$F(n^* - k) = F(n) = \sum_{r=0}^{n} C_{k-1}^{r+k-1} p^k q^r$$

Thus, to find the number of trials n^* to get to the kth success, it is necessary to look up the probability under p, k, and $n = n^* - k$.

Probabilities associated with successive values of n are given for selected combinations of p and k from $p = .05$, $k = 0.1(0.1)0.5$ up to $p = .95$, $k = 2(2)50(10)200$.

Exponential and Weibull Tables

Since the exponential and the Weibull are continuous distributions with an explicit distribution function in closed form:

Exponential: $F(x) = 1 - e^{-x/\mu}$

Weibull: $F(x) = 1 - e^{-[(x-\gamma)/\eta]^{\beta}}$

It is only necessary to obtain tables of e^{-x} to evaluate them. Such tables are available in any mathematical handbook such as the U.S. Department of Commerce Applied Mathematics Series No. 55 (1964). Many hand calculators have such values built in. Appendix Table T3-1 gives selected values of e^{-x}.

Normal Distribution Tables

U.S. Department of Commerce Applied Mathematics Series No. 23 (1953) gives extensive tables of the standard normal distribution showing values of the frequency (density) function

$$f(x) = \frac{1}{\sqrt{2\pi}} e^{-x^2/2}$$

and cumulative probabilities from $-x$ to $+x$ as

$$F(x) - F(-x) = \frac{1}{\sqrt{2\pi}} \int_{-x}^{x} e^{-x^2/2} \, dx$$

for $x = 0(.0001)1(.001)7.800$ and above and also from x to ∞

$$1 - F(x) = \frac{1}{\sqrt{2\pi}} \int_{x}^{\infty} e^{-x^2/2} \, dx$$

for $x = 6(.01)10$

Examples of some of these tables are given in the Appendix; they
include

Appendix table	Distribution	Source	Shows	Range
T3-2	Normal	Burr (1953)	F(z)	z = -3.59(.01)3.59
T3-3	Hypergeometric	Lieberman and Owen (1961)	F(x), f(x)	N = 1(1)10
T3-4	Binomial	Harvard Tables (1955)	1 - F(r - 1)	n = 1(1)33
T3-5	Poisson	Molina (1942)	1 - F(c - 1)	np = .001 to 10.0

If possible, the reader should obtain and use the tables cited or
similar tables in conjunction with work in acceptance sampling. The small
set of tables compiled by Odeh et al. (1977) will be found particularly
useful. Many other tables present values of these and other probability
distributions together with information useful in acceptance sampling.
These include Owen (1962), Beyer (1966), and Burington and May (1970),
among others.

USEFUL APPROXIMATIONS

The complexity of the hypergeometric distribution, and to some extent the
binomial, make it necessary to approximate these distributions at times
with other, more tractable distributions. Fortunately, rules have been
derived which, when adhered to, insure that reasonably good approximations
will be obtained. Naturally, such rules depend upon just how close one
distribution is expected to come to another. A schematic chart showing
some distribution functions approximating the hypergeometric distribution
and the binomial is presented in Fig. 3-9.

The hypergeometric may be approximated by the ordinary p-binomial when
the sample size is less than 10 percent of the population size. When the
sample represents more than 10 percent of the population the so-called
f-binomial may be used for calculations involving a proportion defective,
$p < .10$.

The f-binomial is the standard p-binomial with the sampling proportion
$f = n/N$ used as p and the number of defectives in the population Np used as
the sample size n. The frequency function then becomes

$$f(x) = C_x^{Np}\left(\frac{n}{N}\right)^x\left(1 - \frac{n}{N}\right)^{Np-x}$$

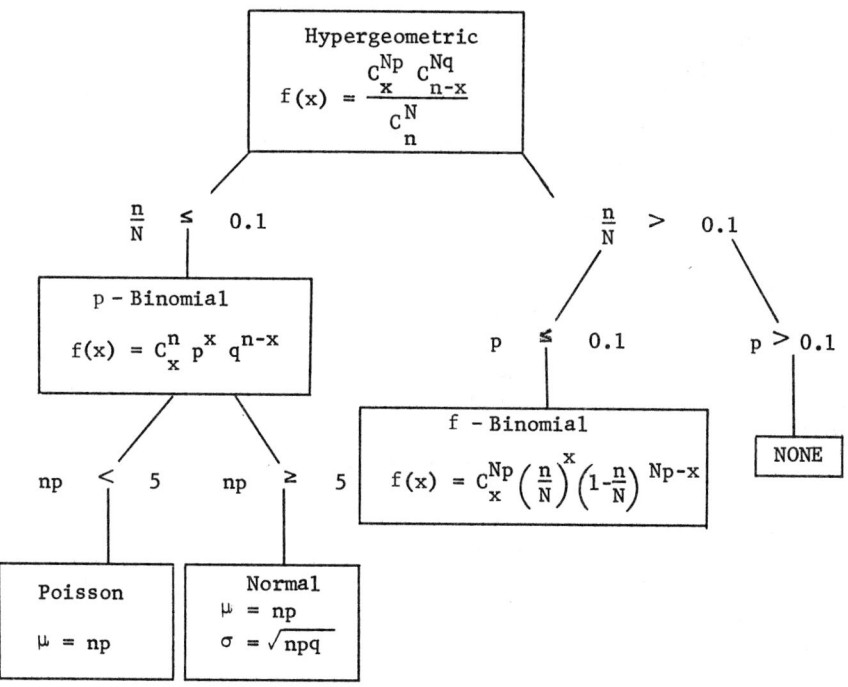

FIG. 3-9 Distributions approximating the hypergeometric and binomial.
(Note: For population of size N containing M defectives: $p = M/N$,
$q = 1 - M/N$.)

Probabilities may be obtained using tables for the standard p-binomial with

$$p = \frac{n}{N} \qquad n = Np$$

The rationale for the f-binomial approximation may be developed as
follows:

$$f(x) = \frac{C_x^{Np} \; C_{n-x}^{N-Np}}{C_n^N}$$

$$= C_x^{Np} \; \frac{(N - Np)!}{(n - x)!(N - Np - n + x)!} \; \frac{n!(N - n)!}{N!}$$

$$= C_x^{Np} \; \frac{(N - n)(N - n - 1) \cdots (N - n - (Np - x) + 1)}{N(N - 1) \cdots (N - Np + 1)} \; \frac{n!}{(n - x)!}$$

$$= C_x^{Np} \left(\frac{N - n}{N} \; \frac{N - n - 1}{N - 1} \cdots \frac{N - n - (Np - x) + 1}{N - (Np - x) + 1} \right) \left(\frac{(n)}{N - (Np - x)} \right.$$

$$\left. \frac{n - 1}{N - (Np - x) - 1} \cdots \frac{n - x + 1}{N - Np + 1} \right)$$

since it can be shown that for $b > a$,

$$\frac{a}{b} > \frac{a - 1}{b - 1}$$

substitute the first ratio, for each succeeding ratio in the first brackets and n/N for each ratio in the second brackets to obtain

$$f(x) < C_x^{Np}\left(\frac{N - n}{N}\right)^{Np-x}\left(\frac{n}{N}\right)^x$$

or

$$f(x) < C_x^{Np}\left(1 - \frac{n}{N}\right)^{Np-x}\left(\frac{n}{N}\right)^x$$

which is the f-binomial.

Note that the approximation could be improved by substituting the ratio $n/(N - Np)$ in the second brackets, but the binomial tables could no longer be used in determining the relevant probabilities.

For proportions defective greater than a tenth when the sampling proportion is greater than 10 percent of the population the hypergeometric itself should be used.

It has been pointed out by William C. Guenther (1973) that the Wise (1954) approximation can be used effectively with binomial tables in the derivation of hypergeometric sampling plans. This approximation to the hypergeometric consists of using the cumulative binomial distribution with

$$p = \frac{2Np - x}{2N - n + 1}$$

to come very close to the hypergeometric values. Details of its use in the development of a sampling plan will be found in Guenther (1977). Another excellent approximation is that of Sandiford (1960).

In turn, the binomial distribution may be approximated by the Poisson distribution for p small and n large (roughly when the product np is less than 5). This is done by looking up Poisson probabilities of x successes when the mean of the Poisson distribution is $\mu = np$.

When the product np is greater than 5, the binomial distribution may be approximated by the normal distribution with

$$\mu = np \qquad \sigma = \sqrt{npq}$$

Note that nq must also be greater than 5. However in acceptance sampling it is usually the case that $q > p$. Here, the normal cumulative probability

is taken over a region corresponding to the number of successes desired.
In approximating a discrete distribution, such as the binomial, with a
continuous distribution, such as the normal, it is necessary to use a
"continuity" correction. Since the probability of a point in a continuous
distribution is zero, it is necessary to approximate each discrete number
of successes by a band on the x axis going out from the number one-half
unit on each side as shown in Fig. 3-10. Thus, the probability of x successes or less would be found as the area up to x + 1/2 under the normal
curve. The probability of x or more successes would be the area above
x - 1/2, and so on.

To illustrate these approximations, let us take a case where the
sampling proportion is equal to a tenth. Suppose lot size is 100, sample
size is 10, p = 0.1, and we desire the probability of 2 or fewer defectives.
Using appropriate formulas and tables we get

Hypergeometric. Owen (N = 100, n = 10, k = 10, x = 2)

$F(2) = .93998$

Binomial. Harvard (n = 10, p = .10, x = 2)

$F(2) = .92981$

Poisson. Molina (np = 10(.1) = 1, x = 2)

$F(2) = .91970$

f-Binomial.

$$F(2) = \sum_{i=0}^{2} C_i^{10} (.1)^i (.9)^{10-i}$$

Harvard (n = 10, p = .10, x = 2)

$F(2) = .92981$

Normal.

$\mu = np = 10(.1) = 1$

$\sigma = \sqrt{npq} = \sqrt{10(.1)(.9)} = .95$

$z = \dfrac{2.5 - 1}{.95} = 1.58$

$F(2) = .9428$

All the approximations were fairly close to the hypergeometric value.
Actually, the normal was used for illustrative purposes only since it
should not usually be used to approximate the binomial when np is less
than 5. This is an indication of the utility of these approximations.

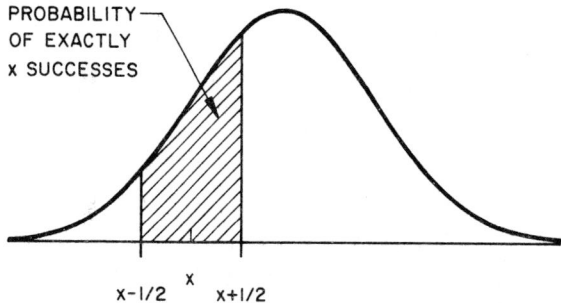

FIG. 3-10 Continuity correction.

TESTS OF FIT

It is not enough to *assume* a distribution to hold in real life applications
of statistics. Statistics is not like mathematics where correct answers
are derived from the assumptions. In statistics, the assumptions must be
correct and must describe the physical situation adequately before correct
answers will be obtained. For this reason it is not enough to assume a
distribution holds to be correct. Enough real data should be analyzed to
assure the assumption is correct.

Frequently, underlying distributions of measurements are characterized
by probability models. In these cases it is necessary to assure that the
data conform to the model used. Methods have been developed to test if
particular distributions are applicable. These include probability plots,
χ^2 tests of goodness of fit, the Kolmogorov-Smirnov test, the Wilk-Shapiro
test, and others. Sample size considerations are quite important in acceptance sampling since large sample sizes are needed to detect aberrations in
the tails of the distribution where the defective material is likely to be
found. It is important that those applying acceptance quality control procedures be familiar with these tests and procedures and apply them to real
data before assuming any distribution shape applies. They are discussed
in most basic texts on applied statistics.

The probability plot is one of the most useful and versatile of the
tests of fit. It involves plotting the ordered observations from a sample
on special paper against the cumulative percentage at which the individual
ordered observations stand in the sample. In this way an empirical cumulative probability distribution plot for the sample is obtained. Estimates
can be made from this plot and its shape can be used as an indication of
the underlying probability distribution which gave rise to the sample.

Special probability papers transform the axis representing cumulative percentage in such a way that if the sample came from the distribution represented by the paper selected, the points will plot roughly in a straight
line. Papers can be obtained to represent a variety of distributions, the
normal and Weibull probably being the most common. Directions for the
construction of normal probability paper have been given by Nelson (1976).

Plotting positions can readily be determined using the formula

$$\hat{P}_{(i)} = \frac{i - 1/2}{n} (100)$$

which gives the approximate probability of obtaining a values less than
$x_{(i)}$. Then, the individual ordered points $x_{(i)}$ are plotted against their
empirical cumulative frequency (usually in percent) estimated by $\hat{P}_{(i)}$. A
straight-line plot is an indication that the underlying distribution of
measurements is that of the paper on which the points are plotted. Substantial departures from a straight-line indicate that the distribution
for the paper may not apply. A straight-line fit through a straight-line
plot can be used to make estimates of the parameters of the underlying
distribution. On normal probability paper, the mean is estimated from the
50th percentile of the empirical plot. Similarly, the standard deviation
can be obtained as half the difference between the 16th and the 84th percentile values.

For example, consider the following data taken from MIL-STD-414. The
specifications for electrical resistance of a certain electrical component
is 650.0 ± 30 ohms. Suppose the values of sample resistance in a sample
of 10 are as follows: 643, 651, 619, 627, 658, 670, 673, 641, 638, 650. A
probability plot for these data appears in Fig. 3-11, which plots the
following points

Order (i)	$x_{(i)}$	$\hat{P}_{(i)}$
1	619	5
2	627	15
3	638	25
4	641	35
5	643	45
6	650	55
7	651	65
8	658	75
9	670	85
10	673	95

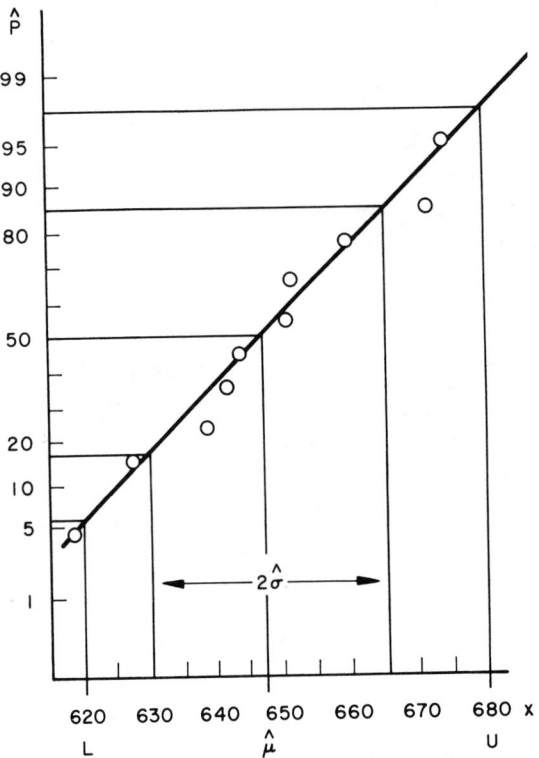

FIG. 3-11   Probability plot.

A straight-line plot is obtained. A line drawn through the points allows
the following estimates to be made (on a plot having a more detailed
resistance scale):

$\hat{\mu}$ = mean = 647.5

$\hat{\sigma}$ = standard deviation = 17.5

P_L = percent below lower spec = 6%

P_U = percent above upper spec = 3%

P_T = total out of spec = 9%

Actually, for these data

$\overline{X}$ = 647.0   s = 17.2

and so the probability plot provided very good estimates of the population
parameters in this case, close to those obtained by the usual computational
methods. Using $\overline{X}$ and s to estimate the percent out of specification limits

$$z_U = \frac{U - \overline{X}}{s} = \frac{680 - 647}{17.2} = 1.92$$

giving 2.74 percent above the upper limit and defining z_L as in MIL-STD-414

$$z_L = \frac{\overline{X} - L}{s} = \frac{647 - 620}{17.2} = 1.57$$

giving 5.82 percent below the lower limit. This gives a point estimate of 8.56 percent out of specifications which is very close to that obtained from the probability plot.

The Weibull distribution is particularly useful in reliability analysis and associated sampling plans. Weibull (1951) first used plots of the Weibull distribution. Later versions and refinements in analysis were developed by Kao (1959), Nelson (1967), and Nelson and Thompson (1971). The reader is referred to these papers for a discussion of the Weibull probability plot and its uses. An extensive book on probability papers has been prepared by King (1971). Shapiro (1980) has prepared an in depth manual on testing normality and other distributional assumptions for the American Society for Quality Control. An excellent introductory text on probability plots has been written by Nelson (1979) and appears in the same series.

PROBLEMS

1. A trial lot of 100 candles is received from a new supplier. Five candles are checked to be sure the wick extends properly above the body. One is found to be defective. What is the probability of one or less defective candles in a sample of 5 if two candles were defective in the lot? How many would you expect? What would be the standard deviation of the number observed in a sample of 5?

2. If the vendor in Prob. 1 maintains a quality of 2 percent defective, what is the probability of one or less defective candles in a sample of 5 from a very large lot? How many would you expect? What would be the standard deviation of the number observed in a sample of 5? Check your answer with the Larson nomograph.

3. A spot welder is expected to produce not more than two defective welds in a shift's production. If the process average were actually two defective welds per shift, what is the probability of obtaining two or fewer bad welds on a given shift? Check your answer with the Thorndyke chart. Why is the answer not one-half? What would be the standard deviation of the number of bad welds per shift?

4. A continuous sampling plan starts by inspecting i successive units.

If no defectives are found a switch to sampling inspection is made. If i = 5, what is the probability of finding a defective on the fifth trial if the proportion defective submitted to the plan is .05? What is the mean number of trials to the second defective?

5. The life of a transistor follows the exponential distribution with mean life μ = 10,000 hrs. What is the probability of a unit failing before 20,000 hrs.? What is the standard deviation of the transistor's lifetime?

6. Express Prob. 5 in terms of the parameters of the Weibull distribution. What is the probability of a lifetime less than or equal to 10,000 hrs.?

7. Bottles are to be filled with one liter of liquid. The amount of fill is normally distributed with standard deviation σ = .01 liter. The mean fill is set at μ = 1.03 liters to minimize the possibility of underfill. To check on the overfill, the contents are poured into a container marked with a "narrow limit" at 1.005 liters. What is the probability of observing a fill less than 1.005 liters when the process mean is actually at 1.03 liters? What is the probability of one such indication in a sample of 3?

8. A new supplier submits a test lot of rods the diameter of which is specified to average 3 ± .001 m. The rods are to be welded together so that average length is important. The distribution of lengths is unknown, but a sample of nine rods yields a mean $\overline{X}$ = 3.001 m. Is such a result likely if the standard deviation of this type of product is σ = .0003 m and the mean is 3 m?

9. Suppose samples of 5 are to be taken from lots of 25 for a simple nondestructive test. If the fraction defective is .08, what approximation is appropriate for the hypergeometric distribution? What is the probability of one or fewer defective in a sample of 5?

10. A probability plot is to be made of the weight of 500 pieces to check for normality. The eightieth ordered observation is 24 while observation 420 is 48. If the fitted line passes through both these points, estimate the standard deviation. Estimate the mean.

REFERENCES

Beyer, W. H. (1966), *Handbook of Tables for Probability and Statistics*, The Chemical Rubber Co., Cleveland, Ohio.

Burington, R. S., and D. C. May (1970), *Handbook of Probability and Statistics with Tables*, 2nd ed., McGraw-Hill, New York.

Burr, I. W. (1953), *Engineering Statistics and Quality Control*, McGraw-Hill, New York.

Defense Systems Department, General Electric Company (1962), *Tables of Individual and Cumulative Terms of the Poisson Distribution*, Van Nostrand, Princeton, New Jersey.

Dodge, H. F., and H. G. Romig (1941), Single Sampling and Double Sampling Inspection Tables, *Bell System Technical Journal*, 20(1): 1-61.

Guenther, W. C. (1973), A Sample Size Formula for the Hypergeometric, *Journal of Quality Technology*, 5(4): 167-173.

Guenther, W. C. (1977), *Sampling Inspection in Statistical Quality Control*, Macmillan, New York.

Harvard University Computing Laboratory (1955), *Tables of the Cumulative Binomial Probability Distribution*, Harvard University Press, Cambridge, Massachusetts.

Kao, J. H. K. (1959), A Graphical Estimation of Mixed Weibull Parameters in Life Testing Electron Tubes, *Technometrics*, 1(4): 389-407.

King, J. R. (1971), *Probability Charts for Decision Making*, Industrial Press, New York.

Larson, H. R. (1966), A Nomograph of the Cumulative Binomial Distribution, *Industrial Quality Control*, 23(6): 270-278.

Lieberman, G. J., and D. B. Owen (1961), *Tables of the Hypergeometric Probability Distribution*, Stanford University Press, Stanford, California.

Molina, E. C. (1942), *Poisson's Exponential Binomial Limit*, Van Nostrand, New York.

Mood, A. M., and F. A. Graybill (1963), *Introduction to the Theory of Statistics*, 2nd ed., McGraw-Hill, New York.

Nelson, L. S. (1967), Weibull Probability Paper, *Industrial Quality Control*, 23(9): 452-453.

Nelson, L. S. (1974), Using Tables of the Cumulative Binomial Distribution, *Journal of Quality Technology*, 6(2): 116-118 [Addendum, 7(1): 49 (1975)].

Nelson, L. S. (1976), Constructing Normal Probability Paper, *Journal of Quality Technology*, 8(1): 56-57.

Nelson, W. B. (1979), *How to Analyze Data with Simple Plots*, Vol. 1, The ASQC Basic References in Quality Control: Statistical Techniques, E. J. Dudewicz, Ed., American Society for Quality Control, Milwaukee, Wisconsin.

Nelson, W. B., and V. C. Thompson (1971), Weibull Probability Papers, *Journal of Quality Technology*, 3(2): 45-50.

Odeh, R. E., D. B. Owen, Z. W. Birnbaum, and L. Fisher (1977), *Pocket Book of Statistical Tables*, Marcel Dekker, New York.

Owen, D. B. (1962), *Handbook of Statistical Tables*, Addison-Wesley, Reading, Massachusetts.

Romig, H. G. (1947), *50-100 Binomial Tables*, John Wiley and Sons, New York.

Sandiford, P. J. (1960), A New Binomial Approximation for Use in Sampling From Finite Populations, *Journal of the American Statistical Association*, 55(292): 718-722.

Schilling, E. G., and P. R. Nelson (1976), The Effect of Non-Normality on the Control Limits of $\overline{X}$-Charts, *Journal of Quality Technology*, 8(4): 183-188.

Shapiro, S. S. (1980), *How to Test Normality and Other Distributional Assumptions*, Vol. 4, The ASQC Basic References in Quality Control: Statistical Techniques, E. J. Dudewicz, Ed., American Society for Quality Control, Milwaukee, Wisconsin.

Shewhart, W. A. (1931), *Economic Control of Quality of Manufactured Product*, Van Nostrand, New York.

Thorndyke, F. (1926), Applications of Poisson's Probability Summation, *Bell System Technical Journal*, 5: 604-624.

United States Department of Commerce (1950), *Tables of the Binomial Probability Distribution*, National Bureau of Standards, Applied Mathematics Series No. 6, Washington, D.C.

United States Department of Commerce (1953), *Tables of Normal Probability Functions*, National Bureau of Standards, Applied Mathematics Series No. 23, Washington, D.C.

United States Department of Commerce (1964), *Handbook of Mathematical Functions*, National Bureau of Standards, Applied Mathematics Series No. 55, Washington, D.C.

Weibull, W. (1951), A Statistical Distribution Function of Wide Applica-
     bility, *Journal of Applied Mechanics*, 18: 293-297.

Williamson, E., and M. H. Bretherton (1963), *Tables of the Negative Binomial
     Probability Distribution*, John Wiley and Sons, New York.

Wise, M. E. (1954), A Quickly Convergent Expansion for Cumulative Hypergeo-
     metric Probabilities, Direct and Inverse, *Biometrika*, 41(3): 317-329.

4

CONCEPTS AND TERMINOLOGY

The fundamental tool for analysis of a sampling plan is the operating characteristic curve. Two types of curves are recognized:

Type A. Sampling from an individual (or isolated) lot, showing probability that the lot will be accepted plotted against lot proportion defective.

Type B. Sampling from a process (such as the producer's process which produced the lot), showing proportion of lots which will be accepted plotted against process proportion defective.

Naturally, the probability distributions utilized in plotting these types of OC curves are inherently different. They also depend upon the measure in which quality is expressed. These include

Attributes. A dichotomous (two class) classification of units into defective and nondefective. For example, number of defective units in a sample of 100 units.

Counting. An enumeration of occurrences of a given characteristic per given number of units counted. For example, number of defects per 100 units in the population.

Variables. The measurement of some characteristic along a continuous scale. For example, diameter of a circular casting as measured in centimeters.

The distinction is made between defect (an imperfection great enough to be counted) and defective (a unit containing one or more defects, which could be rejected for any one of them).

The probability distributions appropriate for the derivation of operating characteristic curves of the two types are shown in Table 4-1. The

TABLE 4-1 Probability Distributions for Operating Characteristic Curves

Characteristic	Type A	Type B
Attribute	Hypergeometric	Binomial
Count	Poisson	Poisson
Measurement	Applicable continuous distribution of measurement involved	

form of these distributions and their properties are shown in Table 3-1.

For variables data the applicable distribution is that of the variable as it would appear to the inspector, that is, including piece to piece variation, measurement error, changes in environmental conditions, and the like. There are means available for separating these sources of error and controlling them. Such methods are addressed in texts on design of experiments, such as Hicks (1973) or Anderson and McLean (1974), and in texts on process quality control, such as Ott (1975).

Differences in the operating characteristic curves associated with these different type plans may be illustrated by the sampling plan

N = lot size = 20

n = sample size = 10

c = acceptance number = 1

Using the methods of Chap. 2, it is possible to compute the different operating characteristic curves. The results are shown in Table 4-2.

We see from the Type A values in Table 4-2 that with the sampling plan n = 10, c = 1, it is impossible to fail if the lot is 5 percent defective when lot size is 20. Why? Because for the lot of 20 to be 5 percent defective it would contain just one defective, and one defective is allowable under the plan. In fact, for a finite lot size, only a limited number of percents defective can be formed, in this case 0, 5, 10, 15, ..., 95, 100.

The Type B operating characteristic curve is not so restricted. In fact, the producer's process could have been running at any percent defective when the lot of 20 was formed. The Type B OC curve views the lot of 20 as a sample from the producer's process and the sample of 10 as a sub-sample of the same process. In this way, it is reasonable to address the probability of acceptance for any percent defective from 0 to 100 when using a Type B operating characteristic curve.

TABLE 4-2 Probabilities of Acceptance for Hypergeometric, Binomial, and Poisson (N = 20, n = 10, c = 1)

Percent defective	Type A hypergeometric	Type B binomial	Defects/100 units Poisson
5	1	.914	.910
10	.763	.736	.736
15	.500	.544	.558
20	.291	.376	.406
25	.152	.244	.287
30	.070	.149	.199
35	.029	.086	.136
40	.010	.046	.092
45	.003	.023	.061
50	.001	.011	.040
55			.027
60			.017
65			.011
70			.007

Finally, if the number of defects are counted in a lot of 20 items the count could easily exceed 20 since one item can have one or more defects. Note that this is not the case for either of the attributes sampling situations which deal with defectives. The number of defectives could not exceed 20, the lot size. A count of defects is often expressed in terms of "defects per 100 units." In this form the measure of quality is analogous to "percent defective," which is also based on 100. However, as noted, defects per hundred units may exceed 100. For instance, if we knew the lot of 20 had 3 defective pieces in it with 4, 10, and 12 defects each, the lot would, in total, contain 26 defects — even though it was composed of only 20 pieces and had only 3 defectives. The mean number of defects per unit in such a situation would be 1.3. The mean number of defects per 10 pieces would be 13, and since a sample size of 10 was specified, the Poisson distribution with a mean $\mu = 13$ would be used to calculate the probability of acceptance in such a case. It should be pointed out that the defects per unit probabilities of acceptance in Table 4-2 were computed using a mean value of

$$\mu = \frac{p \times 10}{100}$$

since the value p is interpreted as defects per hundred units so that p = 5 defects per hundred units implies p/100 = .05 defects per unit, which gives

$$\frac{10p}{100} = 0.5 \text{ defects per 10 units}$$

A plot of the three OC curves is given in Fig. 4-1.

Note how the binomial and Poisson curves are fairly well superimposed for small p and then diverge as p becomes large. This shows the use of the Poisson distribution as an approximation to the binomial when p is small. Also the shape of the Type A (hypergeometric) curve is quite different from

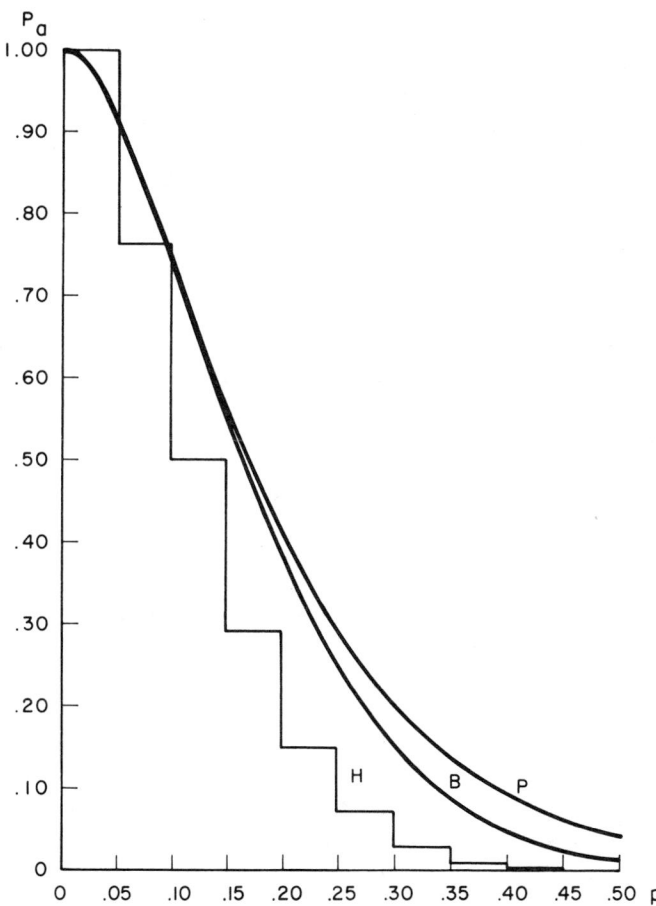

FIG. 4-1 Type A, B, and defects OC curves.

that of the Type B (binomial) and the defects per unit (Poisson) curves, since the sample represents a large proportion (50 percent) of the lot. These curves also illustrate the conservative nature of the approximations since they tend to underestimate for high probability of acceptance and overestimate for low probability of acceptance.

SAMPLE SIZE AND LOT SIZE

Reference to the formulas for the binomial, the Poisson, and the various continuous distributions will indicate that they do not contain any reference to lot size. Only in the hypergeometric distribution will such a parameter be found. Even then, the effect of lot size on the operating characteristic curve is minimal when a small proportion (say less than 10 percent) of the lot is used up in taking the sample. This is illustrated in Table 4-3, which shows the probability of acceptance for the plan n = 10, c = 1 when the lot size is 20, 60, 100, ∞.

A plot of the OC curves would show little difference among the curves except when N = 20. It is apparent that the probabilities change substantially with lot size only when the sample represents a large portion of the lot (say from 1/2 to 1/6). Changes are slight when the sample is a small fraction of the lot (for 1/10 and less). Also note that the values for N = ∞, calculated from the binomial distribution, are conservative in approximating the hypergeometric for smaller lot sizes in that they underestimate P_a for lower percents defective and overestimate for high percents defective. Thus, use of the binomial distribution in constructing a Type A

TABLE 4-3 Type A Probability of Acceptance Using the Hypergeometic Distribution (n = 10, c = 1)

%	N = 20	N = 60	N = 100	N = ∞
5	1	.931	.923	.914
10	.763	.741	.738	.736
15	.500	.533	.538	.544
20	.291	.354	.363	.376
25	.152	.219	.229	.244
30	.070	.126	.136	.149
35	.029	.067	.075	.086
40	.010	.033	.039	.046

TABLE 4-4   Type A Probability of Acceptance Using
the Hypergeometric Distribution (N = 20, c = 1)

%	n = 5	n = 10	n = 15
5	1	1	1
10	.947	.763	.447
15	.860	.500	.140
20	.751	.291	.032
25	.634	.152	.005
30	.517	.070	—
35	.406	.029	—
40	.307	.010	—

operating characteristic curve when the fraction of the lot sampled is
reasonably low will not only give fairly close answers but tends to be
conservative as well. Similar results obtain for the Poisson approximation
to the binomial distribution.

As might be expected, changes in sample size for a given lot size will,
however, have substantial effect on the protection afforded by a plan.
Table 4-4 shows the probability of acceptance for plans with sample size 5,
10, and 15 from a lot of size 20 with acceptance number c = 1 for Type A

TABLE 4-5   Type B Probability of Acceptance Using
the Binomial Distribution (c = 1)

%	n = 5	n = 10	n = 15
5	.977	.914	.829
10	.919	.736	.549
15	.835	.544	.319
20	.737	.376	.167
25	.633	.244	.080
30	.528	.149	.035
35	.428	.086	.014
40	.337	.046	.005

TABLE 4-6 Type A Probability of Acceptance Using
the Hypergeometric Distribution (N = 20, n = 10)

%	c = 0	c = 1	c = 2
5	.500	1	1
10	.237	.763	1
15	.105	.500	.895
20	.043	.291	.709
25	.016	.152	.500
30	.005	.070	.314
35	.002	.029	.175
40	—	.010	.085

(hypergeometric) probabilities.

The effect of sample size is somewhat less pronounced for Type B
(binomial) probabilities; however, Table 4-5 gives an example similar to
Table 4-4 but with infinite lot size.

The most dramatic effect on the probability of acceptance, however
comes with changing acceptance numbers. Even the inherent shape of the
OC curve is changed in going from one acceptance number to another. This
can be seen in Table 4-6 which shows the effect of changing acceptance
numbers for a plan with N = 20, n = 10.

TABLE 4-7 Type B Probability of Acceptance Using
the Binomial Distribution When c/n = .1

%	n = 10, c = 1	n = 20, c = 2	n = 40, c = 4
5	.914	.925	.952
10	.736	.677	.629
15	.544	.405	.263
20	.376	.206	.076
25	.244	.091	.016
30	.149	.035	.003
35	.086	.012	—
40	.046	.004	—

It should be clear, then, that the two principal determinants of the operating characteristics of a sampling plan are acceptance number and sample size. Lot size plays a very minor role in determining protection even when sampling sizable proportions of the lot. This is contrary to intuitive belief and should be constantly borne in mind by the practicing quality control engineer in setting up sampling plans and sampling schemes. Often a relationship of lot size to sample size is specified (even by MIL-STD-105D), but this is for logistic and economic purposes and not primarily for purposes of enhancing the protection afforded by the plan.

We may ask what is the effect of maintaining the acceptance number as a constant proportion of the sample size. Table 4-7 compares three plans which keep the acceptance number at 10 percent of the sample size. These plans are in no sense equivalent. The protection afforded by n = 40, c = 4 is much higher than the other plans. This can be seen by comparing the plans' protection at, say, 20 percent defective. An acceptance number of 4 can, in fact, give more protection than an acceptance number of 2 provided that the sample size is increased accordingly.

In some operations, it has become customary to specify sample size as a proportion of the lot size. Take a 10 percent sample, let us say, usually with an acceptance number c = 0. Since the operating characteristics of a plan are dependent principally on sample size, not lot size, this means that large lots with large samples will be accepted much less often than small lots with small sample sizes at the same percent defective.

TABLE 4-8 Type A Probability of Acceptance Using the Hypergeometric Distribution When Sample Size is Proportionate to Lot Size

%	N = 20, n = 10, c = 0	N = 40, n = 20, c = 0	N = 100, n = 50, c = 0
5	.500	.244	.028
10	.237	.053	.001
15	.105	.010	—
20	.043	.002	—
25	.016	—	—
30	.005	—	—
35	.002	—	—

TABLE 4-9   Effect of Lot Size on Acceptance (Order of 1000)

Lot Size	Number lots	Proportion lots accepted	Expected number lots accepted	Expected pieces accepted
20	50	.50	25	500
40	25	.243	6.075	243
100	10	.028	0.28	28

This is illustrated in Table 4-8 which shows the protection afforded by such a plan. For lots of size 20, 5 percent defective material has a 50 percent chance to be accepted, while for lots of 100 only a 2.8 percent chance of acceptance is provided. An unscrupulous supplier has an incentive to provide small lots as can be seen in Table 4-9. The table shows the results of shipping 5 percent defective material in different lot sizes.

The protection afforded both parties by a plan such as this is clearly dependent upon lot size and is not a rationally determined criterion for protection.

EFFECT OF INSPECTION ERROR

No inspection is perfect all the time. Indeed, it is generally recognized that 100 percent inspection is much less than 100 percent effective in screening out defective items. Studies have indicated that, in the face of monotony and fatigue, only about 80 percent of the defectives will be detected [Juran-Gryna (1970, p. 320)]. The reasons for inspection inaccuracy have been detailed by Juran (1962, pp. 8-25) as follows:

1. Willful errors which include:
    a. Criminal acts such as fraud and collusion.
    b. Falsification for personal convenience of the inspector.
2. Intermediate errors due to bias, rounding off, overzealousness, etc.
3. Involuntary errors due to blunder, fatigue, and other forms of human imperfection.

In particular, flinching, or failure to call a defect when it is close to the specification is a common source of error of what Juran calls the intermediate type.

It should be pointed out that errors can go either way. An overzealous inspector can easily flinch by calling good product bad. Harsh supervision, the mood of the moment, the psychological, and even physical environment can cause marginal and even less than marginal decisions to be incorrectly made.

Sample inspection is also subject to the same type of inspection error. While an advantage of sampling is a reduction of the number of pieces subject to repetitive examination, the same circumstances and motivations exist which may lead to inspector inaccuracy. The result is an inaccurate representation of the quality submitted.

Suppose product is submitted which is of fraction defective p. The inspector misclassifies the product as shown in Table 4-10, where

p_1 = proportion nondefective classified as defectives

p_2 = proportion defective classified as nondefectives

The Statistical Research Group (1948, p. 23) presents the following formula for the apparent level of quality p^* when the true incoming level defective is p.

$$p^* = p_1(1 - p) + p(1 - p_2)$$

This follows from the rightmost column of Table 4-10. When the true fraction defective is small and the proportion of defectives which are missed is not large we have

$$p^* = p_1 + p$$

Similarly when the chance of misclassification of a nondefective item is small

$$p^* = p(1 - p_2)$$

It is often the case that errors will go one way or the other, although both types of misclassification at the same time are possible. For example, suppose due to an error in configuration control the inspector received a print of a symmetric part which was reversed left to right. Unfortunately the written material was on another sheet and the mistake went undetected so that the area of acceptance became that of rejection and vice versa. Then

$$p_1 = 1 \qquad p_2 = 1$$

TABLE 4-10   Proportions Defective Misclassified

Actual condition	Inspector classification	
	Nondefective	Defective
Nondefectives	$1 - p_1$	p_1
Defectives	p_2	$1 - p_2$

and

$$p^* = 1(1 - p) + p(1 - 1) = 1 - p$$

That is, the apparent level of defective material would be the actual proportion nondefective.

The formula works equally well for screening or sampling inspection. In screening it gives the apparent level of quality after 100 percent inspection. In sampling it gives the apparent level of quality as seen by the inspector. The operating characteristic curve can be entered in terms of p^* rather than p to find the probability of acceptance in the face of inspection error. Unfortunately, p_1 and p_2 are rarely known but provide a means for analysis of the possible effect of this type of error. Methods of estimating inspector bias in visual inspection have been discussed by Schilling (1961).

RECTIFICATION

Much of the effect of the imposition of a sampling plan depends upon the disposition of the product after it is inspected. Accepted lots go to the consumer. Rejected lots may be handled in a number of ways as follows:

Destroyed. No effect on overall quality if producer continues to submit at a constant level of quality. Positive effect if quality levels fluctuate nonrandomly from lot to lot.

Resubmitted. No effect on overall quality if producer continues to submit at a constant level of quality

Screened. Quality of rejected lots improved within the limits of inspection error. Properly done 100 percent inspection of rejected lots would transform each rejected lot into a perfect one. As a result the overall level of quality as seen by the consumer would improve.

Acceptance sampling schemes which incorporate 100 percent inspection of rejected lots are called rectification schemes. Formulas are available for calculating the average outgoing quality (AOQ) from such schemes. This is the long-run average quality shipped to the consumer under 100 percent inspection of rejected lots, assuming any defective item found is replaced by a good one. The average is taken over all lots, good and bad, so that assuming no inspection error,

$$AOQ = pP_a\left(1 - \frac{n}{N}\right)$$

since the only defectives transmitted to the consumer would be in the accepted lots (rejected lots having been made perfect). The average proportion defective the consumer would receive then is made up of fraction defective p received a proportion P_a of the time and fraction defective 0 received a proportion $1 - P_a$ of the time. But, for all lots, defective items found in the sampling inspection are also replaced by good ones so that the remaining proportion defective is

$$p\left(\frac{N - n}{N}\right)$$

and

$$AOQ = pP_a\left(\frac{N - n}{N}\right) + 0(1 - P_a)\left(\frac{N - n}{N}\right)$$

$$= pP_a\left(\frac{N - n}{N}\right)$$

$$= pP_a\left(1 - \frac{n}{N}\right)$$

and when the sample size is very small in proportion to the lot size $n/N \sim 0$, so that the formula becomes

$$AOQ = pP_a$$

The maximum value of AOQ over all possible values of fraction defective which might be submitted is called the *average outgoing quality limit* (AOQL). It represents the maximum long term average fraction defective that the consumer can see under operation of the rectification plan.

It is sometimes necessary to determine the average amount of inspection per lot in the application of such rectification schemes, including 100 percent inspection of rejected lots. This average, called the *average total inspection* (ATI), is made up of the sample size n on every lot plus the remaining (N - n) units on the rejected lots, so that

TABLE 4-11   Calculation of AOQ and ATI $(N = 20, n = 10, c = 1)$

p	P_a	$(1 - n/N)$	AOQ	$(1 - P_a)(N - n)$	ATI
.00	1.000	.5	.000	0	10
.05	.914	.5	.023	0.86	10.86
.10	.736	.5	.037	2.64	12.64
.15	.544	.5	.041	4.56	14.56
.20	.376	.5	.038	6.24	16.24
.25	.244	.5	.030	7.56	17.56
.30	.149	.5	.022	8.51	18.51
.35	.086	.5	.015	9.14	19.14
.40	.046	.5	.009	9.54	19.54
.45	.023	.5	.005	9.77	19.77
.50	.011	.5	.003	9.89	19.89
(1)	(2)	(3)	(4) = (1)(2)(3)	(5)	(6) = n + (5)

$$ATI = n + (1 - P_a)(N - n)$$
$$= P_a n + (1 - P_a)N$$

Consider the sampling plan n = 10, c = 1 used on a continuing supply of lots of size 20 from the same producer, that is, in a Type B sampling situation. Clearly rectification plans are meaningless on isolated lots, even though they might be 100 percent inspected if rejected, because there is no long-term average involved. The Type B probabilities of acceptance have already been calculated and are listed in Table 4-11 which shows the calculation of the average outgoing quality (AOQ) and the average total inspection (ATI).

It is apparent that the ATI curve starts at 10, the sample size, when p = 0 since no lots are 100 percent inspected and rises to 20 when p = 1.0 since all lots will be rejected and 100 percent inspected when the lots are completely defective. The ATI curve is shown in Fig. 4-2.

The AOQ curve starts at 0 when p = 0 since no rectification is necessary. It rises to a maximum of around 4.1 percent defective and then declines as more and more 100 percent inspection takes place. When lots are completely defective, they are all rectified and the AOQ is again zero.

The AOQL for this plan can be seen to be around 4.1 percent defective.

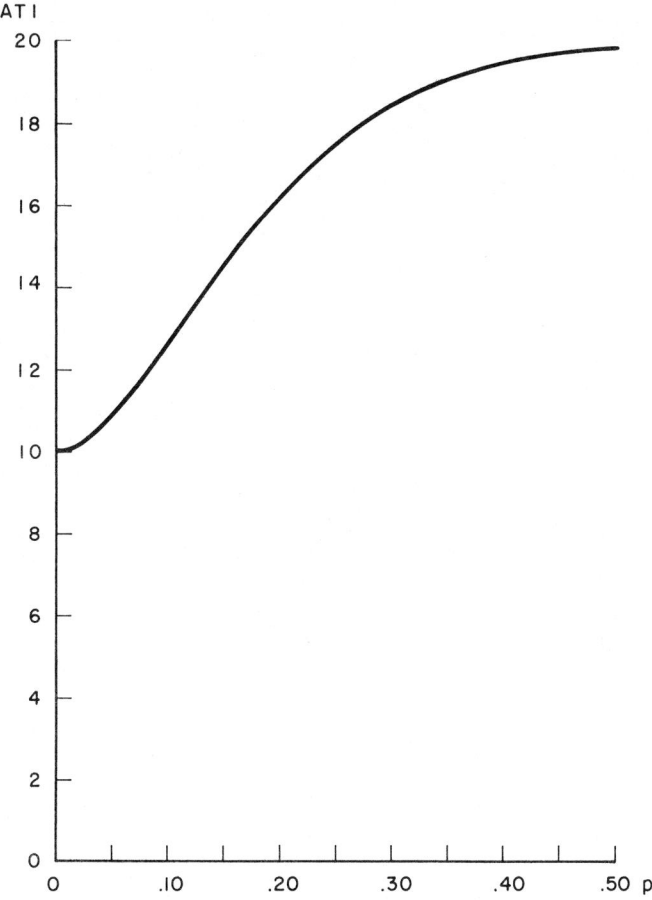

FIG. 4-2 ATI curve (N = 20, n = 10, c = 1).

TABLE 4-12 Determination of AOQL

p	P_a	(1 - n/N)	AOQ
.13	.620	.5	.0403
.14	.582	.5	.0407
.15	.544	.5	.0408
.16	.508	.5	.0406
.17	.473	.5	.0402
(1)	(2)	(3)	(4) = (1)(2)(3)

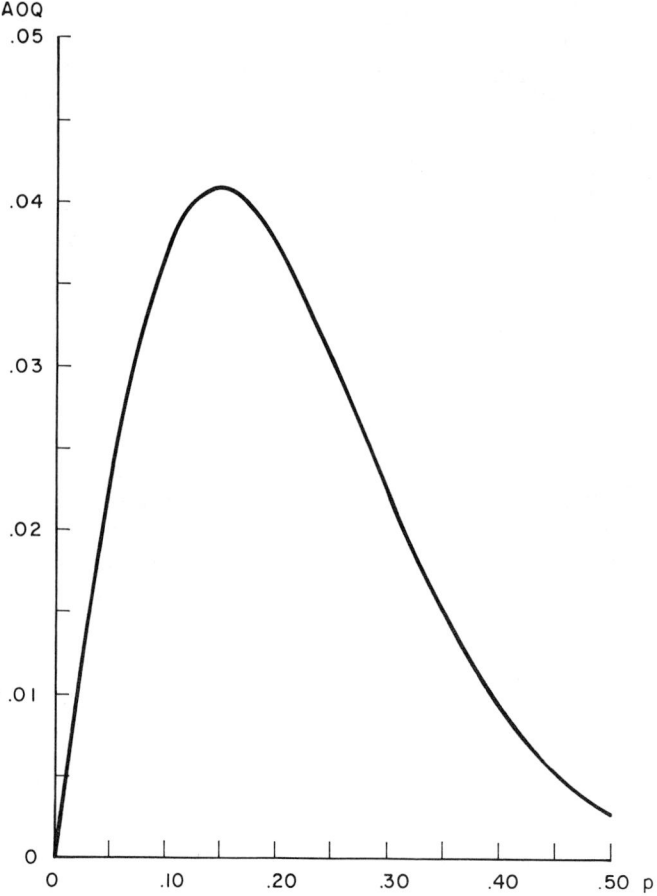

FIG. 4-3 AOQ curve (N = 20, n = 10, c = 1).

We define p_M as the incoming fraction defective at which AOQ reaches its maximum, that is the AOQL occurs when the incoming fraction defective is p_M. Then examining the region close to p = .15 as in Table 4-12 it is apparent that the AOQL is .041 to three-place accuracy and it occurs at p_M = .15. The consumer will never experience a long-term average fraction defective greater than .041, although the average may be considerably higher in the short run. The AOQ curve is given in Fig. 4-3.

CURTAILMENT

Just as there are many procedures for disposing of a lot, there are different ways to treat the sample itself. Consider the following possibilities for a single-sampling plan with sample size n and acceptance number c:

1. Complete inspection. All items in the sample of n are inspected.
2. Semicurtailed inspection. The inspection is stopped when the number of defectives found exceeds the acceptance number. All units are inspected if the lot is accepted.
3. Fully curtailed inspection. The inspection is stopped when the number of defectives found exceeds the acceptance number c or the number of nondefectives is found to exceed n - c. In short, the inspection is stopped once a decision can be made.

Under curtailment, the number of units actually inspected becomes a random variable. There are formulas which can be used to determine the average sample number (ASN) for such procedures. This is the mean number of items inspected per lot. The formulas for a single-sampling plan as given by the Statistical Research Group (1948, p. 212) are

1. Semicurtailed

$$ASN_c = n\ F(c|n) + \frac{c + 1}{p}(1 - F(c + 1|n + 1))$$

2. Fully curtailed

$$ASN_{fc} = \frac{n - c}{q} F(c|n + 1) + \frac{c + 1}{p} (1 - F(c + 1|n + 1))$$

where $F(x|n)$ denotes the probability of x or fewer defectives in a sample of n.

Curtailment of single-sampling plans is usually not recommended because of the difficulty in estimating the process average from such data. Such procedures are quite common with double- or multiple-sampling plans where the first sample may be inspected fully and later samples curtailed.

An unbiased process average proportion defective from fully curtailed single sample data may be estimated using the method of Girshick, et al. (1946) as:

Lot rejected: $\hat{p} = \dfrac{c}{U - 1}$

Lot accepted: $\hat{p} = \dfrac{d}{U - 1}$

where

c = acceptance number

d = number defectives found

U = number units inspected

With semicurtailed inspection, the formula becomes

Lot rejected: $\hat{p} = \dfrac{c}{U - 1}$

Lot accepted: $\hat{p} = \dfrac{d}{U}$

For example, suppose the sampling plan n = 10, c = 1 were to be used with semicurtailed inspection and the second defective was found as the sixth item inspected. Inspection would stop since it is obvious that the lot would be rejected. An estimate of the process average would be

$$\hat{p} = \frac{1}{6 - 1} = .20$$

Calculation of the average sample number if the fraction defective were actually .20 using the binomial distribution gives

$$ASN_c = 10 \ F(1|10) + \frac{2}{.20} \ (1 - F(2|11))$$

$$= 10(.3758) + 10(.3826) = 7.584$$

This indicates that semicurtailment would give an average saving of 2.416 units per inspection at the cost of some precision in estimating the process average.

The concept of average sample number is very useful in determining the average number of samples that will be inspected in using more advanced sampling plans. In double-sampling plans, for example, the second sample is taken only if results from the first sample are not sufficiently defini- tive to lead to acceptance or rejection outright. In such a situation the inspection may be concluded after either one or two samples are taken and so the concept of average sample number (ASN) is necessary to evaluate the average magnitude of inspection in the long run.

TOLERANCE AND CONFIDENCE INTERVALS

Specifications are sometimes written in terms of tolerance intervals. This is particularly true in applications of acceptance sampling in the relia- bility and life-testing areas. Tolerance intervals specify limits which are estimated to contain a specified proportion of the population π with given confidence γ. Confidence intervals merely estimate a range within which a population parameter is expected to lie with a given confidence γ. Both kinds of intervals are available for attributes- and variables-type sampling.

An example of a specification of this type is the following: At least a proportion π of the population must be acceptable with a confidence level (coefficient) of γ. (Or, more simply stated: 100π percent reliability with 100γ percent confidence).

In testing a lot to this type of specification, Type B probabilities are used, since the specification refers to the population produced by the producer's process — not the specific lot. Now just one test of conformance to the specification will be made to accept or reject the lot. The term *confidence* is taken to mean that, whatever method is used, it is to give the correct result in approving a lot, as equaling or exceeding the specified reliability (100π percent), more than 100γ percent of the time in repeated applications. The test on a particular lot will either be correct or incorrect in a single application. But, in the long run, it will accept lots having at least a proportion π nondefective at least a proportion γ of the time. In this sense an accepted lot can be viewed with γ confidence as having π or greater proportion of units conforming to the specification.

To illustrate, consider an example given by Mann, Schafer, and Singpurwalla (1974, p. 374) rephrased as follows: Suppose that n = 20 and the observed number of failures is x = 1. What is the reliability π of the units sampled with 90 percent confidence? Here π is unknown and γ is to be .90. It is necessary to obtain

$$\Pr[q' \geq \pi_0] \geq .90$$

where π_0 is a lower tolerance limit of q', the fraction conforming in the population. For a binomial distribution such as this, the tolerance limit problem resolves itself into finding a lower confidence limit on q' in the population sampled. The binomial tables give

$$\sum_{i=2}^{20} c_i^{20} p^i q^{20-i} = .900+$$

with p = .181 and with corresponding q = .819. So

$$P(q \geq .819) = .90$$

and the lower tolerance limit is π = .819.

An alternate approach is to find the reliability or the confidence

desired directly from the OC curve. This may be done through the use of
the relation

$$\pi = 1 - p \qquad \gamma = 1 - P_a$$

The Type B OC curve for the plan n = 20, c = 1 is shown in Fig. 4-4.
   We see that for

$$P_a = 1 - \gamma = 1 - .90 = .10$$

the corresponding p value from the OC curve is .181. Therefore, as before,
the estimated reliability from the sample is

$$\pi = 1 - p = 1 - .181 = .819$$

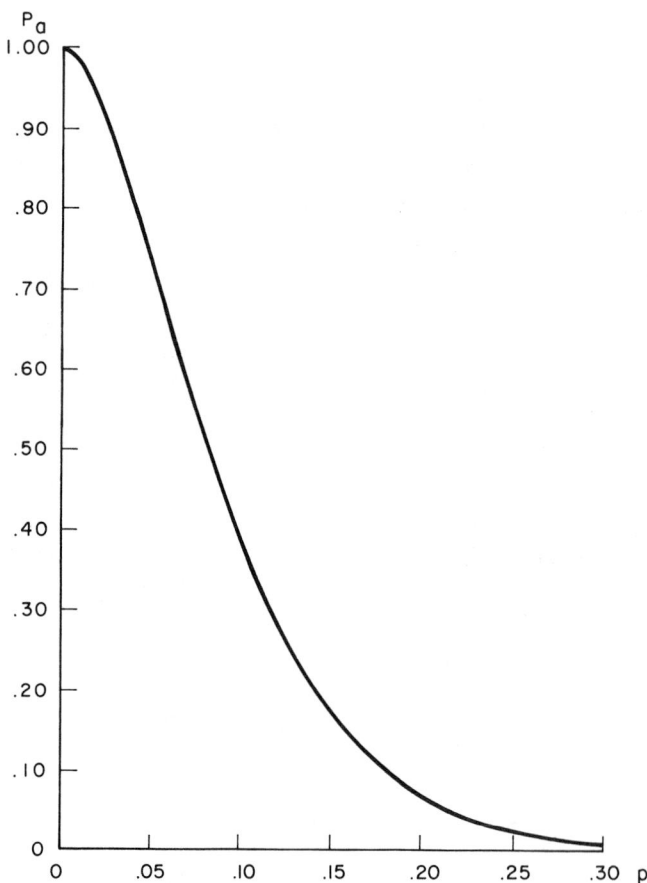

FIG. 4-4   Type B OC curve (n = 20, c = 1).

These relationships can also be employed to find a sampling plan to be used with specifications of the tolerance interval type. Suppose the life of a tire is specified to be such that 87 percent of the population must last more than 20,000 miles with 75 percent confidence. Here we have

$$P_a = 1 - \gamma = 1 - .75 = .25$$

and

$$p = 1 - \pi = 1 - .87 = .13$$

Then it is clear that the plan n = 20, c = 1 would satisfy this requirement since for p = .129 the P_a = .25.

Note that when specifications are stated in terms of tolerance intervals, only one point on the OC curve is specified. Thus, the plan n = 10, c = 0 also satisfies the requirements of the tire example but does not offer the producer as much protection against good lots being rejected.

It should be pointed out that this type of problem may also be solved using measurements and a variables sampling plan with a reduction in sample size. The procedure involved is much the same and will be discussed under variables sampling plans.

LEVELS AND RISKS

It is usually desirable to set up a sampling plan with both the producer's and consumer's interests in mind. This benefits both since their interests are not mutually exclusive and are in fact to a large extent compatible as seen in Table 4-13. While the producer and consumer risks are fairly well defined in terms of good product rejected and bad product accepted,

TABLE 4-13   Producer and Consumer Interests

	Producer	Consumer
Good lots rejected	Good product lost (producer risk)	Potential higher cost
Bad lots accepted	Potential customer dissatisfaction	Paid for bad product (consumer risk)

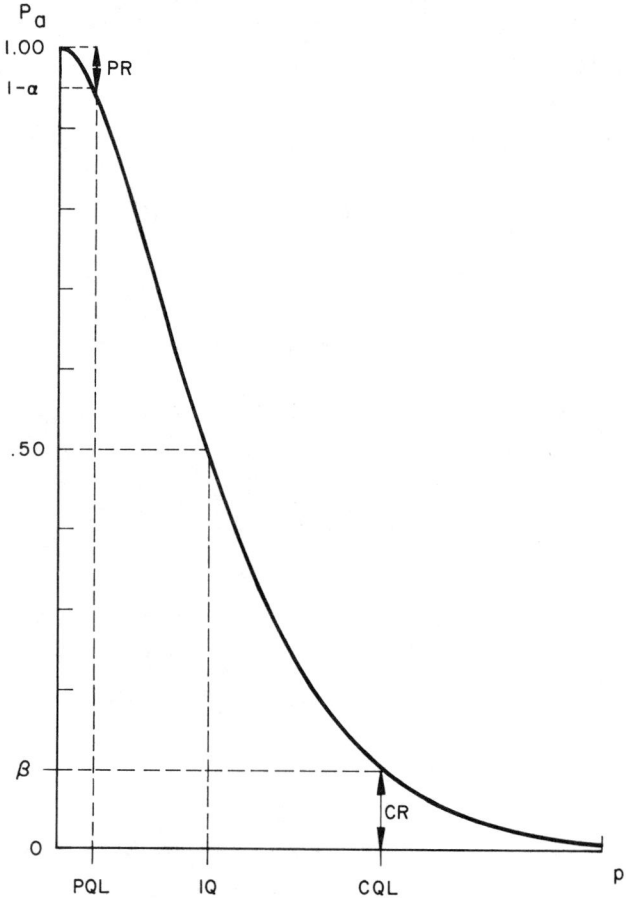

FIG. 4-5   PQL, CQL, PR, CR (n = 20, c = 1).

respectively, each has an interest in having reasonable levels maintained
for the other.

Since specification of two points may be used to define an operating
characteristic curve, it is often desirable to specify these points in
terms of the producer and the consumer. Thus we have

> Producer's quality level (PQL). A level of quality which should
> > be passed most of the time. The state of the art almost
> > always prohibits this from being a fraction defective of zero.
> Producer's risk (PR). The risk of having PQL material rejected by
> > the plan.
> Consumer's quality level (CQL). A level of quality which should
> > be rejected most of the time.

> Consumer's risk (CR). The risk of having CQL material accepted by
> the plan.

It is customary (though not necessary) to designate the producer's risk as
.05 indicating P_a = .95 at the PQL and the consumer's risk as .10 to give
P_a = .10 at the CQL. This value of the CQL in percent is called the *lot
tolerance percent defective* (LTPD) or 10 percent limiting quality of the
plan. Fig. 4-5 shows the location of these points on the OC curve for
n = 20, c = 1. While the PR and CR may take on any values, if the traditional values are taken, we have for n = 20, c = 1: PQL = .018, PR = .05,
CQL = .181, CR = .10.

In addition to these points, a third important point on the curve is
defined as

> Indifference quality level (IQ). The point where the producer and
> the consumer share a 50 percent probability of acceptance.

This point characterizes the plan in the sense of equal risk, although it
is unlikely that the producer will stay in business if 50 percent of the
lots are rejected. Rather, the IQ quantifies the area of vagueness or
indifference between the consumer and the producer.

CLASSIFICATION OF DEFECTS

Defect types are not all of the same concern. Dodge and Torrey (1956)
have pointed out that this is because

1. Defects of different kinds are not all equally serious.
2. Defects of the same kind differ in seriousness according to the
   extent of departure from specified limits or standards.

Accordingly, defects are sometimes classified into groups which reflect
their seriousness. Quality levels for sampling inspection are set accordingly. One such classification has been given by the Statistical Research
Group (1948, p. 82) as

> Major. Will cause failure of the item to function as intended.
> Minor. Will impair the efficiency, shorten the lifetime, or
> otherwise reduce the value of the item.
> Irregularity. A departure from good workmanship not affecting
> the performance or life of an item.

Sometimes an additional category is added reflecting concern for product safety. Using the definition of MIL-STD-414, for example, this may be,

> Critical. Could result in hazardous or unsafe conditions for individ-
> uals using or maintaining the product.

A leak or a flat spot might be a major defect in a tire. A blemish could
be a minor defect while an illegible letter in the brand name may be an
irregularity. Clearly a weak spot or damaged cord that could lead to a
blowout would be a critical defect.

It is essential that any classification of defects be carefully and
explicitly defined before it is used.

CHOOSING QUALITY LEVELS

The choice of quality levels with which to construct a sampling plan must
be made considering the seriousness of the defects to which it is applied,
the operating characteristics of the resulting sampling plan, economic consequences in terms of sample size, the ability of the producer to meet the
levels, and the needs of the consumer which must be met. The construction
of any sampling plan involves a trade-off of these items.

As mentioned earlier, no acceptance control procedure should be instituted without process control information. Necessary and obtainable
quality levels must be determined so that the acceptance sampling scheme
employed is a cost-effective compromise in the interest of both the producer and the consumer. The best tool in choosing quality levels is a
well-designed control chart and possibly process optimization studies to
see what levels can economically be met.

Acceptance control should not be thought of as a policing operation
but rather as the first step toward mutually acceptable process controls
to maximize the cost effectiveness of both operations. This may allow
eventual use of surveillance inspection to detect any departure from
agreed on levels at minimal cost to both parties.

MEASURES OF SAMPLING PLANS: TERMINOLOGY

A distinction may be made between product which is definitely objectionable
to the consumer on the one hand and product which fails to meet specifications on the other. While it is hoped that these categories overlap, they
need not always coincide. This can be seen in Table 4-14. Also specifications imposed on a product, its subassemblies and constituents may be essential to the production operation and its efficiency, but have no relation

TABLE 4-14 Specifications and Defects

Specifications		Product performance	
		Nondefective	Defective
	Met	OK	Need tighter specifications
	Not met	May loosen specifications	OK

to the quality of the product as perceived by the consumer. Thus, parts
may be restricted to certain dimensions for the efficient operation of a
feed mechanism in production but have no relation whatsoever to quality
as measured by the ultimate consumer. For this reason, recent documents
dealing with terminology in quality control have attempted to make a distinction between satisfying the ultimate user and satisfying the specifications. For example, ASQC Standard A2(1978) entitled, "Terms, Symbols and
Definitions for Acceptance Sampling to Control the Percent or Proportion
of Variant Units in a Lot or Batch" defines the following:

Defect. A departure of a quality characteristic from its intended
    level or state that occurs with a severity sufficient to cause
    an associated product or service not to satisfy intended normal,
    or reasonably forseeable, usage requirements.

Defective. A unit of product or service containing at least one de-
    fect, or having several imperfections that in combination cause
    the unit to fail to satisfy intended normal, or reasonably for-
    seeable usage requirements. [The word *defective* is appropriate
    for use when a unit of product or service is evaluated in terms
    of customer usage (as contrasted with conformance to specifica-
    tions.)]

Nonconformity. A departure of a quality characteristic from its
    intended level or state that occurs with a severity sufficient
    to cause an associated product or service not to meet a specifi-
    cation requirement.

Nonconforming unit. A unit of product or service containing at least one nonconformity.

Clearly, the term *defective* is also appropriate for use in the handling of components and materials internal to a production operation since one operation supplying material to another would take on the roles of producer and consumer respectively. In such a situation, a part not meeting specifications would be viewed as a defective by the consuming operation while it may be regarded as a nonconforming unit by the supplier, since the same part might go to a user internally or externally who would find it capable of satisfying usage requirements.

Since acceptance sampling is usually presented in terms of an adversary relationship between a producer and a consumer, and since, in most applications, interest in acceptance sampling is centered on satisfaction of usage requirements on the part of the consumer, the terms *defect* and *defective* will be used here with the understanding that the terms *nonconformity* and *nonconforming unit* should be used when evaluating an item against a specification when no evaluation is being made of its intended use internally or externally to the producer.

In summary, then, a sampling plan may be assessed, at any given incoming proportion defective p, by five basic measures as defined in ASQC Standard A2(1978).

1. <u>Probability of acceptance (P_a)</u>. "The probability that a lot will be accepted under a given sampling plan." A plot of P_a against p comprises the operating characteristic curve. Such curves are of two types:

   Type A. Plots the probability that a lot will be accepted against the proportion defective in the lot inspected.
   Type B. Plots the proportion of lots that will be accepted against the proportion defective in the producer's process which gives rise to the lot inspected.

2. <u>Average sample number (ASN)</u>. "The average number of sample units per lot used for making decisions (acceptance or non-acceptance)." Average sample number is meaningful in Type B sampling situations. A plot of ASN against p is called the *average sample number curve* for the plan.

3. <u>Average outgoing quality (AOQ)</u>. "The expected quality of outgoing product following the use of an acceptance sampling plan for a given value of incoming product quality." This is normally

calculated only when rejected lots are 100 percent inspected since
otherwise AOQ = p for a stream of lots all of incoming product
quality p. Average outgoing quality is meaningful in Type B sampling situations. A plot of AOQ against p is called the *average
outgoing quality curve* of the plan.

4. Average outgoing quality limit (AOQL). "For a given acceptance
    sampling plan, the maximum AOQ over all possible levels of incoming
    quality." This may be seen as the maximum point on the AOQ curve.
    The proportion defective at which the AOQL occurs is denoted as p_M.
    AOQL is sometimes shown as p_L.

5. Average total inspection (ATI). "The average number of units in-
    spected per lot based on the sample size for accepted lots and all
    inspected units in not-accepted lots." Thus, ATI is the total
    average number of units inspected for lots including sample units
    and units involved in 100 percent inspection as required. Average
    total inspection is meaningful in Type B sampling situations. A
    plot of ATI against p is called the *average total inspection curve*
    of the plan.

GRAPHS OF MEASURES

The principal measures of sampling plans are usually presented in the form
of graphs which show at a glance how the plan will perform against various
possible values of proportion defective. Since knowledge of the incoming
fraction defective is usually not available (otherwise there would be no
sense to sample), the graphs allow for rational matching of the plan to the
sampling situation. They portray performance against good and bad quality.
This allows selection of a plan on the basis of its protection and other
measures of performance without knowing the exact fraction defective to
which the plan will actually be applied.

Four such (Type B) curves are illustrated for the plan N = 20, n = 10,
c = 1:

Operating characteristic curve. Shows probability of acceptance plot-
    ted against possible values of proportion defective. It is used
    in assessing the protection afforded by the plan (Fig. 4-6).

Average sample number curve. Shows average sample number plotted
    against possible values of proportion defective. Used with
    plans involving several sampling stages, it shows how average

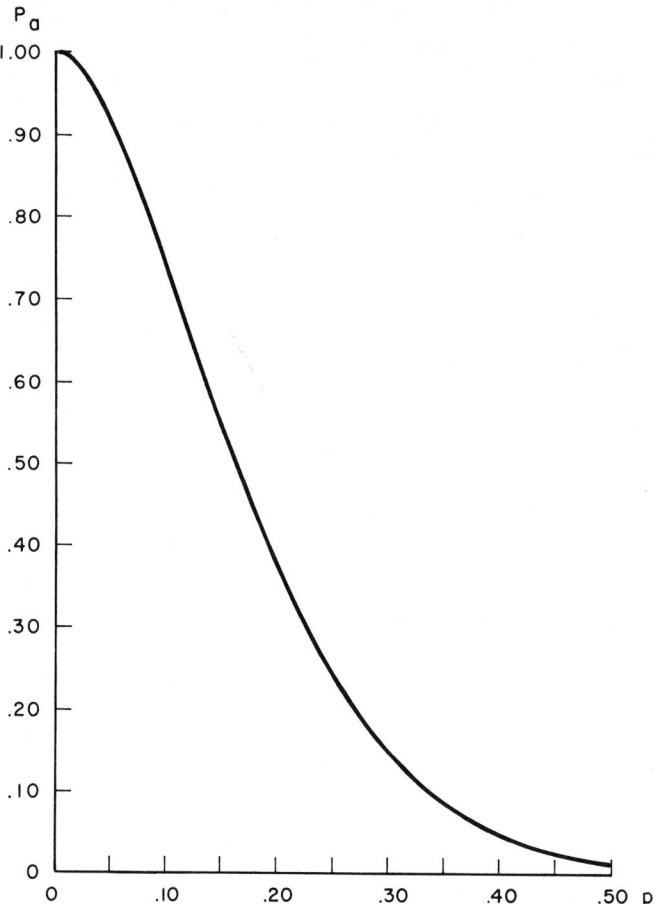

FIG. 4-6 Operating characteristic curve (n = 10, c = 1).

sample size varies as incoming quality changes. It is used in
assessing the inspection requirements for the plan in the
absence of rectification (Fig. 4-7).

Average outgoing quality curve. Shows average outgoing quality plotted against possible values of proportion defective. The AOQ at
the curve's maximum is the AOQL. The proportion defective at
which it occurs is labeled p_M. It is used in evaluating the
effect on average quality going to the consumer after 100 per-

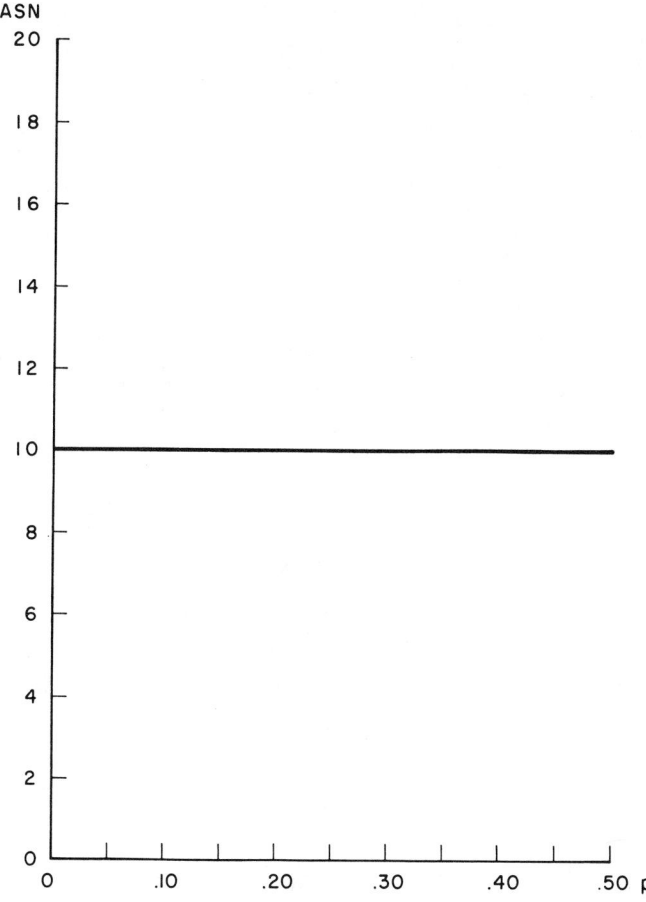

FIG. 4-7 Average sample number curve (n = 10, c = 1).

cent inspection of rejected lots, to determine the level of
assurance afforded to the consumer by a rectification procedure
(Fig. 4-8).

Average total inspection curve. Shows average total inspection
plotted against possible values of proportion defective. Used
with rectification procedures it indicates overall inspection
requirements for the total procedure (Fig. 4-9).

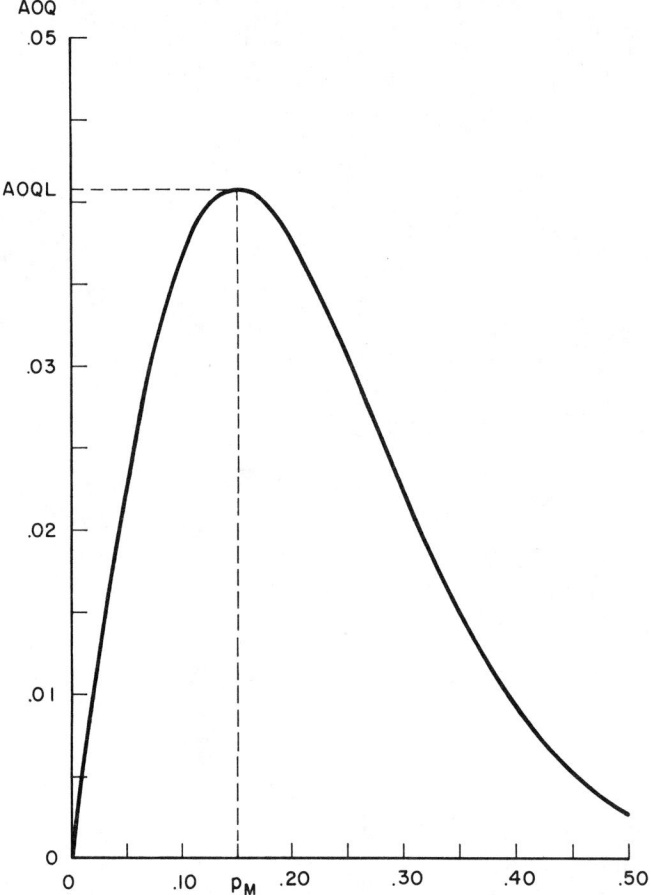

FIG. 4-8   Average outgoing quality curve (N = 20, n = 10, c = 1).

Knowledge of these component measures of sampling plans allows the quality engineer to properly prescribe the plan appropriate to the sampling situation.

SPECIFYING A PLAN

Discriminating use of sampling procedures demands knowledge and specification of the characteristics of the plans to be employed.   A primary con-

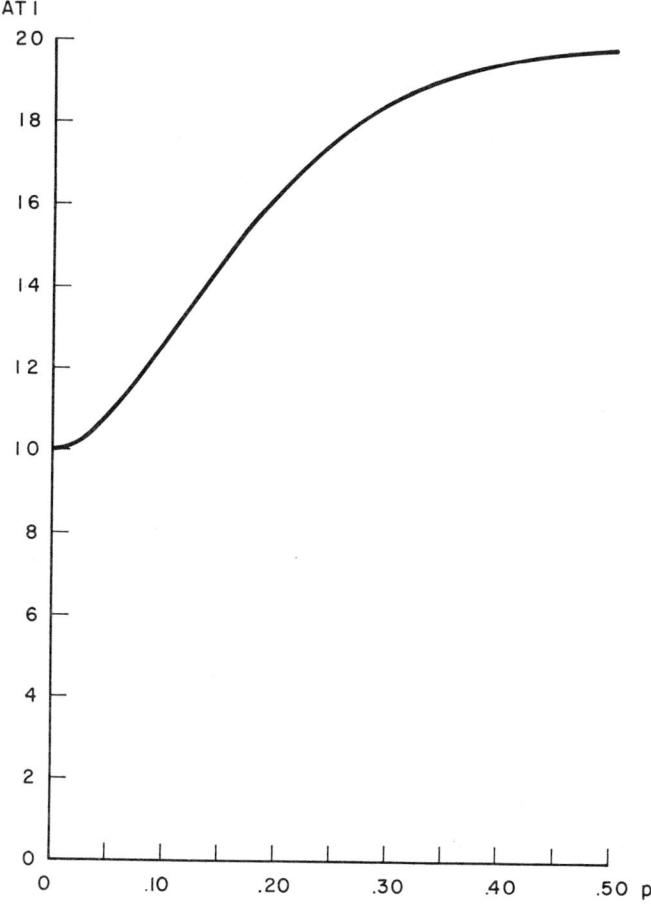

FIG. 4-9 Average total inspection curve (N = 20, n = 10, c = 1).

sideration is the protection afforded to both the producer and consumer.
Since two points may be used to characterize the OC curve, it is customary
to specify:

p_1 = producer quality level

p_2 = consumer quality level

α = producer risk

β = consumer risk

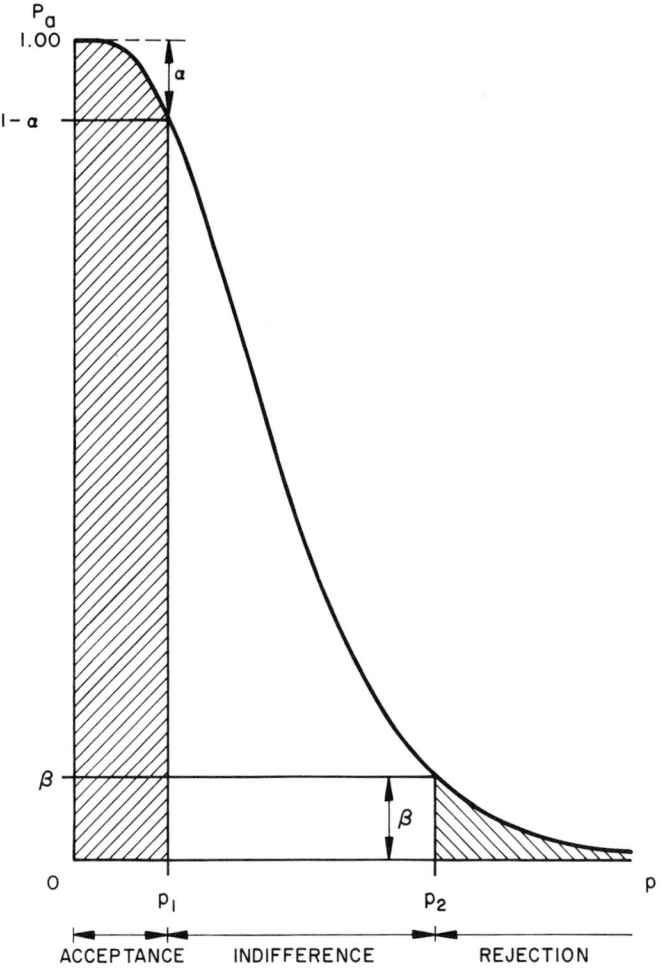

FIG. 4-10   Relation of p_1, p_2, $1 - \alpha$, and β to the OC curve.

For single-sampling attributes plans, $1 - \alpha$ and β can be determined directly from the distribution function of the probability distribution involved.
Fig. 4-10 shows the relation of these quantities to the OC curve. Also
shown are the regions of acceptance, indifference, and rejection defined
by these points. Quality levels of p_1 or better are expected to be accepted most of the time ($\geq 1 - \alpha$) by the plan depicted. Quality levels of p_2
or worse are expected to be rejected most of the time ($\leq 1 - \beta$) while intermediate levels will experience decreasing probability of acceptance
as levels move from p_1 to p_2. Occasionally, only one set of parameters
(p_1, α) or (p_2, β) is specified. When this is done, any plan having an OC

curve passing through the point meets the criterion. A single-sampling attributes plan may be specified by any two of the following: (p_1, α), (p_2, β), n, c. It may also be determined by specifying AOQL and one of the other values listed. The operating ratio $R = p_2/p_1$ is often used to characterize sampling plans. The operating ratio varies inversely with the acceptance number and may be used to derive individual plans for given values of α and β. Unless otherwise stated, α is usually taken to be .05 and β taken to be .10 since these values have become traditional in acceptance sampling having satisfied the test of long-term usage.

The chapters that follow will be primarily devoted to presenting the operating procedure for implementing various sampling plans and schemes together with means for determining P_a, ASN, AOQ, AOQL, and ATI under full and curtailed sampling. In general, except where explicitly stated, Type B measures will be given since they also act as conservative approximations to the Type A results.

PROBLEMS

1. A new firm sends a lot of 100 to qualify as a supplier of circuit boards for use in an electronic assembly. The previous supplier had a process average of 10 percent defective. In inspecting the lot, what type plan should apply: Type A or Type B? What probability distribution is the correct one for constructing the OC curve of the plan selected if specifications are in terms of percent defective? What probability distribution should be used if the specifications are in terms of defects per hundred units?

2. Consider the plan n = 4, c = 0 used on lots of size 8. Draw the Type A and Type B operating characteristic curves. Compute the probability of acceptance at p = .125, .25, .375, .50 as a minimum.

3. In Prob. 2, the lot size is raised to 16 but the plan n = 4, c = 0 is retained. What are the Type A probabilities of acceptance at p = .125, .25, .375, .50? How would these probabilities change if the lot size were made still larger?

4. A special military radio is specified to have less than 1 defect in 50 units. How many defects would be expected in a sample size of 100 units? What is the probability of 0, 1, 2, 3 defects in 100 units if quality were exactly at the specified level? What is the probability of acceptance if two defects were allowed in a sample of 100?

5. The plan n = 5, c = 1 is being used on the inspection of tires for a minor defect on a series of lots. The process average has been 12.5 percent defective for some time. The defect is hard to find and is missed 20 percent of the time. Draw the effective OC curve accounting for the inspection error. Use the points p = .125, .25, .375, .50 at a minimum for the true fraction defective.

6. If rejected lots are 100 percent inspected in Prob. 2, draw the AOQ curve. Should Type A or Type B probabilities be used? Compute AOQ for p = .125, .25, .375, .50 at a minimum.

7. Draw the ATI curve for Prob. 6. Compute ATI for p = .125, .25, .375, .50 at a minimum.

8. The sampling plan n = 15, c = 2 is being used for inspection of gaskets as received. If the incoming process fraction defective is .10 and the plan is curtailed only after finding the third defective gasket, what is the average sample number?

9. If a lot is rejected after 11 units are inspected in Prob. 8, what is the estimated fraction defective?

10. For the plan n = 32, c = 1, 12.2 defects per hundred units will be rejected 90 percent of the time, while 1.66 defects per hundred units will be accepted 90 percent of the time. If PR = CR = .10, what are the PQL and CQL for this plan expressed in defects per unit?

REFERENCES

American Society for Quality Control (1978), *American National Standard: Terms, Symbols, and Definitions for Acceptance Sampling*, ANSI/ASQC A2 (1978), American Society for Quality Control, Milwaukee, Wisconsin.

Anderson, V. L., and R. A. McLean (1974), *Design of Experiments*, Marcel Dekker, New York.

Dodge, H. F., and M. N. Torrey (1956), A Check Inspection and Demerit Rating Plan, *Industrial Quality Control*, 13(1): 5-12.

Girshick, M. A., F. Mosteller, and L. J. Savage (1946), Unbiased Estimates for Certain Binomial Sampling Problems with Applications, *Annals of Mathematical Statistics*, 17: 13-23.

Hicks, C. R. (1973), *Fundamental Concepts in the Design of Experiments*, Holt, Rinehart, and Winston, New York.

Juran, J. M. (1962), *Quality Control Handbook*, 2nd ed., McGraw-Hill, New York.

Juran, J. M., and F. M. Gryna (1970), *Quality Planning and Analysis*, McGraw-Hill, New York.

Mann, N. R., R. E. Schafer, and N. D. Singpurwalla (1974), *Methods for Statistical Analysis of Reliability and Life Data*, John Wiley and Sons, New York.

Ott, E. R. (1975), *Process Quality Control*, McGraw-Hill, New York.

Schilling, E. G. (1961), The Challenge of Visual Inspection in the Electronics Industry, *Industrial Quality Control*, 18(2): 12-15.

Statistical Research Group, Columbia University (1948), *Sampling Inspection*, McGraw-Hill, New York.

5

SINGLE SAMPLING BY ATTRIBUTES

The single-sampling plan is basic to all acceptance sampling. The simplest form of such a plan is single sampling by attributes which relates to dichotomous situations, i.e., those in which inspection results can be classified into only two classes of outcomes. This includes go no-go gauging procedures as well as other classifications, such as measurements in or out of specifications. Applicable to all sampling situations, the attributes single-sampling plan has become the benchmark against which other sampling plans are judged. It is employed in inspection by counting the number of defects found in the sample (Poisson distribution) or evaluating the proportion defective from processes or large lots (binomial distribution) or from individual lots (hypergeometric distribution). Single sampling is undoubtedly the most used of any sampling procedure.

OPERATION

Implementation of an attributes single-sampling plan is very simple. It involves taking a random sample of size n from a lot of size N. The sample

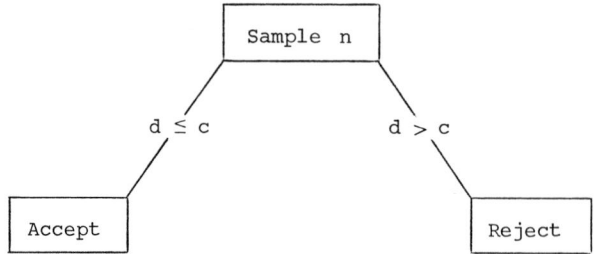

FIG. 5-1 Procedure for single sampling by attributes.

may be intended to represent the lot itself (Type A sampling) or the process used to produce the lot (Type B sampling). The number of defectives
(or defects) d found is compared to an acceptance number c. If the number
found is less than or equal to c, the lot is accepted. If the number found
is greater than c, the lot is rejected. The operation of the plan is illustrated in Fig. 5-1.

SELECTION

Tables of single-sampling attributes plans are available. Perhaps the two
best known sources are military standard MIL-STD-105D (1963) and the Dodge-
Romig tables (1959). These will be discussed in later chapters. The use
of such tables as a collection of individual plans provides ease of selection on the basis of the operating characteristics and other measures
classified therein.

TABLE 5-1 Procedures for Determining Single-Sampling Plans

Type plan	Method	Use
Type B (defectives) (defects)	Tables of Poisson unity values	Tables for derivation of plan given operating ratio R for tabulated values of α, β, and c. Poisson approximation to binomial for defectives. May be used as exact for defects.
Type B (defectives)	Binomial nomograph	Nomograph for derivation of plan given α, β, p_1, p_2. Uses binomial distribution directly. Hence, exact for defectives.
Type A (defectives)	f-Binomial nomograph	Uses binomial nomograph to derive Type A plans given α, β, p_1, p_2 through f-binomial approximation to hypergeometric distribution. Given lot size, gives approximate plan for defectives.
Type B (defects) (defectives)	Thorndyke chart	Procedure for use of Thorndyke chart for Poisson distribution to derive plan given α, β, p_1, p_2. Exact for defects. Approximate for defectives through Poisson approximation to binomial.
Type A (defectives)	Hypergeometric tables	Iterative procedure for derivation of exact hypergeometric plan given N, α, β, p_1, p_2 using Lieberman-Owen tables of hypergeometric distributions.

Procedures are also available for determining so-called two-point single-sampling plans for specified values of

p_1 = producer's quality level

p_2 = consumer's quality level

α = producer's risk

β = consumer's risk

where

$$R = \frac{p_2}{p_1} = \text{operating ratio}$$

Five such procedures will be set forth here. They relate to the derivation of plans as indicated in Table 5-1.

When constructing a plan for defects, rather than defectives, with these procedures use p as the number of defects per unit. The tables of unity values and the Thorndyke chart may then be used directly, where n is simply the sample size, i.e., number of units sampled.

Tables of Poisson Unity Values

Factors for construction of single-sampling plans are available in the literature which are based on the Poisson distribution and which provide excellent approximations to the binomial sampling situation as well. These include the original approach of Peach and Littauer (1946) together with the work of Grubbs (1949) and Cameron (1952) and the tabulations by the U.S. Army, Chemical Corps Engineering Agency (1953). These so-called unity values are expressed as the product np, where n = sample size and p = proportion defective. When dealing with defects, p = defects per unit. The unity values can be easily used to construct and evaluate plans on the basis of the operating ratio. Appendix Tables T5-1 and T5-2 present the values for single-sampling attributes plans developed by Cameron (1952). Additional sets of unity values for matched sets of single, double, and multiple plans have been developed by Schilling and Johnson (1979) and are presented in Appendix Table T6-1. They may be used in the customary situation in which α = .05 and β = .10. Other risk levels associated with p_1 and p_2 will also be found in the Cameron (1952) tables. The values for other risk levels are used in a manner identical to those for the conventional levels of α and β. The theory of construction of unity values is explained by Duncan (1974).

To derive a plan having $\alpha = .05$ and $\beta = .10$, determine the operating ratio $R = p_2/p_1$. Appendix Table T5-1 lists values of R corresponding to various acceptance numbers c and risks α and β. The value of c tabulated closest to the desired value of R for the indicated risks is the acceptance number to be used. Choose a value of R from the table equal to or just less than the value desired, to be conservative, in terms of guaranteeing both risks. To find the sample size n divide np_1 by p_1. Always round up in obtaining sample size.

Appendix Table T5-2 shows probability of acceptance associated with various unity values for the plans and acceptance numbers given in Appendix Table T5-1. This table may be used to evaluate the OC curve of any single-sampling attributes plan. Unity values are shown for various acceptance numbers c tabulated in columns by probability of acceptance P(A). Simply divide the unity values for a given acceptance number by the sample size of the plan to get values of p and find the probability of acceptance for these values from the column headings.

For example, for $\alpha = .05$ and $\beta = .10$, suppose it is desired to have $p_1 = .018$ and $p_2 = .18$ so that $R = 10$. Then for the closest value of R listed (10.96) in Appendix Table T5-1, the acceptance number is shown to be $c = 1$. The sample size is $.355/.018 = 19.7$ which rounds up to 20. The plan is $n = 20$, $c = 1$. If the probability of acceptance is to be evaluated for $P_a = .10$, use Appendix Table T5-2 to find the corresponding $p = 3.89/20 = .194$. The indifference quality for this plan is $1.678/20 = .084$.

In a similar manner, 13 points on the OC curve can be described using Appendix Table T5-2 to obtain the following:

P_a	p	P_a	p
.995	.103/20 = .005	.500	1.678/20 = .084
.990	.149/20 = .007	.250	2.693/20 = .135
.975	.242/20 = .012	.100	3.890/20 = .194
.950	.355/20 = .018	.050	4.744/20 = .237
.900	.532/20 = .027	.025	5.572/20 = .279
.750	.961/20 = .048	.010	6.638/20 = .332
		.005	7.430/20 = .372

The procedure described holds the producer's risk exactly, while the consumer's risk can vary slightly from the specified value when sample size is rounded. It will seldom be possible to hold both risks exactly.

If the consumer's risk is to be held at the expense of the producer's risk, obtain $np_{.10}$ corresponding to P(A) = .10 from Appendix Table T5-2 and divide by p_2 to obtain the sample size. If the result differs from the sample size using p_1, use the larger sample size, or if only one of p_1 and p_2 is of primary interest, use the sample size associated with the value of interest.

Binomial Nomograph

The Larson (1966) nomograph presented earlier as Fig. 3-5 can also be used to derive single-sampling attributes plans. Given p_1, p_2, α and β, plot p_1 and p_2 on the left scale for proportion defective shown as "probability of occurrences on a single trial (p)." Then plot $1 - \alpha$ and β on the right scale for probability of acceptance shown as "probability of c or fewer occurrences in n trials (P)." With a straight edge, connect the points: p_1 with $1 - \alpha$ and p_2 with β. At the intersection of the lines, read the sample size n and the acceptance number c from the grid.

The nomograph can also be used to evaluate the operating characteristic curve of a plan. To do this, plot the point (n,c) on the grid. Locate

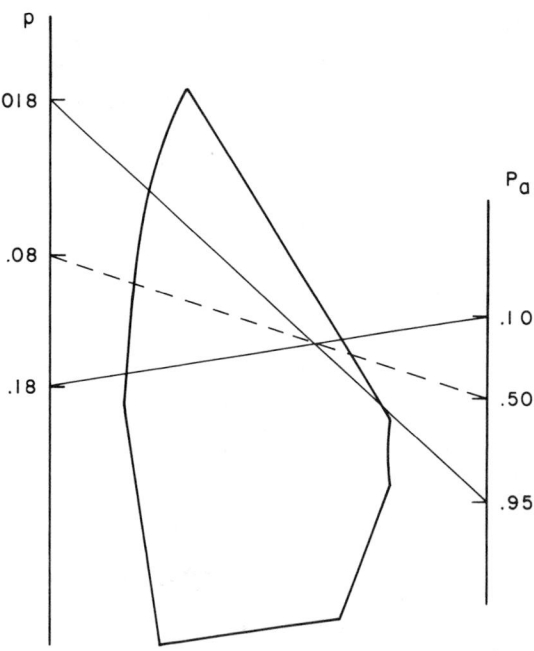

FIG. 5-2   Larson nomograph for n = 20, c = 1.

the position of each value of probability of acceptance to be evaluated on
the right P scale. Then draw a line from P through (n,c) and read the corresponding fraction defective p on the left scale. The procedure can be
reversed to find the value of probability of acceptance for a given value
of proportion defective set on the left scale.

For example, suppose α = .05, β = .10, p_1 = .018, and p_2 = .18. The
derivation of the plan n = 20, c = 1 is shown in Fig. 5-2. The dotted
line shows an indifference quality of p = .08 for this plan.

The Larson nomograph is based on the binomial distribution and so will
allow direct evaluation of Type B plans for fraction defective. It allows
derivation and evaluation of plans for values of probability of acceptance
not shown in the Cameron tables. It provides a reasonable and conservative
approximation for the derivation of plans when the hypergeometric distribution should apply and the binomial approximation to the hypergeometric distribution is appropriate.

f-Binomial Nomograph

Shaul P. Ladany (1971) has provided a method for adapting the Larson binomial nomograph for use in deriving Type A plans for finite lots of size N
when the f-binomial approximation to the hypergeometric distribution
applies. This is when the sampling ratio F = n/N > 0.1 and the fraction
defective p $\leq$ 0.1. Other approximations are listed in Fig. 3-9. A somewhat more complicated direct method for deriving Type A plans using the
Lieberman-Owen (1961) tables is given later in this chapter.

To use the binomial nomograph in this context, for specified p_1, p_2,
α, β, determine two psuedo sample sizes:

$$n_1 = p_1N \qquad n_2 = p_2N$$

and consider two lines, simultaneously drawn, as in Fig. 5-3.

Line A from β on the right scale, through the intersection of c (to
be determined) with the n_2 line on the grid.
Line B from 1 - α on the right scale and the intersection of the
same c with n_1 line on the grid.

When the two lines intersect on the left scale and have identical values
of c at n_2 for line A, and n_1 for line B, respectively, the plan has been
determined. For the value of c specified, the value of the sampling
fraction F = n/N can be read at the point of intersection on the left

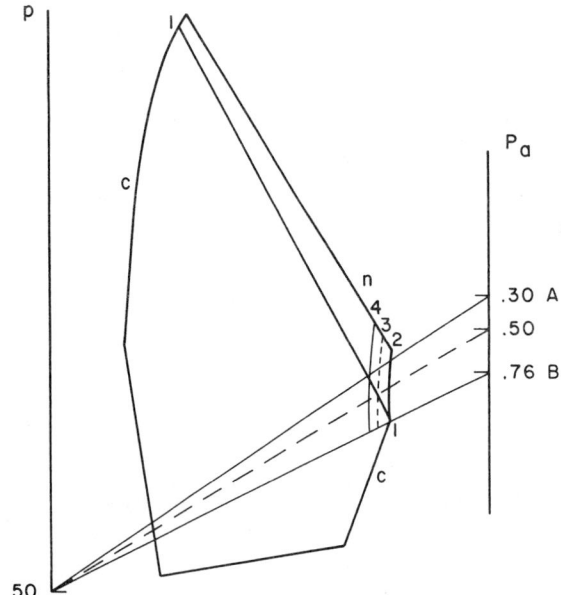

FIG. 5-3   f-Binomial application of nomograph.

scale. Multiplication of this value of F by the lot size will give the
sample size n.

Ladany (1971) suggests the use of a thread or rubber band at β and
1 - α on the right scale which, when looped around a stylus and run up and
down the left scale, would provide a flexible representation of the two
lines. In any event, practice with the method soon makes the use of two
transparent rulers adequate for the purpose.

The procedure may, of course, also be used to evaluate the Type A
probability of acceptance for the plan specified by N, n_1, c, for a given
value of p as follows:

1. Locate the point representing c and n = Np on the grid.

2. Locate the value of n/N on the left (p) axis.

3. Draw a line through these points. The intersection of this line
    with the right (P) scale will indicate the cumulative probability
    of acceptance for that value of p.

Clearly, the procedure could be reversed to find the value of p corresponding to a given value of probability of acceptance. This would involve
drawing a line from P_a on the right scale to n/N on the left. The intersection of the line with the curve for the acceptance constant c involved
gives Np. Division of this value by N gives p.

For example, suppose lot size is N = 20 and it is desired to develop a Type A plan having α = .24, β = .30, p_1 = .10, p_2 = .20. Then we have n_1 = 20(.10) = 2 and n_2 = 20(.20) = 4. The binomial nomograph would appear as in Fig. 5-3.

Line A passes from .30 through (n_2 = 4, c = 1) to p = .50. Line B passes from P = .76 through (n_1 = 2, c = 1) to p = .50. The lines intersect at .50 on the left axis and so the plan has sampling fraction F = n/N = .50. Hence, since N = 20, we find n = 10 and the plan is N = 20, n = 10, c = 1. The indifference quality level may be evaluated using the dotted line in Fig. 5-3 as Np = 3 so that p = 3/20 = .15. This is exactly the value obtained for the plan earlier in Table 4-2 using the tables of Lieberman and Owen (1961).

It should be noted that the discrete nature of the hypergeometric distribution precludes certain fractions defective from occurring. This should be considered throughout in application of the binomial nomograph in this way.

Thorndyke Chart

Although somewhat more complicated than Larson's binomial nomograph, the Thorndyke (1926) chart, as given in Dodge and Romig (1959), may be used to derive a single-sampling attributes plan. This chart, presented earlier as Fig. 3-6, uses cumulative Poisson probabilities on the ordinate and unity values np on the abscissa. Curves for various acceptance numbers are shown. The procedure, adapted from Burgess (1948), is as follows:

1. Project an imaginary horizontal line from β.
2. Place the bottom edge of a piece of paper on the line so that the bottom left corner of the paper lies above np = 1.
3. Project an imaginary horizontal line from 1 - α on the ordinate and mark its intersection with the paper on the vertical left edge.
4. Project an imaginary vertical line from the unity value equal to the operating ratio R desired up to the bottom edge of the paper and mark the paper at the point of intersection.
5. Slide the bottom edge of the paper along the line projected from β until a single c curve most nearly passes through both the 1 - α and the R marks on the paper at the same time. This is the value of c for the sampling plan.

6. For this value of c, read the value of np corresponding to $1 - \alpha$
   or the value of np corresponding to β. Division of either of
   these unity values by p_1 or p_2, respectively, will give the sample
   size for the plan. Unless one risk is specifically to be held,
   use the larger of the two sample sizes or choose an appropriate
   compromise intermediate value.

For use in determining plans for defects, simply substitute the desired
values of defects per unit for p in the above procedure.

To illustrate this method, suppose it is desired to have $\alpha = .05$,
$\beta = .10$, $p_1 = .018$, $p_2 = .18$. The corresponding operating ratio is R = 10.

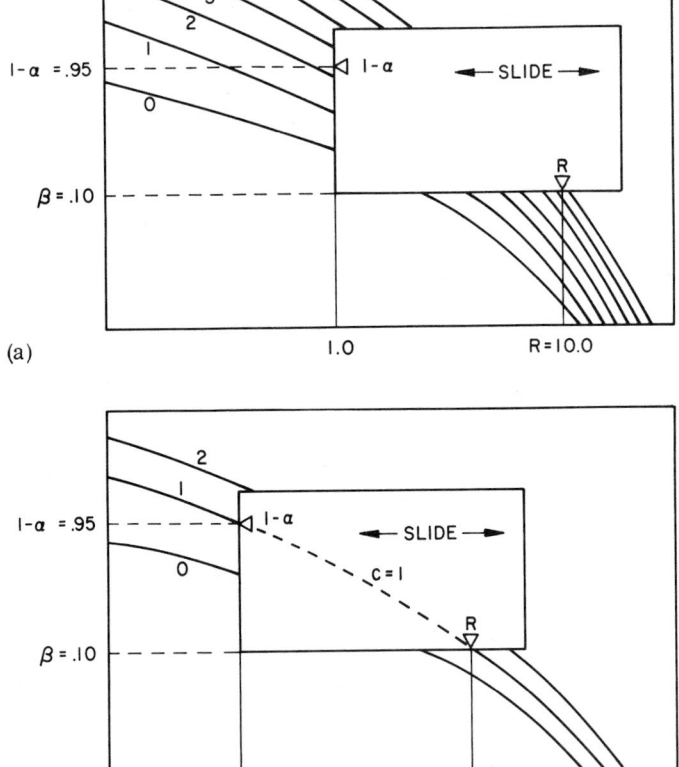

FIG. 5-4 Thorndyke chart to derive plan: $\alpha = .05$, $\beta = .10$, $p_1 = .018$,
$p_2 = .18$. (a) Construction of slide. (b) Derivation of plan.

The resulting Thorndyke chart is shown in Fig. 5-4 and shows the appropriate acceptance number to be $c = 1$. The curve for $c = 1$ shows $np_{.95} = .36$ and $np_{.10} = 3.8$. Dividing these by p_1 and p_2, respectively, we obtain $n = 20$ and $n = 21$. If the producer's risk is to be held, the plan $n = 20$, $c = 1$ would be used.

The Thorndyke chart is based on Poisson probabilities and so like Cameron's tables serves as a good approximation to the binomial distribution for defectives and is exact in dealing with defects per unit. It provides a procedure for determining a plan based on the Poisson distribution for values not tabulated in the tables which employ unity values.

Hypergeometric Tables

When sampling from a single (isolated) lot, the construction of a sampling plan is complicated by the computationally cumbersome hypergeometric distribution. Accordingly, when the situation is such that the approximations shown in Fig. 3-9 are appropriate, they should be used. This may be thought of as substituting the appropriate Type B OC curve for the approximating distribution in place of that of the Type A plan which uses the hypergeometric. Alternatively, Ladany's approach to the use of the binomial nomograph may be employed.

If the approximations are not appropriate or an exact solution is required, an iterative method may be used as outlined below. It requires that tables of the hypergeometric distribution, such as those of Lieberman and Owen (1961), be available. If they are not available, recourse must be made to the computer or in simple problems to hand calculation. The procedure for selecting a Type A plan using the hypergeometric distribution is as follows:

1. Specify PQL = p_1, PR = α, CQL = p_2, CR = β.
2. Use approximations of Fig. 3-9 if appropriate to develop analogous Type B plan.
3. Use hypergeometric tables to obtain plan as follows:
    a. Determine binomial plan n^*, c^* which meets the desired specifications.
    b. For a given lot size, N, determine the number defective in the lot $D_1 = Np_1$ corresponding to the PQL and $D_2 = Np_2$ corresponding to the CQL. Start at $n = n^*$ and $c = c^*$ and iterate through the hypergeometric tables, alternating D_2 and D_1

between successively lower sample sizes and acceptance numbers until a minimum sample size plan meeting the specifications is obtained. This involves simultaneously satisfying the inequalities:

$$F(x) \geq 1 - \alpha \quad \text{at } D_1$$
$$F(x) \leq \beta \quad \text{at } D_2$$

c. Iterate as follows:

N	n	D_2	x	F(x)		N	n	D_1	x	F(x)
N	n*	D_2	c*	β_1	$\rightarrow$	N	n*	D_1	c*	$(1-\alpha)_1$
N	n_a	D_2	c_a	β_2	$\leftarrow$	N	n_a	D_1	c_a	$(1-\alpha)_2$
N	n_b	D_2	c_b	β_3	$\rightarrow$	N	n_b	D_1	c_b	$(1-\alpha)_3$
. . . .						. . . .				
N	n_k	D_2	c_k	$\beta_k \leq \beta$	$\rightarrow$	N	n_k	D_1	c_k	$(1-\alpha)_k \geq 1 - \alpha$

Having started with the binomial plan, which because of its conservative nature assures $\beta_1 \leq \beta$ and $(1-\alpha)_1 \geq 1 - \alpha$, lower the sample size until the inequalities are violated. Next lower the acceptance number and again successively drop the sample size until it is confirmed that the inequalities do not hold. Repeat this process until an acceptance number is found for which a sample size cannot be obtained satisfying the inequalities. The plan identified for the next higher acceptance number is the hypergeometric plan satisfying the specifications. In general, probability of acceptance is lowered by increasing n and lowering c with consequent decrease in β and $1 - \alpha$.

To illustrate the method, suppose a hypergeometric plan is to be selected having $p_1 = .10$, $p_2 = .20$, $\alpha = .24$, and $\beta = .30$ with lot size 20. From the Larson nomogram, the binomial plan satisfying these specifications is n = 17, c = 2. Using the Lieberman-Owen (1961) tables for

$$D_1 = .1(20) = 2$$
$$D_2 = .2(20) = 4$$

the following results are obtained against the specified

$$\beta = .30 \quad 1 - \alpha = .76$$

Step	N	n	D_2	x	F(x)	N	n	D_1	x	F(x)	Plan
1	20	17	4	2	.0877	20	17	2	2	1.0000	
2	20	15	4	2	.2487	20	15	2	2	1.0000	n = 15, c = 2
3	20	14	4	2	.3426	20	14.	2	2	1.0000	
4	20	15	4	1	.0320	20	15	2	1	.4474	
5	20	12	4	1	.1531	20	12	2	1	.6526	
6	20	9	4	1	.3746	20	9	2	1	.8105	
7	20	10	4	1	.2910	20	10	2	1	.7632	n = 10, c = 1
8	20	10	4	0	.0433	20	10	2	0	.2368	
9	20	9	4	0	.0681	20	9	2	0	.2895	
10	20	8	4	0	.1022	20	8	2	0	.3474	

The plan is n = 10, c = 1. This is the same plan which was obtained
by the Ladany f-binomial adaptation of the Larson nomograph. In step 1
the binomial plan was used. Sample size was reduced to obtain the plan
n = 15, c = 2 in step 2, which satisfies the inequalities but which has
not been shown to be optimum in terms of sample size. Step 3 confirms
the plan in step 2. Step 4 lowers the acceptance number, while steps 5
to 8 lead to the plan n = 10, c = 1. Steps 9 and 10 confirm that no plan
exists for the next lower acceptance number.

These results could be obtained using a computer or possibly a program-
mable calculator. Also, the strategy employed can be used in the development of other types of sampling plans. Using a slightly different approach,
Guenther (1969) has outlined a general iterative procedure by which two-
point plans can be obtained from binomial, hypergeometric, or Poisson
tables.

MEASURES

The performance of single-sampling attributes plans may be characterized
by the measures given in Table 5-2. These may be evaluated using the binomial or Poisson distributions as appropriate to the sampling situation.
Care must be exercised in the use of the hypergeometric distribution due to
the depletion of the lot as samples are taken. The formulas should be
modified accordingly. The distributions are listed in Table 3-1. The x
and y values for calculation of the AOQL are explained in Chap. 14 and are
given in Appendix Table T14-1 based on the Poisson model. The approximation
shown for y was developed by the author. The notation $F(c|n)$ indicates the
probability of c or fewer defectives in a sample of n. The frequency

TABLE 5-2 Measures of Single-Sampling Attributes Plans

MEASURE	FORMULA
PROBABILITY OF ACCEPTANCE	$P_a = F(c\|n)$

MEASURE	FORMULA
AVERAGE SAMPLE NUMBER	o FULL INSPECTION $\quad$ASN $= n$ o SEMI-CURTAILED $\quad$ ASN$_c = n\ F(c\|n) + \dfrac{c+1}{p}[1-F(c+1\|n+1)]$ o FULLY CURTAILED $\quad$ ASN$_{fc} = \dfrac{n-c}{1-p}\ F(c\|n+1) + \dfrac{c+1}{p}[1-F(c+1\|n+1)]$
AVERAGE OUTGOING QUALITY	o DEFECTIVES FOUND REPLACED $\quad$ AOQ $= p\ P_a\left(\dfrac{N-n}{N}\right)$ o DEFECTIVES NOT REPLACED $\quad$ AOQ $= p\ P_a\left(\dfrac{N-n}{N-np}\right)$ o APPROXIMATE (n/N small) $\quad$ AOQ $\simeq p\ P_a$
AVERAGE OUTGOING QUALITY LIMIT[*]	o DEFECTIVES FOUND REPLACED WITH GOOD $\quad$ AOQL $= \dfrac{y}{n}\left(\dfrac{N-n}{N}\right)$ o DEFECTIVES NOT REPLACED $\quad$ AOQL $= \dfrac{y}{n}\left(\dfrac{N-n}{N-np}\right)$ o APPROXIMATE (n/N small) $\quad$ AOQL $= \dfrac{y}{n}$ $\quad\quad \simeq \dfrac{.4}{n}\ (1.25\ c+1)$ for $c \leq 5$ o AOQL OCCURS AT $\quad$ $P_M = \dfrac{x}{n}$
AVERAGE TOTAL INSPECTION	ATI $= n + (1-P_a)(N-n)$ $\quad\ = n\ P_a + N(1-P_a)$

[*]
 x,y values given in Appendix Table T14-1.

function $f(c|n)$ is interpreted accordingly.

To illustrate application of these formulas, suppose the measures of the plan n = 20, c = 1 are desired when sampling from a succession of lots of size N = 120. They are to be calculated using Type B (binomial) probabilities when the incoming proportion defective is p = .18.

Probability of acceptance:

$$P_a = F(1|20) = C_0^{20}(.18)^0(.82)^{20} + C_1^{20}(.18)^1(.82)^{19}$$

$$= .019 + .083 = .102$$

Average sample number:

Full inspection, ASN = 20

Semicurtailed

$$ASN_c = 20\ F(1|20) + \frac{1+1}{.18}[1 - F(2|21)]$$

$$= 20\ (.102) + \frac{2}{.18}[1 - .244] = 10.44$$

Fully curtailed

$$ASN_{fc} = \frac{20-1}{1-.18}\ F(1|21) + \frac{1+1}{.18}[1 - F(2|21)]$$

$$= \frac{20-1}{1-.18}(.087) + \frac{1+1}{.18}[1 - .244] = 10.42$$

Average outgoing quality:

Defectives found replaced

$$AOQ = .18(.102)\left(\frac{120-20}{120}\right) = .015$$

Defectives not replaced

$$AOQ = .18(.102)\left(\frac{120-20}{120-20(.18)}\right) = .016$$

Approximate

$$AOQ = .18(.102) = .018$$

Average outgoing quality limit:

Defectives found replaced

$$AOQL = \frac{.8400}{20}\left(\frac{120-20}{120}\right) = .035$$

Defectives not replaced

$$AOQL = \frac{.8400}{20}\left(\frac{120-20}{120-20(.18)}\right) = .036$$

Approximate

$$AOQL = \frac{.8400}{20} = .042$$

AOQL occurs at

$$p_M = \frac{1.62}{20} = .081$$

Average total inspection:

ATI = 20 + (1 - .102)(120 - 20) = 109.8

These calculations illustrate the value of the approximations shown and how little the measures are affected by the variations shown in inspection technique. The formulas could also be evaluated for defects per unit using the Poisson model.

PROBLEMS

1. Construct single-sampling plans to the following specifications given in proportion defective with producer's risk .05 and consumer's risk .10:

   a. PQL = .04, CQL = .21

   b. PQL = .03, CQL = .13

   c. PQL = .02, CQL = .06

   Use the following techniques:

   a. Poisson unity values

   b. Binomial nomograph

   c. Thorndyke chart

   Compare results.

2. Plot the Type B operating characteristic curve for the following single-sampling plans using P_a = .95, .75, .50, .25, .10 at a minimum for plotting positions.

   a. n = 13, c = 1

   b. n = 32, c = 0

   c. n = 125, c = 2

   Use the following techniques:

   a. Poisson unity values

   b. Binomial nomograph

   c. Thorndyke chart

   Compare results.

3. Obtain a Type A plan for lot size 200 with PQL = .025 and CQL = .125 with risks α = .05 and β = .10. Use

   a. Binomial nomograph

   b. Hypergeometric tables (optional)

   Determine the points P_a = .75, .50, .25 on the OC curve.

4. Derive a defects per unit plan having PQL = 1.1 and CQL = 12.2 defects per hundred units for risks α = .05 and β = .05. What is the indifference quality for this plan?

5. If lots are received in quantities of 1000, obtain the AOQ and ATI at the following points for the plans given in Prob. 2 above and find the AOQL of each.

   a. P_a = .95

   b. P_a = .50

   c. P_a = .10

6. Use the binomial nomograph to derive and compare the operating characteristic curves of the plans n = 5, c = 0, 1, at P_a = .95, .50, .10. Plot the AOQ and ATI curves for these plans for lots of size 200. Find their AOQL.

7. Use unity values to derive and compare the operating characteristic curves for sample sizes 5 and 10 for c = 1 at P_a = .95, .50, .10. Plot the AOQ and ATI curves for these plans for lots of size 200. Find their AOQL.

8. Use the Larson nomograph to derive a binomial sampling plan satisfying the specifications p_1 = .03, p_2 = .09, α = .05, β = .10. Then find the appropriate hypergeometric plan for use with a lot of N = 500. This illustrates the importance of considerations of lot size when the sampling fraction is high.

9. Using the Thorndyke chart, derive a plan to satisfy the conditions of Prob. 8 and compare with the results for that problem. Why is the sample size highest using the Thorndyke chart?

10. For even degrees of freedom, it is well known that the complement of the cumulative distribution function for the Poisson can be determined from $\chi^2/2$ with degrees of freedom 2c + 2. Using this fact, derive the unity values for c = 3 for risks α = .05 and β = .05. What is the operating ratio for c = 3? [Hint: See Cameron (1952).]

REFERENCES

Burgess, A. R. (1948), A Graphical Method of Determining A Single Sampling Plan, *Industrial Quality Control*, 4(6): 25-27.

Cameron, J. M. (1952), Tables for Constructing and for Computing the Operating Characteristics of Single Sampling Plans, *Industrial Quality Control*, 9(1): 37-39.

Dodge, H. F., and H. G. Romig (1959), *Sampling Inspection Tables, Single and Double Sampling*, 2nd ed., John Wiley and Sons, New York.

Duncan, A. J. (1974), *Quality Control and Industrial Statistics*, 4th ed., Richard D. Irwin, Homewood, Illinois.

Grubbs, F. E. (1949), On Designing Single Sampling Inspection Plans, *Annals of Mathematical Statistics*, 20: 242-256.

Guenther, W. C. (1969), Use of the Binomial, Hypergeometric and Poisson Tables to Obtain Sampling Plans, *Journal of Quality Technology*, 1(2): 105-109.

Ladany, S. P. (1971), Graphical Determination of Single-Sample Attribute Plans for Individual Small Lots, *Journal of Quality Technology*, 3(3): 115-119.

Larson, H. R. (1966), A Nomograph of the Cumulative Binomial Distribution, *Industrial Quality Control*, 23(6): 270-278.

Lieberman, G. J., and D. B. Owen (1961), *Tables of the Hypergeometric Probability Distribution*, Stanford University Press, Stanford, California.

Peach, P., and S. B. Littauer (1946), A Note on Sampling Inspection, *Annals of Mathematical Statistics*, 17: 81-85.

Schilling, E. G., and L. I. Johnson (1979), Tables for the Construction of Matched Single, Double, and Multiple Sampling Plans with Application to MIL-STD-105D, *Journal of Quality Technology*, 12(4): 220-229.

Thorndyke, F. (1926), Applications of Poisson's Probability Summation, *The Bell System Technical Journal*, 5: 604-624.

United States Department of the Army (1953), *Master Sampling Plans for Single, Duplicate, Double, and Multiple Sampling*, Manual No. 2, Chemical Corps Engineering Agency, Army Chemical Center, Maryland.

United States Department of Defense (1963), *Military Standard, Sampling Procedures and Tables for Inspection by Attributes*, (MIL-STD-105D), Washington, D.C.

6

DOUBLE AND MULTIPLE SAMPLING BY ATTRIBUTES

Double- and multiple-sampling plans reflect the tendency of many experienced inspectors to give a questionable lot an additional chance. Thus, in double sampling if the results of the first sample are not definitive in leading to acceptance or rejection, a second sample is taken which then leads to a decision on the disposition of the lot. This approach makes sense, not only as a result of experience, but also in the mathematical properties of the procedure. For one thing, the average sample number can usually be made to be less for a double-sampling plan than for a single-sampling plan with the same protection.

A natural extension of double sampling is to allow further additional samples to be taken to achieve even more discrimination in the disposition of a lot. Such procedures are called *multiple-sampling plans* when, as with double sampling, the last sample is constructed to force a decision at that point. That is, for a specific last sample (say the kth sample) it is so arranged that $r_k = a_k + 1$, where r_k is the rejection number and a_k is the acceptance number. Thus, double sampling is simply a special case of multiple sampling where $k = 2$.

Multiple-sampling plans allow even more flexibility and still further reduction in average sample size over double-sampling plans, but are often found to be difficult to administer because of the complexity of handling and recording all the samples required. As an example of the reduction in sample size that can be obtained, MIL-STD-105D (Code H, 1.5 AQL, Normal Inspection) shows that for plans matched to the single-sampling plan $n = 50$, $c = 2$, the average sample number at the 95th percentile is

Plan	ASN
Single	50
Double	43
Multiple	35

This is typical of the efficiency in sampling which may be generated by the use of double- and multiple-sampling procedures. Efficiency of this sort may be costly, however, in terms of administration, since there is an increasingly variable workload in going from single to double to multiple sampling. These plans offer an additional dimension to the application of sampling plans, however, by providing increased economy and flexibility when properly applied.

Double- and multiple-sampling plans are said to be matched to single-sampling plans when their OC curves coincide. The inherent shape of a multiple-sampling OC curve is, however, different from that of a single-sampling OC curve. Hence, plans are often matched at two points, usually at $P_{.95}$ and $P_{.10}$.

Inspection is often curtailed, that is, inspection is stopped after reaching a decision, or semicurtailed, that is stopped only on a decision to reject. In either case the first sample is almost always inspected in full so that estimates and records kept on the first sample will have a consistent sample size. Usually the average sample number is assessed at the producer's quality level, since this should be the sustained normal level of the operation of the plan if no problems occur.

OPERATION

Double Sampling

Application of a double-sampling plan requires that a first sample of size n_1 be drawn at random from the lot (usually assumed large). The number of defectives d is counted and compared to the first sample acceptance number a_1 and rejection number r_1.

If $d_1 \leq a_1$, the lot is accepted.

If $d_1 \geq r_1$, the lot is rejected.

If $a_1 < d_1 < r_1$, a second sample is taken.

If needed, a second sample of size n_2 is drawn. The number of defectives

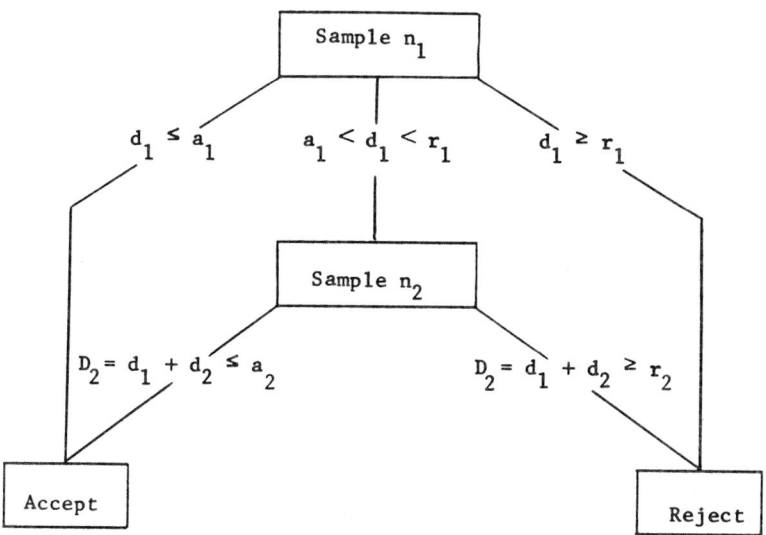

FIG. 6-1 Procedure for double sampling by attributes.

d_2 contained in the second sample is determined. The total number of defectives

$$D_2 = d_1 + d_2$$

is compared to the acceptance number a_2 and the rejection number r_2 for the second sample. In double sampling $r_2 = a_2 + 1$ to insure a decision on the second sample.

If $D_2 \leq a_2$, the lot is accepted.

If $D_2 \geq r_2$, the lot is rejected.

The operation of the plan is shown in Fig. 6-1.

Multiple Sampling

Multiple sampling involves the inspection of specific lots on the basis of from 1 to k successive samples as needed to make a decision. In MIL-STD-105D, k is taken as 7, that is, the multiple-sampling plans contained therein must reach a decision by the seventh sample. Multiple-sampling plans are usually presented in tabular form:

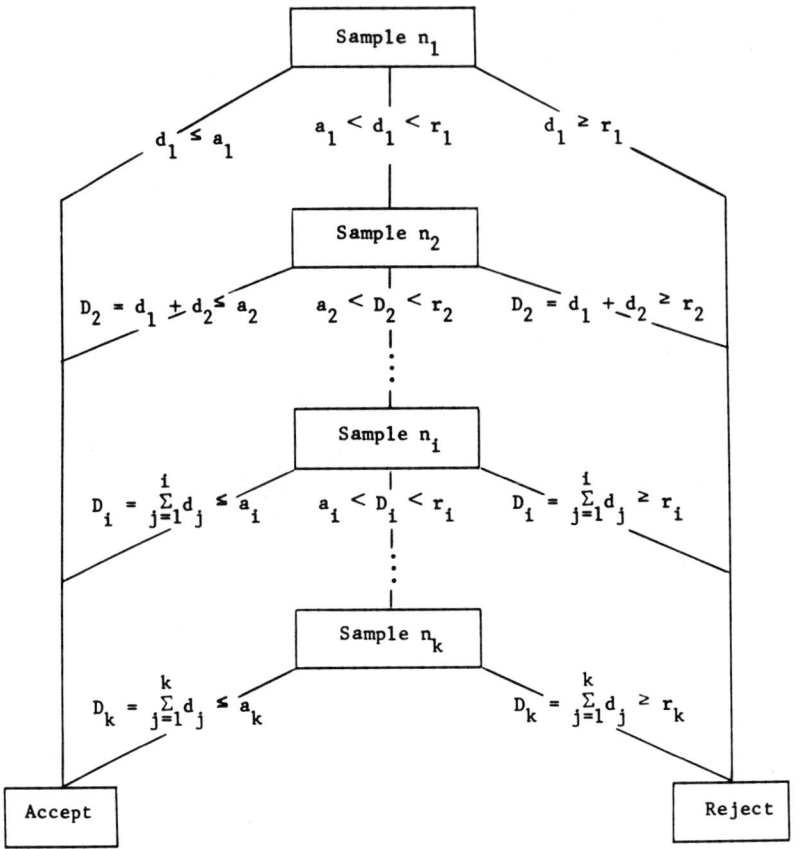

FIG. 6-2   Procedure for multiple sampling by attributes.

Sample	Sample size	Cumulative sample size	Acceptance number	Rejection number
1	n_1	n_1	a_1	r_1
2	n_2	$n_1 + n_2$	a_2	r_2
...	...	...	...	...
k	n_k	$n_1 + n_2 \ldots + n_k$	a_k	$r_k = a_k + 1$

To start the procedure, a sample of n_1 is randomly drawn from a lot of size N and the number of defectives d_1 in the sample is counted.

If $d_1 \leq a_1$, the lot is accepted.

If $d_1 \geq r_1$, the lot is rejected.

If $a_1 < d_1 < r_1$, another sample is taken.

If subsequent samples are needed, the first sample procedure is repeated sample by sample. For each sample, the total number of defectives found at any stage (say the ith)

$$D_i = \sum_{j=1}^{i} d_j$$

is compared with the acceptance number a_i and the rejection number r_i for that stage until a decision is made. Since, for the last (kth) sample $r_k = a_k + 1$, a decision must be made by the kth sample. Sometimes acceptance is not allowed at the early stages of a multiple sampling plan; however, rejection can take place at any stage. When acceptance is not allowed the symbol # is used for the acceptance number. The operation of the plan is shown in Fig. 6-2.

SELECTION

A convenient source of single-, double-, and multiple-sampling plans will be found in the MIL-STD-105D tables. The OC curves and other measures presented in these tables can be used to select an appropriate plan. Matched single- and double-sampling plans are also given in the Dodge-Romig (1959) tables. These sets of tables will be discussed in later chapters.

Procedures are also available for determining double- and multiple-sampling plans using Poisson unity values in a manner similar to single-sampling plans. These require specification of p_1 (producer's quality level), p_2 (consumer's quality level), α (producer's risk), and β (consumer's risk) and calculation of the operating ratio $R = p_2/p_1$. Double- and multiple-sampling plans also require specification of the relationship of the successive sample sizes, that is, a multiple m where for double sampling $n_2 = mn_1$.

Duncan (1974) has provided a compilation of unity values and operating ratios for double and multiple sampling as developed by the U.S. Army Chemical Corps Engineering Agency (1953). The double-sampling plans are for m = 1 and m = 2, respectively, with the rejection numbers constant for both samples. That is,

$$r_1 = r_2 = a_2 + 1$$

The multiple-sampling unity values are presented in terms of plans in which the sample size at each stage is equal, i.e.,

$$n_1 = n_2 = \ldots = n_i = \ldots = n_k$$

The Chemical Corps plans are not in matched sets and do not utilize acceptance numbers corresponding to MIL-STD-105D.

Appendix Table T6-1 was developed by Schilling and Johnson (1979) for the construction and evaluation of matched sets of single-, double-, and multiple-sampling plans. It may be used to derive individual plans to meet specified values of fraction defective and probability of acceptance. It may also be employed to match the scheme performance of the MIL-STD-105D system to that of an individual plan. The tables extend into the range of low probability of acceptance useful in reliability, safety, and compliance testing.

The unity values np listed in Appendix Table T6-1 are based on MIL-STD-105D acceptance and rejection numbers. Values of n are for the first sample sizes; succeeding samples in double and multiple plans are all equal and of size n. The plans are numbered by the corresponding single-sampling acceptance number and a letter (S, D, M) showing the type of plan: single, double, multiple. Two sets of double and multiple plans are included in addition to those from MIL-STD-105D to cover operating ratios R = 33 and R = 20 to facilitate matching an individual plan to the MIL-STD-105D system. No single-sampling plan has an operating ratio in this range. To obtain the operating ratio R = 20 the double-sampling plan requires $n_2 = 5n_1$ and is the only plan in Appendix Table T6-1 in which first and second sample sizes are not equal. Supplementary plans, not in MIL-STD-105D, are included to provide a complete set of plans to match single sampling acceptance numbers from 0 to 15. The table is for $\alpha = .05$ and $\beta = .10$. Its use is similar to that of the Cameron (1952) tables for single sampling including applications to plans to inspect defects per unit.

To construct a given plan,

1. Decide if single, double, or multiple sampling is to be used.
2. Specify

$$p_1 = \text{producer's quality level } (P_a = .95)$$

$$p_2 = \text{consumer's quality level } (P_a = .10)$$

3. Form the operating ratio

$$R = \frac{p_2}{p_1}$$

4. Choose a plan having acceptance numbers associated with an operating ratio just less than or equal to R.

5. Determine the sample size as

$$n = \frac{np_2}{p_2}$$

Round up in determining sample size.

6. The plan consists of the acceptance numbers and sample size chosen.

7. The operating characteristic curve may be drawn by dividing the values of np shown for the plan by the sample size to obtain values of p associated with the values of probability of acceptance listed.

8. The average sample number curve may be drawn by multiplying the values of ASN/n shown by the sample size and plotting the resulting values of ASN against the p values obtained for the corresponding probability of acceptance.

The formula for sample size n is presented showing division by p_2 to ensure the consumer's risk is maintained. Alternatively $n = np_1/p_1$ as in the Cameron tables. The value of np_1 can be found under probability of acceptance .95 for the plan. If values of sample size differ between these two formulas, the probability of acceptance will be exact at the value of p associated with the value of n actually used and approximate for the other value of p. Sometimes a convenient intermediate sample size may be chosen.

For example, suppose a double-sampling plan is desired having 95 percent probability of acceptance at p_1 = .01 and 10 percent probability of acceptance at p_2 = .05.

1. Double sampling is to be used.

2. p_1 = .01 p_2 = .05

3. $R = \frac{.05}{.01} = 5$

4. The operating ratio is R = 4.40 for the plan, giving acceptance numbers

   Ac 1,4   Re 4,5

5. So

$$n_1 = \frac{4.398}{.05} \simeq 87.96 \sim 88$$

This holds p_2 exactly while p_1 = .011. Note that if p_1 is to be held,

$$n_1 = \frac{1.000}{.01} = 100$$

and at this sample size p_2 will be .044.

6. Using n = 88, the plan is

Sample	Sample size	Cumulative sample size	Ac	Re
1	88	88	1	4
2	88	176	4	5

7. The operating characteristic curve is found by dividing the values of np shown by 88 to obtain

P_a	p	P_a	p
.99	.007	.10	.050
.95	.011	.05	.058
.90	.014	.01	.077
.75	.020	.005	.086
.50	.028	.001	.105
.25	.038	.0005	.114
		.0001	.134

8. The average sample number curve is found by multiplying the values of ASN/n_1 shown by 88 and plotting against the values of p obtained for corresponding probabilities of acceptance in step 7 to obtain

p	ASN	p	ASN
.007	94.4	.050	113.8
.011	109.6	.058	106.6
.014	115.8	.077	95.4
.020	125.0	.086	92.7
.028	129.4	.105	89.5
.038	124.4	.114	88.9
		.134	88.3

Appendix Table T6-1 can be used to find matching single (R = 4.89) and multiple (R = 4.67) plans. They are

Sample	Sample size	Cumulative sample size	Ac	Re
Single				
1	134	134	3	4
Multiple				
1	33	33	#	3
2	33	66	0	3
3	33	99	1	4
4	33	132	2	5
5	33	165	3	6
6	33	198	4	6
7	33	231	6	7

where # indicates no acceptance allowed on the first sample. These plans
are matched about as well as those in MIL-STD-105D. Their operating characteristic and average sample number curves are found in a similar manner
to those of the single-sampling plan.

The OC curves of these single-, double-, and multiple-plans are shown
in Fig. 6-3. Their ASN curves are presented in Fig. 6-4.

As a further illustration, suppose a plan is desired having a producer's quality level of 1.78 defects per 100 units and a consumer's quality
level of 19.5 defects per 100 units with risks α = .05 and β = .10. Converting to defects per unit, p_1 = .0178 and p_2 = .195, so that

$$R = \frac{.195}{.0178} = 10.96$$

Appendix Table T6-1 shows that the plans 1S, 1D, and 1M are an appropriate
set of matched plans.

Single sampling: n = 3.89/.195 = 19.95 $\sim$ 20, Ac = 1, and Re = 2.
Double sampling: n = 2.49/.195 = 12.77 $\sim$ 13, Ac = 0,1, and Re = 2,2.
Multiple sampling: n = .917/.195 = 4.70 $\sim$ 5, Ac = #,#,0,0,1,1,2,
                       and Re = 2,2,2,3,3,3,3.

Also, the indifference quality level occurs at the following values of
defects per unit with the associated average sample number shown:

Single sampling: $p_{.50}$ = 1.678/20 = .084, ASN = 1(20) = 20.

Double sampling: $p_{.50}$ = 1.006/13 = .077, ASN = 1.368(13) = 17.8.

Multiple sampling: $p_{.50}$ = .416/5 = .083, ASN = 3.640(5) = 18.2.

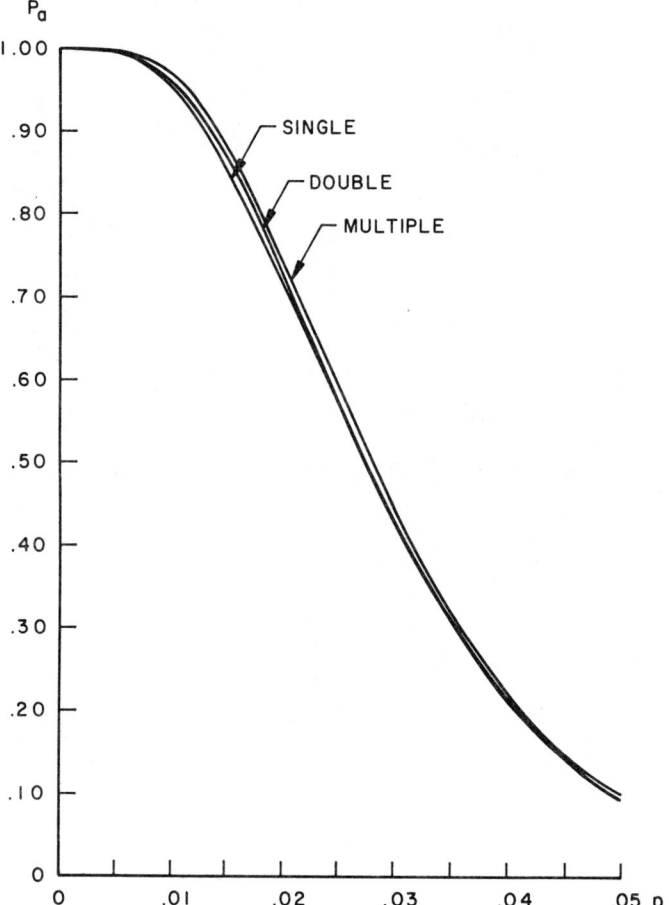

FIG. 6-3 Operating characteristic curves for single,
double, and multiple matched plans.

The producer's and consumer's quality levels for these matched plans are

$$PQL = P_{.95} \qquad CQL = P_{.10}$$

	PQL = $P_{.95}$	CQL = $P_{.10}$
Single	.355/20 = .018	3.890/20 = .195
Double	.207/13 = .016	2.490/13 = .192
Multiple	.103/5 = .021	.917/5 = .183

The plans shown correspond directly to the MIL-STD-105D, Code F, 2.5 AQL
normal plan.

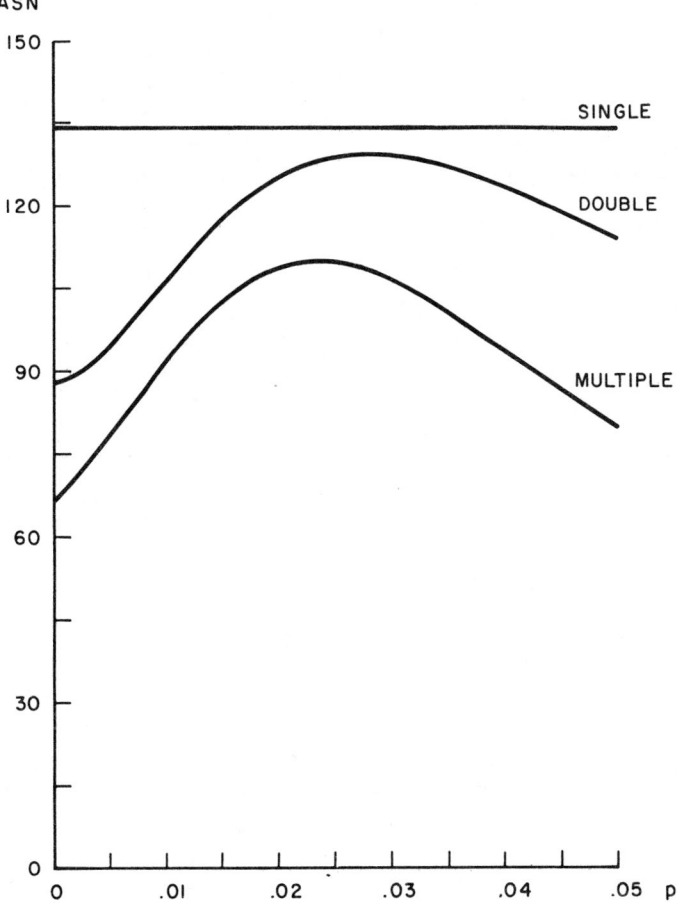

FIG. 6-4 Average sample number curves for single,
double, and multiple matched plans.

MEASURES

Double Sampling

The measures of performance of double-sampling plans are given in Table
6-1. Binomial or Poisson probabilities are appropriate in their evaluation
depending on the sampling situation. These probability distributions are
listed in Table 3-1. As an illustration of application of these formulas,
suppose the measures of the plan $n_1 = n_2 = 13$; $a_1 = 0$, $a_2 = 1$; $r_1 = r_2 = 2$
are to be evaluated when sampling from a succession of lots of size N = 120.
Calculations are to be made using the binomial distribution in a Type B

TABLE 6-1   Measures of Double-Sampling Attributes Plans

MEASURE	FORMULA
PROBABILITY OF ACCEPTANCE	$P_a = P_{a_1} + P_{a_2}$, where P_{a_i} is P_a for i^{th} sample $= F(a_1 \mid n_1) + \sum_{d_1=a_1+1}^{r_1-1} f(d_1 \mid n_1) \, F(a_2-d_1 \mid n_2)$
AVERAGE SAMPLE NUMBER	o FULL INSPECTION $\quad$ ASN $= n_1 + n_2 \left(F(r_1-1 \mid n_1) - F(a_1 \mid n_1) \right)$ o SEMI-CURTAILED $\quad$ ASN$_c = n_1 + \sum_{d_1=a_1+1}^{r_1-1} f(d_1 \mid n_1) \left[\dfrac{r_2-d_1}{p} + n_2 \, F(a_2-d_1 \mid n_2) \right.$ $\quad\quad\quad\quad\quad\quad\quad\quad\quad \left. - \left(\dfrac{r_2-d_1}{p} \right) F(r_2-d_1 \mid n_2+1) \right]$
AVERAGE OUTGOING QUALITY	o DEFECTIVES FOUND REPLACED WITH GOOD $\quad$ AOQ $= p \, P_{a_1} \left(\dfrac{N-n_1}{N} \right) + p \, P_{a_2} \left(\dfrac{N-n_1-n_2}{N} \right)$ o DEFECTIVES NOT REPLACED $\quad$ AOQ $= p \, P_{a_1} \left(\dfrac{N-n_1}{N-n_1 p} \right) + p \, P_{a_2} \left(\dfrac{N-n_1-n_2}{N-n_1 p-n_2 p} \right)$ o APPROXIMATE (n/N small) $\quad$ AOQ $\simeq p \, P_a$
AVERAGE OUTGOING QUALITY LIMIT (Approximate)	o Determine n^* and c^* as follows[†] $\quad$ 1. Average <u>cumulative</u> sample sizes to obtain n^* $\quad$ 2. Average <u>all</u> acceptance and rejection numbers to obtain c^* o Obtain value of y using c^* o AOQL $\simeq \dfrac{y}{n^*} \left(\dfrac{N-n^*}{N} \right)$
AVERAGE TOTAL INSPECTION	ATI $= n_1 \, P_{a_1} + \left(n_1 + n_2 \right) P_{a_2} + N \left(1-P_a \right)$ $= n_1 + n_2 \left(1-P_{a_1} \right) + \left(N-n_1-n_2 \right) \left(1-P_a \right)$

[†]Approximation matching double to single plans given in Schilling, Sheesley and Nelson (1978). Values of y given in Appendix Table T14-1.

sampling situation when the incoming proportion defective is $p = .18$.

Probability of acceptance:

$$P_a = F(0|13) + f(1|13) \ F(0|13)$$

$$= C_0^{13} .18^0 (.82)^{13} + C_1^{13} .18^1 (.82)^{12} [C_0^{13} .18^0 (.82)^{13}]$$

$$= .076 + .216[.076]$$

$$= .076 + .016 \quad \text{(Note: Thus } P_{a_1} = .076 \text{ and } P_{a_2} = .016)$$

$$= .092$$

Average sample number:

Full inspection

$$ASN = 13 + 13 \ (F(1|13) - F(0|13))$$

$$= 13 + 13 \ (.292 - .076)$$

$$= 15.8$$

Semicurtailed

$$ASN_c = 13 + f(1|13)\left[\frac{2-1}{.18} + 13 \ F(0|13) - \frac{2-1}{.18} F(1|14)\right]$$

$$= 13 + .216[5.56 + 13(.076) - 5.56(.253)]$$

$$= 14.1$$

Average outgoing quality:

Defectives found replaced

$$AOQ = .18(.076)\left(\frac{120 - 13}{120}\right) + .18(.016)\left(\frac{120 - 13 - 13}{120}\right)$$

$$= .012 + .002 = .014$$

Defectives not replaced

$$AOQ = .18(.076)\left(\frac{120 - 13}{120 - 13(.18)}\right) + .18(.016)\left(\frac{120 - 13 - 13}{120 - 13(.18) - 13(.18)}\right)$$

$$= .012 + .002 = .014$$

Approximate

$$AOQ \simeq .18(.092) \simeq .017$$

Average outgoing quality limit:

$$n* = \frac{13 + 26}{2} = 19.5 \sim 20$$

$$c* = \frac{0 + 2 + 1 + 2}{4} = 1.25 \sim 1$$

$$y = .8400 \quad \text{(From Appendix Table T14-1)}$$

$$AOQL \simeq \frac{.8400}{20}\left(\frac{120 - 20}{120}\right) \simeq .035$$

Average total inspection:

$$ATI = 13(.076) + (13 + 13)(.016) + 120(1 - .092)$$

$$= .99 + .42 + 108.96 = 110.37$$

TABLE 6-2   Measures of Multiple-Sampling Attributes Plans

MEASURE	FORMULA
PROBABILITY OF ACCEPTANCE	$P_a = \sum_{j=1}^{k} A_j$ where A_j is probability of acceptance on the jth stage. (For explicit formula, see Table 6-3)
AVERAGE SAMPLE NUMBER	o FULL INSPECTION $ASN = \sum_{j=1}^{k} \sum_{m=1}^{j} n_m T_j$ where T_j is probability of termination on the jth stage o SEMI-CURTAILED INSPECTION See Table 6-3
AVERAGE OUTGOING QUALITY	o DEFECTIVES FOUND REPLACED WITH GOOD $AOQ = \sum_{j=1}^{k} \left(\dfrac{N - \sum_{m=1}^{j} n_m}{N} \right) p\, A_j$ o DEFECTIVES NOT REPLACED $AOQ = \sum_{j=1}^{k} \left(\dfrac{N - \sum_{m=1}^{j} n_m}{N - p\sum_{m=1}^{j} n_m} \right) p\, A_j$ o APPROXIMATE (n/N small) $AOQ = p\, P_a$
AVERAGE[†] OUTGOING QUALITY LIMIT (Approximate)	o Determine n^* and c^* as follows: 1. Average cumulative sample sizes to obtain n^* 2. Average all acceptance and rejection numbers to obtain c^*. (Use -1 when acceptance is not allowed at a stage.) o Obtain value of y using c^* o $AOQL \simeq \dfrac{y}{n^*} \left(\dfrac{N - n^*}{N} \right)$
AVERAGE TOTAL INSPECTED	$ATI = N(1 - P_a) + \sum_{j=1}^{k} \sum_{m=1}^{j} n_m A_j$

[†]Approximation matching single to multiple plan given in Schilling, Sheesley and Nelson (1978). Values of y given in Appendix Table T14-1.

These measures are useful in the characterization of this double-sampling plan for p = .18. Repeated calculations for various values of proportion defective would allow construction of the curves describing plan performance.

Multiple Sampling

The measures of performance of multiple-sampling plans are given in Table 6-2. They apply, of course, to double-sampling plans as well. The binomial or Poisson probability distributions are appropriate for their evaluation depending on the sampling situation involved and the degree of approximation desired. These probability distributions are listed in Table 3-1.

As an illustration of the application of these formulas, consider the plan

Stage 1: n_1 = 10 a_1 = # r_1 = 2

Stage 2: n_2 = 10 a_2 = 0 r_2 = 2

Stage 3: n_3 = 10 a_3 = 1 r_3 = 2

where # denotes no acceptance allowed at the first stage. Suppose the plan is to be evaluated at p = .01 for application to lots of size 350. Calculations are to be made using Poisson probabilities as an approximation to the binomial for a Type B sampling situation. Note that for np = (10)(.01) = 0.1,

$f(0|10)$ = .905 $F(0|10)$ = .905 $1 - F(0|10)$ = .095
$f(1|10)$ = .090 $F(1|10)$ = .995 $1 - F(1|10)$ = .005

Probably the best way to portray the evaluation of a multiple-sampling plan is with a probability tree as shown in Fig. 6-5.

The results of Fig. 6-5 give the following probabilities:

Stage	Accept A_j	Reject R_j	Terminate T_j	Indecision I_j
1	0	.005	.005	.995
2	.819	.013	.832	.162
3	.148	.016	.164	0
Total	.967	.034	1.001	X

This listing of the probabilities associated with the tree shows the probability of accepting (A_j), rejecting (R_j), terminating (T_j) and indecision

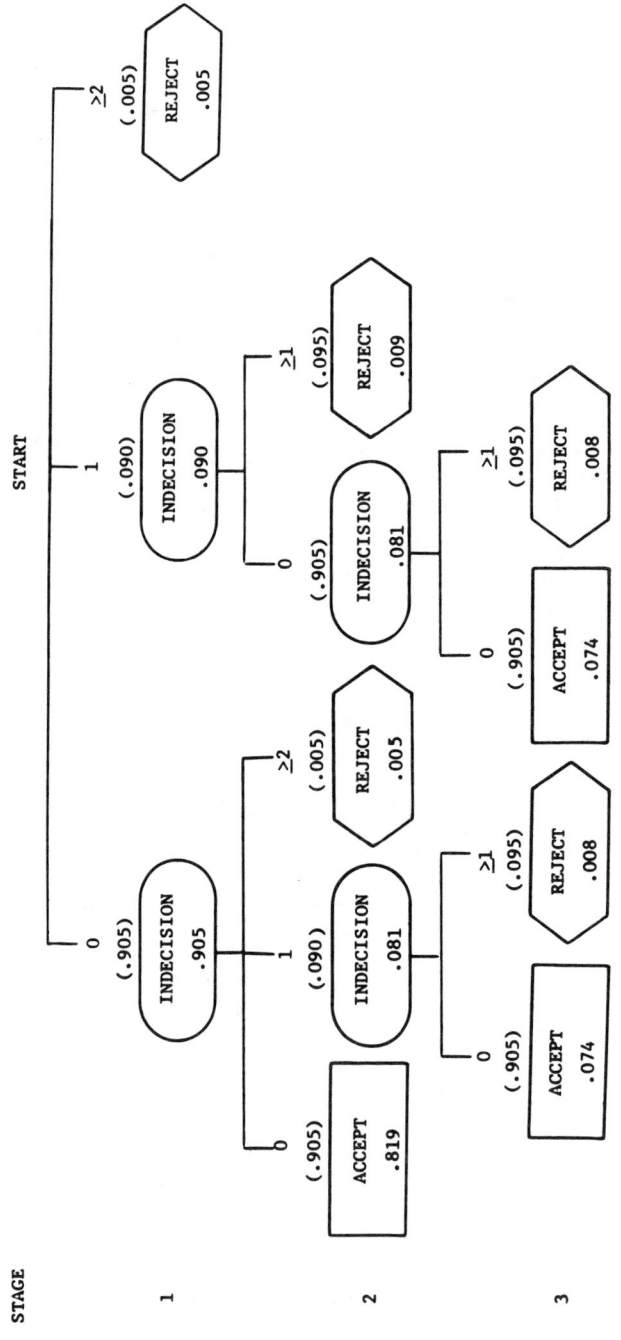

FIG. 6-5   Probability tree evaluating multiple-sampling plan.

(I_j) at each stage. It is constructed simply as the totals of the acceptance, rejection, and indecision probabilities shown at that stage of the tree. Termination is the sum of acceptance and rejection and acts as a check since the termination column must sum to one.

Using the tree and the formulas of Table 6-2 for p = .01 it is possible to obtain

Probability of acceptance:

$$P_a = A_1 + A_2 + A_3$$

$$= 0 + .819 + .148 = .967$$

Average sample number:

$$ASN = n_1 T_1 + (n_1 + n_2) T_2 + (n_1 + n_2 + n_3) T_3$$

$$= 10(.005) + 20(.833) + 30(.164) = 21.6$$

Average outgoing quality:

Defectives found replaced with good,

$$AOQ = p\left[\left(\frac{N - n_1}{N}\right)A_1 + \left(\frac{N - n_1 - n_2}{N}\right)A_2 + \left(\frac{N - n_1 - n_2 - n_3}{N}\right)A_3\right]$$

$$= .01\left[\left(\frac{350 - 10}{350}\right)0 + \left(\frac{350 - 20}{350}\right).819 + \left(\frac{350 - 30}{350}\right).148\right]$$

$$= .01[.772 + .135] = .009$$

Defectives not replaced,

$$AOQ = p\left[\left(\frac{N - n_1}{N - pn_1}\right)A_1 + \left(\frac{N - n_1 - n_2}{N - p(n_1 + n_2)}\right)A_2 + \left(\frac{N - n_1 - n_2 - n_3}{N - p(n_1 + n_2 + n_3)}\right)A_3\right]$$

$$= .01\left[\left(\frac{350 - 10}{350 - .01(10)}\right)0 + \left(\frac{350 - 20}{350 - .01(20)}\right).819\right.$$

$$\left. + \left(\frac{350 - 30}{350 - .01(30)}\right).148\right]$$

$$= .01\left[.773 + .135\right] = .009$$

Approximate,

$$AOQ = pP_a = .01(.967) = .0097$$

Average outgoing quality limit:

$$n* = \frac{10 + 20 + 30}{3} = 20$$

$$c* = \frac{-1 + 0 + 1 + 2 + 2 + 2}{6} = \frac{6}{6} = 1$$

$$AOQL = \frac{y}{n*}\left(\frac{N - n*}{N}\right) = \frac{.8400}{20}\left(\frac{350 - 20}{350}\right) = .040$$

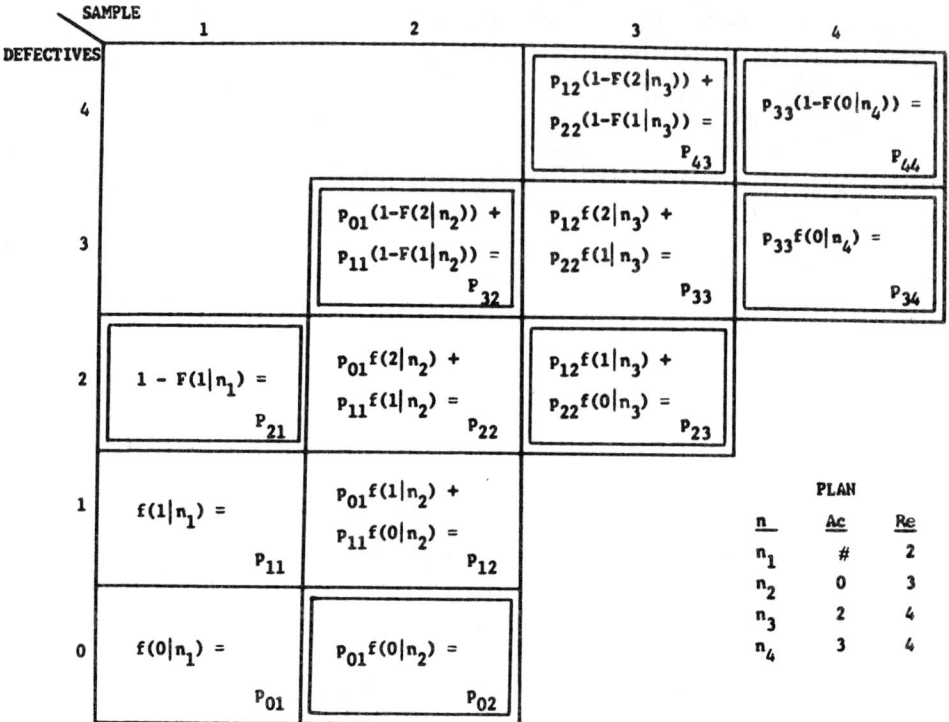

	SAMPLE				
	1	2	3	4	TOTAL
ACCEPT A_j	$A_1=0$	$A_2=P_{02}$	$A_3=P_{23}$	$A_4=P_{34}$	ΣA_j
REJECT R_j	$R_1=P_{21}$	$R_2=P_{32}$	$R_3=P_{43}$	$R_4=P_{44}$	ΣR_j
TERMINATE T_j	$T_1=A_1+R_1$	$T_2=A_2+R_2$	$T_3=A_3+R_3$	$T_4=A_4+R_4$	1.0
INDECISION I_j	$P_{01}+P_{11}$	$P_{12}+P_{22}$	P_{33}	0	X

FIG. 6-6 Control table format.

Average total inspection:

$$ATI = N(1 - P_a) + n_1A_1 + (n_1 + n_2)A_2 + (n_1 + n_2 + n_3)A_3$$
$$= 350(1 - .967) + 10(0) + 20(.819) + 30(.148)$$
$$= 11.55 + 16.38 + 4.44 = 32.37$$

These values measure the performance of this multiple-sampling plan when p = .01. From calculations at other levels of proportion defective the relevant curves characterizing the plan can be drawn.

The operating characteristics and other measures of double- and multiple sampling plans can also be computed using the control table concept originated by the Statistics Research Group (1948). The control table for a plan having acceptance numbers

Sample	Ac	Re
1	#	2
2	0	3
3	2	4
4	3	4

is shown in Fig. 6-6. Sample numbers are listed across the top, while the accumulated number of defectives is shown on the side. The squares represent possible events in the operation of the sampling plan. The boundary of the figure is comprised of the squares leading to acceptance or rejection. If no acceptance decision is possible at a state, the double boundary square is omitted as in the first sample of Fig. 6-6. The top right square in Fig. 6-6, for example, represents an accumulation of at least four defectives on the fourth sample. A double square represents a state at which termination of the plan would occur with acceptance or rejection. Bottom double squares show acceptance, whereas top double squares show rejection. Since top square probabilities indicate rejection, they are cumulative at or exceeding the number shown. They are shown as capital letters (P), as opposed to individual probabilities shown as small letters (p). They are the only squares which use cumulative probabilities.

To fill out the figure proceed as follows:

1. Under sample 1, fill in the appropriate probabilities of each event.

Defective	Probability	Action	Symbol for state
0	$f(0|n_1)$	No decision	P_{01}
1	$f(1|n_1)$	No decision	P_{11}
≥ 2	$1 - F(1|n_1)$	Reject	P_{21}

2. Under sample 2, fill in the appropriate probabilities of each event as the sum of the joint probabilities of events leading to that state.

Defective	Approach		Symbol	Probability	Action
	First Sample	Second Sample			
0	0	0	P_{02} =	$P_{01} f(0\|n_2)$	Accept
1	0	1	P_{12} =	$P_{01} f(1\|n_2)$	No decision
	1	0		$+ P_{11} f(0\|n_2)$	
2	0	2	P_{22} =	$P_{01} f(2\|n_2)$	No decision
	1	1		$+ P_{11} f(1\|n_2)$	
3	0	≥ 3	P_{32} =	$P_{01}(1 - F(2\|n_2))$	Reject
	1	≥ 2		$+ P_{11}(1 - F(1\|n_2))$	

3. Under sample 3, fill in the appropriate probabilities of each event as the sum of the joint probabilities of events leading to that state.

Defective	Approach		Symbol	Probability	Action
	Second Sample	Third Sample			
2	1	1	P_{23} =	$P_{12} f(1\|n_3)$	Accept
	2	0		$+ P_{22} f(0\|n_3)$	
3	1	2	P_{33} =	$P_{12} f(2\|n_3)$	No decision
	2	1		$+ P_{22} f(1\|n_3)$	
4	1	≥ 3	P_{43} =	$P_{12}(1 - F(2\|n_3))$	Reject
	2	≥ 2		$+ P_{22}(1 - F(1\|n_3))$	

4. Under sample 4, fill in the appropriate probabilities of each event as the sum of the joint probabilities of events leading to that state.

	Approach				
Defective	Third Sample	Fourth Sample	Symbol	Probability	Action
3	3	0	P_{34}	$= \quad P_{33}f(0|n_4)$	Accept
4	3	≥ 1	P_{44}	$= \quad P_{33}(1 - F(0|n_4))$	Reject

TABLE 6-3   Explicit Formulas for Measures of Multiple Sampling Plans

MEASURE	FORMULA
PROBABILITY OF ACCEPTANCE	$$P_a = F(a_1|n_1) + \sum_{d_1=a_1+1}^{r_1-1} f(d_1|n_1)\left[F(a_2-d_1|n_2) + \sum_{d_2=a_2-d_1+1}^{r_2-1} f(d_2|n_2)\left[\cdots\right.\right.$$ $$\cdots\left[F(a_j-D_{j-1}|n_j) + \sum_{d_j=a_j-D_{j-1}+1}^{r_j-1} f(d_j|n_j)\left[\cdots\right.\right.$$ $$\cdots\left[F(a_{k-1}-D_{k-2}|n_{k-1}) + \sum_{d_{k-1}=a_{k-1}-D_{k-2}+1}^{r_{k-1}-1} f(d_{k-1}|n_{k-1})\left[F(a_k-D_{k-1}|n_k)\right]\cdots\right]\cdots\right]$$ where $D_j = \sum_{i=1}^{j} d_i$ and cumulative probabilities with negative arguments are taken to be zero. Use $a_i = -1$ if no acceptance is allowed at stage i.
AVERAGE SAMPLE NUMBER	o FULL INSPECTION $$ASN = n_1 + I_1\left[n_2 + I_2\left[n_3 + I_3\left[\cdots\left[n_k + 0\right]\cdots\right]\right]\right]$$ where I_j is probability of no decision at stage j. o SEMI-CURTAILED INSPECTION (equal sample sizes) $$ASN_c = \sum_{j=1}^{k}(nj)T_j + \sum_{j=1}^{k-1}\sum_{i=a_j+1}^{r_j-1}(nj)p_{ij}(1-F(r_{j+1}-i-1|n-1))$$ $$-\sum_{j=2}^{k}\sum_{i=a_{j-1}+1}^{r_{j-1}-1}(nj)p_{i(j-1)}(1-F(r_j-i-1|n-1))$$ $$+\sum_{j=1}^{k-1}\sum_{i=a_j+1}^{r_j-1}p_{ij}\frac{(r_{j+1}-i)}{p}(1-F(r_{j+1}-i|n))$$ where T_j and p_{ij} are taken from the control table and the first sample is always fully inspected. Use $a_i = -1$ if no acceptance is allowed at a stage.

n	Ac	Re
10	#	2
10	0	2
10	1	2

SAMPLE DEFECTIVES	1	2	3
2	$P_{21} = .005$	$P_{01}(1-F(1\|10))+$ $P_{11}(1-F(0\|10))=$ $.905(.005)+$ $.090(.095)=$ $P_{22} = .013$	$P_{12}(1-F(0\|10))=$ $.163(.095)=$ $P_{23} = .015$
1	$P_{11} = .090$	$P_{01}f(1\|10) +$ $P_{11}f(0\|10) =$ $(.905)(.090) +$ $(.090)(.905) =$ $P_{12} = .163$	$P_{12}f(0\|10) =$ $.163(.905) =$ $P_{13} = .148$
0	$P_{01} = .905$	$P_{01}f(0\|10) =$ $(.905)(.905) =$ $P_{02} = .819$	

SAMPLE

	1	2	3	TOTAL
ACCEPT A_j	$A_1=0$	$A_2=.819$	$A_3=.148$	.967
REJECT R_j	$R_1=.005$	$R_2=.013$	$R_3=.015$	.033
TERMINATE T_j	$T_1=.005$	$T_2=.832$	$T_3=.163$	1.000
INDECISION I_j	$I_1=.995$	$I_2=.163$	$I_3=0$	X

FIG. 6-7 Control table illustration.

The probability of acceptance is simply the sum of the probabilities shown in the squares leading to acceptance

$$P_a = P_{02} + P_{23} + P_{34}$$

The probability of rejection can similarly be found as the sum of the probabilities of the squares leading to rejection

$$P(rej) = P_{21} + P_{32} + P_{43} + P_{44}$$

Clearly the probability of acceptance and rejection must add to one.

A summary table similar to the listing of probabilities for the probability tree approach to the evaluation of the multiple plan may be developed for the control table and is shown at the bottom of Fig. 6-6. Using the summary table it is possible to evaluate the measures given in Table 6-2 just as was done with the probability tree. The control table may also be used to evaluate explicit formulas for probability of acceptance and average sample number as given in Table 6-3.

As an example of the control table approach, consider again the multiple plan

n	Ac	Re
10	#	2
10	0	2
10	1	2

The appropriate control table is shown in Fig. 6-7.

The results of the summary table are within rounding error of those for the listing of probabilities obtained from the probability tree, and will give the same results for the measures as given in Table 6-2. Using the control table to evaluate the explicit formulas given in Table 6-3 for this plan, we obtain

Probability of acceptance:

$$P_a = F(\#|n_1) + \sum_{d_1=a_1+1}^{r_1-1} f(d_1|n_1)\Bigg[F(a_2 - d_1|n_2)$$

$$+ \sum_{d_2=a_2-d_1+1}^{r_2-1} f(d_2|n_2)\Big[F(a_3 - D_2|n_3)\Big]\Bigg]$$

$$= F(-1|10) + \sum_{d_1=-1+1}^{2-1} f(d_1|10)\Bigg[F(0 - d_1|10)$$

$$+ \sum_{d_2=0-d_1+1}^{2-1} f(d_2|10)\Big[F(1 - D_2|10)\Big]\Bigg]$$

$$= 0 + \sum_{d_1=0}^{1} f(d_1|10)\Bigg[F(0 - d_1|10) + \sum_{d_2=1-d_1}^{1} f(d_2|10)\Big[F(1 - D_2|10)\Big]\Bigg]$$

$$= 0 + f(0|10)F(0|10) + f(0|10)f(1|10)F(1 - 1|10)$$

$$+ f(1|10)F(-1|10) + f(1|10)f(0|10)F(1 - 1|10)$$

$$+ f(1|10)f(1|10)F(-1|10)$$

$$= 0 + (.905)(.905) + (.905)(.090)(.905) + 0$$

$$+ (.090)(.905)(.905) + 0$$

$$= \underbrace{.819}_{A2} + \underbrace{.074 + .074}_{A3}$$

$$= .967$$

Average sample number:

Full inspection

$$\text{ASN} = n_1 + I_1[n_2 + I_2[n_3]]$$

$$= 10 + .995[10 + .163[10]]$$

$$= 21.57$$

Semicurtailed inspection

$$\text{ASN}_c = \sum_{j=1}^{3} (10j)T_j + \sum_{j=1}^{2} \sum_{i=a_j+1}^{r_j-1} (10j)p_{ij}(1 - F(r_{j+1}-i - 1|9))$$

$$- \sum_{j=2}^{3} \sum_{i=a_{j-1}+1}^{r_{j-1}-1} (10j)p_{i(j-1)}(1 - F(r_j - i - 1|9))$$

$$+ \sum_{j=1}^{2} \sum_{i=a_j+1}^{r_j-1} p_{ij}\left(\frac{r_{j+1}-i}{p}\right)\left(1 - F(r_{j+1}-i|10)\right)$$

$$= \left[10T_1 + 20T_2 + 30T_3\right]$$

$$+ \left[10\, p_{01}(1 - F(1|9)) + 10\, p_{11}(1 - F(0|9))\right.$$

$$\left. + 20\, p_{12}(1 - F(0|9))\right]$$

$$- \left[20\, p_{01}(1 - F(1|9)) + 20\, p_{11}(1 - F(0|9))\right.$$

$$\left. + 30\, p_{12}(1 - F(0|9))\right]$$

$$+ \left[p_{01}\left(\frac{2}{p}\right)(1 - F(2|10)) + p_{11}\left(\frac{1}{p}\right)(1 - F(1|10))\right.$$

$$\left. + p_{12}\left(\frac{1}{p}\right)(1 - F(1|10))\right]$$

$$= [10(.005) + 20(.832) + 30(.163)]$$

$$+ [10(.905)(.0038) + 10(.090)(.0861) + 20(.163)(.0861)]$$

$$- [20(.905)(.0038) + 20(.090)(.0861) + 30(.163)(.0861)]$$

$$+ [(.905)\left(\frac{2}{.01}\right)(.0002) + .090\left(\frac{1}{.01}\right)(.0047) + .163\left(\frac{1}{.01}\right)(.0047)]$$

$$= 21.51$$

FURTHER CONSIDERATIONS

Unity values np presented in Appendix Table T6-1 were derived by Schilling and Johnson (1979) using the theory of unity values as presented by Duncan (1974, pp. 187-188). They are based on the Poisson distribution and can be used to approximate binomial sampling plans where the Poisson approximation to the binomial distribution applies. Since the probability of acceptance and the average sample number can be shown to be a function of np for a given set of acceptance criteria and ratio of subsample sizes, it is possible to vary np_1 and np_2 in such a way that the ratio $R = p_2/p_1$ remains unchanged while the value of n changes. Thus, any member of the set of plans having operating ratio R may be used to generate unity values, values of np when n = 1, by simply dividing the values of np associated with its operating characteristic curve by n. A similar argument holds for the values ASN/n.

PROBLEMS

1. Construct double-sampling plans to the following specifications given in proportion defective with producer's risk .05 and consumer's risk .10.

   a. PQL = .04, CQL = .21

   b. PQL = .03, CQL = .13

   c. PQL = .02, CQL = .06

2. Construct a multiple-sampling plan for the specifications given in Prob. 1.

3. Plot the Type B operating characteristic curve for the following double-sampling plan using P_a = .95, .50, .10 at a minimum for plotting positions

   n: 8, 8

   Ac: 0, 1

   Re: 2, 2

4. Plot the Type B operating characteristic curve for the following multiple-sampling plan using P_a = .95, .50, .10 at a minimum for plotting positions

   n: 3, 3, 3, 3, 3, 3, 3

   Ac: #, #, 0, 0, 1, 1, 2

   Re: 2, 2, 2, 3, 3, 3, 3

5. If lots are received in quantities of 1000, obtain ASN, AOQ, and ATI at the minimum plotting positions for the plan given in Prob. 3 and draw the curves.

6. If lots are received in quantities of 1000, obtain ASN, AOQ, and ATI at the minimum plotting positions for the plan given in Prob. 4 and draw the curves.

7. At present, the single-sampling plan n = 35, c = 3 is being used in acceptance inspection of incoming material from a very good supplier. What double and multiple plans may be substituted? How much would be gained thereby?

8. Use the approximation for determining AOQL to find what single-sampling plan matches the double- and multiple-sampling plans of Probs. 3 and 4. Is this confirmed by the Schilling-Johnson table?

9. What is the average sample number under curtailed inspection at p = .025 for the plan given in Prob. 3? Using the formula for a double-sampling plan, is the curtailment worthwhile at this fraction defective?

10. Regard the plan given in Prob. 3 as a multiple-sampling plan and construct the control table at p = .025 to evaluate the probability of acceptance and the average sample number.

REFERENCES

Cameron, J. M. (1952), Tables for Constructing and for Computing the Operating Characteristics of Single Sampling Plans, *Industrial Quality Control*, 9(1, Part I): 37-39.

Dodge, H. F., and H. G. Romig (1959), *Sampling Inspection Tables, Single and Double Sampling*, 2nd ed., John Wiley and Sons, New York.

Duncan, A. J. (1974), *Quality Control and Industrial Statistics*, 4th ed., Richard D. Irwin, Homewood, Illinois.

Schilling, E. G., and L. I. Johnson (1979), Tables for the Construction of Matched Single, Double, and Multiple Sampling Plans with Application to MIL-STD-105D, *Journal of Quality Technology*, 12(4): 220-229.

Schilling, E. G., J. H. Sheesley, and P. R. Nelson (1978), GRASP: A General Routine for Attribute Sampling Plan Evaluation, *Journal of Quality Technology*, 10(3): 125-130.

Statistical Research Group, Columbia University (1948), *Sampling Inspection*, McGraw-Hill, New York.

United States Department of the Army (1953), *Master Sampling Plans for Single, Duplicate, Double, and Multiple Sampling*, Manual No. 2, Chemical Corps Engineering Agency, Army Chemical Center, Maryland.

7

SEQUENTIAL SAMPLING BY ATTRIBUTES

Single, double, and multiple plans assess one or more successive samples to determine lot acceptability. The most discriminating acceptance sampling procedure involves making a decision as to disposition of the lot or resample successively as each item of the sample is taken. Called *sequential sampling*, these methods may be regarded as multiple-sampling plans with sample size one and no upper limit on the number of samples to be taken. It can be shown that the sequential approach provides essentially optimum efficiency in sampling, that is an average sample number as low as possible [Wald (1947, p. 35)]. For example, in comparing average sample sizes for plans matched to the single-sampling plan n = 50, c = 2, we have at p = .017, chosen as the 95th percentage point of the appropriate MIL-STD-105D plan (Code H, 1.5 percent AQL, Normal Inspection):

Plan	ASN
Single	50
Double	43
Multiple	35
Sequential	33.5

Sequential plans are often applied where sample size is critical so that a minimum sample must be taken. They are somewhat harder to administer than multiple-sampling plans since in specific applications the amount of inspection effort is not determined until after the sample is taken. The possibility of taking one sample at a time must exist; in some operations this would be exceedingly difficult or impossible. Also the operating procedure requires an astute and trusted inspector since it is somewhat more demanding than single, double, or multiple sampling.

OPERATION

Under sequential sampling, samples are taken, one at a time, until a decision is made on the lot or process sampled. After each item is taken a decision is made to (1) accept, (2) reject, or (3) continue sampling. Samples are taken until an accept or reject decision is made. Thus, the procedure is open ended, the sample size not being determined until the lot is accepted or rejected. The average sample number of the plan provides a benchmark as to the expected sample size in any given application.

The plan is often implemented using a chart as shown in Fig. 7-1 in which the cumulative number of defectives found is plotted against the number of individual samples taken where

k = number of sample items taken

d_k = number of defectives found by the kth sample item

Y_2 = sk + h_2 = reject limit at kth sample

Y_1 = sk - h_1 = acceptance limit at kth sample

h_1, h_2 = intercepts

s = slope (not a standard deviation)

When the plot of the cumulative number of defectives found crosses the acceptance or rejection limit lines, the lot is disposed of appropriately. Clearly no acceptance is possible until the acceptance line Y_1 crosses the k axis. The operation of the plan is shown diagrammatically in Fig. 7-2.

The procedure described is called *unit sequential sampling* since items are drawn unit by unit. Occasionally, group sequential procedures are used in which groups of items are successively drawn (say, for example, ten at a time), inspected, and assessed against the acceptance and rejection limits

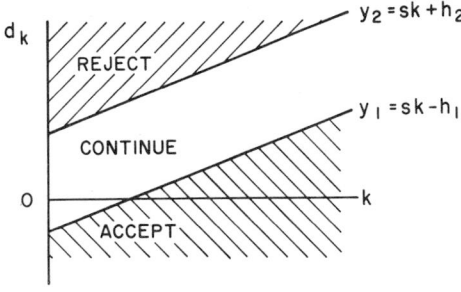

FIG. 7-1 Sequential acceptance plot.

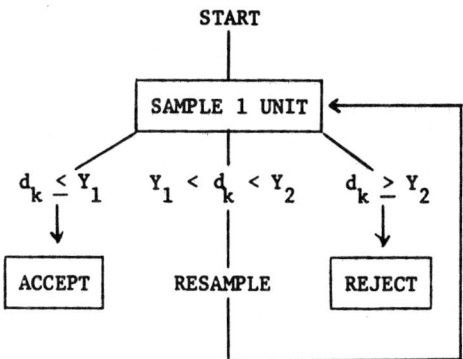

FIG. 7-2 Procedure for sequential sampling by attributes.

at the successive accumulated values of k. This is often done for inspection convenience. When the physical circumstances of the inspection do not
dictate a group size, an expeditious approach, suggested by Cowden (1957)
is to make the group size equal to the number of samples necessary to allow
the first possibility of acceptance. That is, the value of k just beyond
the intersection of the acceptance line Y_1 with the k axis in Fig. 7-1.
Group sequential plans are often listed in the form of a multiple-sampling
plan showing cumulative sample size with acceptance and rejection numbers
for each group. Of course, the listing must remain open ended. When
rounding acceptance and rejection numbers obtained from the sequential
acceptance plot, it is desirable to round the acceptance number upward and
the rejection number downward to minimize the difference between the group
sequential plan and the unit sequential plan from which it is derived. Of
course when the acceptance numbers and rejection numbers are integers for
the unit sequential plan at the successive values of k corresponding to
the multiples of group size, the measures of the group and unit sequential
plans will correspond. This will occur when h_1, h_2, and 1/s are integers.

SELECTION

Sequential sampling plans have been tabulated by the Statistical Research
Group (1945). A table of plans based on their results is given in Appendix Table T7-1 which shows the following characteristics for a variety of
plans when $\alpha = .05$ and $\beta = .10$

p_1 = producer's quality level

p_2 = consumer's quality level

h_1 = acceptance intercept

h_2 = rejection intercept

s = slope

$\bar{n}_0$ = number of samples prior to possibility of acceptance

$\bar{n}_1$ = number of samples prior to possibility of rejection

$\bar{n}_{p_1}$ = ASN at producer's quality level

$\bar{n}_s$ = ASN at proportion defective equal to s

$\bar{n}_{p_2}$ = ASN at consumer's quality level

Using these values, sequential plans for attributes inspection can be readily set up and characterized.

Formulas for the construction and evaluation of sequential plans for arbitrary values of p_1, p_2, α, and β have been derived by Wald (1947) and the Statistical Research Group (1945). The formulas are as follows:

$$h_1 = \frac{\log\left[(1 - \alpha)/\beta\right]}{\log\left(p_2/p_1\right) + \log\left[(1 - p_1)/(1 - p_2)\right]}$$

$$h_2 = \frac{\log\left[(1 - \beta)/\alpha\right]}{\log\left(p_2/p_1\right) + \log\left[(1 - p_1)/(1 - p_2)\right]}$$

$$s = \frac{\log\left[(1 - p_1)/(1 - p_2)\right]}{\log\left(p_2/p_1\right) + \log\left[(1 - p_1)/(1 - p_2)\right]}$$

Either common or natural logarithms can be used in these computations providing they are consistently one or the other.

Then, the acceptance and rejection lines are determined as

$$Y_1 = sk - h_1 \quad \text{(acceptance)} \qquad Y_2 = sk + h_2 \quad \text{(rejection)}$$

and then plotted as shown in Fig. 7-1. These formulas are sometimes expressed as

$$h_1 = \frac{b}{g_1 + g_2} = \frac{b}{G}$$

$$h_2 = \frac{a}{g_1 + g_2} = \frac{a}{G}$$

$$s = \frac{g_2}{g_1 + g_2} = \frac{g_2}{G}$$

for computational convenience where

$$a = \log \frac{1 - \beta}{\alpha}$$

$$b = \log \frac{1 - \alpha}{\beta}$$

$$g_1 = \log \frac{p_2}{p_1}$$

$$g_2 = \log \frac{1 - p_1}{1 - p_2}$$

$$G = g_1 + g_2$$

Appendix Table T7-2 gives values of a and b tabulated for selected values of α and β. Appendix Table T7-3 shows values of g_1 and g_2 for selected values of p_1 and p_2.

For example, suppose a sequential plan is desired having p_1 = .018, p_2 = .18, α = .05, β = .10. Appendix Table T7-1 does not list such a plan so the formulas must be used. Here

$$h_1 = \frac{\log\,[(1 - .05)/.10]}{\log\,(.18/.018) + \log\,[(1 - .018)/(1 - .18)]}$$

$$= \frac{\log\,9.5}{\log\,10 + \log\,1.1976} = 0.907$$

$$h_2 = \frac{\log\,[(1 - .10)/.05]}{\log\,(.18/.018) + \log\,[(1 - .018)/(1 - .18)]}$$

$$= \frac{\log\,18}{\log\,10 + \log\,1.1976} = 1.164$$

$$s = \frac{\log\,[(1 - .018)/(1 - .18)]}{\log\,(.18/.018) + \log\,[(1 - .018)/(1 - .18)]}$$

$$= \frac{\log\,1.1976}{\log\,10 + \log\,1.1976} = 0.0726$$

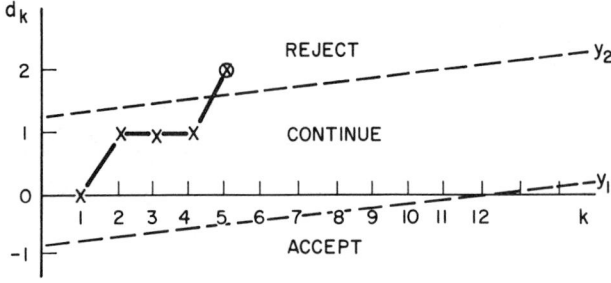

FIG. 7-3 Sequential graph.

so the lines of acceptance and rejection are

$$Y_1 = .0726k - 0.907$$

$$Y_2 = .0726k + 1.164$$

The plot appears as Fig. 7-3 which also shows the plot of sample results if the second and fifth items were defective leading to rejection and cessation of sampling at the fifth item sampled.

MEASURES

Measures of selected values of average sample number are given for the plans in Appendix Table T7-1. This includes $\bar{n}_0$, $\bar{n}_1$, $\bar{n}_{p_1}$, $\bar{n}_s$, and $\bar{n}_{p_2}$. The OC curve may be sketched from the given values of α, β, p_1 and p_2. These are based on Type B sampling as are all of the measures given here.

General formulas exist for probability of acceptance and average sample number. To use these formulas, we employ the auxiliary variable h, where $-\infty < h < \infty$, $h \neq 0$, then for any arbitarily selected value of h, a point (p, P_a) on the operating characteristic curve can be calculated [Wald (1947)] as

$$p = \frac{1 - [(1 - p_2)/(1 - p_1)]^h}{(p_2/p_1)^h - [(1 - p_2)/(1 - p_1)]^h}$$

with

$$P_a = \frac{[(1 - \beta)/\alpha]^h - 1}{[(1 - \beta)/\alpha]^h - [\beta/(1 - \alpha)]^h}$$

Given the combination p, P_a, we have the general formula for ASN

$$ASN = \frac{P_a \log [\beta/(1 - \alpha)] + (1 - P_a) \log [(1 - \beta)/\alpha]}{p \log (p_2/p_1) + (1 - p) \log [(1 - p_2)/(1 - p_1)]}$$

and for large lot size relative to sample size

$$AOQ = pP_a$$

Specifically, when $h = -\infty$, -1, 0, 1, ∞, the formulas given in Table 7-1 are obtained. For example, for the plan $p_1 = .018$, $p_2 = .180$, $\alpha = .05$, $\beta = .10$ where it was found that $h_1 = 0.907$, $h_2 = 1.164$, $s = 0.0726$, the following measures are obtained using Table 7-1:

p	P_a	ASN	AOQ
0	1	12.5	0
.018	.95	14.7	.017
.0726	.562	15.7	.041
.18	.10	8.9	.018
1	0	1.3	0

These values are usually sufficient for a crude sketch of the OC, ASN, and AOQ curves.

SEQUENTIAL SAMPLING FOR DEFECTS PER UNIT

Occasionally, sequential sampling procedures are required for defects per unit. Note that the unit employed need not be an individual piece but may be several pieces considered together. MIL-STD-105D is in terms of "defects per hundred." The following theory applies whatever the unit as long as it is defined beforehand.

A sequential chart for defects per unit is much like the chart for proportion defective in that it plots the sum of the defects found against k, the sample number. The PQL and CQL are, of course, in terms of mean defects per unit, μ_1 and μ_2, respectively, where $\mu_2 > \mu_1$. The parameters

TABLE 7-1 Sequential Sampling by Attributes for Proportion Defective - Points on the OC, ASN, and AOQ Curves

p	P_a	ASN	AOQ
0	1	$\dfrac{h_1}{s}$	0
p_1	$1 - \alpha$	$\dfrac{(1 - \alpha)h_1 - \alpha h_2}{s - p_1}$	$(1 - \alpha)p_1$
s	$\dfrac{h_2}{h_1 + h_2}$	$\dfrac{h_1 h_2}{s(1 - s)}$	$\dfrac{s h_2}{h_1 + h_2}$
p_2	β	$\dfrac{(1 - \beta)h_2 - \beta h_1}{p_2 - s}$	βp_2
1	0	$\dfrac{h_2}{1 - s}$	0

TABLE 7-2   Sequential Sampling by Attributes for
Defects per Unit Points on the OC and ASN Curves*

Mean defects per unit	P_a	ASN
0	1	$\dfrac{h_1}{s}$
μ_1	$1 - \alpha$	$\dfrac{(1 - \alpha)h_1 - \alpha h_2}{s - \mu_1}$
s	$\dfrac{h_2}{h_1 + h_2}$	$\dfrac{h_1 h_2}{s}$
μ_2	β	$\dfrac{(1 - \beta)h_2 - \beta h_1}{\mu_2 - s}$

*The form of the chart and its operation is the
same as that shown for proportion defective.

for the decision lines are as follows using common logarithms:

$$h_1 = \frac{\log\ [(1 - \alpha)/\beta]}{\log \mu_2 - \log \mu_1}$$

$$h_2 = \frac{\log\ [(1 - \beta)/\alpha]}{\log \mu_2 - \log \mu_1}$$

$$s = \frac{\mu_2 - \mu_1}{2.3026(\log \mu_2 - \log \mu_1)}$$

Values of operating parameters for such a plan are given in Table 7-2.

FURTHER CONSIDERATIONS

The theoretical development and application of sequential methods will be
found in Wald (1947) and Wetherill (1975). Proofs associated with these
procedures are fairly straightforward and are developed and presented in
detail in these texts.

Occasionally it is desirable to derive a multiple-sampling plan from
the equations of a sequential plan. A conservative approach to this problem would be to accept and reject only in the appropriate regions of the
sequential plan. This gives rise to the procedure characterized in Fig.
7-4. The acceptance criteria are determined by cutting the sequential
graph at successive samples of n. Here acceptance takes place farther into

the region of acceptance, decreasing α. Similarly β is decreased. Also ASN is increased since it is necessary to wait until n_1 units are inspected to record a defective anywhere in the sample. The multiple plan would be

Cumulative sample size	Ac	Re
n_1	0	d_4
n_2	d_1	d_5
n_3	d_2	d_6
....		

A "middle of the road" approach would be to average premature and delayed decisions so that the overall expectation of ASN and P_a remain the same as for sequential. This approach is shown in Fig. 7-5 in comparison to Fig. 7-4. Here the acceptance and rejection lines split the middle of the steps. This will result in roughly the same average sample number and probability of acceptance as for the sequential plan. The plan would be

Cumulative sample size	Ac	Re
n_1	0	d_4
n_2	d_1	d_5
n_3	d_3	d_6
....		

Incremental sample size for the multiple plan is subject to the same considerations as the size of groups in group sequential plans. Multiple plans require a rejection number one more than the acceptance number on the last sample. Thus, the gap leading to continuance must be closed. This can be achieved by holding the rejection numbers constant in the last few samples. The result is conservative in that it gives more protection to the consumer. This is illustrated for a three-stage plan in Fig. 7-6. The plan is

Cumulative sample size	Ac	Re
n_1	0	d_4
n_2	d_1	d_4
n_3	d_3	d_4

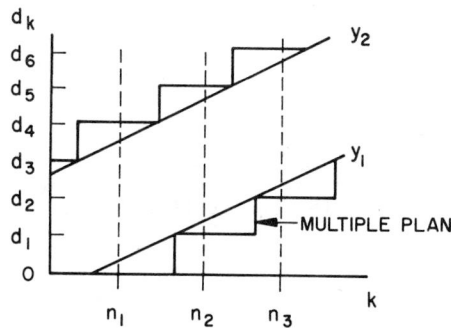

FIG. 7-4 Multiple from sequential –
conservative approach.

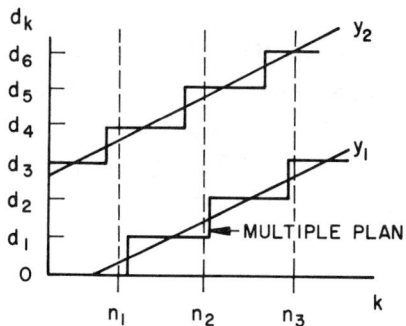

FIG. 7-5 Multiple from sequential –
averaging approach.

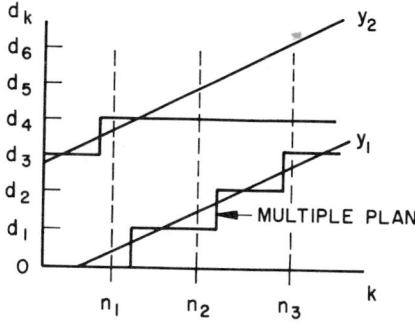

FIG. 7-6 Multiple from sequential –
closing the gap.

The sequential methods presented here are based on the likelihood ratio test of the simple hypothesis

$$H_0: \quad p' = p_1$$
$$H_1: \quad p' = p_2$$

The basic theoretical development will be found in Wald (1947).

PROBLEMS

1. Construct a sequential sampling plan such that $p_1 = .04$ and $p_2 = .20$ with $\alpha = .05$ and $\beta = .10$.

2. Construct a sequential sampling plan such that $p_1 = .07$ and $p_2 = .30$ with $\alpha = .05$ and $\beta = .10$.

3. Construct a sequential sampling plan such that $p_1 = .02$ and $p_2 = .06$ with $\alpha = .05$ and $\beta = .10$.

4. Compare the results for Probs. 1 to 3. What is the effect of increasing the slope? What is the effect of decreasing the slope? What is the effect of increasing h_2? What is the effect of increasing h_1?

5. Plot the Type B operating characteristic curve for the following sequential sampling plan using s and a minimum of two other plotting points.

$$Y_2 = .0723k + 3.8682$$
$$Y_1 = .0723k - 3.0129$$

   What is the ASN at these points?

6. The plan

$$Y_2 = .0656k + 1.9481$$
$$Y_1 = .0656k - 1.5174$$

   has $p_1 = .03$ and $p_2 = .12$ at $1 - \alpha = .95$, $\beta = .10$.
   Compute the ASN and AOQ at these points.

7. MIL-STD-105D, Code L, 1.5 AQL shows $\mu_1 = 1$ defect per hundred units at $P_a = .95$ and $\mu_2 = 5.9$ defects per hundred units at $P_a = .10$. Construct a sequential chart in terms of defects per hundred units which matches this plan.

8. Determine the ASN at μ_1 and μ_2 for the plan developed in Prob. 7.

9. What is the minimum number of samples leading to acceptance in Prob. 7?

10. If samples were assessed two at a time in Prob. 7, stopping after
    three stages to form a multiple sampling plan, what would be the
    acceptance and rejection numbers? How would you recommend closing
    the gap?

REFERENCES

Cowden, D. J. (1957), *Statistical Methods in Quality Control*, Prentice-
Hall, Englewood Cliffs, New Jersey.

Statistical Research Group, Columbia University (1945), *Sequential Analysis
of Statistical Data: Applications*, AMP Report 30.2R, Columbia University, New York.

Wald, A. (1947), *Sequential Analysis*, John Wiley and Sons, New York.

Wetherill, G. B. (1975), *Sequential Methods in Statistics*, Chapman and Hall,
London.

United States Department of Defense (1963), *Military Standard, Sampling
Procedures and Tables for Inspection by Attributes*, (MIL-STD-105D),
Washington, D.C.

VARIABLES SAMPLING FOR PROCESS PARAMETER

Specifications are frequently written in terms of statistical parameters which describe the product to be inspected. For attributes inspection, the parameter to be controlled is, of course, the proportion nonconforming in the lot or process. When specifications are written in terms of measurements, other parameters may be of importance, such as the average (mean) level of a certain characteristic of the process which produced the units to be inspected, or in some instances its standard deviation. This implies Type B sampling. Examples of such specifications are mean life of a lamp, average amount of discharge of an impurity into a stream, average emission of carbon monoxide from cars of a certain make and model, and the standard deviation of an electrical test on transistors for use in a ballistic missile. Specifications of this type are in contrast to those on the individual measurements themselves which relate to individual units of product; variables sampling plans for such specifications will be covered in a later chapter.

It is characteristic of specifications on a process parameter that certain levels are acceptable and should be protected from rejection, while other levels are objectionable and should be rejected by the plan. This was recognized by R. A. Freund (1957) when he distinguished two critical levels:

θ_1: Acceptable process level (APL). A process level which is acceptable and should be accepted most of the time by the plan.

θ_2: Rejectable process level (RPL). A process level which is rejectable and should be rejected most of the time by the plan.

Here, θ is taken to be the parameter specified.

In a manner analogous to attributes plans, the probability of acceptance for each of these levels is defined as

α = probability of rejection at the APL (producer's risk)

1 - α = probability of acceptance at the APL

β = probability of acceptance at the RPL (consumer's risk)

Most variables acceptance sampling plans for process parameter can be specified in terms of these levels and risks.

SINGLE SAMPLING FOR PROCESS PARAMETER

Standard statistical tests of hypotheses form the basis for the methodology of single sampling by variables for process parameters. In fact, such plans are simply tests of hypotheses. Thus, the statistical tests shown in Table 8-1 may be employed as sampling plans in this context. They are used as one-sided or two-sided tests depending on whether the parameter is to be controlled against specifications on one or both sides. The operation of such tests is described in standard statistical texts such as Bowker and Lieberman (1959). Sample sizes are critical in acceptance sampling applications and may be determined from the appropriate power or operating characteristic curves of the test. The power curve shows probability of rejection plotted against hypothetical values of process parameter. Its complement is the operating characteristic curve. These curves for variables plans for process parameter are usually plotted against the standardized displacement of the parameter from the APL (μ_1) such as

$d = (\mu - \mu_1)/\sigma$

TABLE 8-1 Statistical Tests of Hypotheses

Parameter specified	Condition	Test	Statistic
Mean (μ_0)	μ_0 specified, σ known	Normal z-test	$z = \dfrac{\overline{X} - \mu_0}{\sigma_{\overline{X}}} = \dfrac{\overline{X} - \mu_0}{\sigma/\sqrt{n}}$
	μ_0 specified, σ unknown	Student's t-test	$t = \dfrac{\overline{X} - \mu_0}{s_{\overline{X}}} = \dfrac{\overline{X} - \mu_0}{s/\sqrt{n}}$
Standard deviation (σ_0)	σ_0^2 specified	χ^2 test	$\chi^2 = (n - 1)\left(\dfrac{s}{\sigma_0}\right)^2$

or

$$\lambda = \sigma^2/\sigma_1^{\,2}$$

Sample size for two-point plans can be determined from the standard-ized displacement of the RPL (μ_2) from the APL (μ_1). For the tests mention-ed this is as follows:

Test	Displacement
z-test, t-test	$d_0 = \dfrac{|\mu_2 - \mu_1|}{\sigma} = \dfrac{|\text{RPL} - \text{APL}|}{\sigma}$
χ^2 test	$\lambda_0 = \dfrac{\sigma_2^{\,2}}{\sigma_1^{\,2}} = \dfrac{\text{RPL}}{\text{APL}}$

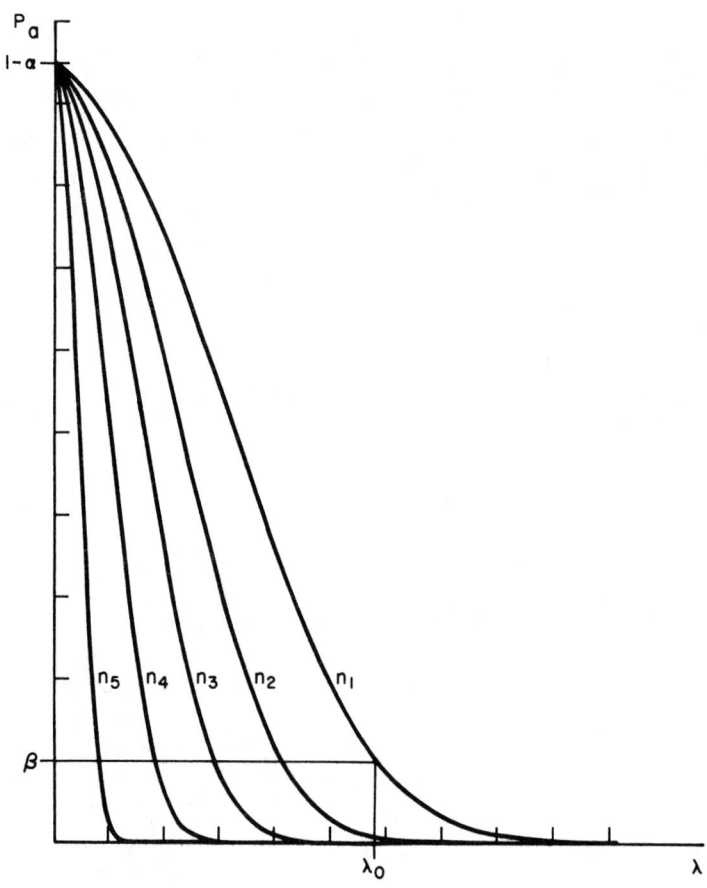

FIG. 8-1 Typical OC curves.

Fig. 8-1 shows typical OC curves. For a set of curves with the specified α risk, the sample size is found from the curve passing through (or
nearest to) the intersection of d_0 or λ_0, plotted on the horizontal axis,
and β, plotted on the vertical axis. If no curve passes through this point,
crude interpolation may be necessary. Frequently a 5 percent producer's
risk and a 10 percent consumer's risk is employed. Operating characteristic
curves are given in the Appendix tables for $\alpha = .05$. They include the
following:

Table	Test
T8-1	One-sided normal z-test
T8-2	Two-sided normal z-test
T8-3	One-sided Student's t-test
T8-4	Two-sided Student's t-test
T8-5	One-sided χ^2 test

For example, suppose mean life of a lamp was specified by the manufacturer as 1000 hr. (APL), while the customer wished to be sure to reject
shipments of lamps having mean life 800 hr. (RPL). The standard deviation
of life is not known but is expected to be in the order of 200 hr. A one-
sided t-test is appropriate since the implied specification on mean life is
one-sided, i.e., μ not less than 1000 hr. The standardized displacement is

$$d_0 = \frac{|RPL - APL|}{\sigma} = \frac{|800 - 1000|}{200} = 1$$

If risks are set at $\alpha = .05$ and $\beta = .10$, the operating characteristic curve
for the one-sided t-test shows that a sample size of n = 10 is required. A
sample of 10 would be selected from the lot and the t-test applied, accepting or rejecting the lot as the null hypothesis is accepted or rejected.

ACCEPTANCE CONTROL CHARTS

A natural extension of a standard test of hypothesis on individual lots is
to plot the results for successive lots in control chart form. This serves
to allow trend and runs analysis on a continuing series of lots and affords
the acceptance control engineer all the advantages of the control chart
technique. The critical value for the test serves as the acceptance control
limit (ACL). Lots which plot inside the acceptance control limit are accepted. Those which plot outside are rejected.

This idea was first proposed by R. A. Freund (1957) in a celebrated

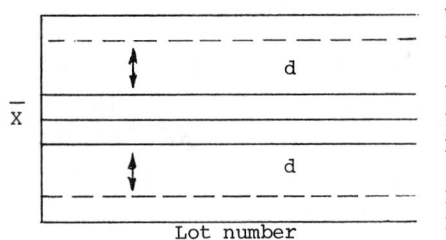

	Upper rejectable process limit, RPL
	Upper acceptance control limit, ACL
	Upper acceptable process limit, APL
	Nominal center line, NCL
	Lower acceptable process limit, APL
	Lower acceptance control limit, ACL
	Lower rejectable process limit, RPL

FIG. 8-2 An acceptance control chart.

paper which later won the Brumbaugh Award from the American Society for
Quality Control as the best technical contribution of the year. A two-
sided chart appears as in Fig. 8-2. A one-sided chart would consist of
the upper half or lower half of the chart shown depending upon the direction in which the mean is to be controlled. Of course, the APL and RPL do
not appear on a chart in application. Only the acceptance control limits
(ACL) and the nominal center line (NCL) are shown in actual use. The
nominal center line is, of course, halfway between the acceptance control
limits in the two-sided case.

The initial work on the acceptance control chart has been primarily
with standard deviation known. Freund (1957, p. 14) points out that: "It
is implied that α and β risks will be selected for the APL and RPL values
respectively and that σ will be known from past experience or estimated in
the usual control chart manner from the $\bar{R}$ or s computed from about 20
samples." Clearly the measure of variability must have been in control
for 20 or more samples to assure the stability implicit in a known standard
deviation application.

The acceptance control chart is easily set up by using appropriate
formulas.

If ACL = acceptance control limit, z_{α} = standard normal deviate cutting
off area of α in upper tail, z_{β} = standard normal deviate cutting off area
of β in upper tail, and d = distance ACL lies from APL in direction of RPL,
then

$$n = \left(\frac{(z_{\alpha} + z_{\beta})\sigma}{\text{RPL} - \text{APL}}\right)^2$$

$$d = \frac{z_{\alpha}}{z_{\alpha} + z_{\beta}} \left|\text{RPL} - \text{APL}\right|$$

where the sign of $\left|\text{RPL} - \text{APL}\right|$ is regarded as always positive. The distance, d, and its relation to the ACL, APL, and RPL is shown in Fig. 8-2.

Freund has derived special factors which facilitate the determination of the acceptance control limits. They allow computation of the limits either from the APL as before or alternatively from the RPL as a baseline. Using γ to represent the risk, the factors are

Factor (A)	Measure of variability (V)
$A_{0,\gamma}$	σ known
$A_{1,\gamma}$	$\bar{\sigma} = \frac{1}{k} \Sigma \sqrt{\dfrac{\Sigma (X - \bar{X})^2}{n}}$ for k lots with n samples
$A_{2,\gamma}$	$\bar{R} = \frac{1}{k} \Sigma R$ for k lots with n samples
$A_{3,\gamma}$	$\bar{s} = \frac{1}{k} \Sigma \sqrt{\dfrac{\Sigma (X - \bar{X})^2}{n - 1}}$ for k lots with n samples

Appendix Table T8-6 gives the Freund A factors for various values of α and β. To use the A factors the sample size must be calculated from the formula as before. For any of the factors A_γ with corresponding measure of variability, V, we have

Lower limit	Upper limit
$ACL = APL - A_\alpha V$	$ACL = APL + A_\alpha V$
$ACL = RPL + A_\beta V$	$ACL = RPL - A_\beta V$

It can be seen directly that $d = A_\alpha V$.

In a two-sided situation, the acceptance control limits may be so close to the nominal value (NCL) as to force consideration of both tails of the distribution of sample means simultaneously. To assess the need for special correction terms in this situation, compute

$$\Delta_1 = \frac{\sqrt{n}(ACL - NCL)}{\sigma} = \text{deviation of upper ACL from NCL in terms of } \sigma_{\bar{X}}$$

or, alternatively,

$$\Delta_2 = \frac{\sqrt{n}(APL - NCL)}{\sigma} = \text{deviation of upper APL from NCL in terms of } \sigma_{\bar{X}}$$

depending on whether the acceptance control limits (ACL) themselves or the acceptable process limit (APL) have already been specified. Apply the correction terms found in Appendix Table T8-7 if the value Δ_1 or Δ_2 calculated is less than the value shown in Table 8-2. If a correction term (CT) is necessary, it can be found in Appendix Table T8-7 corresponding to the value of Δ_1 or Δ_2 calculated. The A factor of Appendix Table T8-6 is then multiplied by the correction term to obtain a new A factor to be used in this two-sided situation. Application of the new A factor proceeds as before. Alternatively, the correction term can be used in the formulas for n and d as follows:

$$n = \left(\frac{[(CT)z_\alpha + z_\beta]\sigma}{RPL - APL}\right)^2$$

$$d = \frac{(CT)z_\alpha}{(CT)z_\alpha + z_\beta}\left|RPL - APL\right|$$

The Freund (1957) paper presents the theory behind acceptance control charts as well as many excellent examples. The following is an adaptation of one such example.

Bottles are filled with 10 cm^3 of a solution. The amount of solution is to be maintained within ± 0.5 cm^3 with less than 0.1 percent of the bottles outside the specification. It is desired to reject if more than 2.5 percent of the bottles are under- or overfilled. A sample is to be taken of each half hour's production to be plotted against an acceptance control chart having $\alpha = .05$ and $\beta = .10$. The standard deviation has been estimated from control charts as $\sigma = .10$ and fill is normally distributed.

Using normal distribution theory to estimate the APL and RPL, we obtain the upper specification:

$$RPL = USL - z_{.025}\sigma = 10.5 - 1.96(.10) = 10.30$$

$$APL = USL - z_{.001}\sigma = 10.5 - 3.09(.10) = 10.19$$

and the lower specification:

$$APL = LSL + z_{.001}\sigma = 9.5 + 3.09(.10) = 9.81$$

$$RPL = LSL + z_{.025}\sigma = 9.5 + 1.96(.10) = 9.70$$

The sample size for $\alpha = .05$, $\beta = .10$ is

$$n = \left(\frac{(z_\alpha + z_\beta)\sigma}{RPL - APL}\right)^2$$

$$= \left(\frac{(1.645 + 1.282).10}{10.30 - 10.19}\right)^2 \simeq 7$$

For a nominal value of 10

$$\Delta_1 = \frac{\sqrt{7}(10.19 - 10.0)}{.10} = 5.03$$

TABLE 8-2   Minimum Δ Values Before Correction Terms Need Be Used

α Risk	Minimum Δ_1	Minimum Δ_2	α Risk	Minimum Δ_1	Minimum Δ_2
.05	2.5	.851	.005	3.2	.619
.01	3.0	.670	.001	3.5	.409

and so no correction term for double specification limits is needed.   Then

$$d = \frac{z_\alpha}{z_\alpha + z_\beta}\left|RPL - APL\right| = \frac{1.645}{1.645 + 1.282}\left|9.70 - 9.81\right| = 0.622$$

Note that, for sample size n = 7, the Freund A factor from Appendix Table T8-6 is 0.622 and d = $A_{0,.05}$ σ = .622(.10) = .062 as it should be.

The acceptance control chart can be set up accordingly and means of samples of size 7 plotted against the acceptance limits to determine acceptance of subsequent lots.   A control chart for variability should also be instituted to detect any change in standard deviation from the known value.

SEQUENTIAL PLANS FOR PROCESS PARAMETER (σ Known)

When sample size must be kept to an absolute minimum, sequential plans also provide an excellent approach in sampling against specified process parameters.   These plans are used on variables data in a manner analogous to sequential plans for attributes.   A greater variety of sequential charts are available for variables, however, since the attributes sequential test is usually limited to a one-sided test against an increase in proportion nonconforming over that specified as the producer's quality level.   Separate sequential tests for variables are available for various parameters against an upper specification limit, a lower specification limit, or double specification limits.   As in attributes testing, a cumulative statistic, Y, is plotted against the number of samples taken.

In testing against an upper specification limit, the decision lines for the sequential plot are of the same form as for attributes, namely the rejection line: $Y_2 = h_2 + sk$ and the acceptance line: $Y_1 = -h_1 + sk$. Formulas for the acceptance constants h_1, h_2, and s for testing the mean and variance against an upper specification limit are given in Table 8-3 together with those for the average sample number at: APL, s, RPL. Formulas for attributes testing are also given in that table for reference, In using the table, all computations should be made consistently in common or natural

TABLE 8-3 Formulas for Single Upper Limit Sequential Plans for Process Parameter (For Common Logs, L = 2.3026; For Natural logs, L = 1)

TEST	PLOT (Y)	h_1	h_2	s	APL	s	RPL
MEAN APL = μ_1 RPL = μ_2	$Y = \Sigma X_i$ = SUM OF OBSERVATIONS	$\dfrac{L\,b\,\sigma^2}{\mu_2 - \mu_1}$	$\dfrac{L\,a\,\sigma^2}{\mu_2 - \mu_1}$	$\dfrac{\mu_2 + \mu_1}{2}$	$P_a = 1 - \alpha$ $ASN = \dfrac{(1-\alpha)h_1 - \alpha h_2}{s - \mu_1}$	$P_a = \dfrac{h_2}{h_1 + h_2}$ $ASN = \dfrac{h_1 h_2}{\sigma^2}$	$P_a = \beta$ $ASN = \dfrac{(1-\beta)h_2 - \beta h_1}{\mu_2 - s}$
VARIANCE APL = σ_1^2 RPL = σ_2^2	$Y = \Sigma(X_i - \mu)^2$ If μ unknown plot $Y' = \Sigma(X_i - \bar{X})^2$ against $k' = k-1$	$\dfrac{2\,L\,b\,\sigma_1^2\,\sigma_2^2}{\sigma_2^2 - \sigma_1^2}$	$\dfrac{2\,L\,a\,\sigma_1^2\,\sigma_2^2}{\sigma_2^2 - \sigma_1^2}$	$\dfrac{L \log\left(\frac{\sigma_2^2}{\sigma_1^2}\right)\sigma_1^2\,\sigma_2^2}{\sigma_2^2 - \sigma_1^2}$	$P_a = 1 - \alpha$ $ASN = \dfrac{(1-\alpha)h_1 - \alpha h_2}{s - \sigma_1^2}$	$P_a = \dfrac{h_2}{h_1 + h_2}$ $ASN = \dfrac{h_1 h_2}{2s^2}$	$P_a = \beta$ $ASN = \dfrac{(1-\beta)h_2 - \beta h_1}{\sigma_2^2 - s}$
PROPORTION DEFECTIVE APL = p_1 RPL = p_2	$Y = \Sigma d_i$ = TOTAL DEFECTIVE	$\dfrac{b}{g_1 + g_2}$	$\dfrac{a}{g_1 + g_2}$	$\dfrac{g_2}{g_1 + g_2}$	$P_a = 1 - \alpha$ $ASN = \dfrac{(1-\alpha)h_1 - \alpha h_2}{s - p_1}$	$P_a = \dfrac{h_2}{h_1 + h_2}$ $ASN = \dfrac{h_1 h_2}{s(1-s)}$	$P_a = \beta$ $ASN = \dfrac{(1-\beta)h_2 - \beta h_1}{p_2 - s}$
DEFECTS PER UNIT APL = μ_1 RPL = μ_2	$Y = \Sigma d_i$ = TOTAL DEFECTS IN k UNITS	$\dfrac{b}{(\log \mu_2 - \log \mu_1)}$	$\dfrac{a}{(\log \mu_2 - \log \mu_1)}$	$\dfrac{\mu_2 - \mu_1}{L(\log \mu_2 - \log \mu_1)}$	$P_a = 1 - \alpha$ $ASN = \dfrac{(1-\alpha)h_1 - \alpha h_2}{s - \mu_1}$	$P_a = \dfrac{h_2}{h_1 + h_2}$ $ASN = \dfrac{h_1 h_2}{s}$	$P_a = \beta$ $ASN = \dfrac{(1-\beta)h_2 - \beta h_1}{\mu_2 - s}$

$$a = \log\left(\frac{1 - \beta}{\alpha}\right) \qquad b = \log\left(\frac{1 - \alpha}{\beta}\right) \qquad g_1 = \log\left(\frac{p_2}{p_1}\right) \qquad g_2 = \log\left(\frac{1 - p_1}{1 - p_2}\right)$$

UPPER SPECIFICATION		LOWER SPECIFICATION	
REJECTION LINE	$Y_2 = h_2 + sk$	REJECTION LINE	$Y_2' = -h_2 + sk$
ACCEPTANCE LINE	$Y_1 = -h_1 + sk$	ACCEPTANCE LINE	$Y_1' = h_1 + sk$

logarithms with the factor L adjusted accordingly. The form of the sequential chart for an upper specification limit is that of Fig. 8-3. Dan J.
Sommers (1979) has pointed out a simple relation between sequential plans
and single-sampling plans for the mean. Given a known standard deviation
single-sampling plan using a sample of n to test μ_1 against an upper limit
μ_2 (i.e., $\mu_2 > \mu_1$) with risks $\alpha = .05$ and $\beta = .10$, the matching sequential
plan has parameters

$$h_1 = .7693 \ \sigma\sqrt{n}$$

$$h_2 = .9877 \ \sigma\sqrt{n}$$

$$s = \mu_1 + 1.4632 \ \frac{\sigma}{\sqrt{n}}$$

or

$$s = \mu_2 - 1.4632 \ \frac{\sigma}{\sqrt{n}}$$

$$ASN(\mu_1) = .4657n$$

$$ASN(s) \ \ = .7598n$$

$$ASN(\mu_2) = .5549n$$

This provides for an immediate assessment of the efficiency of sequential
sampling as against single sampling for the mean.

In testing against a lower specification limit, the acceptance and
rejection regions are, of course, reversed. This gives rise to new decision lines, namely the rejection line: $Y_2' = -h_2 + sk$ and the acceptance
line: $Y_1' = h_1 + sk$ where the values of the acceptance constants are the
same as those given in Table 8-3 for the upper limit. The formulas given
for ASN also remain the same as for the upper limit but must be taken in
absolute value. The results for the decision lines can be seen to come
about from interchanging μ_1 with μ_2 and α with β in the formulas for the
upper limit thus reversing the roles of h_1 and h_2. These changes can also
be made in Sommers' formulas for use with a lower limit ($\mu_2 < \mu_1$) by interchanging μ_1 with μ_2 in s and h_1 with h_2. The formulas for ASN remain the
same. The form of the chart for a lower limit is shown in Fig. 8-4.

In general, when testing the mean against either an upper or a lower
specification limit using the formulas of Table 8-3 for h_1, h_2, and s, it
can be shown that the following relations hold for probability of acceptance and average sample number for any given value of μ except $\mu = s$.

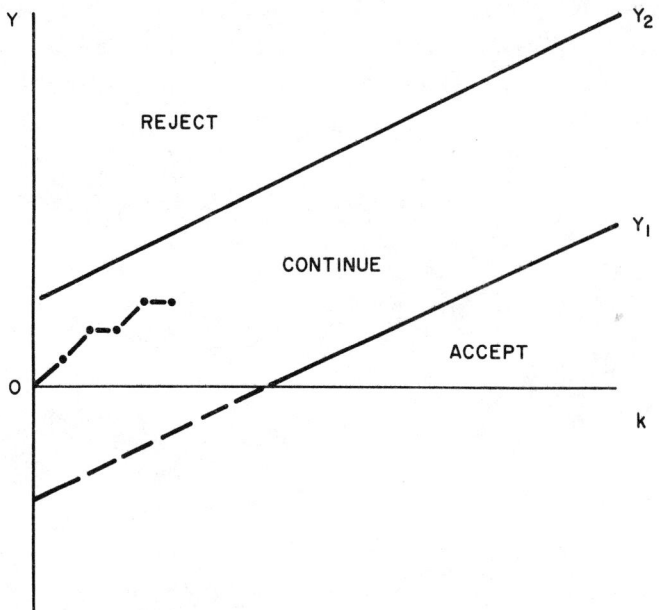

FIG. 8-3 Sequential variables chart: upper specification limit.

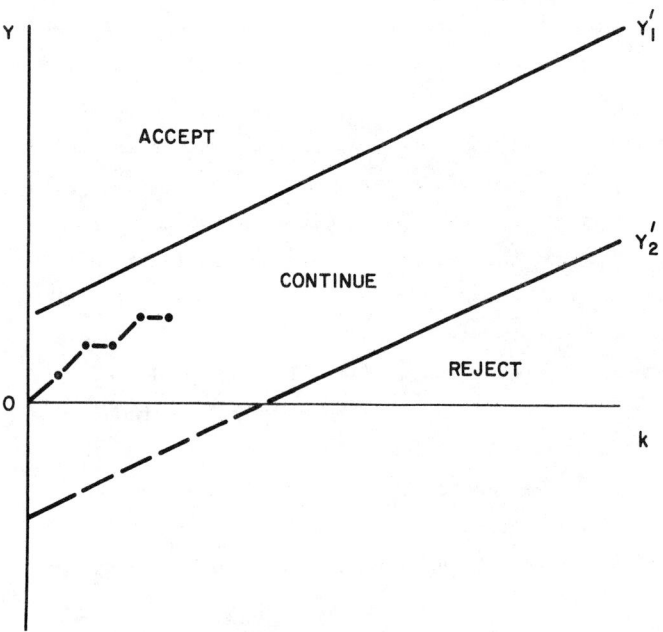

FIG. 8-4 Sequential variables chart: lower specification limit.

$$w = \frac{\mu_2 + \mu_1 - 2\mu}{\mu_2 - \mu_1}$$

so that

$$P_a = \frac{[(1 - \beta)/\alpha]^W - 1}{[(1 - \beta)/\alpha]^W - [\beta/(1 - \alpha)]^W}$$

and

$$ASN = \frac{\left[\dfrac{L\sigma^2 \log[(1-\beta)/\alpha]}{\mu_2 - \mu_1}\right]}{\left[\dfrac{2\mu - \mu_2 - \mu_1}{2}\right]}$$

$$+ \frac{\left[\dfrac{[(1-\beta)/\alpha]^W - 1}{[(1-\beta)/\alpha]^W - [\beta/(1-\alpha)]^W}\right]\left[\dfrac{L\sigma^2 \log[\beta/(1-\alpha)] - L\sigma^2 \log[(1-\beta)/\alpha]}{\mu_2 - \mu_1}\right]}{\left[\dfrac{2\mu - \mu_2 - \mu_1}{2}\right]}$$

Where $L = 1$ when natural logarithms are used and $L = 2.3026$ using common logarithms.

When $\mu = s$,

$$P_a = \frac{\log[(1 - \beta)/\alpha]}{\log[(1 - \beta)/\alpha] + \log[(1 - \alpha)/\beta]}$$

$$ASN = \frac{L^2 \sigma^2 \log[(1 - \beta)/\alpha]\log[(1 - \alpha)/\beta]}{\left(\mu_2 - \mu_1\right)^2}$$

These relations are useful in constructing the OC and ASN curves.

For example, for $\mu_1 = 1$, $\mu_2 = 1.2$, with $\alpha = .025$ and $\beta = .10$, we have the following results for a value of the mean $\mu = 1$ with $\sigma = .1$ using natural logarithms

$$w = \frac{1.2 + 1 - 2(1)}{1.2 - 1} = 1$$

$$P_a = \frac{[.9/.025]^1 - 1}{[.9/.025]^1 - [.1/.975]^1} = .975$$

$$ASN = \frac{\left[\dfrac{1(.01)\log[.90/.025]}{1.2 - 1}\right]}{\left[\dfrac{2 - 1.2 - 1}{2}\right]}$$

$$+ \frac{\left[\dfrac{[.90/.025]^1 - 1}{[.90/.025]^1 - [.10/.975]^1}\right]\left[\dfrac{1(.01)\log[.10/.975] - 1(.01)\log[.90/.025]}{1.2 - 1}\right]}{\left[\dfrac{2 - 1.2 - 1}{2}\right]}$$

$$= \frac{.1792 + [.975][-.2930]}{-.1} = 1.065$$

Occasionally, it may be necessary to plot a linear function of the quality characteristic being measured. This is often done to adjust the slope or scale of the chart to practical proportions. Suppose

$$Y = \Sigma x$$

has been plotted and it is desired to plot

$$Y^* = \Sigma(ax + b)$$

Then the equations for the upper limit decision lines become

$$Y_2^* = ah_2 + (as + b)k$$
$$Y_1^* = -ah_1 + (as + b)k$$

so that, in effect:

$$h_2^* = ah_2$$
$$h_1^* = ah_1$$
$$s^* = as + b$$

For example, if the slope of the chart is too steep, it may be adjusted by subtracting a constant, C, from the points plotted to obtain

$$Y^* = x - C$$

whereupon the equations for the decision lines become

$$Y_2^* = h_2 + (s - C)k$$
$$Y_1^* = -h_1 + (s - C)k$$

For a test of the mean, the constant C is often chosen equal to the APL. This places the APL at $Y = 0$. When $C = s$, the sequential chart will have horizontal limits. When C is less than s, the chart will slope downward. That is, $s^* = s - C$ will be negative. Such an effect will be observed when μ_1, the APL, is subtracted from the individual observations to be plotted on a chart testing against a lower specification limit on the mean. This is because

$$s^* = s - \mu_1$$

$$= \frac{\mu_1 + \mu_2}{2} - \mu_1$$

$$= \frac{\mu_2 - \mu_1}{2}$$

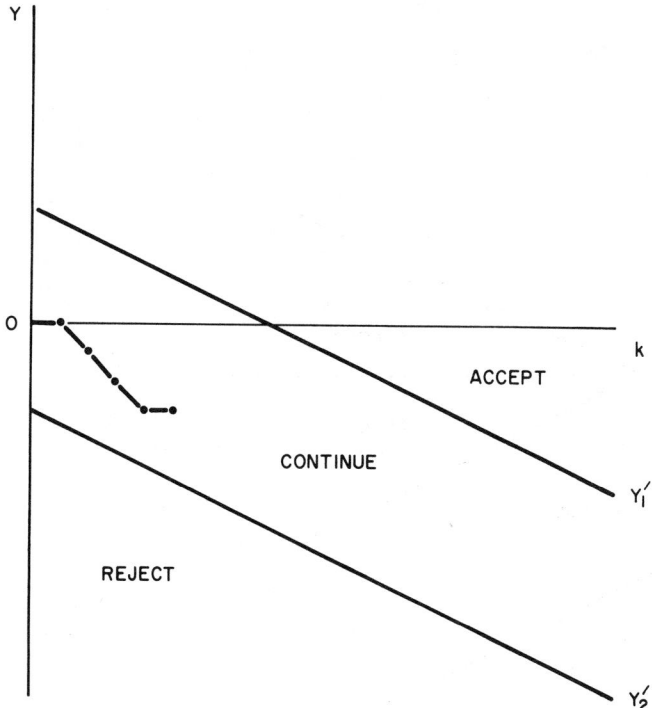

FIG. 8-5 Adjusted sequential variables chart:
lower specification limit.

which will be negative in testing against a lower specification limit and
positive in testing against an upper specification limit. Note that, in
this case, the decision lines for the lower limit case can be found from
the upper limit lines with the same α, β, and $\left| \mu_1 - \mu_2 \right|$ using the relations
rejection line: $Y_2' = -Y_2 = -h_2 - sk$ and acceptance line: $Y_1' = -Y_1 = h_1 - sk$.
The form of such a chart will be seen in Fig. 8-5 for a lower specification
limit where $Y = x - \mu_1$. The form for an upper limit chart using this
adjustment is that of Fig. 8-3 when $Y = x - \mu_1$.

A chart for testing double specification limits on the mean can be
constructed as the superimposition of individual upper and lower specification limit charts. Such a chart is illustrated in Fig. 8-6.

A zero baseline is obtained for plotting the double specification
limits chart by cumulating

$$Y = \Sigma (x - \mu_1)$$

This provides a common abscissa for the constituent upper and lower

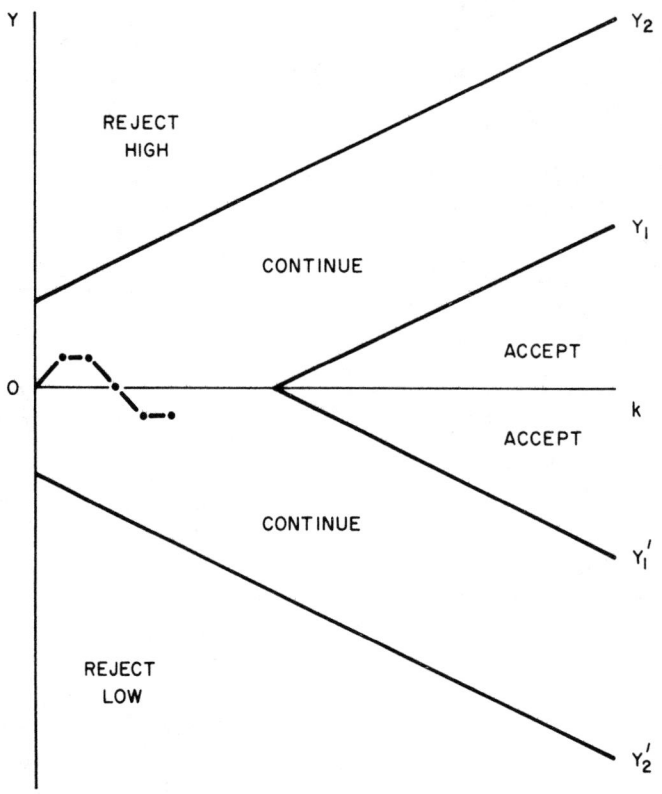

FIG. 8-6 Sequential variables chart: double
specification limits.

specification limit charts. The APL, μ_1, is taken halfway between the upper
and lower RPL so that

$$\mu_1 = \frac{\text{upper RPL} + \text{lower RPL}}{2}$$

and the α risk is apportioned half to each plan so that each is set up
using $\alpha/2$. The formulas must be corrected for subtraction of the constant, μ_1.

As an example of application of such a double-limit sequential plan,
suppose parts are received which are to have a plastic coating of thickness
1 mm ± .2 mm. If it is expensive and possibly destructive to measure the
thickness of the coating, a sequential test is in order. Here there is a
double specification limit involving the construction of two sequential
plans, for the upper and lower limit respectively, which will then be combined into a single chart for application of the plan. Taking the APL to
be halfway between the two RPLs:

$$\text{Lower RPL} = \mu_0 = .8 \text{ mm}$$
$$\text{APL} = \mu_1 = 1 \text{ mm}$$
$$\text{Upper RPL} = \mu_2 = 1.2 \text{ mm}$$

Suppose the standard deviation is known to be $\sigma = 0.1$ mm. The plot must be adjusted by subtracting the APL from each observation so that, C = 1.

If conventional values of $\alpha = .05$, $\beta = .10$ are to be used, the α risk in each plan is $\alpha = .025$. The derivation of the upper limit plan using common logarithms is   .

$$a = \log\left(\frac{1 - \beta}{\alpha}\right) = \log\left(\frac{.9}{.025}\right) = 1.5563$$

$$b = \log\left(\frac{1 - \alpha}{\beta}\right) = \log\left(\frac{.975}{.10}\right) = 0.9890$$

$$h_1 = \frac{Lb\sigma^2}{\mu_2 - \mu_1} = \frac{2.3026(0.9890)(.1)^2}{1.2 - 1.0} = .1139$$

$$h_2 = \frac{La\sigma^2}{\mu_2 - \mu_1} = \frac{2.3026(1.5563)(.1)^2}{1.2 - 1.0} = .1792$$

$$s = \frac{\mu_2 + \mu_1}{2} = \frac{1.2 + 1.0}{2} = 1.1$$

which leads to the limit lines

$$Y_2 = .1792 + (1.1 - 1)k = .1792 + 0.1k$$
$$Y_1 = -.1139 + (1.1 - 1)k = -.1139 + 0.1k$$

For this plan, the operating properties are:
at APL ($\mu = 1$ mm),

$$P_a = 1 - .025 = .975$$

$$\text{ASN} = \frac{(1 - .025)(.1139) - .025(.1792)}{1.1 - 1.0} = 1.07$$

at s ($\mu = 1.1$ mm),

$$P_a = \frac{.1792}{.1139 + .1792} = .6114$$

$$\text{ASN} = \frac{(.1139)(.1792)}{(.1)^2} = 2.04$$

at RPL ($\mu = 1.2$ mm),

$$P_a = .10$$

$$\text{ASN} = \frac{(1 - .1)(.1792) - .1(.1139)}{1.2 - 1.1} = 1.50$$

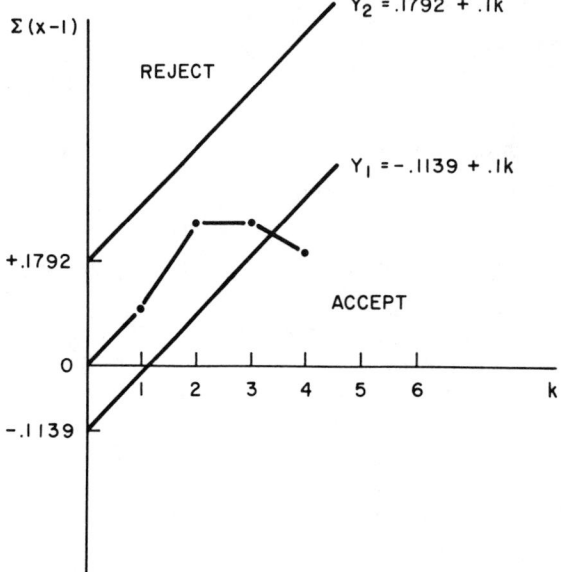

FIG. 8-7   Upper limit chart, $\alpha = .025$, $\beta = .10$.

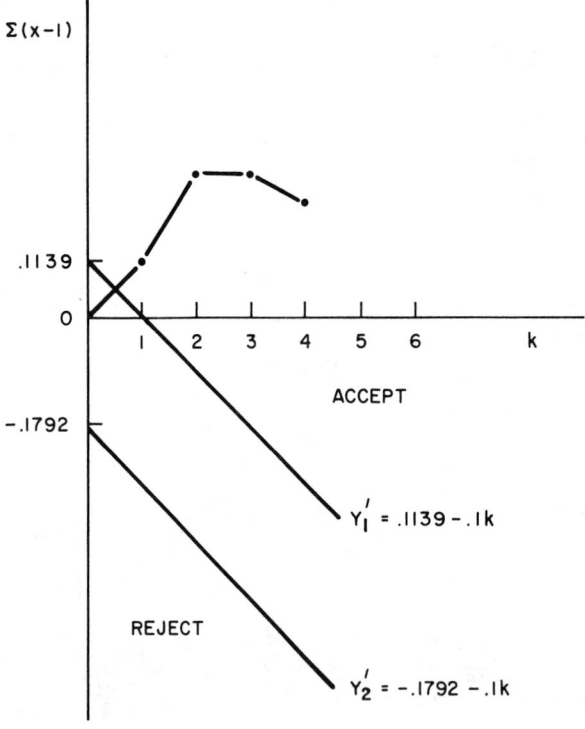

FIG. 8-8   Lower limit chart, $\alpha = .025$, $\beta = .10$.

If successive samples were 1.1, 1.15, 1.0, and 0.95, the sequential chart would appear as in Fig. 8-7 for a test against the upper limit only. Exploiting the symmetry, the relations

$$Y_2' = -Y_2 \qquad Y_1' = -Y_1$$

can be used to obtain the decision lines for testing the lower limit. Using the same data as before, the lower limit chart appears as Fig. 8-8 with

$$Y_1' = .1139 - 0.1k \quad \text{(accept)} \qquad Y_2' = -.1792 - 0.1k \quad \text{(reject)}$$

Clearly, for this chart also, the ASN at the APL is still 1.07 while the ASN at the RPL is, by symmetry, 1.50. For a test against the double specification limits 1.0 ± .2 mm, the charts are superimposed as in Fig. 8-9.

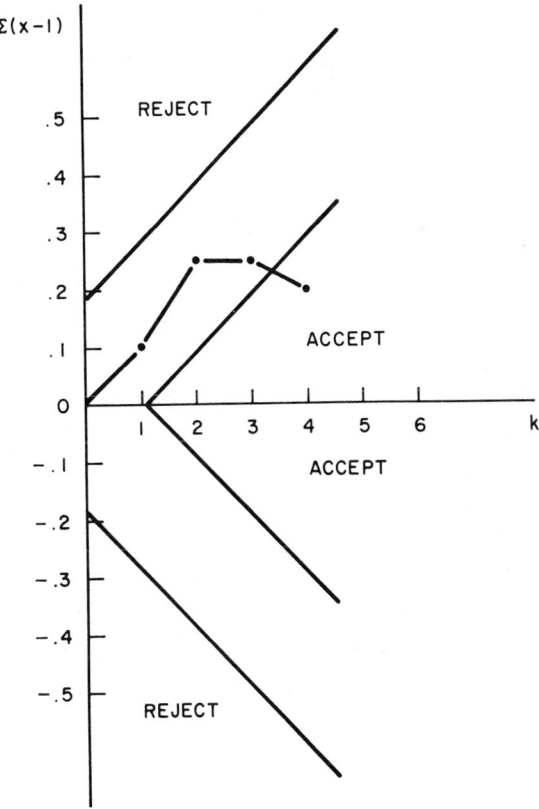

FIG. 8-9 Double specification limit chart,
$\alpha = .05$, $\beta = .10$.

SEQUENTIAL PLANS FOR PROCESS PARAMETER (σ Unknown)

The preceding methods are for the case when the standard deviation is known. When this is not the case, the methods of Barnard (1946) may be employed. Appendix Table T8-8 gives boundary values defining decision lines for a plot against successive values of k of the statistic

$$Y = \frac{\Sigma \ (X - \mu_1)}{\sqrt{\Sigma \ (X - \mu_1)^2}}$$

This is a sequential version of the one-sample t-test against an upper specification limit, that is, where the RPL is greater than the APL. In this case we have $\mu_2 > \mu_1$ and $\mu_2 = \mu_1 + D\sigma$ or $D = (\mu_2 - \mu_1)/\sigma$ where D forms one of the arguments in the table of Barnard's values. The table also gives values of

k_1 = smallest number of values to reach a decision at APL

k_2 = smallest number of values to reach a decision at RPL

$\bar{k}_1$ = average sample number at APL

$\bar{k}_2$ = average sample number at RPL

Values in brackets indicate that no decision is allowed at the given value of k.

For a test of a lower specification limit, that is, when the RPL is less than the APL, proceed as follows:

1. Reverse signs of tabulated boundary values of decision lines Y_1 and Y_2.

2. Reverse acceptance and rejection regions.

Charts for lower and upper specification limits may be combined to test a double specification limit in a manner analogous to that used when the standard deviation was known. Such a chart is illustrated in Fig. 8-10.

As an example of application of the Barnard procedure, consider the previous coating data. Suppose the standard deviation were not known and a chart testing against the upper limit was to be prepared. Assume $\alpha = \beta = .05$ is to be used where

$$\mu_1 = 1.0 \ mm \qquad \mu_2 = 1.2 \ mm$$

The standard deviation is roughly approximated as $\sigma = .1$ so that

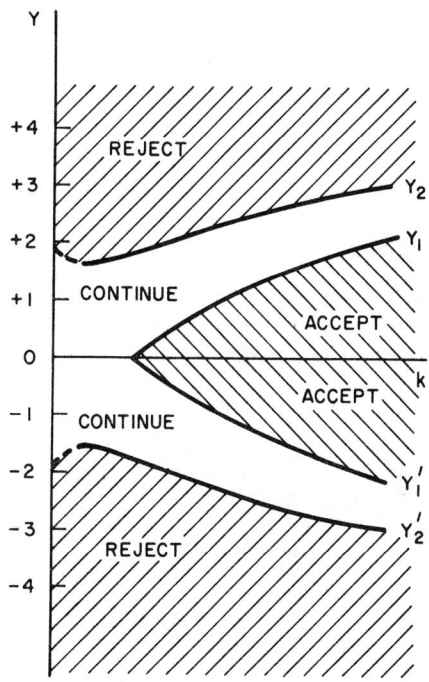

FIG. 8-10   Barnard sequential chart.

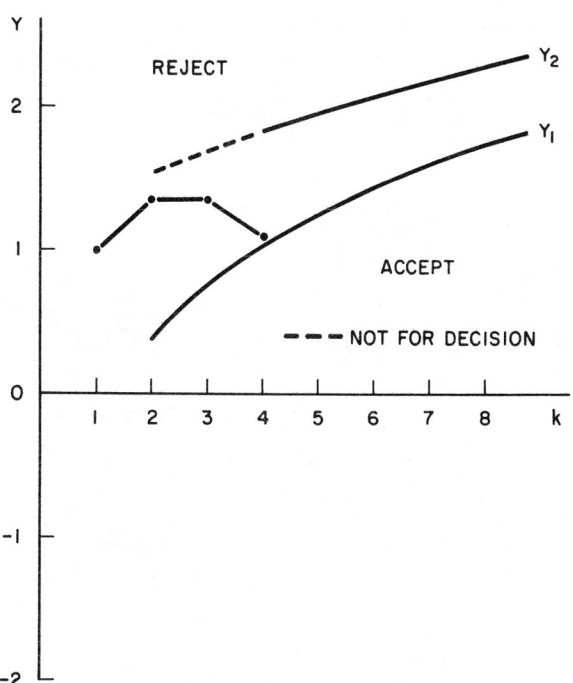

FIG. 8-11   Example of a Barnard sequential
t-test with $\alpha = .05$, $\beta = .05$.

$$D = \frac{1.2 - 1.0}{.1} = 2$$

The necessary calculations are as follows:

(1)	(2)	(3)	(4)	(5)	(6)
k	X	$X - \mu_1$	$\Sigma(X - \mu_1)$	$(X - \mu_1)^2$	$\Sigma(X - \mu_1)^2$
1	1.10	.10	.10	.01	.01
2	1.15	.15	.25	.0225	.0325
3	1.00	0	.25	0	.0325
4	0.95	−.05	.20	.0025	.0350

(7) $\sqrt{\Sigma(X - \mu_1)^2}$	$Y = (4)/(7)$	Decision limits Y_1	Y_2
.10000	1		
.18028	1.387	0.37	(1.56)
.18028	1.387		
.18708	1.069	1.03	1.82

The plot appears as in Fig. 8-11. This test would lead to continued sampling on the fourth sample without a decision.

CUMULATIVE SUM CHARTS

It is sometimes desirable to plot the results of sampling inspection in the form of cumulative sum charts. Originated by Page (1954), their construction has been described by Barnard (1959) and amplified by Johnson and Leone (1962). The charts consist of a sequential plot to which a V-mask is applied point by point to assess the significance of the plot against a specified value of the APL. A typical such mask is shown in Fig. 8-12. Rejection occurs if any of the previous points plotted lie outside the angle defined by the "notch" of the V-mask when the last point plotted is positioned a horizontal distance d from the vertex of the angle of the notch. The mask is determined by two dimensions: the distance d and the angle of the notch 2θ.

N. L. Johnson (1961) has shown that the chart may be regarded "... as (roughly) equivalent to the application of the sequential probability

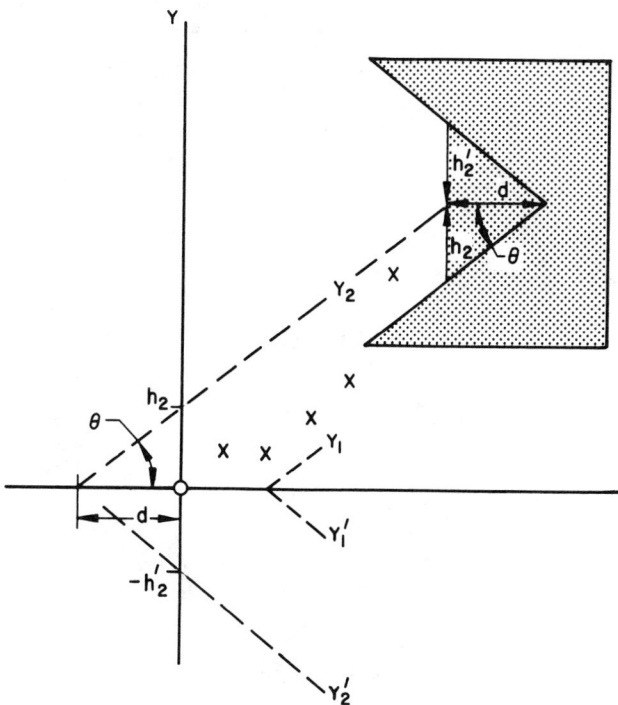

FIG. 8-12 Typical cusum chart.

ratio test in reverse." Thus, the notch of the V-mask corresponds to the
trapezoidal shape formed by the ordinate and the rejection lines of a sequential graph. This is illustrated in Fig. 8-12 which shows the two-sided
sequential graph corresponding to the V-mask.

Using this notion of equivalence, the following relationships may be
employed to convert the sequential parameters given in Table 8-3 to the
dimensions of the V-mask:

$$\tan \theta = s \qquad d = \frac{h_2}{s}$$

Also, the "average run length" of points to a rejection is of interest in
application of the cusum chart. This is analogous to the average sample
number of the sequential procedure. An approximation to the ARL, using
sequential parameters, can also be developed from the results of Johnson
and Leone (1962). We find that at the APL

$$ARL = \frac{h_2}{APL - s}$$

Since a cumulative sum chart cannot "accept" as such, the action rule is
"not to reject" during continuation. Rejection occurs only when the V-mask
is violated. The consumer's risk is taken to be zero for this approximation.
Also, the approximation should not be used when the indicated ARL is 5 or
less.

For example, in testing the mean

$$ARL = \frac{h_2}{\mu_2 - s}$$

so that with some algebra and the formulas of Table 8-3,

$$ARL = \frac{2 \log[(1 - \beta)/\alpha] \sigma^2}{\left(\mu_2 - \mu_1\right)^2}$$

and if β is taken to be zero,

$$ARL = \frac{-2 \log[\alpha]\sigma^2}{\left(\mu_2 - \mu_1\right)^2}$$

So that when $\delta = (\mu_2 - \mu_1)/\sigma = .5$ and $\alpha = .05$, then ARL = 23.97, while the
value calculated by Johnson and Leone (1962) is 24.0.

These results assume the unit length of the vertical and horizontal
scales are plotted 1:1. For a scale using k units of length for the ordinate for one unit length of the abscissa, d remains unchanged; however, the
angle of the mask becomes tan θ = s/k. Naturally, the sequential formulas
for a linear transformation of the points plotted $\Sigma(ax + b)$ may be used to
determine the sequential parameters when a cusum chart is plotted using a
transformed sum for scaling or other purposes.

Another variation on the cumulative sum involves a sequential plot
with horizontal limits. This may be obtained simply by subtracting the
slope s from each point plotted before it is added to the cumulative sum.
The result is a horizontal sequential chart with limits h_2 and $-h_1$ and a
nominal centerline of zero. An interesting procedure for the use of such
a chart in a one-sided test has been given by Kemp (1962). A modification
of his approach using a horizontal one-sided sequential chart derived from
the formulas given in Table 8-3 is as follows:

1. Set an upper limit at h_2 and a lower boundary at 0.
2. Do not cumulate on the chart until an observation has been found
   to exceed s.

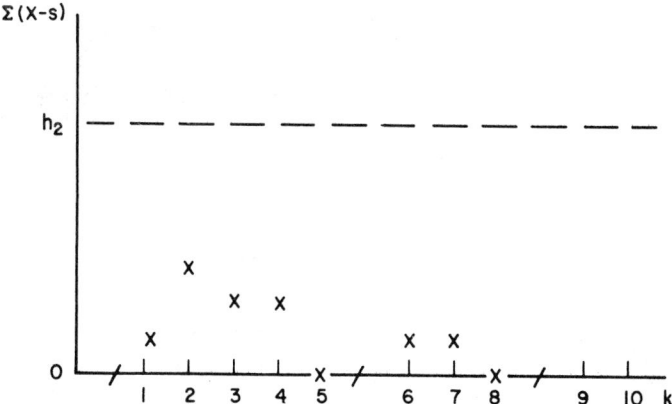

FIG. 8-13 Modified Kemp procedure.

3. When an observation exceeds s, calculate and plot the cumulative
    sum $\Sigma(X - s)$ on the chart until
    a. The cumulative sum $\Sigma(X - s)$ exceeds h_2. In this case, reject
        the process level as significantly greater than the acceptable
        process level.
    b. The cumulative sum $\Sigma(X - s)$ returns to zero. In this case,
        discontinue cumulation and return to step 2.

Such a chart is shown in Fig. 8-13. The action rules for this chart are
the same as those for the cusum chart. Modifications of the procedure
include the corresponding sequential test against a lower limit on the
process level and use of the full chart with both the acceptance and rejection regions defined. Kemp (1962) has also suggested that, for a plot
of $\Sigma(X - s)$ without limits, a significant upward change in process level
is simply indicated when the distance between the lowest point on the plot
and the last point plotted is greater than h_2.

Cumulative sum charts and their variations offer many possibilities
for the quality control engineer in acceptance quality control as well as
process quality control. The reader is referred to Burr (1976), Johnson
and Leone (1977), Duncan (1974), Wetherill (1977), and the literature
cited for more details on this interesting method. An excellent treatment
of the philosophy of application of the cusum chart will be found in Craig
(1969).

A detailed exposition of sequential methods for variables and attributes will be found in Wald (1947) and Wetherill (1975).

PROBLEMS

1. In purchasing high-pressure cylinders, determine the sample size
    needed to assure that the process level does not differ by more than
    1 psi from nominal when the standard deviation is .5 psi.

2. What test would be employed to determine if the variability in lengths
    of leads is at the specified level of 6 cm? Suppose the standard
    deviation of a sample of 15 is 7 cm, should the lot be rejected?

3. An acceptance control chart is to be used in lot acceptance of a
    series of shipments of glass tubes. The tubes are to average not less
    than 90 cm in length. The feed mechanism in the customer's process
    would jam if the tubes are less than 87 cm. The standard deviation
    is known to be 1.5 cm. Construct the appropriate chart. Successive
    lot means are 90, 89, 88, 90, 87, 91, 92, 89. Which lots should be
    rejected?

4. If tubes averaging more than 92 cm in length were also to be rejected
    in Prob. 3, what sample size would be required for an acceptance con-
    trol chart with double limits having α = .005?

5. What is the meaning of seven successive points on one side of the
    nominal center line of an acceptance control chart?

6. The specification on the maximum average weight of a certain construc-
    tion material is 400 lb; however, if the average weight exceeds 408 lb
    the design must be changed. The standard deviation is 8 lb. Set up a
    sequential chart to check the weight of incoming lots of the material.
    Take α = .025, β = .10. Should the lot be accepted if the results
    from a lot are 397, 400, 385, 388, 404, 410, 411, 395, 394, 400?

7. Suppose it is decided that it is important for the material in Prob.
    6 also not to be less than 392 lb. on the average. Construct a two-
    sided sequential chart with α = .05, β = .10. Plot the sequential
    results.

8. It is suspected that the standard deviation in Prob. 6 no longer
    equals 8. If it has increased to 10, the lot should be rejected.
    Using α = .05, β = .10 construct a sequential chart for the variance.
    If successive values of $(X_i - \mu)^2$ are calculated from the data of
    Prob. 6, what is the decision after the last lot shown?

9. Using α = .05, β = .05, what conclusion can be drawn from the data of
    Prob. 6 when the standard deviation is unknown? Draw the sequential
    plot.

10. Convert the parameters of Prob. 7 to those of a cusum V-mask.

REFERENCES

Barnard, G. A. (1946), Sequential Tests in Industrial Statistics, *Journal of the Royal Statistical Society* (Series B), 8: 1-21.

Barnard, G. A. (1959), Control Charts and Stochastic Processes, *Journal of the Royal Statistical Society* (Series B), 21(2): 239-271.

Bowker, A. H., and G. J. Lieberman (1959), *Engineering Statistics*, Prentice-Hall, Englewood Cliffs, New Jersey.

Burr, I. W. (1976), *Statistical Quality Control Methods*, Marcel Dekker, New York.

Craig, C. C. (1969), The $\bar{X}$-and R-Chart and Its Competitors, *Journal of Quality Technology*, 1(2): 102-104.

Duncan, A. J. (1974), *Quality Control and Industrial Statistics*, 4th ed., Richard D. Irwin, Homewood, Illinois.

Freund, R. A. (1957), Acceptance Control Charts, *Industrial Quality Control*, 14(4): 13-23.

Johnson, N. L. (1961), A Simple Theoretical Approach to Cumulative Sum Control Charts, *Journal of the American Statistical Association*, 56: 835-840.

Johnson, N. L., and F. C. Leone (1962), Cumulative Sum Control Charts - Mathematical Principles Applied to Their Construction and Use, *Industrial Quality Control*, Part 1, 18(12): 15-20; Part 2, 19(1): 29-36; Part 3, 19(2): 22-28.

Johnson, N. L., and F. C. Leone (1977), *Statistics and Experimental Design in Engineering and the Physical Sciences*, Vol. 1, 2nd ed., John Wiley and Sons, New York.

Kemp, K. W. (1962), The Use of Cumulative Sums for Sampling Inspection Schemes, *Applied Statistics*, 11: 16-31.

Page, E. S. (1954), Continuous Inspection Schemes, *Biometrika*, 41: 100-115.

Sommers, D. J. (1979), Personal communication with the author.

Wald, A. (1947), *Sequential Analysis*, John Wiley and Sons, New York.

Wetherill, G. B. (1975), *Sequential Methods in Statistics*, Chapman and Hall, London.

Wetherill, G. B. (1977), *Sampling Inspection and Quality Control*, 2nd ed., Chapman and Hall, London.

9

BULK SAMPLING

Most of the literature of acceptance sampling relates to the inspection of discrete units of product. For each unit an associated quality characteristic is determined. This may be either an attribute determination of acceptability (go no-go) or a measurement of some kind taken on each unit in the sample. Another type of product may be distinguished, however. It consists of material in bulk form. The bulk sampling problem has been described by Bicking (1967, p. 96) as follows:

> Bulk materials are essentially continuous and do not consist of populations of discrete, constant, indentifiable, unique units or items that may be drawn into the sample. Rather, the ultimate sampling units must be created, at the time of sampling, by means of some sampling device. The size and form of the units depend upon the particular device employed, how it is used, the nature, condition, and structure of the material, and other factors.

Bulk sampling may address such issues as the inspection of 100 tons of coal, sampling a truck filled with gasoline, or the assessment of the natural gas contained in a particular storage tank. Sampling units might then be a shovel full of coal (whose size depends on the shovel), a sampling bottle full of gasoline (the amount depending on the capacity of the sampling bottle), or a sampling probe delivering gas to a container (of some size and at some pressure). The important point is that the sample is constructed, not gathered up.

The objectives of bulk sampling have been given by Bicking (1978, p. 304) as follows:

1. Characterization of the material in place (as in a natural deposit) as to location, amount, content, or value.

2. Characterization of a material as to grade, need for further processing, or destination.

3. Control during processing.

4. Acceptance on a lot-to-lot basis.

5. Determination of weight or content for purposes of taxation or payment.

6. Determination of properties that must be known so that the end use will be appropriate.

7. Experimentation and analysis to determine future sampling procedures or uses of the material.

Emphasis here will be placed on the fourth objective, that of lot acceptance. As such, bulk sampling may generally be regarded as a form of variables sampling for process parameter.

Various devices have been developed to take samples of bulk materials. These have been aptly described by Bicking (1968, 1978). Their proper use, however, depends upon knowledge of any stratification in the material to be sampled. Sampling approaches in the presence of stratification have also been discussed by Bicking (1967, p. 97). Consider, for example, a shipment of milk contained in a cylindrical tank car. Vertical samples may represent strata disproportionally because of differences in horizontal dimension from top to bottom. The cylinder is wider in the middle. Furthermore, without mixing, it may be very important to be sure the layer of cream is appropriately represented.

In all bulk sampling the population sampled must be appropriately defined. A. J. Duncan (1962) has discussed this in detail. It should be pointed out that it is most advantageous to sample bulk material when it is moving, as on a conveyor belt, in free fall, etc. Steps in developing a standard sampling method as given by Ishikawa (1958) are shown in Table 9-1, taken from Bicking (1967, p. 103).

CONSTRUCTION OF THE SAMPLE

The essential "continuity" of bulk materials allows parts of a sample to be blended or mixed together to form a composite. The composite is then tested once, rather than individual tests on its constituent parts. This

TABLE 9-1 Developing a Standard Sampling Method

---

1. Make clear the purpose of sampling.

    a. What is the population from which the sample will be taken?

    b. What information is required about the population; the mean, the variance, and the precision desired in the estimate?

    c. On what criterion will acceptance of the lot be based?

    d. What action is to be taken to dispose of a rejected lot?

2. Specify the population and investigate the history of a lot.

    a. Is the process that produced the lot in a state of control?

    b. Is the definition of the lot size in conformity with the desires of the producer and the consumer?

    c. Are the methods of handling and storage properly considered in determining the lot size?

3. Study the measurement error.

    a. Separate the measurement error from the sampling error.

    b. Compare the relative sizes of these two sources of error.

4. Estimate the several variances due to the process (within-lots and between-lots).

5. Prepare the sampling instruction, guarding against the following defects:

    a. Lack of clarity in purpose of sampling.

    b. Lack of specific enough instructions for taking increments. ("Take a representative sample" and "take a random sample" are not specific enough.)

    c. Unsuitable containers for the samples.

    d. Failure to provide methods for checking sampling error, reliability, or measurement precision and bias.

    e. Unsuitable methods for handling and reducing the sample in the laboratory.

6. Control the sampling operation.

    a. Train the samplers.

    b. Control the operation of the plan through check samples.

7. Periodically review the sampling instructions to provide for any changes in the process.

Source: Reproduced with permission from C. A. Bicking (1967), The Sampling of Bulk Materials, *Materials Research and Standards*, 7(2): 103. Copyright 1967, American Society for Testing and Materials.

is a *physical* way to average the composited samples. Suppose three samples
of coal were taken from a coal car on a siding as it was being unloaded.
If ash content was to be determined, three separate analyses could be performed. Alternatively, the three samples might be mixed and blended into
one composite sample. An analysis of the composite should yield the same
result as the average of the three distinct samples. Lost, however, would
be any measure of variation.

Often when the population is known to consist of several different
subdivisions which may give different results with respect to the quality
characteristic measured, all the subdivisions, or strata, are deliberately
included in the sample. This is called *stratified sampling*. When the
subdivisions are sampled we have cluster or multistage sampling. In any
such procedure, which deviates from simple random sampling, care must be
taken to properly weight the sample results so that the effect of a result
is proportional to its probability of occurring. This usually involves
proportional allocation. It will be assumed here that all samples are
proportionally allocated. A comparison of types of sampling is shown in
Fig. 9-1 taken from Bicking (1967, p. 99).

In bulk sampling, lots (or populations) of bulk material are regarded
as being composed of mutually exclusive subdivisions or segments. Sometimes obvious segments occur, as when the material comes in boxes or bags.
Sometimes, however, the segments must be artificially created by superim-
posing imaginary grids over the material or by other means of real or syn-
thetic division. Segments may be further subdivided into increments for
sampling within a segment. In sampling theory, segments are often called
primary units, while increments are called *secondary units*.

Segments are treated in a manner similar to the units in discrete
sampling. Their average is considered as an estimate of the average of
the lot and their variability as a measure of variation on which to construct the standard error of the estimate of the lot mean. With bulk
material, however, the possibility for additional sampling within a segment exists. Furthermore, the total variation observed may be broken into
components of variance which estimate the amount of variation that may be
attributed to various stages in the sampling process. Prior estimation of
these components allows for the determination of optimum sample size and
for limits of error in situations in which the sampling strategy precludes
replicate observations. For example, the segments may be sampled giving a
variance between segments. Increments may be taken within segments to

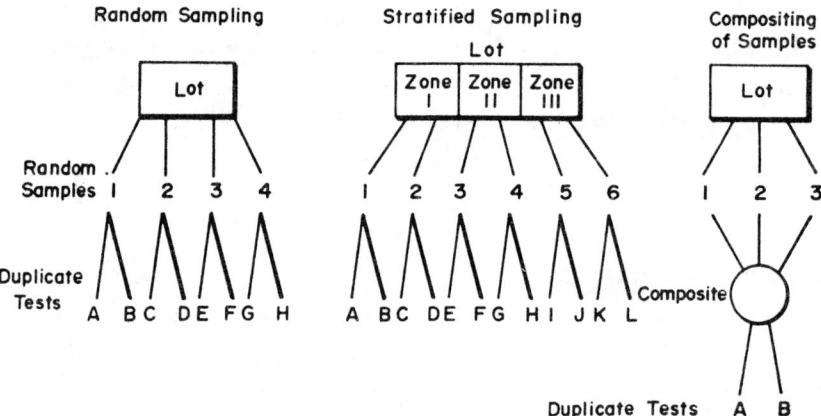

FIG. 9-1 Types of sampling. (From C. A. Bicking (1967), *Materials Research and Standards*, 7(2): 99. Reprinted with permission from the American Society for Testing and Materials, 1916 Race Street, Philadelphia, Pa., copyright 1967.)

give a variance between increments, or sampling variance. The material from each increment or from a composite of increments may then be reduced to the desired particle size by crushing or grinding and the amount of material cut down by quartering to obtain one or more test units of a size just sufficient for laboratory test of the quality characteristic. This gives rise to so-called reduction variance. The tests themselves contribute a variance due to testing. A model[*] for the total variance in the lot as broken into components is

$$\sigma_T^{\,2} = \sigma_1^{\,2} + \sigma_2^{\,2} + \sigma_3^{\,2} + \sigma_4^{\,2}$$

where

$\sigma_T^{\,2}$ = total variance in the lot

$\sigma_1^{\,2}$ = between segment variance

$\sigma_2^{\,2}$ = between increment variance within segments

$\sigma_3^{\,2}$ = testing variance

$\sigma_4^{\,2}$ = reduction variance

[*]The conventional bulk sampling model reverses the roles of $\sigma_3^{\,2}$ and $\sigma_4^{\,2}$; however, the model given provides consistency of enumeration when there is no compositing, i.e., when $\sigma_4^{\,2} = 0$.

The last term in the model will be regarded as also containing variation
from all sources not explicitly shown in the model.

Frequently samples may require no reduction or the reduction variance
σ_4^2 may be assumed so small as to be essentially zero and omitted from the
model. The variation due to reduction will then appear as part of the
testing variance. These components of variance are often assumed constant
across the lot from segment to segment. As with all assumptions in sampling,
however, it is well to check that it is true before setting up a new plan.
This is usually done by control chart.

Duncan (1962) has distinguished two distinct populations which may be
conceptualized and tested through these procedures: populations created by
the act of sampling from what is called *Type A* bulk material with nondistin-
guishable segments, and populations having preexisting elements from what
is called *Type B* bulk material with distinguishable segments. An example
of the former (Type A material) is a pile of coal. An example of the
latter (Type B material) is a lot consisting of 500 bags of fertilizer.
With Type A material, segments and increments must be artificially defined
within the totality of all the product submitted. With Type B material,
the natural segments would be divided into increments for sampling within
segments. Bicking (1967) points out that "Type B materials represent a
transition between piece part sampling and sampling Type A materials."

ESTIMATION

Bulk sampling is primarily used to estimate the lot mean with a given degree of precision. The resulting estimate may be sufficient in itself, or
it may be used to determine lot acceptance. The magnitude of the standard
error of the mean, and hence the precision of the estimate, can, of course,
be controlled by the number of samples taken. If in multistage sampling
of a lot of size N, the number of segments sampled is n_1, the number of
increments taken within a segment is n_2, while n_3 tests are made on each
increment, the variance of the lot sample mean computed from all observations will be composed as follows:

$$\sigma_{\overline{X}}^2 = \frac{\sigma_1^2}{n_1}\left(1 - \frac{n_1}{N}\right) + \frac{\sigma_2^2}{n_1 n_2} + \frac{\sigma_3^2}{n_1 n_2 n_3}$$

where

$\sigma_1{}^2$ = variance between segments

$\sigma_2{}^2$ = variance between increments within segments

$\sigma_3{}^2$ = variance between tests within increments

Now this equation is applicable when increments or even segments are composited. Compositing, however, may lead to an inability to estimate some,
or all, of the components of variance from the sample. It is a useful
device when estimates of these variabilities are not needed, as when the
components of variance are known.

Much is revealed by the partition of the variance of the sample mean.
For example, in sampling homogeneous liquids, $\sigma_2{}^2 = 0$ since the increments
are all equal. Also, if $n_1 = N$, as in stratified sampling, the first term
goes to zero since the population of segments is exhausted. Furthermore,
for given magnitudes of $\sigma_1{}^2$, $\sigma_2{}^2$, and $\sigma_3{}^2$, values of n_1, n_2, and n_3 may be
determined by trial and error to find a combination which will reduce $\sigma_{\bar{X}}^2$
to a desired magnitude. Means of samples of this size will give the desired
precision on $\bar{X}$ whether composited or averaged over individual tests as long
as initial estimates of the magnitudes of the components of variance hold.

For example, suppose bags of argol are to be sampled by a split tube
thief or trier, and it is known that

$$\sigma_1{}^2 = .21 \qquad \sigma_2{}^2 = .31$$

If testing cost is high so that one test is to be made on each increment
and if reduction variance is assumed negligible we have

$$\sigma_{\bar{X}}^2 = \frac{\sigma_1{}^2}{n_1} + \frac{\sigma_2{}^2}{n_1 n_2}$$

where $\sigma_2{}^2$ will now include testing error since the later will not be independently estimated. To determine a plan which will give 95 percent confidence of estimating the mean to within 0.4, using the appropriate z value
of 1.96 from the normal distribution, we must have

$$z\sigma_{\bar{X}} = 0.4$$
$$\sigma_{\bar{X}} = \frac{0.4}{1.96} = 0.2$$
$$\sigma_{\bar{X}}^2 = .04$$

TABLE 9-2 Possible Combinations of n_1 and n_2

n_1	n_2	$\sigma_{\overline{X}}^2$
9	1	.058
	2	.041
10	1	.052
	2	.036
11	1	.047
	2	.033

A few possible combinations of n_1 and n_2 to give the desired $\sigma_{\overline{X}}$ are shown in Table 9-2, where

$$\sigma_{\overline{X}}^2 = \frac{.21}{n_1} + \frac{.31}{n_1 n_2}$$

Of the values shown $n_1 = 10$ and $n_2 = 2$ come closest to the desired $\sigma_{\overline{X}}^2 = .04$ with the smallest number of segments. Note that any combination of n_1 and n_2 giving $\sigma_{\overline{X}}^2 \leq .04$ is acceptable.

It can be shown [Davies (1960, p. 111)] that for a two-stage plan (n_1 segments, n_2 increments) costing c_1 to sample a segment and c_2 to sample an increment from a segment, an economically optimum plan can be developed. For a lot with N segments, where the cost of testing is the same for segments or increments, the most economical sample sizes with which to estimate the lot mean to within $\pm E$ with $1 - \alpha$ confidence are found as

$$n_2 = \sqrt{\frac{c_1 \sigma_2^2}{c_2 \sigma_1^2}}$$

with

$$n_1 = \frac{N\left(\sigma_2^2 + n_2 \sigma_1^2\right)}{N n_2 \left(E/z_{\alpha/2}\right)^2 + n_2 \sigma_1^2}$$

where $z_{\alpha/2}$ is the standard normal deviate associated with the confidence level to be incorporated in the two-sided estimate. For essentially infinite lot sizes the formula for n_1 becomes

$$n_1 = \frac{\sigma_2{}^2 + n_2\sigma_1{}^2}{n_2\left(E/z_{\alpha/2}\right)^2}$$

Using the previous example on argol, if c_1 were known to be \$31 and c_2 were \$21, the sample sizes required would be

$$n_2 = \sqrt{\frac{.31}{.21}\left(\frac{21}{31}\right)} = 1$$

and

$$n_1 = \frac{.31 + 1(.21)}{1(.4/1.96)^2}$$

$$= 12.5 \sim 13$$

The plan to minimize cost in this case is $n_1 = 13$, $n_2 = 1$. The total cost of this plan would be

$$c = \$31(13) + \$21(13)(1) = \$676$$

and result in a standard error of estimate

$$\sigma_{\overline{X}} = \sqrt{\frac{.21}{13} + \frac{.31}{13}} = \sqrt{.04} = .2$$

The plan $n_1 = 10$, $n_2 = 2$ would cost

$$c = \$31(10) + \$21(10)(2) = \$730$$

to produce a standard error of estimate of

$$\sigma_{\overline{X}} = \sqrt{.036} = .19$$

Both plans would meet the desired precision in estimating the lot mean.

Sampling of bulk material can be used most effectively when the components of variance are known. Knowledge of these values can allow estimation of the standard error of the mean of the lot even when extensive compositing results in one test result on the lot. Estimation of these components is straightforward. Let

$\overline{X}$ = lot mean from n_1 segments

$\overline{X}_1$ = segment mean from n_2 increments

$\overline{X}_2$ = increment mean from n_3 tests

X_3 = test result

then the mean squares used in constructing the estimates are

$$MS_1 = \frac{\sum\left(\bar{x}_1 - \bar{x}\right)^2}{n_1 - 1}$$

$$MS_2 = \frac{\sum\left(\bar{x}_2 - \bar{x}_1\right)^2}{n_1(n_2 - 1)}$$

$$MS_3 = \frac{\sum\left(x_3 - \bar{x}_2\right)^2}{n_1 n_2 (n_3 - 1)}$$

with degrees of freedom ν_1, ν_2, and ν_3. Estimates of the components of variance can be determined, regarding N as infinite, as follows:

Estimate of the testing component σ_3^2 is $s_3^2 = MS_3$ with

$$\nu_3 = n_1 n_2 (n_3 - 1).$$

Estimate of the increment within segment component σ_2^2 is

$$s_2^2 = MS_2 - \left(s_3^2/n_3\right) \text{with}$$

$$\nu_2 = \frac{\left(s_2^2\right)^2}{\dfrac{1}{n_1(n_2 - 1)}\left(\dfrac{MS_2}{1}\right)^2 + \dfrac{1}{n_1 n_2 (n_3 - 1)}\left(\dfrac{MS_3}{n_3}\right)^2}$$

Estimate of the between segment component σ_1^2 is

$$s_1^2 = MS_1 - \frac{s_2^2}{n_2} - \frac{s_3^2}{n_2 n_3}$$

with

$$\nu_1 = \frac{\left(s_1^2\right)^2}{\dfrac{1}{n_1 - 1}\left(\dfrac{MS_1}{1}\right)^2 + \dfrac{1}{\nu_2}\left(\dfrac{s_2^2}{n_2}\right)^2 + \dfrac{1}{\nu_3}\left(\dfrac{s_3^2}{n_1 n_2}\right)^2}$$

The above estimates of degrees of freedom are obtained using the Satterthwaite (1946) approximation and will usually be found to be conservative. They should be rounded down to obtain integral values of degrees of freedom. Clearly, when the components of variance are known, they may be regarded as having infinite degrees of freedom.

The mean squares MS_1, MS_2, and MS_3 can be used to construct a nested

TABLE 9-3   Analysis of Variance Table for Nested Sampling

Source	Sum of squares	Degrees of freedom	Mean square	Components of variance estimated by mean square
Between segments	$n_2 n_3 (n_1 - 1) MS_1$	$n_1 - 1$	$n_2 n_3 MS_1$	$n_2 n_3 \sigma_1^2$ $+ n_3 \sigma_2^2 + \sigma_3^2$
Increments within segments	$n_1 n_3 (n_2 - 1) MS_2$	$n_1 (n_2 - 1)$	$n_3 MS_2$	$n_3 \sigma_2^2 + \sigma_3^2$
Tests within increments	$n_1 n_2 (n_3 - 1) MS_3$	$n_1 n_2 (n_3 - 1)$	MS_3	σ_3^2

analysis of variance table to display the variances involved as shown in Table 9-3. The multipliers shown with the mean squares and the components of variance are necessary because analysis of variance is usually performed on observation totals rather than means as shown here. For a discussion of analysis of variance performed using means, with ancillary techniques, see Schilling (1973).

The standard error of the sample mean can be estimated as

$$\sigma_{\overline{X}} = \sqrt{\frac{\sigma_1^2}{n_1}\left(1 - \frac{n_1}{N}\right) + \frac{\sigma_2^2}{n_1 n_2} + \frac{\sigma_3^2}{n_1 n_2 n_3}}$$

when the components of variance are known. When they are estimated the formula for estimation becomes

$$s_{\overline{X}} = \sqrt{\frac{s_1^2}{n_1}\left(1 - \frac{n_1}{N}\right) + \frac{s_2^2}{n_1 n_2} + \frac{s_3^2}{n_1 n_2 n_3}}$$

with approximate degrees of freedom roughly

$$\nu_{\overline{X}} = \frac{\left(s_{\overline{X}}^2\right)^2}{\frac{1}{\nu_1}\left(1 - \frac{n_1}{N}\right)^2\left(\frac{s_1^2}{n_1}\right)^2 + \frac{1}{\nu_2}\left(\frac{s_2^2}{n_1 n_2}\right)^2 + \frac{1}{\nu_3}\left(\frac{s_3^2}{n_1 n_2 n_3}\right)^2}$$

[again obtained from the Satterthwaite (1946) approximation]. This repeated use of the approximation leads to a crude but often useful estimate of the degrees of freedom.

As pointed out by Duncan (1974a), the standard error of the mean can also be obtained directly from the standard deviation of the segment results when they are available. The estimate applies even if some of the segments have been composited to give the results, or in the face of other compositing or reduction within segments. The price of compositing the segments is, of course, a reduction in degrees of freedom. This estimate is usually the only one available when dealing with unique lots. The estimate is

$$s_{\overline{X}} = \frac{1}{\sqrt{n_1}} \sqrt{\frac{\sum(\overline{X}_1 - \overline{X})^2}{n_1 - 1}} = \sqrt{\frac{MS_1}{n_1}}$$

with degrees of freedom $\nu_1 = n_1 - 1$.

It is sometimes necessary to estimate the variability within the lot. This is usually done by taking a sample constructed using n_1 segments, one increment from each segment, one test per increment. It is recommended that at least 10 segments be selected as a sample. No compositing is allowed. The estimate is

$$s_X = \sqrt{\frac{\sum(\overline{X}_1 - \overline{X})^2}{n_1 - 1}}$$

with degrees of freedom $\nu = n_1 - 1$. The value of s_X is then used to characterize the lot with respect to variation in the lot and is also useful in determining sample size. Note that the components of variance associated with this measure are

$$\sigma_X^2 = \sigma_1^2 + \sigma_2^2 + \sigma_3^2$$

As an example of the application of these estimation techniques, consider the data of Table 9-4 from a two-stage bulk sampling plan presented by Tanner and Lerner (1951), which shows a sample taken from a shipment of argol.

Here N is large relative to n_1 so that the finite population correction is not necessary. Also there is only one test per sample, so that σ_2^2 includes the variability of testing and the model for the variance of the lot sample mean is

$$\sigma_{\overline{X}}^2 = \frac{\sigma_1^2}{10} + \frac{\sigma_2^2}{10(2)}$$

TABLE 9-4   Percent Potassium Bitartrate in Shipment of Argol

| | Trierful | | | Standard |
Bag	1	2	Mean	deviation
1	86.37	86.46	86.42	.0636
2	87.50	86.36	86.93	.8061
3	85.75	86.05	85.90	.2121
4	87.09	87.38	87.24	.2051
5	87.31	86.78	87.04	.3748
6	85.85	85.75	85.80	.0707
7	86.46	85.44	85.95	.7212
8	84.62	86.16	85.39	1.0889
9	86.41	86.26	86.34	.1061
10	85.44	86.46	85.95	.7212
Mean	86.28	86.31	86.296	
Standard deviation	.8938	.5350	.6080	

Source: Reproduced with permission from L. Tanner and M. Lerner, Economic Accumulation of Variance Data in Connection with Bulk Sampling, *ASTM STP 114*, p. 9. Copyright 1951, American Society for Testing and Materials.

where σ_3^2 and σ_4^2 are not shown in the model since they cannot be estimated from this sample design.

Now,

$$MS_1 = \frac{(86.42 - 86.296)^2 + (86.93 - 86.296)^2 + \cdots + (85.95 - 86.296)^2}{10 - 1}$$

$$= .36967$$

and

$$MS_2 = \frac{(86.37 - 86.42)^2 + (86.46 - 86.42)^2 + \cdots + (86.46 - 85.95)^2}{10(2 - 1)}$$

$$= .3124$$

so that

$$s_2 = \sqrt{.3124} = .5589$$

with

$$\nu_2 = 10(2 - 1) = 10$$

and

$$s_1 = \sqrt{.36967 - \frac{.3124}{2}}$$

$$= \sqrt{.21347} = .4620$$

with

$$\nu_1 = \frac{(.21347)^2}{\frac{1}{9}\left(\frac{.36967}{1}\right)^2 + \frac{1}{10}\left(\frac{.3124}{2}\right)^2}$$

$$= 2.58 \sim 2$$

The standard error of the mean may then be estimated as

$$s_{\overline{X}} = \sqrt{\frac{s_1^2}{10} + \frac{s_2^2}{10(2)}} = \sqrt{\frac{.2135}{10} + \frac{.3124}{20}}$$

$$= \sqrt{.03697} = .1923$$

with degrees of freedom

$$\nu_{\overline{X}} = \frac{(.03697)^2}{\frac{1}{2}\left(\frac{.21347}{10}\right)^2 + \frac{1}{10}\left(\frac{.3124}{20}\right)^2} = 5.4$$

This is a conservative approximation since

$$s_{\overline{X}} = \sqrt{\frac{s_1^2}{10} + \frac{s_2^2}{20}}$$

$$= \sqrt{\frac{1}{10}\left(\frac{MS_1 - MS_2}{2}\right) + \frac{MS_2}{20}}$$

$$= \sqrt{\frac{MS_1}{10}}$$

and MS_1 has exactly 9 degrees of freedom as can be seen from the formula for its calculation. This estimate can also be obtained directly from the bag (segment) mean as

$$s_{\overline{X}} = \frac{1}{\sqrt{10}} \sqrt{\frac{(86.42 - 86.926)^2 + (86.93 - 86.926)^2 + \cdots}{10 - 1} + \frac{(85.95 - 86.926)^2}{10 - 1}}$$

$$= \frac{1}{\sqrt{10}} \sqrt{MS_1} = .1923$$

with, of course, 9 degrees of freedom. An estimate of this sort would be the only available method for determining the standard error of the mean from a unique lot and is obviously useful regardless of compositing within the segments.

A 95 percent confidence interval for the mean would be

$$\overline{X} \pm ts_{\overline{X}}$$

$$86.3 \pm 2.26(.1923)$$

$$86.296 \pm .43$$

An estimate of the variability in the lot can be obtained using the results of, say, the first trierful. This gives

$$s_X = \sqrt{\frac{(86.37 - 86.28)^2 + (87.50 - 86.28)^2 + \cdots + (85.44 - 86.48)^2}{10 - 1}}$$

$$= .8938$$

with

$$\nu = 9$$

Any estimate of this sort contains bag variation, trier variation within bags, any trier reduction variation and the testing error. In an experiment such as this a number of alternatives would be available for compositing. Some of them are

1. No compositing. In the case of a unique lot or for a pilot study to determine the components of variance to be used in a continuing series of lots, this option provides the most information about the variability involved. This estimate of the mean has standard error

$$\sigma_{\overline{X}} = \sqrt{\frac{\sigma_1^2}{10} + \frac{\sigma_2^2}{20}}$$

   The standard error can be estimated from the sample using the method given in the example above.

2. Composite the trier samples. Here 10 analyses would be required each having a variance

$$\sigma^2 = \sigma_1^2 + \frac{\sigma_2^2}{2}$$

   but the resulting standard error of the mean would be

$$\sigma_{\overline{X}}^2 = \sqrt{\frac{1}{10}\left(\sigma_1^2 + \frac{\sigma_2^2}{2}\right)} = \sqrt{\frac{\sigma_1^2}{10} + \frac{\sigma_2^2}{20}}$$

as before. This standard error could be checked using

$$s_{\overline{X}} = \sqrt{\frac{1}{10} \frac{\sum\left(\overline{x}_i - \overline{x}\right)^2}{10 - 1}}$$

This estimate is useful with unique lots. In that case, this
estimate of the standard error of the mean would provide 9 degrees
of freedom.

3. Composite the odd and even segments (bags) respectively into two
    samples. This would result in two values which would be averaged
    to produce the estimated lot mean. Each of these values would have
    a variance

$$\sigma^2 = \frac{\sigma_1^{\,2}}{5} + \frac{\sigma_2^{\,2}}{10}$$

but the resulting average of the two results would still have

$$\sigma_{\overline{X}} = \sqrt{\frac{1}{2}\left(\frac{\sigma_1^{\,2}}{5} + \frac{\sigma_2^{\,2}}{10}\right)} = \sqrt{\frac{\sigma_1^{\,2}}{10} + \frac{\sigma_2^{\,2}}{20}}$$

However, now the standard error could be checked from the segment
means with 1 degree of freedom as

$$s_{\overline{X}} = \sqrt{\frac{1}{2} \frac{\sum\left(\overline{x}_i - \overline{x}\right)^2}{2 - 1}} = \frac{R}{2}$$

where R is the range of the two readings. This is quite useful
on a continuing series of lots since it provides a check that the
variability has not changed from that predicted from the components
of variance.

4. Composite entire sample. With one analysis this would show just
    one value — the estimated mean of the lot. No estimate of stand-
    ard error would be available from the sample but known components
    of variance could be used to estimate the standard error if avail-
    able and if there were confidence that they had not changed since
    they were obtained. The standard error of the mean would be de-
    termined as

$$\sigma_{\overline{X}} = \sqrt{\frac{\sigma_1^{\,2}}{10} + \frac{\sigma_2^{\,2}}{20}}$$

This procedure can sometimes be used on a continuing series of lots.

Thus, various strategies are available for compositing depending upon the structure of the sample, cost and feasibility constraints, the desired precision of the estimate, the available information, and the ingenuity of the individual designing the procedures. For further discussion, see Davies (1954).

SAMPLING PLANS

The sampling plans that have been suggested for use with bulk materials are essentially variables plans for process parameter. Indeed, if segments are of equal size, the results for the segments can be used with the plans given in Table 8-1 as if the segments were individual units of product. Bulk sampling is, however, somewhat more complicated and is distinguished by exploiting the essential continuity of the basic material in the lot in the development of more complex and informative sampling plans. The method of sampling and compositing must be considered in assessing the overall results. For this reason, tables of bulk sampling plans are not available since the plans must be tailored to the individual sampling situation and the analytical methods used.

Bicking (1970) has enumerated the following steps in setting up a sampling plan:

1. State the problem for which an estimate is desired.
2. Collect information on the relevant properties of the material (average properties, components of variance in the properties).
3. Consider various approaches, taking into account cost, precision, and difficulties.
4. Evaluate these plans in terms of cost of sampling and testing, delay, supervisory time, and convenience.
5. Select a plan.
6. Reconsider the preceding steps.

Consider a test of the mean of a lot against some specified value. The statistic $t = (\overline{X} - U)/s_{\overline{X}}$ would be used for an upper limit where U plays the role of μ_0 in Table 8-1. Similarly L acts as μ_0 when a lower limit is involved. Once the mean of the lot is estimated by $\overline{X}$ and its standard error $s_{\overline{X}}$ determined, the resulting value of t is compared to the relevant upper tail critical value from the t-distribution to determine the disposition of the lot (lower tail for a lower specification limit). Note that the problem of setting special "test" specification limits which take into account measurement error has been addressed by Grubbs and Coon (1954).

In the earlier example, suppose a lower specification limit on the average percent potassium bitartrate in the lot was L = 87 percent with a producer's risk of α = .05. Using the sample results,

$$t = \frac{\overline{X} - L}{s_{\overline{X}}}$$

$$t = \frac{86.3 - 87}{.19} = -3.68$$

Comparison to the critical value of t = -1.83 with 9 degrees of freedom shows -3.68 < -1.83 and the lot should be rejected. This test was made on the lower tail of the t-distribution since a lower specification limit was involved. A diagram showing the application of the test is given in Fig. 9-2 where $\overline{X}$ = L at t = 0. In practice, the consumer's risk involved in such an assessment would be incorporated in determining the sample size. Note that the upper tail of the t-distribution could have been used if the statistic were calculated as t = (L - $\overline{X}$)/$s_{\overline{X}}$.

In discrete sampling, measurements are taken directly on well-defined units of product; however, in bulk sampling, the continuous nature of the bulk within a segment allows for considerable flexibility for sampling within a segment in an attempt to characterize it with respect to the quality characteristic. A wide range of sampling techniques have been, and may be, employed, for example; stratified sampling [see Bennett and Franklin (1954), p. 482], multistage sampling [see Deming (1950), p. 160], ratio estimation [see Deming (1950), p. 183], systematic sampling [see Bicking (1967), p. 111], interpenetrating subsamples from a stream of product [see Duncan (1974), p. 25A-9], among others. It should be pointed out that, in the literature of sampling, segments are referred to as primary units, increments as secondary units, and tests often as tertiary units. Procedures are also available for the assessment of bulk quality

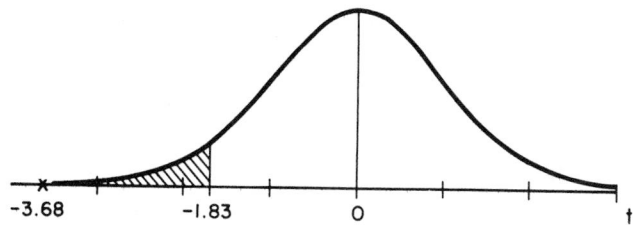

-3.68   -1.83   0   t

FIG. 9-2 The t-Test for purity of liquid.

characteristics in terms of proportions as well as measurements. It is important to caution that chemical *measurements* are frequently expressed in units of proportion or percent but should be analyzed as measurement data; percent carbon monoxide in the exhaust of a car, for example. The statistical analysis of proportions refers to an actual count of a discrete characteristic within a sample of a given size, for example, number of black grains in a sample of 100 grains of sand.

The various sampling techniques available in the literature of sampling theory [see, for example, Cochran (1953), Deming (1950), or Williams (1978)] lead to an estimate of the lot or population mean (or proportion) with its standard error, together with its associated degrees of freedom. These are readily available together with formulas for confidence interval estimation. For use in acceptance sampling, the estimated parameter and its standard error of estimate may be substituted in the criteria of Table 8-1 and used to test conformance to specifications. Furthermore, the procedures of sequential sampling and acceptance control charts given in Chap. 8 can also be used in straightforward fashion once the estimate of the lot mean and its standard error and degrees of freedom are determined. Sequential plans can be used on the segment means to arrive at an early decision on the lot.

An adaptation and modification of the basic procedure of ASTM Standard E-300-73 will be given to illustrate the nature and application of specific bulk sampling plans. While intended for sampling of industrial chemicals, the procedure is easily generalized to other bulk sampling situations. For a complete discussion of the method of ASTM E-300-73 refer to the standard, also see Bicking (1970). Note that this procedure involves sampling for process parameter with α risk of rejection when the process is at the specification limit, which acts as μ_1.

SIMPLE RANDOM SAMPLING OF UNIQUE LOT (Components of Variance Unknown)

Unique lots present a problem in bulk sampling because some or all of the components of variance associated with the inspection of the lot will, in general, be unknown. Further, the variability of the lot will also be unknown, requiring an initial preliminary estimate before sample size can be determined.

Assume a unique lot is to be sampled for low acceptance against a lower specification limit L. The producer's risk is to be $\alpha = .05$, while the consumer's risk is to be $\beta = .10$. Values of θ_1 = acceptable quality level and θ_2 = rejectable quality level are given. The procedure is as follows:

1. Take a preliminary sample of n_1^* segments ($n_1^* \geq 10$) at random from the lot. Use one increment per segment with one test per increment. In other words, use one test unit per segment sampled.

2. Compute

$$\overline{X}^* = \frac{\sum\limits_{i=1}^{n_1^*} X_i}{n_1^*}$$

and

$$s^* = \sqrt{\frac{\sum\limits_{i=1}^{n_1^*} \left(X_i - \overline{X}^*\right)^2}{n_1^* - 1}}$$

3. Calculate

$$d = \frac{L - \theta_2}{s^*}$$

and determine sample size n_1 required from the operating characteristic curve for the t-test.

4. Randomly select an additional $n_1 - n_1^*$ units from the lot and pool them with those of the previous sample. Compute

$$\overline{X} = \frac{\sum\limits_{i=1}^{n_1} X_i}{n_1}$$

and

$$s = \sqrt{\frac{\sum\limits_{i=1}^{n_1} \left(X_i - \overline{X}\right)^2}{n_1 - 1}}$$

so that

$$s_{\overline{X}} = \frac{s}{\sqrt{n_1}}$$

5. Check the adequacy of the sample size selected by recomputing

$$d = \frac{L - \theta_2}{s}$$

and rereading the OC curve to obtain a new estimate of sample size. If this estimate exceeds n_1 by more than 20 percent, obtain additional units as necessary to reach the indicated sample size.

Use the increased sample size as n_1 and return to step 4. Otherwise, proceed to step 6.

6. Using the final estimates of $\overline{X}$ and $s_{\overline{X}}$ calculate

$$t = \frac{L - \overline{X}}{s_{\overline{X}}}$$

and compare this statistic to the upper .05 critical value of the t-distribution with $n_1 - 1$ degrees of freedom. If the calculated value exceeds the critical value, reject the lot. Otherwise, accept.

7. In dealing with an upper specification limit proceed as above using the formulas

$$d = \frac{\theta_2 - U}{s}$$

$$t = \frac{\overline{X} - U}{s_{\overline{X}}}$$

8. For double specification limits, check to be sure that

$$U - L > 6s_{\overline{X}}$$

If not, take additional samples to raise the sample size to

$$n_1 \geq \frac{6s}{U - L}^2$$

Then test the upper and lower limits separately as above, rejecting if either test rejects the lot.

As an example of the application of this procedure let us return to the evaluation of the percent potassium bitartrate given above. Suppose in that case $\theta_1 = L = 87\%$ and $\theta_2 = 86\%$. Using the sample results from Trierful 1 for an initial sample of 10 segments,

$$\overline{X}^* = 86.28 \qquad s^* = .8938$$

with 9 degrees of freedom. Then

$$d = \frac{87 - 86}{.8938} = 1.12$$

The OC curve shows a sample size of 10 is required, so no further samples are required, and

$$s_{\overline{X}} = \frac{.8938}{\sqrt{10}} = .2826$$

so

$$t = \frac{87 - 86.28}{.2826} = 2.55$$

tested against a critical value of 1.83 with 9 degrees of freedom. Since
2.55 > 1.83 the lot would be rejected.

SAMPLING FROM A STREAM OF LOTS

As in discrete sampling, inspection frequently takes place on a steady
stream of product produced by the same supplier. Assuming the process to
be in control, a pilot study on the initial product can be used to estimate
the components of variance. From these estimates, appropriate bulk sampling
plans can be developed. Of course these estimates must be checked during
application of the procedure to be sure they continue to hold.

A procedure for a pilot study to estimate the relevant components of
variance and to assess their stability is suggested in ASTM E-300-73. A
modification and adaptation of the procedure is given here to outline the
detailed analysis necessary before a bulk sampling procedure can be set up
on a stream of lots. The procedure estimates

σ_1^2 = variance between segments (batch variability)

σ_2^2 = variance between increments within segments (sampling
     variability within batches)

σ_3^2 = testing variance (variability between tests)

σ_4^2 = reduction variance (variability introduced by reduction of
     gross sample to test unit size)

Estimation of Testing and Reduction Variances

1. Take 20 increments from each of 5 segments. Make 20 composites
    from the sets of 5 of the first, second, third, ..., 20th incre-
    ments across segments. Make two tests on each composite.

2. Prepare two control charts using standard procedures available in
    any textbook on statistical quality control, such as Burr (1976),
    Duncan (1974), or Grant and Leavenworth (1972) as follows:

    a. Chart I: Range chart on the differences of the two tests on
        each increment to test the stability of the testing variance.

    b. Chart II: Moving range chart on the means of the 20 composites
to test stability of the reduction variance.

3. If both of these charts exhibit a state of control, estimate the
testing and reduction variances. Let $\overline{X}_4$ = composite mean, X_3 =
test result, and $\overline{X}$ = mean of all measurements; then

$$MS_3 = \frac{\sum \sum \left(X_3 - \overline{X}_4\right)^2}{20(2-1)}$$

$$MS_4 = \frac{\sum \left(\overline{X}_4 - \overline{X}\right)^2}{20-1}$$

so that

$$s_3^{\;2} = MS_3$$

$$s_4^{\;2} = MS_4 - \frac{MS_3}{2}$$

These estimates are used in the estimation of the variances between segments
and between increments.

Estimation of Segment and Increment Variances

1. Take 2 increments from each of 25 successive segments produced by
the process.

2. Make a single test on each of the 50 increments under uniform conditions (same time, equipment, operator, etc.).

3. Prepare three control charts as follows using standard procedures:

    a. Chart III: Range chart on difference of results from two
increments from each segment to check stability of within
segment variance.

    b. Chart IV: $\overline{X}$ chart for segment means to check stability of
process from segment to segment with respect to trend, runs,
etc.

    c. Chart V: Moving range chart for segment means to check sta-
bility of variance of segments.

4. If the charts all show evidence of control, without exception, the
components of variance may be determined. As before, let $\overline{X}$ =
grand mean, $\overline{X}_1$ = segment mean, and $\overline{X}_2$ = increment mean; then

$$MS_1 = \frac{\sum \left(\overline{X}_1 - \overline{X}\right)^2}{25-1}$$

$$MS_2 = \frac{\sum\left(\overline{x}_2 - \overline{x}_1\right)^2}{25(2-1)}$$

so that

$$s_2^{\,2} = MS_2 - s_3^{\,2}$$

$$s_1^{\,2} = MS_1 - \frac{MS_2}{2}$$

Note that $s_4^{\,2}$ is not subtracted from $s_2^{\,2}$ since there is no reduction in the sense of compositing in this part of the procedure.

The stability and magnitude of the components of variance having now been determined, it is possible to apply the acceptance procedure to the stream of lots. The procedure suggested by ASTM E-300-73 is based on the results of two composite samples obtained from each lot. The lot is taken to be composed of $N = n_1$ segments all of which are of equal size and sampled to produce a stratified sample of the lot. A sample of n_2 increments is taken from each segment where n_2 is chosen to be an even number. If Type A bulk material is to be sampled, $n_1 n_2$ random increments are taken directly from the lot. The odd and even increments from within segments are separately composited to form two composites A and B. Two tests are made on each composite. The components of variance model for the variance of the mean from this procedure is:

$$\sigma_{\overline{X}}^{\,2} = \frac{\sigma_1^{\,2}}{n_1} + \frac{\sigma_2^{\,2}}{n_1 n_2} + \frac{\sigma_3^{\,2}}{4} + \frac{\sigma_4^{\,2}}{2}$$

In testing against a lower specification limit, L, on the lot mean, the following procedure is employed given values of the acceptable process level $\theta_1 = L$, the rejectable process level θ_2, producer's risk $\alpha = .05$, and consumer's risk $\beta = .10$.

Application of Plan to Stream of Lots

1. Given $n_1 = N$, $n_3 = 2$, $n_4 = 2$, determine n_2 as

$$n_2 = \frac{s_2^{\,2}}{n_1\left(\dfrac{(L - \theta_2)^2}{8.567} - \dfrac{s_1^{\,2}}{n_1} - \dfrac{s_3^{\,2}}{4} - \dfrac{s_4^{\,2}}{2}\right)}$$

where $8.567 = (z_\alpha + z_\beta)^2$. Round up to an even integer. For a test of an upper specification limit, substitute $\theta_2 - U$ for $L - \theta_2$ in the above formula.

2. Perform a check on the validity of the components of variance using two control charts:

   a. Chart VI: Range chart of differences between the two tests made on each composite. Use

   UCL: $3.686\, s_3$
   CL: $1.128\, s_3$
   LCL: 0

   which employ standard control chart factors for the range. This is a continuation of chart I above to check if the testing variance is stable at the estimated level. Proceed if both points plot within the limits and chart exhibits a state of control. Otherwise, revert to the methods for a unique lot.

   b. Chart VII: Range chart of the difference between the mean values of the A and B composites. The chart checks the stability of the other components of variance. Its limits are

   $$\text{UCL:} \quad 3.686 \sqrt{\frac{s_1^2}{n_1} + \frac{2\,s_2^2}{n_1 n_2} + \frac{s_3^2}{2} + s_4^2}$$

   $$\text{Center line:} \quad 1.128 \sqrt{\frac{s_1^2}{n_1} + \frac{2\,s_2^2}{n_1 n_2} + \frac{s_3^2}{2} + s_4^2}$$

   LCL: 0

   Proceed only if the point plots within the limits and the chart exhibits a state of control. Otherwise, revert to the methods for a unique lot.

3. Estimate the standard error of the mean by

$$s_{\overline{X}} = \sqrt{\frac{s_1^2}{n_1} + \frac{s_2^2}{n_1 n_2} + \frac{s_3^2}{4} + \frac{s_4^2}{2}}$$

4. Accept for single specification limits if: (a) for lower specification limit $\overline{X} \geq L - 1.645\, s_{\overline{X}}$ or (b) for upper specification limit $\overline{X} \leq U + 1.645\, s_{\overline{X}}$. For double specification limits the acceptance procedure is as follows. If

$$(U - L) \leq 6 \sqrt{\frac{s_1^2}{n_1} + \frac{s_2^2}{n_1 n_2} + \frac{s_3^2}{4} + \frac{s_4^2}{2}}$$

discontinue inspection since specification limits are too close to be assessed at this sample size. Otherwise, proceed to test both upper and lower specification limits separately. Reject if either test fails. Accept if both pass.

To illustrate application of this procedure, consider the concentration of an ingredient in shipments of a certain chemical compound. The average level of the ingredient is not to exceed 10 percent. Shipments consist of 6 bags each containing 50 pounds of the material. The customer does not wish to accept any material if the process average from which the sample was taken exceeds 12 percent. It is known that

$\sigma_1^2 = 1.0 = $ variance of bags

$\sigma_2^2 = 1.9 = $ variance of samples within bags

$\sigma_3^2 = 0.8 = $ variance of testing

$\sigma_4^2 = 0 = $ reduction variance (assumed zero)

The plan is applied as follows:

1. Given $n_1 = 6$, $n_3 = 2$ with the reduction variance assumed negligible. Then the number of increments needed from each bag will be

$$n_2 = \frac{1.9}{6\left(\frac{(12 - 10)^2}{8.567} - \frac{1.0}{6} - \frac{0.8}{4}\right)}$$

$$n_2 = 3.16 \sim 4$$

From a shipment of six bags, four increments are taken from each. The first and third increments from each of the bags are composited into composite A while the second and fourth increments from each of the bags are composited into composite B. Two tests are made on composite A and two tests on composite B. Results are

	Composite A	Composite B
Test 1	8.3	8.8
Test 2	8.2	8.7

with an overall mean $\overline{X} = 8.5$.

2.   The differences

$$R_A = 8.3 - 8.2 = .1$$
$$R_B = 8.8 - 8.7 = .1$$

are plotted on a range chart to check the stability of the testing variance. Similarly, the mean values of the two composites

$$\overline{X}_A = 8.25 \qquad \overline{X}_B = 8.75$$

are used to obtain the range of composite means

$$R_{\overline{X}} = 8.75 - 8.25 = 0.5$$

which is plotted on a range chart to check the stability of the other components of variance using an upper control limit of

$$3.686\sqrt{\frac{1.0}{6} + \frac{2(1.9)}{6(4)} + \frac{0.8}{2} + 0} = 3.14$$

a center line of

$$1.128\sqrt{\frac{1.0}{6} + \frac{2(1.9)}{6(4)} + \frac{0.8}{2} + 0} = 0.96$$

and a lower control limit of 0. Both tests are in control.

3.   The standard error of the mean is

$$s_{\overline{X}} = \sqrt{\frac{1.0}{6} + \frac{1.9}{6(4)} + \frac{0.8}{4} + 0} = .67$$

4.   Since

$$\overline{X} < 12 + 1.645(.67)$$
$$8.5 < 13.1$$

the lot is accepted.

PROBLEMS

A shipment of crushed raw material is received in five special railroad cars, each with two compartments believed filled separately, which dump from the bottom. It is to be tested for an impurity which is specified to be less than five percent. Levels of seven percent or more cannot be tolerated by the customer's process.

1.   If the components of variance were unknown, how might the preliminary sample be taken?

2.   If in the preliminary sample $\overline{X} = 5.0$ percent and $s = 3$ percent, what additional sample size is necessary? How should these be taken?

3. Final estimates are $\overline{X}$ = 5.5 percent and s = 2 percent. Should the shipment be accepted?

4. Additional information was gathered on 20 increments from the first 5 segments (compartments) in an effort to estimate the testing and reduction variances. The results were MS_3 = .7 and MS_4 = .45. What are the estimates of the testing and reduction variances?

5. Successive shipments are made. After 25 compartments have each been sampled twice, control charts confirmed the stability of the data. The segment and increment mean squares were MS_1 = 4.75 and MS_2 = 2.2. Estimate the segment and increment variances.

6. Present the mean squares given in Prob. 5 in the form of an analysis of variance table.

7. A shipment of eight railroad cars is received. On the basis of results from Probs. 4 and 5, how many increments should be taken from each compartment if odd and even increments from each compartment are composited and two tests are made on each composite?

8. If the grand mean of the results from the eight cars was $\overline{X}$ = 5.9, should the shipment be accepted?

9. What would be the standard deviation of a single observation from the shipment from Prob. 8? Construct a 95 percent confidence interval for the lot mean in Prob. 8.

10. If the lot mean is to be estimated within ± 1 percent in Prob. 7, when the cost of sampling a segment is equal to that of an increment, what are the most cost effective segment and increment sample sizes disregarding any testing or reduction variance?

REFERENCES

American Society for Testing and Materials (1974), *Standard Recommended Practice for Sampling Industrial Chemicals*, E-300-73, ASTM Standards, Part XXX General Test Methods, Philadelphia, Pennsylvania.

Bennett, C. A., and N. L. Franklin (1954), *Statistical Analysis in Chemistry and the Chemical Industry*, John Wiley and Sons, New York.

Bicking, C. A. (1967), The Sampling of Bulk Materials, *Materials Research and Standards*, 7(2): 95-116.

Bicking, C. A. (1968), Sampling, *Encyclopedia of Chemical Technology* (Kirk-Othmer, ed.), 2nd ed., Vol. 17, John Wiley and Sons, New York, pp. 744-762.

Bicking, C. A. (1970), ASTM E-105-58 and ASTM E-300-69 Standards for the Sampling of Bulk Materials, *Journal of Quality Technology*, 2(3): 165-173.

Bicking, C. A. (1978), Principles and Methods of Sampling, *Treatise on Analytical Chemistry* (I. M. Kolthoff and P. J. Elving, eds.), 2nd ed., Vol. 1, Part I, Sec. B, Chap. 6, John Wiley and Sons, New York, pp. 299-359.

Bowker, A. H., and G. J. Lieberman (1955), *Handbook of Industrial Statistics*, Prentice-Hall, Englewood Cliffs, New Jersey.

Burr, I. W. (1976), *Statistical Quality Control Methods*, Marcel Dekker, New York.

Cochran, W. G. (1953), *Sampling Techniques*, John Wiley and Sons, New York.

Davies, O. L., ed. (1954), *Statistical Methods in Research and Production*, Hafner Publishing Co., New York.

Davies, O. L., ed. (1960), *Design and Analysis of Industrial Experiments*, Hafner Publishing Co., New York.

Deming, W. E. (1950), *Some Theory of Sampling*, John Wiley and Sons, New York.

Duncan, A. J. (1962), Bulk Sampling Problems and Lines of Attack, *Technometrics*, 4(3): 319-344.

Duncan, A. J. (1974), *Quality Control and Industrial Statistics*, 4th ed., Richard D. Irwin, Homewood, Illinois.

Duncan, A. J. (1974a), Bulk Sampling, *Quality Control Handbook* (J. M. Juran, ed.), 3rd ed., Sec. 25A, McGraw-Hill, New York, pp. 25A: 1-14.

Grant, E. L., and R. S. Leavenworth (1972), *Statistical Quality Control*, 4th ed., McGraw-Hill, New York.

Grubbs, F. E., and H. J. Coon (1954), On Setting Test Limits Relative to Specification Limits, *Industrial Quality Control*, 10(5): 15-20.

Ishikawa, K. (1958), How to Rationalize the Physical Material Sampling in Plants, *Reports of Statistical Applied Research*, Japanese Union of Scientists and Engineers, 5(2): 15.

Satterthwaite, F. E. (1946), An Approximate Distribution of Estimates of Variance Components, *Biometrika Bulletin*, 2: 110-114.

Schilling, E. G. (1973), A Systematic Approach to Analysis of Means, Part I. Analysis of Treatment Effects, *Journal of Quality Technology*, 5(3): 93-108.

Tanner, L., and M. Lerner (1951), *Economic Accumulation of Variance Data in Connection with Bulk Sampling*, ASTM STP 114, American Society for Testing and Materials, Philadelphia, Pennsylvania, pp. 8-12.

Williams, W. H. (1978), *A Sampler on Sampling*, John Wiley and Sons, New York.

SAMPLING BY VARIABLES FOR PROPORTION NONCONFORMING

The distinction between discrete and continuous variables involves good grammar as well as good statistics. We state how *many* we have of a discrete variable and how *much* when the variable is continuous. We may be interested in how many cans of soup were underweight by as much as a milligram; or how many rivets were off center by as much as 0.5 mm. These statements imply that continuous (measurement) variables can be subjected to an attributes (go no-go) type test simply by counting the number of items in a sample beyond some limit. Thus, attributes sampling plans could be applied in these two cases.

Alternatively, if the shape of the underlying distribution of individual measurements were known, acceptance sampling could be performed directly on the measurements themselves. Such procedures form the basis for variables sampling plans for proportion nonconforming and, when applicable, provide a considerable savings in sample size.

The basic idea of variables sampling for proportion nonconforming is to show that the sample results are sufficiently far within the specification limit(s) to assure the acceptability of the lot with reasonable probability.

Variables plans involve comparing a statistic, such as the mean $\overline{X}$, with an acceptance limit A in much the same way that the number nonconforming, d, is compared to an acceptance number, c, in attributes plans. A comparison of the procedures involved in variables and attributes sampling plans is shown in Fig. 10-1.

Some of the advantages of variables sampling are as follows:

1. Same protection with smaller sample size than attributes
2. Feedback of data on process which produced the units

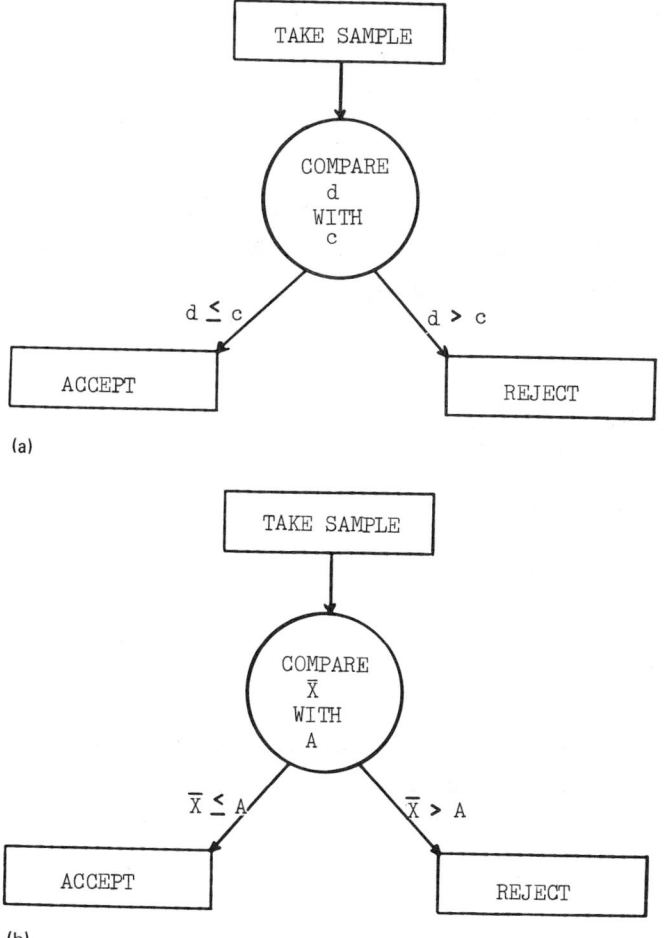

FIG. 10-1 Comparison of attributes and variables
sampling. (a) Attributes, single sampling; (b)
Variables, single sampling (upper specification
limit). (Schilling, 1974a, p. 16. Reprinted by
permission.)

3. More data available in waiver situations

4. Extent of conformity of each unit given weight in application of
    the plan

5. Increased likelihood of errors in measurment being detected

Some of the disadvantages of variables sampling are as follows:

1. Dependence of results on correctness of assumption of shape of
    underlying distribution of measurements

2. Applicable to one (only) characteristic at a time

3. Higher inspection cost per unit

4. Higher clerical cost per unit

5. Possibility of no nonconforming unit found to show to producer after rejection

The principal advantage of variables plans over attributes is reduction in sample size. For example, in comparing average sample sizes for plans matched to the single-sampling attributes plan $n = 50$, $c = 2$ we have, for a single specification limit:

Plan	ASN
Single attributes	50
Double attributes	43
Multiple attributes	35
Sequential attributes	33.5
Variables (σ unknown)	27
Variables (σ known)	12

SPECIFICATION LIMITS

Specification limits can be of two types. A single specification limit implies only one boundary value for acceptability, either upper U or lower L. Thus, a measurement does not conform to the specification limit if

$$X > U$$

for an upper specification limit, or if

$$X < L$$

for a lower specification limit. Double specification limits place both upper and lower boundary values on the acceptability of a measurement. That is, the measurement X is acceptable if and only if

$$L \leq X \leq U$$

ASSUMPTIONS AND THEORY

Probably the most important consideration in applying variables sampling plans is the requirement that the shape of the underlying distribution of measurements to which the plan is to be applied must be known and stable. This means that probability plots or statistical tests on past data must

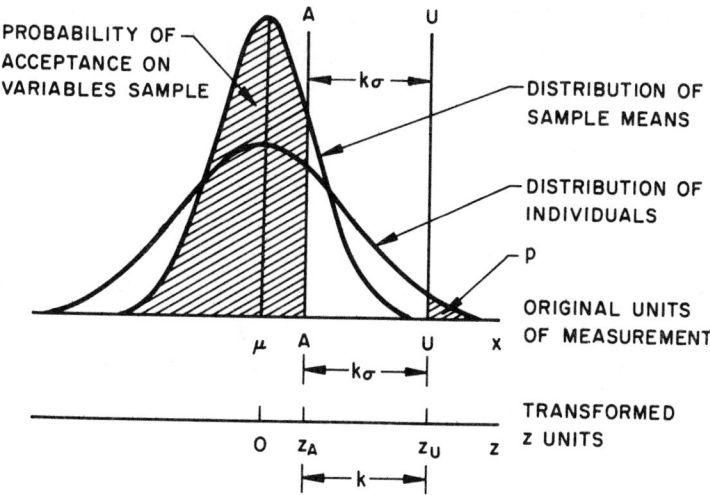

FIG. 10-2 Distributions in variables sampling.
(Schilling, 1974a, p. 17. Reprinted by permission.)

show that the distribution of measurements involved actually is that assumed
by the plan. Control chart evidence also is desirable to indicate its stability. For a known distribution, it is the underlying distribution of
measurements which relates the proportion of units outside the specification
limit to a fixed position of the population mean of the measurements. Variables plans for process parameter may then be used to confirm or deny that
the population mean is in the proper position. In a crude way, this is how
variables sampling works. In fact, some plans are devised in just this way.
It must be emphasized that the underlying distribution must be known to be
that assumed by the plan for variables sampling to be properly applied.

The basic theoretical nature of variables acceptance sampling plans is
illustrated in Fig. 10-2, which involves an upper specification limit and
assumes the underlying distribution of individual measurements to be normal.
If the procedure of Fig. 10-1 is applied, the mean $\bar{X}$ of a sample of n measurements is compared to an acceptance limit A and the lot accepted if $\bar{X}$ is
not greater than A. Figure 10-2 shows that if the distribution of individual measurements is as shown, with σ known, a proportion p of the product
above the upper specification limit U implies the mean of the distribution
must be fixed at the position indicated by μ. Sample means of size n then
would be distributed about μ as shown and so the probability of obtaining
an $\bar{X}$ not greater than A is as indicated by the shaded area in the diagram.

Published plans for known standard deviation often are given in terms of sample size and k, the distance in units of the standard deviation, between the (upper) specification limit U and the acceptance limit A. From Fig. 10-2 we see

$$k = \frac{U - A}{\sigma} = z_U - z_A$$

for the distribution of individual measurements positioned as shown where the z values are taken from the standard normal table. The situation is analogous, but reversed, for a lower specification limit.

Using Fig. 10-2 and normal probability theory, the probability of acceptance P_a can be calculated for various possible values of p, the

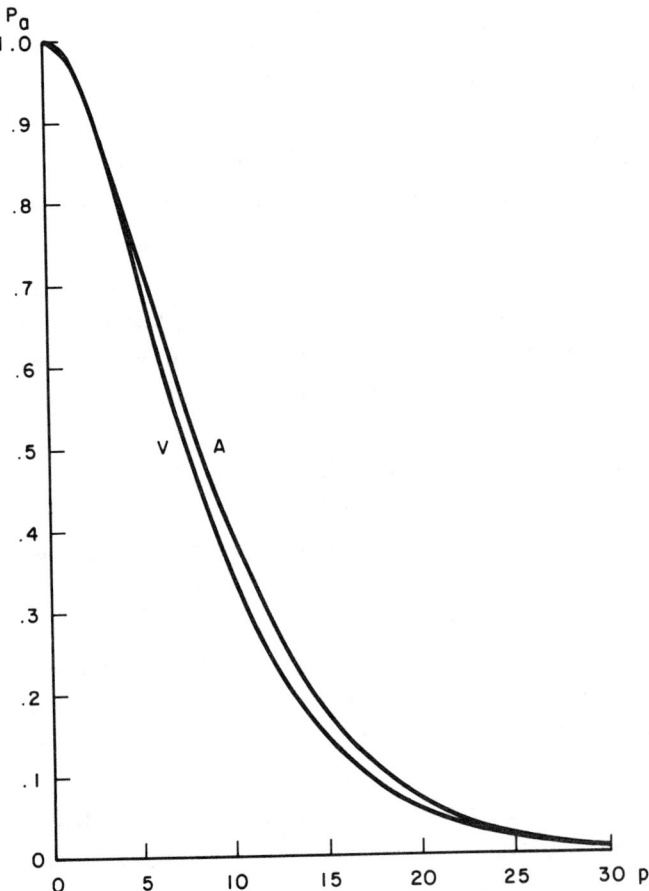

FIG. 10-3 Comparison of operating characteristic curves. For variables, V, n = 7 and k = 1.44. For attributes, A, n = 20 and c = 1.

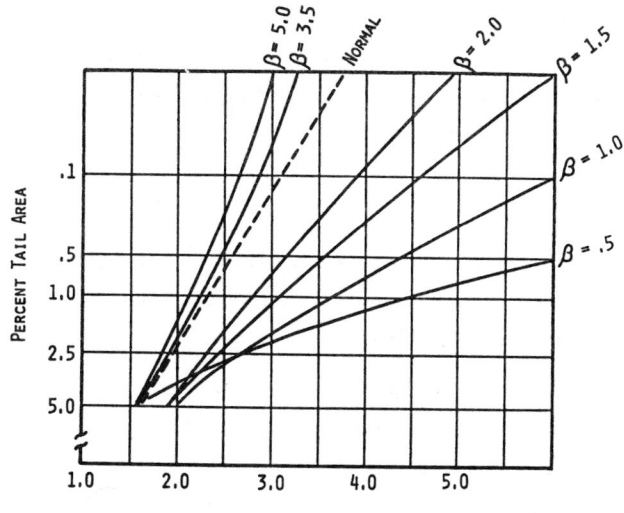

FIG. 10-4 Curves of upper tail areas of several Weibull
distributions. (Schilling, 1974a, p. 18. Reprinted
with permission.)

proportion nonconforming. Figure 10-3 shows the operating characteristic
curve of the variables plan n = 7, k = 1.44 for known standard deviation
compared to that of the attributes plan n = 20, c = 1. Operating characteristic curves of variables plans are generally considered to be Type B.

It can be seen that the two OC curves are well matched, that is they
give about the same protection. The variables plan, however, uses only
about a third as large a sample size as the attributes plan. Thus, the
variables plan appears superior. It must be remembered, however, that the
superiority of the variables plan rests on assumption of the normality of
the underlying distribution of the measurements. If this assumption cannot
be justified, the variables plan may give unreliable results and recourse
must be either to an attributes or to a mixed variables-attributes plan.

The danger involved in using a variables plan which assumes normality
when, in fact, the underlying distribution of individual measurements is
actually nonnormal is illustrated by Fig. 10-4. This shows the proportion
of the product beyond z standard deviation units from the mean to be
heavily dependent on the shape of the distribution. A variety of distributions is represented by various shape parameters for the family of Weibull
distributions. Note that the tail area beyond three standard deviations is
over 2 percent for a Weibull distribution with shape parameter 0.5, while

it is 0.13 percent for a normal distribution.

It cannot be overemphasized that the shape of the underlying distribution must be that assumed by the plan for variables acceptance sampling for proportion nonconforming to be properly applied.

AREAS OF APPLICATION

Certain areas of product inspection are better suited to application of variables plans for proportion nonconforming and defective than others. As pointed out by the author in the Juran (1974) handbook:

> Since some fundamental knowledge of the process pro-
> ducing the product is required for proper application
> of variables sampling plans for percent defective, a
> natural area of application of such plans is to in-
> house inspections, e.g., process control or final
> inspection. Use of these plans in incoming inspection
> should be restricted to product from known and trusted
> suppliers with a confirmed history of a reasonably
> stable process steadily producing product with a known
> shape of distribution. Process history should be ini-
> tially developed under an attributes sampling plan. A
> switch to variables may be considered after a plot on
> probability paper or a goodness-of-fit test ... indi-
> cates that the distribution of product is as assumed.
> A control chart for at least 20 lots is useful to con-
> firm process stability. There are times when the
> inspection situation demands the use of fewer lots.
> In such situations, appropriate limits for a 10-lot
> control chart have been developed by Hillier (1969).
> Variables plans are inappropriate for inspection of
> single lots unless the sample size is large enough to
> allow for meaningful goodness-of-fit tests to confirm
> that the shape of the underlying distribution of
> measurements is as assumed.*

*Reprinted by permission from *Quality Control Handbook*, 3rd ed. (J. M. Juran, Frank M. Gryna, Jr. and Richard S. Bingham, eds.), McGraw-Hill, New York, Section 25, E. G. Schilling, Sampling by Variables, p. 25-4. Copyright 1974, 1962 by McGraw-Hill, Inc. All rights reserved.

TABLE 10-1 $\overline{X}$ Method

Lower specification limit	Upper specification limit
1. Set A = L + kσ.	1. Set A = U - kσ.
2. Select a random sample of size n.	2. Select a random sample of size n.
3. Compute $\overline{X}$.	3. Compute $\overline{X}$.
4. If $\overline{X} \geq A$, accept the lot; if $\overline{X} < A$, reject the lot.	4. If $\overline{X} \leq A$, accept the lot; if $\overline{X} > A$, reject the lot.

OPERATION

$\overline{X}$ Method

The simplest application of variables plans for proportion nonconforming is when a single specification limit is involved and the standard deviation is known. In this case, a straightforward procedure, which we shall call the $\overline{X}$ method, may be employed. It requires that the sample size and an acceptance constant k be specified and that σ be known. An acceptance limit A for $\overline{X}$ is set a distance kσ within the specification limit. The procedure, then, is as shown in Table 10-1.

k Method

The $\overline{X}$ method is actually a special case of what is called the k method. The procedure involved in the k method is shown in Table 10-2. The more general k method may be used when the standard deviation is not known simply by substituting the sample standard deviation

$$s = \sqrt{\frac{\sum \left(x - \overline{x}\right)^2}{n - 1}}$$

for σ in the known standard deviation procedure and choosing an appropriate value of k and sample size n for the unknown standard deviation case. It may be applied in two alternative but equivalent ways as shown in Table 10-2.

It can be seen that the $\overline{X}$ method is a special case of the k method, since for a lower specification limit acceptance would occur if

$$\overline{X} - k\sigma \geq L$$
$$\overline{X} \geq L + k\sigma$$
$$\overline{X} \geq A$$

TABLE 10-2   k Method (Given n,k)

Lower specification limit	Upper specification limit
1. Select a random sample of size n.	1. Select a random sample of size n.
2. Compute $z = (\overline{X} - L)/\sigma$, for σ known or $z = (\overline{X} - L)/s$, for σ unknown.	2. Compute $z = (U - \overline{X})/\sigma$, for σ known or $z = (U - \overline{X})/s$, for σ unknown.
3. If $z \geq k$, accept the lot; if $z < k$, reject the lot.	3. If $z \geq k$, accept the lot; if $z < k$, reject the lot.
4. Equivalently, if $\overline{X} - k\sigma \geq L$, accept the lot; if $\overline{X} - k\sigma < L$, reject the lot. If σ is unknown, use appropriate values of n and k with $\overline{X} - ks$ as above.	4. Equivalently, if $\overline{X} + k\sigma \leq U$, accept the lot; if $\overline{X} + k\sigma > U$, reject the lot. If σ is unknown, use appropriate values of n and k with $\overline{X} + ks$ as above.

Also, note that A is a fixed constant only if σ is known and so the k method
is the only real alternative for the case of unknown standard deviation.
The $\overline{X}$ method, however, offers a simpler approach for lot acceptance when it
is applicable. It can be directly presented diagrammatically as in Fig.
10-1. A diagrammatic representation of the relationship between the $\overline{X}$ and
k methods when σ is known as presented by the author in the Juran (1974)
Quality Control Handbook is given in Fig. 10-5.

Double Specification Limits

When double specification limits are involved, the procedure for implementing variables sampling plans becomes somewhat more complicated. This is
because, when variability is large relative to the distance between the
lower and upper specification limits, it is possible to have a significant
proportion of product outside both specification limits at the same time.
Clearly, if the specification limits are sufficiently far apart, two separate single specification limit plans may be used since the occurrence of
product outside either of the limits will be mutually exclusive. That is,
product may be outside one or the other of the specification limits, but
not both.

     When the standard deviation is known, a simple procedure, modified
from that suggested by Duncan (1974) may be used to determine if two separate single specification limit plans may be used. Suppose a plan is to be

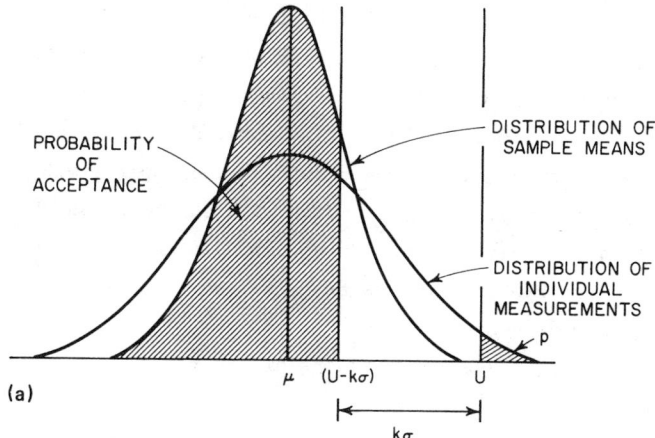

(a)

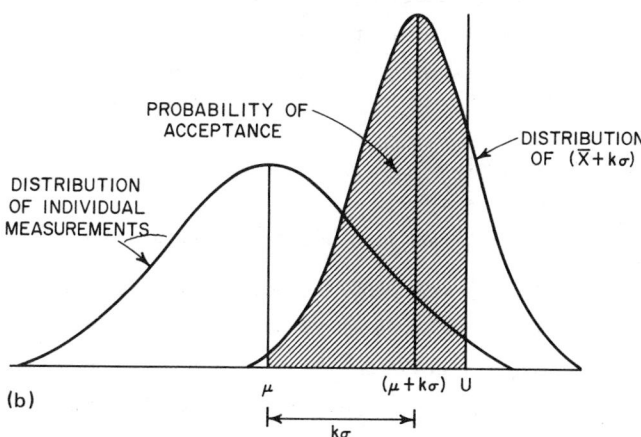

(b)

FIG. 10-5 $\overline{X}$ and k methods compared.   (a) U – kσ method.
(b) $\overline{X}$ + kσ method. Reprinted by permission from *Quality*
Control Handbook, 3rd ed. (J. M. Juran, Frank M. Gryna, Jr.,
and Richard S. Bingham, eds.), McGraw-Hill, New York, 1974,
Section 25, E. G. Schilling, Sampling by Variables, p.
25-5. Copyright 1974, 1962 by McGraw-Hill, Inc. All rights
reserved.

instituted with producer's quality level PQL = p_1 and consumer's quality
level CQL = p_2. The method is as follows:

1. Compute z_p = (U – L)/2σ.
2. Find p* from the normal table as the upper tail area correspond-
    ing to z_p. This is the minimum proportion nonconforming outside
    one of the specification limits.
3. Criteria:

a. If $2p^* \leq p_1/2$, use two single-limit plans.

b. If $p_1/2 < 2p^* \leq p_1$, the specifications may be too close to pre-
    vent nonconformities on both sides when the distribution is
    centered. Using normal probability theory, determine the
    split of proportion nonconforming outside the upper limit p_U
    and outside the lower limit p_L, which will sum to p_1 as the
    distribution is moved between the specifications. Use the
    larger of these two proportions as p_1 in two single-limit
    plans together with specified p_2.

c. If $p_1 < 2p^* < p_2$, the specifications of the plans must be
    reconsidered.

d. If $2p^* \geq p_2$, the product should be rejected outright.

For example, suppose a plan is desired to check on the resistance of
a certain electrical device. The specifications are $U = 100 \ \Omega$ and $L = 90 \ \Omega$
with $p_1 = .01$ and $p_2 = .05$. The standard deviation is known to be $1.5 \ \Omega$.
Then

1. $z_p = (100 - 90)/2(1.5) = 3.33$
2. From the normal table $p^* = .0004$
3. Since $.0008 < p_1/2 = .005$, two single-sided plans are appropriate.

When the standard deviation is unknown, the double specification limit
problem becomes still more difficult since there are two random quantities
to be considered in the acceptance decision: the mean $\overline{X}$ and the standard
deviation s. In such a situation, it is customary to check the sample
standard deviation against the maximum value before proceeding to check
against two separate single-limit plans. The so-called maximum standard
deviation becomes part of the acceptance procedure. It may be approximated
as follows from a procedure suggested by Wallis (1950):

1. Find the upper tail normal area p^{**} corresponding to $z_p^* = k$.
2. Find z_p^{**} corresponding to a normal upper tail area of $p^{**}/2$.
3. The maximum standard deviation is MSD $\simeq (U - L)/2z_p^{**}$.

The acceptance criteria for double specification limits then adds the
following initial check to the procedure for two single-limit plans:

Check s against MSD:

a. If $s \leq MSD$, use two separate single specification limit plans.

b. If $s > MSD$, reject the lot since the standard deviation is too
    large to be consistent with the acceptance criteria.

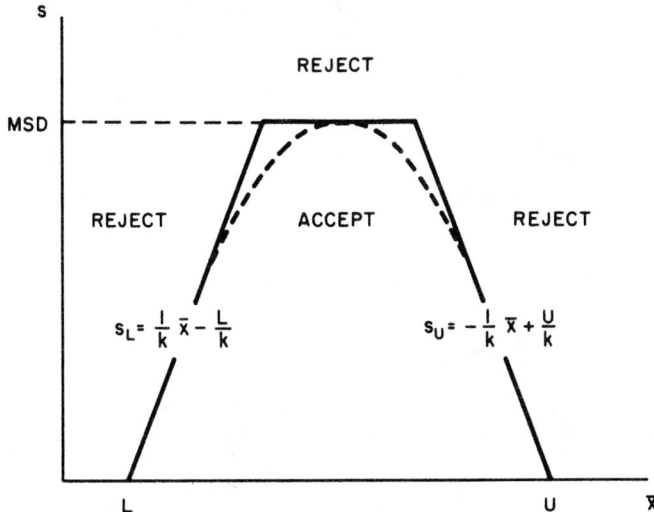

FIG. 10-6 Acceptance polygon.

The idea can be expressed graphically as shown in Fig. 10-6. The sample standard deviation is plotted against the sample mean. The x-axis, the MSD, and the two single-sided acceptance sampling criteria ($\overline{X} + ks = U$ and $\overline{X} - ks = L$) form an acceptance polygon. If the point ($\overline{X}, s$) plots within the polygon, the lot should be accepted. If not, the lot is rejected.

Actually, the polygon shown is an approximation of a more accurate acceptance region, the development of which was attributed by Wallis (1950) to Kenneth J. Arnold. The region is defined by the points ($\overline{X}', \overline{Y}'$) such that for any two proportions p_0' and p_0'' summing to p^{**} corresponding to $z_p^* = k$:

$$\overline{X}' = \frac{U z_{p_0'} + L z_{p_0''}}{z_{p_0'} + z_{p_0''}}$$

$$s' = \frac{U - L}{z_{p_0'} + z_{p_0''}}$$

Given a point ($\overline{X}', s'$) on one side of the "polygon," of course the symmetric point for a given s' is

$$\overline{X}'' = U + L - \overline{X}'$$

Such a polygon is shown in Fig. 10-6 in dotted lines. The dotted curve will intersect the straight sides of the original polygon at approximately

$$s = \frac{U - L}{3 + k}$$

It can be seen that the first polygon approximation to the acceptance region is slightly loose in that it overstates the acceptance region. The solution for s' at $\overline{X}' = (U + L)/2$ results in the approximation for MSD given above.

Wallis (1950) outlined the method for determining the more accurate acceptance region as follows:

1. Determine n and k from the usual one sided procedures (given in the next section).

2. Find the indifference quality p**, which is the probability that a standard normal deviate will exceed $z_p^* = k$.

3. Divide p** into two parts p_1' and p_2' such that $p_1' + p_2' = p^{**}$. Each pair p_1' and p_2' leads to a point on the acceptance region boundary.

4. Find z_1 and z_2 as normal deviates corresponding to the upper tail areas p_1' and p_2' respectively.

5. Substitute each pair, z_1 and z_2, into the equation

$$\overline{X}' = \frac{U\,z_1 + L\,z_2}{z_1 + z_2}$$

$$s' = \frac{U - L}{z_1 + z_2}$$

6. Plot enough points to define the acceptance region.

For example, consider the unknown standard deviation plan

$n = 13 \quad k = 1.44$

to be applied against the previous specifications for resistance, $U = 100\ \Omega$, $L = 90\ \Omega$. The acceptance polygon would appear as in Fig. 10-7.

The polygon is constructed as follows:

1. The two acceptance lines are

$$s_L = \frac{1}{k}\,\overline{X} - \frac{L}{k} = .694\,\overline{X} - 62.5$$

$$s_U = -\frac{1}{k}\,\overline{X} + \frac{U}{k} = -.694\,\overline{X} + 69.4$$

2. The maximum standard deviation is determined as

   a. $z_p^* = 1.44$; so $p^{**} = .0749$.

   b. $\dfrac{p^{**}}{2} = .0375$; so $z_p^{**} = 1.78$.

   c. MSD $= (U - L)/2z_p^{**} = 10/[2(1.78)] = 2.809$.

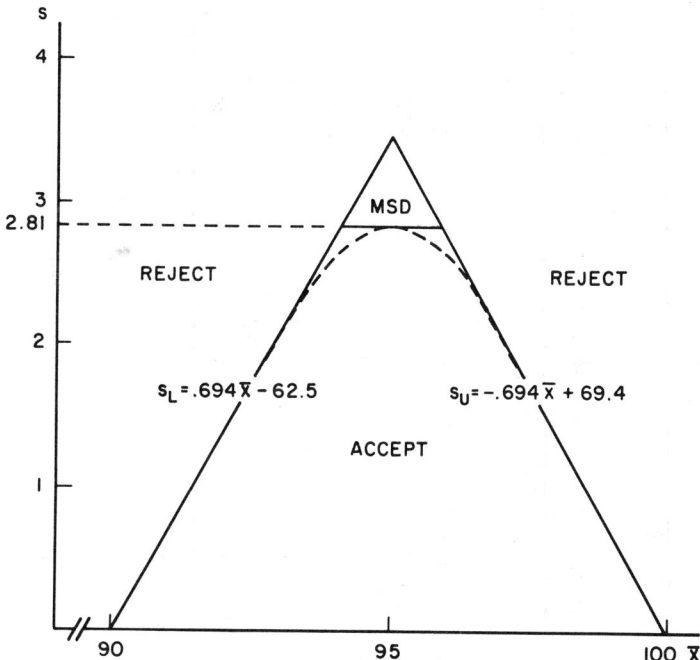

FIG. 10-7   Acceptance polygon: example.

3.   The two single specification lines will intersect when

$$\overline{X} = \frac{U + L}{2} = \frac{100 + 90}{2} = 95$$

which corresponds to a height (s) of

$$s = -.694(95) + 69.4 = 3.47$$

4.   The lines are drawn to obtain the polygon.

Using Wallis' method the more accurate acceptance region can be obtained using the tabulation shown in Table 10-3 given $p^{**} = .075$.

SELECTION

Tables

Extensive tables of variables plans for proportion nonconforming and defective will be found in the well-known military standard MIL-STD-414. The OC curves presented therein can be used to select a plan appropriate to the sampling situation. MIL-STD-414 will be discussed in a later chapter. Procedures for the development of variables plans assuring LTPD or AOQL

TABLE 10-3   Polygon Derived by Wallis Method

(1) p_1'	(2) p_2'	(3) z_1	(4) z_2	(5) (3) + (4)	(6) (3) − (4)	(7) (6)/(5)	(8) 2/(5)	$\bar{x} = \frac{U+L}{2} + \frac{U-L}{2}$ (7)	$s = \frac{U-L}{2}$ (8)
.0375	.0375	1.78	1.78	3.56	0	0	.5618	95.00	2.809
.0325	.0425	1.85	1.72	3.57	0.13	.0364	.5602	95.18	2.801
.0275	.0475	1.92	1.67	3.59	0.25	.0696	.5571	95.35	2.786
.0225	.0525	2.00	1.62	3.62	0.38	.1050	.5525	95.52	2.762
.0175	.0575	2.11	1.58	3.69	0.53	.1436	.5420	95.72	2.710
.0125	.0625	2.24	1.53	3.77	0.71	.1883	.5305	95.94	2.652
.0075	.0675	2.43	1.49	3.92	0.94	.2398	.5102	96.20	2.551
.0025	.0725	2.81	1.46	4.27	1.35	.3162	.4684	96.58	2.342
.0001	.0749	3.89	1.44	5.33	2.45	.4597	.3752	97.30	1.876

protection have been developed by H. G. Romig (1939) and are presented in his Ph.D. dissertation.

Appendix Table T10-2, computed by Sommers (1981), gives acceptance criteria for single-sampling variables plans as well as matched double-sampling variables plans (discussed later). Sample sizes are shown for standard deviation known (n_σ) and unknown (n_s) for a given acceptance constant (k). The table is indexed by producer's and consumer's quality level with $\alpha = .05$ and $\beta = .10$. Plans were derived using the computational formulas given below. The Wallis approximation was used for standard deviation unknown plans.

The selection of p_1 and p_2 values used by Sommers was made to be the same as those used by the Statistical Research Group (1947) in a similar tabulation to facilitate a comparison. As an example of the use of Appendix Table T10-2, it will be seen that for $p_1 = .01$ and $p_2 = .05$ with $\alpha = .05$ and $\beta = .10$ the following plans are given:

    Known standard deviation: n = 19, k = 1.94
    Unknown standard deviation: n = 54, k = 1.94

Matching binomial attributes and narrow limit plans have been tabulated by Schilling and Sommers (1981) for the same selection of p_1 and p_2 values and appear with the single-sampling variables plans in Appendix Table T13-3.

Formulas

The acceptance criteria for variables plans may be readily determined from computational formulas for n and k. In these formulas the standard normal deviates, z, represent

$$z_{p_1} = \text{area of } p_1 \text{ in upper tail}$$

$$z_{p_2} = \text{area of } p_2 \text{ in upper tail}$$

$$z_\alpha = \text{area of } \alpha \text{ in upper tail}$$

$$z_\beta = \text{area of } \beta \text{ in upper tail}$$

The values of k and n are obtained from the following formulas. It will be seen that the formula for n depends upon the state of knowledge of the standard deviation. Results should always be rounded *up*.

$$k = \frac{z_{p_2} z_\alpha + z_{p_1} z_\beta}{z_\alpha + z_\beta}$$

for σ known

$$n = \left(\frac{z_\alpha + z_\beta}{z_{p_1} - z_{p_2}}\right)^2$$

for σ unknown

$$n = \left(\frac{z_\alpha + z_\beta}{z_{p_1} - z_{p_2}}\right)^2 \left(1 + \frac{k^2}{2}\right)$$

The latter formula is due to Wallis (1947) and corrects the sample size found for σ known by the factor $(1 + k^2/2)$ obtained from the non-central t-distribution. Although this is an approximation, it is surprisingly accurate and certainly adequate for practical purposes. It can be shown to be extremely accurate when compared to the exact values obtained using the non-central t-distribution. This can be seen from Appendix Table T10-3 prepared by the Columbia Statistical Research Group (1947, p. 65) for unknown standard deviation plans where both approximate and exact producer and consumer quality levels are given.

Suppose $p_1 = .018$ and $p_2 = .18$. For $\alpha = .05$ and $\beta = .10$, the formulas give

$$k = \frac{0.92(1.64) + 2.10(1.28)}{1.64 + 1.28} = 1.44$$

for σ known

$$n = \left(\frac{1.64 + 1.28}{2.10 - 0.92}\right)^2 = 6.12 \sim 7$$

and for σ unknown

$$n = \left(\frac{1.64 + 1.28}{2.10 - 0.92}\right)^2 \left(1 + \frac{1.44^2}{2}\right)$$

$$= 6.12(2.04) = 12.5 \sim 13$$

Jacobson Nomograph for Plan Selection

A nomograph also exists for determining variables plans for proportion nonconforming. Due to Jacobson (1949) it can be used in a manner similar to that of Larson (1966) for attributes plans. It is based on the Wallis formula. It is shown in Fig. 10-8. To use the nomograph to derive a plan, given p_1, p_2, α, and β, proceed as follows for the case of σ unknown:

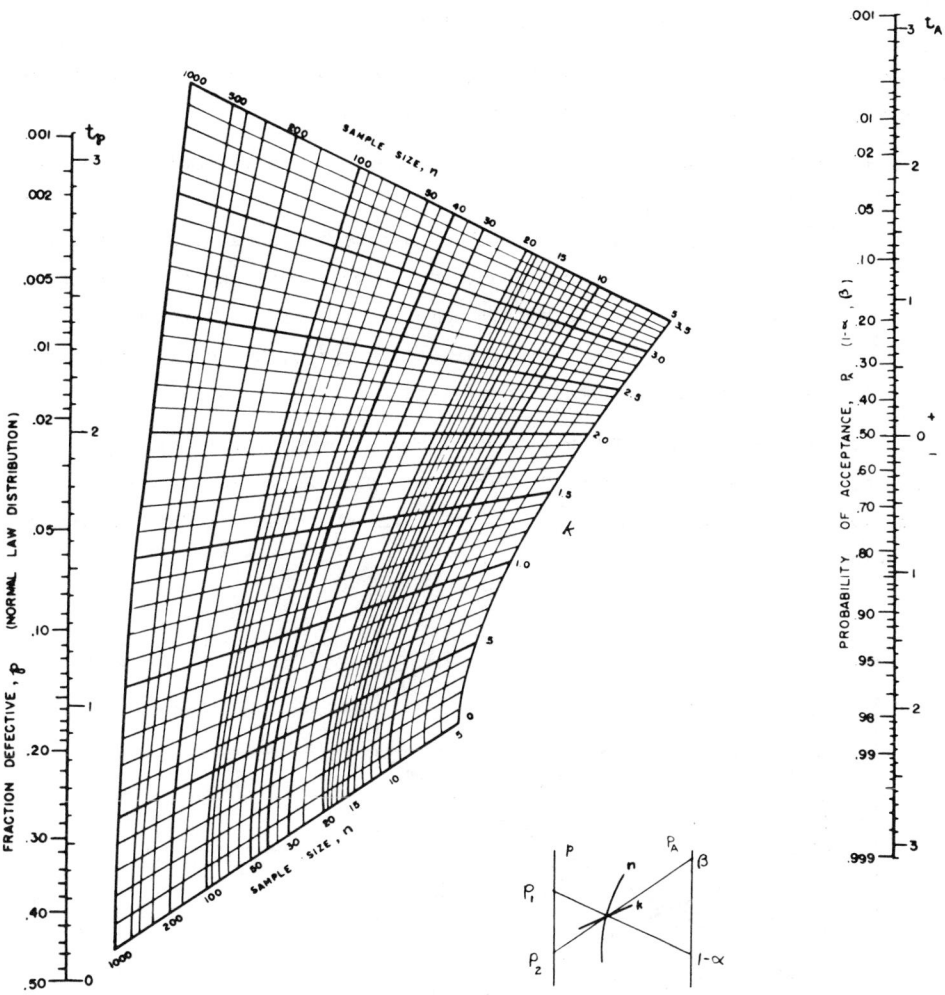

FIG. 10-8 Jacobson nomograph for variables plans for proportion defective. (From Jacobson, 1949, p. 42. Reprinted by permission.)

1. Connect p_1 on the left "fraction defective" axis with $(1 - \alpha)$
    on the right "probability of acceptance" axis.

2. Connect p_2 on the left fraction defective axis with β on the
    right probability of acceptance axis.

3. From the point of intersection of the two lines, read the sample
    size n and the acceptance constant k.

When the standard deviation is known, the author has found that the nomograph can be employed to derive a plan as follows:

1. Draw the two lines and obtain the point of intersection as above
    and read the value of k.

2. Draw a line through the point of intersection parallel to the left
    and right vertical axes and read the value of n at the intersection
    of this line with the *bottom* sample size scales on the chart (i.e.,
    where k = 0). This follows since for k = 0, the Wallis formula
    reduces to that of the known standard deviation plan.

Figure 10-9 shows the derivation of the σ known plan k = 1.44, n = 7
when p_1 = .018, p_2 = .18, α = .05, β = .10. The dotted line shows the
location of the sample size of 13 for the plan when the standard deviation
is unknown.

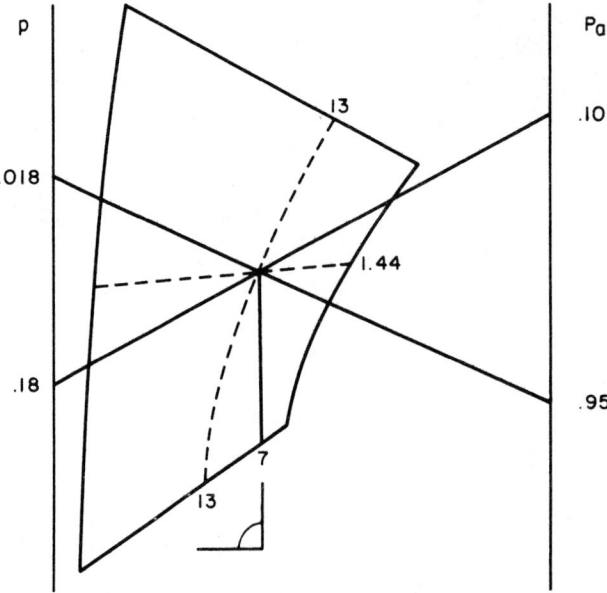

FIG. 10-9 Application of Jacobson nomograph.

MEASURES

Jacobson Nomograph for Operating Characteristics

The Jacobson nomograph shown in Fig. 10-8 may be used to derive the operating characteristic curve for a variables plan for proportion nonconforming.
The procedure differs slightly between the case of standard deviation known
and that of standard deviation unknown. The method is as follows for a plan
specified by n and k.

Standard deviation unknown

1. Locate the point (n,k).
2. Locate the proportion nonconforming p on the left fraction
    defective axis.
3. The intersection of a line through the two points with the right
    axis gives the probability of acceptance.
4. The process may be reversed to give the proportion nonconforming
    associated with a given probability of acceptance.

Standard deviation known

1. Locate the sample size, on the *bottom* sample size axis (i.e.,
    where k = 0).
2. Draw a line through the sample size parallel with the right and
    left vertical axes.
3. The intersection of the line drawn with the appropriate curve
    for k gives the point (n,k) for the known standard deviation case.
4. Follow the unknown standard deviation procedure from step 2.

It will be observed that the two lines drawn in Fig. 10-9 may be regarded as representative of the procedure for the unknown standard deviation
plan n = 13, k = 1.44 or for the known standard deviation plan n = 7, k =
1.44, respectively. They show the 10th and 95th percentage points of the
OC curve.

Calculation: σ Known

When the standard deviation is known, calculation of operating characteristic curves for variables plans may be performed using the normal distribution. Referring to Fig. 10-2, which describes the $\overline{X}$ method, we see that
for any given proportion nonconforming p, the population mean μ must be a
fixed distance $z_U \sigma$ from the upper specification limit. Also, the distance

from the population mean to the acceptance limit A is $z_A \sigma$, where

z_U = standard normal deviate for the distribution of individual measurements corresponding to proportion nonconforming in the upper tail

z_A = standard normal deviate for the distribution of individual measurements corresponding to the acceptance limit A

As noted previously, $k = z_U - z_A$; so $z_A = z_U - k$.

To find the probability of acceptance, the probability of obtaining a mean below A must be determined. But, means are distributed according to the distribution of sample means which is also shown in Fig. 10-2. To get the probability of acceptance, the distance between A and μ must be found in terms of the standard deviation of the distribution of sample means. Recall

$$\sigma_{\overline{X}} = \frac{\sigma}{\sqrt{n}}$$

so

$$\sqrt{n}\sigma_{\overline{X}} = \sigma$$

and the distance z_A between μ and A in terms of the standard deviation of individuals may be used to obtain the distance in terms of the standard deviation of means since

$$z_A \sigma = \sqrt{n}\, z_A \frac{\sigma}{\sqrt{n}} = \overline{z}_A \sigma_{\overline{X}}$$

where the conventional bar denoting average is used to indicate that $\overline{z}_A$ is from the distribution of sample means. We then have

$$\overline{z}_A = \sqrt{n}\, z_A \quad \text{or} \quad \overline{z}_A = \sqrt{n}(z_U - k)$$

Thus, for any given value of p, the probability of acceptance can be determined as follows for an upper specification limit

1. Determine z_U from p.
2. Obtain $z_A = z_U - k$.
3. Convert z_A to the distribution of sample means as $\overline{z}_A = \sqrt{n}\, z_A$.
4. The probability of a normal variate exceeding $\overline{z}_A$ is the probability of rejection. Its complement, the probability of a result less than $\overline{z}_A$, is the probability of acceptance.

For a lower specification limit, this becomes

1. Determine z_L from p.

TABLE 10-4   Calculation of Probability of Acceptance: n = 7, k = 1.44

PROPORTION NONCONFORMING p	z_U	$z_A = z_U - k$	$\bar{z}_A = \sqrt{n}\, z_A$	$P_r = (1-P_a)$	$P_a = (1-P_r)$	PROBABILITY OF ACCEPTANCE n=20, c=1
.0075	2.43	0.99	2.62	.0044	.9956	.99
.018	2.10	0.66	1.75	.0401	.9599	.95
.027	1.93	0.49	1.30	.0968	.9032	.90
.048	1.66	0.22	0.58	.2810	.7190	.75
.083	1.39	-0.05	-0.13	.5517	.4483	.50
.129	1.13	-0.31	-0.82	.7939	.2061	.25
.181	0.91	-0.53	-1.40	.9192	.0808	.10
.216	0.78	-0.66	-1.75	.9599	.0401	.05
.289	0.56	-0.88	-2.33	.9901	.0099	.01

2. Obtain $z_A = z_L + k$.

3. Convert z_A to the distribution of sample means as $\bar{z}_A = \sqrt{n}\, z_A$.

4. The probability of a normal variate equal to or exceeding $\bar{z}_A$ is the probability of acceptance. Its complement, the probability of a result less than $\bar{z}_A$ is the probability of rejection.

For example, consider the plan n = 7, k = 1.44; Table 10-4 shows the computation of the probability of acceptance and compares the results to the attributes plan n = 20, c = 1. A plot of both OC curves was given in Fig. 10-3.

Calculation: σ Unknown

As suggested by Wallis (1950), the OC curve for an unknown standard deviation plan, specified by k and n_s, can be approximated by using the known standard deviation procedure above with

$$n = \frac{n_s}{1 + k^2/2}$$

or, using a slightly more accurate form of the Wallis (1947) approximation to relate the sample sizes of known (n_σ) and unknown (n_s) standard deviation plans

$$n_\sigma = \frac{n_s}{1 + \left(k^2 n_s / 2(n_s - 1)\right)}$$

So that, if z_{1-P_a} denotes the upper tail standard normal deviate corresponding to the probability of rejection $(1 - P_a)$, Wallis (1947) gives the following relation:

$$z_A = z_U - k = z_{1-P_a} \sqrt{\frac{1}{n_s} + \frac{k^2}{2(n_s - 1)}}$$

Note that, the standard normal deviate corresponding to the probability of acceptance is

$$z_{P_a} = -z_{1-P_a}$$

When the standard deviation is unknown, exact calculation of the OC curve becomes less straightforward. The statistic (shown here for an upper specification limit)

$$t = \frac{U - \overline{X}}{s}$$

has a Student's t-distribution only for 50 percent nonconforming. For all other values of proportion nonconforming, the statistic is distributed by the non-central t-distribution, the distribution involved in calculating the OC curve when the standard deviation is unknown.

For a variate t which can be expressed as

$$t = \frac{z + \delta}{\sqrt{w}}$$

with

   z: distributed standard normal $(\mu = 0, \ \sigma = 1)$
   w: distributed χ^2/f with f degrees of freedom independent of z
   δ: a constant

The non-central t-probability distribution function has been expressed by Resnikoff and Lieberman (1957) as

$$P(f,\delta,t_0) = \frac{f!}{2^{(f-1)/2} \Gamma(f/2) \sqrt{\pi f}} \int_{-\infty}^{t_0} e^{-(1/2)(f\delta^2)/(f+t^2)}$$

$$\left(\frac{f}{(f+t^2)}\right)^{(f+1)/2} \int_0^\infty \frac{v^f}{f!} e^{-(1/2)\left(v - \delta t/\sqrt{f+t^2}\right)^2} dv \, dt$$

x \ p	.2500	.1500	.1000	.0650	.0400	.0250	.0100	.0040	.0025	.0010
− 0.50	0.0000	0.0000	0.0000	0.0000	0.0000	0.0000	0.0000	0.0000	0.0000	0.0000
− 0.45	0.0001	0.0000	0.0000	0.0000	0.0000	0.0000	0.0000	0.0000	0.0000	0.0000
− 0.40	0.0001	0.0000	0.0000	0.0000	0.0000	0.0000	0.0000	0.0000	0.0000	0.0000
− 0.35	0.0002	0.0000	0.0000	0.0000	0.0000	0.0000	0.0000	0.0000	0.0000	0.0000
− 0.30	0.0004	0.0000	0.0000	0.0000	0.0000	0.0000	0.0000	0.0000	0.0000	0.0000
− 0.25	0.0006	0.0000	0.0000	0.0000	0.0000	0.0000	0.0000	0.0000	0.0000	0.0000
− 0.20	0.0010	0.0000	0.0000	0.0000	0.0000	0.0000	0.0000	0.0000	0.0000	0.0000
− 0.15	0.0017	0.0000	0.0000	0.0000	0.0000	0.0000	0.0000	0.0000	0.0000	0.0000
− 0.10	0.0029	0.0000	0.0000	0.0000	0.0000	0.0000	0.0000	0.0000	0.0000	0.0000
− 0.05	0.0047	0.0000	0.0000	0.0000	0.0000	0.0000	0.0000	0.0000	0.0000	0.0000
0.00	0.0075	0.0001	0.0000	0.0000	0.0000	0.0000	0.0000	0.0000	0.0000	0.0000
0.05	0.0119	0.0002	0.0000	0.0000	0.0000	0.0000	0.0000	0.0000	0.0000	0.0000
0.10	0.0184	0.0004	0.0000	0.0000	0.0000	0.0000	0.0000	0.0000	0.0000	0.0000
0.15	0.0279	0.0007	0.0000	0.0000	0.0000	0.0000	0.0000	0.0000	0.0000	0.0000
0.20	0.0413	0.0012	0.0000	0.0000	0.0000	0.0000	0.0000	0.0000	0.0000	0.0000
0.25	0.0594	0.0022	0.0001	0.0000	0.0000	0.0000	0.0000	0.0000	0.0000	0.0000
0.30	0.0832	0.0039	0.0002	0.0000	0.0000	0.0000	0.0000	0.0000	0.0000	0.0000
0.35	0.1134	0.0067	0.0004	0.0000	0.0000	0.0000	0.0000	0.0000	0.0000	0.0000
0.40	0.1503	0.0110	0.0009	0.0000	0.0000	0.0000	0.0000	0.0000	0.0000	0.0000
0.45	0.1939	0.0177	0.0016	0.0001	0.0000	0.0000	0.0000	0.0000	0.0000	0.0000
0.50	0.2435	0.0273	0.0030	0.0002	0.0000	0.0000	0.0000	0.0000	0.0000	0.0000
0.55	0.2983	0.0407	0.0052	0.0004	0.0000	0.0000	0.0000	0.0000	0.0000	0.0000
0.60	0.3567	0.0587	0.0088	0.0009	0.0000	0.0000	0.0000	0.0000	0.0000	0.0000
0.65	0.4173	0.0820	0.0143	0.0016	0.0001	0.0000	0.0000	0.0000	0.0000	0.0000
0.70	0.4784	0.1109	0.0223	0.0030	0.0002	0.0000	0.0000	0.0000	0.0000	0.0000
0.75	0.5383	0.1456	0.0336	0.0052	0.0005	0.0000	0.0000	0.0000	0.0000	0.0000
0.80	0.5958	0.1858	0.0488	0.0087	0.0010	0.0001	0.0000	0.0000	0.0000	0.0000
0.85	0.6496	0.2310	0.0685	0.0141	0.0018	0.0002	0.0000	0.0000	0.0000	0.0000
0.90	0.6992	0.2803	0.0930	0.0217	0.0032	0.0004	0.0000	0.0000	0.0000	0.0000
0.95	0.7439	0.3326	0.1227	0.0323	0.0055	0.0008	0.0000	0.0000	0.0000	0.0000
1.00	0.7837	0.3866	0.1572	0.0464	0.0091	0.0015	0.0000	0.0000	0.0000	0.0000
1.05	0.8185	0.4412	0.1963	0.0644	0.0143	0.0028	0.0001	0.0000	0.0000	0.0000
1.10	0.8487	0.4952	0.2394	0.0866	0.0217	0.0047	0.0002	0.0000	0.0000	0.0000
1.15	0.8745	0.5476	0.2856	0.1133	0.0317	0.0077	0.0003	0.0000	0.0000	0.0000
1.20	0.8964	0.5976	0.3341	0.1442	0.0447	0.0120	0.0006	0.0000	0.0000	0.0000
1.25	0.9148	0.6445	0.3839	0.1792	0.0612	0.0182	0.0012	0.0001	0.0000	0.0000
1.30	0.9302	0.6879	0.4341	0.2177	0.0813	0.0265	0.0021	0.0001	0.0000	0.0000
1.35	0.9429	0.7276	0.4836	0.2593	0.1052	0.0373	0.0035	0.0002	0.0001	0.0000
1.40	0.9534	0.7635	0.5319	0.3031	0.1329	0.0510	0.0056	0.0004	0.0001	0.0000
1.45	0.9621	0.7956	0.5781	0.3484	0.1641	0.0679	0.0087	0.0008	0.0002	0.0000
1.50	0.9691	0.8240	0.6220	0.3944	0.1984	0.0881	0.0131	0.0014	0.0004	0.0000
1.55	0.9749	0.8491	0.6629	0.4405	0.2356	0.1115	0.0190	0.0024	0.0007	0.0001
1.60	0.9796	0.8709	0.7009	0.4859	0.2749	0.1382	0.0267	0.0038	0.0013	0.0001
1.65	0.9834	0.8900	0.7357	0.5300	0.3159	0.1679	0.0365	0.0059	0.0021	0.0003
1.70	0.9865	0.9064	0.7674	0.5725	0.3579	0.2004	0.0486	0.0089	0.0034	0.0005
1.75	0.9890	0.9205	0.7959	0.6129	0.4003	0.2352	0.0633	0.0129	0.0052	0.0008
1.80	0.9911	0.9326	0.8215	0.6509	0.4426	0.2719	0.0805	0.0182	0.0078	0.0013
1.85	0.9927	0.9429	0.8443	0.6864	0.4842	0.3099	0.1005	0.0250	0.0113	0.0021
1.90	0.9941	0.9517	0.8645	0.7192	0.5248	0.3489	0.1230	0.0335	0.0159	0.0033
1.95	0.9952	0.9592	0.8824	0.7495	0.5639	0.3884	0.1480	0.0440	0.0219	0.0050
2.00	0.9961	0.9655	0.8980	0.7771	0.6013	0.4278	0.1753	0.0564	0.0293	0.0073
2.05	0.9968	0.9709	0.9117	0.8021	0.6367	0.4667	0.2047	0.0711	0.0384	0.0104
2.10	0.9974	0.9754	0.9237	0.8248	0.6701	0.5049	0.2359	0.0879	0.0493	0.0144
2.15	0.9978	0.9792	0.9341	0.8452	0.7013	0.5419	0.2685	0.1069	0.0622	0.0194
2.20	0.9982	0.9825	0.9431	0.8634	0.7302	0.5776	0.3022	0.1280	0.0770	0.0258
2.25	0.9985	0.9852	0.9510	0.8797	0.7569	0.6116	0.3367	0.1511	0.0939	0.0335
2.30	0.9988	0.9875	0.9577	0.8942	0.7815	0.6439	0.3716	0.1761	0.1127	0.0427
2.35	0.9990	0.9894	0.9636	0.9071	0.8040	0.6744	0.4066	0.2028	0.1335	0.0535
2.40	0.9992	0.9910	0.9686	0.9184	0.8245	0.7030	0.4414	0.2310	0.1561	0.0661
2.45	0.9993	0.9924	0.9730	0.9285	0.8431	0.7297	0.4756	0.2603	0.1804	0.0803

FIG. 10-10 Resnikoff-Lieberman table. Reprinted from G. J. Resnikoff and G. J. Lieberman, *Tables of Non-Central t-Distribution*, Stanford University Press, Stanford, California, 1957, p. 327, with the permission of the publishers. Copyright 1957 by the Board of Trustees of the Leland Stanford Junior University.

where

f = degrees of freedom in t

δ = non-centrality parameter

t = random variate

Resnikoff and Lieberman (1957) have extensively tabulated the non-central t-distribution. A sample page is shown as Fig. 10-10. For a non-centrality parameter $\delta = \sqrt{f + 1}\ K_p$, they give values of the distribution function

$$\Pr(t/\sqrt{f} \leq x)$$

tabulated by p and x, where

K_p = standard normal deviate exceeded with probability p

To use the non-central t-distribution to obtain the probability of acceptance when the standard deviation is unknown, the acceptance criterion may be expressed as

$$\frac{U - \overline{X}}{s} \geq k$$

$$\sqrt{n}\left(\frac{U - \overline{X}}{s}\right) \geq \sqrt{n}\ k$$

and

$$\left(\frac{\sqrt{n}(U - \mu)}{\sigma} - \frac{\sqrt{n}(\overline{X} - \mu)}{\sigma}\right)\frac{\sigma}{s} \geq \sqrt{n}\ k$$

The left-hand side of the inequality is distributed non-central t with

$$f = n - 1$$

$$\delta = \frac{\sqrt{n}(U - \mu)}{\sigma} = \sqrt{n}\ z_U$$

so that the probability of acceptance is simply

$$P_a = \Pr(t \geq \sqrt{n}\ k)$$

and since tables are for $t/\sqrt{f}$ we have*

$$P_a = 1 - \Pr\left(\frac{t}{\sqrt{n - 1}} \leq \sqrt{\frac{n}{n - 1}}\ k\right)$$

$$= 1 - P\left(n - 1,\ \sqrt{n}\ z_U,\ \sqrt{\frac{n}{n - 1}}\ k\right)$$

*Note that this relation allows the Jacobson nomograph to be used in reverse to obtain approximate values for the non-central t-distribution by using $n = f + 1$; $k = t\sqrt{n - 1}/n$; $t_p = \delta/\sqrt{f + 1}$; $1 - P(f,\delta,t) = P_a$.

Hence, using the Resnikoff-Lieberman tables, proceed as follows to evaluate the operating characteristics of an upper specification limit plan specified by n and k for proportion nonconforming p.

1. Degrees of freedom are f = n - 1; select the table of the probability integral with f degrees of freedom.
2. Compute $\sqrt{n/(n-1)}\ k$; this is x in the table.
3. For the value of p given and x calculated, obtain the probability of rejection P(R) from the table.
4. The complement is the probability of acceptance $P_a = 1 - P(R)$.

For example, consider the plan n = 13, k = 1.44 to be evaluated at p = .025

1. f = 13 - 1 = 12
2. $x = \sqrt{\dfrac{13}{12}}(1.44) = 1.50$
3. P(R) = .0881
4. P_a = .9119

This value may be obtained from Fig. 10-10.

For reasons of symmetry, evaluation of the operating characteristic curve for a lower specification limit plan is analogous. It will be seen that the specification limit (upper or lower) does not appear in the steps for determining probability of acceptance. The procedure will work for either.

Double Specification Limits

Evaluation of the probability of acceptance in the two-sided specification limit case is analogous to the single specification limit procedure when the standard deviation is known. This amounts to an evaluation of the probability of rejection and the proportion nonconforming outside each of the specification limits over various fixed positions of the population mean. Their sum gives values of p and P_a which may be plotted as the operating characteristic curve.

Consider the earlier double specification limit example in which U = 100 Ω, L = 90 Ω, and suppose σ = 2.0. The plan n = 7, k = 1.44 is to be applied to both specification limits. The double specification limit analog to Table 10-3 would appear as does Table 10-5, where p_U and p_L simply represent the proportion nonconforming outside U and L, respectively.

TABLE 10-5 Calculation of Probability of Acceptance: Double Specification Limits
(Known Standard Deviation $\sigma = 2$; $n = 7$, $k = 1.44$)

μ	$z_U = \dfrac{U-\mu}{\sigma}$	P_U	$z_A = z_U - k$	$\bar{z} = \sqrt{n}\, z_A$	$P(R)$	$z_L = \dfrac{\mu-L}{\sigma}$	P_L	$z_A = z_L - k$	$\bar{z} = \sqrt{n}\, z_A$	$P(R)$	p	$P(R)$	$P(A)$
90	5.0	0	3.56	9.42	0	0	.5000	-1.44	-3.81	.9999	.5000	.9999	.0001
91	4.5	0	3.06	8.10	0	0.5	.3085	-0.94	-2.49	.9936	.3085	.9936	.0064
92	4.0	0	2.56	6.77	0	1.0	.1587	-0.44	-1.16	.8770	.1587	.8770	.1230
93	3.5	.0002	2.06	5.45	0	1.5	.0668	0.06	0.16	.4364	.0670	.4761	.5239
94	3.0	.0013	1.56	4.13	0	2.0	.0228	0.56	1.48	.0694	.0241	.0694	.9306
95	2.5	.0062	1.06	2.80	.0026	2.5	.0062	1.06	2.80	.0026	.0124	.0052	.9948
96	2.0	.0228	0.56	1.48	.0694	3.0	.0013	1.56	4.13	0	.0241	.0694	.9306
97	1.5	.0668	0.06	0.16	.4364	3.5	.0002	2.06	5.45	0	.0670	.4761	.5239
98	1.0	.1587	-0.44	-1.16	.8770	4.0	0	2.56	6.77	0	.1587	.8770	.1230
99	0.5	.3085	-0.94	-2.49	.9936	4.5	0	3.06	8.10	0	.3085	.9936	.0064
100	0	.5000	-1.44	-3.81	.9999	5.0	0	3.56	9.42	0	.5000	.9999	.0001

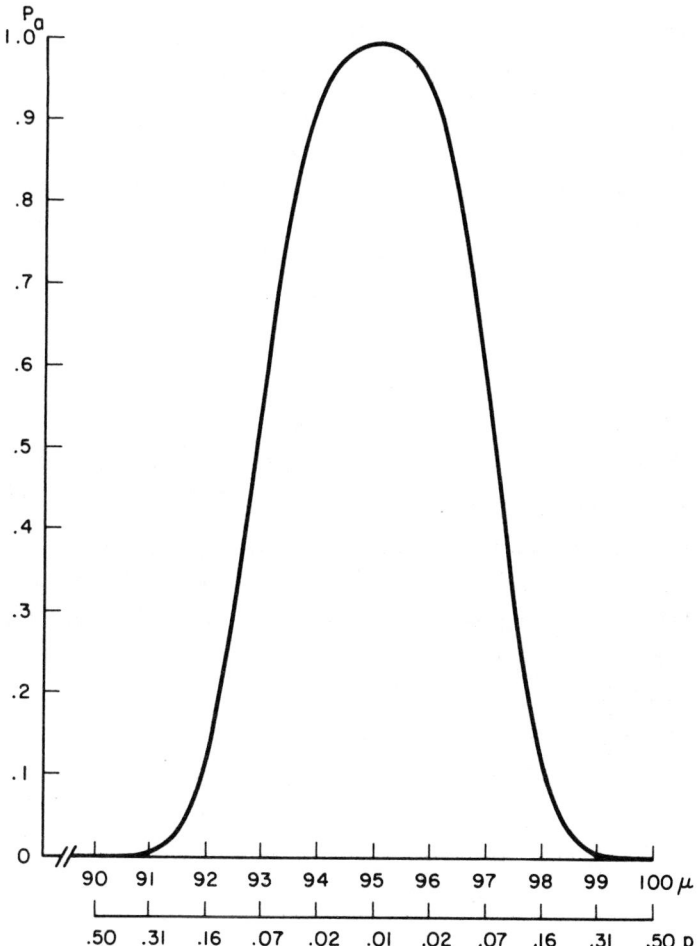

FIG. 10-11 OC curve: double specification limits, σ known.

The OC curve would appear as Fig. 10-11.

When the standard deviation is unknown complications in deriving the operating characteristic curve arise. The curve becomes a narrow band of possible values for probability of acceptance. A special procedure has been developed, however, utilizing a minimum variance unbiased estimate of the proportion nonconforming in the lot. Called the *M method* it is the only procedure recommended by MIL-STD-414 for use with double specification limits. The procedure and the corresponding operating characteristics will be discussed later in the chapter.

Measures of Performance

In addition to the probability of acceptance, there are other measures of performance of variables plans for proportion nonconforming, such as AOQ and ATI. Since these measures are functions based on the operating characteristic curve, the formulas for average outgoing quality and average total inspection remain the same as in attributes inspection as shown in Table 5-1,

$$AOQ \simeq pP_a$$

and

$$ATI = nP_a + N(1 - P_a)$$

Average outgoing quality limit (AOQL) must, however, be evaluated from the AOQ curve. A crude approach to finding the AOQL of a variables plan would be to use that of a matching attributes single-sampling plan. The match must necessarily be very good in the region of the AOQL. Because of the difference in the inherent shape of the OC curves of variables and attributes plans however such an approach would have to be regarded as only a very rough approximation.

As an example, consider the known standard deviation plan $n = 7$, $k = 1.44$. For $p = .18$, we have $P_a = .08$. Hence, for lots of size $N = 120$

$$AOQ \simeq pP_a = .18(.08) = .014$$

while

$$ATI = nP_a + N(1 - P_a)$$
$$= 7(.08) + 120(.92) = 110.96$$

Further, since the plan is matched to the attributes plan $n = 20$, $c = 1$, a crude measure of the AOQL would be that of the attributes plan $p_M = .036$. Actually, calculation of AOQ over the range of Table 10-4 would indicate that this is not far from the actual value.

M METHOD

Occasionally, it is desirable to base lot acceptance on estimates of the proportion nonconforming in the lot. This provides those administering the inspection with ancillary information which is meaningful to those not familiar with statistical methods. An estimate of this sort can be made under attributes inspection simply by dividing the number of nonconformances or defectives d found by the sample size n, to obtain an unbiased estimate

	Attributes	Variables
k Method	$d \leq c$	$z = \dfrac{U - \overline{X}}{\sigma} \geq k$
M Method	$p = \dfrac{d}{n} \leq \dfrac{c}{n} = M$	$Q = \dfrac{U - \overline{X}}{\sigma} \Rightarrow p \leq M$

FIG. 10-12 Equivalent criteria for acceptance (sample of n).
(Based on Schilling, 1974a, p. 19. By permission.)

of the proportion nonconforming in the lot. This estimate would then be
compared to the constant c/n to determine lot acceptance. When variables
procedures are employed more sophisticated methods of estimation must be
used. Such a procedure has been developed by Lieberman and Resnikoff (1955)
which involves use of a uniform minimum variance unbiased estimate of the
lot proportion nonconforming p in the acceptance sampling criteria. Using
$\overline{X}$ and s (or σ) a standard normal deviate Q is obtained which is then adjusted and employed to estimate p. A comparison of the equivalent k and
M methods for variables and attributes is shown in Fig. 10-12.

The procedure for application of the M method when the standard deviation is known is given in Table 10-6.

When the standard deviation is known, M may be found as

TABLE 10-6 M Method Standard Deviation Known (Given n,M)

LOWER SPECIFICATION LIMIT	UPPER SPECIFICATION LIMIT
1. Select a random sample of size n	1. Select a random sample of size n
2. Compute $\overline{X}$	2. Compute $\overline{X}$
3. Compute $Q_L = \dfrac{\overline{X}-L}{\sigma} \sqrt{\dfrac{n}{n-1}}$ and obtain $$\hat{P}_L = \int_{Q_L}^{\infty} \frac{1}{\sqrt{2\pi}} e^{-\frac{t^2}{2}} dt$$ from tables. This gives a minimum variance unbiased estimate of p.	3. Compute $Q_U = \dfrac{U-\overline{X}}{\sigma} \sqrt{\dfrac{n}{n-1}}$ and obtain $$\hat{P}_U = \int_{Q_U}^{\infty} \frac{1}{\sqrt{2\pi}} e^{-\frac{t^2}{2}} dt$$ from tables. This gives a minimum variance unbiased estimate of p.
4. If $\hat{P}_L \leq M$ accept $\hat{P}_L > M$ reject	4. If $\hat{P}_U \leq M$ accept $\hat{P}_U > M$ reject

$$M = \int_{k\sqrt{n/(n-1)}}^{\infty} \frac{1}{\sqrt{2\pi}} e^{-t^2/2} \, dt$$

so that M is simply the upper normal tail for $z_M = k\sqrt{n/(n-1)}$. Note also, that the estimated proportion nonconforming is simply the upper normal tail area corresponding to Q_U or Q_L.

When the standard deviation is unknown, the non-central t-distribution is involved in the estimation procedure, which leads to incorporation of values from the incomplete beta function in the estimate of p. The *incomplete beta function* is defined as

$$I_x(a,b) = \frac{\Gamma(a+b)}{\Gamma(a)\Gamma(b)} \int_0^x v^{a-1}(1-v)^{b-1} \, dv \qquad \begin{cases} 0 \leq x \leq 1 \\ a > 0 \\ b > 0 \end{cases}$$

and may be evaluated using special tables, such as those by Pearson (1968) or, in special cases by its relation to the binomial distribution

$$I_x(a,b) = 1 - F_{bin}(a - 1 | p = x, n = a + b - 1)$$

or

$$F_{bin}(y | p,n) = I_{x=p}(a = y + 1, b = n - y)$$

An example of the Pearson (1968) tables for the cumulative function $I_x(p,q)$ is shown in Fig. 10-13. For example, we have from the table

$$I_{.31}(14,14) = .0191640$$

When the standard deviation is unknown, the basic procedure for the M method is the same as that shown for σ known in Table 10-6 using s in the denominator of Q so that

$$Q_U = \frac{U - \overline{X}}{s} \qquad Q_L = \frac{\overline{X} - L}{s}$$

The estimation procedure then becomes

$$\hat{p}_U = I_x(a,b)$$

where

$$x = \max\left\{0, \frac{1}{2} - \frac{1}{2} Q_U \frac{\sqrt{n}}{n-1}\right\}$$

$$a = b = \frac{n}{2} - 1$$

and

	$p = 14$	$p = 15$	$p = 16$	$p = 17$	$p = 18$	$p = 19$
$B(p,q) =$	$\cdot3561\ 0481 \times \frac{1}{10^8}$	$\cdot1780\ 5241 \times \frac{1}{10^8}$	$\cdot9209\ 6072 \times \frac{1}{10^9}$	$\cdot4911\ 7905 \times \frac{1}{10^9}$	$\cdot2693\ 5625 \times \frac{1}{10^9}$	$\cdot1515\ 1289 \times \frac{1}{10^9}$
x						
·10	·0000 001					
·11	·0000 002					
·12	·0000 006	·0000 001				
·13	·0000 015⁻	·0000 004	·0000 001			
·14	·0000 036	·0000 009	·0000 002	·0000 001		
·15	·0000 083	·0000 023	·0000 006	·0000 002		
·16	·0000 179	·0000 053	·0000 015⁺	·0000 004	·0000 001	
·17	·0000 362	·0000 114	·0000 035⁻	·0000 010	·0000 003	·0000 001
·18	·0000 699	·0000 232	·0000 075⁻	·0000 023	·0000 007	·0000 002
·19	·0001 289	·0000 451	·0000 153	·0000 051	·0000 016	·0000 005⁺
·20	·0002 285⁺	·0000 840	·0000 300	·0000 105⁻	·0000 036	·0000 012
·21	·0003 905⁻	·0001 505⁺	·0000 565⁻	·0000 207	·0000 074	·0000 026
·22	·0006 454	·0002 603	·0001 022	·0000 391	·0000 146	·0000 054
·23	·0010 346	·0004 357	·0001 786	·0000 714	·0000 279	·0000 107
·24	·0016 129	·0007 079	·0003 024	·0001 261	·0000 514	·0000 205⁺
·25	·0024 500⁻	·0011 186	·0004 972	·0002 157	·0000 915⁻	·0000 380
·26	·0036 333	·0017 227	·0007 954	·0003 585⁻	·0001 580	·0000 682
·27	·0052 692	·0025 906	·0012 406	·0005 799	·0002 651	·0001 188
·28	·0074 840	·0038 098	·0018 895⁻	·0009 149	·0004 333	·0002 011
·29	·0104 244	·0054 872	·0028 145⁺	·0014 097	·0006 907	·0003 317
·30	·0142 565⁺	·0077 498	·0041 060	·0021 247	·0010 758	·0005 338
·31	·0191 640	·0107 453	·0058 737	·0031 364	·0016 390	·0008 395⁺
·32	·0253 448	·0146 415⁻	·0082 480	·0045 398	·0024 458	·0012 918
·33	·0330 071	·0196 246	·0113 810	·0064 503	·0035 789	·0019 470
·34	·0423 632	·0258 962	·0154 452	·0090 047	·0051 404	·0028 777
·35	·0536 230	·0336 688	·0206 321	·0123 619	·0072 539	·0041 748
·36	·0669 863	·0431 604	·0271 494	·0167 023	·0100 652	·0059 503
·37	·0826 346	·0545 876	·0352 165⁺	·0222 258	·0137 436	·0083 385⁺
·38	·1007 226	·0681 578	·0450 585⁻	·0291 489	·0184 800	·0114 979
·39	·1213 695⁻	·0840 603	·0568 991	·0376 996	·0244 858	·0156 106
·40	·1446 518	·1024 577	·0709 528	·0481 117	·0319 886	·0208 816
·41	·1705 958	·1234 768	·0874 151	·0606 167	·0412 272	·0275 361
·42	·1991 729	·1472 002	·1064 535⁻	·0754 351	·0524 450⁻	·0358 154
·43	·2302 954	·1736 584	·1281 977	·0927 668	·0658 810	·0459 706
·44	·2638 151	·2028 244	·1527 306	·1127 809	·0817 611	·0582 549
·45	·2995 240	·2346 087	·1800 799	·1356 048	·1002 864	·0729 146
·46	·3371 573	·2688 580	·2102 116	·1613 152	·1216 229	·0901 777
·47	·3763 986	·3053 548	·2430 256	·1899 290	·1458 901	·1102 430
·48	·4168 872	·3438 207	·2783 531	·2213 963	·1731 506	·1332 674
·49	·4582 276	·3839 219	·3159 570	·2555 957	·2034 010	·1593 544
·50	·5000 000⁶	·4252 770	·3555 356	·2923 324	·2365 648	·1885 428
·51	·5417 724	·4674 668	·3967 279	·3313 385⁺	·2724 881	·2207 979
·52	·5831 128	·5100 463	·4391 231	·3722 780	·3109 378	·2560 042
·53	·6236 014	·5525 576	·4822 716	·4147 531	·3516 034	·2939 618
·54	·6628 427	·5945 434	·5256 976	·4583 149	·3941 031	·3343 861
·55	·7004 760	·6355 607	·5689 143	·5024 763	·4379 922	·3769 115⁺
·56	·7361 849	·6751 941	·6114 385⁻	·5467 265⁻	·4827 758	·4210 989
·57	·7697 046	·7130 675⁺	·6528 057	·5905 478	·5279 236	·4664 475⁻
·58	·8008 271	·7488 543	·6925 850⁺	·6334 320	·5728 872	·5124 096
·59	·8294 042	·7822 851	·7303 914	·6748 975⁺	·6171 186	·5584 088
·60	·8553 482	·8131 542	·7658 968	·7145 044	·6600 889	·6038 596
·61	·8786 305⁺	·8413 213	·7988 385⁻	·7518 685⁻	·7013 066	·6481 886
·62	·8992 774	·8667 127	·8290 244	·7866 721	·7403 338	·6908 548
·63	·9173 654	·8893 184	·8563 352	·8186 725⁻	·7768 004	·7313 691
·64	·9330 137	·9091 878	·8807 239	·8477 057	·8104 145⁺	·7693 114
·65	·9463 770	·9264 229	·9022 118	·8736 881	·8409 698	·8043 435⁻
·66	·9576 368	·9411 698	·9208 825⁺	·8966 138	·8683 479	·8362 190
·67	·9669 929	·9536 104	·9368 734	·9165 484	·8925 170	·8647 875⁺
·68	·9746 552	·9639 519	·9503 658	·9336 209	·9135 271	·8899 950⁺
·69	·9808 360	·9724 173	·9615 740	·9480 132	·9315 008	·9118 786
·70	·9857 435⁻	·9792 367	·9707 346	·9599 475⁻	·9466 222	·9305 579

FIG. 10-13 Tables of the incomplete β function. (Reproduced with permission from E. S. Pearson, *Tables of the Incomplete Beta-Function*, 2nd ed., Cambridge University Press, London, 1968, p. 296.)

$$\hat{p}_L = I_x(a,b)$$

where

$$x = \max \left\{0, \frac{1}{2} - \frac{1}{2} Q_L \frac{\sqrt{n}}{n-1}\right\}$$

$$a = b = \frac{n}{2} - 1$$

For single specification limits, the M and k methods can be shown to be equivalent for a given sample size with

$$k = \frac{(n-1)(1-2B_M)}{\sqrt{n}}$$

$$B_M = \frac{1}{2}\left(1 - k\frac{\sqrt{n}}{n-1}\right)$$

where B_M is defined such that

$$I_{B_M}\left(\frac{n-2}{2}, \frac{n-2}{2}\right) = \frac{M}{100}$$

for M expressed in percent.

The M method is unique for double specification limits since the sum of the estimates of p above and below specification limits give the total estimated proportion nonconforming, that is,

$$\hat{p}_L + \hat{p}_U = \hat{p}$$

This total estimate $\hat{p}$ can be compared to M to determine acceptance of the lot. Under the uniform minimum variance unbiased estimation technique, borders for an acceptance polygon can be found by finding a k equivalent to M and using the formulas for the k method. The acceptance region itself will be found to be slightly different than that given by Wallis and is defined by the points $(\overline{x}', s')$ resulting from the simultaneous solution of

$$\overline{X} = \frac{U\left(1 - 2B_{p_0'}\right) - L\left(2B_{p_0''} - 1\right)}{2\left(1 - B_{p_0'} - B_{p_0''}\right)}$$

$$s = \frac{(U-L)}{2\left(1 - B_{p_0'} - B_{p_0''}\right)}\left(\frac{\sqrt{n}}{n-1}\right)$$

where

$$M = p^{**} = p_0' + p_0''$$

When the k method is to be used with double specification limits without the benefit of a polygon, a more refined estimate of the maximum standard deviation (MSD) may be obtained using the method of Lieberman and Resnikoff (1955). We have, for a given value of M,

$$MSD = \frac{\sqrt{n}}{2(1 - 2B_{M/2})(n - 1)}(U - L)$$

To obtain $B_{M/2}$, a value of x from the incomplete β distribution must be found such that

$$I_x\left(\frac{n - 2}{2}, \frac{n - 2}{2}\right) = \frac{M}{2}\left(\frac{1}{100}\right)$$

Then $B_{M/2} = x$ and the above formula for MSD can be evaluated. This is the method used in MIL-STD-414.

As an example to show the relation of the k and M methods, consider the plan n = 30, k = 2.00 which is listed in MIL-STD-414 as the Code J, 0.65 AQL normal plan. Using the relation given above, M is such that

$$B_M = \frac{1}{2}\left(1 - k\frac{\sqrt{n}}{n - 1}\right)$$
$$= \frac{1}{2}\left(1 - (2.00)\frac{\sqrt{30}}{29}\right) = .311$$

It is then necessary to evaluate the incomplete β distribution to obtain the value of M since

$$I_{B_M}\left(\frac{n - 2}{2}, \frac{n - 2}{2}\right) = \frac{M}{100}$$
$$I_{.311}(14,14) = .0197$$

by interpolation from Fig. 10-13. Using the binomial relation

$$I_{.311}(14,14) = 1 - F_{bin}(13|p = .311, n = 27)$$
$$= 1 - .9803 = .0197$$

by linear interpolation in the Department of Commerce (1950) binomial tables. So

$$\frac{M}{100} = .0197 \qquad M = 1.97$$

MIL-STD-414 shows M = 1.98. The difference is due to rounding.

Furthermore, the maximum standard deviation associated with this plan is

$$MSD = \frac{\sqrt{n}}{2(1 - 2B_{M/2})(n - 1)}(U - L)$$

For M = 1.98, it is necessary to find $I_x(14,14)$. By linear interpolation from Fig. 10-13, or using the binomial relation

$$I_{.288}(14,14) \simeq .0099$$

hence,

$$\text{MSD} = \frac{\sqrt{30}}{2(1 - 2(.288))(29)}(U - L) = .223(U - L)$$

and MIL-STD-414 gives this value for the MSD of this plan.

PLANS BASED ON SAMPLE RANGE

In practice it is often desirable to substitute the sample range for the sample standard deviation in applying variables plans for proportion nonconforming or process parameter. The range has several desirable properties and is used extensively in quality control. Among them are

1. Easy and inexpensive to compute
2. Quick to compute
3. Easy to explain
4. Easy to verify

The chief disadvantage of the range is the loss of efficiency resulting from its use. This can be compensated for by increasing the sample size. Use of the average range of m random subgroups of size n_R taken from the original sample also improves the efficiency somewhat. The d_2^* factor developed by Duncan (1955) can be used to estimate the standard deviation as

$$\hat{\sigma} = \frac{\overline{R}}{d_2^*}$$

This estimate has the same bias as s. Accordingly, the following procedure can be used if the average range is to be substituted for s in a given variables sampling plan which requires a sample size of n_s. Normality of the underlying observations is assumed.

1. Select the subgroup size to be used. (Subgroups of size 5 are often used, although 8 is considered to be optimum.)
2. Determine the equivalent sample size for the range plan. If the original plan using s had sample size n_s, the number of subgroups m of size n_R in the range plan will be approximately

$$m \simeq \frac{n_s - 1}{.9(n_R - 1)}$$

This value of m is conservative when rounded upward.

3. Substitute $\bar{R}/d_2^*$ for s in the original sampling plan, where

$$d_2^* \simeq d_2 \left(1 + \frac{0.2778}{m(n_R - 1)}\right)$$

and d_2 is the standard control chart factor for subgroups of size n_R.

4. Use the original sampling plan as given with the decision criteria, using the statistic and sample size as modified above.

The above approximation to Duncan's d_2^* factor from Schilling (1973) has been found to be quite sufficient for practical purposes. The formula for m uses Ott's (1955) approximation

$$\nu_R = .9m(n_R - 1)$$

for degrees of freedom of the range, ν_R. More accurate values for d_2^* and the degrees of freedom associated with the average range estimate of s as given by Nelson (1975) will be found in Appendix Table T10-1. The values of the constant difference (c.d.) found at the bottom of Nelson's table can be used to determine degrees of freedom for numbers of samples not listed as

$$\nu' = \nu + (c.d.)(k' - k)$$

where degrees of freedom ν' are required for k' samples of n, but only ν for k samples of n is listed. Thus, using k = 20, the degrees of freedom for 25 samples of 5 is

$$\nu' = 72.7 + 3.62(25 - 20) = 90.8 \sim 91.$$

Ott's approximation gives

$$\nu_R = .9(25)(5 - 1) = 90$$

EXAMPLE. Consider the following example taken from MIL-STD-414.

The specifications for electrical resistance of a certain electrical component is 650.0 ± 30 Ω. A lot of 100 items is submitted for inspection ... with AQL = 2.5 percent ... . Suppose the values of sample resistance in the order reading from left to right are as follows:

643, 651, 619, 627, 658, 670, 673, 641, 638, 650

For these data

$$\overline{X} = 647$$

$$s = \sqrt{\frac{\Sigma(X - \overline{X})^2}{n - 1}} = 17.22$$

Let us assume the standard deviation was unknown and it was desired to institute a plan based on average range. The first step in determining an average range plan would be to develop the appropriate plan for variability unknown using the sample standard deviation. For this example, take $p_2 = .215$. Then, for the standard deviation plan,

$p_1 = .025 \qquad \alpha = .05$
$p_2 = .215 \qquad \beta = .10$

so

$$k = \frac{z_{p_2} z_\alpha + z_{p_1} z_\beta}{z_\alpha + z_\beta} = \frac{0.79(1.645) + 1.96(1.282)}{1.645 + 1.282} = 1.30$$

$$n_s = \left(\frac{z_\alpha + z_\beta}{z_{p_1} - z_{p_2}}\right)^2 \left(1 + \frac{k^2}{2}\right) = \left(\frac{1.645 + 1.282}{1.96 - 0.79}\right)^2 \left(1 + \frac{1.30^2}{2}\right) = 11.5 \sim 12$$

and to find MSD

a. p^{**} corresponding to $z_p^* = k = 1.30$ is $p^{**} = .0968$
b. z_p^{**} corresponding to $p^{**}/2 = .0484$ is $z_p^{**} = 1.66$
c. so MSD $\simeq (U - L)/2z_p^{**} = (680 - 620)/(2(1.66)) = 18.07$

In applying this plan, for $s = 17.22$ and $\overline{X} = 647$ as indicated in the example,

1. $s = 17.22 \leq$ MSD $= 18.07$
2. $T_U = (U - \overline{X})/s = (680 - 647)/17.22 = 1.92 > k = 1.30$
3. $T_L = (\overline{X} - L)/s = (647 - 620)/17.22 = 1.57 > k = 1.30$

and since all three acceptance criteria are met, the lot would be accepted. To convert the above plan for use of the average range,

1. Use subgroup size $n_R = 5$ (arbitrarily).
2. Number of subgroups m:

$$m = \frac{n_s - 1}{.9(n_R - 1)} = \frac{10 - 1}{.9(5 - 1)} = \frac{9}{3.6} = 2.5 \sim 3$$

3. Substitute $\bar{R}/d_2^*$ for s in above acceptance criteria, where

$$d_2^* \simeq d_2\left(1 + \frac{.2778}{m(n_R - 1)}\right) = 2.326\left(1 + \frac{.2778}{3(5 - 1)}\right) = 2.38$$

Note that Nelson's table (Appendix Table T10-1) gives $d_2^* = 2.38$ with 11.1 degrees of freedom in contrast to the 9 degrees of freedom that would have been obtained under the standard deviation plan with five fewer observations.

The average range plan then requires a sample of three random subgroups of size 5 each. Assume the following values are obtained:

643	670	651
651	673	627
619	641	670
627	638	641
658	650	650

Then

$$\bar{X} = 647.27$$

$$\bar{R} = \frac{39 + 35 + 43}{3} = 39$$

so the acceptance criteria become

1. $\dfrac{\bar{R}}{d_2^*} \le MSD$

$\bar{R} \le d_2^* \; MSD = MAR$

$39 \le 2.38(18.07) = 43.0$

where MAR is the maximum allowable range which serves the same purpose as the maximum standard deviation.

2. $T_U = \dfrac{U - \bar{X}}{\bar{R}/d_2^*} = \dfrac{680 - 647.27}{39/2.38} = 2.00 > k = 1.30$

3. $T_L = \dfrac{\bar{X} - L}{\bar{R}/d_2^*} = \dfrac{647.27 - 620}{39/2.38} = 1.66 > k = 1.30$

and the lot is accepted.

Many production situations demand the simplicity and utility of the use of the range. Even hand calculators which can calculate s directly will not supplant the intuitive understanding and familiarity which operators and inspectors have for the range as a measure of spread. It is

important that users of acceptance sampling techniques have an understanding of the basic approach. The range can contribute much in this regard.

DOUBLE SAMPLING BY VARIABLES

As in the case of attributes sampling plans, savings in average sample size
can be achieved by using double-sampling plans. Indications are that a
reduction in average sample number to about 80 percent of a single-sampling
unknown standard deviation plan is possible for single sample sizes greater
than 25. Such plans are presented by Bowker and Goode (1952) and follow the
same procedure as in attributes double sampling with variables acceptance
criteria in place of the familiar first and second sample acceptance and rejection numbers.

A variables double-sampling plan is specified for the case of standard
deviation unknown by

n_1 = first sample size

n_2 = second sample size

k_a = first sample acceptance constant

k_r = first sample rejection constant

k_t = second sample acceptance constant

Primes are added to the acceptance constants to give k_a', k_r', k_t' for use
when the standard deviation is known.

To apply a double-sampling plan to a single specification limit,

1. Draw the first sample of n_1 and calculate $\overline{X}_1$ and s_1 from the data.
2. Compute

$$T_{U_1} = \frac{U - \overline{X}_1}{s_1}$$

for an upper specification limit, and

$$T_{L_1} = \frac{\overline{X}_1 - L}{s_1}$$

for a lower specification limit.

3. Test first sample

$$T_{U_1} \geq k_a \quad \text{or} \quad T_{L_1} \geq k_a \quad \text{accept}$$

$$T_{U_1} \leq k_r \quad \text{or} \quad T_{L_1} \leq k_r \quad \text{reject}$$

$$k_r < T_{U_1} < k_a \quad \text{or} \quad k_r < T_{L_1} < k_a \quad \text{resample}$$

4. Draw the second sample of n_2 from the lot. Combine the data with
    that from the first sample to obtain the total sample. Calculate

$$\bar{X}_t = \frac{\sum X_1 + \sum X_2}{n_1 + n_2}$$

$$s_t = \sqrt{\frac{(n_1 + n_2)\left(\sum X_1^2 + \sum X_2^2\right) - \left(\sum X_1 + \sum X_2\right)^2}{(n_1 + n_2)(n_1 + n_2 - 1)}}$$

5. Compute

$$T_{U_t} = \frac{U - \bar{X}_t}{s_t}$$

for an upper specification limit, and

$$T_{L_t} = \frac{\bar{X}_t - L}{s_t}$$

for a lower specification limit

6. Test combined total sample

$$T_{U_t} \geq k_t \quad \text{or} \quad T_{L_t} \geq k_t \quad \text{accept}$$

$$T_{U_t} < k_t \quad \text{or} \quad T_{L_t} < k_t \quad \text{reject}$$

When the standard deviation is known, the procedure is the same with
s_1 and s_t replaced by σ. In the case of double specification limits, test
against both specification limits as above rejecting if indicated by any
of the tests. In addition, calculate maximum standard deviation values
from k_r and k_t, respectively, using the method of Wallis to obtain MSD_1
and MSD_t. Reject if $s_1 > MSD_1$ or $s_t > MSD_t$.

Bowker and Goode (1952) have tabulated double-sampling plans for
standard deviation known and unknown. They also give information on the
operating characteristics and AOQL of the plans. Note that in their tabulation, Bowker and Goode define AQL to be the 95th percentile of the
operating characteristic curve.

The following is an example of application given by Bowker and Goode
(1952) in presenting their double-sampling plans. A manufacturer purchases
stud bolts which are to have a minimum tensile strength of 125,000 lbs.
per square inch. The plan

$$n_1 = 8 \qquad n_2 = 16$$

TABLE 10-7   Tensile Strength of Stud Bolts

Item	Ultimate strength, in pounds per square inch	Item	Ultimate strength, in pounds per square inch
First sample		Second sample	
1	129,500	9	129,500
2	131,000	10	131,000
3	128,500	11	129,500
4	126,500	12	128,000
5	130,000	13	129,000
6	130,500	14	127,000
7	127,500	15	132,500
8	129,000	16	130,500
		17	129,000
		18	130,000
		19	133,000
		20	129,000
		21	131,500
		22	130,000
		23	132,000
		24	128,500

Reproduced with permission from A. H. Bowker and H. P. Goode, *Sampling Inspection by Variables*, McGraw-Hill, New York, 1952, p. 95.

$$k_a = 3.041 \qquad k_r = 1.344 \qquad k_t = 2.245$$

is to be used with standard deviation unknown. The measurements for a first and second sample are shown in Table 10-7. Applying the plan

1. $\bar{X}_1 = 129062 \qquad s_1 = 1522$

2. $T_{L_1} = \dfrac{129062 - 125000}{1522} = 2.67$

3. $2.67 < k_a = 3.041 \qquad$ cannot accept

    $2.67 > k_r = 1.344 \qquad$ cannot reject

    Must resample.

4. $\bar{X}_t = 129688 \qquad s_t = 1647$

5. $T_{L_t} = \dfrac{129688 - 125000}{1647} = 2.85$

6. $2.85 > k_t = 2.245 \qquad$ accept

D. J. Sommers (1981) has obtained two-point double-sampling variables plans which provide minimum average sample number when the proportion non-conforming is at the producer's quality level. The plans are given in Appendix Table T10-2 and cover the values of p_1 and p_2 for $\alpha = .05$ and $\beta = .10$ which were tabulated by the Statistical Research Group (1947). The plans presented are for $n_1 = n_2$ and $k_t = k_r$; hence only n_1, k_a, and k_r are shown. Sample sizes are given as n_σ and n_s for known and unknown standard deviation, respectively. Average sample numbers are represented in a similar manner. For known standard deviation plans, $k_a' = k_a$ and $k_t' = k_r' = k_r$. Given these constraints, Sommers used an adaptation of the Wallis approximation together with an iterative procedure to minimize ASN at the specified p_1. Appendix Table T10-2 also presents a set of matched single-sampling plans for each p_1 and p_2. For example, when $p_1 = .01$, $p_2 = .05$, $\alpha = .05$, $\beta = .10$,

Known standard deviation: $n_1 = 13$, $n_2 = 13$, $k_a' = 2.09$, $k_r' = k_t' = 1.87$
Unknown standard deviation: $n_1 = 36$, $n_2 = 36$, $k_a = 2.09$, $k_r = k_t = 1.87$

The average sample numbers for these plans are 14.9 and 41.5 for known and unknown standard deviation, respectively. The matched single-sampling plans have sample size 19 and 54. This is indicative of the type of average sample size reduction possible through double-sampling variables plans.

TOLERANCE INTERVALS AND VARIABLES PLANS FOR PERCENT NONCONFORMING

The use of tolerance intervals in attributes sampling was discussed earlier. Tolerance intervals constructed from variables data can be used in a similar way as a medium for lot acceptance and reliability assessment.

The form in which reliability specifications are written, i.e., requirements of π reliability with γ confidence, fosters the use of tolerance intervals in this regard. This is natural since a tolerance interval guarantees that at least a stated proportion π of the population is contained within the limits of the interval with λ confidence. Therefore, a tolerance interval, constructed from the data, which is entirely within the specification limits shows the requirements have been met with the corresponding confidence. A tolerance interval which overlaps the specification limits is evidence that the requirements have not been met provided

a. The π and γ used are reasonably exact (not gross inequalities).
b. Sample size has been chosen to take the producer's risk into account.

Unless these conditions are checked, judgment should be withheld as to whether the lot should be rejected.

As shown earlier, specifications in terms of reliability can be converted into the usual quality control notation through the relations

$$\pi = 1 - p \qquad \gamma = 1 - P_a$$

Thus, the tolerance interval approach has found application in acceptance sampling as well as reliability.

We shall be concerned here with tolerance intervals on measurements. An underlying normal distribution is assumed. It should be pointed out that when all population parameters are known, a tolerance interval having $\gamma = 1$ can be obtained from the normal distribution itself. For example, a resistor having $\mu = 10 \ \Omega$ and $\sigma = 1 \ \Omega$ will have 95 percent of the population within

$$\mu \pm 1.96 \ \sigma$$

or

$$8.04 \text{ to } 11.96 \ \Omega$$

This tolerance interval is constructed with 100 percent confidence, so that if the specification limits are 8.0 to 12.0 Ω and a reliability of 95 percent is required, the product should be accepted. This is true even if less confidence, say $\gamma = 90$ percent was originally specified. Specification of a confidence value always means *at least* the stated amount for lot acceptance.

When population parameters are unknown, 100 percent confidence can rarely be achieved short of 100 percent inspection. Even then we cannot often be 100 percent confident of the inspection procedure. Estimates must be substituted for population parameters, confidence levels set, and more sophisticated procedures, often based on the non-central t-distribution, employed. When parameters are unknown, a typical one-sided variables tolerance interval is of the form

$$\overline{X} + ks$$

for an upper tolerance limit, or

$$\overline{X} - ks$$

for a lower tolerance limit. A two-sided interval may be expressed as

$$\overline{X} \pm ks$$

It will be recalled that the acceptance criteria for the k method in the variables procedure were precisely

$$\overline{X} + ks < U$$

or

$$\overline{X} - ks > L$$

for one-sided plans, with corresponding criteria for the two-sided case.

An extensive set of tables of tolerance limit factors and associated criteria has been published by Robert E. Odeh and Donald B. Owen (1980). The resulting tabulation is useful in acceptance sampling and relaibility applications. The contents of the tables is described in Table 10-8.

Odeh-Owen Tables 1, 3, and 7 are primarily intended for tolerance and confidence interval estimation. Their Table 7 provides confidence limits for the tail areas of the normal distribution using the procedure of Owen and Hua (1977). The Odeh-Owen Tables 8, 9, and 10 are useful in implementing a screening strategy in which the proportion of variable Y above a lower specification limit L is improved by screening on a related variable X. By selecting out a proportion β of the population in which $X > \mu_x - z_\beta \sigma_x$ the proportion of Y above L is raised from γ to δ. Of course, the effectiveness of the procedure depends on the strength of the correlation ρ between X and Y. However, it presents a useful alternative for screening when tests are destructive as, for example, in life tests. This type of screening strategy is described in Owen, McIntire, and Seymore (1975).

Tables 2, 5, and 6 are intended to be used directly in acceptance sampling. Table 2 presents one-sided k factors and sample sizes for specified α, β, p_1, p_2. Table 5, reproduced here as Appendix Table T10-4, shows two-sided equal-tailed k values for specified $P = p_2$ and $\beta = .10$. To use the table with specified producer's risk (say at $\alpha = .05$), it is necessary to use the Wallis formula to determine sample size and then improve the approximation by selecting k from the Odeh-Owen table.

For example, suppose a plan is desired having $p_1 = .005$, $p_2 = .10$, $\alpha = .05$, $\beta = .10$. It is to be used to check the length of Kovar leads which are specified to be 9 cm $\pm$ .05 mm. Using the Wallis formula

$$k = \frac{z_{\alpha/2} z_2 + z_\beta z_1}{z_{\alpha/2} + z_\beta}$$

$$= \frac{1.96(1.28) + 1.28(2.58)}{1.96 + 1.28} = 1.79$$

TABLE 10-8 Content of Odeh-Owen Tables of Tolerance Limits

TABLE	CONTENT	APPLICATION
1	Factors for one sided tolerance limits (k by γ, n, P = π)	o One sided variables plans with one risk specified o One sided tolerance intervals for reliability estimation
2	Sample size for one sided sampling plans (n, k by α, β, P_1, P_2)	o One sided variables plans with both risks specified
3	Two sided (central) tolerance limits to control both tails equally (k by γ, n, P = π)	o Two sided tolerance intervals for estimation with equal tails
4	Two sided (non-central) tolerance limits to control tails separately (k by γ, n, P = π)	o Two sided tolerance intervals for estimation with non-equal tails
5	Two sided sampling plan factors to control equal tails (k by γ = .90, n, p)	o Two sided equal tails variables plans with β = .10 (only) specified o May be used with two risks by approximating n with Wallis formula (use $\alpha/2$). Then find k from Table 5

6	Two sided sampling plan factors to control tails separately (k by γ = .90, n, p)	o Two sided unequal tails variables plans with β = .10 (only) specified o May be used with two risks by approximating n with Wallis formula (use α/2). Then find k_L and k_U for lower and upper limits separately from Table 6
7	Confidence limits for proportion in tail of normal distribution	o Lower confidence limit on proportion of product above lower specification limit. Shows proportion above L tabulated by confidence = η, n, and K = $(\bar{X} - L)/s$
8	Screening proportion - population parameters known (β by δ, γ, ρ)	o Proportion measurement Y above L is γ. Y to be screened on concomitant variable X. μ_X, μ_Y, σ_X, σ_Y, ρ known. Proportion Y above L may be raised from γ to δ by selecting all X above $\mu_X - z_\beta \sigma_X$ where β is proportion of population which will be selected
9	Screening proportion - population parameters unknown (β by f, δ, γ, ρ)	o Same as Table 8 for γ, δ, ρ known and $\bar{X}$, s_x used as estimates from preliminary sample with f degrees of freedom
10	Confidence limits on the correlation coefficient	o Confidence limits for ρ. Shows upper, lower, and two sided confidence limits given n, sample R = r, and risk α o Used with Table 9 when ρ is unknown

$$n = \left(\frac{z_{\alpha/2} + z_{\beta}}{z_1 - z_2}\right)^2 \left(1 + \frac{k^2}{2}\right)$$

$$= \left(\frac{1.96 + 1.28}{2.58 - 1.28}\right)^2 \left(1 + \frac{1.79^2}{2}\right)$$

$$= 16.2 \sim 17$$

The Odeh-Owen Table 5 shows that for $p_2 = .10$ and sample size 17, a more accurate value of k would be 1.95. This will hold the consumer's quality level at .10. The sampling plan is

    Sample size: n = 17
    Accept if: $8.95 \leq \overline{X} - 1.95$ s   and $\overline{X} + 1.95s \leq 9.05$
    Reject if: $\overline{X} - 1.95$ s < 8.95 or $9.05 < \overline{X} + 1.95s$

Sometimes the consumer's quality level is specified to be different for the lower and upper tails. In this case the Odeh-Owen Table 6 reproduced here as Appendix Table T10-5 is used for specified p_L and p_U in the lower and upper tails respectively; here also $\beta = .10$. More extensive values are given in Odeh-Owen Table 1. Again the Wallis formula is used to obtain the approximate sample size associated with a specified producer's risk α.

Suppose longer leads could be tolerated by the process better than shorter leads in the above example so that the consumer's quality level was broken into two parts, $p_L = .025$ against the lower specification and $p_U = .05$ against the upper specification, still with the producer's quality level $p_1 = .005$ with risks $\alpha = .05$ and $\beta = .10$. Using the Wallis formula with z_L and z_U corresponding to p_L and p_U, k and n are calculated for the lower and upper specification limits shown in Table 10-9.

Since only one sample size can be taken, it will be necessary to take a sample size of approximately 94. This might be rounded to 100 for administrative purposes. The Odeh-Owen Table 6 shows for n = 100; $k_L = 2.203$ and $k_U = 1.861$. The sampling plan is

    Sample size: n = 100
    Accept if: $8.95 \leq \overline{X} - 2.20$ s   and $\overline{X} + 1.86$ s ≤ 9.05
    Reject if: $\overline{X} - 2.20$ s < 8.95 or $9.05 < \overline{X} + 1.86$ s

This procedure may be used as a substitute for the M method and is somewhat simpler for inspectors to understand and use. Of course, the Odeh-Owen Table 2 can be used for one-sided specification limits.

The theory of the use of tolerance limits in sampling inspection has

TABLE 10-9   Calculation of k and n for Lower and Upper
Specification Limits

Lower specification limit	Upper specification limit

$$k = \frac{z_{\alpha/2} z_L + z_\beta z_1}{z_{\alpha/2} + z_\beta}$$ $$k = \frac{z_{\alpha/2} z_U + z_\beta z_1}{z_{\alpha/2} + z_\beta}$$

$$= \frac{1.96(1.96) + 1.28(2.58)}{1.96 + 1.28}$$ $$= \frac{1.96(1.64) + 1.28(2.58)}{1.96 + 1.28}$$

$$= 2.20$$ $$= 2.01$$

$$n = \left(1 + \frac{k^2}{2}\right)\left(\frac{z_{\alpha/2} + z_\beta}{z_1 - z_L}\right)^2$$ $$n = \left(1 + \frac{k^2}{2}\right)\left(\frac{z_{\alpha/2} + z_\beta}{z_1 - z_U}\right)^2$$

$$= \left(1 + \frac{2.2^2}{2}\right)\left(\frac{1.96 + 1.28}{2.58 - 1.96}\right)^2$$ $$= \left(1 + \frac{2.01^2}{2}\right)\left(\frac{1.96 + 1.28}{2.58 - 1.64}\right)^2$$

$$= 93.4$$ $$= 35.9$$

been described in detail by Owen (1964, 1967), Owen and Frawley (1971), and
Owen, Rao, and Subrahamaniam (1972). Earlier tables of tolerance factors
in the quality control literature include Lieberman (1958) and Zobel (1958).
Tables of tolerance limits based on the range have been given by Bingham
(1962) and Owen, Frawley, Kapadia, and Rao (1971). L. S. Nelson (1977) has
discussed tolerance intervals in which the mean and standard deviation are
estimated by separate samples.

SEQUENTIAL PLANS FOR PROPORTION NONCONFORMING

When the standard deviation is known, a simplified relationship between
sequential plans for the mean and single sampling variables plans, given by
Sommers (1979), can be used to easily obtain sequential variables plans
for proportion nonconforming with $\alpha = .05$ and $\beta = .10$ for specified p_1 and
p_2. Once a known standard deviation variables plan for proportion nonconforming has been obtained to these specifications, its parameters n_σ, k can
be converted to those of a sequential plan for an upper specification limit
by using the relations*:

$$h_1 = .7693\ \sigma\sqrt{n_\sigma}$$

$$h_2 = .9877\ \sigma\sqrt{n_\sigma}$$

*The formula for s is obtained from that given by Sommers (1979) by using
the relation $\mu_1 = U - z_1\sigma = U - (k + z_\alpha/\sqrt{n})\sigma$.

$$s = U - k\sigma - .1818 \frac{\sigma}{\sqrt{n}}$$

with

$$\text{ASN}(p_1) = .4657\, n_\sigma \qquad \text{ASN}(p_2) = .5549\, n_\sigma$$

Thus, the sequential plan matching the known standard deviation plan $n_\sigma = 7$, $k = 1.44$ has parameters

$$h_1 = 2.04\, \sigma \qquad h_2 = 2.61\, \sigma \qquad s = U - 1.51\, \sigma$$

with

$$\text{ASN}(p_1 = .018) = 3.26 \qquad \text{ASN}(p_2 = .180) = 3.88$$

This approach may be applied to a lower specification limit using the procedures outlined in Chap. 8, where s can be calculated from the lower specification limit as

$$s = L + k\sigma + .1818 \frac{\sigma}{\sqrt{n}}$$

FURTHER CONSIDERATIONS

Derivation of n, k Formulas

When the standard deviation is known, the formulas which give n and k as a function of fixed p_1, p_2, α, and β are easily derived. Figure 10-14 will be used as motivation for the algebra involved. It supposes an upper specification limit U and is based on the $\overline{X}$ method.

To obtain n, if we regard the z values as representing upper tail probability points

$$U - A = k\sigma$$
$$= \mu_1 + z_1\sigma - (\mu_1 + z_\alpha \sigma_{\overline{X}}) \tag{1}$$
$$= \mu_2 + z_2\sigma - (\mu_2 - z_\beta \sigma_{\overline{X}}) \tag{2}$$

Since (1) and (2) equal each other,

$$z_1\sigma - z_\alpha \frac{\sigma}{\sqrt{n}} = z_2\sigma + z_\beta \frac{\sigma}{\sqrt{n}}$$

$$(z_\alpha + z_\beta) \frac{1}{\sqrt{n}} = (z_1 - z_2)$$

$$n = \left(\frac{z_\alpha + z_\beta}{z_1 - z_2} \right)^2$$

To obtain k,

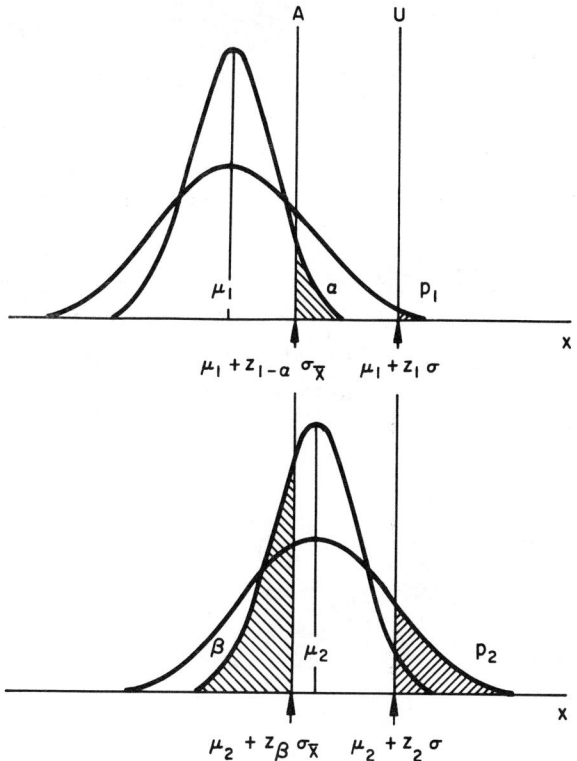

FIG. 10-14   Derivation of variables plan.

$$k\sigma = -\mu_1 + z_1\sigma - \left(\mu_1 + z_\alpha \frac{\sigma}{\sqrt{n}}\right)$$

$$k = z_1 - \frac{z_\alpha}{\sqrt{n}} \quad \text{or} \quad k = z_2 + \frac{z_\beta}{\sqrt{n}}$$

So carrying further

$$\frac{k\sqrt{n}}{z_\alpha} = \frac{z_1\sqrt{n}}{z_\alpha} - 1$$

and

$$\frac{k\sqrt{n}}{z_\beta} = \frac{z_2\sqrt{n}}{z_\beta} + 1$$

so, adding

$$\frac{k\sqrt{n}}{z_\alpha} + \frac{k\sqrt{n}}{z_\beta} = \frac{z_1\sqrt{n}}{z_\alpha} + \frac{z_2\sqrt{n}}{z_\beta}$$

$$k\left(\frac{z_\alpha + z_\beta}{z_\alpha z_\beta}\right) = \frac{z_1 z_\beta + z_2 z_\alpha}{z_\alpha z_\beta}$$

$$k = \frac{z_1 z_\beta + z_2 z_\alpha}{z_\alpha + z_\beta}$$

The formulas may be developed equally well using a lower specification limit.

The derivation of the formulas when the standard deviation is not known is more complicated and is given by Wallis (1947) who developed the formulas. An excellent discussion of approximations of the type proposed by Wallis, their application, and their efficiency in relation to indifference quality has been given by Hamaker (1979).

Need for Normality

The plans discussed in this chapter are based on the assumption of normality of the underlying measurements. While Jennett and Welch (1939) have argued that they may give roughly correct results when applied to distributions other than normal, extreme care must be taken in such circumstances. Certainly when dealing with small values of proportion nonconforming, i.e., the tails of the distribution, the validity of the assumption of normality is critical.

Plans may be derived for underlying distributions other than normal. D. B. Owen (1969) has presented an excellent paper in this regard.

PROBLEMS

1. Given the following specifications

   $p_1 = .006$ $1 - \alpha = .95$
   $p_2 = .057$ $\beta = .10$

   derive the following variables plans for percent nonconforming

   a. $\overline{X}$ method - σ known, $\sigma = .08$
   b. k method - σ unknown

   by formula and from the Jacobson nomograph.

2. The upper specification limit for the response time for a certain emergency warning device is 7 sec. They are known to be normally distributed. A sample of 10 such devices yields the following results: 6.82, 6.85, 6.70, 6.73, 6.89, 6.95, 6.96, 6.80, 6.79, 6.85. Apply the

plan obtained in Prob. 1 to these data.

3. Convert the results of Prob. 1a to the M method and apply the plan to the results of Prob. 2.

4. Determine the value of M to be used with Prob. 1b. What is the estimated proportion nonconforming if $U = 1000$, $\bar{X} = 973$, $s = 9$, $n = 30$? Should the lot be accepted?

5. Convert the plan of Prob. 1b to a plan using the range with subgroups of 5.

6. Maximum and minimum specifications for hardness of a certain material are 69 and 60 when read on a certain scale. Construct an unknown standard deviation plan to the following specifications:

$p_1 = .01$ $1 - \alpha = .95$

$p_2 = .14$ $\beta = .10$

7. If the results of sampling a lot from Prob. 6 yielded 63.0, 64.5, 64.0, 62.5, 63.0, 70.0, 71.0, 61.0, 60.0, 67.5, 66.5, 64.0, 68.0, should the lot be accepted?

8. Using the tolerance interval approach, assess the results of Prob. 7.

9. Would the double-sampling plan $n_1 = 10$, $n_2 = 10$, $k_a = 2.51$, $k_r = 1.58$, $k_t = 2.05$ lead to a second sample on the basis of Prob. 2 results?

10. Draw the OC curve for the sampling plan given in Prob. 1a. What is the indifference quality level?

REFERENCES

Bingham, R. S. (1962), Tolerance Limits and Process Capability Studies, *Industrial Quality Control*, 19(1): 36-39.

Bowker, A. H., and H. P. Goode (1952), *Sampling Inspection by Variables*, McGraw-Hill, New York.

Duncan, A. J. (1955), The Use of Ranges in Comparing Variabilities, *Industrial Quality Control*, 11(5): 18-22.

Duncan, A. J. (1974), *Quality Control and Industrial Statistics*, 4th ed., Richard D. Irwin, Homewood, Illinois.

Hamaker, H. C. (1979), Acceptance Sampling for Percent Defective by Variables and by Attributes, *Journal of Quality Technology*, 11(3): 139-148.

Hillier, F. S. (1969), $\bar{X}$- and R-Chart Control Limits Based on Small Number of Subgroups, *Journal of Quality Technology*, 1(1): 17-26.

Jacobson, L. J. (1949), Nomograph for Determination of Variables Inspection Plan for Fraction Defective, *Industrial Quality Control*, 6(3): 23-25.

Jennett, W. J., and B. L. Welch (1939), The Control of Proportion Defective as Judged by a Single Quality Characteristic Varying on a Continuous Scale, *Supplement to the Journal of the Royal Statistical Society* (Series B), 6: 80-88.

Juran, J. M., ed. (1974), *Quality Control Handbook*, 3rd ed., McGraw-Hill, New York.

Larson, H. R. (1966), A Nomograph of the Cumulative Binomial Distribution, *Industrial Quality Control*, 23(6): 270-278.

Lieberman, G. J. (1958), Tables for One-Sided Statistical Tolerance Limits, *Industrial Quality Control*, 14(10): 7-9.

Lieberman, G. J., and G. J. Resnikoff (1955), Sampling Plans for Inspection by Variables, *Journal of the American Statistical Association*, 50: 457-516.

Nelson, L. S. (1975), Use of the Range to Estimate Variability, *Journal of Quality Technology*, 7(1): 46-48.

Nelson, L. S. (1977), Tolerance Factors for Normal Distributions, *Journal of Quality Technology*, 9(4): 198-199.

Odeh, E., and D. B. Owen (1980), *Tables for Normal Tolerance Limits, Sampling Plans, and Screening*, Marcel Dekker, New York.

Ott, E. R. (1967), Analysis of Means — A Graphical Procedure, *Industrial Quality Control*, 24(2): 101-109.

Owen, D. B. (1964), Control of Percentages in Both Tails of the Normal Distribution, *Technometrics*, 6(4): 377-387 [Errata, 8(3): 570 (1966)].

Owen, D. B. (1967), Variables Sampling Plans Based on the Normal Distribution, *Technometrics*, 9(3): 417-423.

Owen, D. B. (1969), Summary of Recent Work on Variables Acceptance Sampling with Emphasis on Non-Normality, *Technometrics*, 11(4): 631-637.

Owen, D. B., and W. H. Frawley (1971), Factors for Tolerance Limits Which Control Both Tails of the Normal Distribution, *Journal of Quality Technology*, 3(2): 69-79.

Owen, D. B., W. H. Frawley, C. H. Kapadia, and J. N. K. Rao (1971), Tolerance Limits Based on Range and Mean Range, *Technometrics*, 13(3): 651-656.

Owen, D. B., and T. A. Hua (1977), Tables of Confidence Limits on the Tail Area of a Normal Distribution, *Communications in Statistics*, B6: 285-311.

Owen, D. B., D. D. McIntire, and E. Seymore (1975), Tables Using One or Two Screening Variables to Increase Acceptable Product Under One-Sided Specifications, *Journal of Quality Technology*, 7(3): 127-138.

Owen, D. B., J. N. K. Rao, and K. Subrahamaniam (1972), Effect of Non-Normality on Tolerance Limits Which Control Percentages in Both Tails of the Normal Distribution, *Technometrics*, 14(3): 571-575.

Pearson, E. S. (1968), *Tables of the Incomplete Beta-Function*, 2nd ed., Cambridge University Press, London.

Resnikoff, G. J., and G. J. Lieberman (1957), *Tables of the Non-Central t-Distribution*, Stanford University Press, Stanford, California.

Romig, H. G. (1939), *Allowable Average in Sampling Inspection*, Ph.D. Dissertation submitted to Columbia University, New York.

Schilling, E. G. (1973), A Systematic Approach to Analysis of Means, Part I. Analysis of Treatment Effects, *Journal of Quality Technology*, 5(3): 93-108.

Schilling, E. G. (1974), Sampling by Variables, *Quality Control Handbook* (J. M. Juran, ed.), 3rd ed., Section 25, McGraw-Hill, New York, pp. 25.1-25.41.

Schilling, E. G. (1974a), Variables Sampling and MIL-STD-414, *Quality Progress*, 7(5): 16-20.

Schilling, E. G., and D. J. Sommers (1981), Two-Point Optimal Narrow Limit Plans with Applications to MIL-STD-105D, *Journal of Quality Technology*, 13(2): 83-92.

Sommers, D. J. (1979), Personal communication with the author.

Sommers, D. J. (1981), Two-Point Double Variables Sampling Plans, *Journal of Quality Technology*, 13(1): 25-30.

Statistical Research Group, Columbia University (1947), *Techniques of Statistical Analysis*, McGraw-Hill, New York.

United States Department of Commerce (1950), *Tables of the Binomial Probability Distribution*, National Bureau of Standards, Applied Statistics Series No. 6, U.S. Government Printing Office, Washington, D.C.

United States Department of Defense (1957), *Military Standard, Sampling*

Procedures and Tables for Inspection by Variables for Percent Defective (MIL-STD-414), U.S. Government Printing Office, Washington, D.C.

Wallis, W. A. (1947), Use of Variables in Acceptance Inspection for Percent Defective, *Techniques of Statistical Analysis*, (C. Eisenhart, M. Hastay, and W. A. Wallis, eds.), Statistical Research Group, Columbia University, McGraw-Hill, New York, Chap. 1, pp. 3-93.

Wallis, W. A. (1950), Lot Quality Measured by Proportion Defective, *Acceptance Sampling — A Symposium*, American Statistical Association, Washington, D.C., pp. 117-122.

Zobel, S. P. (1958), One-Sided Statistical Tolerance Limits, *Industrial Quality Control*, 15(4): 35.

ATTRIBUTES SAMPLING SCHEMES : MIL-STD-105D

Sampling plans are frequently used in consort to produce levels of protection not attainable by any of the component plans individually. Such combinations of plans are called *sampling schemes* or *sampling systems*. I. D. Hill (1962, p. 31) has defined a *sampling plan* as "the specification of the rules to be followed in sentencing any particular batch of articles," while sampling schemes involve procedures for the use of a set of specific plans. According to Hill (1962, p. 31), a *sampling scheme* is "an over-all strategy specifying the way in which sampling plans are to be used." H. F. Dodge (1963) has used the term *system* in a manner analogous to the word *scheme* to achieve a somewhat less sinister description of the procedure. Stephens and Larson (1967, p. 310) viewed "a *sampling system* as an assigned grouping of two or three sampling plans and the rules for using (i.e., switching between) these plans for sentencing lots or batches of articles to achieve a blending of the advantageous features of each of the sampling plans." Sampling plans are the basic elements of sampling schemes, while sampling systems may be considered to involve a grouping of one or more sampling schemes.

Attributes sampling schemes include the tables of AOQL plans prepared by H. F. Dodge and H. G. Romig (1959), which result in a stated AOQL with minimum total inspection when used as directed with 100 percent inspection. These will be discussed later. Many schemes, however, are included in AQL systems. AQL refers to the acceptable quality level, i.e., what has been called the producer's quality level for a single plan. These systems are intended to be applied to a stream of lots. Such plans specify an upper limit on quality, the AQL, not to be exceeded by the producer without the penalty of an excessive number of rejected lots. That is, for levels of

quality less than the AQL, rejections will be relatively infrequent, say
less than 1 in 10, while for levels of quality in excess of the AQL, rejections will be more frequent, say more than 1 in 10. This is achieved by
switching back and forth between the plans included in the system. Tighter
plans are used when quality levels are shown to be poor, while looser plans
involving smaller sample sizes are utilized when quality is shown to be
good. Over a continuing supply, schemes can be devised to incorporate the
best properties of the plans included as elements. Frequently, schemes
are selected within a system in relation to the lot size involved.

Military Standard 105D (p. 3) defines the AQL by the following:

> 4.2 DEFINITION. The AQL is the maximum percent defective
> (or the maximum number of defects per hundred units) that,
> for purposes of sampling inspection, can be considered
> satisfactory as a process average ...
>
> 4.3 NOTE ON THE MEANING OF AQL. When a consumer des-
> ignates some specific value of AQL for a certain defect
> or group of defects, he indicates to the supplier that
> his (the consumer's) acceptance sampling plan will accept
> the great majority of the lots or batches that the sup-
> plier submits, provided the process average level of
> percent defective (or defects per hundred units) in
> these lots or batches be no greater than the designated
> value of AQL. Thus, the AQL is a designated value of
> percent defective (or defects per hundred units) that
> the consumer indicates will be accepted most of the time
> by the acceptance sampling procedure to be used. The
> sampling plans provided herein are so arranged that the
> probability of acceptance at the designated AQL value
> depends upon the sample size, being generally higher for
> large samples than for small ones, for a given AQL. The
> AQL alone does not describe the protection to the con-
> sumer for individual lots or batches but more directly
> relates to what might be expected from a series of lots
> or batches, provided the steps indicated in this publi-
> cation are taken. It is necessary to refer to the
> operating characteristic curve of the plan, to determine
> what protection the consumer will have.

It also contains the following limitation:

> 4.4 LIMITATION. The designation of an AQL shall not
> imply that the supplier has the right to supply knowingly
> any defective unit of product.

Military Standard 105D (1963) is not a sampling plan. It is a sampling
system. As such, it combines several individual sampling plans in schemes
constructed to employ economic, psychological and operational means to motivate the producer to sustain acceptable quality levels. The procedure for
switching between plans is essential to the system; it is so designed as to
exert pressure on the producer to take corrective action when quality falls
below prescribed levels and to provide rewards, in terms of reduced sample
size, for quality improvement.

The standard ties together sets of three attributes sampling plans,
each at a different level of severity, into a unified procedure for lot
acceptance through the use of its switching rules. These action rules determine the level of severity to be employed depending on the level of
quality previously submitted. Thus, inspection of a succession of lots is
intended to move among the specified set of tightened, normal and reduced
sampling plans as quality levels degenerate or improve. Switching between
tightened and normal plans is made mandatory by the standard, while use of
reduced plans is optional.

The MIL-STD-105D system, as such, does not allow for application of
individual plans without use of the switching rules, since such an approach
can lead to a serious loss of protection from that achieved when the system
is properly applied. Quality levels are specified in terms of acceptable
quality level (AQL) for the producer, while consumer protection is afforded
by the switching rules which lead to tighter plans when quality is poor.
The operation of MIL-STD-105D has been described in detail by Hahn and
Schilling (1975), and is the subject of several military and international
handbooks including the classic MIL-HDBK-53 (1965).

When an isolated lot is to be inspected, special tables of limiting
quality are presented in the standard. In such instances, MIL-STD-105D
merely represents a convenient collection of individual plans indexed by
the limiting quality table. In no sense, however, is this the use for
which the MIL-STD-105D system was designed.

Unfortunately, the standard may sometimes be misused, particularly in
nonmilitary applications, through the selection and use of normal plans

only — disregarding the tightened and reduced plans, and the switching rules. This deprives the consumer of the protection provided by the tightened plan when quality is poor, and it foregoes the advantage to the producer of smaller sample sizes and slightly increased protection afforded by the reduced plan when quality is good.

The operation of MIL-STD-105D is straightforward. Lot sizes are linked to sample size by a system of code letters. Matched sets of single, double, and multiple plans provide a complete choice among these types of plans in application. The average sample size of double and multiple plans can be arrived at from average sample number curves which are given. MIL-STD-105D also contains tables presenting the AOQL resulting from the use of its normal plans together with 100 percent inspection of rejected lots. Complete sets of OC curves and probability points of the normal and tightened plans are contained in the standard.

The standard is written in terms of inspection for defectives (expressed in percent defective) and also for defects (expressed in defects per 100 units). The approach and operation of the scheme is the same for both and so they will be used interchangeably here for economy of presentation. Their measures of performance, however, are based on different probability distributions (binomial and Poisson) and so they must be addressed separately where operating characteristics and other measures are concerned.

The structure of MIL-STD-105D is shown in Fig. 11-1.

OPERATION

Proper use of the MIL-STD-105D sampling system demands close attention and adherence to the rules for switching among the sets of three plans (tightened, normal, and reduced) which are presented. In so doing the producer receives adequate protection against excessive rejections when quality is better than the AQL, while the consumer receives increased protection when quality is running worse than the AQL. The operation of the switching rules is shown in Fig. 11-2.

A MIL-STD-105D scheme always starts with the normal inspection plan. The plan continues to be used until sufficient evidence is generated to indicate that a switch to the tightened or the reduced plan is appropriate. Note that MIL-STD-105D makes use of the reduced plan optional, although for full economic benefit of the procedure, it should be utilized where possible.

A switch to tightened inspection involves moving to the acceptance

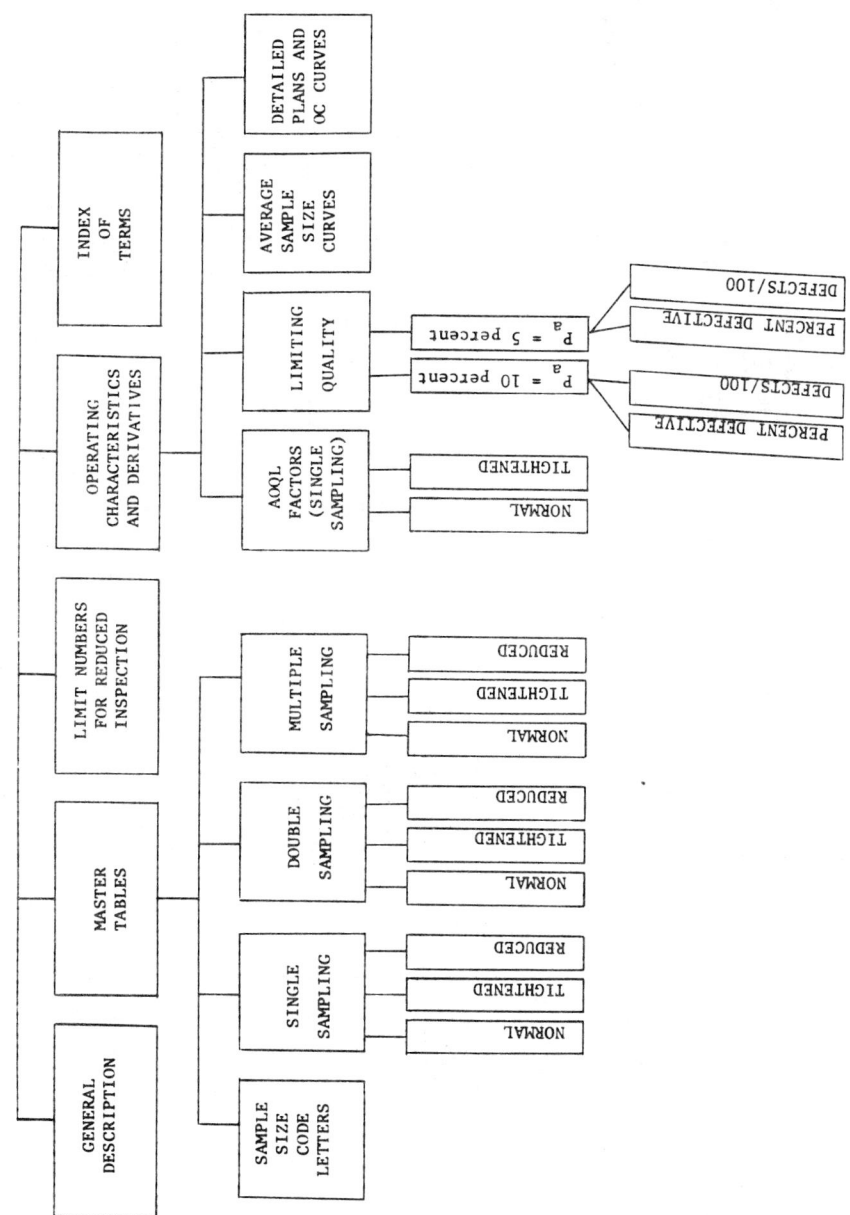

FIG. 11-1   Structure of MIL-STD-105D.

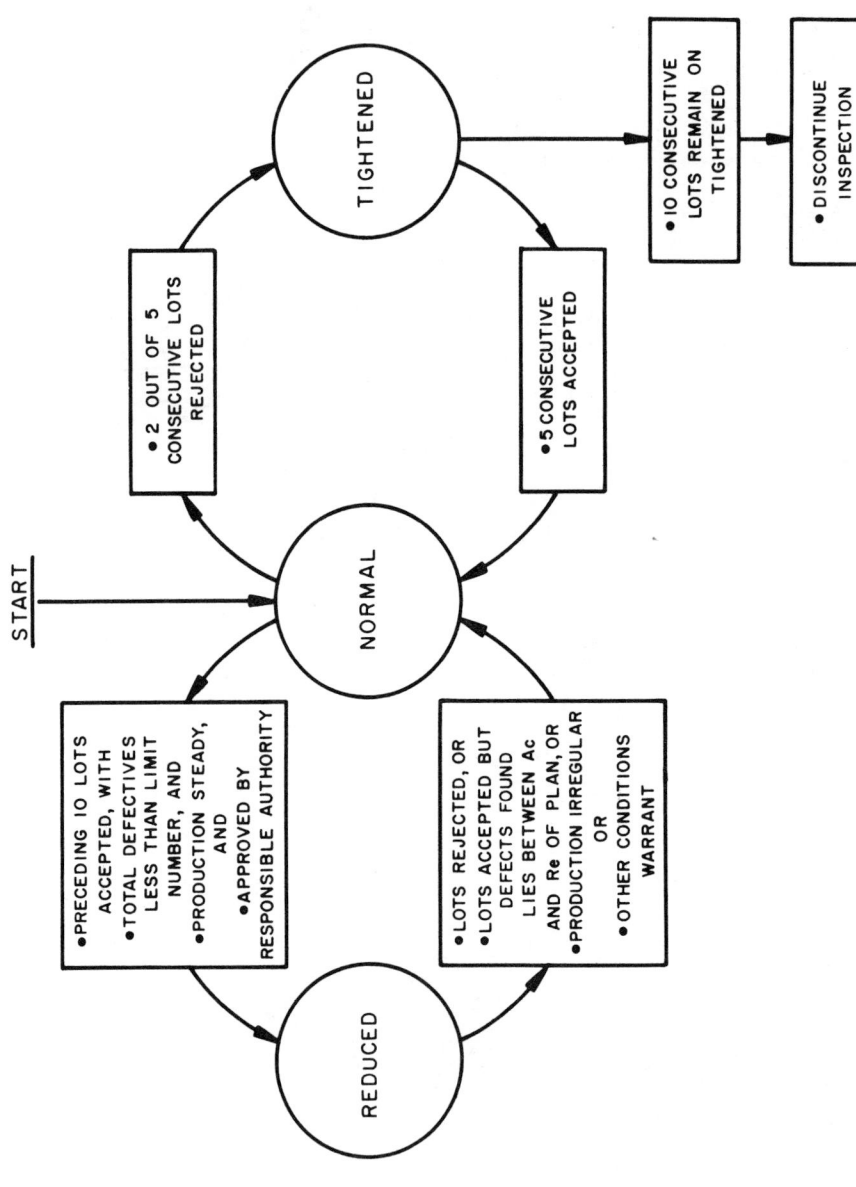

FIG. 11-2 Switching rules for MIL-STD-105D. (Based on
Schilling and Sheesley, 1978, p. 77. Reprinted by permission.)

criteria of the next lower AQL category while retaining the sample size
used in the normal plan. This results in a more stringent plan with less
consumer's risk at the expense of increased producer's risk. *Tightened
inspection* is imposed when two out of five consecutive lots are rejected
on original inspection. *Normal inspection* is reinstated when five consecutive lots are accepted on original inspection.

A switch to reduced inspection involves changing both the sample size
and the acceptance numbers. Sample size is reduced two sample size code
letter categories below that originally used for normal inspection. The
final acceptance and rejection numbers are separated by a gap. If the
number of defectives found falls in the gap, the lot is accepted but the
scheme reverts to the normal plan on the next lot. The gap is used to pre-
vent rejection of a lot on reduced inspection when it might be accepted
under normal inspection. Otherwise the acceptance and rejection numbers
are used in the conventional manner on reduced inspection. A shift is made
to *reduced inspection* when

1. The preceding 10 lots have been accepted on original inspection
   under normal sampling.
2. The total number of defectives from the preceding 10 lots is less
   than or equal to the limit numbers given in Table VIII of the
   standard. Results from all samples (not just first samples) should
   be used if double or multiple sampling is employed.
3. Production must be steady.
4. Reduced inspection is considered desirable by responsible authority.

MIL-STD-105D Table VIII is reproduced here as Appendix Table T11-1. To use
the table, the accumulated sample size from the last 10 lots is entered and
the limit number read from the AQL. When the accumulated sample size is not
sufficient for reduced inspection, additional lots must be taken until a
limit number can be obtained from the table. Clearly, the additional lots
must be from the same uninterrupted sequence.

Normal inspection must be reinstated from reduced when

1. A lot is rejected.
2. The results of inspection of a lot falls in the gap between the
   reduced acceptance and rejection numbers.
3. Production becomes irregular or delayed.
4. Other conditions warrant.

Application of the switching rules is much simplified by use of a

LEGEND

n = NUMBER CONSECUTIVE LOTS INSPECTED UNDER FORM OF INSPECTION (R, N, NC, T)

Rn = REDUCED INSPECTION

Nn = NORMAL INSPECTION

NCn = NORMAL CONDITIONAL INSPECTION (1 LOT PREVIOUSLY REJECTED)

Tn = TIGHTENED INSPECTION

F = PREVIOUS LOT FAILED UNDER TIGHTENED

Ai = PRECEDING i LOTS ACCEPTED UNDER TIGHTENED INSPECTION

PAR 8.4

INSPECTION SHOULD BE DISCONTINUED OR 100% INSPECTION INSTITUTED.

DEFINITIVE CORRECTIVE ACTION TAKEN TO IMPROVE PROCESS AND/OR QUALITY OF SUBMITTED MATERIAL.

WHEN OBJECTIVE EVIDENCE OF ABOVE IS AVAILABLE AND VERIFIED, RETURN TO T_1

INSTRUCTIONS

1. START AT "ENTER" (E), CIRCLE CODE N], OR ON APPROPRIATE CODE BASED ON EXISTING QUAL HISTORY OF THE UNIT OF PRODUCT IN ACCORDANCE WITH SWITCHING CRITERIA.

2. INSPECT MATERIAL TO SAMPLING TABLE INDICATED BY CIRCLE CODE (R, N, T).

3. MOVE TO NEXT APPLICABLE CIRCLE BY FOLLOWING ARROW BASED ON LOT DECISION TO ACCEPT OR REJECT.

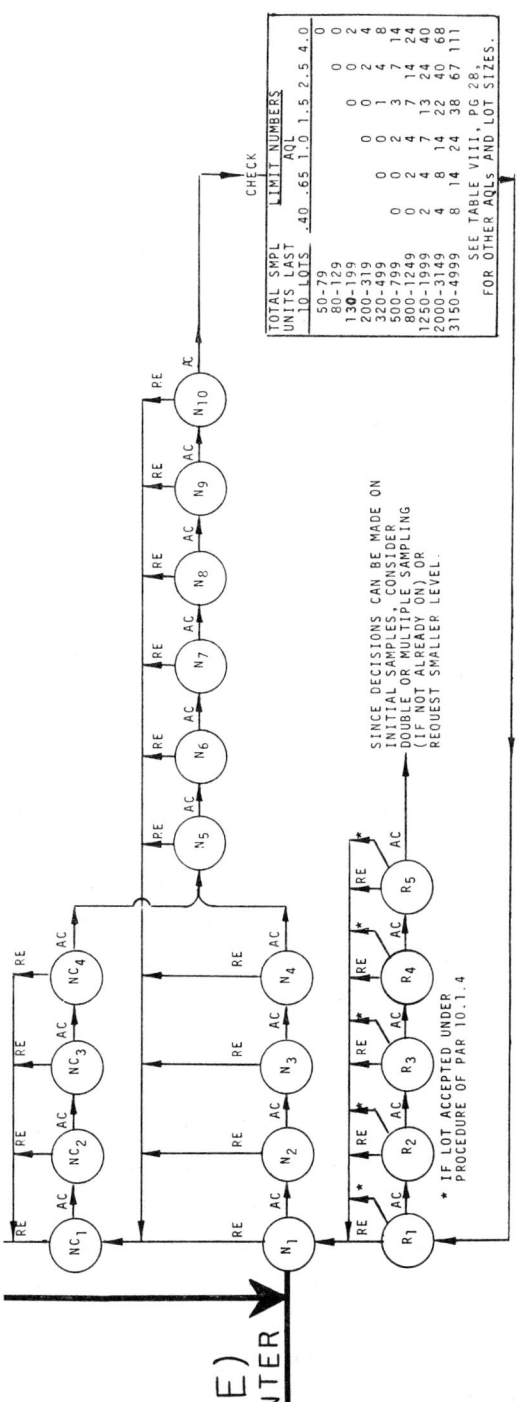

FIG. 11-3 Instantaneous switching procedure for MIL-STD-105D. (From Kaplan and MacDonald, 1969, p. 77. Reprinted by permission.)

chart prepared by Kaplan and MacDonald (1969, p. 172). From their experience,

> It provides a practically foolproof system for switching
> that requires less than *one minute* of training, even for
> a new employee. The inspector has only to follow the
> three simple steps noted at the right side of the chart,
> and switching from tightened, normal or reduced inspec-
> tion becomes automatic, adequate and accurate.

The chart is shown in Fig. 11-3. It is used checkerboard fashion as
each lot is inspected. It may be posted in the inspection area or copies
appropriately marked may be kept with the inspection records to provide an
immediate update when the next lot comes in. The legend at the left of the
chart interprets the symbols used. A somewhat simpler switching diagram
with an associated efficient and informative record keeping procedure has
been given by Whittingham (1977).

MIL-STD-105D was intended to be used with a continuing series of lots
or batches. However, occasionally specific plans may be selected from the
standard and used without the switching rules. This is not the intended
application of MIL-STD-105D and its use in this way should not be referred
to as inspection under MIL-STD-105D. When employed in this manner, the
standard simply represents a repository for a collection of individual plans
indexed by AQL. In this sense, AQL has no operational meaning and the
operating characteristics and other measures of a plan so chosen must be
assessed individually for that plan from the tables of performance provided
in MIL-STD-105D. It is a convenience to the user that tables are provided
to be used in this way. They are described in a later section.

SELECTION

The selection of a set of tightened, normal, and reduced plans from MIL-STD-
105D is fairly straightforward. The key elements in the selection of a plan
are lot size and AQL. The definition of a lot is often governed by the
operational situation and the available information. Lots may be composed
of the material delivered at one time, or produced in the same time period
(a day or a month), or that made under a particular set of operating conditions (raw material, operator, etc.). The standard suggests that "as far
as practicable, each lot should consist of units of product manufactured
under essentially the same conditions and at essentially the same time."

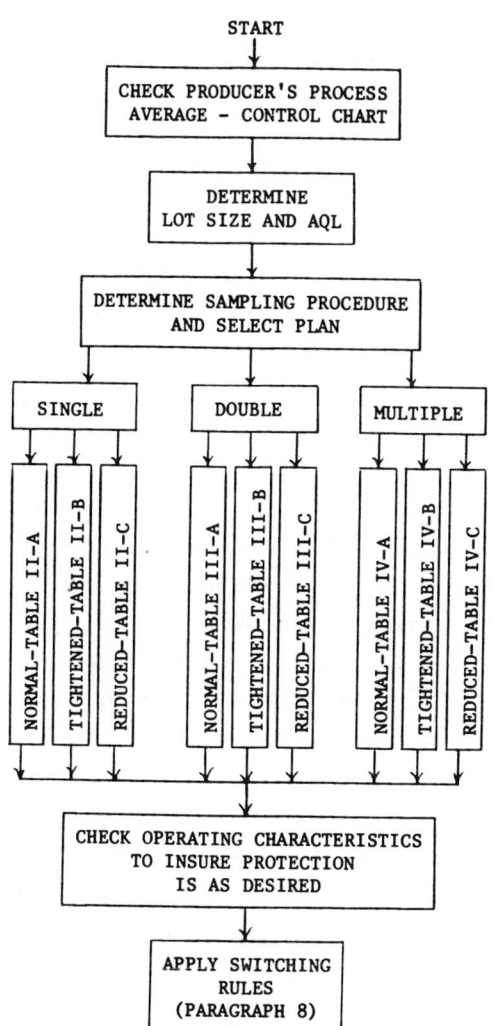

FIG. 11-4 Check sequence for selecting
a plan from MIL-STD-105D.

A quest for homogeneity tends toward small lots, while large lots are
desirable in allowing for larger sample sizes with greater discrimination
between good and bad quality. Thus, determination of lot size is often a
compromise frequently settled by practical considerations.

The AQL is central to the entire MIL-STD-105D system. It must be set
with due consideration for the producer's process capability and the consumer's need for a reasonable quality level relative to the state of the
art. The ideal AQL would be set in terms of process capability studies to

determine reasonable levels and costs of quality for the producer's process
as well as the tolerance of the consumer to changes in quality level and
the associated costs. An excellent discussion of process capability studies
will be found in Mentch (1980).

A further consideration in the determination of AQLs is defect class.
MIL-STD-105D uses the defect classification:

Critical: likely to result in hazardous or unsafe conditions

Major: likely to result in failure or to reduce usability
            materially

Minor: not likely to reduce usability materially

AQLs are usually assigned to each category with each defect type in the
category counted against the category AQL.

Ultimately the AQL to be used in sampling must be determined by negotiation between the producer and the consumer with due consideration of the
tradeoffs in both a physical and economic sense. The sequence of steps
involved in the selection of a set of plans from MIL-STD-105D is shown in
Fig. 11-4.

Once the lot size and AQL have been determined, a set of sampling plans
can be found. The lot size is used to enter Table I of the standard, reproduced here as Appendix Table T11-2. A sample size code letter is then
obtained appropriate to the inspection level to be used. Inspection Level
II is normally used unless some other inspection level is specified. Inspection Levels I and III allow for control of discrimination (lower or higher)
depending on past history and operating circumstances. The special inspection levels, S-1 through S-4, are generally used with expensive or destructive tests where sample size is at a premium and more extensive inspection
is not economic or not warranted on the basis of past history and the intent
of application of the plan.

The advantages and disadvantages of single-, double-, and multiple-
sampling plans have previously been discussed. Single-sampling plans are
easy to administer and understand. Double-sampling plans allow for a reduction in average sample size at the expense of the possibility of taking an
additional sample. Multiple-sampling plans are somewhat difficult to administer, but provide the greatest economy in terms of average sample size.
The choice, again, depends on the operating situation and the experience
and reliability of the inspection personnel involved. MIL-STD-105D does
provide average sample number curves to help in the allocation of inspection

STEP	TIGHTENED	NORMAL	REDUCED
PREPARATORY	MIL-STD-105D TABLE I DETERMINE SAMPLE SIZE CODE LETTER FROM LOT SIZE AND INSPECTION LEVEL (II)		
DETERMINE CRITERIA Single Sampling	MIL-STD-105D Table IIB	MIL-STD-105D Table IIA	MIL-STD-105D Table IIC
Double Sampling	MIL-STD-105D Table IIIB	MIL-STD-105D Table IIIA	MIL-STD-105D Table IIIC
Multiple Sampling	MIL-STD-105D Table IVB	MIL-STD-105D Table IVA	MIL-STD-105D Table IVC
DECISION RULES	• Accept if total defects $\leq$ Ac • Reject if total defects $\geq$ Re	• Accept if total defects $\leq$ Ac • Reject if total defects $\geq$ Re	• Accept if total defects $\leq$ Ac • Reject if total defects $\geq$ Re • Accept lot but switch if total defects on final sample fall in gap
SWITCHING	See Fig 11-2 for switching rules		

FIG. 11-5 Application of MIL-STD-105D.

effort when double- and multiple-sampling plans are involved.

Once the sampling procedure has been selected, a set of plans is found in the appropriate tables of the standard. Figure 11-5 shows how the sampling plans are selected and gives the decision rules to be applied in application of the plan.

The master tables for plan selection are reproduced here in the Appendix as follows:

Appendix table	MIL-STD-105D table	Content
T11-1	Table VIII	Limit number for reduced inspection
T11-2	Table I	Sample size code letters
T11-3	Table IIA	Single sampling, normal inspection
T11-4	Table IIB	Single sampling, tightened inspection
T11-5	Table IIC	Single sampling, reduced inspection
T11-6	Table IIIA	Double sampling, normal inspection

T11-7	Table IIIB	Double sampling, tightened inspection
T11-8	Table IIIC	Double sampling, reduced inspection
T11-9	Table IVA	Multiple sampling, normal inspection
T11-10	Table IVB	Multiple sampling, tightened inspection
T11-11	Table IVC	Multiple sampling, reduced inspection

These tables completely specify the plans included in MIL-STD-105D. In
selection of specific plans from the tables, however, it must be emphasized
that the vertical arrows direct the user to a completely new set of acceptance criteria; that is, both the sample size and acceptance number of the
indicated plan must be used in satisfying the intent of the arrow.

MEASURES

MIL-STD-105D contains detailed tables showing measures of performance of
individual plans. This includes

Appendix table	MIL-STD-105D table	Content
T11-12	Table VA	AOQL for normal plans
T11-13	Table VB	AOQL for tightened plans
T11-14	Table VIA	Limiting quality in percent defective for P_a = 10 percent
T11-15	Table VIB	Limiting quality in defects/100 units for P_a = 10 percent
T11-16	Table VIIA	Limiting quality in percent defective for P_a = 5 percent
T11-17	Table VIIB	Limiting quality in defects/100 units for P_a = 5 percent
T11-18	Table IX	Average sample number curves for double and multiple sampling
T11-19	Table X	OC curves and probability points for plans by code letter (Code F only)

These tables are presented here as Appendix Tables T11-12 through T11-19
as indicated above.

The distinction between percent defective and defects per 100 units
is particularly important when dealing with operating characteristics and
other measures of performance. Since Type B operating characteristics are
involved, the binomial distribution is exact in assessing percent defective,
while the Poisson distribution is employed in determining defects per 100
units. This is carried through MIL-STD-105D and appears explicitly in

Table X. The Poisson distribution, however, is used as an approximation to the binomial except for AQLs of 10.0 or less and sample sizes of 80 or less. This simplifies the presentation somewhat with little loss of accuracy. The AOQL tables are based on the Poisson distribution as are the average sample number curves.

The tables of operating characteristics and other measures are fairly well self-explanatory, however certain features should be pointed out.

1. AOQL factors are approximate in Table V and can be corrected by multiplying by $\left(1 - \dfrac{\text{sample size}}{\text{lot or batch size}}\right)$

2. ASN curves in Table IX are selected by the single sample size acceptance number c. The vertical axis is then interpreted in proportion to the single sample size n at the top, and the horizontal axis in terms of unity values np. As a result a comparison of single, double, and multiple average sample sizes can be made for any matched set in the standard. For example, for Code F, 2.5 percent AQL, the single-sampling plan is n = 20, c = 1. Using the c = 1 set of curves, the vertical axis becomes 0, 5, 10, 15, 20 and the horizontal axis 0, .05, .10, .15, .20. The double and multiple plans, then, have approximately the same average sample size at a proportion defective of about p = .10, that is, where np = 2. The arrows show the position of the AQL, obviously at proportion defective .025 for the plan n = 20, c = 1. This is made possible by the fact that for a given single-sampling acceptance number, the product (n × AQL) is constant for all sample sizes.

The operating characteristic curves and other measures of performance given in MIL-STD-105D relate to the performance of the constituent individual plans and so can be used to assess its operation at any given stage or to determine how the plan will perform in moving from normal to tightened or reduced inspection. This is helpful in determining AQLs. Unfortunately, the standard does not give measures of performance of the system as a whole, including the switching rules. Detailed tables of scheme performance patterned after the MIL-STD-105D tables cited above have been prepared by Schilling and Sheesley (1978). They are based on the work of Stephens and Larson (1967) and of Burnett (1967), who did much to develop the theory of evaluation of scheme characteristics.

A representation of the Schilling-Sheesley tables is given in the

Appendix as follows:

Appendix table	MIL-STD-105D table	Content
T11-20	V	MIL-STD-105D scheme AOQL
T11-21	VI	MIL-STD-105D scheme limiting quality for P_a = 10 percent
T11-22	VII	MIL-STD-105D scheme limiting quality for P_a = 5 percent
T11-23	IX, X	MIL-STD-105D scheme P_a, ASN, AOQ, ATI (Code F only)

The first three tables correspond directly to those given only for the individual normal, and tightened plans in MIL-STD-105D. The fourth is an example of the complete listing of measures by code letter and provides values for examples to follow. They characterize the performance of the standard when it is properly used, with the switching rules.

In application of the MIL-STD-105D system, it is intended that a switch to tightened inspection with possible discontinuation of inspection will, in the case of poor quality, provide a psychological and economic incentive for the producer to improve the level of quality submitted. In actual application, this may or may not be the case. When used in internal inspection to take advantage of the increased protection and economy afforded by the switching procedure a scheme may be used with no intention to discontinue inspection. Further, in early stages of process development producers may expect to have a large proportion of lots rejected and it may be impossible to improve the process given the state of the art. MIL-HDBK-53 (1965) points out that when inspection is discontinued, "If the supplier otherwise has an excellent quality history for similar products, the specified AQL should be investigated." Thus, the AQL and not the process may be changed. It is quite possible, however, as pointed out by Stephens and Larson (1967, p. 313) that "the actual behavior of the process under the influence of the sampling procedure may be ... very dynamic."

In discussing the problem of evaluating the performance of a sampling system, such as MIL-STD-105D, which may itself induce such process changes, Stephens and Larson

> ... adopt a somewhat simpler model which is tractable
> and which permits relative comparisons to be made be-
> tween different plans or ... different sets of plans ...
> which allows an evaluation of the operating behavior of

the system of plans for different values of fraction
defective. This is the same type of approach taken
in the presentation of an ordinary OC curve for a
sampling plan.

The same approach has also been used by Pabst (1963) and by Dodge (1965).
A producer would not usually be expected to operate at the limiting quality
level of any simple sampling plan or complex sampling scheme for very long
without taking action of one kind or another. However, ordinary OC curves
do not reflect such actions.

The Stephens-Larson model as evaluated by Schilling and Sheesley (1978)
does not incorporate considerations of possible process changes resulting
from psychological pressures inherent in the use of the switching rules or
discontinuation of inspection. After discontinuation, the Schilling-
Sheesley tables essentially assume restart under tightened inspection with
no change in fraction defective. Thus, the term *scheme performance*, as used
with respect to the scheme operating characteristic curve, has a very special meaning. It refers to how the MIL-STD-105D system of switching rules
would operate at a given process level under the assumption that the process
stays at that level even after discontinuation of inspection. It should be
noted that this gives a conservative "worst case" description of the performance of a scheme in the sense that, if psychological pressures were
operative, the probability of acceptance at low levels of fraction defective
would be increased while probability of acceptance at high levels of fraction defective would be decreased relative to the values given by Schilling
and Sheesley.

The compilation of complete tables of measures of scheme performance
allows the following approximate procedure to be used when the stream of
consecutive lots, on which MIL-STD-105D is based, is broken to produce an
isolated lot (known not to be part of the stream) or a short sequence of
lots of a unique character.

1. Obtain the LQ for the scheme at P_a = 10 percent from Schilling-
    Sheesley (Appendix Table T11-21) using the appropriate AQL and
    sample size code letter.
2. Select the plan from the MIL-STD-105D LQ table with P_a = 10 percent
    (Appendix Table T11-14, T11-15) which has the LQ of the scheme at
    the AQL listed.

This procedure will guarantee about the same protection on the isolated lot as would have been obtained under use of the switching rules with the continuing series of lots. A more refined approach is given later.

IMPLEMENTATION OF MIL-STD-105D

The implementation of MIL-STD-105D is probably best explained by example. Suppose the producer and the consumer agree on an AQL of 2.5 percent and lot sizes are expected to be N = 100. Inspection Level II will be used since no other inspection level was agreed upon. Using the lot size of 100 and Inspection Level II, the sample size code letter table (Appendix Table T11-2) gives Code F. Using the master tables for tightened, normal, and reduced inspection, we find the following set of matched single, double, and multiple plans to apply.

Code F, 2.5 AQL	Tightened			Normal			Reduced		
	n	Ac	Re	n	Ac	Re	n	Ac	Re
Single	32	1	2	20	1	2	8	0	2
Double	20	0	2	13	0	2	5	0	2
	20	1	2	13	1	2	5	0	2
Multiple	8	#	2	5	#	2	2	#	2
	8	#	2	5	#	2	2	#	2
	8	0	2	5	0	2	2	0	2
	8	0	3	5	0	3	2	0	3
	8	1	3	5	1	3	2	0	3
	8	1	3	5	1	3	2	0	3
	8	2	3	5	2	3	2	1	3

The plans under tightened inspection are found by use of the arrow which directs the user to the next set of sample sizes and acceptance numbers. The symbol # in multiple sampling indicates that no acceptance decision can be made at that stage of the sampling plan. Notice that the final acceptance and rejection numbers under reduced inspection differ by more than 1, thus showing the gap that can lead to lot acceptance with a switch to normal inspection.

Suppose single sampling were used. The first lot would be inspected using the plan n = 20, c = 1. This plan would continue in use on subsequent lots until a switch was called for. At that time the plan n = 32, Ac = 1, Re = 2 or n = 8, Ac = 0, Re = 2 would be used depending on whether the switch was to tightened or reduced inspection. For example, consider the following sequence of lot acceptance and rejection.

A A R A A A R R A R A A A A A A A A A A A A A A A A A R

Inspection would start using the normal plan. At the second rejection a switch to tightened inspection is instituted since 2 out of 5 lots are rejected under normal inspection. Tightened inspection continues until the 15th lot signals a switch to normal. A switch to reduced is called for after 10 lots are accepted under normal inspection. However, the total number of defectives in the last 10 lots must be less than the limit number of 2 found in MIL-STD-105D Table VIII for 200 accumulated sample units. If one defective was found the switch would be made, only to revert back to normal inspection with the rejection at the end of the sequence. This, of course, assumes the other conditions for switching were met. This sequence can be followed on the Kaplan-MacDonald diagram (Fig. 11-3).

The measures of performance of these individual plans are easily found from the tables given in the Appendix. They are

AOQL (normal) = 4.2 percent (3.4 percent corrected)

AOQL (tightened) = 2.6 percent (1.8 percent corrected)

10% LQ (normal, percent) = 18 percent

10% LQ (normal, defects) = 20 defects per 100 units

 5% LQ (normal, percent) = 22 percent

 5% LQ (normal, defects) = 24 defects per 100 units

ASN at AQL (double) $\simeq$ 15

ASN at AQL (multiple) $\simeq$ 18

MIL-STD-105D Table X for Code F reproduced here as Appendix Table T11-19 shows probability points for tightened and normal inspection for percent defective. The plan n = 8, c = 1 was similarly evaluated for reduced inspection. Appendix Table T11-23 shows the probability points of the resulting scheme as a whole as computed by Schilling and Sheesley (1978). They may be compared as follows:

P_a	Normal	Tightened	Reduced	Scheme	P_a	Normal	Tightened	Reduced	Scheme
.99	0.75	0.475	2.00	0.978	.25	12.9	8.19	30.3	8.21
.95	1.80	1.13	4.64	1.85	.10	18.1	11.6	40.6	11.6
.90	2.69	1.67	6.88	2.47	.05	21.6	14.0	47.1	14.0
.75	4.81	3.01	12.1	3.66	.01	28.9	19.0	58.8	19.0
.50	8.25	5.19	20.1	5.40					

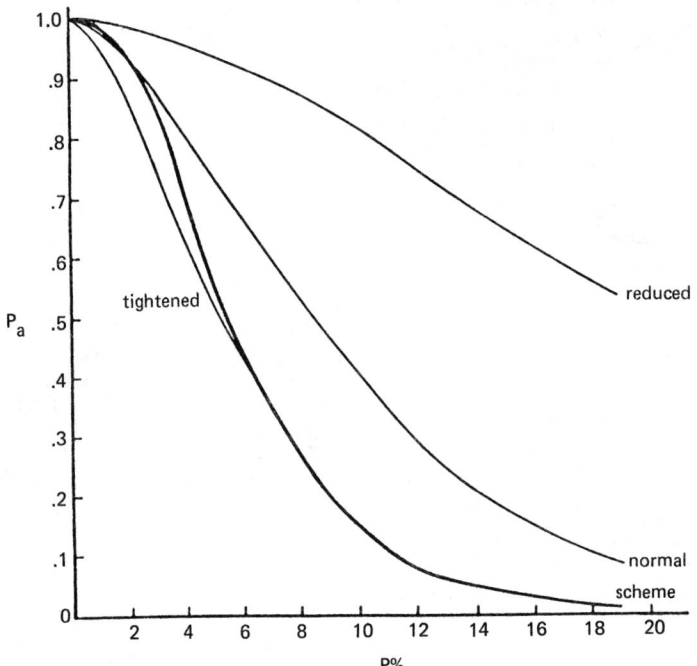

FIG. 11-6   Scheme operating characteristic curves, Code F,
2.5 percent AQL. (From Schilling and Sheesley, 1978, p. 79.
Reprinted by permission.)

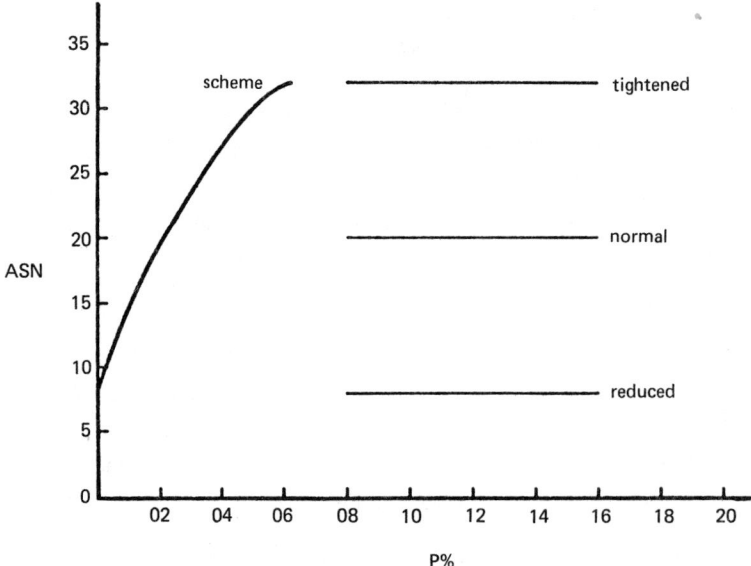

FIG. 11-7   Scheme average sample number curves, Code F,
2.5 percent AQL. (From Schilling and Sheesley, 1978, p. 80.
Reprinted by permission.)

Notice that scheme performance is slightly looser than the normal inspection plan for levels of quality well below the AQL but is much tighter
for levels of quality above the AQL. In fact, scheme performance is close
to that of the tightened plan at and below the indifference quality level.
This illustrates the advantage of using the scheme over any of its individual component plans. This can also be seen in the composite OC curve shown
in Fig. 11-6.

Since different sample sizes are involved in the plans constituting
the scheme for a given code letter and AQL, the sample size for the scheme
can only be represented as an expected value. This is the average sample
number (ASN) for the scheme. Although the scheme OC curve shows minimal
increase in probability of acceptance over that for the normal plan alone
when quality is good, the reduction in the average sample number in that
region is substantial because of the possibility of going to reduced inspection. This may be seen in the average sample number curve for Code F,
2.5 percent AQL, shown in Fig. 11-7. The sample sizes for the component
plans are also indicated.

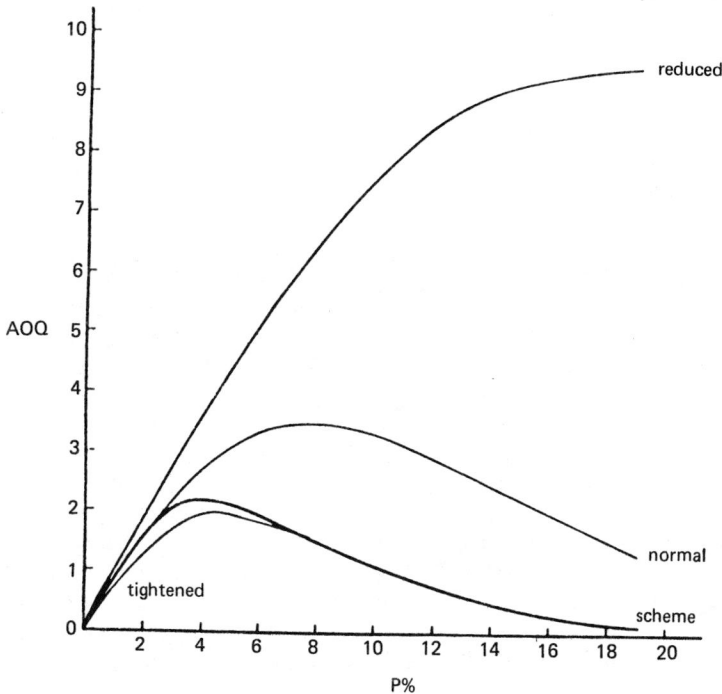

FIG. 11-8 Scheme average outgoing quality curves, Code F,
2.5 percent AQL. (From Schilling and Sheesley, 1978, p. 80.
Reprinted by permission.)

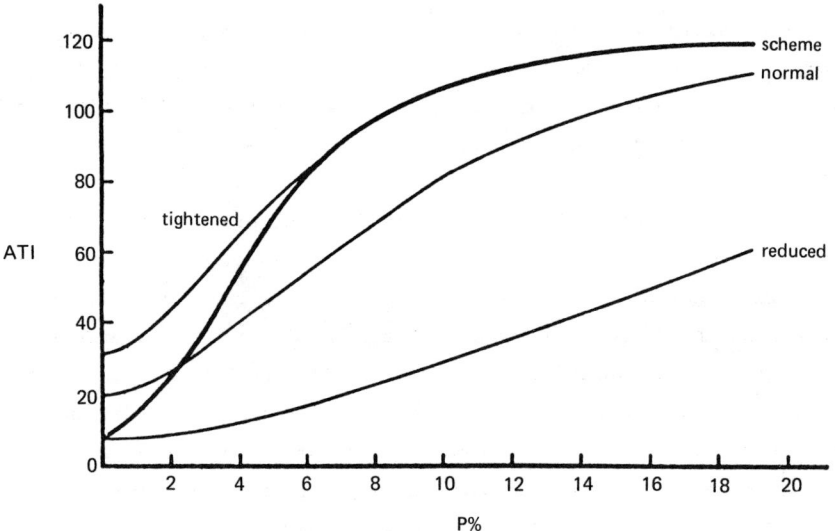

FIG. 11-9 Scheme average total inspection curves, Code F,
2.5 percent AQL. (From Schilling and Sheesley, 1978, p. 80.
Reprinted by permission.)

TABLE 11-1 Scheme P_a and ASN Compared to Normal Plan Only for Single, Double, and
Multiple Sampling (Code F, 2.5 Percent AQL)

	Probability of Acceptance						Average Sample Number					
	Scheme			Normal Only			Scheme			Normal Only		
p	SINGLE	DOUBLE	MULTIPLE	SINGLE	DOUBLE	MULTIPLE	SINGLE	DOUBLE	MULTIPLE	SINGLE	DOUBLE	MULTIPLE
0.978	99	98.5	99.6	98.4	98.0	98.9	14.6	11.7	8.7	20	14.5	16.4
1.85	95	93.2	96.4	94.8	93.5	96.1	19.1	15.9	16.3	20	15.5	17.4
2.47	90	87.2	92.1	91.4	89.4	93.1	21.5	18.3	19.0	20	16.1	17.9
3.66	75	70.6	77.6	83.5	80.3	85.5	26.2	23.3	24.0	20	17.0	18.7
5.40	50	46.7	49.2	70.6	66.1	71.9	30.8	26.9	28.2	20	17.7	19.1
8.21	25	23.9	22.4	50.3	45.4	49.4	32.0	26.4	25.1	20	18.0	18.5
11.6	10	10.4	8.1	30.8	27.1	28.2	32.0	24.5	20.4	20	17.5	16.7
14.0	5	5.7	3.8	20.8	18.3	18.1	32.0	23.2	17.8	20	16.9	15.2
19.0	1	1.6	0.8	8.4	7.7	6.6	32.0	21.4	14.1	20	15.6	12.4

From E. G. Schilling and J. H. Sheesley (1978, p. 82). Reprinted by permission.

When rectification is employed, the average outgoing quality level for
the scheme is much improved over levels reported for the normal inspection
plan, although they are not as low as that given for the tightened plan.
This can be seen for Code F, 2.5 percent AQL in the AOQ curve shown in Fig.
11-8.

Average total inspection is given as a guide to determine the inspection effort requirements when rectification is used with a MIL-STD-105D
scheme. A plot of the ATI curve for Code F, 2.5 percent AQL, is given in
Fig. 11-9.

Curves such as those shown can easily be constructed from the tables
presented by Schilling and Sheesley (1978).

A comparison of single, double, and multiple plans for Code F, 2.5
percent AQL, indicates that the operating characteristic curves of the
schemes using these sampling procedures are about as well matched as those
of the constituent individual plans. This can be seen in Table 11-1. It
also suggests savings in sample size may result when using double or multiple plans with the switching procedure, particularly below AQL levels of
percent defective.

Thus, there are significant advantages in the use of the switching
rules to achieve operating characteristics and other measures of performance
not attainable through individual sampling plans.

MATCHING INDIVIDUAL SAMPLING PLANS TO MIL-STD-105D SYSTEM PERFORMANCE

Occasionally, it is necessary or economically desirable to abandon the
scheme aspect of MIL-STD-105D in favor of a single-sampling plan. This may
be because a unique or isolated lot and not a continuing stream is to be
inspected. It may be that the stream of lots is too short to provide effective use of the MIL-STD-105D switching rules. When this is the case, the
sampling plan for normal inspection is sometimes incorrectly selected to be
used alone without the switching rules. This results in less consumer protection than would be afforded by use of the overall scheme. A more conservative approach for the consumer would be to use the tightened plan
alone, but this can result in an objectionably high level of rejection for
the producer of quality at or better than the AQL. For example, for Code
H, 1.5 percent AQL, the scheme will accept 1.59 percent defective quality
95 percent of the time and reject 7.56 percent defective 90 percent of the
time. The normal plan will accept 1.59 percent defective 81 percent of the
time while the tightened plan will reject 7.56 percent defective 73 percent

of the time.

It is possible to use the unity values developed by Schilling and Johnson (1980) shown in Appendix Table T6-1 to derive a unique individual sampling plan to match the scheme performance of MIL-STD-105D. The plan obtained will usually require a larger sample size than that given for the normal plan since protection under the scheme is better than under the normal plan taken alone. The difference in sample size reflects the advantage in protection obtained by using the switching rules.

Appendix Table T11-24 from Schilling and Johnson (1980) shows values of the operating ratio R for the AQL code letter combinations of the MIL-STD-105D system. It was derived from the tabulations of MIL-STD-105D scheme performance by Schilling and Sheesley (1978) and includes switching among tightened, normal, and reduced plans. Appendix Table T11-21, from Schilling and Sheesley (1978), shows the lot tolerance percent defective, LTPD (LQ = 10 percent) associated with code letter-AQL combinations of the MIL-STD-105D system. To use Tables T11-24 and T11-21 to obtain a unique plan having the performance of a MIL-STD-105D scheme:

1. Decide if single, double, or multiple sampling is to be used.
2. Use Table T11-24 to obtain the appropriate MIL-STD-105D scheme operating ratio.
3. Use Table T11-21 to obtain the LTPD resulting from the use of the MIL-STD-105D scheme.
4. Use this operating ratio and a value of p_2 equal to the LTPD to determine a matching individual plan from Appendix Table T6-1.

For example, if MIL-STD-105D specifies Code F, 2.5 percent AQL, a matching individual sampling plan must have R = 6.70 and LTPD = 12.2 percent, when the Poisson approximation to the binomial is used. Either a single, a double, or a multiple plan may be used. Application of Appendix Table T6-1 produces the following possibilities using plans 2S, 2D, and 2M.

	Sample	Sample size	Ac	Re		Sample	Sample size	Ac	Re
Single	1	44	2	3	Multiple	1	12	#	2
						2	12	0	3
						3	12	0	3
Double	1	28	0	3		4	12	1	4
	2	28	3	4		5	12	2	4
						6	12	3	5
						7	12	4	5

Any of these plans will provide scheme performance protection equivalent to the MIL-STD-105D, Code F, 2.5 percent AQL scheme. It should be noted that the average sample size for the MIL-STD-105D scheme at the AQL is about 22, while the single-sampling plan to match the scheme has sample size 44. This illustrates the advantage of use of the switching rules which are incorporated in MIL-STD-105D. Also, use of the normal inspection plan alone, without the switching rules, would result in an operating ratio of 10.96, and in considerably less consumer protection than that of the scheme or of the plan derived above to match the scheme.

Appendix Table T11-25 may also be used in reverse to find a MIL-STD-105D sampling scheme to match an individual sampling plan. The procedure is as follows:

1. Find the operating ratio of the individual plan.

2. Find the LTPD of the individual plan.

3. Locate the diagonal of Table T11-24 showing operating ratios just less than or equal to that of the given plan.

4. On the corresponding diagonal in Table T11-21, find the sample size-code letter combination which has the desired LTPD for the MIL-STD-105D scheme.

5. Use this MIL-STD-105D scheme, *with the switching rules*, in lieu of the individual plan.

For example, the plan n = 20, c = 2 has an operating ratio of 6.5 with 26.6 percent LTPD, using the Poisson approximation to the binomial. Table T11-24 shows values of R close to 6.5 on the second diagonal. The second diagonal of Table T11-21 gives LTPD = 19.4 for Code E, 4.0 AQL, which is closest to that desired. Use of this code letter-AQL combination, with the switching rules, will given an average sample size of about 15 at the AQL with the same scheme performance as the plan n = 20, c = 2.

Occasionally the acceptance criteria of a MIL-STD-105D plan must be altered to meet operating conditions. Suppose it is necessary to destructively sample 13 units under Code E, 1.0 AQL, normal inspection, from a shipment of 84 units packed a dozen to a box. The units are to be resold and it is desirable to reduce the sample size to 12 so that 7 full cartons will remain after sampling. To assess the effect of a sample size of 12:

1. The original normal plan from MIL-STD-105D is n = 13, c = 0.

2. The operating ratio of the normal plan is R = 44.9 with p_1 = .004 and p_2 = .177.

3. If sample size 12 is used with the same acceptance number, we
    have from Appendix Table T6-1

$$p_1 = \frac{.0513}{12} = .004$$

and

$$p_2 = \frac{2.303}{12} = .192$$

If the slight degradation in consumer protection can be tolerated by the
consumer a switch to the plan n = 12, c = 0 may be reasonable.

Thus, since matched MIL-STD-105D criteria are used in the Schilling-
Johnson (1980) tables they can be employed to assess the effect of any
changes from the nominal sample sizes given in that standard to other values
made necessary by operating conditions, or to compensate for such changes.
Individual sampling plans can also be derived to match MIL-STD-105D scheme
performance for use under conditions in which switching is difficult or
impossible. These tables also provide unity values for very low probability
of acceptance for use in reliability, safety, and compliance sampling. Sufficient values are given to allow the operating characteristic and average
sample number curves to be evaluated as necessary.

FURTHER CONSIDERATIONS

The background of MIL-STD-105D and its development out of the 105 series is
given in an excellent paper by Pabst (1963). It explains some of the intri-
cacies of the system and its development. The theory behind the structure
of the MIL-STD-105D tables is well presented in a paper by Hill (1973). A
detailed explanation of the procedural aspects of the use of the standard
is given by Hahn and Schilling (1975). An extensive and informative investigation of the properties of MIL-STD-105D schemes is presented in a paper
by Stephens and Larson (1967). Scheme properties are also investigated by
Schilling and Sheesley (1978) and measures of performance tabulated.

Several important handbooks and references on MIL-STD-105D operation
and performance have been prepared. These include Military Handbook H-53
(1965) and the International Organization for Standardization document
ISO/DIS 3319 which serves as a guide to the international version of the
standard, ISO 2859. A set of plans indexed by limiting quality and compatible with MIL-STD-105D (same lot size classes and sample sizes) has
been proposed by Duncan, Mundel, Godfrey and Partridge (1980). The proposed table, which simplifies the selection of a limiting quality plan, can

be used independently or in conjunction with MIL-STD-105D and associated standards.

Based on his work with M. N. Torrey on continuous sampling plans, H. F. Dodge (1965) has pointed out that the scheme OC curve resulting from the combination of two plans into a scheme using the MIL-STD-105D normal-tightened (only) switching rules is easily determined. Consider a normal plan N and tightened plan T so combined. Then at a given proportion defective with associated probabilities of acceptance P_{a_N} and P_{a_T} respectively, the system probability of acceptance is determined by calculating

$$a = \frac{2 - P_{a_N}^4}{\left(1 - P_{a_N}\right)\left(1 - P_{a_N}^4\right)}$$

$$b = \frac{1 - P_{a_T}^5}{\left(1 - P_{a_T}\right)P_{a_T}^5}$$

Then $a/(a + b)$ represents the proportion of the time the plan will be on normal inspection and $b/(a + b)$ represents the proportion of time the plan will be on tightened inspection, so that

$$P_a = \frac{aP_{a_N} + bP_{a_T}}{a + b}$$

(Those familiar with CSP-2 will recognize: $a = fv$, $b = u$, $i = 5$, $k = 4$.)

For example, if under Code F, 2.5 AQL, the reduced procedure is not used, the scheme probability of acceptance at the AQL may be determined from the plans

$$N: n = 20 \quad c = 1 \quad P_{a_N} = .9118$$
$$T: n = 32 \quad c = 1 \quad P_{a_T} = .8097$$

as

$$a = \frac{2 - .9118^4}{(1 - .9118)(1 - .9118^4)} = 48.05$$

$$b = \frac{1 - .8097^5}{(1 - .8097).8097^5} = 9.844$$

$$P_a = \frac{48.05(.9118) + 9.844(.8097)}{48.05 + 9.844} = .894$$

Note that the Dodge formula gives scheme performance of MIL-STD-105D when only the normal and tightened plans are used and reduced inspection is omitted.

PROBLEMS

1. MIL-STD-105D 1.0 percent AQL is specified and a lot of 390 pieces is to be inspected. Find the associated set of normal, tightened, and reduced plans for
   a. Single sampling
   b. Double sampling
   c. Multiple sampling

2. The "exact" AOQL for the scheme represented in Prob. 1 is 0.93 percent. What is the AOQL of the constituent
   a. Normal plan?
   b. Tightened plan?
   What does this suggest as a rough measure of the AOQL of the tightened plan? Of the scheme?

3. What is the limiting quality for $P_a = 10$ percent for the tightened and normal plans of Prob. 1 for
   a. Percent defective?
   b. Defects per hundred units?

4. Which type of plan (single, double, multiple) gives minimum average sample size at the AQL for the tightened plan of Prob. 1?

5. What action should be taken if, after a switch, the sixth lot is the first lot rejected and the switch was to
   a. Normal inspection?
   b. Tightened inspection?
   c. Reduced inspection?

6. For the scheme, Code F, 4.0 AQL, what are the following properties of the scheme for defects per hundred units:
   a. Probability of acceptance at the AQL?
   b. Average sample number at the AQL?
   c. AOQL?
   d. Average total inspection for lots of size 120 at the AQL?

7. What is the probability of having a succession of 10 lots on tightened inspection after a switch is made if the process is running at
   a. The indifference quality level of the tightened plan?
   b. The LTPD of the tightened plan?

8. The reduced plan for Code C, 10 percent AQL is n = 2, Ac = 0, Re = 2. What is the probability of simultaneously accepting a lot but switching back to normal inspection if the producer's process is running at 10 percent defective?

9. The sample sizes in MIL-STD-105D are in a geometric progression with ratio $10^{1/5}$. What would be the next single sample size after S in the tightened table if Code T were added? What would be the acceptance number for Code T and 0.015 AQL? What would be the approximate AOQL?

10. A contract requires MIL-STD-105D, 4.0 AQL. A single isolated lot of size N = 140 is to be inspected. Derive a single-sampling plan which will match the performance of the MIL-STD-105D scheme specified by the contract. What MIL-STD-105D scheme will afford performance equivalent to the plan n = 14, c = 3?

REFERENCES

Burnett, T. L. (1967), *Markov Chains and Attribute Sampling Plans*, IBM Technical Report No. 67-825-2175, IBM Federal Systems Division, Oswego, New York.

Dodge, H. F. (1963), Choosing Acceptance Sampling Plans, *Industrial Quality Control*, 20(2): 40-43.

Dodge, H. F. (1965), *Evaluation of a Sampling Inspection System Having Rules for Switching Between Normal and Tightened Inspection*, Technical Report No. 14, The Statistics Center, Rutgers-The State University, New Brunswick, New Jersey.

Dodge, H. F., and H. G. Romig (1959), *Sampling Inspection Tables, Single and Double Sampling*, 2nd ed., John Wiley and Sons, New York.

Duncan, A. J., A. B. Mundel, A. B. Godfrey, and V. A. Partridge (1980), LQL Indexed Plans that are Compatible with the Structure of MIL-STD-105D, *Journal of Quality Technology*, 12(1): 40-46.

Hahn, G. R., and E. G. Schilling (1975), An Introduction to the MIL-STD-105D Acceptance Sampling Scheme, *ASTM Standardization News*, 3(9): 20-30.

Hill, I. D. (1962), Sampling Inspection and Defense Specification DEF-131, *Journal of the Royal Statistical Society* (Series A), 125(1): 31-73.

Hill, I. D. (1973), The Design of MIL-STD-105D Sampling Tables, *Journal of Quality Technology*, 5(2): 80-83.

International Organization for Standardization (1974), *Sampling Procedures and Tables for Inspection by Attributes* (ISO 2859), Geneva, Switzerland.

International Organization for Standardization (1974), *A Guide to the Use of ISO 2859 "Sampling Procedures and Tables for Inspection by Attributes,"* (ISO 3319), Geneva, Switzerland.

Kaplan, A., and E. MacDonald (1969), Instantaneous Switching Procedure for MIL-STD-105D, *Journal of Quality Technology*, 1(3): 172-174.

Mentch, C. C. (1980), Manufacturing Process Quality Optimization Studies, *Journal of Quality Technology*, 12(3): 119-129.

Pabst, W. R., Jr. (1963), MIL-STD-105D, *Industrial Quality Control*, 20(5): 4-9.

Schilling, E. G., and L. I. Johnson (1980), Tables for the Construction of Matched Single, Double, and Multiple Sampling Plans with Applications to MIL-STD-105D, *Journal of Quality Technology*, 12(4): 220-229.

Schilling, E. G., and J. H. Sheesley (1978), The Performance of MIL-STD-105D Under the Switching Rules, *Journal of Quality Technology*, Part 1, 10(2): 76-83; Part 2, 10(3): 104-124.

Stephens, K. S., and K. E. Larson (1967), An Evaluation of the MIL-STD-105D System of Sampling Plans, *Industrial Quality Control*, 23(7): 310-319.

United States Department of Defense (1965), *Guide to Sampling Inspection, Quality and Reliability Assurance Handbook* (MIL-HDBK-53), Office of the Assistant Secretary of Defense (Installations and Logistics), U.S. Department of Defense, Washington, D.C.

United States Department of Defense (1963), *Military Standard, Sampling Procedures and Tables for Inspection by Attributes* (MIL-STD-105D), U.S. Government Printing Office, Washington, D.C.

Whittingham, P. R. B. (1977), Visual Guide to Switching Rules for MIL-STD-105D, *Journal of Quality Technology*, 9(1): 33-37.

VARIABLES SAMPLING SCHEMES : MIL-STD-414

Like MIL-STD-105D, Military Standard 414 is an acceptable quality level
(AQL) sampling system. It was devised by the military, as a consumer, to
be used to assess percent defective beyond contractual limits. Since it is
an AQL system, it incorporates switching rules to move from normal to
tightened or reduced inspection and return to achieve consumer protection.
These switching rules must be used if the standard is to be properly applied.
The switching rules differ somewhat from those used in MIL-STD-105D. The
standard assumes underlying normality of the distribution of the measurements to which it is applied and is intended to be used with a steady stream
of lots.

MIL-STD-414 allows for use of three alternative measures of variability:
known standard deviation (σ), estimated standard deviation (s), or average
range of subsamples of five ($\overline{R}$). If the variability of the process producing
the product is known and stable, it is profitable to use σ. The choice between s and $\overline{R}$ when σ is unknown is an economic one. The range requires
larger sample sizes but is easier to compute. OC curves given in the standard are based on use of s, the σ and $\overline{R}$ plans having been matched as closely
as possible to those using s.

The basic statistic to be calculated in applying MIL-STD-414 may be
considered to be the standardized distance from the sample mean to the specification limit. For an upper specification limit U,

$$t_U = \frac{U - \overline{X}}{\sigma}$$

When σ is unknown,

$$t_U = \frac{U - \overline{X}}{s}$$

or

$$t_U = \frac{U - \bar{X}}{\bar{R}}$$

is substituted depending on the measure of variability chosen. A comparison of t_U to the acceptance constant k will show whether the sample mean is or is not in the region of acceptance.

MIL-STD-414 offers an alternate procedure to use of the acceptance constant k, the M method discussed in Chap. 10. This involves using a statistic similar to those above to estimate proportion defective in the lot and is referred to in the standard as Form 2. The k method, involving simple comparison of t to k to determine acceptability, is called Form 1. Form 2 is the preferred procedure since the switching rules cannot be applied unless the fraction defective $\hat{p}$ of each lot is estimated from the sample.

MIL-STD-414 is complex. It consists of sections indexed by measure of variability, type of specification (single or double) and form number of

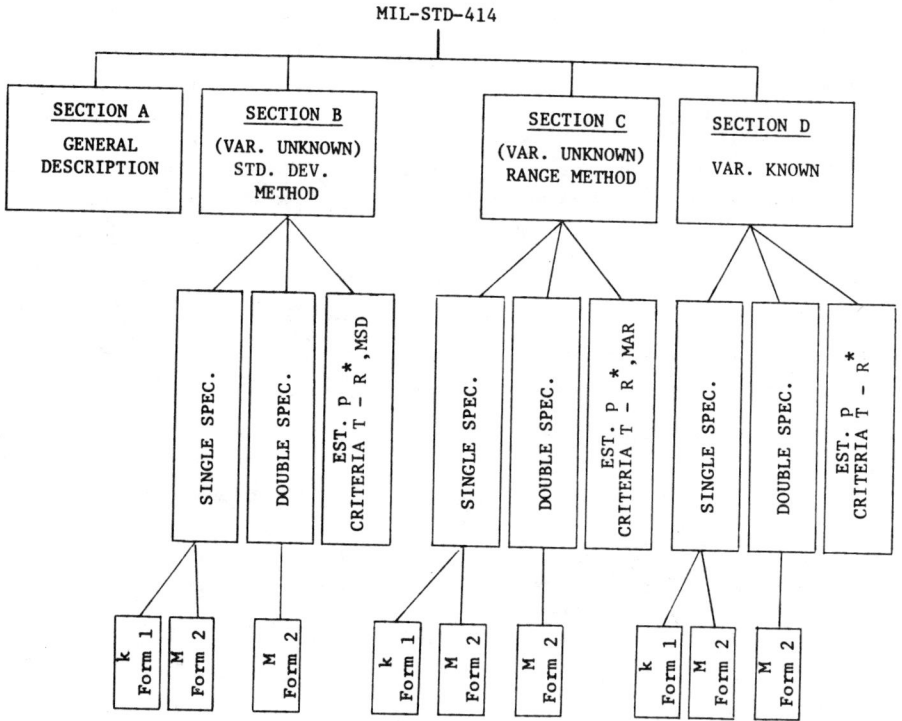

FIG. 12-1 Content of MIL-STD-414. (Asterisk (*) indicates tables for estimating p and criteria for tightened and reduced inspection.) (From Schilling, 1974a, p. 19. Reprinted by permission.)

the acceptance procedure. Only Form 2 is officially available for the case
of double specification limits. The standard's structure is shown in Fig.
12-1.

Application of MIL-STD-414 follows the pattern of MIL-STD-105D. Note
that MIL-STD-414 and MIL-STD-105D plans are not matched. The classification
of defects used in MIL-STD-414 is the same as that used in MIL-STD-105D:
critical, major, and minor. Sample sizes are determined from lot size and
AQL and, after choosing the measure of variability to be used and the form
of acceptance procedure, appropriate acceptance constants are obtained from
the standard.

MIL-STD-414 has a liberal supply of excellent examples. The reader
should refer to the standard for detailed numerical examples of its application.

The necessary assumption of a known, stable underlying normal distribution of individual measurements inherent in the MIL-STD-414 variables plans
is a serious limitation in their application. Use of MIL-STD-414 plans
without investigating the true nature of the underlying distribution is
foolhardy, for the results can be very bad indeed.

Nonetheless, sensible evaluation of the nature of the underlying distribution and implementation of prudent procedures to insure stability can
provide sufficient justification for use of MIL-STD-414. This is particularly true for in-process and final inspections where the distribution of
the process producing the product is not beyond the control or investigation
of those applying the plan. The rewards for painstaking, thorough analysis
are great in terms of sample size and worthwhile information on the process
involved.

OPERATION

Since it is a sampling system, proper use of MIL-STD-414 requires diligent
use of the switching rules. It is with this procedure that protection is
afforded both the producer and the consumer through tightening and relaxing
the severity of inspection consistent with the demonstrated performance of
the producer. The operation of the switching rules is shown schematically
in Fig. 12-2.

A MIL-STD-414 sampling scheme always starts on normal inspection, which
is continued until a switch to tightened or reduced inspection is warranted.
Normal inspection is reinstituted when the conditions justifying tightened
or reduced inspection can be shown to no longer apply. The switching rules

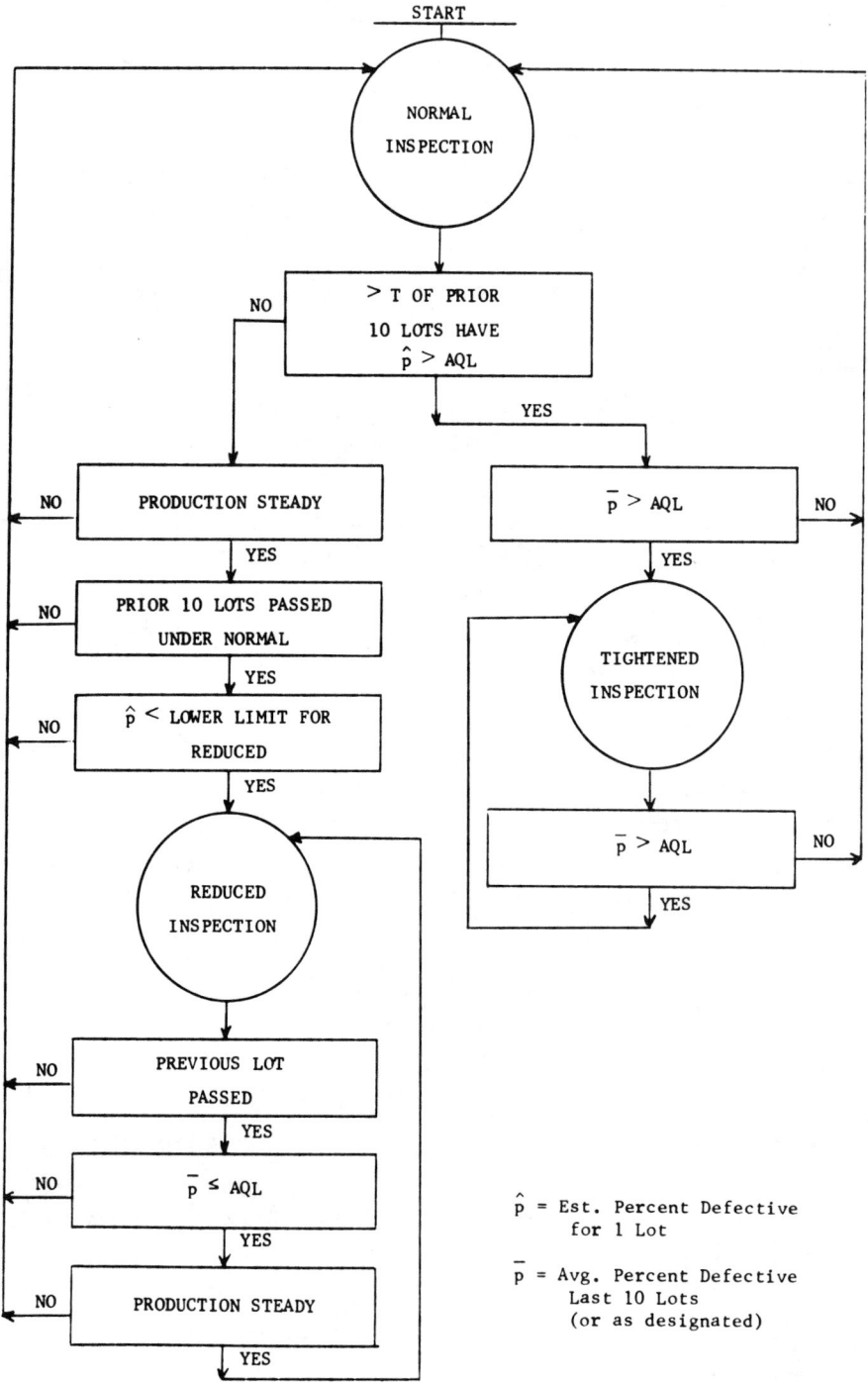

FIG. 12-2 MIL-STD-414 normal-tightened-reduced. (From Schilling, 1974a, p. 18. Reprinted by permission.)

of MIL-STD-414 are such that the probability of switching from normal to
tightened or reduced inspection respectively is less than .005 when quality
is running at the level of the AQL.

An important part of the switching procedure is the estimated percent
defective in each lot, obtained using the M method of Chap. 10 from tables
given as part of Form 2 of the standard. The estimated process average,
which is the mean of these percents defective, is also employed in switching.

A switch to tightened inspection involves changing the acceptance criterion to the next lower AQL category, while retaining the sample size
associated with the code letter involved. This leads to decreased consumer's
risk at the expense of an increase in the producer's risk. *Tightened inspection is instituted* under the following conditions:

1. More than T of the last 10 lots (or such other number of lots as
   designated) have estimates of percent defective, obtained through
   use of the M method (Form 2), exceeding the AQL.

2. The process average obtained from the estimated percents defective
   of the last 10 lots (or such other number of lots as designated)
   is greater than the AQL.

Values of T are given in each of the sections (see Fig. 12-1) for
application against the last 5, 10, or 15 lots, whichever has been designated.
If the sample size code letter is not the same for all the previous lots, the
table of T is entered using the code letter of the smallest sample size involved. As an example MIL-STD-414 Table B-6, which gives T values for the
standard deviation known section, is presented as Appendix Table T12-1.

Normal inspection is reinstated from tightened when

> The estimated process average of the last 10 lots (or such other
> number of lots as designated) is equal to or less than the AQL.

A switch to reduced inspection involves changing both the sample size
and the acceptance criteria to obtain a reduction in the sample size. This
reduction is typically around 40 percent. The producer's risk is decreased
slightly thereby while the consumer's risk is increased. *Reduced inspection
is instituted* when

1. Production is at a steady rate.

2. The preceding 10 lots (or such other number of lots as designated)
   have been accepted under normal inspection.

3. The estimated percent defective for *each* of the preceding 10 lots
   (or such other number of lots as designated) is less than the

applicable lower limit number tabulated. Or, for certain plans
having small sample size and low AQL, the estimated lot percent
defective must be zero for a specified number of consecutive lots.

Values of the lower limit number (or number of consecutive lots) are given
in each of the sections (see Fig. 12-1) for application against the preceding 5, 10, or 15 lots, whichever has been designated. As an example, MIL-
STD-414 Table B-7 showing limit numbers for reduced inspection is presented
in Appendix Table T12-2.

Normal inspection is reinstated from reduced when

1. A lot is rejected.
2. The estimated process average from the last 10 lots (or such other
    number of lots as is designated) is greater than the AQL.
3. Production becomes irregular or delayed.
4. Other conditions warrant that normal inspection should be instituted.

These switching rules are somewhat more complicated than those of MIL-STD-105D
and are patterned after those used in MIL-STD-105A. Nevertheless, their use
is economically effective in reducing sample size with increased protection
over that which could be achieved by use of single plans alone.

The *process average* is defined as the average percent defective, based
upon a group of lots submitted for original inspection. It is constructed
using estimates of percent defective from a specified number of preceding
lots from first submissions only. Product known to have been produced under
atypical conditions is excluded from the estimated process average. Normally, it is computed as the arithmetic mean of the estimated percents
defective from the last 10 lots unless some other number of lots has previously been designated.

SELECTION

The selection of a set of plans for normal, tightened, and reduced inspection
is more complicated in MIL-STD-414 than in MIL-STD-105D in that MIL-STD-414
offers complete sets of plans and procedures for each of three methods for
estimating variability. In fact, MIL-STD-414 could easily be separated into
three self-contained standards each based on its own measure of variability,
$\hat{\sigma}$. As seen in Fig. 12-1, they are

Standard deviation method (Sec. B):

$$\hat{\sigma} = s = \sqrt{\frac{\sum (x - \overline{x})^2}{n - 1}}$$

Range method (Sec. C):

$$\hat{\sigma} = \frac{\overline{R}}{d_2^*}$$

Variability known (Sec. D):

$$\hat{\sigma} = \sigma$$

Section A applies to each measure of variability and presents a general description of the sampling plans, gives AQL ranges to be covered by the AQLs in the standard, supplies sample size code letters, and presents OC curves.

Obtaining a plan from MIL-STD-414 involves more than selection of a measure of variability, however. The sequence for selection of a set of plans is given in Fig. 12-3.

First, the underlying distribution of measurements to which the plan is to be applied should be checked for normality. This involves probability plots, statistical goodness-of-fit tests, control charts, and other statistical procedures as appropriate. MIL-STD-414 assumes a normal distribution of measurements and this assumption needs to be constantly verified during application of the standard.

As in application of MIL-STD-105D, the lot size and AQL must be determined. If the AQL chosen is not one of those used to index MIL-STD-414 plans, Table A-1 of Sec. A allows for conversion to one of the specified values. Fortunately, the AQL values given in MIL-STD-414 are the same as those used in MIL-STD-105D. As in MIL-STD-105D, lot size is used to determine sample size code letter using Table A-2 of Sec. A. Five inspection levels are given. Since assignment of code letters to lot size ranges is primarily based on economic and other nonstatistical considerations in Type B sampling situations, the alternative inspection levels in both systems provide some flexibility in this regard. Unlike MIL-STD-105D, Inspection Level IV is normally used unless some other inspection level is specified. Also the lot size ranges are not the same in the two standards. Tables A-1 and A-2 of MIL-STD-414 are reproduced here in the Appendix as Table T12-3 and T12-4, respectively.

The choice between the k method (Form 1) and the M method (Form 2) is an important initial decision. These methods were described in Chap. 10.

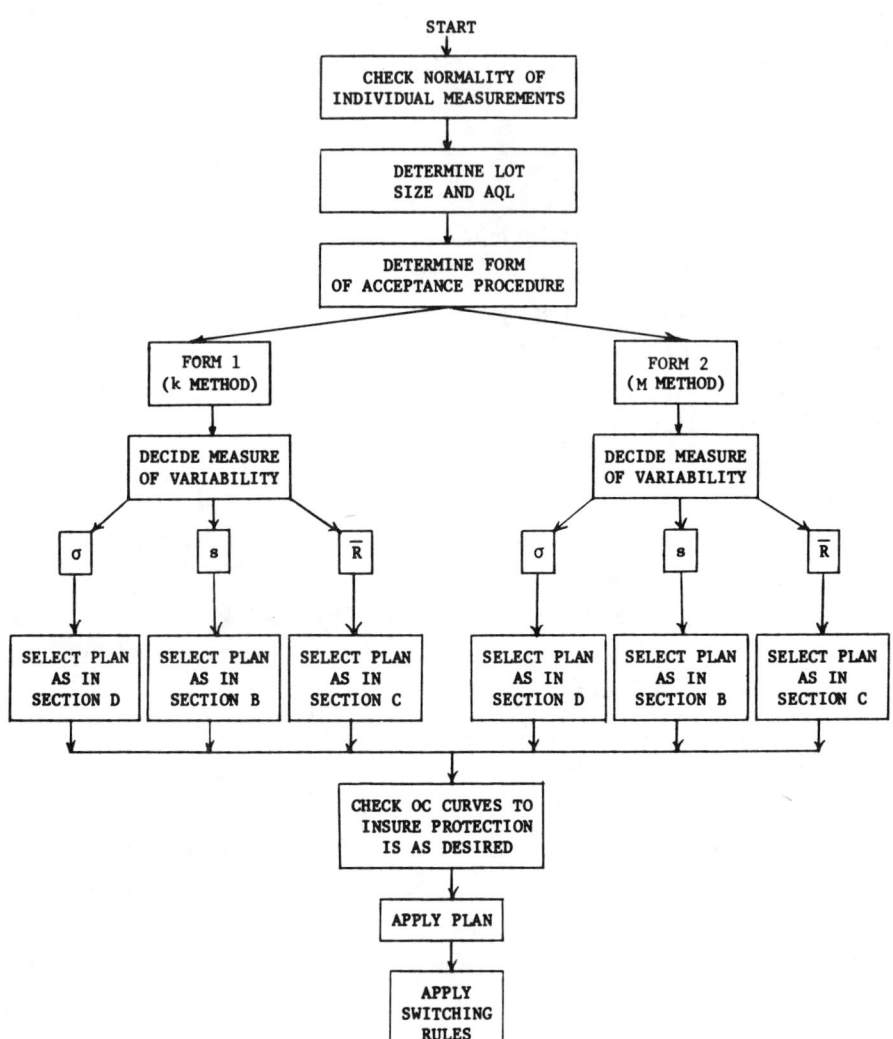

FIG. 12-3 Check sequence for selecting a plan from MIL-STD-414.
(Reprinted by permission from *Quality Control Handbook*, 3rd ed. (J. M.
Juran, Frank M. Gryna, Jr., and Richard S. Bingham, eds.), McGraw-Hill,
New York, 1974, Section 25, Sampling by Variables by E. G. Schilling,
p. 25-15. Copyright 1974, 1962 by McGraw-Hill, Inc. All rights
reserved.)

For single specification limits Form 1 is more straightforward. Values of
the maximum standard deviation (MSD) and maximum allowable range (MAR) are
provided in the standard for use of the k method with double specification
limits. These are useful when a plan is to be plucked out of the standard
to be used singly and not as part of a MIL-STD-414 scheme. However, MIL-
STD-414 recognizes only the M method (Form 2) when the standard is to be
applied to double specification limits. Furthermore, the switching rules
cannot be used unless Form 2 tables and procedures are used. Therefore,
Form 2 is to be recommended for use with a MIL-STD-414 scheme if only for
reasons of economy of effort. Form 1 is easier to explain and administer,
however, and the associated sampling plans are recommended if individual
plans are to be taken from the standard and used not as MIL-STD-414, but
separately as individual variables plans for proportion nonconforming.

It is, of course, necessary to select a measure of variability to be
used. If control charts have confirmed the existence and consistency of a
known standard deviation, variability known (Sec. D) will be the most economic source of sampling plans. If the standard deviation is unknown (but
the distribution shape consistently stays normal), plans using the standard
deviation method (Sec. B) or the range method (Sec. C) may be chosen. The
range plans are easier to explain and calculate than those using sample
standard deviation but are also less efficient. Hand calculators, when
available, have facilitated the computation of the standard deviation. The
choice should be made in keeping with the sampling situation and the competence of the inspection personnel.

Having made these decisions, the specific set of normal, tightened,
and reduced plans is selected from the standard. Vertical arrows shown in
the table are used in the same manner as those in MIL-STD-105D. Figure
12-4 from Schilling (1974) shows how the plans are implemented once they
have been selected. The sample size and acceptance criteria are obtained
from appropriate tables. The statistics associated with a specific plan
are then computed. These are listed in Fig. 12-4 and are explained by
worked examples in each section of MIL-STD-414. The statistic is compared
directly to the acceptance criteria in the manner of the k method when
using Form 1. Form 2 requires estimation of the percent defective in the
lot to obtain the percent estimated to be above an upper specification
limit p_U or below a lower limit p_L. This is done from special tables. The
total estimated percent defective is then compared to M taken from the Form
2 table for acceptance or rejection.

STEP	SECTION	FORM 1	FORM 2
PREPARATORY	—	OBTAIN k AND n FROM APPROPRIATE TABLES	OBTAIN M AND n FROM APPROPRIATE TABLES
DETERMINE CRITERIA	SECTION B (s)	$z_U = \dfrac{U-\bar{x}}{s}$ $z_L = \dfrac{\bar{x}-L}{s}$	$Q_U = \dfrac{U-\bar{x}}{s}$ $Q_L = \dfrac{\bar{x}-L}{s}$
	SECTION C (R)	$z_U = \dfrac{U-\bar{x}}{\bar{R}}$ $z_L = \dfrac{\bar{x}-L}{\bar{R}}$	$Q_U = \dfrac{(U-\bar{x})c}{\bar{R}}$ $Q_L = \dfrac{(\bar{x}-L)c}{\bar{R}}$
	SECTION D (σ)	$z_U = \dfrac{U-\bar{x}}{\sigma}$ $z_L = \dfrac{\bar{x}-L}{\sigma}$	$Q_U = \dfrac{(U-\bar{x})v}{\sigma}$ $Q_L = \dfrac{(\bar{x}-L)v}{\sigma}$
ESTIMATION	—	—	ENTER TABLE WITH n AND Q_U OR Q_L TO GET P_U OR P_L
ACTION	SINGLE SPEC.	ACCEPT IF $z_U \geq k$ OR $z_L \geq k$	ACCEPT IF $P_U \leq M$ OR $P_L \leq M$
	DOUBLE SPEC.	ACCEPT IF[*] $z_U \geq k,\ z_L \geq k$ AND $s \leq$ MSD OR $\bar{R} \leq$ MAR	ACCEPT IF $P_U + P_L \leq M$
c = SCALE FACTOR		$v = \sqrt{\dfrac{N}{N-1}}$	[*]NOT OFFICIAL PROCEDURE

FIG. 12-4 Application of MIL-STD-414. (From Schilling, 1974a, p. 20. Reprinted by permission.)

Detailed examples of the selection of plans from MIL-STD-414 and their operation are given in later sections of this chapter.

MEASURES

Only the operating characteristic curves are given as measures of the plans contained in MIL-STD-414. These are for individual plans and not for the scheme as a whole. Since the plans for the standard deviation, range, and variability known methods are matched, and the k and M methods are equivalent for single specification limits, only one set of OC curves is given. These are for the standard deviation method. The others are assumed sufficiently well matched to be represented by those shown. The operating characteristic curve of the plan Code F, 2.5 percent AQL is shown in Fig. 12-5.

The operating characteristic curves of MIL-STD-414 may be used to select individual plans to be used outside the MIL-STD-414 sampling system. In this case, the operating characteristic curve desired is found and the acceptance criteria determined from the corresponding table of normal plans. In no sense should the resulting plan be referred to as a MIL-STD-414 plan, since MIL-STD-414 implies full use of the sampling system based on the switching rules. Nevertheless, it is a natural compendium of variables plans for proportion nonconforming and can be used to effectively select individual plans for special applications.

IMPLEMENTATION OF FORM 2

Implementation of MIL-STD-414 is best shown by example. Since Form 2, the M method, is the preferred procedure in that the switching rules are based on its estimates, it will be presented first. Also, the standard deviation method is used since the range method and variability known involve only slight modifications of the procedure (see Fig. 12-4). Double specification limits are shown since the single specification limit procedure follows from that given. Consider the following example, adapted from MIL-STD-414 (p. 69):

> The specifications for electrical resistance of a certain
> electrical component is 650.0 ± 30 ohms. A lot of 100
> items is submitted for inspection with AQL = 2.5% for the
> upper and AQL = 1% for the lower specification limits.
> Suppose the values of sample resistances are as follows:
> 643, 651, 619, 627, 658, 670, 673, 641, 638, 650

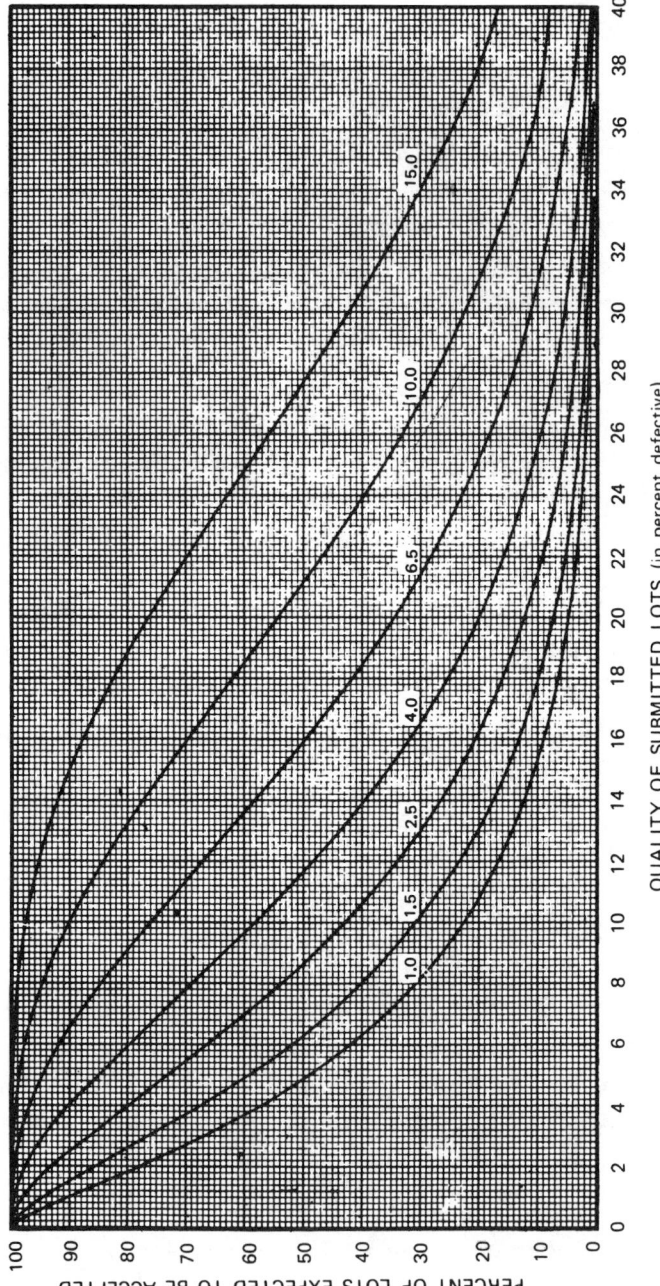

FIG. 12-5 MIL-STD-414 Operating characteristic curve for Code F, standard deviation method. (MIL-STD-414, Table A-3, p. 11.)

Assume the electrical resistances of this device have been shown to be normally distributed. In fact, these data were plotted on normal probability paper in Chap. 3. The sample size code letter table (Table T12-4) shows Code F to apply under Inspection Level IV, which is used unless some other level is specified. The master table for normal and tightened inspection (Table T12-5) and the master table for reduced inspection (Table T12-6) give the following criteria for the plans involved in the MIL-STD-414 sampling scheme

	1% AQL	2.5% AQL
Tightened	n = 10, M = 2.17	n = 10, M = 4.77
Normal	n = 10, M = 3.26	n = 10, M = 7.29
Reduced	n = 4, M = 5.50	n = 4, M = 16.45

Since the scheme starts on normal inspection, we will illustrate application of the normal plan. We find

$$\bar{X} = 647 \qquad s = 17.2$$

The quality indices are

$$Q_L = \frac{(\bar{X} - L)}{s} = \frac{647 - 620}{17.2} = 1.57$$

$$Q_U = \frac{(U - \bar{X})}{s} = \frac{680 - 647}{17.2} = 1.92$$

Using the table for estimating the lot percent defective (Table T12-7), the values of Q_L and Q_U and the sample size are cross tabulated to give estimates of percent defective of

Lower specification: p_L = 4.92
Upper specification: p_U = 1.62
Overall: $p = p_L + p_U$ = 6.54

These estimates are compared against the respective critical values of M, to obtain

$$p_L = 4.92 > M_L = 3.26 \qquad \text{Reject}$$
$$p_U = 1.62 < M_U = 7.29 \qquad \text{Accept}$$
$$p = 6.54 < M_U = 7.29 \qquad \text{Accept}$$

The lot is rejected since the lower estimated percent defective does not meet the acceptance criterion M. When there are different AQLs for the lower and upper specifications, it is necessary to test the upper and lower estimated percent defective separately. Furthermore, the total estimated percent defective p is tested against the value of M associated with the larger AQL.

These estimates of percent defective are slightly smaller than those obtained using the probability plot of Chap. 4, due to estimation by the minimum variance-unbiased technique. This procedure is somewhat more complicated than the procedure required if both specification limits had the same AQL. Suppose the AQL of 1.0 percent applied to both specification limits. Then the acceptance procedure would simply be to compute the total estimated percent defective and compare it to the value of M for 1.0 percent AQL. That is,

$$p = 6.54 > M = 3.26$$

and the lot would be rejected. The OC curve for such a plan will be found in Fig. 12-5 labeled 1.0 percent AQL. The OC curves for different AQLs on the lower and upper specification limits are not given since they would depend upon the split of percent defective between the specifications.

Having rejected the lot, the estimated value of p = 6.54 would be entered into the process average to be utilized in the switching procedure.

IMPLEMENTATION OF FORM 1

As an illustration of Form 1, a single specification limit will be used, since the standard does not advocate use of the k method with double specification limits. However, if a Form 1 plan is to be used with double specification limits, the maximum standard deviation, MSD, can be found from the table of values of F for maximum standard deviation (Table T12-10). It gives values of F which are used to compute the maximum standard deviation as

$$MSD = F(U - L)$$

where U and L are the upper and lower specification limits. The plan is applied to each specification limit separately if $s \leq MSD$. Of course the lot is rejected if $s > MSD$.

Consider the following example, adapted from MIL-STD-414 (p. 69):

> The specification for minimum electrical resistance of a
> certain electrical component is 620 ohms. A lot of 100
> items is submitted for inspection with an AQL = 1.0%.
> Suppose the values of sample resistances are as follows:
> 643, 651, 619, 627, 658, 670, 673, 641, 638, 650

Assume the electrical resistances of this component have been shown to be normally distributed. The sample size code letter table (Table T12-4) shows Code F to apply using Inspection Level IV, which is used unless some other level is specified. The master tables for normal and tightened inspection (Table T12-8) and the master table for reduced inspection (Table T12-9) give the following criteria for the plans involved in the MIL-STD-414 sampling scheme, Form 1.

> Tightened: n = 10, k = 1.84
>
> Normal: n = 10, k = 1.72
>
> Reduced: n = 4, k = 1.34

The switching rules begin with normal inspection, and so the normal inspection plan will be illustrated here. In application of the normal plan we have

$$\overline{X} = 647 \qquad s = 17.2$$

$$t_L = \frac{(\overline{X} - L)}{s} = \frac{647 - 620}{17.2} = 1.57$$

and since $1.57 < 1.72$, the lot is rejected. The MSD is, of course, not used with single specification limits.

To use the switching rules, the estimated percent defective must be determined. This is done using the Form 2 criteria and tables under the M method. From the previous example of implementation of Form 2, we have $p_L = 4.92$ percent with the same data. This is the value that would be entered into the computations of the process average and compared to appropriate criteria for application of the switching rules.

IMPLEMENTATION OF PLANS FOR RANGE AND VARIABILITY KNOWN

Implementation of Form 1 and Form 2 under the range method or variability known is very much like that under the standard deviation method. The principle change is in the statistic to be compared to the acceptance criteria. For Form 1 the statistic remains essentially the same as that for the standard deviation method with $\overline{R}$ or σ substituted for s (see Fig. 12-4). When Form 1 is to be used with double specification limits, the maximum allowable range (MAR), which serves the same purpose as the maximum standard deviation (MSD), is calculated from f factors given in the standard to obtain

$$MAR = f(U - L)$$

in a manner similar to the procedure used with s. For Form 2, the statistic
is changed by the addition of a constant so that $\overline{R}/c$ or σ/v is substituted
for s (see Fig. 12-4). Here

$$c = d_2^* \qquad v = \sqrt{\frac{n}{n-1}}$$

where d_2^* is the adjusted d_2 factor developed by Duncan (1955). This is
necessary in order to obtain the minimum variance unbiased estimate of p
characteristic of the M method.

In all range plans, $\overline{R}$ is the average range of subsamples of 5 in the
sample of n. Units are assigned to the subsamples in the order in which
they are drawn (assuming random sampling). Naturally, for small samples
not divisible by 5 the range of the full sample is used (i.e., 3, 4, 7).

Aside from these changes, the procedures for implementation of plans
for the three measures are essentially the same.

MATCH BETWEEN MIL-STD-414 AND MIL-STD-105D

In 1976 the American National Standards Institute Committee Z-1 on Quality
Assurance recommended that a revision of the ANSI version of MIL-STD-414
be made incorporating some of the suggestions made by Gascoigne (1976)
resulting from his work on British Defence Standard (05-30/1) and with the
International Organization for Standardization. Principal among these was
a method for adjusting the code letter of the ANSI version of MIL-STD-414
to make its operating characteristic curves roughly match those of the ANSI
version of MIL-STD-105D (ANSI Z1.4(1971)) at the adjusted code letter and
AQL. Revision of the ANSI version of MIL-STD-414 (ANSI Z1.9) was accomplished by the American Society for Quality Control Standards Group and it
now appears as ANSI Z1.9(1980). Table 12-1 shows the match between the
revised ANSI Z1.9(1980) code letter, the MIL-STD-414 code letter, and the
corresponding code letter of MIL-STD-105D. Plans with these code letters
are roughly matched and will allow switching between variables and attributes plans within the code letters shown at a given AQL. To preserve the
match, MIL-STD-414 AQLs 0.04, 0.065, and 15.00 should not be used and were
dropped from ANSI Z1.9(1980). For example, MIL-STD-105D, Code J, 1.5 AQL
is roughly matched to MIL-STD-414, Code K, 1.5 AQL which matches ANSI Z1.9
(1980), Code J, 1.5 AQL.

Other changes in ANSI Z1.9(1980) from the earlier version identical
to MIL-STD-414 included an update of terminology and changes in the

TABLE 12-1 Matching Code Letters[*]

MIL-STD-105D (ANSI Z1.4(1971)) Code Letter	MIL-STD-414 Code Letter	ANSI Z1.9(1980) Code Letter
B	B	B
C	C	C
D	D	D
E	E	E
F	F	F
G	G	G
H	H	H
H	I	I
J	K	J
K	M	K
L	N	L
M	O	M
N	P	N
P	Q	P

[*]Delete MIL-STD-414 AQLs: 0.04, 0.065, 15.00.

switching rules, inspection levels and other features to match MIL-STD-105D. Standards Z1.4 and Z1.9 may be obtained from the American National Standards Institute.

FURTHER CONSIDERATIONS

An excellent description of the theory behind MIL-STD-414 has been given by Lieberman and Resnikoff (1955) in the *Journal of the American Statistical Association*. Much of this material was later presented in a detail technical report on MIL-STD-414 published by the Assistant Secretary of Defense (1958). These works give a detailed technical description of the background of the standard. A classic review of MIL-STD-414 was undertaken by Kao (1971) and appears in the *Journal of Quality Technology*. In a two part series, Duncan (1975) and Bender (1975) described the history and matching of MIL-STD-414 to other national and international standards including MIL-STD-105D.

PROBLEMS

1. MIL-STD-414, 1.0 percent AQL is specified and a lot of 390 pieces is to be inspected. Find the associated set of single-sided Form 1 normal, tightened, and reduced plans when the standard deviation is unknown and estimated by s.

2. If the upper specification limit was 130, determine the acceptability of a lot for the plans of Prob. 1 if $\overline{X}$ = 110, s = 10.

3. MIL-STD-414, 1.0 percent AQL is specified and a lot of 390 pieces is to be inspected. Find the associated set of two-sided Form 2 normal, tightened, and reduced plans when the standard deviation is unknown and estimated by s.

4. If the upper and lower specification limits are 130 and 90, respectively, determine acceptance under the plans found in Prob. 3 if $\overline{X}$ = 110, s = 10.

5. If Form 1 is to be used with the double specification limits of Prob. 4, what is the MSD? Would s = 10 pass the MSD?

6. What is the LTPD of the plan Code F, 0.4 AQL? What is its indifference quality?

7. What action should be taken under Code G, 4.0 AQL normal inspection if 7 of 10 lots have estimated percent defective greater than the AQL and the process average of the last 10 lots exceeds the AQL? What would be the minimum possible process average under these circumstances? Is it possible to switch to reduced inspection under these conditions?

8. What MIL-STD-414 plan would roughly match MIL-STD-105D, Code K, 0.65 AQL?

9. Suppose range plans are substituted for the standard deviation plans of Prob. 4. The criteria for the normal range plan is n = 30, M = 2.81, c = 2.353. Compute Q_U when $\overline{X}$ = 110, $\overline{R}$ = 23.53. If the corresponding p_U = 1.88, should the lot be accepted?

10. Suppose the standard deviation was known and the inspection of Prob. 4 was to be applied. The criteria for the normal standard deviation known plan is n = 9, M = 2.59, v = 1.061. If $\overline{X}$ = 110, σ = 10.61, should the lot be accepted? Use upper tail percentages of the normal distribution as estimates of p_U and p_L.

REFERENCES

American National Standards Institute (1980), *American National Standard: Sampling Procedures and Tables for Inspection by Variables for Percent Nonconformance*, ANSI/ASQC Standard Z1.9-1980, American Society for Quality Control, Milwaukee, Wisconsin.

Bender, A. (1975), Sampling by Variables to Control the Fraction Defective: Part II, *Journal of Quality Technology*, 7(3): 139-143.

Duncan, A. J. (1955), The Use of Ranges in Comparing Variabilities, *Industrial Quality Control*, 11(5): 18-22.

Duncan, A. J. (1975), Sampling by Variables to Control the Fraction Defective: Part I, *Journal of Quality Technology*, 7(1): 34-42.

Gascoigne, J. C. (1976), *Future International Standards on Sampling by Variables*, American Society for Quality Control Technical Conference Transactions, Toronto, Ontario, pp. 472-478.

Kao, J. H. K. (1971), MIL-STD-414 Sampling Procedures and Tables for Inspection by Variables for Percent Defective, *Journal of Quality Technology*, 3(1): 28-37.

Lieberman, G. J., and G. J. Resnikoff (1955), Sampling Plans for Inspection by Variables, *Journal of the American Statistical Association*, 50: 457-516.

Schilling, E. G. (1974), Variables Sampling and MIL-STD-414, *Quality Progress*, 7(5): 16-20.

United Kingdom Ministry of Defence (1974), *Sampling Procedures and Charts for Inspection by Variables* (Defence Standard 05-30), Ministry of Defence, Directorate of Standardization, London.

United States Department of Defense (1950), *Military Standard, Sampling Procedures and Tables for Inspection by Attributes* (MIL-STD-105A), U.S. Government Printing Office, Washington, D.C.

United States Department of Defense (1963), *Military Standard, Sampling Procedures and Tables for Inspection by Attributes* (MIL-STD-105D), U.S. Government Printing Office, Washington, D.C.

United States Department of Defense (1957), *Military Standard, Sampling Procedures and Tables for Inspection by Variables for Percent Defective* (MIL-STD-414), U.S. Government Printing Office, Washington, D.C.

United States Department of Defense (1958), *Mathematical and Statistical Principles Underlying MIL-STD-414*, Technical Report, Office of the Assistant Secretary of Defense (Supply and Logistics), Washington, D.C.

SPECIAL PLANS AND PROCEDURES

A variety of plans and procedures has been developed for special sampling
situations involving both measurements and attributes. Only a few of them
can be shown here. Each is tailored to do a specific job under prescribed
circumstances. They range from a simplified variables approach involving
no calculations to a more technically complicated combination of variables
and attributes sampling in a so-called mixed plan. They provide useful
options in the application of acceptance sampling plans to unique sampling
situations.

NO-CALC PLANS

Since variables plans for percent nonconforming usually assume an underlying
normal distribution of measurements, probability plots would seem to be a
natural tool for acceptance sampling. Such plots can provide a visual check
on the normality of the distribution involved, while at the same time affording an opportunity to estimate the fraction nonconforming in the lot. This
was shown in Chap. 3. Probability plots can also be used directly for lot
acceptance. Such a plan has been developed by Chernoff and Lieberman (1957).
It assumes underlying normality of individual measurements. While its
results are approximate, it requires no calculations and can be used in inspection situations where mathematical calculation is out of the question.
The authors of the plan point out (p. 5) that "No-Calc is not a replacement
for the usual variables procedures when a contract between two parties
exists and calls for inspection by variables." Nevertheless, it is particularly useful for internal in-process acceptance inspections and the like.

The No-Calc procedure is matched to MIL-STD-414. Plans are identified
by code letter and AQL. The OC curves of MIL-STD-414 approximate those of

No-Calc and can be used to select a plan. Sample sizes are, of course, limited to the MIL-STD-414 sequence which appears in the No-Calc tables. For a given sample size, the No-Calc procedure is as follows:

1. Plot the sample results on normal probability paper using the No-Calc plotting positions of Appendix Table T13-1 when $n \leq 20$; when $n > 20$ use the approximation

$$\hat{P}_{(i)} = \frac{i - \frac{1}{2}}{n} \times 100$$

2. If the points do not plot roughly in a straight line, discontinue the procedure on the grounds that the underlying population may not be normal.

3. Estimate the underlying normal distribution by drawing a straight line through the points.

4. Locate the specification limits on the x-axis and use the straight line to estimate the percent nonconforming beyond the single or double specification limits. Call this estimate $\hat{p}$.

5. Obtain the critical value of p^* from Appendix Table T13-2.

6. If $\hat{p} \leq p^*$, accept the lot; if $\hat{p} > p^*$, reject the lot.

Clearly p^* plays the role of M in MIL-STD-414, while $\hat{p}$ acts as p_L, p_U or p_T.

To illustrate the application of the No-Calc plan, consider the following example:

> The specification for minimum electrical resistance of a certain electrical component is 620 ohms. A lot of 100 items is submitted for inspection with an AQL = 1.0 percent. A 10 percent Limiting Quality of 15 percent is desired. Suppose values of sample resistances are as follows: 643, 651, 619, 627, 658, 670, 673, 641, 638, 650.

A search through the OC curves of MIL-STD-414 shows Code F, 1.0 percent AQL is closest to the specifications of the plan. Its 10 percent limiting quality is just about 15 percent, while Code E and Code G differ substantially from that at 1.0 percent AQL. Reference to Appendix Table T13-2 shows that a sample size of 10 should be taken with a critical value of $p^* = 3.88$. Plotting positions are obtained from Appendix Table T13-1 and associated with the observations as follows:

Order (i)	$X_{(i)}$	$P_{(i)}$
1	619	4.4
2	627	16.4
3	638	26.2
4	641	35.8
5	643	45.3
6	650	54.7
7	651	64.2
8	658	73.8
9	670	83.6
10	673	95.6

The probability plot appears as Fig. 13-1. It estimates that 6 percent of the underlying distribution is below the lower specification limit of 620 ohms. Since

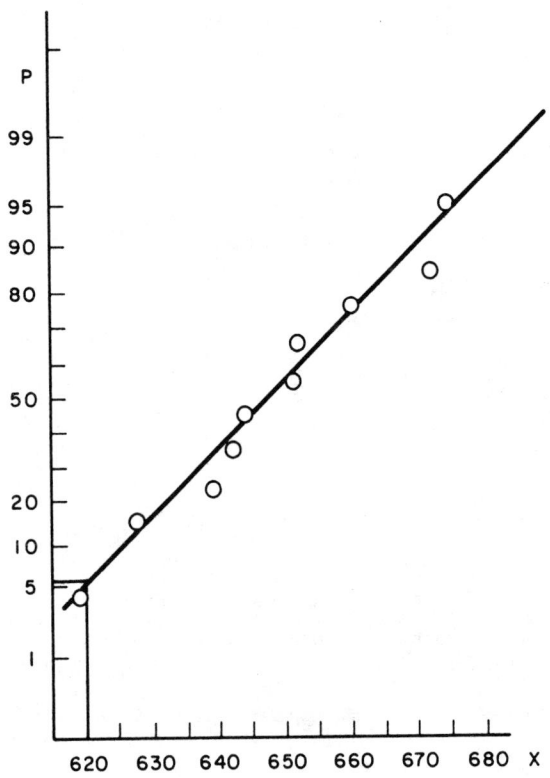

FIG. 13-1 Probability plot.

$\hat{p} = 6.0 > p* = 3.88$

the lot is rejected. It is interesting to note that this is the same estimate obtained from the probability plot of Chap. 3 and illustrates how good
the approximation

$$\hat{p}_{(i)} = \frac{i + \frac{1}{2}}{n}$$

is even with a reduced amount of data.

The No-Calc plan can easily be implemented in the shop by drawing a
vertical decision line at the specification limit (in the example 620). On
this line mark the critical value p* (3.88 here). Label the line below p*
as "accept" and above p* as "reject." Take the action indicated by the
intersection of the probability plot line with the decision line. Care
should be taken so that the decision line does not prejudice drawing the
probability plot line.

LOT PLOT PLANS

Probably no acceptance sampling procedure has more intuitive appeal for
inspectors than the Lot Plot method, developed by Dorian Shainin at the
Hamilton Standard Division of United Aircraft Company. Shainin (1950)
published an extensive introduction and description of the plan in *Industrial Quality Control*. The reader is well advised to study his paper for
the details of the method. Its wide acceptance attests to its appeal and
value in practical applications. Detailed examples of its use in various
companies have also been given by Shainin (1952).

Lot Plot uses a constant sample size of 50 observations. It is based
on construction of the histogram of the sample. The mean and the standard
deviation are estimated and the resulting "3σ" limits used as an acceptance
criterion when compared to the specification limits. It serves as a particularly useful tool in the introduction of statistical acceptance sampling
techniques and in applications where more sophisticated methods are inappro-
priate or not likely to be well received. As Shainin (1952) has pointed
out, with the Lot Plot method, " ... it was possible to bring the method of
analysis down to where anyone who can read a micrometer can be taught in
less than a week to analyze Lot Plots completely."

The original procedure utilizes the Pearson-Tippett method to estimate
the standard deviation from 10 ranges of five obtained from subsamples of
the 50 observations. We shall present a variation of Lot Plot, due to

LOT PLOT SAMPLING INSPECTION

Vendor _Beacon Screw Co_   P.O. Number _19530_
Part No. _10-16799_   Name _Adjusting Screw_   Quantity _600_
R.F. or Contract _11392_   Insp.Oper. _After Steelblast_   Date _5-22-51_
Spec. _Rockwell C46-56_   Inspector _937_

		1-20	FX	X	F	X²	FX²	
	+10			+10		100		ΣFX = +6
	+9			+9		81		
	+8			+8		64		.02ΣFX = X̄ = +.12
	+7			+7		49		
	+6			+6		36		ΣFX² = 102
56	+5			+5		25		
55	+4	✓ (ULL)	4	+4	1	16	16	.02ΣFX² = 2.04
54	+3	✓✓	6	+3	2	9	18	
53	+2	✓✓✓✓✓	10	+2	5	4	20	3σ = 4.2
52	+1	✓✓✓✓✓✓✓✓✓✓	10	+1	10	1	10	
51	0	✓✓✓✓✓✓✓✓✓✓✓✓✓✓✓ (X̄)	0	0	15	0	0	Cell Interval = 1
50	-1	✓✓✓✓✓✓✓✓✓	-10	-1	10	1	10	
49	-2	✓✓✓✓✓✓✓	-14	-2	7	4	28	
48	-3			-3		9		
47	-4	(L.L.L)		-4		16		
46	-5			-5		25		
	-6			-6		36		
	-7			-7		49		
	-8			-8		64		
	-9			-9		81		
	-10			-10		100		

Accept	✓		Spec	%	Extent		Remarks	
Rework			Beyond H.L.					
Mat'l Review			Beyond L.L.					
Screen 100 %								

FIG. 13-2 Example of completed modified Lot Plot form. (From Ashley, 1952, p. 30. Reprinted by permission.)

Ashley (1952), which was applied at the Bendix Aviation Corporation. This method calculates the standard deviation directly from the frequency distribution itself, thus avoiding use of the range. This preserves all the qualities intended by Shainin but leads to somewhat more rapid calculation, possible computerization, and no need to order the observations into subgroups as taken.

A completed modified Lot Plot form appears as Fig. 13-2. After completing the heading, the form is filled in from left to right as follows:

1. A sample of 50 is taken.

2. The mean of the first five observations is used to locate the center of the distribution. Enter this value in the leftmost column next to the value of 0 in the second column, suitably rounded to obtain a "nice" starting point.

3. Mark off the cells above and below the center value. Individual measurements (cell width of 1) are desirable but not necessary. If cells of width other than 1 are to be used, enter the cell midpoints above and below the middle cell. Shainin suggests a

cell width roughly equal to one-fourth the range of the first
five observations.

4. Each observation is tallied by a check mark in the space provided.
   This will automatically provide a histogram of the sample.

5. If the histogram appears to be obviously nonnormal, stop and investigate the cause.

6. The tally for each row is recorded in the F column. The numbers
   at the top of the grid facilitate the count.

7. The FX column is filled in as the product of the F and X values
   shown for each row.

8. The FX^2 column is filled in as the product of the F and X^2 values
   shown for each row.

9. The sum of the FX column is recorded in the upper right box.
   When the sum is multiplied by 0.02 the result is a coded $\overline{X}$ recorded in the second box.

10. The sum of the FX^2 column is recorded in the third box on the
    right. When the value is multiplied by 0.02 the result is recorded
    in the fourth box.

11. Table 13-1 is then used to estimate 3σ. The coded $\overline{X}$ (second box)
    is entered at the top and $.02 \, \Sigma \, FX$ (fourth box) is entered at the
    side. The resulting closest tabulated value estimates 3σ and is
    entered in the fifth box. If the table does not cover the values
    obtained, the estimate of 3σ of the X's can be calculated from
    the formula

$$3\sigma = 3 \sqrt{.02 \, \Sigma \, FX^2 - \overline{X}^2}$$

12. The cell width w is entered in the sixth box. If it is necessary
    to estimate the mean $\hat{\mu}$ and standard deviation $\hat{\sigma}$ in units of the
    original measurement, use

$$\hat{\mu} = w \, \overline{X}$$

$$\hat{\sigma} = \frac{w(3\sigma)}{3} = w \sqrt{.02 \, \Sigma \, FX^2 - \overline{X}^2}$$

To assess the acceptability of the lot, the upper lot limit (ULL) and lower
lot limit (LLL) as well as the specifications are drawn on the chart. To
draw these limits,

1. Using the X column (second and twenty-fourth columns) draw a
   horizontal line at the coded $\overline{X}$. This is an estimate of the mean

TABLE 13-1   3σ Values for Lot Plot

X Values to Nearest Tenth

√ΣX²/20	.0	.1	.2	.3	.4	.5	.6	.7	.8	.9	1.0	1.1	1.2	1.3	1.4	1.5	1.6	1.7	1.8	1.9	2.0	2.1	2.2	2.3	2.4	2.5
.0																										
.5	2.1	2.1	2.0	1.9	1.7	1.5	1.1	.3																		
1.0	3.0	3.0	2.9	2.9	2.7	2.6	2.4	2.1	1.8	1.3																
1.5	3.7	3.7	3.6	3.6	3.5	3.4	3.2	3.0	2.8	2.5	2.1	1.6	.7													
2.0	4.2	4.2	4.2	4.1	4.1	4.0	3.8	3.7	3.5	3.3	3.0	2.7	2.2	1.7	.6											
2.5	4.7	4.7	4.7	4.7	4.6	4.5	4.4	4.3	4.1	3.9	3.7	3.4	3.1	2.7	2.2	1.5										
3.0	5.2	5.2	5.2	5.1	5.1	5.0	4.9	4.8	4.6	4.4	4.2	4.0	3.7	3.4	3.1	2.6	2.0	1.0								
3.5	5.6	5.6	5.6	5.5	5.5	5.4	5.3	5.2	5.1	4.9	4.7	4.5	4.3	4.0	3.7	3.4	2.9	2.3	1.5							
4.0	6.0	6.0	6.0	5.9	5.9	5.8	5.7	5.6	5.5	5.4	5.2	5.0	4.8	4.6	4.3	4.0	3.6	3.2	2.6	1.9						
4.5	6.4	6.4	6.3	6.3	6.2	6.2	6.1	6.0	5.9	5.8	5.6	5.4	5.2	5.0	4.8	4.5	4.2	3.8	3.4	2.8	2.1	.9				
5.0	6.7	6.7	6.7	6.6	6.6	6.5	6.5	6.4	6.3	6.1	6.0	5.8	5.7	5.5	5.2	5.0	4.7	4.4	4.0	3.5	3.0	2.3	1.2			
5.5	7.0	7.0	7.0	7.0	6.9	6.9	6.8	6.7	6.6	6.5	6.4	6.2	6.0	5.9	5.6	5.4	5.1	4.8	4.5	4.1	3.7	3.1	2.4	1.4		
6.0	7.3	7.3	7.3	7.3	7.3	7.2	7.1	7.0	6.9	6.8	6.7	6.6	6.4	6.2	6.0	5.8	5.6	5.3	5.0	4.6	4.2	3.8	3.2	2.5		
6.5	7.6	7.6	7.6	7.6	7.6	7.5	7.4	7.4	7.3	7.2	7.0	6.9	6.7	6.6	6.4	6.2	6.0	5.7	5.4	5.1	4.7	4.3	3.9	3.3	1.5	1.5
7.0	7.9	7.9	7.9	7.9	7.8	7.8	7.7	7.7	7.6	7.5	7.3	7.2	7.1	6.9	6.7	6.5	6.3	6.1	5.8	5.5	5.2	4.8	4.4	3.9	2.6	2.6
7.5	8.2	8.2	8.2	8.2	8.1	8.1	8.0	7.9	7.9	7.8	7.6	7.5	7.4	7.2	7.1	6.9	6.7	6.4	6.2	5.9	5.6	5.3	4.9	4.5	3.3	3.4
8.0								8.2	8.1	8.0	7.9	7.8	7.7	7.5	7.4	7.2	7.0	6.8	6.5	6.3	6.0	5.7	5.3	5.0	4.0	4.0
8.5											8.2	8.1	8.0	7.8	7.7	7.5	7.3	7.1	6.9	6.6	6.4	6.1	5.7	5.4	4.5	4.5
9.0													8.3	8.1	8.0	7.8	7.6	7.4	7.2	7.0	6.7	6.4	6.1	5.8	5.0	5.0
9.5															8.2	8.1	7.9	7.7	7.5	7.3	7.0	6.8	6.5	6.2	5.4	5.4
10.0																	8.2	8.0	7.8	7.6	7.3	7.1	6.8	6.6	5.8	5.8
10.5																		8.3	8.1	7.9	7.6	7.4	7.1	6.8	6.2	6.2
11.0																				8.2	7.9	7.7	7.4	7.2	6.5	6.5
11.5																					8.2	8.0	7.7	7.5	6.9	6.9
12.0																						8.3	8.0	7.8	7.2	7.2
12.5																							8.3	8.1	7.5	7.5
13.0																								8.3	7.8	7.8
13.5																									8.1	8.1

(From Ashley, 1952, p. 31.   Reprinted by permission.)

of the distribution and can be decoded simply by reading the
corresponding value from the cell midpoints recorded (first column).

2. Mark a distance 3σ (fifth box) in terms of the X column above and
    below the coded $\overline{X}$. These are the lot limits, LLL and ULL. They
    may be read in terms of the original measurements simply by extend-
    ing them to the cell midpoints (column 1) and reading off the
    appropriate values.

3. Draw the specification limits on the chart in terms of the cell
    midpoints (column 1).

The acceptance criteria are

1. Lot limits within specification limits, accept.

2. Lot limits outside specification limits

    a. Count the number of X spaces by which the lot limit exceeds
        the specification limit. Call this E.

    b. Compute

        $$Z = 3\left(1 - \frac{E}{3\sigma}\right)$$

        where the 3σ value is taken from the fifth box.

    c. Estimate the proportion of product out of specification as the
        upper normal tail area corresponding to Z. This may be done
        graphically by the inspector simply by providing normal proba-
        bility paper having a straight line drawn which corresponds to
        the standard normal distribution. A suitable table could also
        be provided.

    d. If the estimated value is less than a predetermined allowable
        value, accept the lot.

    e. Otherwise, reject the lot.

The method suggested here for estimating the proportion out of specification
has been found by the author to be accurate enough for most practical
purposes.

NARROW LIMIT GAUGING

The predominance of attributes type data in industry attests to the economic
advantages of collecting go no-go data over recording specific variables
data. Gauging is often to be preferred over measurement. This is because
it takes less skill to gauge properly, is faster, less costly, and has
become something of a tradition in certain industries. As put by Ladany
(1976, p. 225),

Variables sampling plans have the known advantage, over
sampling plans for attributes, of requiring a much smaller
sample size ... This is due to the possibility of utilizing more effectively quantitative data as opposed to
qualitative data. The statistical advantage may be out-
weighed by economic considerations, since the cost of
inspecting a unit, using a simple go-no-go gage, is often
much lower than the cost of determining the exact value
of the critical characteristic variable by a measuring
instrument.

Narrow limit sampling plans (sometimes called *compressed limit plans*)
effectively bridge the gap between variables and attributes procedures by
utilizing go no-go gauges set up on the principles of variables inspection.
Originated in England by Dudding and Jennett (1944), they were introduced
into the United States by Mace (1952). Ott and Mundel (1954) did much to
extend the theory and application of the procedure. The narrow limit plans
were initially regarded as a process control device as evidenced by the
title, "Quality Control Chart Technique When Manufacturing to a Specification," used by Dudding and Jennett. Nevertheless, narrow limit plans provide an excellent technique for acceptance sampling in that they are based
on the same assumptions as known standard deviation variables plans for
proportion nonconforming but require little calculation and are easier to
use.

The basic idea is a simple one. Since the sample size required by an
attributes plan is related inversely to the size of the proportion nonconforming it is required to detect, a pseudo-specification, or narrow limit,
is set inside the specification limits. The sampling plan is set up on the
number of items failing the narrow limit rather than the specification
limit itself. Since the relationship between the pseudo and the actual
proportions nonconforming is strictly monotonically increasing, the one can
be used to control the other. This is then done by using the narrow limit.
These plans assume the standard deviation σ to be known and the underlying
distribution of measurements to be normal. Of course, when the specification limits are more than 6σ apart, individual narrow limit plans can be
applied on each side of double specification limits.

Using the notation of Ott and Mundel (1954), narrow limit gauge plans are specified by three quantities:

  n = the sample size

  c = the acceptance number for units allowed outside the narrow limit
      gauge

  t = the compression constant, the narrow limit is set to inside the
      specification limit

To implement a plan,

1. Check to be sure the underlying distribution of measurements is
   consistently normal using probability plots, tests of fit, control
   charts, etc., on past data.

2. For a single upper specification limit U or a lower specification
   limit L set the narrow limit gauge at

        $U - t\sigma$ or $L + t\sigma$

   When double specification limits are at least 6σ apart, individual
   narrow limit plans can be applied to each of the specification
   limits separately.

3. Take a random sample of size n.

4. Gauge to the narrow limit. Items outside the narrow limit are
   treated as nonconforming to the narrow limit gauge. Items inside
   the narrow limit are treated as conforming.

5. Accept if the number nonconforming to the narrow limit gauge is
   less than or equal to c; otherwise, reject.

It should be noted that changes in the criteria for acceptance affects the
operating characteristic curve of the narrow limit gauging procedure in
different ways:

  n increased $\longrightarrow$ plan tightened

  c increased $\longrightarrow$ plan loosened

  t increased $\longrightarrow$ plan tightened

A large value of t can lead to rejections even when p = 0. Ott and Mundel
have found a compression constant of

        $t = 1$

to be very good in practice with moderately small sample sizes. Operating
characteristic curves for several plans having t = 1 are shown in Fig. 13-3
taken from Ott and Mundel (1954).

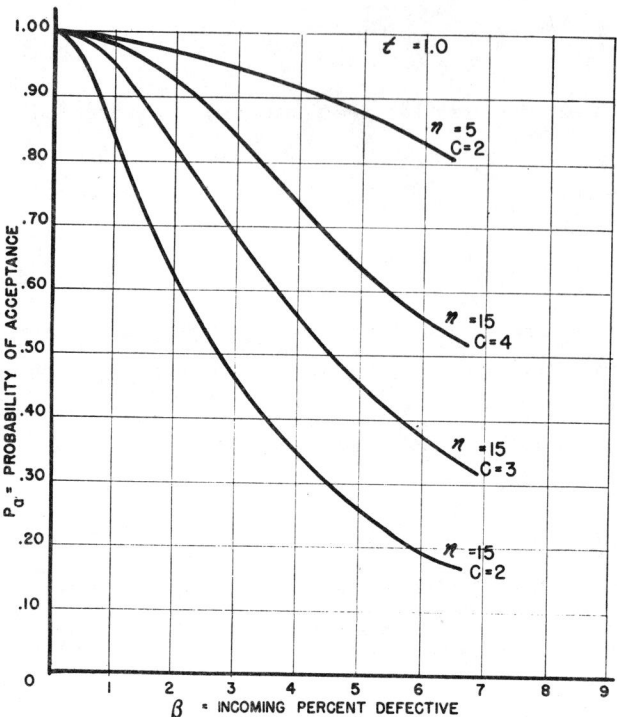

FIG. 13-3 OC curves for NLG plans with t = 1.
(From Ott and Mundel, 1954, p. 25. Reprinted
by permission.)

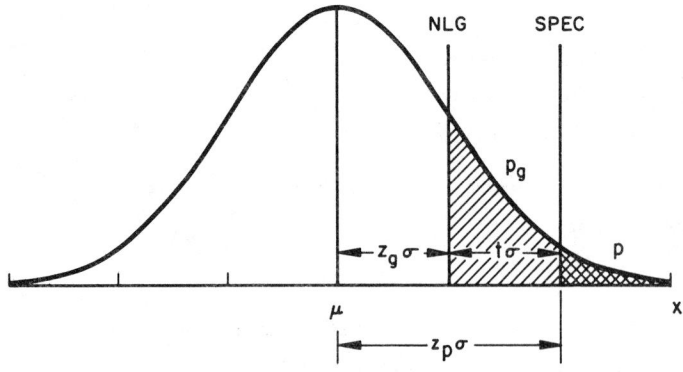

FIG. 13-4 Narrow limit distribution.

The operating characteristic curve of a narrow limit plan is relatively easy to compute. Figure 13-4 shows diagrammatically the principle behind its computation.

Assuming an underlying normal distribution of measurements, each value of proportion nonconforming p will be associated with a fixed position of the mean μ. If we let

Z_γ = standard normal deviate having area γ in the upper tail

then the specification limit will be a distance $Z_p\sigma$ from the mean. Also, the narrow limit gauge will be a distance $t\sigma$ from the specification limit, or a distance $Z_g\sigma$ from the mean. So we have

$$Z_g = Z_p - t$$

The proportion of units outside the narrow limit p_g is the upper tail normal area cut off by Z_g. The sampling plan will then be applied to a proportion p_g when there is proportion p out of the specification limit. The calculations are summarized in Table 13-2 which illustrates finding the probability of acceptance for the plan n = 15, c = 2, t = 1 using Poisson probabilities to approximate the binomial. Care should be taken to be sure the Poisson approximation applies. If not, the binomial distribution should be used directly. It will be seen that these values are shown on the OC curve of Fig. 13-3.

The following procedure may be utilized to derive narrow limit gauge plans when the Poisson approximation applies. Refer to Fig. 13-4 for a diagrammatic representation of the procedure.

Given

P_1 = producer's quality level

P_2 = consumer's quality level

TABLE 13-2   Calculation of Probability of Acceptance for n = 15, c = 2, t = 1

Proportion out of spec. p	σ Units Z_p	$Z_g = Z_p - t$	Proportion out of NLG p_g	$n\, p_g$	Poisson P_a
.005	2.58	1.58	.057	0.86	.94
.03	1.88	0.88	.189	2.84	.46
.08	1.405	0.405	.341	5.12	.12

α = producer's risk

β = consumer's risk

t = compression constant

1. Determine Z_{p_1} and Z_{p_2}.

2. Compute $Z_{g_1} = Z_{p_1} - t$ and $Z_{g_2} = Z_{p_2} - t$.

3. Obtain upper tail areas p_{g_1} and p_{g_2}.

4. Compute the operating ratio

$$R = \frac{p_{g_2}}{p_{g_1}}$$

5. Determine standard acceptance sampling plan n, c with risks α, β, and operating ratio R.

6. The narrow limit plan is specified as n, c, t.

For example, suppose the following plan is desired:

p_1 = .005 α = .05   t = 1.0

p_2 = .08 β = .10

1. $Z_{.005}$ = 2.576 and $Z_{.08}$ = 1.405

2. Z_{g_1} = 1.576   and Z_{g_2} = 0.405

3. p_{g_1} = .0575   and p_{g_2} = .3427

4. $R = \frac{.3427}{.0575} = 5.96$

5. Use of the table of unity factors gives c = 2, n = 14.2 $\sim$ 15

6. The plan is n = 15, c = 2, t = 1

It can be confirmed that the desired characteristics were essentially obtained by reference to Table 13-2 which was used to compute the operating characteristic curve of this plan. Use of the unity values requires that the Poisson approximation to the binomial apply to both values of np_g.

It is frequently desirable to obtain an optimum narrow limit plan with regard to sample size. For p_1, p_2, α, β, specified as before, Ladany (1976) has developed an iterative procedure for the construction of such a plan. It utilizes a special nomograph based on the Larson (1966) nomograph for the binomial distribution. The nomograph is shown in Fig. 13-5. It

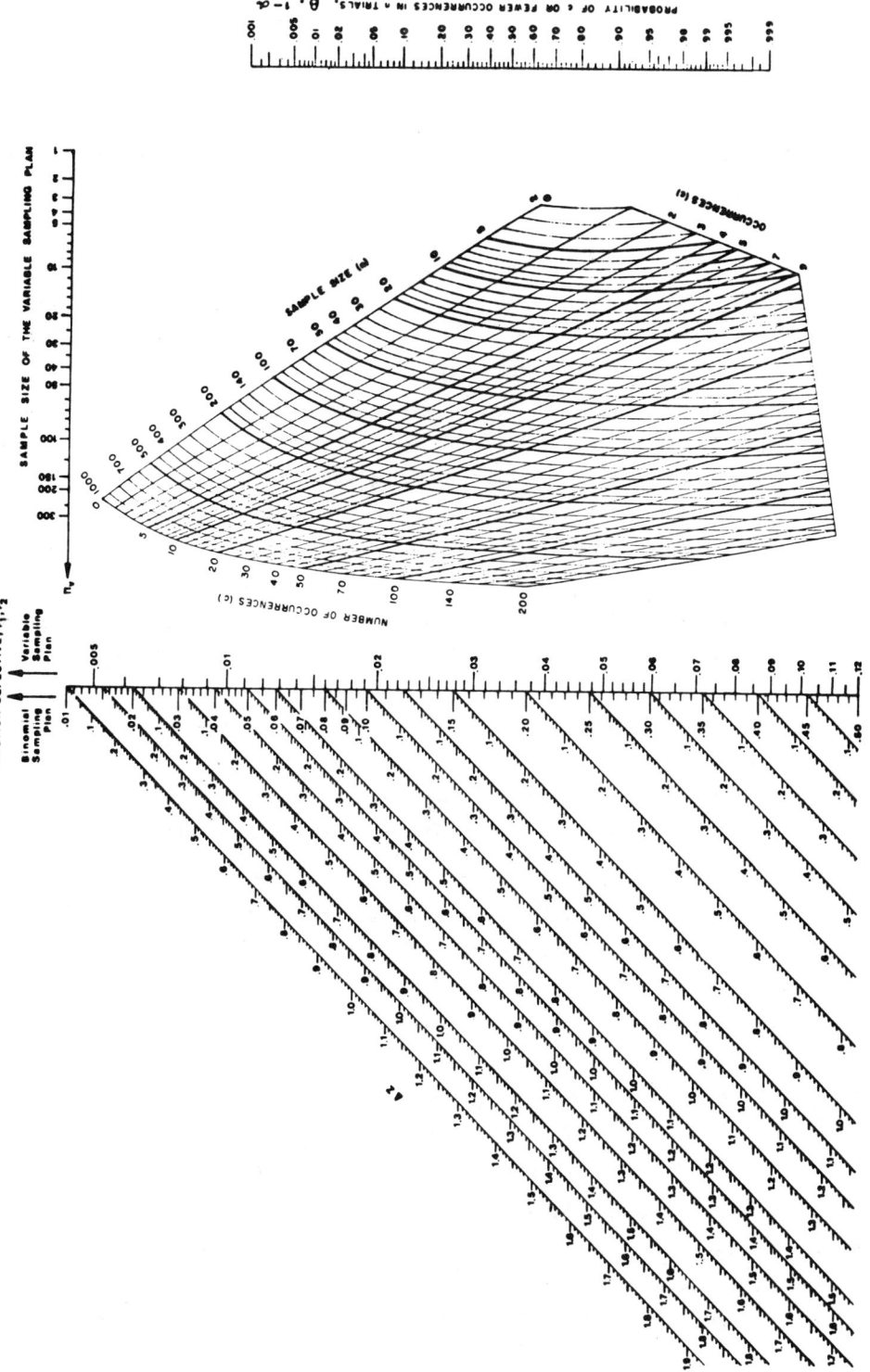

FIG. 13-5 Ladany nomograph of narrow limit gauging sampling plans. (From Ladany, 1976, p. 227. Reprinted by permission.)

should be noted that in Ladany's notation

$t = \Delta Z$

Steps in the application of the procedure are as follows:

1. Connect p_1 and p_2 on the variables sampling plan axis (middle axis, right side) with $(1 - \alpha)$ and β respectively on the probability axis (right half) using two straight lines.

2. Read the corresponding σ known variables plan sample size, from the horizontal axis on top, directly above the point of intersection of the two lines. This serves as an extreme lower bound for the narrow limit sample size.

3. Locate p_1 and p_2 on the binomial sampling plan axis (middle axis, left side) and connect with $(1 - \alpha)$ and β respectively on the probability axis (right side). Using the Larson nomograph read the sample size n_0 and acceptance number c_0 from the grid. This is the plan that would apply without the narrow limit. That is, when $t = 0$.

4. Select a trial value of t, say t_1. From p_1 move down the slanted $\Delta Z (=t)$ axis a distance of t_1. Read over horizontally to obtain p_{g_1} on the binomial sampling plan axis (middle axis, left side). Similarly, move down from p_2 on the slanted $\Delta Z (=t)$ axis a distance t and read over to get p_{g_2}. For example, moving from $p_2 = .08$ a distance $t = 1.0$ on the diagonal scale and reading across gives $p_{g_2} = .343$ which is the same value obtained in Table 13-2.

5. Connect p_{g_1} and p_{g_2} on the binomial sampling plan scale with $1 - \alpha$ and β, respectively, on the probability scale. The intersection of these two lines gives the value of n and c which will provide the desired risks with $t = t_1$.

6. Select another value of t and determine values of n and c for it as in steps 4 and 5.

7. Continue the iterative procedure until the last derived narrow limit plan starts to increase in sample size, with no indication that sample size may be further reduced, or until p_{g_2} exceeds 0.50.

The σ known sample size provides a rough indication of how close the iterative procedure is to optimum. Naturally, the variables sample size will never be reached, however the narrow limit gauge should reduce the sample size by roughly 80 percent of the difference between attributes and variables.

It is best to keep a running table of the results of the iterations.
Note that this table could also be developed by changing t in the tabular
method presented earlier. Ladany (1976) gives the following example:

$$p_1 = .02 \quad \alpha = .05$$
$$p_2 = .08 \quad \beta = .10$$

The nomograph for the example is shown in Fig. 13-6. We see that

1. The initial lines are shown dotted from .02 on the variables
    sampling plan scale to .95 on the probability scale and from .08
    to .10. They cross at point 0.

2. The σ known sample size is n = 20.

3. The points .02 and .08 on the binomial axis are connected to .95
    and .10 on the probability axis to give the plan n = 98, c = 4
    when t = 0.

4. A trial value of t = 0.3 is selected. Moving from .02 on the
    binomial axis down the diagonal ΔZ(=t) axis a distance 0.3 and
    reading over gives p_{g_1} = .04. Similarly, moving from .08 on the
    binomial axis down the diagonal ΔZ(=t) axis a distance 0.3 and
    reading over gives p_{g_2} = .135.

5. Connecting p_{g_1} and p_{g_2} with α and β on the probability axis and
    reading the Larson grid gives n = 67, c = 5 as the plan when
    t = 0.3.

6. The next value of t selected is t = 0.65 and the procedure starts
    again.

7. The nomogram is used iteratively to produce Table 13-3. The
    procedure stops at t = 1.405 since at that value p_{g_2} = .50.

Hence the optimum narrow limit plan for these conditions is n = 31, c = 11,
t = 1.405.

Use of the Larson nomograph constrains the Ladany procedure to values
of $p_g \leq .50$. Schilling and Sommers (1981) have computed tables of optimal
narrow limit plans based on the binomial distribution through an iterative
procedure not subject to this constraint. Appendix Table T13-3 shows narrow limit plans which have minimum sample size tabulated by producer's
quality level p_1 and consumer's quality level p_2 for fixed α = .05 and
β = .10. Also shown are matched binomial attributes plans and the single
variables plans which appear in Appendix Table T10-2. All the plans in
Appendix Tables T10-2 and T13-3 are matched and tabulated using the same

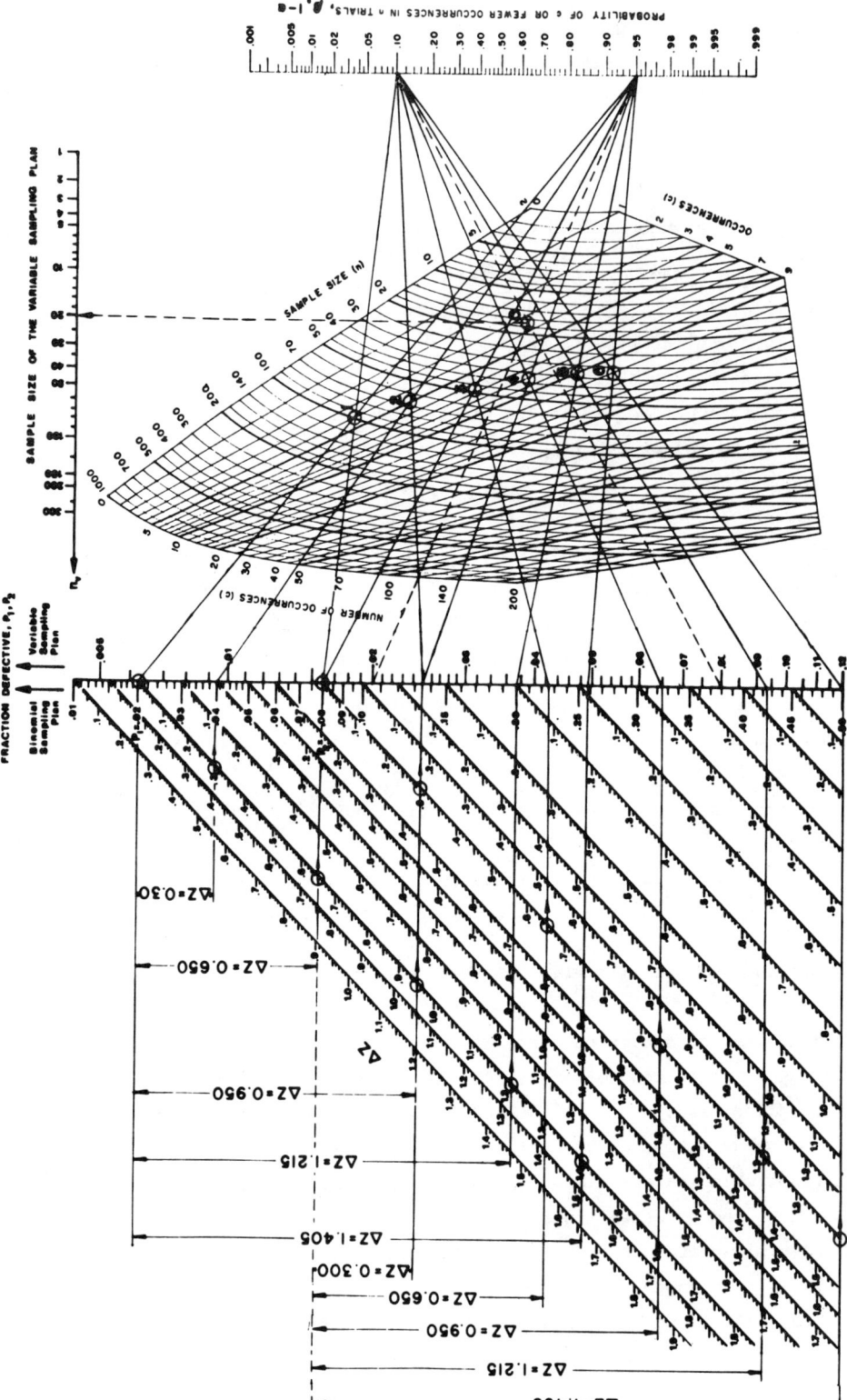

FIG. 13-6 Use of the Ladany nomograph of narrow limit gauging sampling plans for solution of example. (From Ladany, 1976, p. 229. Reprinted by permission.)

TABLE 13-3   Iterative Use of Ladany Nomogram

Iteration	t	p_{g_1}	p_{g_2}	n	c	Point number in Fig. 13-6
0	0	.02	.08	98	4	1
1	.30	.04	.135	67	5	2
2	.65	.08	.225	48	7	3
3	.95	.135	.325	38	8	4
4	1.215	.200	.425	33	10	5
5	1.405	.258	.500	31	11	6

values of p_1 and p_2. In assessing narrow limit plans, comparison should be made with known standard deviation variables plans since the standard deviation is assumed known for both procedures.

As an example of the use of Appendix Table T13-3, consider the example used with the Ladany nomograph

$$p_1 = .02 \quad \alpha = .05$$
$$p_2 = .08 \quad \beta = .10$$

These specifications result in the following plans:

Attributes:	n = 97, c = 4
Narrow limit:	n = 31, c = 15, t = 1.69
Variables (σ known):	n = 21, k = 1.69
Variables (σ unknown):	n = 50, k = 1.69

This narrow limit plan differs slightly from that given by Ladany since for the plan developed from the nomogram the α and β risks are not held exactly because of the constraint $p_{g_2} \leq .50$. For the Ladany plan $\alpha = .079$ and $\beta = .075$ whereas, using the Schilling-Sommers table $\alpha = .052$ and $\beta = .101$.

In this example the narrow limit plan affects a two-thirds reduction in sample size relative to attributes, compared to an 80 percent reduction using the variables plan. Advantages of narrow limit plans over variables are:

1. No calculations for the inspector
2. Ease and accuracy in collecting data
3. Ease of understanding and use

Of course less information is generated by the narrow limit plans for possible feedback in acceptance control.

Schilling and Sommers (1981) found that a simple heuristic approximation can be used to develop an optimal narrow limit plan from the known
standard deviation variables plan having the same p_1, p_2, α, β. If the
variables plan has sample size n_v and acceptance constant k, the parameters

$$n = 1.5n_v \qquad\qquad t = k$$
$$c = .75n_v - .67$$

provide an excellent approximation to the optimal narrow limit plan. This
can be confirmed from the results of the preceding example. Using this
procedure, an approximation of the known standard deviation plan n = 21,
k = 1.69 is

$$n = 31.5 \sim 32 \qquad\qquad t = 1.69$$
$$c = 15.08 \sim 15$$

which is very close to the optimal narrow limit plan.

Tables of optimal narrow limit plans matching the MIL-STD-105D normal,
tightened, and reduced tables are also presented by Schilling and Sommers
(1981). This allows use of these narrow limit plans as substitutes for
the attributes plans given in the standard when the assumptions of narrow
limit gauging are met. Use with the MIL-STD-105D AQL system and its
switching rules allows for significant reductions in scheme sample size.

The tables of optimal narrow limit plans matching MIL-STD-105D are
given here as follows:

    Appendix Table T13-4: Tightened inspection optimal narrow limit
                                       plans for MIL-STD-105D

    Appendix Table T13-5: Normal inspection optimal narrow limit
                                       plans for MIL-STD-105D

    Appendix Table T13-6: Reduced inspection optimal narrow limit
                                       plans for MIL-STD-105D

The OC curves of the resulting plans closely follow those of the counter-
part attributes plans from MIL-STD-105D. Thus, when substituted for the
attributes plans the operating characteristics and other measures of the
narrow limit plans are essentially the same as those given in that standard.
The following tables from MIL-STD-105D can be used directly to assess their
properties.

    MIL-STD-105D Table V-A: Average outgoing quality limit factors
                                       for normal inspection

MIL-STD-105D Table V-B: Average outgoing quality limit factors
                                       for tightened inspection

MIL-STD-105D Table VI-A: Limiting quality for which P_a = 10 percent

MIL-STD-105D Table VII-A: Limiting quality for which P_a = 5 percent

MIL-STD-105D Table X-A: Tables for sample size code letter.

The MIL-STD-105D average sample size Table IX is not represented among
these tables since average sample sizes using narrow limit plans will be
much less than those given in MIL-STD-105D and, further, the narrow limit
plans shown are for single sampling only. When the AQL sampling scheme
which MIL-STD-105D represents is properly used (with the switching rules)
the average sample number for the overall scheme using the narrow limit
plans can be computed. This has been tabulated using the approach of
Schilling and Sheesley (1978) for the overall tightened-normal-reduced
scheme except that the limit numbers for switching to reduced inspection
were not utilized in the tabulation. The resulting average sample sizes
are shown in Appendix Table T13-7 for the case when the process is running
at the AQL. Probabilities of acceptance for the scheme at the AQL are
also shown in Appendix Table T13-7.

When the MIL-STD-105D system is applied using narrow limit plans, the
switching rules and other procedures may be used directly. Use of the
limit numbers in switching to reduced inspection poses a problem, however,
in that gauging is to the narrow limit and not to the specification limit.
Accordingly, units not conforming to the narrow limit would have to be
regauged to determine the number of defectives (or nonconformances to the
specification limit) in the sample to compare to the limit numbers. Also,
the sample sizes are reduced to such an extent by using narrow limit plans
that it would take considerably more than 10 lots to accumulate a sample
large enough to use the limit numbers for reduced inspection in Table VIII
of MIL-STD-105D. It is recommended that the limit numbers be dropped from
the switching procedure. As stated by Schilling and Sheesley (1978), "The
effect of the limit numbers for reduced inspection on the operating
characteristics is minimal. Yet they serve as an impediment to easy use
of the switching rules."

Use of the switching rules with narrow limit plans can result in a
significant decrease in average sample size. For example, with Code M,
1.5 percent AQL, the sample size for attributes plans drops from 315 for
the normal plan alone to 268 for the scheme with the switching rules.

TABLE 13-4   Narrow Limit Plans Substituted for Attribute Plans in
MIL-STD-105D, Code F, 2.5 AQL

	Attributes as given	Narrow limit counterparts
Normal	n = 20, Ac = 1, Re = 2	n = 9, Ac = 4, Re = 5, t = 1.43
Tightened	n = 32, Ac = 1, Re = 2	n = 11, Ac = 5, Re = 6, t = 1.67
Reduced	n = 8, Ac = 0, Re = 2	n = 6, Ac = 1, Re = 4, t = 1.07

When narrow limit plans are substituted in the scheme the average sample
size drops even further from 79 for the normal plan alone to 50.5 when the
switching rules are used.

As an example of the use of narrow limit plans in the MIL-STD-105D
system, consider the plan Code F, 2.5 percent AQL. A comparison of MIL-
STD-105D attributes plans with their narrow limit counterparts is shown in
Table 13-4. Here, the scheme average sample size is 21.5 at the AQL using
the attributes plans and 8.6 when the narrow limit plans are substituted.

It should be noted that the acceptance criteria for reduced plans
under MIL-STD-105D show a gap between the acceptance and rejection numbers.
Sample results falling in this gap initiate a switch back to normal inspection although the lot itself is accepted under the reduced plan. When the
tables for the narrow limit plans were prepared, the plan at the attributes
rejection number was matched at P_a = .95 and P_a = .10 and made optimum.
The plan for the corresponding attributes acceptance number was then matched
as closely as possible at P_a = .10 using the sample size, n, and compression
constant, t, from the plan derived from the rejection number.

The assumption of normality upon which the narrow limit plans presented
are based is an important consideration in application. Preliminary investigation by Schilling and Sommers (1981) showed increasing sensitivity to
the assumption with small p (large t). The risks may differ considerably
from those specified by the plan depending on the degree of nonnormality.
The standard deviation must, of course, be known and stable.

As an extreme illustration, suppose the plan n = 31, c = 15, t = 1.69
was set up to be used with a standard normal distribution of product. This
implies p_1 = .02 has .95 probability of acceptance and p_2 = .08 has .10
probability of acceptance. If the distribution subsequently changed to
that of a t-distribution with one degree of freedom (i.e., the symmetrical
thick-tailed Cauchy distribution) with an appropriate location parameter

and an interquartile range the same as the assumed normal distribution, p = .103 would have .95 probability of acceptance while p = .139 would have .10 probability of acceptance. Thus, it is very important that the normal assumption be verified and monitored in the use of narrow limit plans.

With variables sampling, the sampling data currently obtained could be used to set up control charts for checking on known variability and the continued validity of the normality assumption. Control charts using gauging techniques have been discussed by Ott and Mundel (1954) and by Stevens (1948).

Properly used in applications in which a normal distribution is assured and where σ has been accurately estimated, narrow limit plans provide an excellent vehicle for sample size reduction. Their use with the MIL-STD-105D scheme switching rules can lead to still further reductions in sample size and utilization of that standard in situations in which the attributes sample sizes required by the standard would be prohibitive. They provide a useful and viable alternative in a continuing effort to attain maximum quality at minimum costs. In the words of Ott and Mundel (1954), "The advantages which are inherent in a program of quality control require an appreciation of its philosophy, and understanding of its techniques, and provision for competent management of the program." Used in such an environment, narrow limit plans are an excellent tool for quality assessment and control.

MIXED VARIABLES - ATTRIBUTES PLAN

The choice between acceptance sampling by attributes and by variables has commonly been considered a first step in the application of sampling plans to specific problems in industry. The dichotomy is more apparent than real, however, since other alternatives exist in the combination of both attributes and variables results to determine the disposition of the lot. One such procedure is the so-called *mixed* variables-attributes sampling plan. It is, in essence, a double-sampling procedure involving variables inspection of the first sample and subsequent attributes inspection if the variables inspection of the first sample does not lead to acceptance.

As early as 1932, Dodge (1932) suggested that variables criteria be used in the first stage (only) of a double-sampling plan "... for judging the results of a first sample and for determining when a second, substantially larger sample should be inspected before rejecting the lot." Such

procedures are now called *mixed* or *variables-attributes* sampling plans.
The double-sampling feature distinguishes these plans from single-sampling
plans using both variables and attributes criteria as proposed by Woods
(1960) and Kao (1966).

Mixed variables-attributes sampling differs from the ordinary double-
sampling procedure in the sense that only acceptance can take place as a
result of the application of the variables plan to the first sample. If
acceptance is not indicated, a second sample is drawn, acceptance or rejection then being determined on an attributes basis. Use of variables on the
first sample with attributes on the second sample combines the economy of
variables for quick acceptance on the first sample with the broad nonparametric protection of attributes sampling when a questionable lot requires
a second sample. Schematically, the procedure is shown in Fig. 13-7.

Mixed plans are of two types, so-called *independent* and *dependent*
plans. Independent mixed plans do not incorporate first sample results in
the assessment of the second sample, that is, decisions on the two samples
are kept independent. Dependent mixed plans combine the results of the
first and second samples in making a decision if a second sample is necessary; thus the second sample decision is dependent on first sample results.
In describing mixed variables-attributes plans, Bowker and Goode (1952,
p. 8) indicate that "Under this procedure, a sample is drawn and inspected
on a variables basis ... if the action indicated is rejection an additional
sample is drawn. This additional sample is inspected on an attribute basis,
and the final decision concerning disposal of the lot is made on the basis
of the attribute plan."

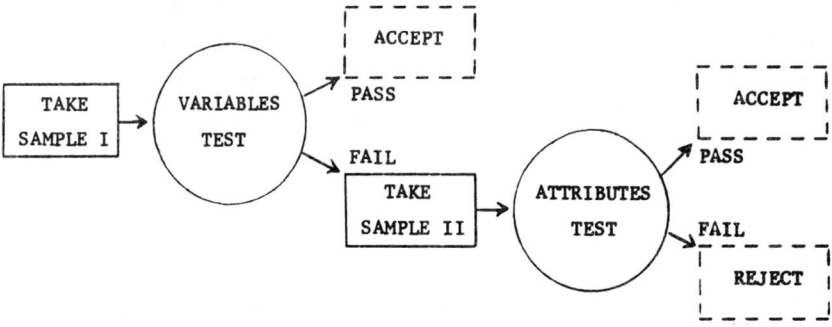

FIG. 13-7 Operation of a mixed plan. (From Schilling, 1966, p. 83.)

Their discussion of mixed plans is, for the most part, limited to the independent case; that is, to plans in which the attributes procedure, when
called for, is applied to the results of the second sample only. This keeps
the probabilities of acceptance of the variables and attributes components
of the plan independent. Schilling (1966) has provided procedures for deriving independent mixed plans given two points on the OC curve. In contrast, the so-called dependent mixed plan is one in which attributes data
arising from both the first and second samples are combined for testing
when the attributes procedure is employed. This makes the probabilities
of acceptance of the variables and attributes parts of the plan dependent.
Dependent plans have been examined by Gregory and Resnikoff (1955), Savage
(1955), and by Schilling and Dodge (1969).

Independent mixed plans maintain stochastic independence between the
probabilities of the variables and attributes constituents of the procedure.
Bowker and Goode (1952) suggest that independent plans have conventionally
been carried out as follows:

1. Obtain first sample.

2. Test first sample against a given variables acceptance criterion
    and

    a. Accept if the test meets the variables criterion.

    b. Resample if the test fails to meet the variables criterion.

3. Obtain a second sample if necessary (per 2b).

4. Test the second sample (only) against a given attributes criterion
    and accept or reject as indicated by the test.

Dependent mixed plans are those in which the probabilities of the
variables and attributes constituents of the procedure are made dependent.
The dependent procedure, as proposed by Savage (1955), can be summarized
as follows:

1. Obtain first sample.

2. Test first sample against a given variables acceptance criterion
    and

    a. Accept if the test meets the variables criterion.

    b. If the test fails to meet the variables criterion,

        i.   Reject if the number nonconforming in the first
             sample exceeds a given attributes criterion.

        ii. Otherwise resample.

3. Obtain a second sample if necessary (per 2bii).

4. Test the results for the first and second samples taken together
    against the given attributes criterion and accept or reject as
    indicated by the test.

Note that this procedure can be generalized by providing for the use
of different attributes criteria in steps 2 and 4. Such a generalized
dependent mixed plan has been presented by Schilling and Dodge (1967).

The dependent plan provides the optimal procedure in terms of the
size of average sample number (ASN) associated with the plan. Attention
will be directed here to mixed plans for the case of single specification
limit, known standard deviation, when a normal distribution of product is
assumed. Gregory and Resnikoff (1955) have examined the case of dependent
plans with standard deviation unknown, while Bowker and Goode (1952) provide an approximation useful in estimating the operating characteristic
curve of such plans. Adams and Mirkhani (1976) have derived an approach
to standard deviation unknown when c = 0 and examine the effect of non-
normality on combined variables-attributes plans.

Advantages and Disadvantages of Mixed Plans

The assumption of normality inherent in most variables acceptance procedures
has proved to be both their strength and their undoing. Perturbations in
the production process or screening of nonconforming product may make otherwise normally distributed product anything but normal. Whatever the potential source of nonnormality, the possibility of submission of such product
to standard variables plans is a serious consideration weighing against their
use except under conditions where normality is well assured. Nonetheless,
the reduction in sample size attendant with variables plans makes them particularly inviting.

The mixed variables-attributes plan achieves some of the reduction of
sample size associated with a variables plan without some of the related
disadvantages. The mixed procedure appeals to the psychology of inspectors
by giving a questionable lot a second chance. In rejecting lots it is also
often a decided psychological or legal advantage to be able to show actual
defectives to the producer, a feature which can be had only by rejecting on
an attributes basis. Truncated and nonnormal distributions cannot be rejected for poor variables results alone, but only on the basis of defective
or nonconforming units found in the attributes sample. Furthermore, with
regard to acceptance-rejection decisions, the effect of changes in shape of
distribution can be minimized by accepting only on variables evidence so

good as to be practically beyond question for most distributions which might reasonably be presented to the plan. Thus, mixed plans provide a worthwhile alternative to variables plans used alone.

The principal advantage of a variables-attributes plan over attributes alone is a reduction in sample size for the same protection. The variables aspect of the mixed plan also allows for a far more careful analysis of the distribution of product presented to the plan than would be possible with attributes inspection alone. Variables control charts kept on this data can provide information on the variability and stability of product from lot to lot. Control charts should normally be used in conjunction with acceptance sampling procedures involving variables inspection.

With small first samples, the mixed plan provides an excellent form of surveillance inspection on product which is generally expected to be of good quality but which may, at times, show degradation. A small variables first sample can be employed to accept at relatively low values of proportion nonconforming and the second attributes sample is then used to provide a definitive criterion for disposition of the lot if it is not accepted on the first sample.

Unfortunately, mixed plans do not provide the same protection against nonnormality for acceptance as they do for rejection, since product is accepted at the first stage of the plan on a variables basis. It is possible, however, to minimize this disadvantage for product well within specification by designing the plan in such a way as to accept on a variables basis only product with distribution located far enough from the specification limit so that reasonable changes in the shape of the distribution will not cause appreciable changes in proportion nonconforming. In this way a tight variables criterion could be employed to minimize the effect of changes in shape of distribution on the operating characteristic curve of the plan. See Schilling (1967).

In application, it is also conceivable that mixed plans might be more difficult to administer than either variables plans or attributes plans alone. As with all plans using variables criteria, a separate mixed plan must be developed for each characteristic to which it is applied. Any increase in complexity would, however, probably be compensated for by the advantages of the mixed procedure.

Generalized Mixed Dependent Procedure

Given an upper specification limit, the inspection procedure for application of a single specification limit U, known standard deviation σ, dependent

mixed plan has been generalized by Schilling and Dodge (1969) by allowing
for two acceptance numbers. Symmetry obviates the necessity for parallel
consideration of a lower specification limit. The first acceptance number
c_1 is applied to the attributes results of the first sample after rejection
by variables and before a second sample is taken. The second acceptance
number c_2 is applied to the combined first and second sample attributes
results. As a special case, the two acceptance numbers may be made the same;
this is the plan proposed by Savage (1955). Providing for the use of different acceptance numbers increases the flexibility and potential of the dependent mixed plan. Of the several methods of specifying the variables constituent of known standard deviation variables plans, the $\overline{X}$ method involving
designation by sample size n_1 and acceptance limit on the sample average A
is used here since it simplifies the notation somewhat. Note that $A = U - k\sigma$
for upper specification limit and standard variables acceptance factor k.

Let

N = lot size

n_1 = first sample size

n_2 = second sample size

A = acceptance limit on sample mean ($\overline{X}$)

c_1 = attributes acceptance number on first sample

c_2 = attributes acceptance number on second sample

Then the generalized plan would be carried out in the following manner:

1. Determine the parameters of the mixed plan: n_1, n_2, A, c_1, c_2.
2. Take a random sample of n_1 from the lot.
3. If the sample average $\overline{X} \leq A$, accept the lot.
4. If the sample average $\overline{X} > A$, examine the first sample for the
   number of defectives d_1 therein.
5. If $d_1 > c_1$, reject the lot.
6. If $d_1 \leq c_1$, take a second random sample of n_2 from the lot and
   determine the number of defectives d_2 therein.
7. If in the combined sample of $n = n_1 + n_2$, the total number of
   defectives $d = d_1 + d_2$ is such that $d \leq c_2$, accept the lot.
8. If $d > c_2$, reject the lot.

When semicurtailed inspection is employed the procedure remains the
same, except that, if c_2 is exceeded at any time during the inspection of
the second sample, inspection is stopped at once and the lot rejected.

Semicurtailed inspection involves stopping inspection of the second sample only upon rejection of the lot.

Measures: Independent Mixed Plan

The four principal curves which describe the properties of an acceptance sampling plan for various proportions nonconforming are the operating characteristic or OC curve, the average sample number or ASN curve, the average total inspection or ATI curve, and the average outgoing quality or AOQ curve. The operation of mixed plans cannot be properly assessed until these curves, for given values of the true proportion nonconforming, are defined. In particular, attention will be directed here to Type B OC curves (i.e., sampling from a process).

Let

n_1 = first sample size

n_2 = second sample size

V' = probability of acceptance under variables plan (with sample size n_1)

A' = probability of acceptance under attributes plan (with sample size n_2)

p = proportion nonconforming in process.

Then, the probability of acceptance and other measures of independent mixed plans can be developed by analogy to attributes sampling (Schilling, 1966) for a lot of size N as

$$P_a = V' + (1 - V')A'$$
$$ASN = n_1 + (1 - V')n_2$$
$$ASN_c = n_1 + (1 - V')ASN_c^*$$
$$AOQ \simeq pP_a$$
$$ATI = n_1V' + (n_1 + n_2)(1 - V')A' + (N)(1 - V')(1 - A')$$

where ASN_c^* is the average sample number under semicurtailed inspection for the attributes plan.

It is important to note that these equations for independent plans hold whatever the nature of the variables or attributes sampling plans involved. *Any* variables plan (using range or standard deviation) can be combined with *any* attributes plan (single or multiple) using the independent procedure, provided, of course, that the underlying assumptions of the two plans are appropriate to the situation to which the mixed plan is to be applied. Also, the probabilities of acceptance, V' and A', are usually readily available since they can be read directly from the OC curves of the variables and attributes plans used.

The assumption of a known underlying distribution inherent in variables sampling would seem to indicate sufficient knowledge of the underlying process to allow use of known standard deviation variables plans in most applications. The possibility of a process generating product with a distribution of constant shape but frequent changes in variability suggests that unknown standard deviation plans may sometimes be in order. The appropriate selection should, of course, be made subsequent to investigation of the stability of the distribution from lot to lot as revealed by a control chart and by examinations of the shape of the distribution and its constancy. As with variables plans, mixed plans should not be used "in the blind" with product of unknown history. Unknown standard deviation plans are easily derived and measures determined for the independent case using the above procedure. A method for assessing the operating measures of unknown standard deviation dependent mixed plans when $c = 0$ is given by Adams and Mirkhani (1976).

For example, consider the independent variables-attributes plan:

Variables: $n_1 = 7$ $k = 1.44$
Attributes: $n_2 = 20$ $c = 1$

The probability of acceptance has previously been calculated for the two constituents of the independent mixed plan (see Chap. 5 and Chap. 10). For example, when $p = .18$, it was found for lot size 120 that

Measure	Variables plan	Attributes plan
Probability of acceptance	$V' = .08$	$A' = .10$
Average sample number	7	20
Average sample number (semicurtailed)	–	10.44
Average outgoing quality (approximate)	.014	.018
Average total inspection (N = 120)	111.0	109.8
Average outgoing quality limit	.036	.036

So for the independent mixed plan

$$P_a = .08 + (1 - .08).10 = .172$$
$$ASN = 7 + (1 - .08)20 = 25.4$$
$$ASN_c = 7 + (1 - .08)10.44 = 16.6$$
$$AOQ = pP_a = .18(.172) = .031$$
$$ATI = 7(.08) + (7 + 20)(1 - .08)(.10) + (120)(1 - .08)(1 - .10) = 102.4$$

Thus, while probability of acceptance and average sample numbers are higher for this proportion nonconforming, the other measures AOQ and ATI are improved over the attributes plan taken alone.

Measures: Dependent Mixed Plan

Formulas for the measures of the generalized dependent mixed plan as given by Schilling and Dodge (1969) are shown in Table 13-5, where

$$P(Y) = \text{probability of Y}$$
$$P_n(Y,W) = \text{probability of Y and W in a sample of n}$$
$$P_n(Y|W) = \text{probability of Y given W in a sample of n}$$
$$P(i;n) = \text{probability of i defectives in a sample of n}$$
$$p = \text{population (process) fraction defective}$$

Since σ is assumed known, it is possible to evaluate the expressions shown in Table 13-5 using tables of $P_n(i, \overline{X} > A)$ for a standard normal universe, i.e., $\mu = 0$, $\sigma = 1$. Such values are given in Appendix Table T13-8 for first sample size $n_1 = 5$. To accomplish this, the value of $P_n(i, \overline{X} > A)$

TABLE 13-5 Formulas for Measures of Dependent Mixed Plans*

Measure	Formula
P_a	$P_a = P(\overline{X} \leq A) + \sum\limits_{i=0}^{c_1} \sum\limits_{j=0}^{c_2-1} P_{n_1}(i, \overline{X} > A)\, P(j;n_2)$
ASN	$ASN = n_1 + n_2 \sum\limits_{i=0}^{c_1} P_{n_1}(i, \overline{X} > A)$
ASN_c	$ASN_c = n_1 + \sum\limits_{i=0}^{c_1} P_{n_1}(i, \overline{X} > A)\left[\dfrac{c_2-i+1}{p} \sum\limits_{k=c_2-i+2}^{n_2+1} P(k;n_2+1)\right.$ $\left. + n_2 \sum\limits_{j=0}^{c_2-1} P(j;n_2)\right]$
ATI	$ATI = ASN + (N-n_1) \sum\limits_{i=c_1+1}^{n_1} P_{n_1}(i, \overline{X} > A) +$ $(N-n_1-n_2)\left(1-P_a - \sum\limits_{i=c_1+1}^{n_1} P_{n_1}(i, \overline{X} > A)\right)$
AOQ	$AOQ = \dfrac{p}{N}\left[P(\overline{X} \leq A)(N-n_1) + (P_a - P(\overline{X} \leq A))(N-n_1-n_2)\right]$

*Except for ASN, all formulas are the same with or without curtailed inspection.

From Schilling and Dodge (1969, p. 344). Reprinted by permission.

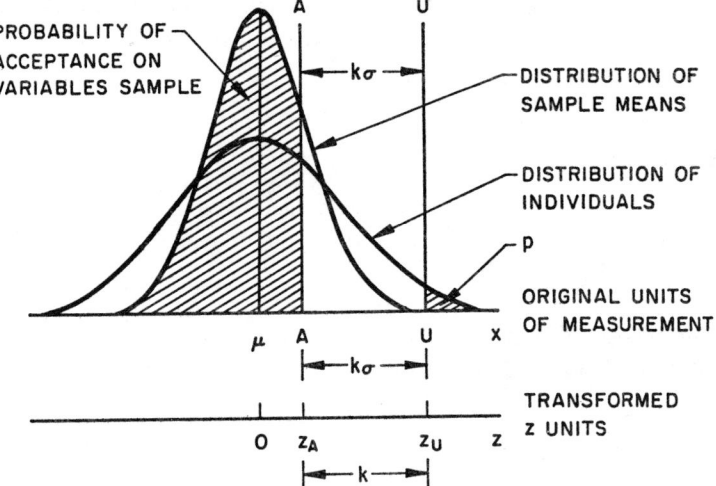

FIG. 13-8 Relationship of k and A. (From Schilling and
Dodge, 1969, p. 346. Reprinted by permission.)

for a particular application can be found by transforming the variates involved to standard normal deviates by use of the familiar z transformation.
This expresses the departure of given values from the population mean in
units of the (known) standard deviation. Thus, an upper specification
limit U is expressed as z_U, where

$$z_U = \frac{U - \mu}{\sigma}$$

and μ is the population mean of a normal distribution such that fraction
defective p of the said distribution exceeds the upper specification limit
U (see Fig. 13-8). Thus

$$P_n(i, \ \overline{X} > A) = P_n(i, \ \overline{z} > z_A)$$

where $\overline{z}$ and z_A are standard normal deviates such that

$$\overline{z} = \frac{\overline{X} - \mu}{\sigma} \qquad z_A = \frac{A - \mu}{\sigma}$$

The tables in the appendix are entered with these values for the mean and
the acceptance limit.

The values shown in the appendix were calculated using the method
given in Schilling and Dodge (1966). Similar tables for sample sizes 4 to
10 are presented in Schilling and Dodge (1967a) and for sample sizes 11 to
20 when c = 0 in Schilling and Dodge (1967b).

Figure 13-8 shows the relationship of k and A for a given distribution of product with mean μ associated with fraction nonconforming p. It also displays the role of the transformed variables z_A and z_U.

Mixed plans have been discussed in terms of the $\overline{X}$ method mentioned above since this simplifies the notation somewhat. Variables plans specified in terms of the other methods can be converted to the $\overline{X}$ method using the k method

$$A = U - k\sigma$$

or the M method

$$A = U - \sqrt{\frac{n-1}{n}}\, k\sigma,$$

k such that $\displaystyle\int_k^\infty \frac{1}{\sqrt{2\pi}}\, e^{-t^2/2}\, dt = \frac{M}{100}$ in the notation of MIL-STD-414,

respectively.

In combining any two variables and attributes plans in a dependent mixed plan, the formulas of Table 13-5 define the probability of acceptance, or OC curve, and associated measures of the combined plan. Note that the formulas simplify greatly when $c_1 = 0$.

To illustrate the inspection procedure to be followed and the methods to be used in determining the properties of a mixed-acceptance sampling plan, consider the following example.

Suppose the plan

$$n_1 = 5 \qquad k = 1.5$$
$$n_2 = 20 \qquad c_1 = 1,\ c_2 = 2$$

is to be applied to the lot-by-lot acceptance inspection of a particular kind of device. The characteristic to be inspected is the "operating temperature" of the device, for which there is a specified upper limit of $209.0^\circ F$. For this characteristic, the standard deviation of the process is known to be $4.0^\circ F$, based on past experience substantiated by a control chart. What inspection procedure should be followed and what are the properties of this plan?

For this example,

$$U = 209.0^\circ F$$
$$\sigma = 4.0^\circ F$$
$$A = U - k\sigma = 209.0 - 1.5(4.0) = 203.0^\circ F$$

The procedure would be carried out as follows:

Step	Result
1. Determine parameters of plan	$n_1 = 5$, $n_2 = 20$, $A = 203.0$, $c_1 = 1$, $c_2 = 2$.
2. Take sample of $n_1 = 5$ from lot.	First sample results: 205, 202, 208, 198, 207.
3. If $\overline{X} \leq A$, accept the lot.	$\overline{X} = 204$; not $\leq A = 203.0$, so go to next step.
4. If $\overline{X} > A$, examine first sample for number of defectives d_1 therein.	No sample value $> U = 209.0$; so $d_1 = 0$.
5. If $d_1 > c_1$, reject the lot.	$d_1 = 0$, not $> c_1 = 1$; so go to next step.
6. If $d_1 \leq c_1$, take second sample of $n_2 = 20$ and determine number nonconforming or defective d_2 therein.	Second sample results: 3 nonconforming in $n_2 = 20$; so $d_2 = 3$.
7. If in combined sample, total nonconforming or defective $d = d_1 + d_2 \leq c_2$, accept the lot.	$d = d_1 + d_2 = 0 + 3 = 3$, not $\leq c_2 = 2$; so go to next step.
8. If $d > c_2$, reject the lot.	$d = 3 > c_2 = 2$; reject the lot.

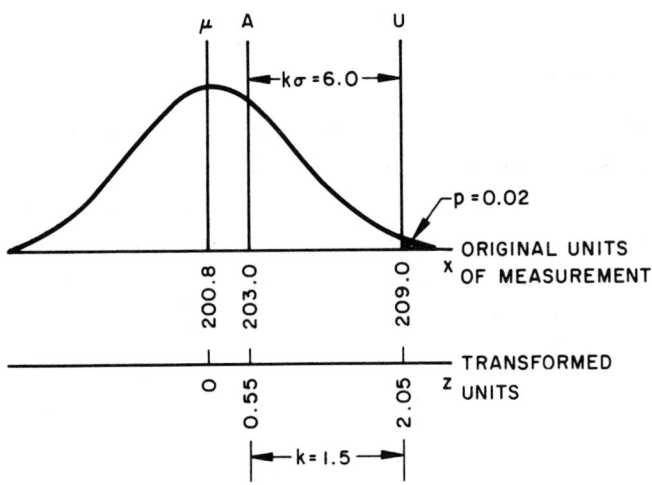

FIG. 13-9 Distribution of individuals when $p = .02$, known $\sigma = 4.0$. (From Schilling and Dodge, 1969, p. 349. Reprinted by permission.)

Suppose the probability of acceptance P_a and associated measures are to be calculated for fraction nonconforming $p = .02$. Then, since the distribution is normal, $p = .02$ implies the distribution of individuals will be as indicated in Fig. 13-9, and from a normal probability table we find $z_U = 2.05$ for $p = .02$. Thus

$$z_A = z_U - k = 2.05 - 1.5 = .55$$

The following, then, are the probability of acceptance and associated measures of the plan given above.

1. Probability of acceptance (at $p = .02$)

$$P_a = P(\overline{X} \le A) + \sum_{i=0}^{c_1} \sum_{j=0}^{c_2-i} P_{n_1}(i, \overline{X} > A) P(j; n_2)$$

$$= P(z \le \sqrt{n_1}\, z_A) + P_5(0, \overline{z} > z_A) \sum_{j=0}^{2} P(j; 20)$$

$$+ P_5(1, \overline{z} > z_A) \sum_{j=0}^{1} P(j; 20)$$

$$= P(z \le \sqrt{5}\,(0.55)) + P_5(0, \overline{z} > 0.55) \sum_{j=0}^{2} P(j; 20)$$

$$+ P_5(1, \overline{z} > 0.55) \sum_{j=0}^{1} (j; 20)$$

$$= .8907 + .0693(.9929) + .037(.9401)$$

$$= .9943$$

2. Average sample number (at $p = .02$)

$$ASN = n_1 + n_2 \sum_{i=0}^{c_1} P_{n_1}(i, \overline{X} > A)$$

$$= 5 + 20 \sum_{i=0}^{1} P_5(i, \overline{z} > z_A)$$

$$= 5 + 20 \sum_{i=0}^{1} P_5(i, \overline{z} > 0.55)$$

$$= 5 + 20[.0693 + .037]$$

$$= 7.126$$

3. Average sample number under semicurtailed inspection (at $p = .02$)

$$ASN_c = n_1 + \sum_{i=0}^{c_1} P_{n_1}(i, \bar{X} > A)$$

$$\left[\frac{c_2 - i + 1}{p} \sum_{k=c_2-i+2}^{n_2+1} P(k; n_2 + 1) + n_2 \sum_{j=0}^{c_2-i} P(j; n_2) \right]$$

$$= 5 + \sum_{i=0}^{1} P_5(i, \bar{z} > 0.55)$$

$$\left[\frac{2 - i + 1}{.02} \sum_{k=2-i+2}^{20+1} P(k; 20 + 1) + 20 \sum_{j=0}^{2-i} P(j; 20) \right]$$

$$= 5 + .0693 \left[\frac{3 - 0}{.02} \sum_{k=4-0}^{21} P(k; 21) + 20 \sum_{j=0}^{2-0} P(j; 20) \right]$$

$$+ .037 \left[\frac{3 - 1}{.02} \sum_{k=4-1}^{21} P(k; 21) + 20 \sum_{j=0}^{2-1} P(j; 20) \right]$$

$$= 5 + .0693[150(.0007) + 20(.9929)] + .037[100(.0081)$$

$$+ 20(.9401)]$$

$$= 7.109$$

4. Average total inspection, for lot size $N = 1000$ (at $p = .02$)

$$ATI = ASN + (N - n_1) \sum_{i=c_1+1}^{n_1} P_{n_1}(i, \bar{X} > A) + (N - n_1 - n_2)$$

$$\left(1 - P_a - \sum_{i=c_1+1}^{n_1} P_{n_1}(i, \bar{X} > A) \right)$$

$$= 7.126 + (1000 - 5) \sum_{i=1+1}^{5} P_5(i, \bar{z} > 0.55) + (1000 - 5 - 20)$$

$$\left(1 - .9943 - \sum_{i=1+1}^{5} P_5(i, \bar{z} > 0.55) \right)$$

but

$$\sum_{i=2}^{5} P_5(i, \bar{X} > A) = P(\bar{X} > A) - \sum_{i=0}^{1} P_5(i, \bar{X} > A)$$

$$\sum_{i=2}^{5} P_5(i, \bar{z} > 0.55) = P(z > \sqrt{5}(.055)) - \sum_{i=0}^{1} P_5(i, \bar{z} > 0.55)$$

$$= .1093 - (.0693 + .037)$$

$$= .003$$

so

ATI = 7.126 + 995(.003) + 975(1 - .9943 - .003)

   = 7.126 + 2.985 + 2.632

   = 12.743

5. Average outgoing quality, for lot size N = 1000 (at p = .02)

$$AOQ = \frac{p}{N}[P(\overline{X} \le A)(N - n_1) + (P_a - P(\overline{X} \le A))(N - n_1 - n_2)]$$

$$= \frac{.02}{1000}[P(z \le \sqrt{n_1}\, z_A)(1000 - 5)$$

$$+ (.9943 - P(z \le \sqrt{n_1}\, z_A))(1000 - 5 - 20)]$$

$$= .00002[.8907(995) + (.9943 - .8907)(975)]$$

$$= .0197$$

MIL-STD-414 Dependent Mixed Plans

Dependent mixed variables-attributes plans are, in fact, specified for use
in MIL-STD-414 in paragraphs A9.2.2 to A9.4.2 reproduced as follows:

A9.2.2 Mixed Variables-Attributes Inspection. Mixed
variables-attributes inspection is inspection of a
sample by attributes, in addition to inspection by
variables already made of a previous sample, before a
decision as to acceptability or rejectability of a lot
can be made.

A9.3 Selection of Sampling Plans. The mixed variables-
attributes sampling plan shall be selected in accordance
with the following:

A9.3.1 Select the variables sampling plan in accordance
with Section B, C, or D.

A9.3.2 Select the attributes sampling plan from MIL-
STD-105, paragraph 10, using a single sampling plan and
tightened inspection. The same AQL value(s) shall be
used for the attributes sampling plan as used for the
variables plan of paragraph A9.3.1.

(Additional sample items may be drawn, as necessary,
to satisfy the requirements for sample size of the
attributes sampling plan. Count as a defective each
sample item falling outside of specification limit(s).)

9.4 Determination of Acceptability. A lot meets the
acceptability criterion if one of the following conditions is satisfied:

Condition A. The lot complies with the appropriate variables acceptability criterion of Section
B, C, or D.

Condition B. The lot complies with the acceptability criterion of paragraph 11.1.2 of MIL-STD-105.

A9.4.1   If Condition A is not satisfied, proceed in accordance with the attributes sampling plan to meet Condition B.

A9.4.2   If Condition B is not satisfied, the lot does not meet the acceptability criterion.

To illustrate the method for determining the OC curve of a combination of two such plans, suppose the following two plans are combined after the manner of MIL-STD-414:

MIL-STD-414, Code F (AQL = 4.0) : n = 5, k = 1.20

MIL-STD-105D, Code F (AQL = 4.0 tightened) : n = 20, c = 1

Note that in combining these published plans in the manner of MIL-STD-414, the second sample size is n_2 = 15 in the calculations since 5 units are contributed by the first sample to the attributes determination.

Let $c_1 = c_2 = 1$. The combined Type B OC curve would be derived as follows:

1. The formula is

$$P_a = P(\overline{X} \le A) + \sum_{i=0}^{1} \sum_{j=0}^{1-i} P_5(i, \overline{X} > A) P(j;15)$$

2. Computation then proceeds in the same manner as the example above. For example, to obtain the probability of acceptance when p = .05,

$$P_a = P(z \le \sqrt{n_1} z_A) + P_5(0, \overline{z} > z_A) \sum_{j=0}^{1} P(j;15) + P_5(1, \overline{z} > z_A) P(0;15)$$

$$= P(z \le \sqrt{5}(0.44)) + P_5(0, \overline{z} > 0.44) \sum_{j=0}^{1} P(j;15)$$

$$+ P_5(1, \overline{z} > 0.44) P(0;15)$$

$$= .8374 + .0649(.8290) + .079(.4633)$$

$$= .928$$

Comparison of Independent and Dependent Mixed Plans

A comparison can be made between independent and dependent plans which have "essentially" the same OC curve. A criterion for comparison is the average sample number of the two plans. The probability of acceptance and average sample number of an independent mixed plan can be calculated as

$$P_a = P_{n_1} (\overline{X} \leq A) + P_{n_1} (\overline{X} > A) \sum_{j=0}^{c_2} P(j;n_2)$$

$$ASN = n_1 + n_2 P_{n_1} (\overline{X} > A)$$

Now, it can be shown (Schilling and Dodge, 1967) that if the two plans have the same first stage variables plan and attributes acceptance number c_2 (where for the dependent plan $c_1 \leq c_2$), the second sample size of the independent plan will be greater than that of the dependent plan.

Therefore, for the same probability of acceptance, i.e., the same OC curve, the independent plan requires a larger second sample size. But even if the second sample size of the dependent plan is kept the same as that of the independent plan, the ASN of the dependent plan will be lower since

$$ASN \text{ (independent)} \geq ASN \text{ (dependent)}$$

$$n_1 + P(\overline{X} > A)n_2 \geq n_1 + n_2 \sum_{i=0}^{c_1} P_{n_1} (i, \overline{X} > A)$$

$$P(\overline{X} > A) \geq \sum_{i=0}^{c_1} P_{n_1} (i, \overline{X} > A)$$

Thus, the dependent plan is superior to the independent plan in terms of the same protection with a smaller sample size.

The difference in average sample number can become quite large if particularly bad quality is submitted to the plan and if, as seems customary, the independent plan has no provision for rejection on an attributes basis immediately after taking the first sample and before taking the second sample. Thus, in the event of poor quality the attributes plan is utilized to a greater extent in the independent scheme than in the dependent procedure with further possible increase in the average sample number.

As an example of the superiority of dependent plans, consider the following:

$$n_1 = 5 \quad k = 2 \quad n_2 = 20 \quad c_1 = c_2 = 0$$

The probability of acceptance and average sample numbers were calculated for the specified mixed plan, assuming it to be carried out in dependent and independent form. A comparison of the results for the dependent and independent procedures is shown in Table 13-6.

TABLE 13-6   Comparison of P_a and ASN for a Specified
Mixed Plan Applied in Dependent and Independent Form:
$n_1 = 5$, $n_2 = 20$, $k = 2$, $c_1 = c_2 = 0$

p	Dependent P_a	ASN	Independent P_a	ASN
.005	.980	6.7	.991	6.9
.01	.931	8.9	.958	9.6
.02	.794	12.5	.849	14.1
.05	.415	16.4	.493	20.8
.10	.119	15.8	.169	23.9
.15	.032	13.6	.054	24.7
.20	.008	11.5	.016	24.9

Comparison of Mixed with Other Type Plans

As an indication of the relative merit of mixed plans, variables plans and
single- and double-sampling attributes plans were matched as closely as
possible by Schilling and Dodge (1967) to the same dependent mixed plan

$$n_1 = 5 \qquad k = 2$$
$$n_2 = 20 \qquad c_1 = c_2 = 0$$

at the two points

$$p_1 = .008 \qquad P_a = .953$$
$$p_2 = .107 \qquad P_a = .098$$

TABLE 13-7   Comparison of Various Plans to Match: $p_1 = .008$, $P_a = .953$;
$p_2 = .107$, $P_a = .098$

Plan	Criteria	Probability of acceptance p = .008	p = .107	Average sample number p = .008	p = .107
Dependent mixed	$n_1 = 5$, $k = 2$ $n_2 = 20$, $c_1 = c_2 = 0$	.953	.098	8.1	15.5
Variables	$n = 6$, $k = 1.75$	.947	.106	6	6
Attributes (single)	$n = 37$, $c = 1$	.965	.083	37	37
Attributes (double)	$n_1 = 21$, $c_1 = 0$ $n_2 = 42$, $c_2 = 1$	.947	.095	27.0	30.8

which lie on the OC curve of the mixed plan. Because of inherent differences in the shape of the various OC curves, exact matches could not be obtained; however, all the plans obtained show probability of acceptance within ± 0.015 of the mixed plan at these points. The results are shown in Table 13-7.

Comparison of the average sample number at these points for the various plans gives a rough indication of the advantages of mixed plans against either single- or double-sampling attributes plans. Also, it would appear that for low percents nonconforming the average sample number for the mixed plan approaches that of the variables plan as illustrated in the following tabulation:

	Probability of acceptance			Average sample number		
	$p \sim 0$	$p = .005$	$p = .01$	$p \sim 0$	$p = .005$	$p = .01$
Dependent mixed	1.0	.980	.931	5.0	6.7	8.9
Variables	1.0	.979	.922	6	6	6
Single attributes	1.0	.985	.947	37	37	37
Double attributes	1.0	.977	.922	21.0	25.0	28.2

This is reasonable, since if "perfect" product (within the constraint of the assumption of normality) were submitted to both plans, it would be accepted on the first stage of the mixed procedure resulting in an average sample number of 5 compared to the variables average sample number of 6.

PHILIPS STANDARD SAMPLING SYSTEM

The Philips Standard Sampling System was originated by Dr. Hugo C. Hamaker and his associates, "For the practical execution of sampling inspection by unskilled factory personnel...." Developed at the Philips Works in Holland, the plans are presented in a single table which is simple, straightforward and easy to use. The table appears as Table 13-8. Hamaker, Taudin Chabot, and Willemze (1950) have provided a detailed account of the theory, application, and documentation of the system.

The plans are indexed by the so-called point of control p_0 or indifference quality. That is, the percent defective having 50 percent probability of acceptance. This makes the producer and the consumer share equally in the risks of the system at the point of control. Single sampling is used for lot size up to 1000 and double sampling thereafter. The second sample size in double sampling is always twice the first and the rejection number on the first and second samples is the same, $c_2 + 1$.

TABLE 13-8 Philips Standard Sampling System

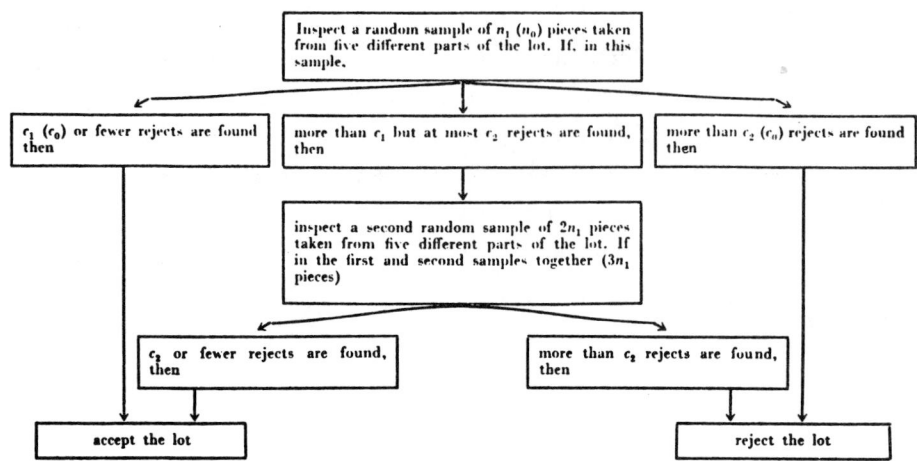

N.B. Where the letter A is given under n_0 this means that no sample is to be taken but that the whole lot has to be inspected.

Single-sampling plans

LOT SIZE \ POINT OF CONTROL	$\frac{1}{4}\%$		$\frac{1}{2}\%$		1%		2%		3%		5%		7%		10%	
	n_0	c_0	n_0	c_0	n_0	c_0	n_0	c_0	n_0	c_0	n_0	c_0	n_0	c_0	n_0	c_0
20-50	A	—	A	—	A	—	30	0	20	0	13	0	10	0	7	0
51-100	A	—	A	—	60	0	30	0	20	0	13	0	10	0	7	0
101-200	A	—	100	0	60	0	35	0	55	1	35	1	25	1	17	1
201-500	175	0	100	0	135	1	75	1	55	1	35	1	40	2	25	2
501-1000	225	0	225	1	150	1	85	1	85	2	55	2	55	3	35	3

Double-sampling plans

LOT SIZE	$\frac{1}{4}\%$			$\frac{1}{2}\%$			1%			2%			3%			5%			7%			10%		
	n_1	c_1	c_2	n_1	c_1	c_2	n_1	c_1	c_2	n_1	c_1	c_2	n_1	c_1	c_2	n_1	c_1	c_2	n_1	c_1	c_2	n_1	c_1	c_2
1001-2000	330	0	1	150	0	1	110	0	2	55	0	2	45	0	3	25	0	3	30	1	5	22	1	5
2001-5000	425	0	2	200	0	2	135	0	3	70	0	3	70	1	5	45	1	5	55	2	10	40	2	10
5001-10000	525	0	3	260	0	3	220	1	5	110	1	5	125	2	10	75	2	10	75	3	15	55	3	15
10000-20000	875	1	5	440	1	5	380	2	10	190	2	10	180	3	15	110	3	15	100	4	20	70	4	20
20000-50000	1500	2	10	750	2	10	540	3	15	270	3	15	240	4	20	140	4	20	120	5	25	85	5	25
50000-100000	2200	3	15	1100	3	15	700	4	20	390	4	20	290	5	25	175	5	25	145	6	30	105	6	30

From Hamaker, Taudin Chabot, and Willemze (1950, p. 365). Reprinted by permission.

The Philips approach is centered on the point of control p_0 and the relative slope h_0 of the OC curve at the point of control. It is well known from the work of Campbell (1923) that, for the Poisson distribution, the following approximations hold:

$$np_0 = c + \frac{2}{3}$$

and

$$\frac{\pi}{2} h_0^2 = np_0$$

Of course, the absolute slope of the OC curve at any fraction defective is dP_a/dp the relative slope at p_0 being

$$h_0 = \left| \frac{dP_a/dp}{P_a/p} \right|_{p = p_0}$$

where h_0 is taken in absolute value. Note that h_0 is dimensionless.

Since Poisson probabilities depend on np, a change in sample size simply results in a change of scale along the p-axis of the OC curve. But such a transformation will not alter h_0, which is independent of the absolute size of the sample.

The Philips Standard Sampling System plans are selected from the relations

$$c = \frac{\pi}{2} h_0^2 - .73 \qquad n = \frac{c + .67}{p_0}$$

which employs an empirical correction term developed by Hamaker in the formula for c. Note that, in contrast to Campbell,

$$c + \frac{2}{3} = \frac{\pi}{2} h_0^2 - .06$$

is used to make the approximation more effective for low values of c.

The plans incorporate a steeper operating characteristic (higher h_0) with larger lot sizes and/or increasing points of control. The functional relationship between h_0, p_0, and lot size was adjusted as appropriate to the practical needs of the factory and incorporated into the Philips Table. Hamaker, Taudin Chabot, and Willemze (1950, p. 364) point out that for any plan to be used on the factory floor it must be presented in a form that is quick and easy to employ. "Only when presented in this form will a sampling table be acceptable in the factory where simplicity of manipulation is a paramount requirement and where even the simplest calculations should preferably be avoided." The theory of the Philips System is presented in an excellent paper by Hamaker (1950).

PROBLEMS

1. A Code H, 1.0 AQL, MIL-STD-414 plan is to be used in in-process inspection. If a No-Calc plan is to be substituted, what is the plotting position of the largest value in the sample? If the largest value is exactly at the upper specification limit, should the lot be accepted?

2. A modified lot plot form was drawn up for 50 observations of coating weight of instrument pins. It was found that $\hat{\mu} = 15$ mg and $\hat{\sigma} = 1$ mg. The lot limit exceeded the specification limit by 1 space. Estimate the fraction nonconforming. Cell width is 1. Should the lot be accepted if it is important to have less than .01 nonconforming?

3. The plan $n = 10$, $c = 0$, $t = 1.5$ is applied to an upper specification limit $U = 110$. The standard deviation is known to be $\sigma = 6$. If the largest value in the sample is 102, should the lot be accepted?

4. Sketch the OC curve of the narrow limit plan $n = 5$, $t = 1.92$, $c = 2$ through the points for $p = .0025$, .034, and .109. Compare to MIL-STD-105D, Code F, .65 AQL. Use Poisson and binomial probabilities. Why are the two closer using binomial probabilities?

5. Use the Ladany nomogram to obtain an "optimum" plan when $p_1 = .03$, $p_2 = .12$, $\alpha = .05$, $\beta = .10$.

6. Find tightened, normal, and reduced optimum narrow limit plans which match those for the MIL-STD-105D, Code H, 1.0 AQL scheme.

7. Suppose MIL-STD-414 normal and MIL-STD-105D tightened unknown standard deviation plans for Code F, 4.0 AQL, are to be combined in a mixed sampling procedure. The plans are

     MIL-STD-414: $n = 10$, $k = 1.23$
     MIL-STD-105D: $n = 20$, $Ac = 1$, $Re = 2$

   Calculate the following measures for $p = .04$ when the plans are combined to form an independent mixed plan:

   a. Probability of acceptance
   b. Average sample number
   c. Average outgoing quality

8. To conform to the procedure for combining mixed plans recommended in MIL-STD-414, a dependent mixed plan should be used. Compute the measures of Prob. 7 for $p = .05$ when the Code G, 0.65 AQL, known

standard deviation plan n = 5, k = 1.88, from MIL-STD-414 is combined with the corresponding tightened plan from MIL-STD-105D, n = 32, c = 0, in a dependent mixed procedure. Note that, for the MIL-STD-414 method the total combined second sample size would be 32 so that n_2 = 27.

9. What single-sampling plan from the Philips System would roughly correspond to the MIL-STD-105D, 0.65 AQL, normal plan when lot size is 125?

10. Using the relationship developed by Campbell,

$$np_0 = c + \frac{2}{3}$$

compute the indifference quality level for n = 100, c = 0,1,2,3,4,5. Compare with the values obtained from the Schilling-Johnson table.

REFERENCES

Adams, R. M., and K. Mirkhani (1976), *Combined Variables/Attributes Plans — Sigma Unknown*, American Society for Quality Control Annual Technical Conference Transactions, Toronto, pp. 292-300.

Ashley, R. L. (1952), Modification of the Lot Plot Method, *Industrial Quality Control*, 8(5): 30-31.

Bowker, A. H., and H. P. Goode (1952), *Sampling Inspection by Variables*, McGraw-Hill, New York.

Campbell, G. A. (1923), Probability Curves Showing Poisson's Exponential Limit, *Bell System Technical Journal*, 2(1): 95-113.

Chernoff, H., and G. J. Lieberman (1957), Sampling Inspection by Variables with No Calculations, *Industrial Quality Control*, 13(7): 5-7.

Dodge, H. F. (1932), Statistical Control in Sampling Inspection, *American Machinist*, October: 1085-1088; November: 1129-1131.

Dudding, B. P., and W. J. Jennett (1944), *Control Chart Technique When Manufacturing to a Specification*, British Standards Institution, London.

Gregory, G., and G. J. Resnikoff (1955), *Some Notes on Mixed Variables and Attributes Sampling Plans*, Technical Report No. 10, Applied Mathematics and Statistics Laboratory, Stanford University, Stanford, California.

Hamaker, H. C. (1950), The Theory of Sampling Inspection Plans, *Philips Technical Review*, 11(9): 260-270.

Hamaker, H. C., J. J. M. Taudin Chabot, and F. G. Willemze (1950), The Practical Application of Sampling Inspection Plans and Tables, *Philips Technical Review*, 11(12): 362-370.

Kao, J. H. K. (1966), *Single-Sample Attri-Vari Plans for Item-Variability in Percent Defective*, American Society for Quality Control Annual Technical Conference Transactions, New York, pp. 743-758.

Ladany, S. P. (1976), Determination of Optimal Compressed Limit Gaging Sampling Plans, *Journal of Quality Technology*, 8(4): 225-231.

Larson, H. R. (1966), A Nomograph of the Cumulative Binomial Distribution, *Industrial Quality Control*, 23(6): 270-278.

Mace, A. E. (1952), The Use of Limit Gauges in Process Control, *Industrial Quality Control*, 8(4): 24-31.

Ott, E. R., and A. B. Mundel (1954), Narrow Limit Gaging, *Industrial Quality Control*, 10(5): 2-9.

Savage, I. R. (1955), *Mixed Variables and Attributes Plans: The Exponential Case*, Technical Report No. 23, Applied Mathematics and Statistics Laboratory, Stanford University, Stanford, California.

Schilling, E. G. (1966), *Mixed Variables-Attributes Sampling, The Independent Case*, Transactions of the 18th Annual Conference on Quality Control at Rutgers - The State University, New Brunswick, New Jersey, pp. 82-89.

Schilling, E. G. (1967), *A General Method for Determining the Operating Characteristics of Mixed Variables-Attributes Sampling Plans, Single Sided Specification, Standard Deviation Known*, Ph.D. Dissertation, Rutgers - The State University, New Brunswick, New Jersey.

Schilling, E. G., and H. F. Dodge (1966), *On Some Joint Probabilities Useful in Mixed Acceptance Sampling*, Technical Report No. N-26, Rutgers - The State University Statistics Center, New Brunswick, New Jersey.

Schilling, E. G., and H. F. Dodge (1967), *Dependent Mixed Acceptance Sampling Plans and Their Evaluation*, Technical Report No. N-27, Rutgers - The State University Statistics Center, New Brunswick, New Jersey.

Schilling, E. G., and H. F. Dodge (1967a), *Tables of Joint Probabilities Useful in Evaluating Mixed Acceptance Sampling Plans*, Technical Report No. N-28, Rutgers - The State University Statistics Center, New Brunswick, New Jersey.

Schilling, E. G., and H. F. Dodge (1967b), *Supplement to Tables of Joint Probabilities*, Technical Report No. N-29, Rutgers - The State University Statistics Center, New Brunswick, New Jersey.

Schilling, E. G., and H. F. Dodge (1969), Procedures and Tables for Evaluating Dependent Mixed Acceptance Sampling Plans, *Technometrics*, 11(2): 341-372.

Schilling, E. G., and J. H. Sheesley (1978), The Performance of MIL-STD-105D Under the Switching Rules, *Journal of Quality Technology*, Part 1, 10(2): 76-83; Part 2, 10(3): 104-124.

Schilling, E. G., and D. J. Sommers (1981), Two-Point Optimal Narrow Limit Plans with Applications to MIL-STD-105D, *Journal of Quality Technology*, 13(2): 83-92.

Shainin, D. (1950), The Hamilton Standard Lot Plot Method of Acceptance Sampling by Variables, *Industrial Quality Control*, 7(1): 15-34.

Shainin, D. (1952), Recent Lot Plot Experiences Around the Country, *Industrial Quality Control*, 8(5): 20-29.

Stevens, W. L. (1948), Control by Gauging, *Journal of the Royal Statistical Society*, Series B, 10(1): 54-108.

United States Department of Defense (1963), *Military Standard, Sampling Procedures and Tables for Inspection by Attributes* (MIL-STD-105D), U.S. Government Printing Office, Washington, D.C.

United States Department of Defense (1957), *Military Standard, Sampling Procedures and Tables for Inspection by Variables for Percent Defective* (MIL-STD-414), U.S. Government Printing Office, Washington, D.C.

Woods, W. M. (1960), *Variables Inspection Procedures Which Guarantee Acceptance of Perfectly Screened Lots*, Technical Report No. 47, Applied Mathematics and Statistics Laboratory, Stanford University, Stanford, California.

14

SERIES OF LOTS : RECTIFICATION SCHEMES

While it is impossible to "inspect quality into the product," it is possible to use 100 percent inspection or screening operations in such a way as to insure with known probability that levels of quality in lots outgoing from an inspection station will not, either individually or on the average, exceed certain levels. Often this is done with minimum average total inspection (ATI). Schemes which utilize this concept are

> LTPD schemes: Specify LTPD protection on each lot. Assuming screening of rejected lots, the sampling plan is selected to make ATI
> a minimum at a projected process average level of percent defective.
> AOQL schemes: Specify AOQL protection for the lots. Assuming screening of rejected lots, the sampling plan is selected to make ATI a minimum at the projected process average level of percent defective.

In all rectification schemes 100 percent inspection of rejected lots with replacement of defective units with good ones, or equivalent screening, is assumed.

The LTPD plans are useful when the producer desires LTPD protection on individual lots with the intention of screening rejected material in a large sequence of lots. The purpose is to minimize the total amount of inspection, including the screening, by using a plan with the lowest possible average total inspection. This is particularly suitable for internal in-process sampling where the costs of screening can be borne by the component responsible for producing the product. Also, on final inspection these plans provide LTPD protection for the consumer while the producer, who may do the screening, keeps the overall testing costs to a minimum.

The AOQL plans are concerned with the series of lots as a whole. They do not focus on individual lots, but guarantee that the average outgoing quality limit (AOQL), will not exceed a certain specified amount. Thus the consumer is assured that the average quality level received will not exceed the AOQL in the long run. The producer, or the inspection agency, achieves minimum sample size. Again, these plans are useful in in-process or incoming inspections to guarantee outgoing quality levels regardless of the quality coming into the inspection station.

Note that a specification of LTPD is much more severe than that of AOQL. When operating at the LTPD level of percent defective, the average outgoing quality (AOQ) will be about 10 percent of the LTPD specified. Thus, a 4 percent LTPD is much tighter than a 4 percent AOQL. With plans of this type, the actual AOQ should generally be less than half to two-thirds the value of the AOQL. A 1.5 percent AOQL implies that the producer must hold about a 1.0 percent process level to avoid excessive rejections and consequent screening.

Special procedures have been developed for application lot-by-lot to provide LTPD or AOQL protection. Like the AQL in the MIL-STD-105D or MIL-STD-414 AQL schemes, the AOQL or LTPD becomes the index for such rectification schemes. Two of these, the Dodge-Romig scheme and the Anscombe Rectifying Inspection Scheme will be presented here. The selection of a simple AOQL plan will first be discussed.

SINGLE-SAMPLING AOQL PLAN

It is sometimes desirable to quickly select and compare AOQL plans without regard to process average levels or minimization of inspection. For a single-sampling attributes plan, the AOQL can easily be determined from a graph developed by Altman (1954).

The Altman diagram, which assumes sample size to be small relative to lot size (less than 10 percent), is shown in Fig. 14-1. It is based on the Poisson distribution. The diagram allows the comparison of n, c, and AOQL for various plans to achieve the combinations desired. For specified AOQL and sample size, the diagram gives the appropriate acceptance number, c. For example, suppose limitations on inspection staff are such that a sample size of about 20 is deemed feasible while an AOQL of 4 percent is desired. Lot size is large. Cross-reference of these two criteria on the graph indicates that an acceptance number of c = 1 is appropriate. The plan becomes n = 20, c = 1.

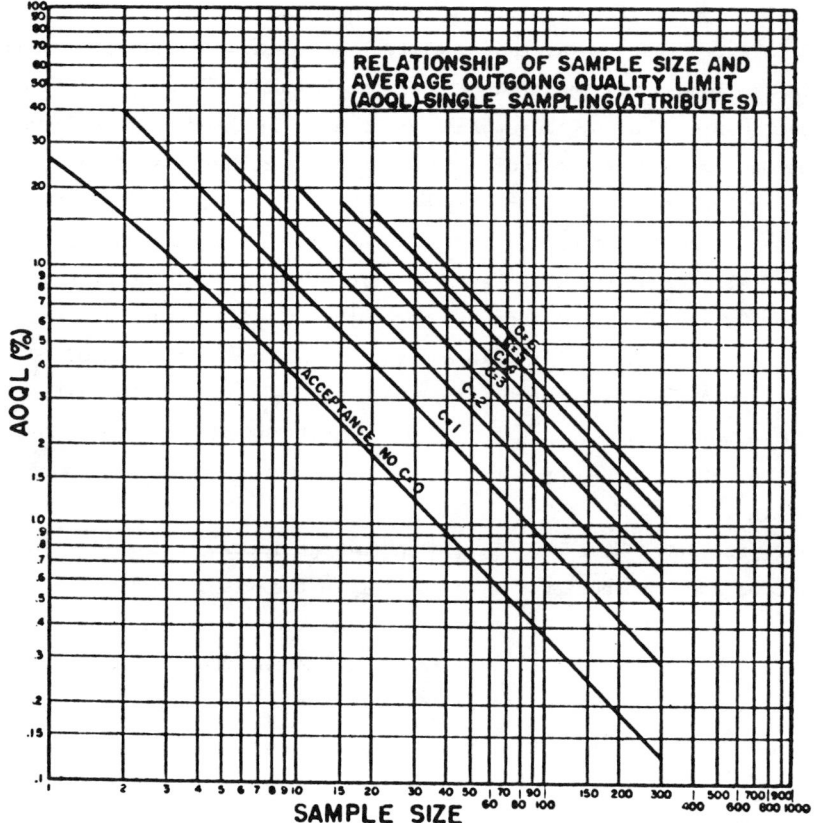

FIG. 14-1 Altman AOQL diagram. (From Altman, 1954, p. 30.
Reprinted by permission.)

DODGE-ROMIG SAMPLING SCHEME

Rectification schemes stand among the earliest examples of sampling schemes
as such. They precede the development of AQL schemes by well over a decade.
The celebrated Dodge-Romig (1941) tables are an excellent example of such
early efforts. Developed by Harold F. Dodge and Harry G. Romig at Bell
Telephone Laboratories in the late 1920s and through the 1930s, they were
first published in the *Bell System Technical Journal* in the early 1940s.
The tables were published in book form in 1944 with the revised second
edition coming forth in 1959.

There are two sets of tables:

LTPD single- and double-sampling plans which minimize ATI for values
    of LTPD = 0.5%, 1.0%, 2.0%, 3.0%, 4.0%, 5.0%, 7.0%, 10.0%

> AOQL single- and double-sampling plans which minimize ATI for values
> of AOQL = 0.1%, 0.25%, 0.5%, 0.75%, 1.0%, 1.5%, 2.0%, 2.5%, 3.0%,
> 4.0%, 5.0%, 7.0%, 10.0%

The LTPD tables are set up to minimize ATI based on Type B probabilities while maintaining LTPD protection (shown as p_t%) determined from Type A probabilities, since the lot size is specified. The AOQL tables utilize Type B probabilities in determining both ATI and AOQL. Thus, the same plan may appear in different lot size ranges in the AOQL and LTPD tables. The disparity represents the difference in use of the scheme in protecting individual lots (Type A) or providing protection on the process producing the lots (Type B). The Type A probability used applies to the middle of the lot size range shown and so is exact for that value only. LTPDs are always calculated using Type A probabilities in the Dodge-Romig scheme.

Both sets of tables require knowledge of the process average percent defective to achieve an optimum plan to minimize ATI. This implies the producer, or the incoming inspection station, must keep adequate records and control charts to properly assess the process average.

The tables are set up such that if the process average is not known, they can be entered at the highest level of process average percent defective shown in the table. In such a situation, the protection desired will be guaranteed with a somewhat less than optimum plan until the necessary information can be developed.

In general, with rectification sampling plans, larger lot sizes result in less overall inspection. Too large a lot size, of course, may preclude

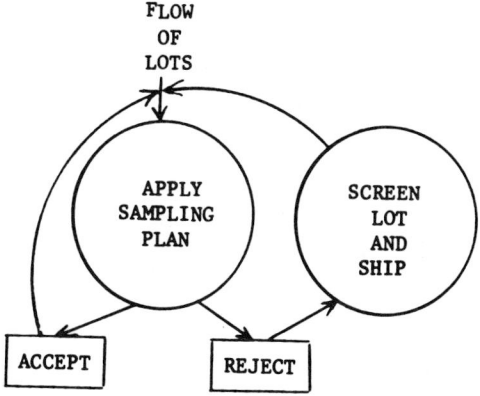

FIG. 14-2 Operation of Dodge-Romig plans.

effective random sampling. Also such plans may actually provide an incentive to improve quality by forcing the producer to incur screening costs on lots of poor quality.

It is important to note that the Dodge-Romig tables contain an appendix with an excellent collection of binomial OC curves for some of the most commonly used plans, including

$$c = 0, \ n = 2(1)15, \ 16(2)34, \ 35(5)50, \ 50(10)100, \ 100(20)200, \ 200(50)500$$
$$c = 1, \ n = 3(1)20, \ 20(2)50, \ 50(5)95, \ 90(10)200, \ 200(20)500$$
$$c = 2, \ n = 5(1)20, \ 20(2)36, \ 30(5)100, \ 100(10)160, \ 160(20)280, \ 250(50)500$$
$$c = 3, \ n = 8(1)19, \ 20(2)36, \ 35(5)75, \ 70(10)200, \ 200(20)300, \ 300(50)500$$

where $X(Y)Z$ indicates the curves start at X and progress in increments of Y up to Z. Operating characteristic curves of the AOQL single- and double-sampling plans specified in the tables are also given.

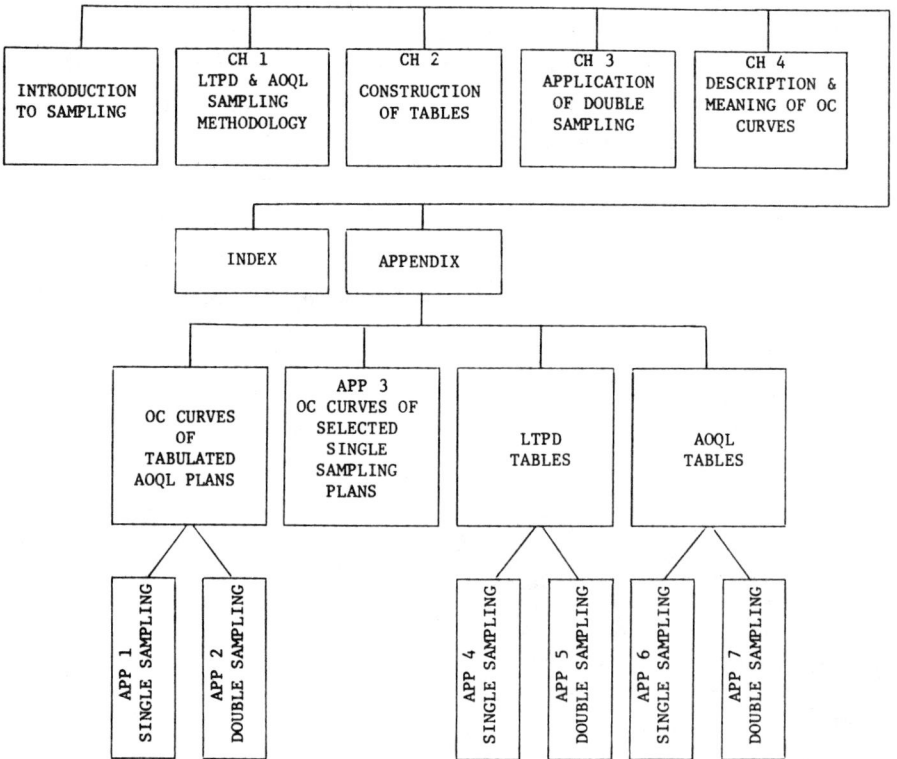

FIG. 14-3   Structure of Dodge-Romig tables.

Operation

All the plans contained in the Dodge-Romig tables assume 100 percent inspection of rejected lots. For plans indexed by LTPD or AOQL, the operation of the scheme is as indicated in Fig. 14-2.

The Dodge-Romig work is more than just tables. It describes the mathematical development behind the plans presented together with much practical material on the application of sampling plans and the meaning of operating characteristic curves and other measures. The content is structured as indicated in Fig. 14-3.

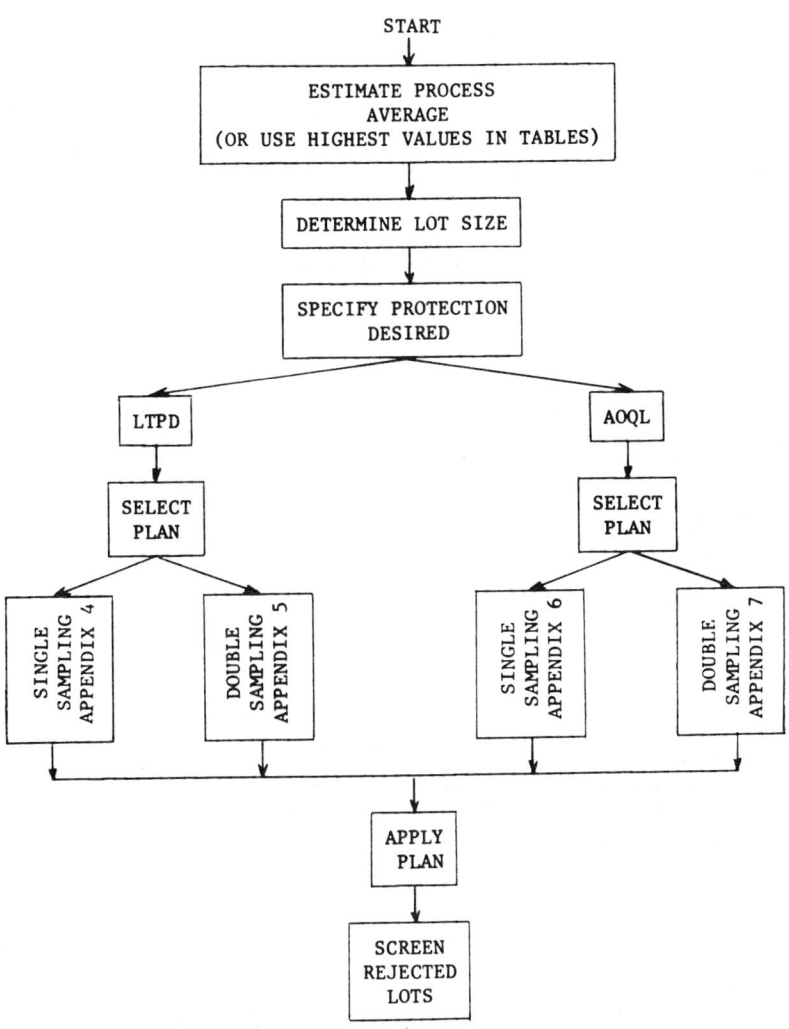

FIG. 14-4 Check sequence for selecting Dodge-Romig plan.

Selection

The tables are indexed in two sets by LTPD or AOQL respectively. Plans are
selected on the basis of lot size and process average. The sets of tables
are entered at the specific value of LTPD or AOQL and the plan determined
by cross-referencing process average and lot size. If the process average
is not known, the largest value of process average appearing in the table
is used until adequate information can be developed. Plans are available
for single or double sampling. The tables give the sample sizes and acceptance criteria and also show the value of the measure not specified, i.e.,
AOQL if LTPD is specified and vice versa. A check sequence for the selection of a Dodge-Romig plan is given in Fig. 14-4.

Examples of the Dodge-Romig (1959) tables and OC curves are shown here
as follows:

Table	Dodge-Romig table	Content
Table 14-1	Appendix 6	4% AOQL single-sampling plans
Table 14-2	Appendix 7	4% AOQL double-sampling plans
Table 14-3	Appendix 4	4% LTPD single-sampling plans
Table 14-4	Appendix 5	4% LTPD double-sampling plans

To exemplify use of these tables and figures, suppose a plan is desired having 4% AOQL with lot size 250 and process average percent defective
1.6%.

Tables 14-1 and 14-2 give the following plans which will guarantee AOQ
less than 4 percent defective with minimum average total inspection.

AOQL single-sampling (Table 14-1)

$$n = 20 \quad c = 1 \quad LTPD = 19\%$$

AOQL double-sampling (Table 14-2)

Sample size	Cumulative sample size $(n_1 + n_2)$	Acceptance number	Rejection number
$n_1 = 16$	16	$c_1 = 0$	$c_2 + 1 = 3$
$n_2 = 18$	34	$c_2 = 2$	$c_2 + 1 = 3$
	$LTPD = 17.4\%$		

The rejection number for both samples in double-sampling is always
one more than the acceptance number for the second sample, c_2. Also, the

TABLE 14-1   Dodge-Romig Single Sampling Table for Average Outgoing Quality Limit (AOQL) = 4%

Lot Size	Process Average 0 to 0.08%			Process Average 0.09 to 0.80%			Process Average 0.81 to 1.60%			Process Average 1.61 to 2.40%			Process Average 2.41 to 3.20%			Process Average 3.21 to 4.00%		
	n	c	$p_t\%$	n	c	$p_t\%$	n	c	$p_t\%$	n	c	$p_t\%$	n	c	$p_t\%$	n	c	$p_t\%$
1–10	All	0	—	All	0	—	All	0	—	All	0	—	All	0	—	All	0	—
11–50	8	0	23.0	8	0	23.0	8	0	23.0	8	0	23.0	8	1	23.0	8	0	23.0
51–100	8	0	24.0	8	0	24.0	8	0	24.0	8	0	24.0	17	1	21.5	17	1	21.5
101–200	9	0	22.0	9	0	22.0	19	1	20.0	19	1	20.0	19	1	20.0	19	1	20.0
201–300	9	0	22.5	9	0	22.5	20	1	19.0	20	1	19.0	31	2	16.8	31	2	16.8
301–400	9	0	22.5	20	1	19.1	20	1	19.1	32	2	16.2	32	2	16.2	43	3	15.2
401–500	9	0	22.5	20	1	19.1	20	1	19.1	32	2	16.3	32	2	16.3	44	3	14.9
501–600	9	0	22.5	20	1	19.2	20	1	19.2	32	2	16.3	45	3	14.6	60	4	12.9
601–800	9	0	22.5	20	1	19.2	33	2	15.9	33	2	15.9	46	3	14.3	60	4	13.0
801–1000	9	0	22.5	21	1	18.3	33	2	16.0	46	3	14.3	60	4	13.0	75	5	12.2
1001–2000	9	0	22.5	21	1	18.4	34	2	15.6	47	3	14.1	75	5	12.2	105	7	11.0
2001–3000	9	0	22.5	21	1	18.4	34	2	15.6	60	4	13.2	90	6	11.3	125	8	10.4
3001–4000	21	1	18.4	21	1	18.4	48	3	13.8	65	4	12.2	110	7	10.7	155	10	9.8
4001–5000	21	1	18.5	34	2	15.7	48	3	13.9	80	5	11.6	110	7	10.8	175	11	9.5
5001–7000	21	1	18.5	34	2	15.7	48	3	13.9	80	5	11.6	125	8	10.4	210	13	9.0
7001–10,000	21	1	18.5	34	2	15.7	65	4	12.3	95	6	11.1	145	9	9.8	245	15	8.6
10,001–20,000	21	1	18.5	34	2	15.7	65	4	12.3	110	7	10.8	195	12	9.0	340	20	7.9
20,001–50,000	21	1	18.5	49	3	13.6	80	5	11.6	145	9	9.8	250	15	8.5	460	26	7.4
50,001–100,000	21	1	18.5	49	3	13.6	95	6	11.1	165	10	9.6	310	18	8.0	540	30	7.1

n = sample size; c = acceptance number
"All" indicates that each piece in the lot is to be inspected
P_t = lot tolerance per cent defective with a Consumer's Risk (P_C) of 0.10

From H. F. Dodge and H. G. Romig, *Sampling Inspection Tables*, 2nd ed., 1959. Copyright © 1944,
Bell Telephone Laboratories. Reprinted by permission of John Wiley and Sons, Inc.

TABLE 14-2 Dodge-Romig Double Sampling Table for Average Outgoing Quality Limit (AOQL) = 4.0%

Lot Size	Process Average 0 to 0.08%						Process Average 0.09 to 0.80%						Process Average 0.81 to 1.60%					
	Trial 1		Trial 2			p_t %	Trial 1		Trial 2			p_t %	Trial 1		Trial 2			p_t %
	n_1	c_1	n_2	n_1+n_2	c_2		n_1	c_1	n_2	n_1+n_2	c_2		n_1	c_1	n_2	n_1+n_2	c_2	
1-10	All	0	-	-	-	-	All	0	-	-	-	-	All	0	-	-	-	-
11-50	8	0	-	-	-	23.0	8	0	-	-	-	23.0	8	0	-	-	-	23.0
51-100	12	0	7	19	1	22.0	12	0	7	19	1	22.0	12	0	7	19	1	22.0
101-200	13	0	8	21	1	21.0	13	0	8	21	1	21.0	15	0	17	32	2	18.0
201-300	13	0	9	22	1	20.5	16	0	18	34	2	17.4	16	0	18	34	2	17.4
301-400	14	0	8	22	1	20.0	16	0	19	35	2	17.0	18	0	28	46	3	15.5
401-500	14	0	8	22	1	20.0	16	0	19	35	2	17.0	19	0	28	47	3	15.3
501-600	16	0	19	35	2	17.0	16	0	19	35	2	17.0	19	0	29	48	3	15.1
601-800	16	0	20	36	2	16.7	16	0	20	36	2	16.7	19	0	30	49	3	14.9
801-1000	16	0	20	36	2	16.7	16	0	20	36	2	16.7	20	0	45	65	4	13.8
1001-2000	17	0	19	36	2	16.6	19	0	31	50	3	14.8	21	0	44	65	4	13.6
2001-3000	17	0	19	36	2	16.6	19	0	31	50	3	14.8	21	0	44	65	4	13.6
3001-4000	17	0	20	37	2	16.5	19	0	31	50	3	14.8	22	0	58	80	5	13.0
4001-5000	17	0	20	37	2	16.5	19	0	31	50	3	14.8	22	0	58	80	5	13.0
5001-7000	17	0	20	37	2	16.5	19	0	31	50	3	14.8	22	0	58	80	5	13.0
7001-10,000	17	0	20	37	2	16.5	19	0	36	55	3	14.6	23	0	57	80	5	12.7
10,001-20,000	17	0	20	37	2	16.5	21	0	44	65	4	13.6	23	0	72	95	6	12.0
20,001-50,000	17	0	20	37	2	16.5	21	0	44	65	4	13.6	43	1	92	135	8	10.6
50,001-100,000	17	0	20	37	2	16.5	23	0	62	85	5	12.5	44	1	106	150	9	10.3

From H. F. Dodge and H. G. Romig, *Sampling Inspection Tables*, 2nd ed., 1959. Copyright © 1944, Bell Telephone Laboratories. Reprinted by permission of John Wiley and Sons, Inc.

TABLE 14-3 Dodge-Romig Single Sampling Table for Lot Tolerance Percent Defective (LTPD) = 4.0%

Lot Size	Process Average 0 to 0.04% n	c	AOQL %	Process Average 0.05 to 0.40% n	c	AOQL %	Process Average 0.41 to 0.80% n	c	AOQL %	Process Average 0.81 to 1.20% n	c	AOQL %	Process Average 1.21 to 1.60% n	c	AOQL %	Process Average 1.61 to 2.00% n	c	AOQL %
1–35	All	0	0	All	0	0	All	0	0	All	0	0	All	0	0	All	0	0
36–50	34	0	0.35	34	0	0.35	34	0	0.35	34	0	0.35	34	0	0.35	34	0	0.35
51–100	44	0	0.47	44	0	0.47	44	0	0.47	44	0	0.47	44	0	0.47	44	0	0.47
101–200	50	0	0.55	50	0	0.55	50	0	0.55	50	0	0.55	50	0	0.55	50	0	0.55
201–300	55	0	0.57	55	0	0.57	85	1	0.71	85	1	0.71	85	1	0.71	85	1	0.71
301–400	55	0	0.58	55	0	0.58	90	1	0.72	120	2	0.80	120	2	0.80	145	3	0.86
401–500	55	0	0.60	55	0	0.60	90	1	0.77	120	2	0.87	150	3	0.91	150	3	0.91
501–600	55	0	0.61	95	1	0.76	125	2	0.87	125	2	0.87	155	3	0.93	185	4	0.95
601–800	55	0	0.62	95	1	0.78	125	2	0.93	160	3	0.97	190	4	1.0	220	5	1.0
801–1000	55	0	0.63	95	1	0.80	130	2	0.92	165	3	0.98	220	5	1.1	255	6	1.1
1001–2000	55	0	0.65	95	1	0.84	165	3	1.1	195	4	1.2	255	6	1.3	315	8	1.4
2001–3000	95	1	0.86	130	2	1.0	165	3	1.1	230	5	1.3	320	8	1.4	405	11	1.6
3001–4000	95	1	0.86	130	2	1.0	195	4	1.2	260	6	1.4	350	9	1.5	465	13	1.6
4001–5000	95	1	0.87	130	2	1.0	195	4	1.2	290	7	1.4	380	10	1.6	520	15	1.7
5001–7000	95	1	0.87	130	2	1.0	200	4	1.2	290	7	1.5	410	11	1.7	575	17	1.9
7001–10,000	95	1	0.88	130	2	1.1	230	5	1.4	325	8	1.5	440	12	1.7	645	19	1.9
10,001–20,000	95	1	0.88	165	3	1.2	265	6	1.4	355	9	1.6	500	14	1.8	730	22	2.0
20,001–50,000	95	1	0.88	165	3	1.2	295	7	1.5	380	10	1.7	590	17	2.0	870	26	2.1
50,001–100,000	95	1	0.88	200	4	1.3	325	8	1.6	410	11	1.8	620	18	2.0	925	29	2.2

From H. F. Dodge and H. G. Romig, *Sampling Inspection Tables*, 2nd ed., 1959. Copyright © 1944, Bell Telephone Laboratories. Reprinted by permission of John Wiley and Sons, Inc.

TABLE 14-4 Dodge-Romig Double Sampling Table for Lot Tolerance Percent Defective (LTPD) = 4.0%

Lot Size	0.81–1.20% Trial 1 n_1	c_1	Trial 2 n_2	n_1+n_2	c_2	AOQL %	1.21–1.60% Trial 1 n_1	c_1	Trial 2 n_2	n_1+n_2	c_2	AOQL %	1.61–2.00% Trial 1 n_1	c_1	Trial 2 n_2	n_1+n_2	c_2	AOQL %
1–35	All	0	—	—	—	0	All	0	—	—	—	0	All	0	—	—	—	0
36–50	34	0	—	—	—	0.35	34	0	—	—	—	0.35	34	0	—	—	—	0.35
51–75	40	0	—	—	—	0.43	40	0	—	—	—	0.43	40	0	—	—	—	0.43
76–100	50	0	25	75	1	0.46	50	0	25	75	1	0.46	50	0	25	75	1	0.46
101–150	55	0	30	85	1	0.55	55	0	30	85	1	0.55	55	0	30	85	1	0.55
151–200	60	0	55	115	2	0.68	60	0	55	115	2	0.68	60	0	55	115	2	0.68
201–300	60	0	65	125	2	0.75	60	0	90	150	3	0.84	60	0	90	150	3	0.84
301–400	65	0	95	160	3	0.86	65	0	95	160	3	0.86	65	0	120	185	4	0.92
401–500	65	0	100	165	3	0.92	65	0	130	195	4	0.96	105	1	140	245	6	1.0
501–600	65	0	135	200	4	1.0	105	1	145	250	6	1.1	105	1	175	280	7	1.1
601–800	65	0	140	205	4	1.0	105	1	185	290	7	1.2	105	1	210	315	8	1.2
801–1000	110	1	155	265	6	1.2	110	1	210	320	8	1.2	145	2	230	375	10	1.3
1001–2000	110	1	195	305	7	1.3	150	2	240	390	10	1.5	180	3	295	475	13	1.6
2001–3000	110	1	260	370	9	1.4	185	3	305	490	13	1.6	220	4	410	630	18	1.7
3001–4000	150	2	255	405	10	1.5	185	3	340	525	14	1.6	285	6	465	750	22	1.8
4001–5000	150	2	285	435	11	1.6	185	3	395	580	16	1.7	285	6	520	805	24	1.9
5001–7000	150	2	320	470	12	1.6	185	3	435	620	17	1.7	320	7	585	905	27	2.0
7001–10,000	150	2	325	475	12	1.7	220	4	460	680	19	1.9	320	7	645	965	29	2.1
10,001–20,000	150	2	355	505	13	1.7	220	4	495	715	20	1.9	350	8	790	1140	35	2.2
20,001–50,000	150	2	420	570	15	1.7	255	5	575	830	24	2.0	385	9	895	1280	40	2.3
50,001–100,000	150	2	450	600	16	1.8	255	5	665	920	27	2.1	415	10	985	1400	44	2.4

From H. F. Dodge and H. G. Romig, *Sampling Inspection Tables*, 2nd ed., 1959. Copyright © 1944, Bell Telephone Laboratories. Reprinted by permission of John Wiley and Sons, Inc.

second sample size is not kept at a constant ratio of the first sample size. It varies. In MIL-STD-105D, $n_1 = n_2$.

LTPD plans afford more protection on individual lots and so require larger sample sizes. If 4 percent LTPD were specified (rather than 4 percent AOQL) with lot size 250 and process average percent defective 1.6 percent, we would have from Table 14-3 and 14-4:

LTPD single-sampling (Table 14-3)

$n = 85$ $c = 1$ AOQL = 0.71%

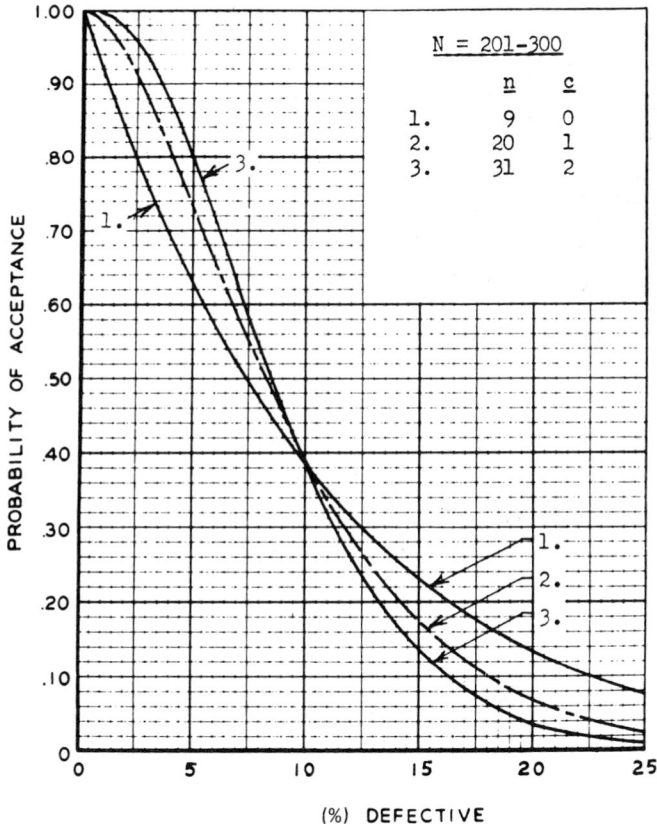

FIG. 14-5 Operating characteristic curves, single-sampling plans: average outgoing quality limit, AOQL = 4.0%. (From H. F. Dodge and H. G. Romig, *Sampling Inspection Tables*, 2nd ed., 1959, p. 94. Copyright Ⓒ 1944, Bell Telephone Laboratories. Reprinted by permission of John Wiley and Sons, Inc.)

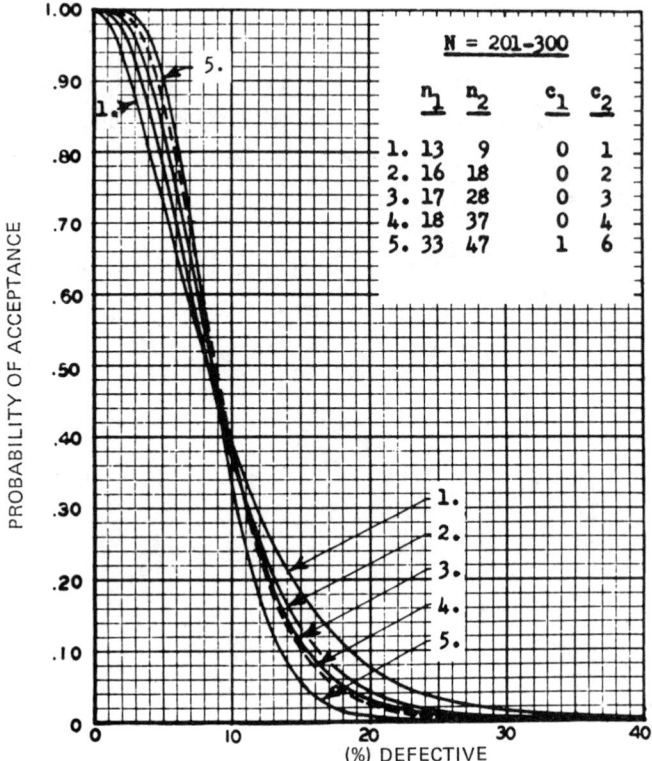

FIG. 14-6 Operating characteristic curves, double-sampling
plans: average outgoing quality limit, AOQL = 4.0%. (From
H. F. Dodge and H. G. Romig, *Sampling Inspection Tables*,
2nd ed., 1959, p. 151. Copyright (c) 1944, Bell Telephone
Laboratories. Reprinted by permission of John Wiley and
Sons, Inc.)

LTPD double-sampling (Table 14-4)

	Sample size	Cumulative sample size	Acceptance number	Rejection number
$n_1 = 60$		60	$c_1 = 0$	$c_2 + 1 = 4$
$n_2 = 90$		150	$c_2 = 3$	$c_2 + 1 = 4$

$$AOQL = 0.84\%$$

This dramatically shows the difference in sample size which can result
from specifying AOQL or LTPD. It is vital to select the proper measure for
the sampling situation when applying rectification schemes of this type as
with all sampling plans.

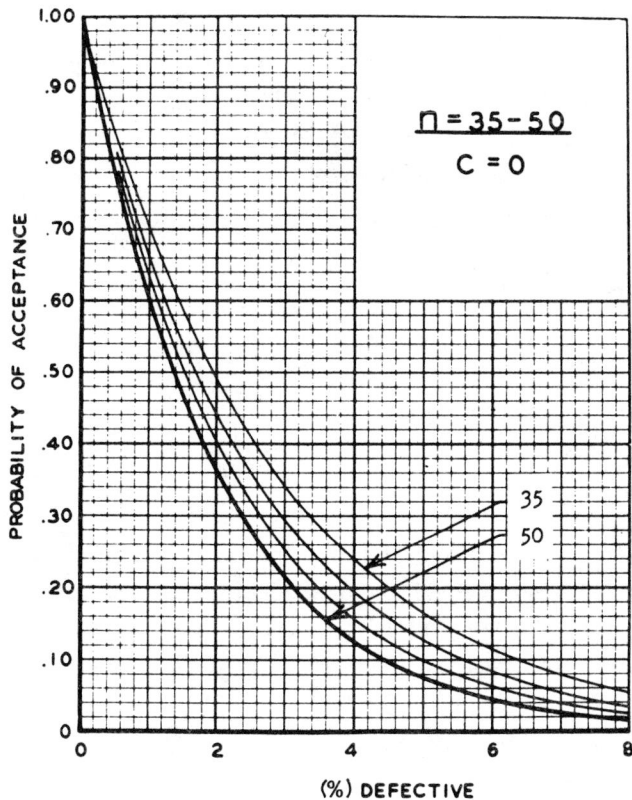

FIG. 14-7 Operating characteristic curves, single-sampling plans: acceptance number, c = 0. (From H. F. Dodge and H. G. Romig, *Sampling Inspection Tables*, 2nd ed., 1959, p. 173. Copyright Ⓒ 1944, Bell Telephone Laboratories. Reprinted by permission of John Wiley and Sons, Inc.)

Measures

Operating characteristic curves are given for all AOQL single- and double-sampling plans. The binomial OC curves for selected single-sampling plans are also shown for reference. To illustrate the curves, Figs. 14-5 and 14-6 show the OC curves for the AOQL single- and double-sampling plans of the previous example. Figure 14-7 gives an example of the OC curves from the collection of binomial OC curves given by Dodge and Romig.

Further Considerations

Sampling plans meeting the Dodge-Romig criterion for minimum average total inspection can be derived using procedures developed by Dodge and Romig

(1929) for LTPD plans and by Dodge and Romig (1941) for AOQL plans. These
papers provide the basis and proofs underlying the technical development
of the Dodge-Romig (1959) tables. The technical background of these plans
is interesting and informative. The Dodge-Romig plans minimize average
total inspection for both the LTPD and AOQL plans.

For the LTPD plans

$$ATI = n + (N - n)\left(1 - \sum_{i=0}^{c} \frac{e^{-n\bar{p}}(n\bar{p})^i}{i!}\right)$$

is minimized subject to

$$.10 = \sum_{i=0}^{c} \frac{C_i^{Np_t} C_{n-i}^{N(1-p_t)}}{C_i^N}$$

Thus, the LTPD is calculated using (Type A) hypergeometric probabilities
(or approximations thereto) since LTPD is on individual lots. The AOQL,
however, is determined using the (Type B) Poisson approximation to the
binomial distribution since AOQL has meaning only in terms of a series of
lots from a process.

For the AOQL plans, it is necessary to maximize AOQ to find the

$$AOQL = p_L = \max\left[p\frac{(N - \bar{I})}{N}\right]$$

which will occur when $\bar{I}$, the average number of units inspected in a lot,
is at a minimum. Substituting the formula for ATI in place of $\bar{I}$

$$AOQL = \max\left[p - \frac{p}{N}\left[n + (N - n)\left(1 - \sum_{i=0}^{c} \frac{e^{-n\bar{p}}(n\bar{p})^i}{i!}\right)\right]\right]$$

which can be shown to be

$$AOQL = \max\left[p\left(\frac{N - n}{N}\right)\left(\sum_{i=0}^{c} \frac{e^{-n\bar{p}}(n\bar{p})^i}{i!}\right)\right]$$

and it is obvious that all the probabilities in this calculation involve
the Poisson approximation to the binomial (Type B). Differentiating and
setting the results equal to zero gives

$$AOQL = x\left(\frac{N - n}{Nn}\right)\sum_{i=0}^{c} \frac{e^{-x}x^i}{i!}$$

where $x = np_M$ and p_M represents the value of p at which the AOQL occurs.

Then

$$AOQL = y\left(\frac{1}{n} - \frac{1}{N}\right)$$

where

$$y = x\sum_{i=0}^{c}\frac{e^{-x}x^{i}}{i!}$$

which is shown by Dodge and Romig to equal

$$y = \frac{e^{-x}x^{c+2}}{c!}$$

and finally

$$n = \frac{yN}{Np_{L} + y}$$

It follows that

$$np_{L} = y\left(1 - \frac{n}{N}\right)$$

Values of x and y are given in Appendix Table T14-1. Use of these values with the acceptance number which gives minimum $\bar{I}$ (Fig. 14-11) forms the basis of the Dodge-Romig AOQL plans.

Constructing LTPD Plan With Minimum ATI

To find an LTPD plan which will achieve minimum average total inspection I_{min} proceed as follows:

1. Given lot size (N), LTPD, process average proportion defective $(\bar{p})$. Define p_{t} = LTPD/100 = tolerance fraction defective.
2. Enter Fig. 14-8, with the ratio $\bar{p}/p_{t}$ on the x-axis and the product $p_{t}N$ on the y-axis, to find the acceptance number region which will make ATI a minimum. Use this acceptance number c.
3. Enter Fig. 14-9 with the product $p_{t}N$ on the x-axis and read the product $(p_{t}n)$ on the y-axis from the appropriate curve for c.
4. Divide $(p_{t}n)$ by p_{t} to obtain n.
5. The plan n, c will give the LTPD protection desired on each lot with minimum ATI.
6. The minimum ATI can be found from Fig. 14-10 by finding the point corresponding to $\bar{p}/p_{t}$ on the x-axis and $p_{t}N$ on the y-axis and interpolating between the closest curves to obtain $p_{t}(ATI) = p_{t}I_{min}$

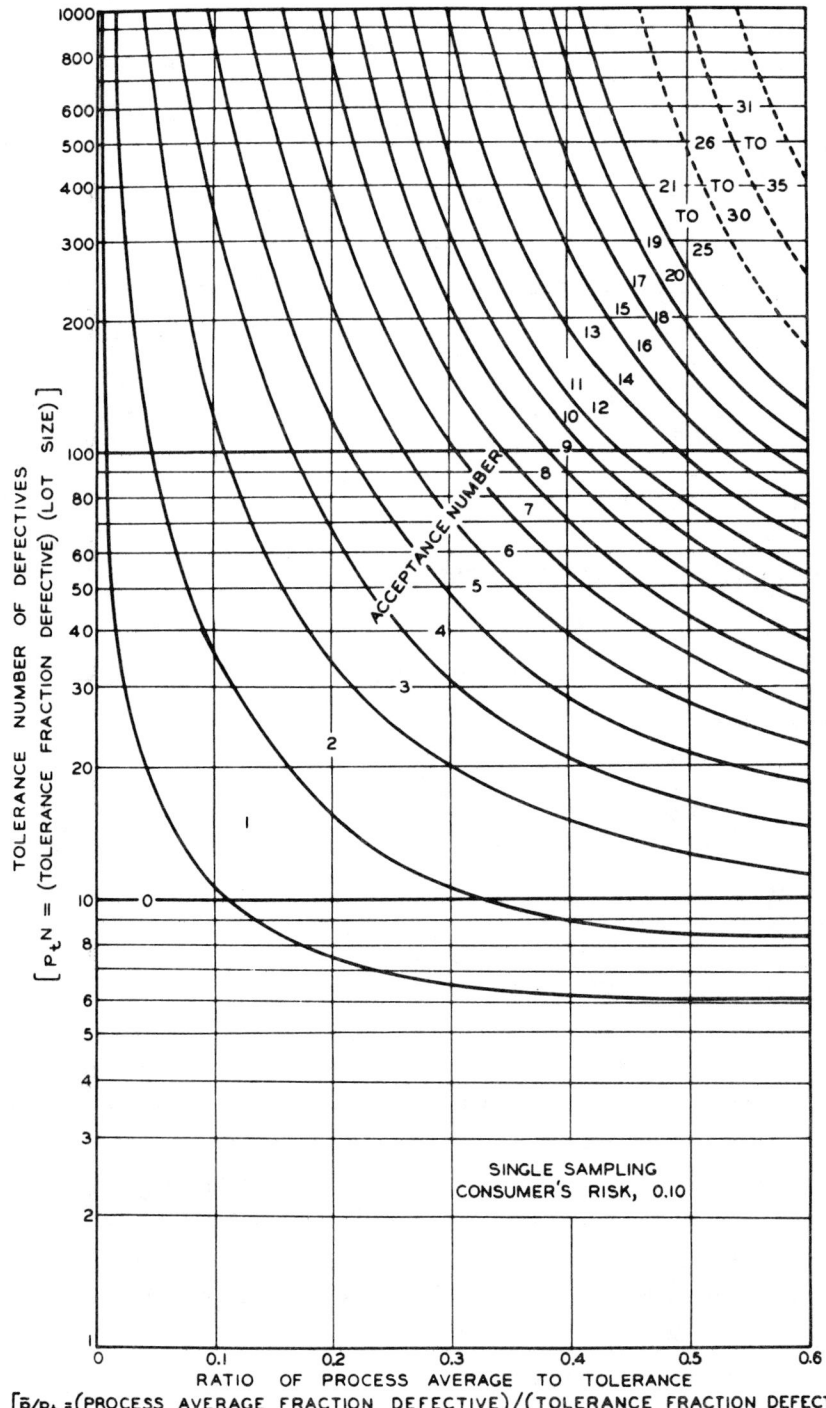

FIG. 14-8 Dodge-Romig curves for finding the acceptance number. (From H. F. Dodge and H. G. Romig, *Sampling Inspection Tables*, 2nd ed., 1959, p. 14. Copyright © 1944, Bell Telephone Laboratories. Reprinted by permission of John Wiley and Sons, Inc.)

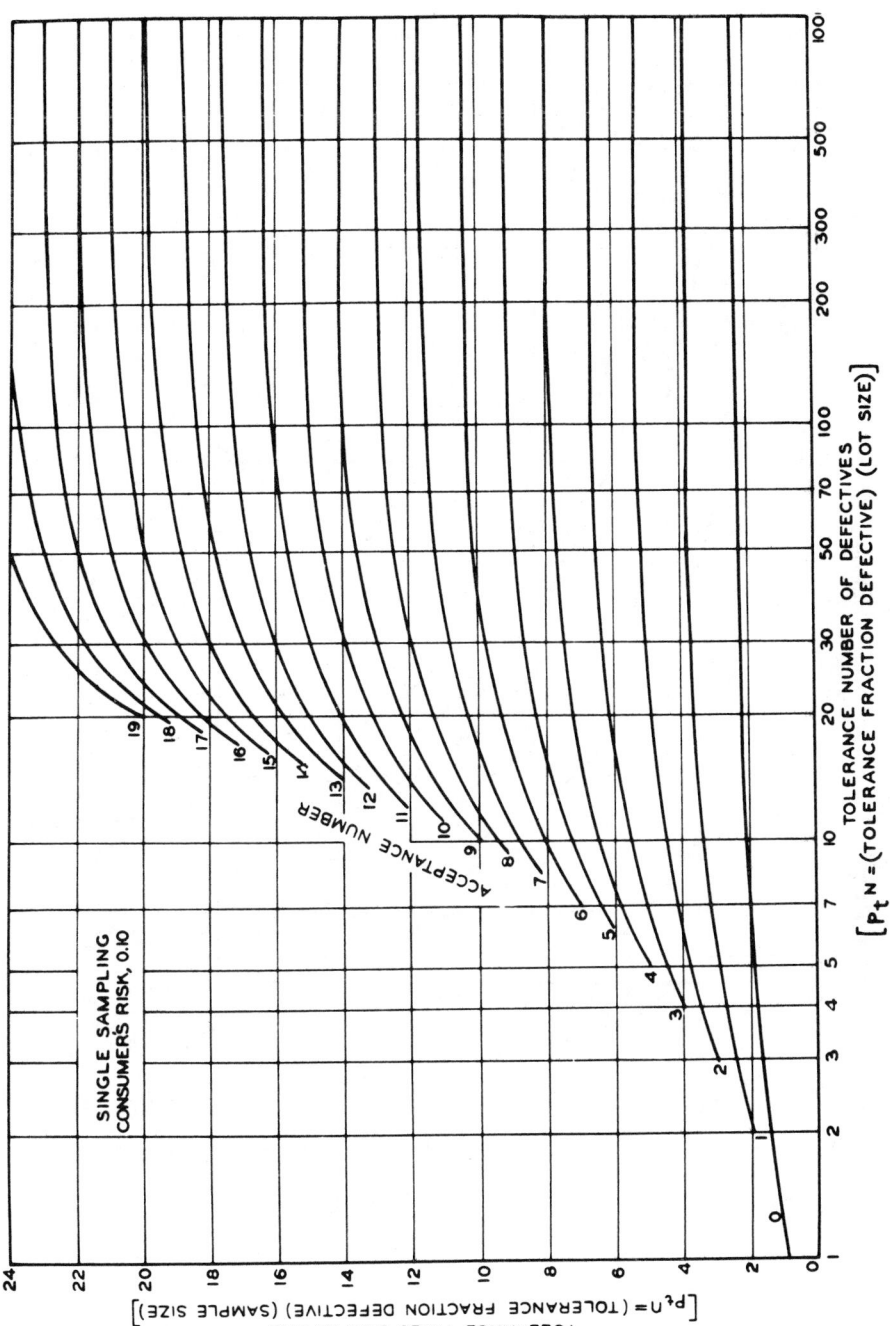

FIG. 14-9 Dodge-Romig curves for finding the size of the sample. (From
H. F. Dodge and H. G. Romig, *Sampling Inspection Tables*, 2nd ed., 1959,
p. 15. Copyright (c) 1944, Bell Telephone Laboratories. Reprinted by
permission of John Wiley and Sons, Inc.)

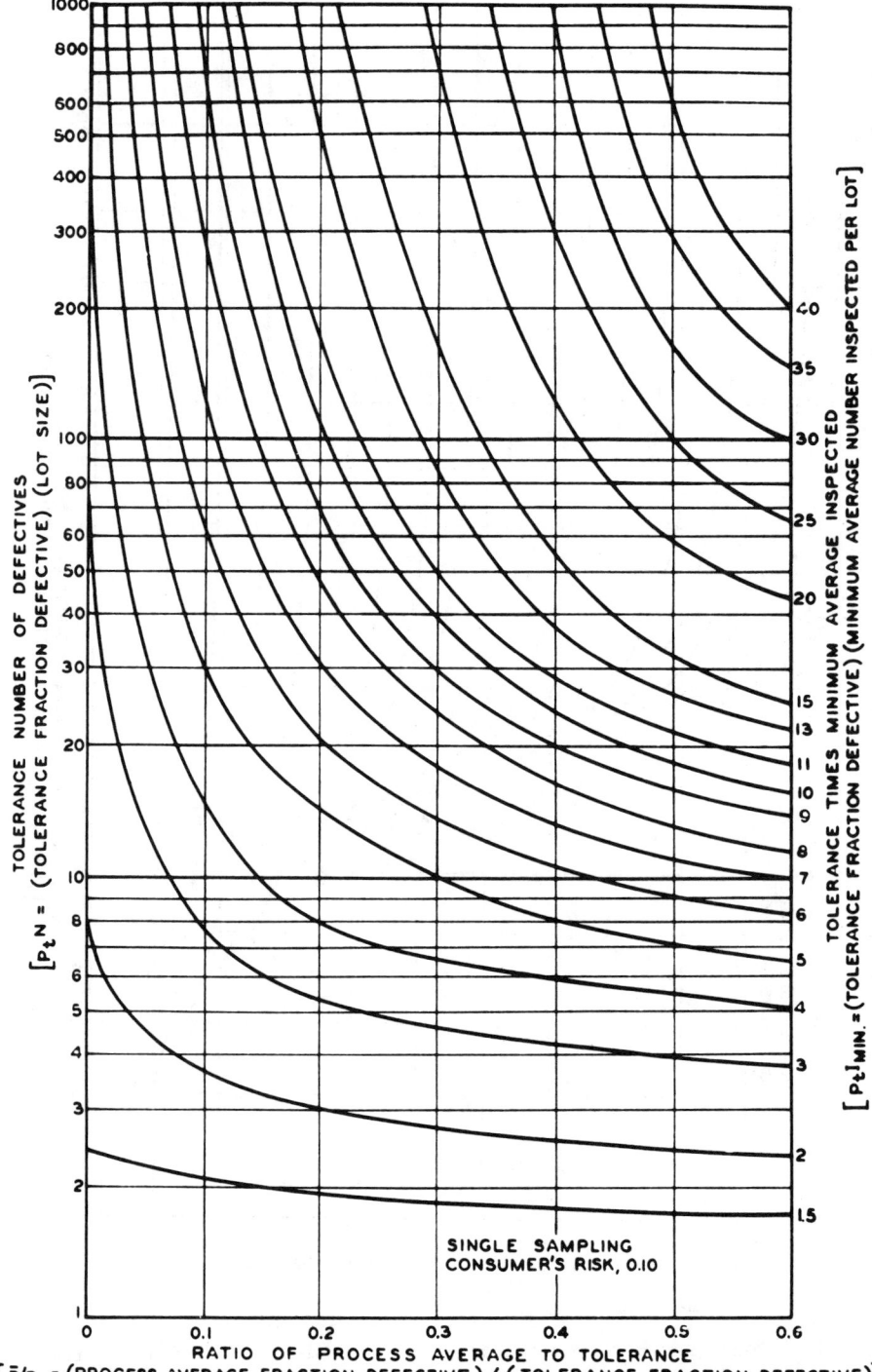

FIG. 14-10 Dodge-Romig curves for finding the minimum amount of inspection per lot. (From H. F. Dodge and H. G. Romig, *Sampling Inspection Tables*, 2nd ed., 1959, p. 16. Copyright ©️ 1944, Bell Telephone Laboratories. Reprinted by permission of John Wiley and Sons, Inc.)

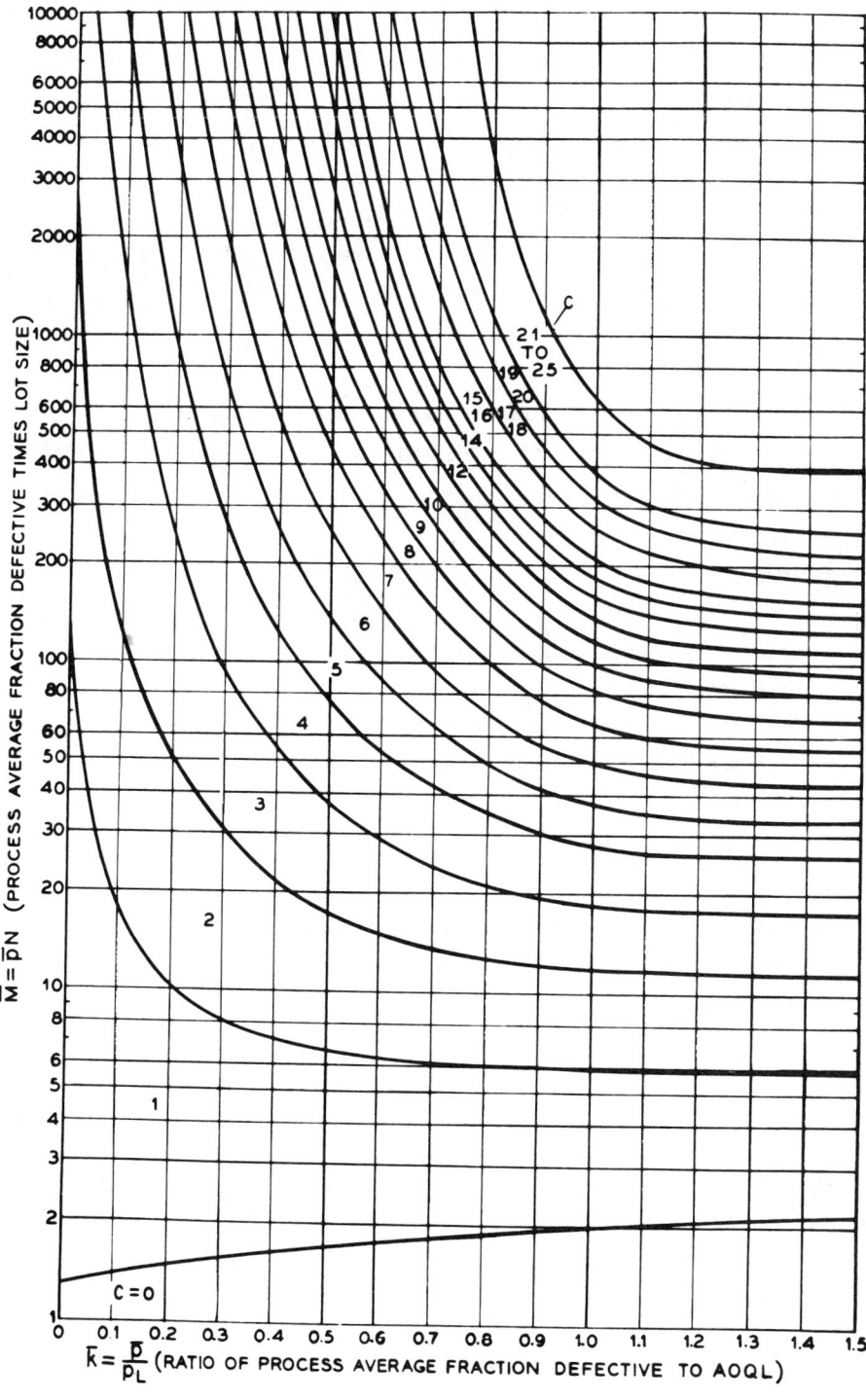

FIG. 14-11 Dodge-Romig curves for determining the acceptance number, c;
AOQL protection. (From H. F. Dodge and H. G. Romig, *Sampling Inspection
Tables*, 2nd ed., 1959, p. 40. Copyright © 1944, Bell Telephone Labora-
tories. Reprinted by permission of John Wiley and Sons, Inc.)

on the right axis. Division of p_t(ATI) by p_t gives the ATI that
will minimize average total inspection.

For example, suppose N = 250, $\bar{p}$ = .016, and LTPD = 4 percent; so
p_t = .04. Then $\bar{p}/p_t$ = 0.4 and $p_t N$ = 10, so that Fig. 14-8 shows c = 2.
Entering Fig. 14-9 with $p_t N$ = 10 and reading from c = 2 gives $p_t n$ = 4.5 so
that n = 4.5/.04 = 112.5. The plan is n = 113, c = 2. Entering Fig.
14-10, the point (0.4, 10) is nearest the curve for p_t(ATI) = 5.7 (interpolating) and so ATI$_{min}$ = 5.7/.04 = 142.5 $\sim$ 143.

Constructing AOQL Plan With Minimum ATI

To find an AOQL plan which minimizes average total inspection of accepted
samples plus 100 percent inspection of rejected lots, proceed as follows:

1. Given lot size (N), AOQL (p_L), process average proportion defective ($\bar{p}$).
2. Calculate the ratio $\bar{k}$ = $\bar{p}/p_L$ and the product $\bar{M}$ = $\bar{p}N$.
3. Enter Fig. 14-11 with $\bar{k}$ on the x-axis and M on the y-axis to find
   the acceptance number region which will make ATI a minimum. Use
   the acceptance number c.
4. Use Appendix Table T14-1 to find the values of x and y specified
   for the c value obtained in step 3.
5. Compute

$$n = \frac{yN}{p_L N + y}$$

6. The plan n, c will give the AOQL specified with minimum ATI. The
   AOQL will occur at

$$np_M = x$$

   so that

$$p_M = \frac{x}{n}$$

For example, suppose we take AOQL = 4 percent; so p_L = .04, lot size
N = 250, and process average $\bar{p}$ = .016. Then $\bar{k}$ = .016/.04 = 0.4 and
$\bar{M}$ = .016(250) = 4. Hence, from Fig. 14-11, c = 1. Appendix Table T14-1
shows y = 0.84 and x = 1.62,

$$n = \frac{.84(250)}{.04(250) + .84} = 19.4 \sim 20$$

The plan is n = 20, c = 1, and the AOQL will occur at

$$P_M = \frac{1.62}{20} = .08$$

at $\bar{p}$ = .016, the ATI will be

$$ATI = nP_a + N(1 - P_a)$$
$$= 20(.959) + 250(.041)$$
$$= 29.4$$

ANSCOMBE RECTIFYING INSPECTION PROCEDURE

F. J. Anscombe (1949) has presented an adaptable, easy to use, inspection procedure which is appropriate in guaranteeing an LTPD and an AOQL in a sequence of inspections. The method does not rely on 100 percent inspection of rejected lots. Rather, successive samples are taken on each lot until a stopping rule is satisfied and the lot accepted, or until the lot is exhausted. The stopping rule is set in such a way that poor lots will, in general, be extensively sampled while good lots will require minimal sampling. It assumes that any defective items found will be replaced with effective ones. The LTPD is guaranteed with minimum ATI.

Operation

The procedure, in the words (notation slightly modified[*]) of Anscombe (1949, p. 193), is as follows:

> From a batch of N articles, a first sample of f_1N articles
> is inspected, and then further samples of f_2N articles
> each. Defective articles found are removed or replaced by
> good ones. Inspection ceases after the first sample if
> no defectives have been found, or after the second sample
> if altogether one defective has been found, or, generally,
> after the (r + 1)th sample if altogether r defectives have
> been found. Inspection is continued until either this
> stopping rule operates or the whole batch is inspected.

This ingenious procedure may be summarized as follows:

---

[*] Anscombe's original α and β are given here as f_1 and f_2 after the manner of Duncan (1974).

Sample (i)	Sample size	Acceptance number ($c_i = i - 1$)
1	$f_1 N$	0
2	$f_2 N$	1
3	$f_2 N$	2
...	...	...
k	$f_2 N$	k - 1

Selection

The parameters of these plans are f_1 and f_2. These quantities and associated measures have been tabulated exactly by Anscombe (1949) and are exemplified in Table 14-5, which is part of the original Table IV of the Anscombe paper. Anscombe's notation compares to that of this book as follows:

Anscombe	Present notation
Z_t	$p_t N$
ε	β
α	f_1
β	f_2
A	ASN
N	N
Y	$\bar{p}N$
AOQL	(N)AOQL
Y*(last column)	$(N)p_M$

In the notation of this book, the heading of Anscombe's first table would appear as follows:

$$Np_t = 5, \beta = .10$$

Scheme   ASN/N for $N\bar{p}$ equal to

f_1 f_2 0 1 2 3 4 5 6 8 10 12   (N)AOQL $(N)p_M$

Tables are given for $\beta = .10$ and $\beta = .01$ by Anscombe; however, only the table for $\beta = .10$ (i.e., LTPD protection) is shown here.

Anscombe's tables are indexed by limiting quality $Z_t = Np_t$ and associated risk $\varepsilon = \beta$. They show the ratio of average sample number to lot size, that is ASN/N for various values of $Y = \bar{p}N$. The value of ASN/N which appears in bold type (denoted by asterisk here) is the minimum ratio

TABLE 14-5 Anscombe Rectifying Inspection Schemes for Lot Tolerance $Z_t = p_t N$ with Risk $\epsilon = \beta = .10$.
Note: Asterisk indicates minimum ratio for column.

Scheme		Average Sample Size (A/N) for Y equal to										AOQL	Y*
α	β	0	1	2	3	4	5	6	8	10	12		
					$Z_t = 5$, $\epsilon = 0.10$								
.3690	.1900	.369*	.439*	.536	.674	.650	.743	.737	.890	.916	.935	1.4	3
.4238	.0982	.424	.465	.515*	.576*	.626*	.677	.709	.800	.844	.877	1.8	5
.4773	.0639	.477	.508	.542	.581	.635	.670*	.706*	.769	.815	.849	1.9	6
.5241	.0459	.524	.548	.574	.603	.652	.678	.713	.760*	.802	.835	2.0	7
.5642	.0348	.564	.584	.605	.627	.670	.691	.723	.761	.799	.828	2.1	8
.5986	.0275	.599	.615	.632	.651	.688	.705	.734	.765	.798*		2.2	9
.6283	.0223	.628	.642	.657	.672	.705	.719	.745	.771			2.2	10
.6540	.0185	.654	.666	.679	.692	.721	.733					2.2	11
.6767	.0155	.677	.687	.698	.709							2.2	12
					$Z_t = 10$, $\epsilon = 0.10$								
.2057	.1694	.206*	.241*	.287	.352	.444	.569	.677	.820	.897	.940	2.8	5
.2337	.0967	.234	.256	.283*	.316*	.356*	.406	.469	.526	.673	.708	3.7	7
.2669	.0686	.267	.285	.306	.330	.357	.389*	.427*	.495	.584	.639	4.3	9
.3002	.0530	.300	.316	.334	.353	.375	.400	.427*	.492*	.556	.610	4.6	10
.3323	.0429	.332	.347	.362	.379	.397	.418	.440	.499	.549*	.601	4.9	11
.3625	.0357	.363	.375	.389	.404	.420	.438	.456	.510	.552	.599*	5.1	13
.3908	.0304	.391	.403	.415	.429	.443	.458	.474	.523	.559	.602	5.2	14
.4170	.0262	.417	.428	.439	.452	.464	.478	.492	.537	.568		5.3	15
.4414	.0229	.441	.452	.462	.473	.485	.497	.510				5.4	16

From Anscombe (1949, p. 202). Reprinted by permission.

for the column. That is, minimum ATI, since for these plans

        ASN = ATI

Also shown are values of N(AOQL) and Np_M for each plan. The original
tables are in terms of number defective in the lot, and not proportion
defective. Accordingly, the values shown must be suitably transformed by
multiplying or dividing by the lot size to obtain the conventional values.

    To find a plan having a desired AOQL, compute N(AOQL) and search the
second last column of the Anscombe tables for the desired value. This will
not guarantee minimum ATI for the AOQL given.

    To use Table 14-5 to guarantee a specified LTPD = p_t for estimated
fraction defective $\bar{p}$, proceed as follows:

1. Enter the table with $Z_t = Np_t$ and $\varepsilon = \beta = .10$.
2. Find the column corresponding to $Y = N\bar{p}$.
3. The value of ASN/N in bold type (denoted by asterisk here) indi-
    cates the row for minimum ATI. Use the values of $\alpha = f_1$ and
    $\beta = f_2$ obtained from that row.
4. Multiply ASN/N by N to obtain ASN. Divide the value of AOQL
    shown by N to put it in terms of fraction defective. Similarly,
    divide the corresponding value of Y by N to obtain p_M.

    For example, if a plan is desired having 4 percent LTPD with lot size
250 for process average percent defective 1.6 percent, we have

1. $Z_t = 250(.04) = 10$ and $\varepsilon = \beta = .10$
2. $Y = 250(.016) = 4$
3. Bold (asterisk) ASN/N = .356; so $f_1 = .2337$ and $f_2 = .0967$
4. Measures are as follows:

$$ASN = .356(250) = 89$$
$$AOQL = 3.7/250 = .015$$
$$p_M = 7/250 = .028$$

Sample sizes will be

$$f_1(N) = .2337(250) = 58.4 \sim 59$$
$$f_2(N) = .0967(250) = 24.2 \sim 25$$

Application of this plan would involve a first sample of 59 followed by
successive samples of 25. Inspection would be terminated if at any time
the accumulated number of defectives is less than the number of samples

minus one, or when the lot is exhausted. The plan would appear as follows:

Sample	Sample size	Cumulative sample size	Cumulative acceptance number
1	59	59	0
2	25	84	1
3	25	109	2
...	...	...	...
k	25	$59 + 25(k - 1)$	$k - 1$

Measures

The following approximate measures of scheme performance have been given by Anscombe. They may be used whenever

$$p < \frac{1 - f_1}{f_2 N}$$

1. Average sample number

$$ASN = N\left[\frac{f_1}{1 - pNf_2} - \frac{f_1 f_2^2 pN}{(1 - pNf_2)^3}\right]$$

2. Average outgoing quality at p

$$AOQ = p\left[1 - \frac{f_1}{1 - pNf_2} + \frac{f_1 f_2}{(1 - pNf_2)^3}\right]$$

3. Average outgoing quality limit

$$AOQL = \frac{1}{N}\left[\frac{(1 - \sqrt{f_1})^2}{f_2} + \frac{1}{\sqrt{f_1}} - 1\right]$$

which is attained at

$$p_M = \frac{1}{N}\left[\frac{1 - \sqrt{f_1}}{f_2} + \frac{3}{2\sqrt{f_1}} - 1\right]$$

For example, for the plan derived above where LTPD = 4 percent, N = 250, $\bar{p}$ = .016, we obtained f_1 = .2337, f_2 = .0967. Hence at $\bar{p}$ = .016,

$$ASN = 250\left[\frac{.2337}{1 - .016(250)(.0967)} - \frac{.2337(.0967)^2(.016)(250)}{(1 - (.016)(250)(.0967))^3}\right]$$

$$= 85.8$$

$$AOQ = .016\left[1 - \frac{.2337}{1 - .016(250)(.0967)} + \frac{.2337(.0967)}{(1 - (.016)(250)(.0967))^3}\right]$$

$$= .011$$

$$AOQL = \frac{1}{250}\left[\frac{(1 - \sqrt{.2337})^2}{.0967} + \frac{1}{\sqrt{.2337}} - 1\right]$$

$$= .015$$

$$P_M = \frac{1}{250}\left[\frac{1 - \sqrt{.2337}}{.0967} + \frac{3}{2\sqrt{.2337}} - 1\right]$$

$$= .030$$

PROBLEMS

1. Using the Altman diagram, find the AOQLs associated with the following plans:

   a. $n = 50$, $c = 1$

   b. $n = 80$, $c = 1$

   c. $n = 20$, $c = 0$

2. From the Altman diagram, derive a plan for AOQL = 5 percent when sample size must be restricted to 30 or less.

3. Find Dodge-Romig single- and double-sampling plans for AOQL = 4.0 percent for the lot sizes and process average percents defective shown.

   a. $N = 125$, $\bar{p} = 1$ percent

   b. $N = 1250$, $\bar{p} = 1$ percent

   c. $N = 5500$, $\bar{p}$ unknown

4. Find Dodge-Romig LTPD plans for LTPD = 4 percent meeting the specifications given in Prob. 3.

5. Construct an LTPD plan for $N = 1250$, $p_t = 4$ percent when the process average is at one percent defective. What is the minimum average total inspection for this plan?

6. Construct an AOQL plan for $N = 1250$, $p_L = 4$ percent when the process average is at one percent defective. At what fraction defective will the AOQL occur?

7. Observing that $c/y \simeq .7$, use the Campbell relationship $np_0 = c + .67$ to obtain an estimate of the AOQL from the indifference quality, p_0.

8. A lot of size 125 is to be screened using the Anscombe procedure. LTPD protection of $p_t = .04$ is desired, while the process average has been $\bar{p} = .008$. Construct the appropriate Anscombe scheme. What is its average sample number, the AOQL, and the point at which the AOQL occurs?

9. What Dodge-Romig single-sampling plan corresponds to the Anscombe plan
    developed in Prob. 8? Find the minimum amount of inspection per lot
    from Fig. 14-10 and compare to the ASN of the Anscombe plan.

10. Using f_1 = .300 and f_2 = .053, verify the values given in the table
    for Z_t = 10, ε = .10 when $\overline{Np}$ = 0.

REFERENCES

Altman, I. B. (1954), Relationship Between Sample Size and AOQL for Attri-
    butes Single Sampling Plans, *Industrial Quality Control*, 10(4): 29-30.

Anscombe, F. J. (1949), Tables of Sequential Inspection Schemes to Control
    Fraction Defective, *Journal of the Royal Statistical Society* (Series A),
    112, Part II: 180-206.

Dodge, H. F., and H. G. Romig (1929), A Method of Sampling Inspection, *The
    Bell System Technical Journal*, 8(10): 613-631.

Dodge, H. F., and H. G. Romig (1941), Single Sampling and Double Sampling
    Inspection Tables, *The Bell System Technical Journal*, 20(1): 1-61.

Dodge, H. F., and H. G. Romig (1959), *Sampling Inspection Tables, Single
    and Double Sampling*, 2nd ed., John Wiley and Sons, New York.

Duncan, A. J. (1974), *Quality Control and Industrial Statistics*, 4th ed.,
    Richard D. Irwin, Homewood, Illinois.

CONTINUOUS SAMPLING PLANS

In the sampling of some processes, lots are not clearly defined. In a
sense, lot size is N = 1, since units are produced item by item. Examples
might be cars coming off an assembly line, soft drink bottles from a continuous glass ribbon machine, or welded leads emanating from a welding
operation. Yet AOQL and perhaps some form of LTPD protection may be desired. Sometimes in such situations, it is possible to artificially define
a lot, such as the production of an hour, a shift, a day, or a week. This
is often quite arbitrary, however, and other alternatives exist.

Emanating from the original CSP-1, published by H. F. Dodge (1943),
several different continuous sampling plans have been developed to deal
with this situation, usually with AOQL protection. They apply to a steady
stream of individual items from the process and require sampling of a specified fraction, f, of the items in order of production, with 100 percent
inspection of the flow at specified times. Several such plans have been
described in detail by K. S. Stephens (1980) in a manual prepared for the
American Society for Quality Control.

Special measures of performance apply to continuous plans, they include

AOQ = Average outgoing quality

AOQL = Average outgoing quality limit

P_a = Average fraction of production accepted under sampling

AFI = Average fraction of production inspected

The AOQ and AOQL are previously defined. The symbol P_a is used to denote
the average fraction of production accepted under sampling since in concept P_a implies the probability of an item being accepted on a sampling
basis (whether included in the sample or not). In this sense P_a will be
seen to be analogous to the lot-by-lot probability of acceptance under

rectification. AFI indicates the average fraction of product actually inspected including items inspected during sampling or in screening. Then

$$P_a = \frac{1 - AFI}{1 - f}$$

DODGE CONTINUOUS PLANS

Dodge CSP-1

The most celebrated continuous sampling plan and the plan which undoubtedly
has received the most application is also the original — the Dodge CSP-1
plan. It is carried out on a stream of product, with items inspected in
order of production. The procedure is as follows:

1. Specify f = sampling frequency
          i = clearing interval.
2. Begin 100 percent inspection.
3. After i units in succession have been found without a defective,
   start sampling inspection.
4. Randomly inspect a fraction f of the units.
5. When a defective is found, revert to 100 percent inspection (step 2).

A diagrammatic representation of CSP-1 will be found in Fig. 15-1.

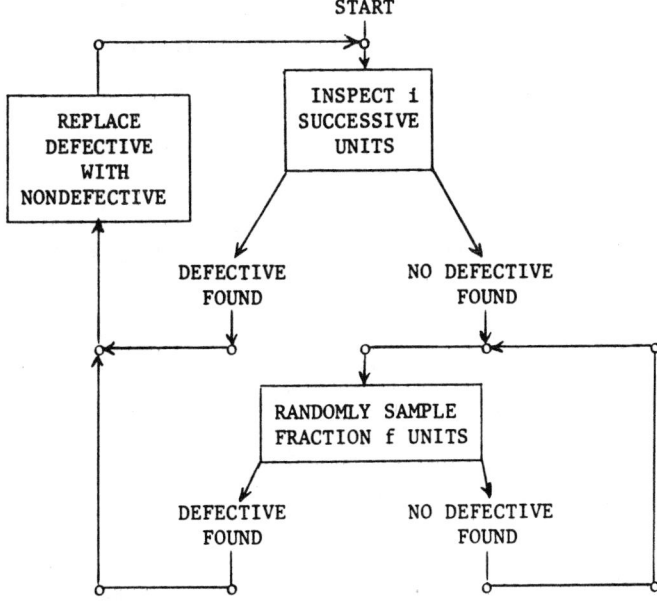

FIG. 15-1 Dodge CSP-1 procedure.

A detailed discussion of the Dodge CSP-1 plan and its relation to other approaches for the sampling of a steady stream of product is given in Dodge (1947).

The Dodge CSP-1 plan and its later modifications CSP-2 and CSP-3 (see below) use a special measure to evaluate protection against spottiness, that is, surges of highly defective product. Based on a finite production run of 1000 units p_t% is the percent defective in a consecutive run of 1000 units for which the probability of remaining under sampling is 10 percent for a sample of f. It shows the percent defective which will result in a 90 percent chance of reverting to 100 percent inspection within a run of 1000 consecutive units.

The notation here is somewhat deceptive, in that p_t% is not the LTPD of the continuous plan. Burr (1976) has shown that, if f is the fraction sampled in a CSP-1 plan,

$$.10 = \left(1 - \frac{p_t\%}{100}\right)^{1000f}$$

which gives

$$p_t\% = 100\left[1 - (.10)^{\frac{1}{1000f}}\right]$$

a function of f only.

If the LTPD is to be determined in terms of a 10 percent probability of acceptance P_a as defined above for continuous plans, we find, for CSP-1, when P_a = .10

$$LTPD = 100\left[1 - \left(\frac{f}{9+f}\right)^{1/i}\right]$$

which is a function of both f and i.

Both these equations can be solved using logarithms. For example, for the CSP-1 plan f = .1, i = 38, and p_t% = 2.3% while LTPD = 11.2%.

A particular CSP-1 plan is determined by the values of f and i selected. The choice of f and i, of course, depends on the value of AOQL and possibly the protection against spotty quality desired for the plan. Dodge presented a simple diagram, shown in Fig. 15-2, to be used in setting these quantities. Values of f are given on the left axis while values of p_t% are given on the right. Values of i are shown on the abscissa. The curves represent various levels of AOQL in percent so that any curve defines alternative f and i to obtain the AOQL. The corresponding p_t% can be read from the right axis.

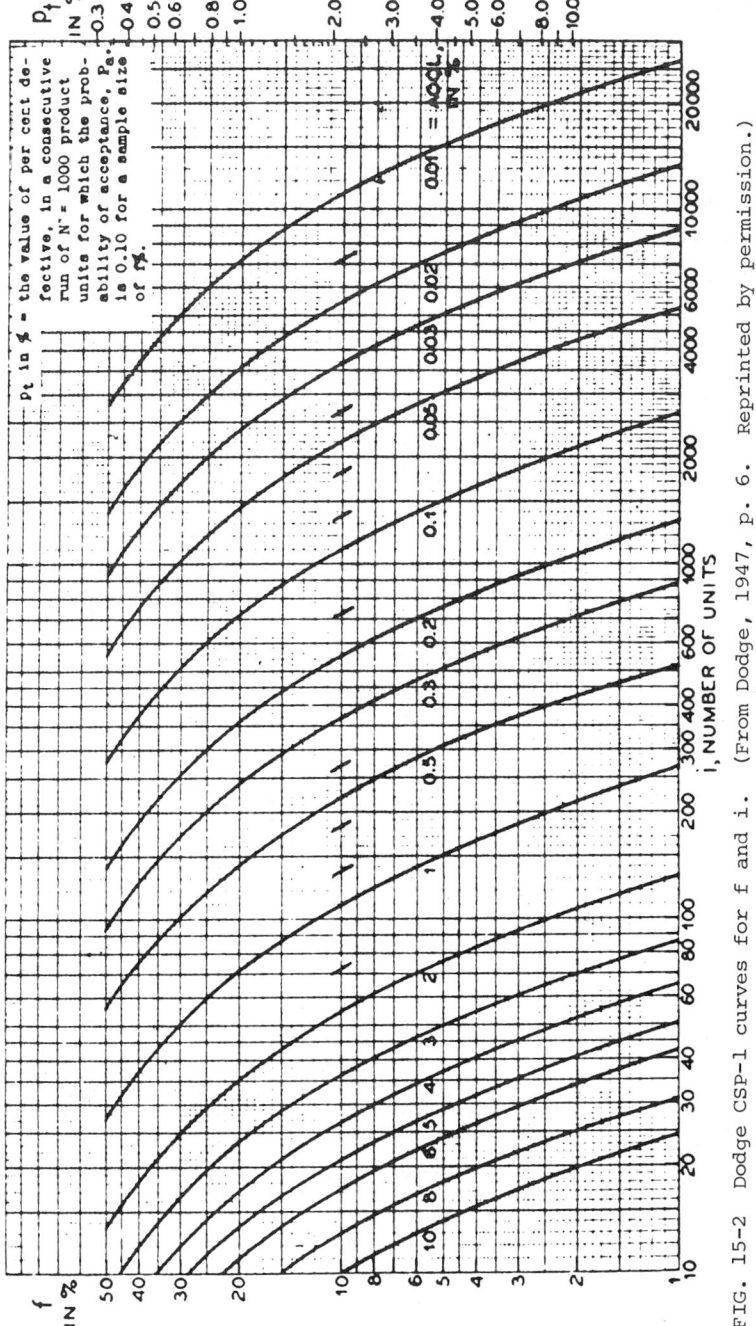

FIG. 15-2 Dodge CSP-1 curves for f and i. (From Dodge, 1947, p. 6. Reprinted by permission.)

For example, if an AOQL of 2.9 percent is desired with protection against spotty quality of 2.3 percent we have, from the diagram, i = 38 and f = .10. This would result in 100 percent inspection until 38 units are found good in succession with random sampling of 10 percent of the units thereafter until a defective is found.

The Dodge CSP-1 plan provides AOQL protection for most practical conditions; however, when the process is not in a state of control, instances may be found where the AOQL may be exceeded. This was pointed out by Wald and Wolfowitz (1945). Subsequently, Lieberman (1953) showed that the Dodge CSP-1 procedure absolutely guarantees an unlimited AOQL, or UAOQL, of

$$UAOQL = \frac{(1/f) - 1}{(1/f) + i}$$

when defective items are replaced with good items.

For the plan f = .1, i = 38, this becomes

$$UAOQL = \frac{(1/.1) - 1}{(1/.1) + 38} = .188$$

which is considerably worse than the nominal AOQL of 2.9 percent. Under this highly conservative approach, i would have to be 301 to guarantee a 2.9 percent UAOQL. Fortunately, the nominal AOQL of 2.9 percent will hold in most practical situations and so can be safely used to characterize the plan. When defective items are not replaced, the UAOQL guaranteed becomes

$$UAOQL = \frac{1 - f}{f(i - 1) + 1}$$

as given by Banzhaf and Brugger (1970). This would give UAOQL = .191 for the plan f = .1, i = 38 when defectives are not replaced.

Dodge-Torrey CSP-2 and CSP-3

H. F. Dodge and M. N. Torrey later improved upon CSP-1 somewhat, particularly with regard to the occurrence of an occasional stray random defective (CSP-2) and, in addition, with regard to "spotty" quality (CSP-3). That is, short bursts of bad quality. In the Dodge-Torrey (1951) paper, they proposed to extend the CSP-1 procedure by changing the steps given above for CSP-1 as follows:

For CSP-2,

> Step 5. When a defective is found, continue sampling for k successive sample units. If no defective is found in the k

samples, continue sampling on a normal basis (step 4). If a defective is found in the k samples, revert to 100 percent inspection immediately (step 2).

For CSP-3,

Step 5. Same as CSP-2, except, in addition, begin step 5 as follows: When a defective is found, inspect the next 4 units, if an additional defective is found revert to 100 percent inspection (step 2); otherwise, continue sampling for k ...

The other steps remain the same in each procedure. A flow chart for the

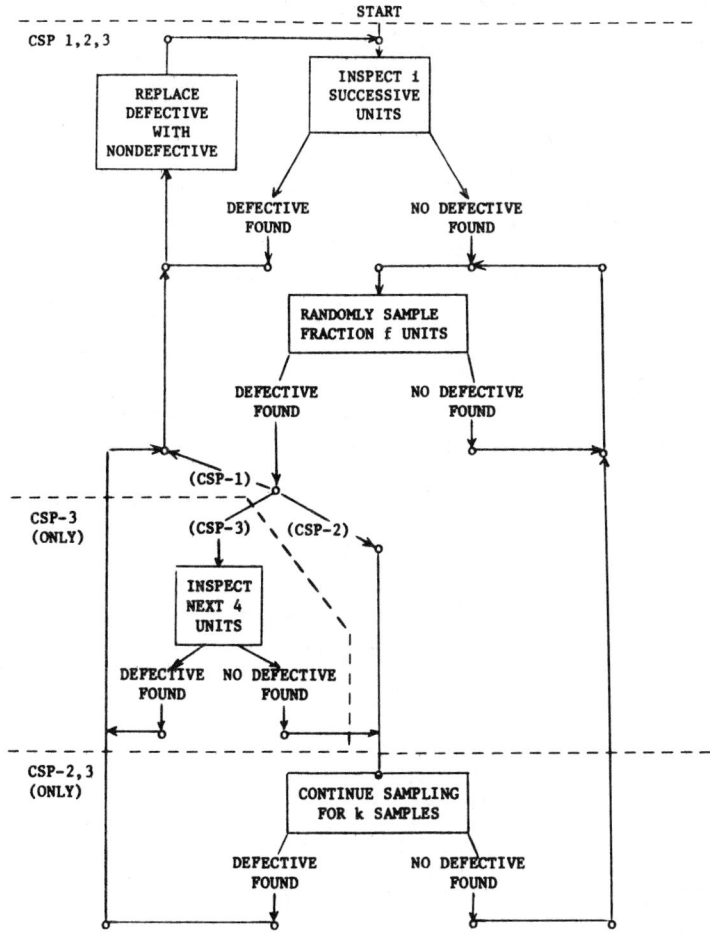

FIG. 15-3 Dodge CSP-1,2,3 procedure.

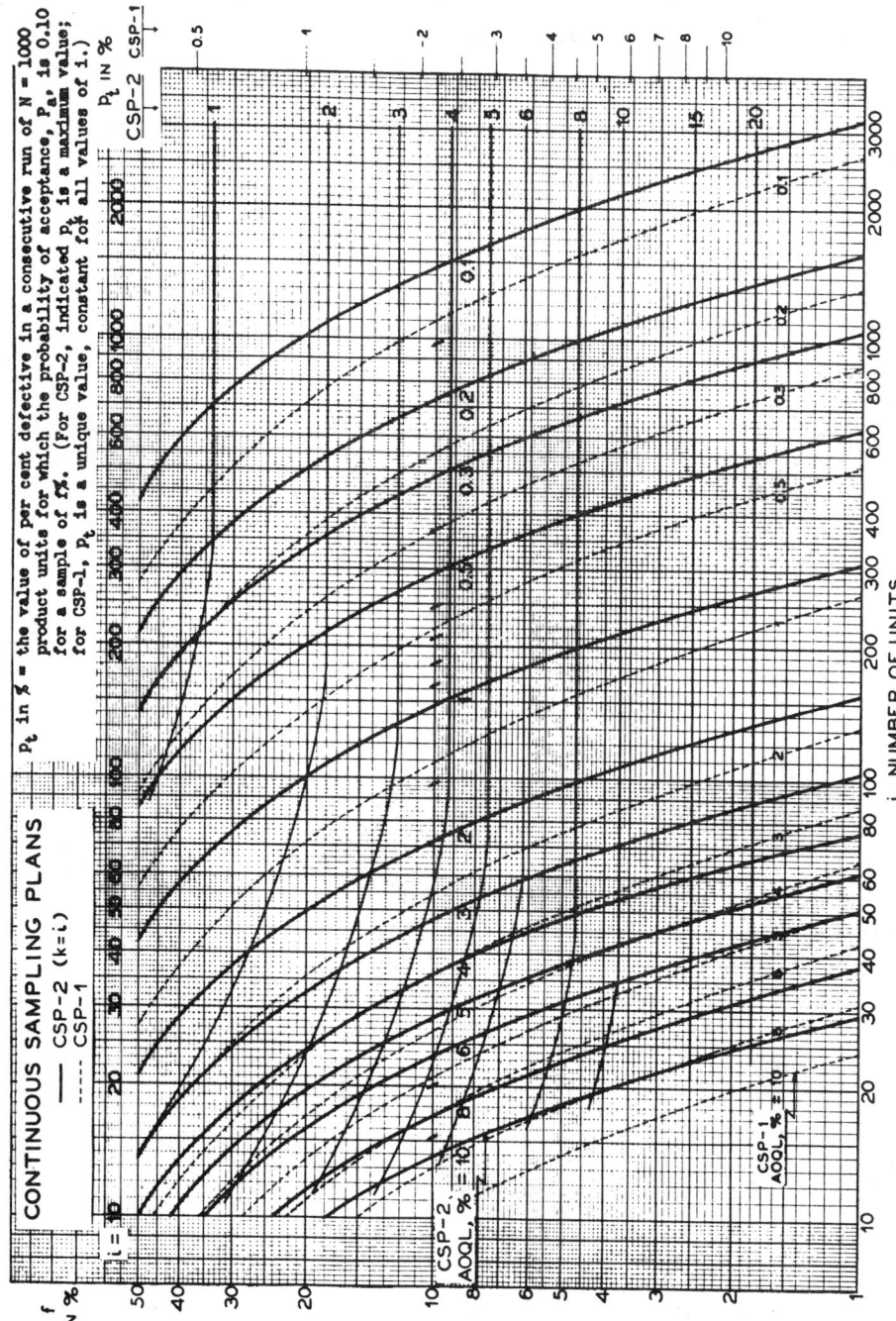

FIG. 15-4 Dodge–Torrey CSP-2 curves for f and i. (From Dodge and Torrey, 1951, p. 8. Reprinted by permission.)

CSP-1, 2, and 3 procedures is given in Fig. 15-3 which highlights the differences between them.

Since CSP-3 is a very slight modification of CSP-2, the curves for finding f and i for CSP-2 are used also for CSP-3, as a very good approximation. A set of curves is presented in Dodge and Torrey (1951) for CSP-2 for the case when k = i. These curves are given in Fig. 15-4. They are employed in a manner identical to those for CSP-1. In fact the CSP-1 curves are superimposed on the diagram as dotted lines. We find the CSP-2 plan f = .1, i = 50 gives an AOQL of 2.9 percent as did the CSP-1 plan previously discussed. The value of p_t% for this plan is about 4.2 percent, higher than the value of 2.3 percent from the CSP-1 plan because of the increased difficulty of reverting to 100 percent inspection under CSP-2.

CSP-2 can also be shown to guarantee an AOQL even when the process is not in a state of control. The upper limit on AOQL, as given by Banzhaf and Brugger (1970) when defective items are not replaced is

$$\text{UAOQL} = \frac{2(1 - f)}{if + 2(1 - f)}$$

This formula may also be used as an upper limit on the AOQL of CSP-3.

The operating characteristic curves of continuous plans are expressed in terms of the percent of total production accepted on a sampling basis (100 P_a) plotted against incoming values of percent defective. A set of such curves from the Dodge-Torrey (1951) paper is given in Fig. 15-5.

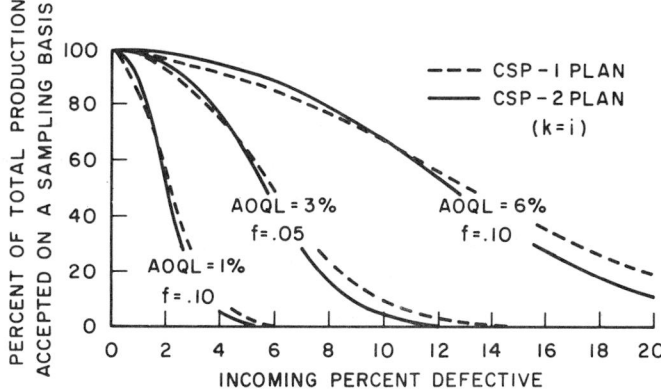

FIG. 15-5 Operating characteristic curves of three CSP-1 and CSP-2 plans. (From Dodge and Torrey, 1951, p. 9. Reprinted by permission.)

Measures of CSP-1, 2, 3

Formulas are available to more precisely determine the various measures of CSP plans for incoming proportion defective p, where $q = 1 - p$ and

u = average number of units inspected on a 100 percent inspection basis

$$u = \frac{1 - q^i}{pq^i} \qquad \text{(CSP-1, CSP-2)}$$

v = average number of units passed during sampling inspection

$$v = \frac{1}{fp} \qquad \text{(CSP-1)}$$

$$v = \frac{2 - q^k}{fp(1 - q^k)} \qquad \text{(CSP-2)}$$

The formulas are as follows:

Average fraction inspected

F = average fraction of total product inspected in the long run,

$$F = \frac{u + fv}{u + v} \qquad \text{(CSP-1, CSP-2)}$$

Specifically,

$$F = \frac{f}{f + q^i(1 - f)} \qquad \text{(CSP-1)}$$

$$F = \frac{f(1 - q^i)(1 - q^k) + q^i f(2 - q^k)}{f(1 - q^k)(1 - q^i) + q^i(2 - q^k)} \qquad \text{(CSP-2)}$$

$$F = \frac{f}{f(1 - q^i)^2 + q^i(2 - q^i)} \qquad \text{(CSP-2, k = i)}$$

$$F = \frac{f(1 - q^i)(1 - q^{k+4}) + fq^i + 4\,fpq^i + fq^{i+4}(1 - q^k)}{f(1 - q^i)(1 - q^{k+4}) + q^i + 4\,fpq^i + q^{i+4}(1 - q^k)} \qquad \text{(CSP-3)}$$

Average outgoing quality (defectives replaced by good)

P_A = AOQ = average outgoing quality

AOQ = $p[1 - F]$ (defectives replaced with good)

Specifically,

$$AOQ = p\left[\frac{(1 - f)q^i}{f + (1 - f)q^i}\right] \qquad \text{(CSP-1)}$$

$$AOQ = p\left[\frac{(1 - f)q^i(2 - q^k)}{f(1 - q^i)(1 - q^k) + q^i(2 - q^k)}\right] \quad \text{(CSP-2)}$$

$$AOQ = p\left[\frac{(1 - f)q^i(2 - q^i)}{f + (1 - f)q^i(2 - q^i)}\right] \quad \text{(CSP-2, } k = i\text{)}$$

$$AOQ = p\left[\frac{(1 - f)q^i(1 + q^4 - q^{k+4})}{f(1 - q^i)(1 - q^{k+4}) + q^i(1 + q^4 - q^{k+4}) + 4\,fpq^i}\right] \quad \text{(CSP-3)}$$

Average outgoing quality (defectives removed but not replaced by good)

$$p_A' = AOQ' = \text{average outgoing quality}$$

$$AOQ' = \frac{p(1 - F)}{(1 - pF)} = \frac{AOQ}{q + AOQ}$$

Specifically,

$$AOQ' = p\left[\frac{(1 - f)q^i}{fq + (1 - f)q^i}\right] \quad \text{(CSP-1)}$$

$$AOQ' = p\left[\frac{(1 - f)q^i(2 - q^i)}{fq + (1 - f)q^i(2 - q^i)}\right] \quad \text{(CSP-2, } k = i\text{)}$$

The AOQL may be found by the differentiation-iteration technique used by Dodge (1943) or by trial and error. The relationship between p_L, the AOQL, and the point at which it occurs p_M is

$$AOQL = p_L = \frac{(i + 1)p_M - 1}{i}$$

so that

$$p_M = \frac{1 + ip_L}{i + 1}$$

These formulas can be used to calculate specific measures for a given CSP plan. As an example, consider the CSP-1 plan $f = .1$, $i = 38$ evaluated at $p = .054$. We have

$$q = .946$$

$$u = \frac{1 - .946^{38}}{(.054)(.946^{38})} = 134.148$$

$$v = \frac{1}{(.1)(.054)} = 185.185$$

Average fraction inspected

$$F = \frac{134.148 + .1(185.185)}{134.148 + 185.185} = .478$$

or alternatively,

$$F = \frac{.1}{.1 + .946^{38}(.9)} = .478$$

Average outgoing quality (with replacement)

$$AOQ = .054(1 - .478) = .028$$

or alternatively,

$$AOQ = .054\left[\frac{(.9)(.946^{38})}{.1 + .9(.946^{38})}\right] = .028$$

Average outgoing quality (without replacement)

$$AOQ' = \frac{.054(1 - .478)}{(1 - .054(.478))} = \frac{.028}{.946 + .028} = .029$$

or alternatively,

$$AOQ' = .054\left[\frac{.9(.946^{38})}{.1(.946) + .9(.946^{38})}\right] = .029$$

The AOQL will occur at

$$P_M = \frac{1 + 38(.029)}{38 + 1} = .054$$

giving, of course

$$AOQL = \frac{(39)(.054) - 1}{38} = .029$$

Now consider the CSP-2 plan such that f = .1, i = 50, k = 50 evaluated at p = .054. We have

q = .946

$$u = \frac{1 - .946^{50}}{.054(.946^{50})} = 278.682$$

$$v = \frac{2 - .946^{50}}{.1(.054)(1 - .946^{50})} = 382.676$$

Average fraction inspected

$$F = \frac{278.682 + .1(382.676)}{278.682 + 382.676} = .479$$

or alternatively,

$$F = \frac{.1(1 - .946^{50})(1 - .946^{50}) + .946^{50}(.1)(2 - .946^{50})}{.1(1 - .946^{50})(1 - .946^{50}) + .946^{50}(2 - .946^{50})} = .479$$

or when $k = i$,

$$F = \frac{.1}{.1(1 - .946^{50})^2 + .946^{50}(2 - .946^{50})} = .479$$

Average outgoing quality (with replacement)

$$AOQ = .054(1 - .479) = .028$$

or alternatively,

$$AOQ = .054\left[\frac{.9(.946^{50})(2 - .946^{50})}{.1(1 - .946^{50})(1 - .946^{50}) + .946^{50}(2 - .946^{50})}\right] = .028$$

or when $k = i$,

$$AOQ = .054\left[\frac{.9(.946^{50})(2 - .946^{50})}{.1 + .9(.946^{50})(2 - .946^{50})}\right] = .028$$

Average outgoing quality (without replacement)

$$AOQ' = \frac{.054(1 - .479)}{1 - .054(.479)} = \frac{.028}{.946 + .028} = .029$$

or when $k = i$,

$$AOQ' = .054\left[\frac{.9(.946^{50})(2 - .946^{50})}{.1(.946) + .9(.946^{50})(2 - .946^{50})}\right] = .029$$

The AOQL for this plan should also be .029. Hence it should occur at $p_M = .054$ as before for the CSP-1 plan.

Finally, evaluating the corresponding CSP-3 plan $f = .1$, $i = 50$, $k = 50$, we have

$$F = \frac{\begin{pmatrix}.1(1 - .946^{50})(1 - .946^{54}) + .1(.946^{50}) \\ + 4(.1)(.054)(.946^{50}) + .1(.946^{54})(1 - .946^{50})\end{pmatrix}}{\begin{pmatrix}.1(1 - .946^{50})(1 - .946^{54}) + .946^{50} \\ + 4(.1)(.054)(.946^{50}) + .946^{54}(1 - .946^{50})\end{pmatrix}}$$

$$= .508$$

and with replacement of defectives

$$AOQ = .054(1 - .508) = .027$$

or alternatively,

$$AOQ = .054 \left[\frac{.9(.946^{50})(1 + .946^4 - .946^{54})}{.1(1 - .946^{50})(1 - .946^{54}) + .946^{50}(1 + .946^4 - .946^{54}) + 4(.1)(.054)(.946^{50})} \right]$$

$$= .027$$

The AOQL for this plan should also be .029. This too should occur at $p_M = .054$.

Stopping Rules and Selection of CSP-1 Plans

Occasionally, it may become obvious that the process level of fraction defective has moved upward from nominal levels. This is exhibited by excessively long sequences of 100 percent inspection. It is for such an eventuality that stopping rules, rules which indicate when the process should be stopped for corrective action, were devised. Such rules have been extensively investigated by Murphy (1959). The rules studied by Murphy are summarized in Table 15-1.

Typical of the stopping rules is the so-called rule r which involves stopping as soon as r defective units are found in any screening sequence. We shall consider only this rule here.

To uniquely determine r for a given CSP-1 plan having

TABLE 15-1 Stopping Rules for CSP-1 Plans

Rule ($n^* - i$)	Stop as soon as a defective unit is found in any one screening sequence after the sequence has exceeded $n^* - i$ units.
Rule (r)	Stop as soon as a specified number r of defective units are found in any one screening sequence.
Rule (N,R)	Stop as soon as a specified number R of defective units are found in any block of a specified number N of inspected units. (Blocks do not overlap.)
Rule (n^*)	Stop as soon as a specified number n^* of units have been inspected in any one screening sequence without ending it.

From Murphy (1959, p. 10). Reprinted by permission.

$\quad\quad\quad\quad$ f = fraction inspected

$\quad\quad\quad\quad$ i = clearing interval

$\quad\quad\quad\quad$ A = AOQL = p_L

$\quad\quad\quad\quad$ p_M = proportion defective at AOQL

it is necessary to specify

$\quad\quad\quad\quad$ E = average number of units produced between successive stops

$\quad\quad\quad\quad\quad\quad$ when p = p_M

recalling

$$p_M = \frac{1 + i(AOQL)}{i + 1}$$

Note that for large values of i (say greater than 100)

$$p_M \simeq \frac{1}{i + 1} + AOQL$$

Algebraically manipulating the formulas given by Murphy we find that to assure an interval between stops of E at proportion defective p_M it is necessary to set

$$r = \frac{-\log (1 - f + EFA)}{\log ((F - f)/F)}$$

using any base for the logarithms employed.

$\quad\quad$ Thus, for the plan f = .1, i = 38 with AOQL = .029, we have

$\quad\quad\quad\quad$ f = .1

$\quad\quad\quad\quad$ i = 38

$\quad\quad\quad\quad$ A = .029

$\quad\quad\quad\quad$ p_M = .054

$\quad\quad\quad\quad$ F = .478

so for an interval between stops of say 2i = 76, so that E = 76 and

$$r = \frac{-\log(1 - .1 + 76(.478)(.029))}{\log((.478 - .1)/.478)}$$

$$r = \frac{-\log 1.9535}{\log .7908} = \frac{-.2908}{-.1019} = 2.85 \sim 3$$

$\quad\quad$ As we have seen, the Dodge charts give a wide variety of choice of f and i for a given AOQL. Murphy (1959a) has presented a way to uniquely define a CSP-1 plan given

$\quad\quad$ A = AOQL

$\quad\quad$ p´ = producer's nominal quality level

$\quad\quad$ F´ = fraction inspected at producer's nominal quality level

For a given AOQL this allows the producer to specify a quality level which is to have minimal inspection. P' is chosen to be a fraction defective which is to require a reasonably small fraction inspected F' when quality is at the specified level. The procedure given by Murphy is as follows for plans where $(P' < A)$

1. Calculate

$$B = \frac{A - P'}{2.3\sqrt{(1 - A)(1 - P')}}$$

and

$$H = \frac{(1 - F')(1 - P')B}{AF'}$$

2. Using the graph given by Murphy (Fig. 15-6) find the value of C corresponding to H.

3. Then $i = C/B$.

4. Find f from the Dodge chart for CSP-1 (Fig. 15-2).

Murphy gives, as an example, the selection of a plan having A = .10, P' = .05, and F' = .10. We have

FACTOR FOR DETERMINING i

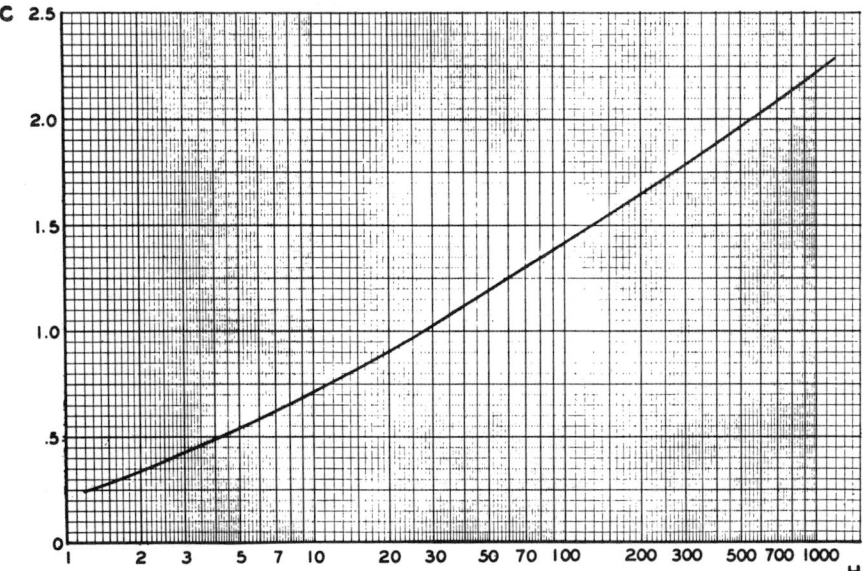

FIG. 15-6 C factor for determining i. (From Murphy, 1959a, p. 20. Reprinted by permission.)

$$B = \frac{.10 - .05}{2.3\sqrt{(0.90)(0.95)}} = .0235$$

and

$$H = \frac{(.90)(.95)(.0235)}{(.10)(.10)} = 2.01$$

$$C = 0.34 \qquad \text{(from Fig. 15-6)}$$

$$i = \frac{0.34}{.0235} = 14.5 \sim 15$$

$$f = .043 \qquad \text{(from Fig. 15-2)}$$

So the plan having the desired characteristics is

$$f = .043 \qquad i = 15$$

Note that for a plan selected in this way, it is possible to compute a value of r for the stopping rule which will give an interval between stops of E' when quality is at the producer's nominal quality level P' by using the formula

$$r = \frac{-\log(1 + E'P'F'(1 - P')^i)}{\log(1 - (1 - P')^i)}$$

Any base logarithms may be used. For the plan just selected in the example, if $E' = 10,000$, we have

$$r = \frac{-\log(1 + 10,000(.05)(.10)(1 - .05)^{15})}{\log(1 - (1 - .05)^{15})}$$

$$= \frac{-\log(24.1646)}{\log(.5367)}$$

$$= \frac{-1.3832}{-.2703} = 5.12 \sim 5$$

MULTI-LEVEL PLANS

In an ingenuous extension of the Dodge CSP-1 concept, Lieberman and Solomon (1955) conceived the idea of sampling fewer items as quality gives increasing evidence of being acceptable. This notion resulted in the so-called multi-level plan, which reduces the sampling frequency as successively more product is passed without finding a defective. This involves less inspection than CSP-1 under certain conditions to achieve the same AOQL. The AOQL given for the multi-level plans assumes that the production process is in control as in the Dodge plans.

Based on a Markov chain approach, the plans thus produced may be characterized theoretically as a random walk with reflecting barriers.

The procedure, allowing the possibility of infinite levels, is as
follows:

1. Specify

    i = clearing interval.

    f = initial sampling frequency.

    k_0 = maximum number of levels to be used.

2. Set $k = 1$ and begin 100 percent inspection.

3. After i units in succession have been found without a defective,
    start sampling at a rate of f^k.

4. If i sampled units are found free of defects, increase k by one
    and go to step 3. However, k must not exceed k_0, that is
    $k \leq k_0$.

5. If a defective is found, decrease k by one and go to step 3. If
    $k = 0$, go to step 2.

While the number of levels in a multi-level plan may be unrestricted, that
is $k_0 = \infty$, it is often desirable to stop the progression of levels at a
certain number of stages. For this reason, a value of k_0 may be specified at the outset, which k is not allowed to exceed. Thus we have two-
level plans ($k_0 = 2$), three-level plans ($k_0 = 3$) and so forth. It should
be noted that when $k_0 = 1$, the multi-level plan reduces to the Dodge CSP-1
plan.

A schematic representation of the multi-level plan is presented in
Fig. 15-7.

Lieberman and Solomon have provided charts, similar to those of CSP-1,
to determine values of f and i for specified AOQL. Figure 15-8 shows curves
for the infinite-level plans ($k_0 = \infty$) as solid lines, as contrasted with
the Dodge CSP-1 equivalent ($k_0 = 1$) as dotted lines. Figure 15-9 gives the
AOQL curves for a two-level plan ($k_0 = 2$).

For example, it can be seen from Fig. 15-9 that the two level multi-
level plan $i = 38$, $f = .10$ has an AOQL of 4 percent compared to a 2.9 percent AOQL for a CSP-1 plan with the same f and i. From Fig. 15-8 an
infinite-level plan having 4 percent AOQL and a clearing interval of $i = 38$
would require $f = .27$.

For an infinite-level plan with defective units replaced by good
items, measures can be determined as follows at fraction defective p for
a plan having AOQL = A.

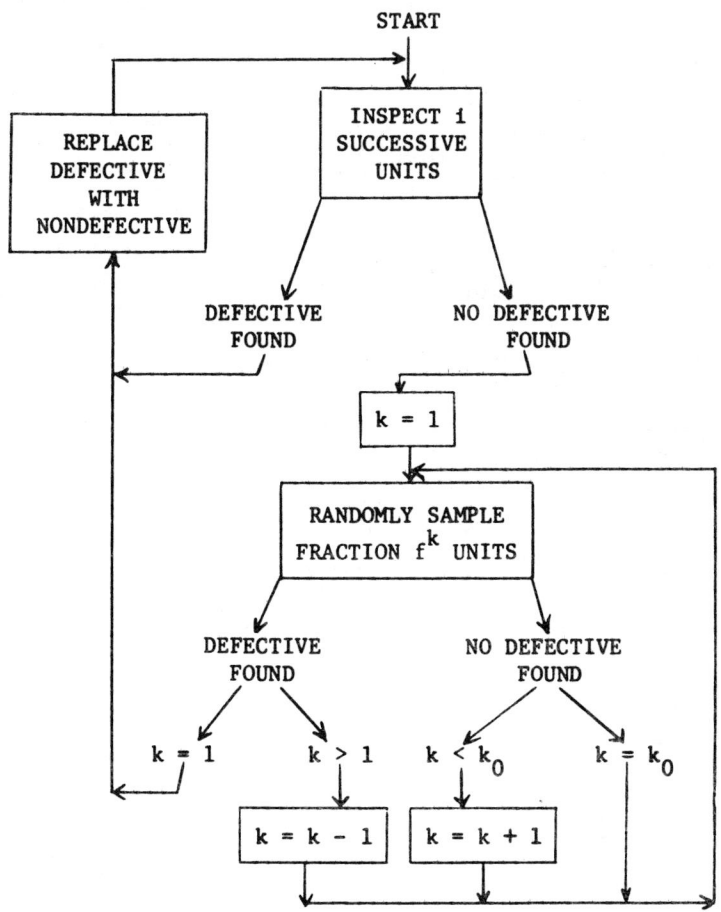

FIG. 15-7 Multi-level procedure.

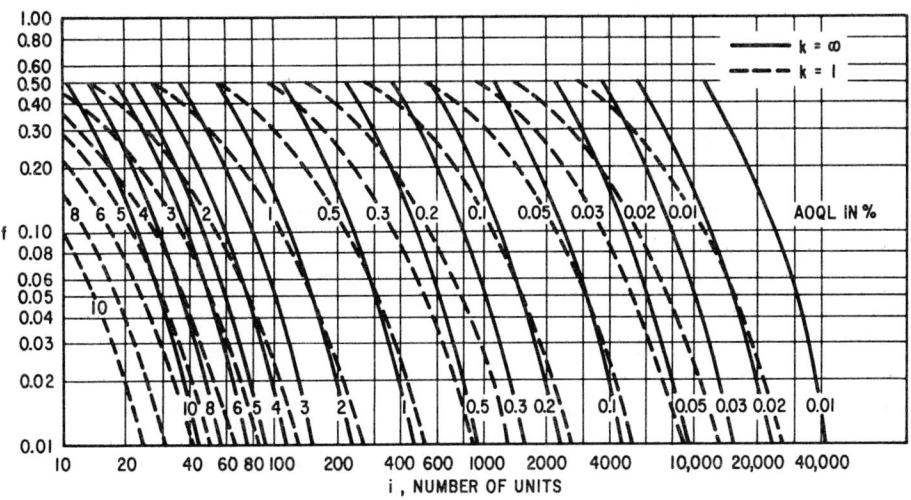

FIG. 15-8 Multi-level AOQL curves for $k_0 = 1, \infty$. (From Lieberman and Solomon, 1955, p. 696. Reprinted by permission.)

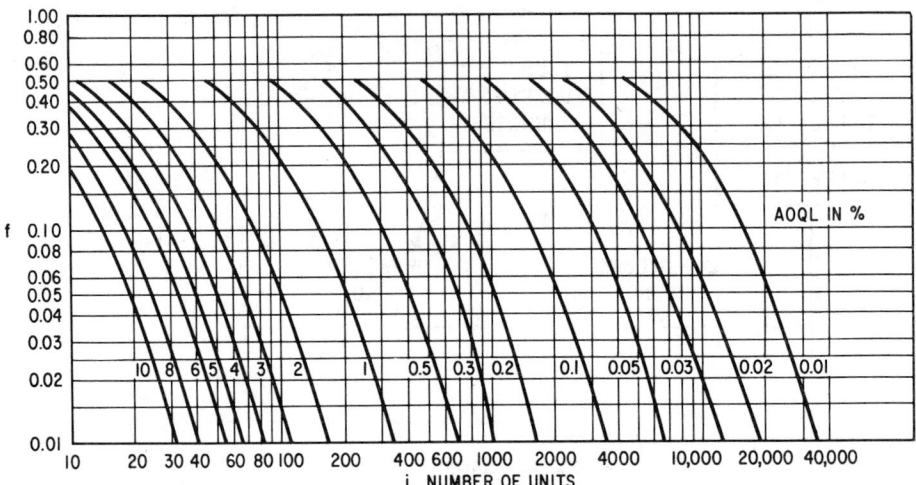

FIG. 15-9 Multi-level AOQL curves for two-level plan, $k_0 = 2$. (From
Lieberman and Solomon, 1955, p. 698. Reprinted by permission.)

Initial sampling frequency

$$f = \frac{(1 - A)^i}{1 - (1 - A)^i}$$

Average outgoing quality limit

$$AOQL = 1 - \left(\frac{f}{1 + f}\right)^{1/i}$$

Fraction defective p_M at which AOQL occurs

$$p_M = AOQL$$

Average fraction inspected

$$F_\infty = \frac{\left(\frac{1 - A}{1 - p}\right)^i - 1}{\left(\frac{1 - A}{1 - p}\right)^i - 2(1 - A)^i} \qquad\qquad p > A$$

$$F_\infty = 0 \qquad\qquad p \leq A$$

Average outgoing quality

$$AOQ_\infty = p(1 - F_\infty)$$

Thus, for the infinite level plan i = 38, f = .27 having AOQL = .04, we
have

$$f = \frac{(1 - .04)^{38}}{1 - (1 - .04)^{38}} = .27$$

with

$$AOQL = 1 - \left(\frac{.27}{1 + .27}\right)^{1/38} = .04$$

which occurs at

$$p_M = .04$$

The average fraction inspected at fraction defective $p = .041$ is

$$F_\infty = \frac{\left(\frac{1 - .04}{1 - .041}\right)^{38} - 1}{\left(\frac{1 - .04}{1 - .041}\right)^{38} - 2(1 - .04)^{38}}$$

$$= \frac{1.0404 - 1}{1.0404 - .4240} = .066$$

and

$$AOQ_\infty = .041(1 - .066) = .038$$

Note that the average fraction inspected at $p = AOQL = .04$ is

$$AFI = F = 0$$

hence

$$AOQ_\infty = .04(1 - 0) = .04$$

The mathematical development of multi-level plans is described in the Lieberman-Solomon (1955) paper. In discussing the advantages of the multi-level procedure the authors state (p. 686) that their purpose was "... to consider an extension of Dodge's first plan which (a) allows for smoother transition between sampling inspection and 100% inspection, (b) requires 100% inspection only when the quality submitted is quite inferior, and (c) allows for a minimum amount of inspection when quality is definitely good." Burr (1976) has pointed out, however, that f must be fairly large "... in order to avoid extremely low fractions on higher powers of f. This makes the saving small at f or even f^2, and the multi-levels make scheduling of workloads difficult."

In any event, the multi-level procedure provides a useful alternative in the application of continuous sampling plans and has found application, "to a variety of products, ranging from EAM cards to very complicated equipments ... with substantial savings," as reported by Ireson and Biedenbender (1958).

TIGHTENED MULTI-LEVEL PLANS

A set of tightened multi-level plans has been developed by Derman, Littauer, and Solomon (1957). They offered "three generalizations of MLP, accomplished by altering the manner in which the transition can occur ...". One of these, the simplest, will be discussed here. It is the tightest and has been labeled MLP-T.

The MLP-T plan is simply a multi-level plan which requires a switch all the way back to 100 percent inspection, at any level, whenever a defective unit is found. This provides quick rectification in the event of a shift in quality.

Measures for the MLP-T infinite level plan when defectives are replaced with effective units are

Average fraction inspected

$$F = \frac{1 - \frac{(1 - p)^i}{f}}{1 - (1 - p)^i} \qquad\qquad f > (1 - p)^i$$

$$f = \infty \qquad\qquad f \leq (1 - p)^i$$

Average outgoing quality

$$AOQ = \frac{p(1 - p)^i}{1 - (1 - p)^i}\left(\frac{1 - f}{f}\right) \qquad\qquad f > (1 - p)^i$$

$$AOQ = p \qquad\qquad f \leq (1 - p)^i$$

Average outgoing quality limit

$$AOQL = 1 - f^{1/i}$$

with

$$p_M = AOQL$$

For the infinite level plan i = 38, f = .27 at p = .04, we have

$$f = .27 > (1 - p)^i = .21$$

so

$$F = \frac{1 - \frac{(1 - .04)^{38}}{.27}}{1 - (1 - .04)^{38}} = .273$$

$$AOQ = \frac{.04(1 - .04)^{38}}{1 - (1 - .04)^{38}}\left(\frac{1 - .27}{.27}\right) = .029$$

$$\text{AOQL} = 1 - .27^{1/38} = .034$$

and

$$p_M = .034$$

Here we see the average fraction inspected is increased over the corresponding multi-level plan while the AOQ at $p = .04$ and the AOQL are decreased by the quick return to 100 percent inspection.

BLOCK CONTINUOUS PLANS

Both the multi-level and the Dodge continuous plans basically assume a steady flow of production with no attempt to segregate the product into lots or segments. While either of these procedures may be carried on by sampling at random a fraction f units from successive segments of a given size, special plans have been designed for this purpose and may be characterized as block continuous plans. These plans divide the sequence of production into successive blocks, taking a prescribed sample from each block. A specified proportion of the block is screened once a signal for 100 percent inspection is given. Block continuous plans are easily adapted to the inspection of successive lots and are useful when the process itself generates natural segments. It should be noted that while the Dodge plans essentially fix both i and f, the multi-level plans fix i but allow the fraction inspected to vary with the level of the plans. The block continuous plans typically sample one unit from a group of items of specified size (that is f fixed) and allow the clearing interval to vary. Action is usually taken on the cumulative number of defectives found.

Wald-Wolfowitz Plan

The first block continuous plans were proposed by Wald and Wolfowitz (1945). Of the three plans they proposed, only one will be discussed here, their SPC plan. They divide the production flow into segments of size N_0 which are sampled by taking one item from groups of size $1/f$ to achieve a sampling frequency of f. The plan is applied as follows for specified N_0, M^*, and f.

1. Demark the flow of production into segments of fixed size N_0.
2. Break each segment into fN_0 groups of size $k = 1/f$.
3. Start with partial inspection of one item from each group.
4. Continue accumulating the sum of defectives found, Σd, until

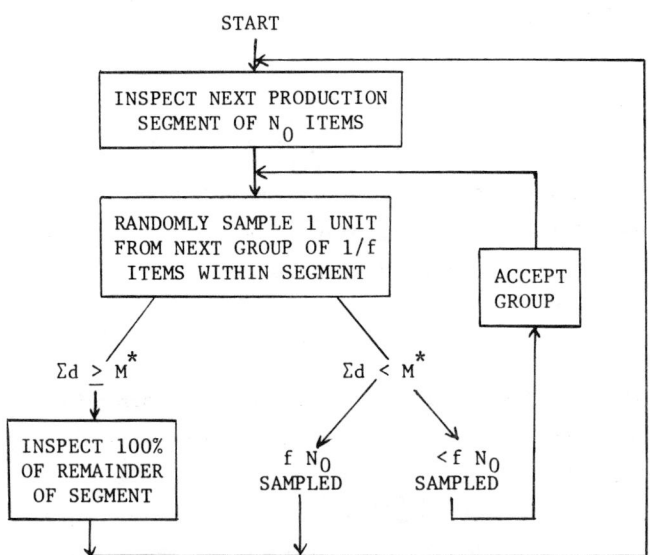

FIG. 15-10 Wald-Wolfowitz SPC procedure.

    a. M* defectives are found, then begin 100 percent inspection

        of the remainder of the segment starting with the next group.

    b. The segment has been completely partially inspected.

  5. Repeat the procedure anew on the next segment.

A schematic representation of the procedure is given in Fig. 15-10.

    This procedure can be adapted to the inspection of a sequence of lots
by substituting the word *lot* for the word *segment* in the above steps. In
such applications it is often convenient to draw successive samples from
the entire lot rather than from individual groups. In this case a sample
of fN_0 items is taken successively at random from the lot. As soon as the
number of defectives found equals M* sampling is discontinued. At this
point, if the M*th defective occurred on the N'th sample, an additional
$(N_0 - (1/f)N')$ units are 100 percent inspected and the lot is released. If
fN_0 items are sampled without reaching M* defectives the lot is also released. This procedure guarantees the same UAOQL as the procedure for
sampling from groups.

    The Wald-Wolfowitz procedure guarantees the AOQL regardless of the
state of control of the process. For this plan, when defectives are replaced with good items:

$$UAOQL = \frac{M^*(1 - f)}{fN_0}$$

so that, given UAOQL = A, N_0, and f, a plan can be set up using

$$M^* = \frac{AfN_0}{(1 - f)}$$

Other measures of these plans at fraction defective p are

Average fraction inspected

$$AFI = F = 1 - \frac{A}{p} + \frac{(1 - f)}{pfN_0} \sum_{i=0}^{M^*-1} (M^* - 1) \binom{fN_0}{C_i} p^i (1 - p)^{fN_0-i}$$

Average outgoing quality

$$AOQ = A \left[1 - \frac{1}{M^*} \sum_{i=0}^{M^*-1} (M^* - i) \binom{fN_0}{C_i} p^i (1 - p)^{fN_0-i} \right]$$

As an example of application, consider setting up a plan which is to have UAOQL = A = .029 and f = .1 for production segments of N_0 = 310. Then

$$M^* = \frac{.029(.1)310}{(1 - .1)} = .999 \sim 1$$

The plan would be applied to segments of 310 units. From each segment one unit would be sampled from each of .1(310) = 31 groups of size 1/.1 = 10. As soon as a defective was found in a segment, the remaining groups in the segment would be 100 percent inspected before starting afresh with the next segment. For this plan at fraction defective p = .054

$$AFI = 1 - \frac{.029}{.054} + \frac{(1 - .1)}{.054(.1)310}(1) \binom{31}{C_0} .054^0 (1 - .054)^{31}$$

$$= 1 - .537 + .538(.179) = .559$$

$$AOQ = .029 \left[1 - \frac{1}{1}(1) \binom{31}{C_0} .054^0 (1 - .054)^{31} \right]$$

$$= .029(1 - .179) = .024$$

Girshick Plan

M. A. Girshick (1954) has provided a modification of the Wald-Wolfowitz approach which avoids the necessity for segmenting production, but which achieves essentially the same result. The procedure is as follows for specified f, m, N:

1. Divide the flow of production up into groups of size 1/f.
2. Start with partial inspection of one item from each group.
3. Cumulate the number of defectives Σd and the number of samples

taken, n.

4. When the cumulative number of defectives equals m, compare the number of samples inspected to the integer N.

5. a. If $n \geq N$, product previously inspected is confirmed as good.

   b. If $n < N$, 100 percent inspect the next $N - n$ groups [that is $(N - n)(1/f)$ units] replacing defectives with good.

6. Start anew.

The Girshick procedure guarantees

$$\text{UAOQL} \leq \frac{(1 - f)m}{N}$$

regardless of the state of control of the process. It will be seen that this is essentially a modification of the Wald-Wolfowitz SPC plan with $N = fN_0$ and $m = M^*$. To set up such a plan, N should be small enough that 100 percent inspection can reasonably be performed if necessary. Then for a given UAOQL = A, f, and N,

$$m = \frac{NA}{(1 - f)}$$

If UAOQL = .29, f = .1 and N = 310, the plan is essentially the same as the Wald-Wolfowitz example above if

$$m = \frac{310(.029)}{(1 - .1)} \simeq 10$$

without the necessity for setting up arbitrary divisions on the flow of production. The Girshick (1954) monograph presents the mathematical characterization and measures of the procedure.

Based on the work of Girshick, Minneapolis-Honeywell has utilized an adaptation of the sequential sampling chart to perform both process control and continuous inspection. The sequential chart appears as in Fig. 15-11 and is divided into regions which are used to signal the state of the process to the operators (and management) through a set of three colored lights. This gives early warning of the need for assembly of a screening crew as well as a signal to the operator of the need for corrective action. The approach is described in detail by Albrecht, Gulde, MacLean and Thompson (1955).

The associated continuous acceptance sampling procedure employed for AOQL protection is as follows:

1. For specified p_1, p_2, α, and β, set up a sequential sampling plan

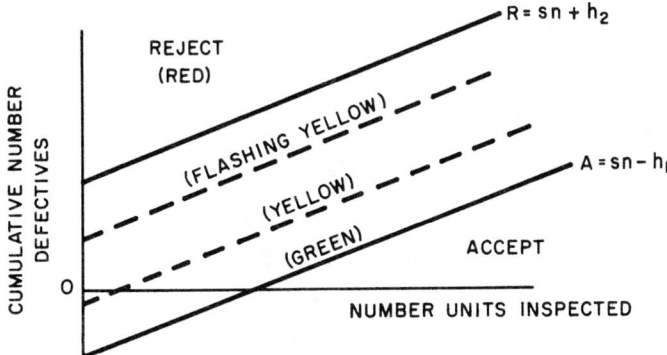

FIG. 15-11 Minneapolis-Honeywell sequential chart for
continuous inspection. (From Albrecht et al. 1955,
p. 7. Reprinted by permission.)

with associated decision lines

$$R = sn + h_2$$
$$A = sn - h_1$$

2. Start with partial inspection on one unit out of each group of
    $1/f$ units. Pass the material while the cumulative number of
    defectives is between the decision lines.

3. Continue until

    a. The cumulative number of defectives falls below the acceptance
        line. This indicates the product passed to this point is
        acceptable.

    b. The cumulative number of defectives exceeds the rejection
        line. This indicates line quality is unacceptable. In this
        case a lot of size i of perfect material must be produced to
        dilute the poor material, where $i = h_2/sf$. Nothing is passed
        until the i units have been produced, checked and passed.

4. Return to partial inspection starting over with a new sequential
    chart after either 3a or 3b has occurred.

This plan can be set up for specified UAOQL, f, and i as follows:

$$s = \frac{UAOQL}{1 - f} \qquad h_2 = isf$$

Since Albrecht et al. (1955) recommended equal risks, $\alpha = \beta$, in this case
take $h_1 = h_2$ if an acceptance line is to be utilized in the procedure.

The protection guaranteed by the plan will be as follows:

Average outgoing quality limit

UAOQL = A = (1 - f)s

Average outgoing quality at p

$$AOQ = \frac{sp(1 - f)E}{sE + (1 - P_a)h_2}$$

where, from the formulas for sequential sampling, E = ASN = average
sample number and P_a = probability of acceptance at p.

For example, if p_1 = .018, p_2 = .18, α = .05, β = .10 as in the example of
Chap. 7, we have

R = .0726 n + 1.164

A = .0726 n - 0.907

and at p = .018, P_a = .95 and ASN = 14.7. Then, if f = .1

UAOQL = (1 - .1).0726 = .065

and at p = .018,

$$AOQ = \frac{.0726(.018)(1 - .1)(14.7)}{.0726(14.7) + (1 - .95)(1.164)} = \frac{.0173}{1.1254} = .0154$$

This compares to an AOQ of approximately .017 for the conventional sequential plan, which, of course, would require inspection of every unit of a
rejected lot.

Considering block continuous plans in general, it would seem that the
Wald-Wolfowitz plans are particularly well suited where lot inspection is
involved or when the production stream is naturally divided into segments
of a given size. The Girshick plan would appear to be quite good for a
continuous flow of product where product might be set aside as required
for screening without the need for immediate 100 percent inspection as
would be required in the Dodge or multi-level plans. Finally, the Minneapolis-Honeywell method provides an interesting alternative for combined
process control and acceptance inspection.

MIL-STD-1235A

Military Standard 1235A entitled, "Single and Multi-level Continuous
Sampling Procedures and Tables for Inspection by Attributes" is a collection of continuous sampling plans indexed by AQL. The standard takes care
to point out in its definition of acceptable quality level that *"For*

continuous sampling plans, the AQL is an index to the plans, and has no other meaning." The AQL index is used to tie the standard to contractual levels of protection incorporated in contracts involving MIL-STD-105D, MIL-STD-414, and other such sampling plans. It is not an AQL plan and has no switching rules.

Since continuous sampling plans are usually used, specified, and indexed by AOQL, the AOQL of the plans given in MIL-STD-1235A is always shown together with the AQL index. In fact, the plans included were chosen to match, as well as possible, representative values of the *scheme* AOQLs of the MIL-STD-105D plans having the AQL index shown. Reference to Appendix Table T11-20 shows the *scheme* AOQLs range from 2.9 to 3.2 over the 2.5 AQL column for nonzero acceptance numbers[*]. Thus, the continuous plans indexed under 2.5 AQL in MIL-STD-1235A show AOQL values of 2.9 percent. In this way the results of the MIL-STD-1235A plans correspond to the results of the MIL-STD-105D system when MIL-STD-105D is used to guarantee AOQLs per paragraphs 11.3 and 11.4 of that standard.

Five different types of continuous plans are given in MIL-STD-1235A. The user has the option of selecting the plan which is most suitable for the inspection situation involved. The plans included are CSP-1, CSP-2, CSP-F, CSP-T, and CSP-V. These plans are characterized in Table 15-2.

Of course, CSP-1 and CSP-2 are the standard plans used in the standard way. The CSP-F plan is intended for use with short production runs, short periods of production within a production interval (defined by the standard to be a period of homogeneous quality such as a shift, but at most a day). The criteria are adjusted to account for a finite period of production, N items in length. The CSP-T plans are tightened multi-level plans incorporating three levels. They are modified from those of Lieberman and Solomon (1955) and Derman, Littauer, and Solomon (1957) after the manner of Guthrie and Johns (1958) in that the sampling frequency is cut in half from level to level, rather than by powers of f as in the conventional multi-level plans. As a multi-level plan, CSP-T allows a reduction in sampling frequency as quality improves, reducing the amount of sampling necessary. It is sometimes desirable to cut the clearing interval rather than sampling frequency with improved quality, particularly when the sampling inspector

---

[*]The MIL-STD-105D system AOQLs were calculated using tightened-normal switching only and so correspond only roughly to the Schilling-Sheesley (1978) values given in Appendix Table T11-20 which incorporates switching to reduced inspection also.

TABLE 15-2   Type and Purpose of MIL-STD-1235A Plans

Section	Plan	Type	Purpose
2	CSP-1	Standard CSP-1 plan	Simple, popular, easy to use and administer
3	CSP-F	CSP-1 procedure with parameters modified for application to sequence of specified length	Smaller clearing intervals for short production runs or when long clearing intervals are impractical
4	CSP-2	Standard CSP-2 plan	Provides against stray defectives and gives warning that screening crew may be needed
5	CSP-T	Tightened 3-stage multi-level plan with modified sampling frequencies	Permits reduction of sampling frequency (f) when superior quality is demonstrated
6	CSP-V	Modified CSP-1 procedure with shortened clearing interval if previous i units free of defectives	Permits reduction of clearing interval (i) when superior quality is demonstrated

cannot be switched to other work. The screening crew will then have less to do. The CSP-V plans are designed to do just this by reducing the clearing interval if evidence of superior quality exists.

The structure of MIL-STD-1235A is shown in Fig. 15-12. Tables of f and i are provided for each type of plan. In addition, except for CSP-F, tables of S values are also given as criteria to allow termination of excessively long periods of screening. The stopping rule employed is the "rule $n^* - i$" of Murphy (1959). That is, clearing is stopped as soon as a defective is found in any one screening sequence exceeding S units. Clearing is started anew at the beginning of the clearing interval after corrective action has been taken.

MIL-STD-1235A also provides for the possibility of a check inspection of screened lots. If the check inspector finds one defective, the customer is to be notified and corrective action taken on the screening crew. If two defectives are found, product acceptance may be suspended.

A diagrammatic representation of the application of MIL-STD-1235A is presented in Fig. 15-13 which gives a check sequence for the operation of the standard. This can be used to insure that the standard is properly employed.

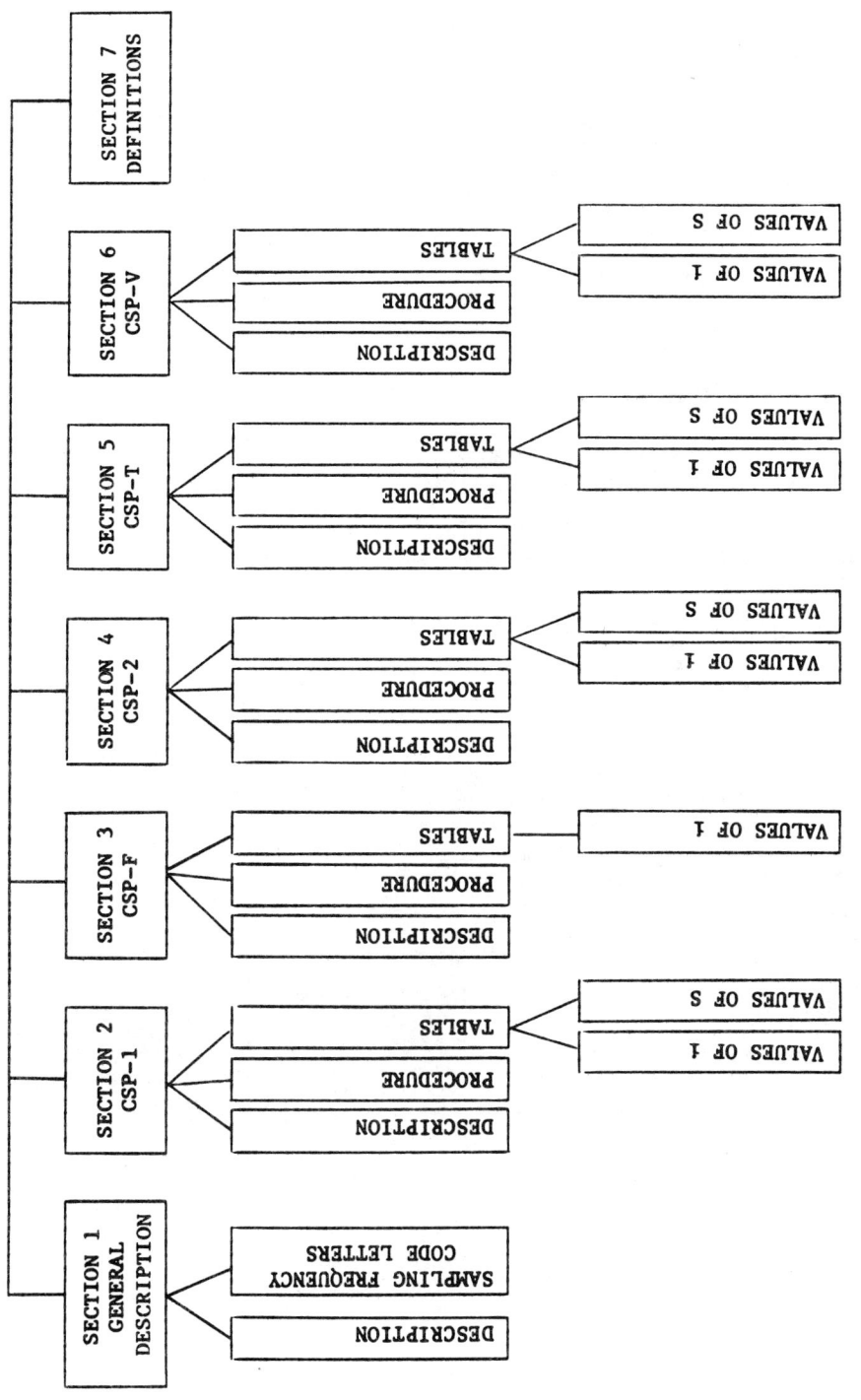

FIG. 15-12 Structure of MIL-STD-1235A.

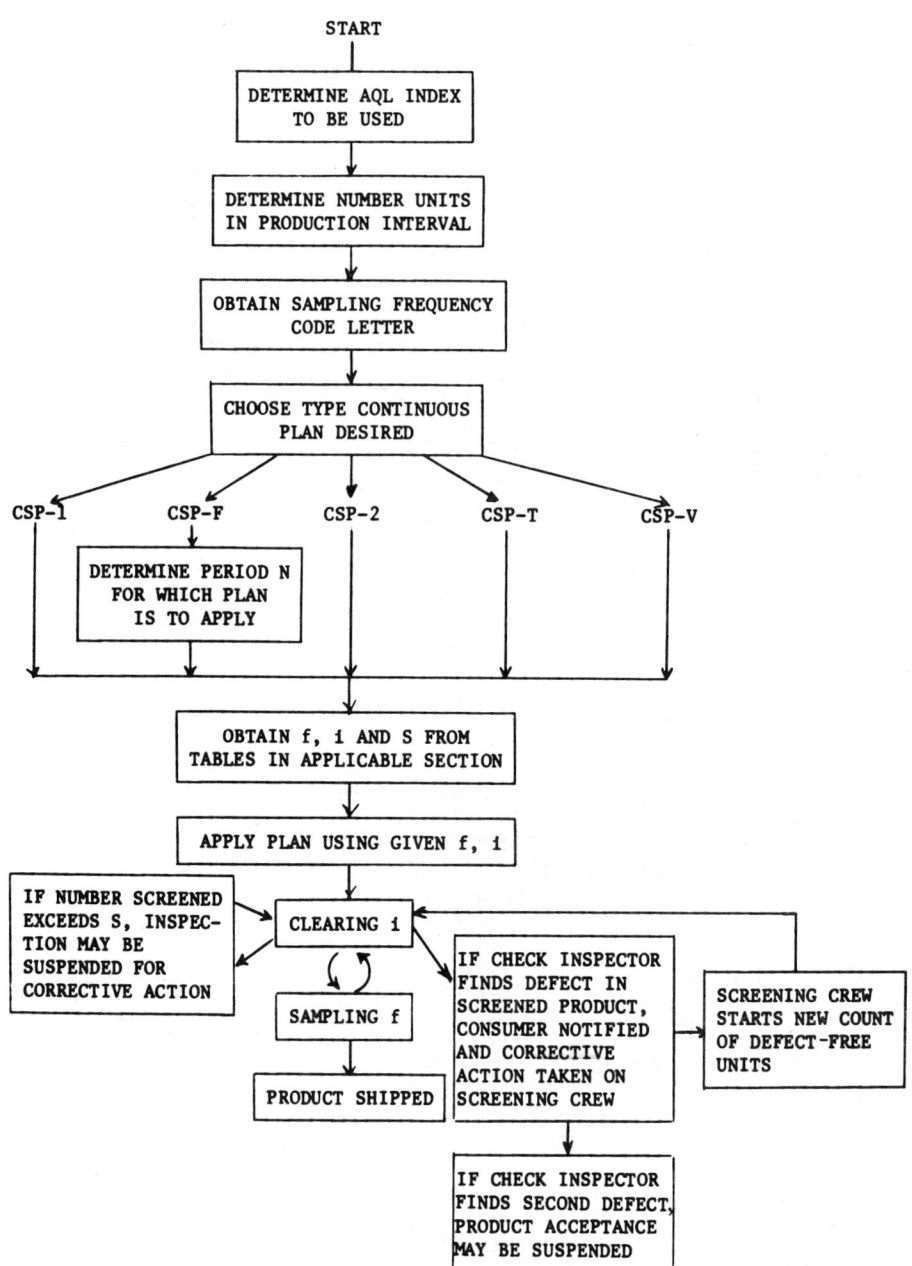

FIG. 15-13 Check sequence for applying MIL-STD-1235A.

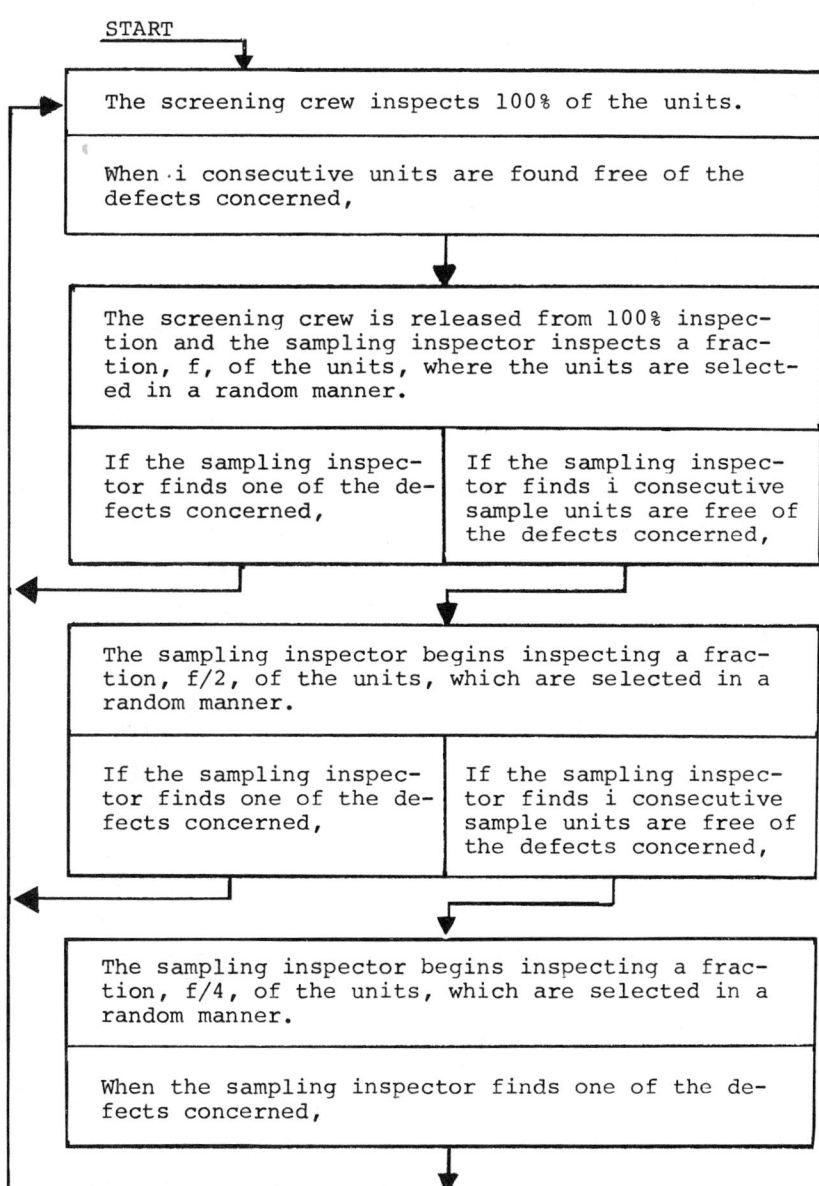

FIG. 15-14 Procedure for CSP-T plans. (From MIL-STD-1235A, p. 41.)

TABLE 15-3 Sampling Frequency Code Letters

Number of Units in Production Interval	Permissible Code Letters
2-8	A,B
9-25	A through C
26-90	A through D
91-500	A through E
501-1200	A through F
1201-3200	A through G
3201-10,000	A through H
10,001-35,000	A through I
35,001-150,000	A through J
150,001-up	A through K

From MIL-STD-1235A, p. 7.

To illustrate the application of MIL-STD-1235A, consider its use to obtain a CSP-T plan to be employed on a contract specifying an AQL of 0.65 percent with the production interval expected to be about 5000 units. The operation of the CSP-T plan is illustrated in Fig. 5-A of the standard and given here as Fig. 15-14. Table 1 of the standard, shown here as Table 15-3, indicates that Code H may be used. Values of f, i, and S can be obtained from Tables 5-A and 5-B of MIL-STD-1235A which correspond to Tables 15-4 and 15-5. They show that the plan to be employed should be

$f = 1/25$

$i = 217$

$S = 1396$

It should be emphasized that AQL is used here as an index only. In fact, the plan given has an AOQL equal to 0.79 percent. Note that, if AOQL protection is desired, an appropriate plan can be selected from the tables simply by using the AOQL listed with any desired value of f to find the corresponding value of i. The value of S for such a plan can also readily be located if a stopping rule is to be employed. For example, the MIL-STD-105D, Code F, 2.5 percent AQL system has an AOQL of 2.9 percent from Appendix Table T11-20. If a CSP-T plan is to be employed having equal AOQL protection with a sampling frequency of, say, $f = .2$, we find the plan $f = .2$, $i = 29$ to be appropriate. For this plan $S = 93$.

The theoretical development of the standard and particularly the CSP-F, CSP-T, and CSP-V plans was largely accomplished by R. A. Banzhaf and R. M. Brugger of the U.S. Army, Armament Procurement and Supply Agency, Product

TABLE 15-4 Values of i for CSP-T Plans

Samp Freq Code Ltr	f	AQL* in %							
		0.40	0.65	1.0	1.5	2.5	4.0	6.5	10.0
A	1/2	87	58	38	25	16	10	7	5
B	1/3	116	78	51	33	22	13	9	6
C	1/4	139	93	61	39	26	15	11	7
D	1/5	158	106	69	44	29	17	12	8
E	1/7	189	127	82	53	35	21	14	9
F	1/10	224	150	97	63	41	24	17	11
G	1/15	266	179	116	74	49	29	20	13
H	1/25	324	217	141	90	59	35	24	15
I	1/50	409	274	177	114	75	44	30	19
J,K	1/100	499	335	217	139	91	53	37	23
		0.53	0.79	1.22	1.90	2.90	4.94	7.12	11.46
					AOQL in %				

*AQL's are provided as indices to simplify use of this table, but have no other meaning relative to the plans.

From MIL-STD-1235A, p. 42.

TABLE 15-5 Values of S for CSP-T Plans

Samp Freq Code Ltr	f	AQL* in %							
		0.40	0.65	1.0	1.5	2.5	4.0	6.5	10.0
A	1/2	159	117	77	52	34	22	13	12
B	1/3	256	197	128	80	59	35	25	18
C	1/4	379	253	167	103	78	43	38	24
D	1/5	444	320	210	130	93	54	43	30
E	1/7	725	460	289	188	137	81	59	34
F	1/10	857	619	398	261	189	104	88	58
G	1/15	1254	900	584	368	376	152	126	84
H	1/25	1885	1396	923	545	421	235	198	122
I	1/50	3283	2477	1604	1013	764	408	374	223
J,K	1/100	5753	4541	2948	1754	1341	708	653	391
		0.53	0.79	1.22	1.90	2.90	4.94	7.12	11.46
		AOQL in %							

*AQL's are provided as indices to simplify use of this table, but have no other meaning relative to the plans.

Quality Evaluation Division, using a Markov chain approach; their work on the original MIL-STD-1235 standard has already been cited (Banzhaf and Brugger, 1970). The theoretical background of the Dodge CSP-1 and CSP-2 plans has already been given. CSP-F developed out of the Markov chain approach to CSP-1 formula derivation presented by Roberts (1965) and the study by Lasater (1970) of the theory and performance of CSP-1 when applied to a finite number of units.

Measures of the plans incorporated in MIL-STD-1235A are presented in a companion document, MIL-STD-1235A-1. All plans included in MIL-STD-1235A are represented except for the CSP-F plans. For each plan, curves are given for AOQ, AFI, and OC. These are defined by MIL-STD-1235A-1 as

> The *average outgoing quality* (AOQ) for a particular process average is the long run expected percentage of defective material in the accepted material, if the associated sampling plan is followed faithfully.

> The *average fraction inspected* (AFI) is the fraction of product that will be inspected over the long run if the process average is a particular value.

> The *operating characteristics* (OC) of a continuous sampling plan describe the percent of product accepted during the sampling phases of the plan over the long run if the process average is a particular value.

The standard states that "Curves for CSP-F are not provided, since exact methods for their determination have not been developed."

Illustrations of the MIL-STD-1235A-1 curves are given here for the multi-level plan CSP-T: Code D, 2.5 percent AQL ($f = 1/5$, $i = 29$).

> Fig. 15-15 AOQ curve for CSP-T, Code D, 2.5 percent AQL
> Fig. 15-16 AFI curve for CSP-T, Code D, 2.5 percent AQL
> Fig. 15-17 OC curve for CSP-T, Code D, 2.5 percent AQL

PROBLEMS

1. Construct a CSP-1 plan for AOQL = 4 percent which will have a sampling frequency of about 10 percent. What is the UAOQL for this plan when defectives are replaced by good units?

2. Find a CSP-2 plan with $k = i$ which will afford about the same protection as the plan in Prob. 1. What is its UAOQL when defectives are not replaced with good items?

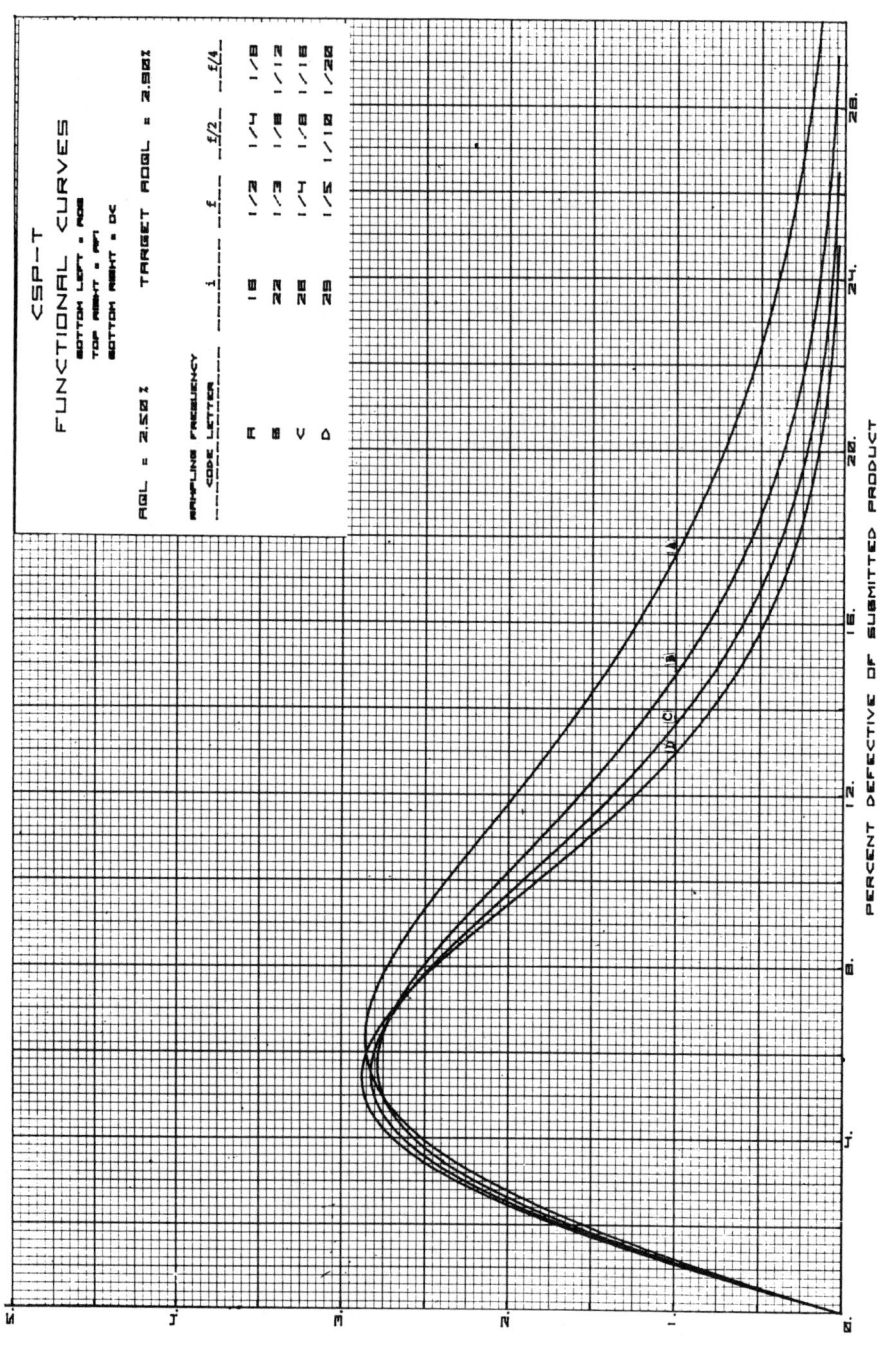

FIG. 15-15 MIL-STD-1235A-1 AOQ curves - CSP-T. (From MIL-STD-1235A-1, p. 178.)

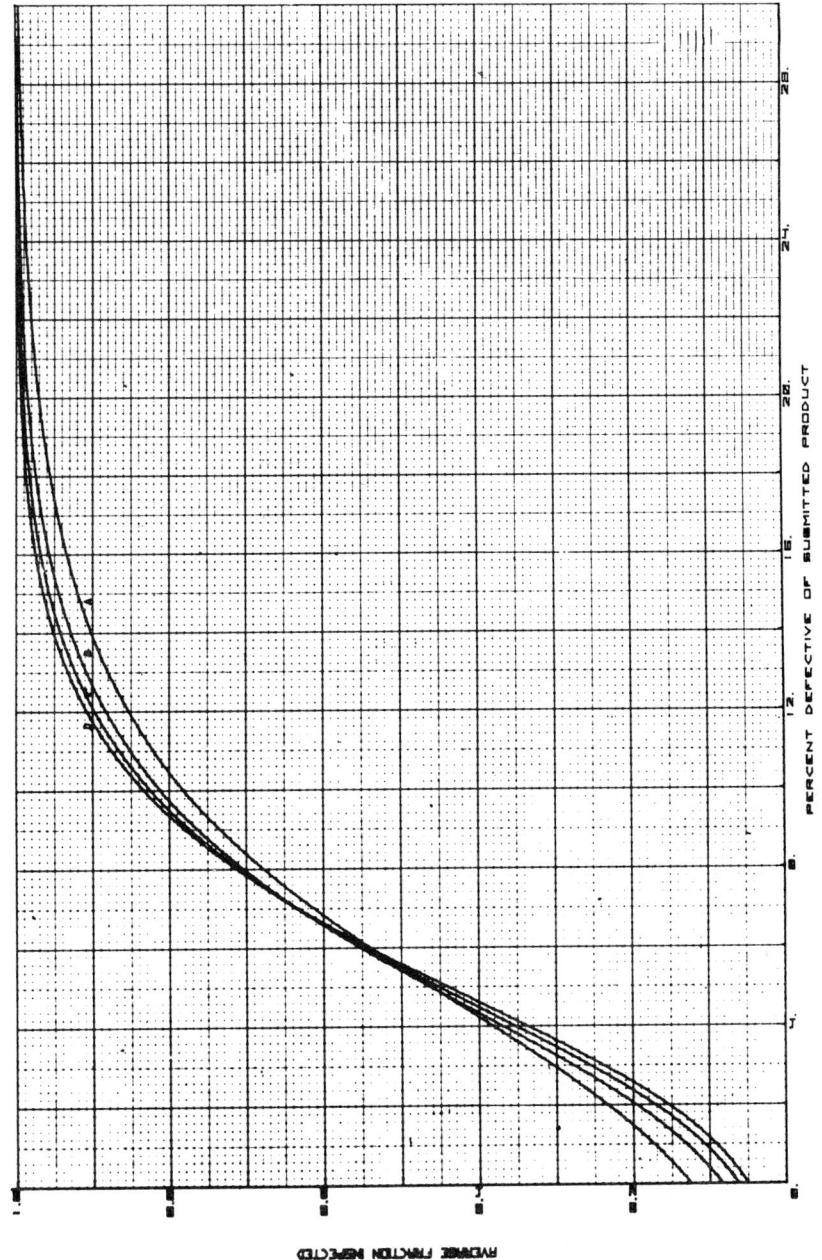

FIG. 15-16 MIL-STD-1235A-1 AFI curves - CSP-T (From MIL-STD-1235A-1, p. 179.)

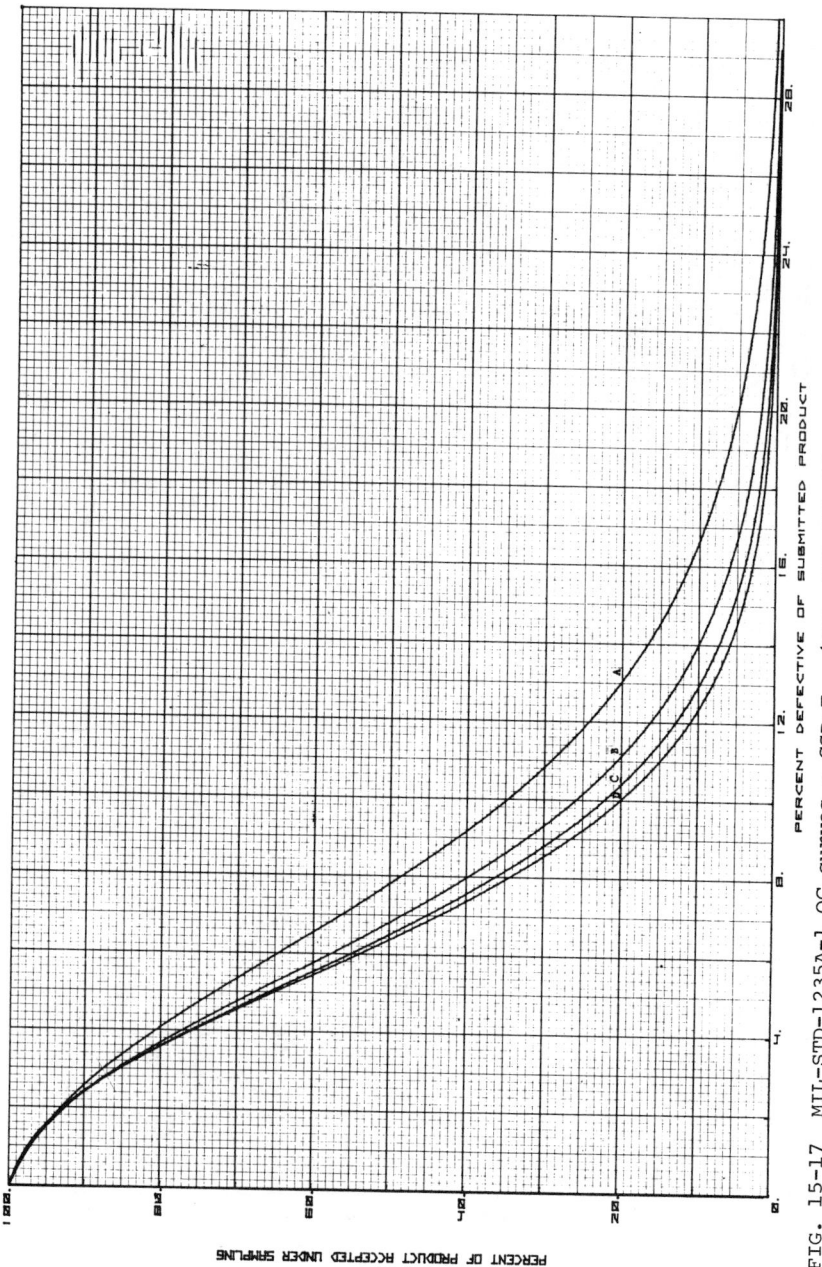

FIG. 15-17 MIL-STD-1235A-1 OC curves – CSP-T. (From MIL-STD-1235A-1, p. 179.)

3. What is the average fraction inspected when the proportion defective submitted is .08 for

    a. Prob. 1

    b. Prob. 2

4. Stopping rule r is to be instituted on the plan of Prob. 1. Find r if 50 units may be produced between successive stops when the process proportion defective is such that the AOQL is realized (i.e., $p = p_M$).

5. Use Murphy's procedure to find a CSP-1 plan which will have a 5 percent AOQL with 25 percent of the product inspected when the process average is expected to be 1 percent defective.

6. Find an infinite multi-level plan which will have an AOQL of 4 percent with an initial sampling frequency of about 10 percent. What is the AOQL if it is used as a tightened multi-level plan?

7. Production is boxed in crates of 24 units with 24 crates to a skid. An AOQL of 6 percent is desired. Construct the appropriate Wald-Wolfowitz plan.

8. What would be the parameters of a Girshick plan corresponding to Prob. 7?

9. A MIL-STD-105D scheme is being used with screening of rejected lots to provide AOQL protection. For the scheme, AQL = 6.5 with lots of size 550. A process change mandates a change to continuous sampling. At present a tightened/normal sample size of 80 is being used which implies a sample size-lot size ratio of .145 ~ 1/7. What MIL-STD-1235A, CSP-T, plan should be used? What should be the maximum number screened before clearing is stopped?

10. Compare the code letters of MIL-STD-1235A to those of MIL-STD-105D. Are they comparable? That is, if Code J is being used on MIL-STD-105D should J be used on MIL-STD-1235A?

REFERENCES

Albrecht, L., H. Gulde, A. MacLean, and P. Thompson (1955), Continuous Sampling at Minneapolis-Honeywell, *Industrial Quality Control*, 12(3): 4-9.

Banzhaf, R. A., and R. M. Brugger (1970), MIL-STD-1235 (ORD), Single and Multi-Level Continuous Sampling Procedures and Tables for Inspection by Attributes, *Journal of Quality Technology*, 2(1): 41-53.

Burr, I. W. (1976), *Statistical Quality Control Methods*, Marcel Dekker, New York.

Derman, C., S. Littauer, and H. Solomon (1957), Tightened-Multi-Level Continuous Sampling Plans, *Annals of Mathematical Statistics*, 28: 395-404.

Dodge, H. F. (1943), A Sampling Plan for Continuous Production, *Annals of Mathematical Statistics*, 14(3): 264-279.

Dodge, H. F. (1947), Sampling Plans for Continuous Production, *Industrial Quality Control*, 4(3): 5-9.

Dodge, H. F., and M. N. Torrey (1951), Additional Continuous Sampling Inspection Plans, *Industrial Quality Control*, 7(5): 7-12.

Girshick, M. A. (1954), *A Sequential Inspection Plan for Quality Control*, Technical Report No. 16, Applied Mathematics and Statistics Laboratory, Stanford University, Stanford, California.

Guthrie, D., and M. Johns (1958), *Alternative Sequences of Sampling Rates for Tightened Multi-Level Continuous Sampling Plans*, Technical Report No. 36, Applied Mathematics and Statistics Laboratory, Stanford University, Stanford, California.

Ireson, W. G., and R. Biedenbender (1958), Multi-level Continuous Sampling Procedures and Tables for Inspection by Attributes, *Industrial Quality Control*, 15(4): 10-15.

Lasater, H. A. (1970), *On the Robustness of a Class of Continuous Sampling Plans Under Certain Types of Process Models*, Ph.D. Dissertation, Rutgers - The State University, New Brunswick, New Jersey.

Lieberman, G. J. (1953), A Note on Dodge's Continuous Sampling Plan, *Annals of Mathematical Statistics*, 24: 480-484.

Lieberman, G. J., and H. Solomon (1955), Multi-Level Continuous Sampling Plans, *Annals of Mathematical Statistics*, 26: 686-704.

Murphy, R. B. (1959), Stopping Rules with CSP-1 Sampling Inspection Plans, *Industrial Quality Control*, 16(5): 10-16.

Murphy, R. B. (1959a), A Graphical Method of Determining a CSP-1 Sampling Inspection Plan, *Industrial Quality Control*, 16(6): 20-21.

Roberts, S. W. (1965), *States of Markov Chains for Evaluating Continuous Sampling Plans*, Transactions of the 17th Annual All-Day Conference on Quality Control, Metropolitan Section ASQC and Rutgers University, New Brunswick, New Jersey, pp. 106-111.

Schilling, E. G., and J. H. Sheesley (1978), The Performance of MIL-STD-105D Under the Switching Rules, *Journal of Quality Technology*, Part 1, 10(2): 76-83; Part 2, 10(3): 104-124.

Stephens, K. S. (1980), *How to Perform Continuous Sampling*, Vol. 2, The
    ASQC Basic References in Quality Control: Statistical Techniques (E.
    J. Dudewicz, ed.), American Society for Quality Control, Milwaukee,
    Wisconsin.

United States Department of Defense (1963), *Military Standard, Sampling
    Procedures and Tables for Inspection by Attributes* (MIL-STD-105D),
    U.S. Government Printing Office, Washington, D.C.

United States Department of Defense (1957), *Military Standard, Sampling
    Procedures and Tables for Inspection by Variables for Percent Defective*,
    (MIL-STD-414), U.S. Government Printing Office, Washington, D.C.

United States Department of Defense (1974), *Military Standard, Single and
    Multi-Level Continuous Sampling Procedures and Tables for Inspection
    by Attributes* (MIL-STD-1235A), U.S. Government Printing Office,
    Washington, D.C.

United States Department of Defense (1974), *Military Standard, Single and
    Multi-Level Continuous Sampling Procedures and Tables for Inspection
    by Attributes — Functional Curves of the Continuous Sampling Plans*
    (MIL-STD-1235A-1), U.S. Government Printing Office, Washington, D.C.

Wald, A., and J. Wolfowitz (1945), Sampling Inspection Plans for Continuous
    Production Which Insure a Prescribed Limit on the Outgoing Quality,
    Annals of Mathematical Statistics, 16: 30-49.

16

CUMULATIVE RESULTS PLANS

Except for continuous sampling plans, the acceptance sampling plans discussed so far have been applied on an individual lot-by-lot basis. The AQL schemes incorporated in MIL-STD-105D and MIL-STD-414 do, in fact, utilize the results from the most recent 10 lots as part of the switching rules, but the acceptance criteria applied to any one lot does not specifically incorporate the results of the inspection of the immediately preceding lots.

The continuous sampling plans discussed in the last chapter require knowledge of the results from the immediately preceding samples as part of the action rule for any sample inspected. This is particularly evident in CSP-2 and CSP-3, but applies to all the plans discussed. Thus, continuous sampling plans are a member of the class of, so-called, cumulative results plans. Other members include skip-lot plans, chain sampling plans, and the Cone-Dodge cumulative results plan. Cumulative results plans, however, usually involve lot-by-lot inspection of a stream of product. It is the purpose of this chapter to examine these plans as a means of dealing with the frequent problem of minimizing sample size because of economic constraints while still affording a reasonable amount of protection.

In general, cumulative results plans require certain assumptions to be met about the nature of the inspection. As described by Dodge (1955) in introducing chain sampling plans, these are

1. The lot should be one of a continuing series of supply.
2. Lots should normally be expected to be of the same quality.
3. The consumer should have no reason to believe that the lot to be inspected is poorer than any of the immediately preceding lots.
4. The consumer must have confidence in the supplier, in that

advantage would not be taken of a good record to slip in a sub-
standard lot.

Under these conditions, it is reasonable to use the record of previous inspections as a means of reducing the amount of inspection required on any
given lot.

SKIP-LOT SAMPLING PLANS

SkSP-1

Continuous sampling plans are intended to be applied on individual units
produced in sequence from a continuing source of supply. The principles of
continuous sampling can, however, be applied to individual lots received in
a steady stream from a trusted supplier. Just as units are "skipped" during
the sampling phase of a continuous sampling plan, so lots may be skipped
(and passed) under an analogous skip-lot plan. It is surprising, but for-
tuitous, that skip-lotting can actually increase protection per unit sampled.

The first skip-lot plan, SkSP-1, was introduced by Dodge (1955a) as an
adaptation of CSP-1 to the inspection of raw materials purchased regularly
from a common source. Materials are often inspected using bulk sampling
procedures with an output of one laboratory determination per lot. Disposition of the lot is in accord with whether or not the laboratory determination conforms to specification requirements. Thus, by regarding each lot
of raw material as a single "unit" — either conforming or not conforming
to specifications — continuous sampling plans are readily applied. In
this case, however, the AOQL provides an upper bound on the average percentage of accepted *lots* that will be nonconforming. Similarly, other measures
of continuous sampling plans may be interpreted as referring to "lots"
rather than units. Dodge (1955a) contrasts CSP-1 and SkSP-1 for 2.0 percent
AOQL in tabular form, shown here as Table 16-1.

The skip-lot procedure may be represented schematically as in Fig.
16-1 which presents the skip-lot concept.

In such applications, rejected lots are not usually 100 percent inspected or replaced by known good material. They are simply rejected and
disposed of. When this is the case, Dodge (1955a) points out, i should be
increased by one in CSP-1 plans to maintain the protection guaranteed.

While any continuous sampling plan may be used in this application,
Dodge (1955a) proposes a CSP-1 plan with 2 percent AOQL for general use:

TABLE 16-1   Comparison of CSP-1 and SkSP-1

CSP-1 (product units)	SkSP-1 (lots of a raw material)
Series of units	Series of lots (or batches)
Inspect a unit	Make lab analysis of a sample of material
Defective unit (a unit which fails to meet the applicable specification requirement)	Nonconforming lot (a lot whose sample fails to meet the applicable specification requirement)
Units in succession found clear of defects	Lots in succession found conforming
Incoming % defective: % of incoming units that are defective	Incoming % defective: % of incoming lots that are nonconforming
Meaning of 2% AOQL: an average of not more than 2% of accepted units will be defective for the characteristics under consideration.	Meaning of 2% AOQL: an average of not more than 2% of accepted lots will be nonconforming for the characteristic under consideration.

From Dodge (1955a, p. 4).   Reprinted by permission.

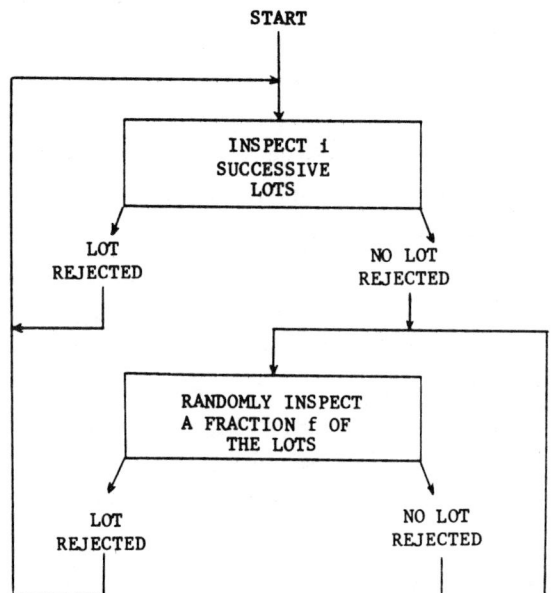

FIG. 16-1   Skip-lot procedure.

   Procedure A-1   (Each nonconforming lot corrected or replaced by a
      conforming lot.)   i = 14, f = 1/2

   Procedure A-2   (Each nonconforming lot rejected and not replaced by a
      conforming lot.)   i = 15, f = 1/2

Other choices of f and i can be made from Fig. 15-2 for CSP-1 or by using
appropriate means for other continuous sampling plans.

   SkSP-1 plans are implemented as follows, illustrated using procedure
A-2 on lots which are rejected and not replaced.

   1.   Apply plan separately to each characteristic under consideration.
        When several characteristics are involved, try to test at least
        one characteristic in each lot. That is, lots skipped for one
        characteristic should be tested for another so that each lot
        receives tests on some of the characteristics.

   2.   Start by testing each lot consecutively as received until 15 lots
        in succession are found conforming.

   3.   When 15 lots in succession are found conforming, test only half
        the lots at random. Accept the lots not tested.

   4.   When a lot is rejected, revert to step 2.

This will guarantee that on the average a maximum of only 2 percent of the
accepted lots will be nonconforming. Procedure A-1, for use when nonconforming lots are replaced with conforming lots, is the same with i = 14
rather than i = 15 as above.

   Other skip-lot plans of this sort can easily be devised using the
procedures for the continuous sampling plans of Chap. 15.

SkSP-2

While SkSP-1 was intended to be used in circumstances leading to a simple
and absolute go no-go decision on each lot, the continuous sampling approach
to skipping lots may be utilized when a standard sampling plan is applied to
each lot. When sampling plans are used, a lot is accepted or rejected with
an associated producer's or consumer's risk. These risks have been factored
into the skip-lot procedure by Dodge and Perry (1971) in their development
of SkSP-2. These plans are intended to be applied to a series of lots or
batches of discrete items which are sampled using a standard "reference"
sampling plan. Two stages of sampling are distinguished:

   Normal inspection: Use of the reference plan on every lot as received.

> Skipping Inspection: Use of the reference plan on a randomly selected fraction f of the lots. Skipped lots are accepted.

CSP-1 is applied to the inspection results to determine whether normal or skipping inspection is to be applied. The procedure is as follows:

1. Start with normal inspection of each lot using the reference plan.
2. When i consecutive lots have been accepted, switch to skipping inspection, inspecting a fraction f of the lots at random as received.
3. When a lot is rejected, switch to normal inspection, step 1.
4. Screen each rejected lot replacing nonconforming units with conforming units.

Thus, an SkSP-2 plan is specified by

1. The reference sampling plan applied to each lot.
2. i, the clearing interval.
3. f, the sampling frequency.

Application of the plan n = 20, c = 1, i = 4, f = .20 would proceed as follows:

1. Inspect consecutive lots using the reference plan n = 20, c = 1 on each under normal inspection.
2. When i = 4 lots in succession have passed the reference plan, go to skipping inspection. Inspect at random only a fraction f = .20 of the lots using the reference plan on each. Pass the lots not inspected.
3. When a lot is rejected, revert to normal inspection, step 1.
4. Screen all rejected lots.

Measures of SkSP-2 are analogous to SkSP-1. Let

P = probability of acceptance of reference plan

$$U = \frac{1 - P^i}{P^i(1 - P)} = \text{expected number of lots during normal inspection}$$

$$V = \frac{1}{f(1 - P)} = \text{expected number of lots during skipping inspection}$$

Then, measures of the SkSP-2 plan are as follows as derived by Perry (1970) using both a power series approach, as in the derivation of CSP-1, and Markov chains.

P_a = probability of acceptance (long-run proportion of lots accepted)

$$= \frac{(1 - f)P^i + fP}{(1 - f)P^i + f}$$

F = average fraction lots inspected (long-run average fraction of lots inspected)

$$= \frac{U + fV}{U + V} = \frac{f}{(1 - f)P^i + f}$$

ASN_{Sk} = average sample number (long-run average sample size over lots inspected)

$$ASN_{Sk} = F(ASN_R)$$

where

ASN_R = average sample number of reference plan

For a single sampling plan (n,c) this becomes

$$ASN_{Sk} = Fn$$

$AOQL_1$ = Unit AOQL. Upper bound on the long-run average proportion of outgoing product units that are defective.

$$AOQL_1 = \frac{Y}{n}$$

where selected values of Y from Perry (1970) are given in Appendix Table T16-1 for single sampling reference plans for various values of c, f, and i (assume Type B sampling).

$AOQL_2$ = Lot AOQL. Upper bound on the long-run average proportion of outgoing lots that are nonconforming, i.e., lots which fail the reference plan

$$AOQL_2 = AOQL \text{ of CSP-1 plan having same f and i.}$$

Values of the lot $AOQL_2$ have been tabulated by Perry (1973) and are shown in Table 16-2. Consider the SkSP-2 plan n = 20, c = 1, i = 4, f = .25. Assume an incoming proportion defective of p = .05. For the reference plan, at this fraction defective, P = .736. Also the AOQL = .042. The measures of the skip-lot plan are

TABLE 16-2 Values of Lot $AOQL_2$ Given f and i

	f						
i	2/3	3/5	1/2	2/5	1/3	1/4	1/5
4	.034	.044	.060	.081	.098	.126	.148
6	.024	.030	.042	.057	.069	.089	.105
8	.018	.023	.032	.044	.053	.069	.081
10	.015	.019	.026	.035	.043	.056	.066
12	.013	.016	.022	.030	.037	.047	.056
14	.011	.014	.019	.026	.031	.041	.048

From Perry (1973, p. 130). Reprinted by permission.

Probability of acceptance

$$P_a = \frac{(1 - .25)(.736)^4 + .25(.736)}{(1 - .25)(.736)^4 + .25} = .860$$

Average fraction lots inspected

$$F = \frac{.25}{(1 - .25)(.736)^4 + .25} = .532$$

Average sample number

$$ASN_{Sk} = .532(20) = 10.6$$

Unit AOQL

$$AOQL_1 = \frac{.9861}{20} = .049$$

Lot AOQL

$$AOQL_2 = .126$$

Thus, at this fraction defective, the probability of acceptance is higher under skip-lotting, from .736 to .860, while the AOQL is increased only slightly, from .042 to .049.

The operating characteristic curve for the plan n = 20, c = 1, i = 4, f = .25 is given in Fig. 16-2 taken from Perry (1973). Also included are the OC curves for a similar plan with i = 10 and for the reference plan. Note that skip-lotting swells the shoulder of the OC curve, improving the producer's risk, but leaves the LTPD essentially unchanged from the reference plan. A typical set of ASN curves is illustrated in Fig. 16-3. Note the substantial reductions in ASN in regions of good quality ($P_a > .5$).

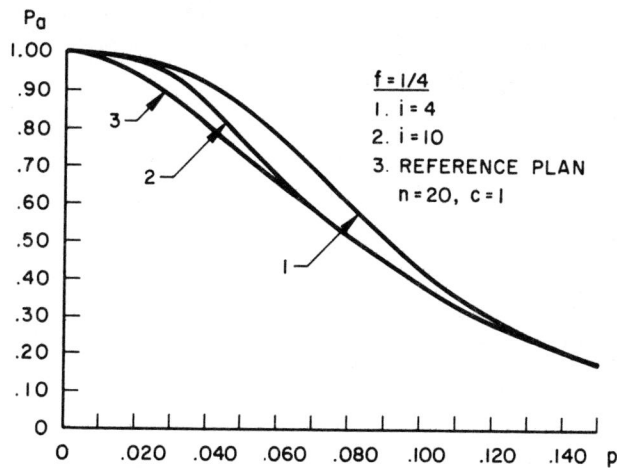

FIG. 16-2   OC curves for some skip-lot plans.
(From Perry, 1973, p. 125.   Reprinted by permission.)

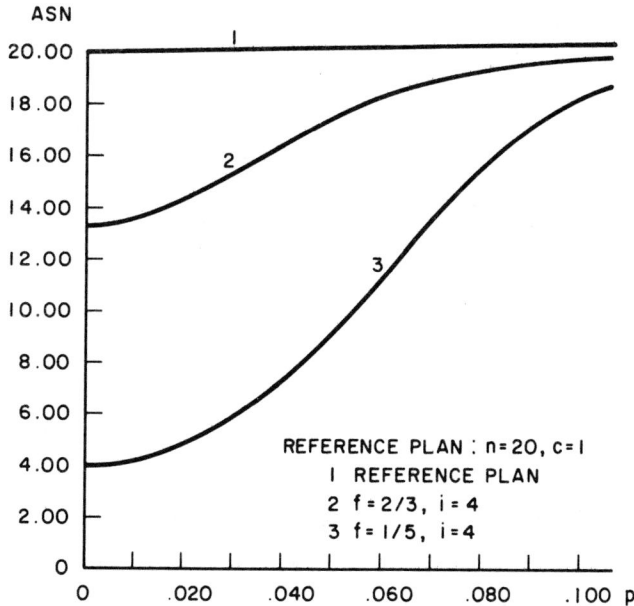

FIG. 16-3   ASN curves for some skip-lot plans.
(From Perry, 1973, p. 127.   Reprinted by permission.)

Using unity values from the Poisson distribution, Dodge and Perry (1971) developed a table which can be used to easily derive skip-lot plans to match single-sampling plans having acceptance numbers from 2 to 10. It is presented here as Appendix Table T16-2. It shows alternate skip-lot plans having a given operating ratio. Of course, the operating ratios of skip-lot plans cover a wide range of possible values and are not restricted to those of the single-sampling plans. Additional unity values have been given by Perry (1970).

To use the Dodge-Perry table, given the operating ratio desired

1. Find the OR listed closest to the operating ratio desired, where OR = $p_{.10}/p_{.95}$.

2. Find the corresponding single-sampling plan for that operating ratio by using the associated acceptance number c and sample size n found by dividing the value $np_{.95}$ given for the single-sampling plan by the value of $p_{.95}$ used to compute the operating ratio.

3. To find the matched SkSP-2 plan

   a. Pick convenient values of f and i listed for the operating ratio given.

   b. The reference plan will have the value of c listed and sample size found by dividing the value of $np_{.95}$ for the SkSP-2 plan by the value of $p_{.95}$ used to obtain the operating ratio.

For example, suppose a plan is desired such that $p_{.95} = .01$ and $p_{.10} = .04$. The operating ratio is

$$OR = \frac{.04}{.01} = 4$$

1. The closest operating ratio for a single-sampling plan is 4.057.

2. The single-sampling plan closest to OR = 4 has c = 4 and

$$n = \frac{1.97}{.01} = 197$$

3. A corresponding SkSP-2 plan would have f = .5, i = 4, with a reference plan having c = 3 and a sample size

$$n = \frac{1.645}{.01} = 164.5 \sim 165$$

The table shows the ratio of these two sample sizes to be .830.

Of course, Appendix Table T16-2 can be used in a number of ways to derive and evaluate SkSP-2 plans. For example, a SkSP-2 plan matching the

single-sampling plan n = 200, c = 3, has f = .2, i = 14, c = 2 and a
reference plan sample size 73.1 percent of the matched single-sampling plan,
or

$$n = .731(200) = 146.2 \sim 147$$

This utilizes the column of ratios of reference sample sizes of SkSP-2 plans
to matched single-sampling plans. From this column it is apparent that, in
matching single-sampling plans, not only are some lots skipped, but the
sample size and acceptance number of the reference plan applied to the lots
inspected will be less than that of the matched single-sampling plan. This
is because, in SkSP-2 inspection, a tight reference plan is used a fraction
of the time to achieve the same result as consistent application of a looser
matched single-sampling plan. Sizable savings in average sample size can
be achieved by using SkSP-2 plans. Of course, the gain is achieved by not
inspecting all the lots, which may, at times, be a serious disadvantage.

The skip-lot concept has been extended by Perry (1973a) to achieve
greater flexibility in application by using two stages. Since the skip-lot
plans may be derived from any continuous sampling procedure, not just CSP-1,
three procedures are proposed by Perry based on other continuous plans:

> Plan 2L.1. Two-stage plan based on the multi-level plan of Lieberman
> and Solomon(1955).

> Plan 2L.2. Two-stage plan based on the tightened multi-level plan of
> Derman, Littaurer, and Solomon (1957) as extended by Gutherie and
> Johns (1958).

> Plan 2L.3. A unique two-stage plan developed by Perry which determines
> the sampling rate on the basis of the number of consecutive lots
> accepted.

The plans presented allow for any combination of sampling rates to be used,
and thus are more general than those of the conventional multi-level plans
which prescribe a geometric relationship between sampling rates. The details of these plans, together with an exposition of their properties are
presented in the Perry (1973a) paper.

CHAIN SAMPLING PLANS

A prime example of the use of cumulative results to achieve a reduction of
sample size while maintaining or even extending protection can be found in
the chain sampling plans introduced by Dodge (1955). These plans were
originally conceived to overcome the problem of lack of discrimination in

c = 0 sampling plans. The procedure was developed to "chain" together the
most recent inspections in a way that would build up the shoulder of the
OC curve of c = 0 plans. This is especially desirable in situations in
which small samples are demanded because of the economic or physical difficulty of obtaining a sample.

ChSP-1

The original chain sampling inspection procedure as developed by Dodge
(1955) is as follows:

    1. From each lot, select a sample of n units.

    2. Accept if

        a. No defectives are found in the sample.

        b. One defective is found in the sample, but no defectives were
            found in the previous i samples of n.

    3. Reject otherwise.

Specification of n and i completely determines a ChSP-1 plan. The ChSP-1
procedure is illustrated schematically in Fig. 16-4.

    The operating characteristic curve of the ChSP-1 procedure is shown
by Dodge (1955) to be determined by

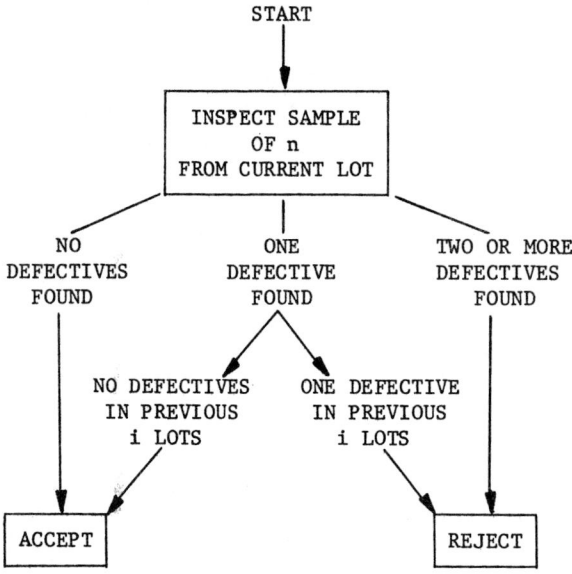

FIG. 16-4 ChSP-1 procedure.

$$P_a = p(0) + p(1) [p(0)]^i$$

where

$p(x)$ = probability of x defectives in a sample of n

Clearly, when $i = \infty$, ChSP-1 reverts to the single-sampling plan having $c = 0$. Thus, for the plan $n = 10$, $i = 2$ when $p = .10$, we have

$$p(0) = .3487 \qquad p(1) = .3874$$

so

$$P_a = .3487 + .3874(.3487)^2 = .3958$$

Of course, average outgoing quality can be found by

$$AOQ = pP_a$$

which gives

$$AOQ = .10(.3958) = .04$$

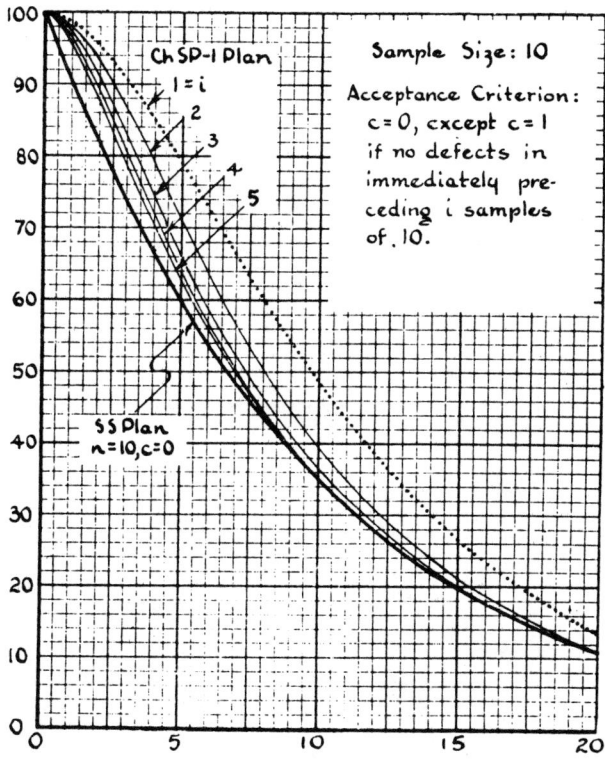

FIG. 16-5 OC curve for ChSP-1 ($n = 10$; $i = 1, \ldots,$ 5; ∞). (From Dodge, 1955, p. 11. Reprinted by permission.)

A comparison of ChSP-1 plans for n = 10 and several values of i is shown in Fig. 16-5 from Dodge (1955). The solid line represents the single-sampling plan n = 10, c = 0. It illustrates the shoulder built up on the c = 0 operating characteristic curve when the chain sampling criterion is imposed. The curve for i = 1 is shown dotted since its use is not recommended by Dodge. Note that in the region of low probability of acceptance the OC curves for various values of i seem to coincide with the exception of the curve for i = 1.

Chain sampling plans are easily evaluated using Poisson unity values developed by Soundararajan (1978). Table 16-3 shows values of np corresponding to various probabilities of acceptance for values of i from 1 to 6 and for i = ∞, which is simply the single-sampling plan with c = 0. Notice that with values of low probability of acceptance the unity values are those of the corresponding single-sampling plan having c = 0. When divided by sample size, the unity values give the proportion defective corresponding to the probability of acceptance shown. Thus for the plan n = 10, i = 3, we have

P_a	p
.99	.0057
.95	.0139
.50	.0768
.10	.2303
.05	.2996
.01	.4605

TABLE 16-3   Unity Values for Evaluation of ChSP-1 Operating Characteristic Curves

P_a \\ i	0.99	0.95	0.50	0.10	0.05	0.01
1	0.086	0.207	1.0066	2.490	2.996	4.605
2	0.067	0.162	0.8399	2.325	2.996	4.605
3	0.057	0.139	0.7675	2.303	2.996	4.605
4	0.051	0.124	0.7325	2.303	2.996	4.605
5	0.046	0.114	0.7135	2.303	2.996	4.605
6	0.042	0.106	0.7034	2.303	2.996	4.605
∞	0.010	0.051	0.693	2.303	2.996	4.605

From Soundararajan (1978, p. 56). Reprinted by permission.

A table for constructing ChSP-1 plans has also been given by Soundararajan (1978) and is presented here as Appendix Table T16-3. Based on
Poisson unity values, it allows determination of a ChSP-1 plan from the
desired operating ratio p_2/p_1. Values of np are given at .95 and .10 probability of acceptance for i from 1 to 10 and i = ∞. Also the AOQL of the
ChSP-1 procedure can be found from sample size using nAOQL values or from
the producer's quality level p_1 using values of $AOQL/p_1$. Further, the proportion defective p_M at which the AOQL occurs can be determined from values
of np_M.

For example, we find for the plan n = 10, i = 3, the following properties given by Appendix Table T16-3.

$$P_a = .95 \text{ at } p = .0139$$

$$P_a = .10 \text{ at } p = .2303$$

$$p_2/p_1 = 16.568$$

$$AOQL = 2.798(.0139) = .0389$$

$$p_M = .0902$$

It is sometimes desirable to construct a ChSP-1 plan having a specified AOQL. For this purpose, Table 16-4 has been developed by Soundararajan (1978), showing values of sample size n and cumulative results criterion i. We see that an AOQL of .04 is achieved by the plan n = 10, i = 3.

While Poisson unity values provide an excellent device for constructing
plans as an approximation to the binomial distribution and are exact when
dealing with defects, it is sometimes desirable to have exact tables for
the selection of plans based on the binomial distribution itself. Appendix
Table T16-4 from Soundararajan (1978a) gives ChSP-1 plans indexed by AQL
($p_{.95}$ value) and LTPD ($p_{.10}$ value). It shows that, using the binomial distribution, for $p_{.95}$ = .015 and $p_{.10}$ = .220 the plan n = 10, i = 3 would be
appropriate. In addition, plans may be constructed for a given AQL/AOQL
combination (AQL = $p_{.95}$) using Appendix Table T16-5 given by Soundararajan
(1978a). For an AQL/AOQL combination of .015/.035 Appendix Table T16-5
gives the plan n = 15, i = 1, which should give protection roughly equivalent to the plan n = 20, c = 1 which has $p_{.95}$ = .018 and AOQL = .035. Note
the obvious saving in sample size.

Two-Stage Plans

Two-stage chain sampling plans generalizing ChSP-1 have been the subject of
extensive work by H. F. Dodge and K. S. Stephens. These plans provide a

TABLE 16-4   ChSP-1 Plans Having Given AQQL

i.	0.10	0.25	0.50	0.75	1.0	1.5	2.0	2.5	3.0	3.5	4.0	4.5	5.0	6.0	7.0	8.0	9.0	10.0
1.	504	202	101	68	51	34	26	21	17	15	13	12	11	9	8	7	6	5
2.	420	168	89	56	42	28	22	17	14	12	11	10	9	7	6	6	5	5
3.	389	156	78	52	39	26	20	16	13	12	10	9	8	7	6	5	5	4
4.	377	151	76	51	38	26	19	16	13	11	10	9	8	7	6	5	5	4
5.	372	149	74	50	38	25	19	15	13	11	10	9	8	7	6	5	5	4
6.	369	148	74	50	37	25	19	15	13	11	10	9	8	7	6	5	4	4
7.	369	148	74	50	37	25	19	15	13	11	10	9	8	7	6	5	4	4
8.	368	148	74	49	37	25	19	15	13	11	10	9	8	7	6	5	4	4
9.	368	148	74	49	37	25	19	15	13	11	10	9	8	7	6	5	4	4
10.	368	148	74	49	37	25	19	15	13	11	10	9	8	7	6	5	4	4
∞	368	148	74	49	37	25	19	15	13	11	10	9	8	7	6	5	4	4

AQQL in per cent

From Soundararajan (1978, p. 58).   Reprinted by permission.

generalization of ChSP-1 plans in that two stages for the implementation of
the plan are defined.

1. Restart procedure. The period during which the chain sampling procedure is started or immediately following a rejection. During
this phase, samples of n_1 are chained with a cumulative results
criterion of C_1 allowable defectives in the cumulative results.
When k_1 lots have been accepted, the normal procedure is instituted.

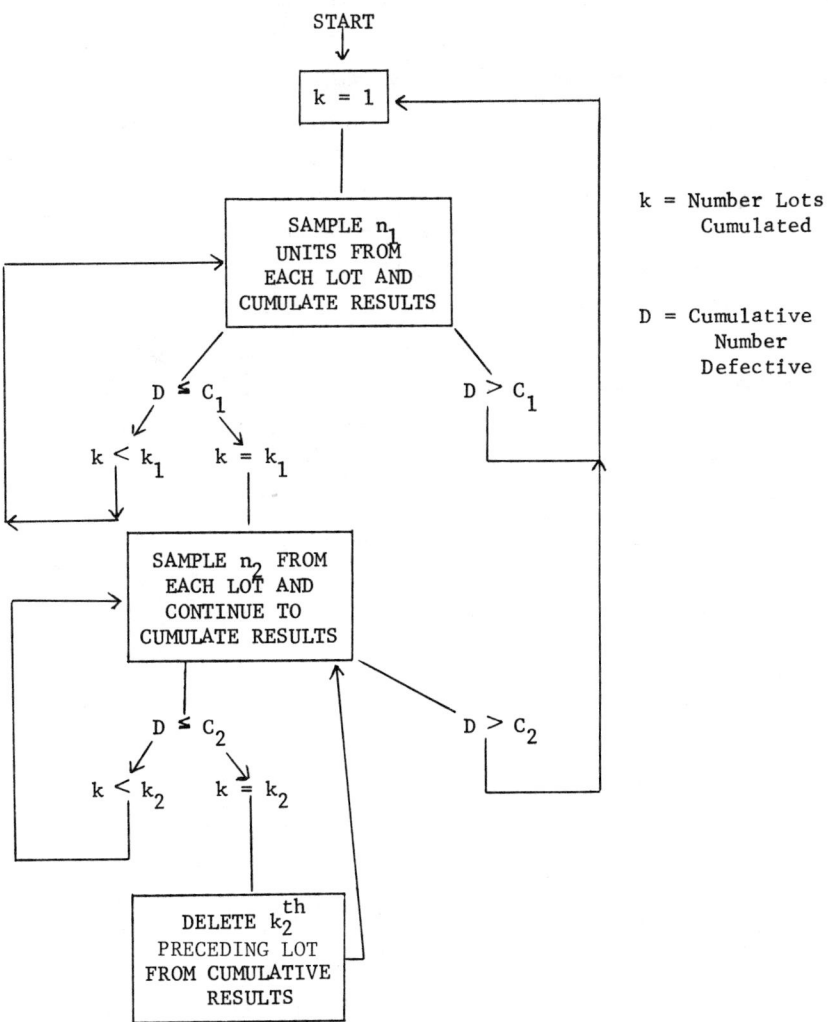

FIG. 16-6 Two-stage chain sampling procedure.

2. Normal procedure. After k_1 lots have been accepted, additional lots are chained until a running total of k_2 lots is reached and maintained. During this period, samples of n_2 are taken from each lot using a cumulative results criterion of C_2 allowable defectives in the cumulative results. The restart procedure is initiated as soon as a lot is rejected.

This approach as introduced by Dodge and Stephens (1966) can be represented schematically as in Fig. 16-6.

The solution of the operating characteristic problem of the general family of chain sampling plans is described by Stephens and Dodge (1974). It involves imbedding a Markov chain in the chain sampling process by an appropriate definition of states. The two-stage plans have been designated by Stephens and Dodge (1976a) as $\text{ChSP}(n_1, n_2) - C_1, C_2$ with k_1, k_2 separately specified. The original ChSP-1 plan is equivalent to $\text{ChSP}(n,n) - 0,1$ with $i = k_1 = k_2 - 1$. The first two-stage plans by Dodge and Stephens (1966) maintained a constant sample size in both the restart and normal procedures. These plans will be found designated $\text{ChSP-}C_1, C_2$ in the literature with n, k_1, k_2 separately specified. The advantages of greater generality in the selection of chain sampling parameters are greater flexibility in matching and use, and of course, improved discrimination through the use of the generalized two-stage procedure.

Stephens and Dodge (1976) have provided a comparison of ChSP-1 and two-stage chain plans against single- and double-sampling plans. For example, they have found the following to be matched using $k_1 = 1$, $k_2 = 2$, $n = n_1 = n_2 = 50$. Double sampling rejection numbers are $c_2 + 1$ on both samples.

Chain sampling	Single sampling	Double sampling	
ChSP-1 $i = 1$, $n = 50$	$n = 85$, $c = 1$	$n_1 = 40$, $n_2 = 80$ $c_1 = 0$, $c_2 = 1$	ASN = 53.2
ChSP-0,2, $n = 50$	$n = 105$, $c = 2$	$n_1 = 72$, $n_2 = 144$ $c_1 = 1$, $c_2 = 3$	ASN = 79.2
ChSP-0,3, $n = 50$	$n = 120$, $c = 3$	$n_1 = 54$, $n_2 = 108$ $c_1 = 0$, $c_2 = 4$	ASN = 79.6
ChSP-0,4, $n = 50$	$n = 135$, $c = 5$	$n_1 = 69$, $n_2 = 138$ $c_1 = 1$, $c_2 = 6$	ASN = 75.5

The values of ASN shown are for a proportion defective of .005 which had

greater than .92 probability of acceptance under the double-sampling plan.
Since for fractions defective greater than .005 the ASN of the double-
sampling plans were much higher, the comparison seems favorable to the
chain sampling plans shown.

DEFERRED SENTENCING SCHEMES

Deferred sentencing schemes were among the earliest of cumulative results
plans. They trace their origins to the British Ministry of Supply when
H. B. Spalding, P. Halliday, and E. H. Sealey applied the method during
World War II under the name of "rational sentencing." The term *sentencing*
was regularly used by the Government Inspectorates in connection with
acceptance of ammunition. Deferred sentencing involves delay of disposition
of questionable lots until subsequent lots have been inspected.

The scheme has been aptly described and evaluated by Anscombe, Godwin,
and Plackett (1947), who presented several approaches to this type of sentencing. Their simplest scheme has been the most fully investigated and is
described by them (p. 199) in the following:

> Sentencing rule. - The product, as it leaves the line,
> is divided into small lots, and one item is selected
> from each for test. D and n being given integers,
> whenever n defective items are encountered out of D or
> fewer consecutive lots tested, all the lots consecu-
> tively from that giving the first to that giving the
> nth defective in the cluster are rejected. Lots not
> rejected by this rule are accepted.

This simple deferred sentencing scheme is represented diagrammatically in
Fig. 16-7.

The basic idea is to defer sentencing lots after a defect is found
until it can be shown that the subsequent D - 1 lots are of acceptable quality, that is they give less than n - 1 defectives. When this is shown,
lots are released up to the next defective and the process is repeated.
This approach may be used as a continuous sampling plan on individual
items, which does not require screening and so may be employed with destructive tests. Deferred sentencing is primarily intended, however, for use
with lots of product sufficiently small that one test per lot is reasonable.
Note that it is particularly well suited to bulk sampling applications.

The selection of n and D is facilitated by a table of the percentage

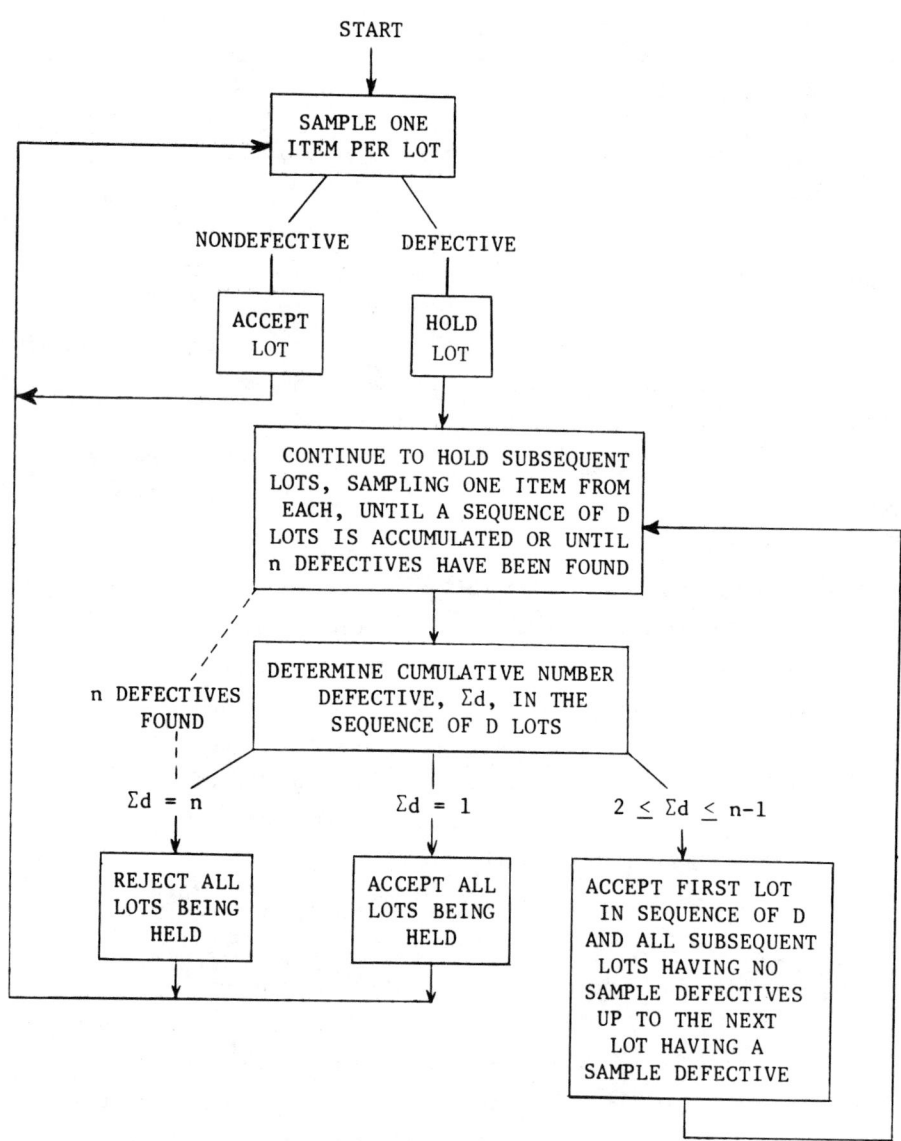

FIG. 16-7 Simple deferred sentencing scheme.

TABLE 16-5 Values of Dp Having Specified Probability of Acceptance

n	99 percent	90 percent	Percent Output Accepted 50 percent	10 percent	1 percent
3	.35	.89	2.20	4.5	7.3
4	.63	1.34	2.84	5.3	8.1
5	.97	1.84	3.54	6.1	9.0
6	1.36	2.39	4.28	7.0	10.0
7	1.78	2.96	5.04	7.9	11.0
10	3.25	4.85	7.43	10.8	14.2

From Anscombe, Godwin, and Plackett (1947, p. 200). Reprinted by permission.

points of the product Dp presented by Anscombe, Godwin, and Plackett (1947) shown here as Table 16-5.

Using Lagrangian interpolation to determine the $p_{.95}$ values, the operating ratios $R = p_{.10}/p_{.95}$ for these plans are approximately

n	R
3	7.5
4	5.7
5	4.4
6	3.8
7	3.4
10	2.7

A graphical representation of Table 16-5 has also been given by Anscombe, Godwin, and Plackett (1947) which is shown in Fig. 16-8. It gives curves for n plotted by Dp and the percentage of product accepted. Note that the latter corresponds roughly to P_a and can be used to construct an operating characteristic curve for these schemes.

To use the table, or the figure, n may be chosen to correspond with one of the operating ratios given. If it is desired to hold a value of LTPD, say $100p_t$, the value of Dp corresponding to a proportion of output accepted of 10 percent, is divided by p_t. This gives the value of D to be used in the plan. For example, if a deferred sentencing scheme is to be determined having about the same protection as the plan n = 50, c = 5, which has an operating ratio R = 3.6 and an LTPD = 18.6 percent, we have

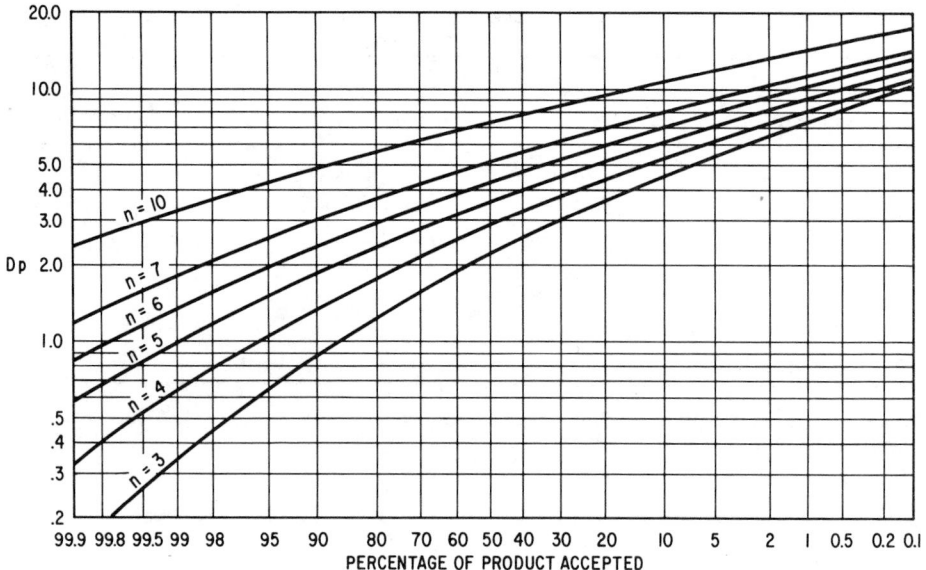

FIG. 16-8 Operating characteristics of simple deferred sentencing schemes.
(From Anscombe, Godwin, and Plackett, 1947, p. 201. Reprinted by permission.)

$$n = 7$$

$$D = \frac{Dp}{P_t} = \frac{7.9}{.186} = 42.5 \sim 43$$

The plan is implemented by taking one unit from each lot. As long as
no defectives are found, the lots are passed. As soon as a defective is
obtained, lots are held. If 6 or more defectives are found in the next 42
lots, all lots are rejected up to and including that providing the seventh
defective. If five or less defectives are forthcoming, the first lot is
passed together with all subsequent lots up to the next lot showing a defective. From that point, the procedure is applied again.

The OC curve for this scheme may be obtained by dividing the values
of Dp given in Table 16-5 for $n = 7$ by $D = 43$.

P_a	Deferred sentencing p	Single sampling $n = 50, c = 5$
.99	.041	.036
.90	.069	.063
.50	.117	.113
.10	.184	.186
.01	.256	.262

The chart shown in Fig. 16-8 is also very useful for this purpose.

There are many variations possible for the deferred sentencing scheme. Some are given by Anscombe, Godwin, and Plackett (1947). In one procedure the lots held for deferred sentencing may extend from the lot in which a defect is found, forward and back for a number of lots. A double-sampling approach has been given by Hill, Horsnell, and Warner (1959) which incorporates into the scheme samples of size greater than one from a lot. Deferred sentencing is attractive in certain applications where time is not at a premium and lots can be put aside. However, it suffers from the delay inherent in holding lots for disposition for any period of time.

DEMERIT RATING PLAN

The check inspection and demerit rating plan used extensively by Western Electric and described by Dodge and Torrey (1956) is an audit plan intended to characterize quality levels and to provide a check inspection with relatively small samples. It supplies management with a demerit rating of defects on specific products and a demerit index of quality across defect and product types. At the same time, the inspection results necessary for surveillance of quality are used in product acceptance. The plan provides continuing surveillance through control charts. H. F. Dodge (1962) has recommended its use in conjunction with the cumulative results criterion (CRC) plan discussed below. Its successful application in this regard has been described by Cone and Dodge (1963).

The demerit rating plan is initiated by taking small samples at regular intervals (by shift, day, or week) across the product and product types to be included. Sample sizes are chosen with regard to the homogeneity of the universe of product to be sampled and its complexity. For a given quantity of output N to be represented by the sample, the sample size is chosen as in Table 16-6. These samples sizes were arrived at empirically and represent the result of 15 years or more experience.

Recognizing differences in the variety and nature of defect types, a classification of defects was developed as follows:

Class A (very serious)	Will surely cause an operating failure.
Class B (serious)	Will probably cause an operating failure.
Class C (moderately serious)	May possibly cause an operating failure.
Class D (not serious)	Minor defect which will not affect operation, maintenance, or life.

TABLE 16-6   Sample Size for Check Inspection and Demerit Rating

	Universe	
	Heterogeneous	Homogeneous
Product	Product comprising a variety of types of different construction	Product of a specific type or a group of types of similar construction
Complex. Construction subject to variations in adjustment	$n = 2.5 \sqrt{2N}$	$n = 1.5 \sqrt{2N}$
Simple. Construction non-adjustable or stable	$n = 2 \sqrt{2N}$	$n = \sqrt{2N}$

From Dodge and Torrey (1956, p. 7).   Reprinted by permission.

These classes are assigned demerits for use in constructing a demerit rating and demerit index in the operation of the plan. These demerits are

Class A: 100 demerits
Class B: 50 demerits
Class C: 10 demerits
Class D: 1 demerit

Table 16-7 as given by Dodge and Torrey (1956) describes the classification further.

The sample size having been determined, samples are taken periodically over the time period for which the demerit rating is to be constructed. If the number of defects observed in any sample exceeds the nonconformance criteria given in Table 16-8, a second sample twice as large is taken. If the combined number of defectives in the first and second samples exceeds the nonconformance criteria, the lot or batch represented by the samples is rejected subject to action by the proper authority.

The nonconformance criteria are set at three standard deviations distant from nonconformance levels (NL) which roughly correspond to AQLs for the class of defects involved in the sense that, according to Dodge and Torrey (1956), "... if products are maintained at acceptable quality levels, the chances of the criteria being exceeded are very remote." These are

TABLE 16-7 Important Aspects of Classification of Defects

Defect Class	Demerit Weight	Cause Pers. Injury	Cause Operating Failure	Cause Intermit. Op.Trouble Difficult to Locate in Field	Cause Subst'd Per-form-ance	Involve Incr'd Maint'ce or Decr'd Life	Cause Increase in Instal. Effort by Customer	Appear'ce Finish or Work'ship Defects
A	100	Liable to	Will Surely*	Will Surely	—	—	—	—
B	50	—	Will Surely** Will Probably	—	Will Surely	Will Surely	Major Increase	—
C	10	—	May Possibly	—	Likely to	Likely to	Minor Increase	Major
D	1	—	Will not	—	Will not	Will not	—	Minor

*Not readily corrected in the field.
**Readily corrected in the field.

From Dodge and Torrey (1956, p. 8). Reprinted by permission.

Defect class	Nonconformance level (NL) proportion defective
A	.0025
B	.005
C	.02
D	.04

Thus, the limits become

$$n(NL) + 3 \sqrt{n(NL)}$$

using Poisson limits for the number of defects.

After the samples for the period have been collected, a demerit rating is calculated as

$$D = w_A d_A + w_B d_B + w_C d_C + w_D d_D$$

where the weights w_K are simply the demerits assigned to defects of class

TABLE 16-8   Nonconformance Criteria

No. of Units in Sample, n	Max. No. of Defects in Sample			
	Class A	Class B	Class C	Class D
1-2	0	0	0	0
3-4	0	0	0	1
5-8	0	0	1	1
9-16	0	0	1	2
17-18	0	0	2	3
19-25	0	1	2	3
26-31	0	1	2	4
32-36	0	1	3	4
37-48	1	1	3	5
49-50	1	1	3	6
51-65	1	1	4	6
66-75	1	2	5	7
76-90	1	2	5	8
91-100	1	2	6	9
Over 100	**	**	**	**

**Class A = $0.0025n + .150 \sqrt{n}$.
  Class B = $0.0050n + .212 \sqrt{n}$.
  Class C = $0.0200n + .424 \sqrt{n}$.
  Class D = $0.0400n + .600 \sqrt{n}$.

From Dodge and Torrey (1956, p. 9). Reprinted by permission.

K and the number of defects found in that class is d_K. Demerits per unit may be calculated as

$$U = \frac{D}{n}$$

where n is the sample size collected for the period. These values are plotted on control charts with limits set to reflect a "standard quality level" which represents engineering estimates of what quality at delivery should be, taking into account considerations of quality and cost. For defect classes A, B, C, and D, these are represented by μ_A, μ_B, μ_C, μ_D, each in terms of defects per unit. A given unit of product would then have a standard quality level U_S, in defects per unit, determined as

$$U_S = w_A \mu_A + w_B \mu_B + w_C \mu_C + w_D \mu_D$$

This linear combination of Poisson variates has a standard variance factor

$$C_S = w_A^2 \mu_A + w_B^2 \mu_B + w_C^2 \mu_C + w_D^2 \mu_D$$

Hence, limits for a control chart showing the sample value of demerits per unit, D, plotted for samples of n taken each period are

$$U_S \pm 3 \sqrt{\frac{C_S}{n}}$$

When products or dissimilar product types are to be combined to give a quality index for a line, a department, or a plant, the types included should each be weighted to represent the number produced or other salient considerations. In this case, the overall demerit index I_0 is

$$I_0 = \frac{\sum w_i U_i / U_{S_i}}{\sum w_i}$$

with

$$\sigma_{I_0} = \sqrt{\sum \frac{1}{n_i} \left(\frac{w_i^2}{\left(\sum w_i\right)^2} \frac{C_{S_i}}{U_{S_i}^2} \right)}$$

where n_i represents the sample size for the period for the ith type. The values I_0 are plotted on a control chart with limits

$$1 \pm 3\sigma_{I_0}$$

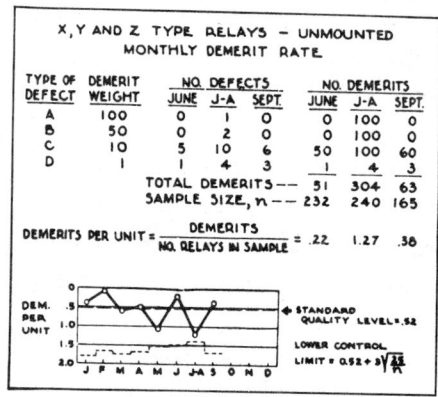

FIG. 16-9 Example of summary of
demerits per unit and control chart.
(From Dodge and Torrey, 1956, p. 10.
Reprinted by permission.)

since the expected demerits per unit of the index calculated in this way
is 1. When all types included in the index have the same standard value,
$W_i = 1$, and the index degenerates into a simple "demerit index", I.

An example of the calculation and display of demerits per unit as
shown by Dodge and Torrey (1956) is given in Fig. 16-9.

CUMULATIVE RESULTS CRITERION PLAN (CRC)

The necessity for small samples when tests are costly or difficult to
administer often reduces protection to the consumer. In cases of audit
inspection and demerit rating, check samples in production, small lots and
destructive tests, sample sizes of 5 or 10 or even less are common. Under
such circumstances, a cumulative results criterion can be used to increase
the effectiveness of the inspection in protecting the consumer. A. F. Cone

and H. F. Dodge (1963) outlined such a plan which has had successful application at the Sandia Corporation. The procedure as described by Dodge (1962)
is as follows:

1. For a given quality characteristic, choose a standard quality level
   SQL which estimates what quality should be at delivery considering
   costs and needs of service.

2. Choose a standard acceptance sampling plan for lot acceptance to
   be used regardless of whether the cumulative results criterion plan
   is also applied.

3. When a lot fails the standard acceptance sampling plan, reject the
   lot and advise the supplier that the cumulative results criterion
   will apply to subsequent lots.

4. Cumulate the results of the standard acceptance sampling plan over
   subsequent lots and compare the results for each lot to the cumulative results criterion CRC_1. A stated moving cumulative sample
   size m shall be maintained once attained.

5. If at any lot, the cumulative results fails to meet the CRC, the
   immediate lot is rejected and the process is also declared nonconforming.

6. Declaration of the process as nonconforming entails:
   a. Ceasing inspection until the supplier submits written evidence
      that corrective action has been taken.
   b. Starting a new sequence of cumulative results when inspection
      is resumed.
   c. If inspection is stopped a second time during the period in
      which the cumulative results criterion is in force, it is not
      resumed until evidence, satisfactory to higher authority, has
      been furnished.

7. The cumulative results criterion is continued until a succession
   of m units has been found to have results equal to or better than
   the criterion for discontinuance CRC_2. At that time, the supplier
   is notified that the cumulative results criterion has been removed.

To determine the cumulative results criterion, let

   Y = statistic generated by standard acceptance sampling plan
      (p, c, or $\overline{X}$, etc.)

cum-Y = cumulative results

CRC_1 = cumulative results action criterion

CRC_2 = cumulative results discontinuance criterion

Y_S = standard quality level for Y

σ_{Y_n} = standard error of Y for given Y_S and cumulative sample size n

A specific CRC plan is determined by three constants.

m = maximum moving sample size

Z_1 = multiple of standard deviation for action on the cumulative results criterion

Z_2 = multiple of standard deviation for discontinuance of the cumulative results criterion

then

$$CRC_1 = Y_S + Z_1 \, \sigma_{Y_n}$$

and

$$CRC_2 = Y_S + Z_2 \, \sigma_{Y_n}$$

Usually Z_1 is taken to be 1.65, 2, or 3, while Z_2 may be 0 or 1. In application at Sandia the constants

$$Z_1 = 3 \qquad Z_2 = 0 \qquad m = 100$$

have been found very effective. Cone and Dodge (1963) describe the favorable experience generated at Sandia over a period of more than two years.

It should be emphasized that Y, the statistic cumulated, can take many forms. For example,

Y = p   (fraction defective)

Y = u   (defects per unit)

Y = U   (demerits per unit)

$Y = \bar{X}$ (sample mean)

In all cases, the sampling distributions involved will be those of a known universe with parameters specified by the standard quality level employed. Thus, the procedure can be adapted to a wide variety of sampling situations.

J. R. Troxell (1972) has made an extensive investigation of types of suspension systems for small sample inspections exemplified by the CRC plan, together with applications of the procedure to MIL-STD-105D.

PROBLEMS

1. The plan n = 50, c = 3 is being used in lot-by-lot inspection. Derive a SkSP-2 plan that will afford the same protection.

2. The SkSP-2 plan n = 165, c = 3, i = 4, f = .5 is being used in sampling inspection of a continuing series of lots. The reference plan has $p_{.95}$ = .008. For this plan evaluate the following when the process level is p = .008:

   a. Probability of acceptance

   b. Average fraction lots inspected

   c. Average sample number

   d. Unit AOQL

   e. Lot AOQL

3. Find a single-sampling plan that roughly matches the SkSP-2 plan n = 143, c = 6, i = 12, f = .5. Also, find matching (a) double- and (b) multiple-sampling plans. Compare the ASN of the single, double, and multiple plans to that of the skip-lot plan for $p_{.95}$. (Hint: Use the Dodge-Perry and the Schilling-Johnson tables.)

4. Draw the OC curves for the ChSP-1 plan i = 3 and n = 20. What is its AOQL? Evaluate the formula for P_a when p = .10.

5. Find a ChSP-1 plan having an AOQL of 6 percent where, for administrative purposes, i should be no greater than 2.

6. The MIL-STD-105D system for Code C, 2.5 AQL has an overall operating ratio R = 20.14 with an LTPD = 28.8 percent. Find a ChSP-1 plan which will give this protection. What is $p_{.95}$ for this plan? What is its AOQL and at what process average does it occur?

7. Find a deferred sentencing plan matching the single-sampling plan n = 50, c = 3. The OC curves should match as closely as possible at the LTPD. What is the indifference quality for the plan?

8. A certain simple component used in one specific product is made at the rate of 10,000 units per month. What should be the sample size per month to be used in a demerit rating plan?

9. Using the demerit weight given by Dodge and Torrey together with the nonconformance levels of the defect classes, in terms of defects per hundred units, compute the standard quality level U_S and standard variance factor C_S. What would be the control limits on a chart for n = 1000? Would a signal result if for classes A, B, C, D, there were found 0, 2, 1, 4 defectives, respectively, in a sample of 1000?

10. A cumulative results criterion for p is set up on the attributes inspection plan n = 10, c = 1, where Z_1 = 3, Z_2 = 0, and m = 100. The standard quality level is Y_S = .02. After 10 lots have been inspected under the criterion, one defect has been found, what action should be taken?

REFERENCES

Anscombe, F. J., H. J. Godwin, and R. L. Plackett (1947), Methods of Deferred Sentencing in Testing the Fraction Defective of a Continuous Output, *Supplement to the Journal of the Royal Statistical Society*, 9: 198-217.

Cone, A. F., and H. F. Dodge (1963), *A Cumulative Results Plan for Small Sample Inspection*, American Society for Quality Control Technical Conference Transactions, Chicago, Illinois, pp. 21-30. [Also published in *Industrial Quality Control*, 21(1): 4-9.]

Derman, C., S. Littauer, and H. Solomon (1957), Tightened Multi-Level Continuous Sampling Inspection Plans, *Industrial Quality Control*, 7(5): 7-12.

Dodge, H. F. (1955), Chain Sampling Inspection Plan, *Industrial Quality Control*, 11(4): 10-13.

Dodge, H. F. (1955a), Skip-Lot Sampling Plan, *Industrial Quality Control*, 11(5): 3-5.

Dodge, H. F. (1962), *A Cumulative-Results Sampling Plan for Small Sample Inspection*, Technical Report No. 11, Rutgers - The State University Statistics Center, New Brunswick, New Jersey.

Dodge, H. F., and R. L. Perry (1971), *A System of Skip-Lot Plans for Lot by Lot Inspection*, American Society for Quality Control Technical Conference Transactions, Chicago, Illinois, pp. 469-477.

Dodge, H. F., and K. S. Stephens (1966), Some New Chain Sampling Inspection Plans, *Industrial Quality Control*, 23(2): 61-67.

Dodge, H. F., and M. N. Torrey (1956), A Check Inspection and Demerit Rating Plan, *Industrial Quality Control*, 13(1): 5-12.

Gutherie, D., and M. Johns (1958), *Alternative Sequences of Sampling Rates for Tightened Multi-Level Continuous Sampling Plans*, Technical Report No. 36, Applied Mathematics and Statistics Laboratory, Stanford University, Stanford, California.

Hill, I. D., G. Horsnell, and B. T. Warner (1959), Deferred Sentencing Schemes, *Applied Statistics*, 8(2): 76-91.

Lieberman, G. J., and H. Solomon (1955), Multi-Level Continuous Sampling Plans, *Annals of Mathematical Statistics*, 28: 686-704.

Perry, R. L. (1970), A System of Skip-Lot Sampling Plans for Lot Inspection, Ph.D. Dissertation, Rutgers - The State University, New Brunswick, New Jersey.

Perry, R. L. (1973), Skip-Lot Sampling Plans, *Journal of Quality Technology*, 5(3): 123-130.

Perry, R. L. (1973a), Two-Level Skip-Lot Sampling Plans - Operating Characteristic Properties, *Journal of Quality Technology*, 5(4): 160-166.

Soundararajan, V. (1978), Procedures and Tables for Construction and Selection of Chain Sampling Plans (ChSP-1) - Part 1, *Journal of Quality Technology*, 10(2): 56-60.

Soundararajan, V. (1978a), Procedures and Tables for Construction and Selection of Chain Sampling Plans (ChSP-1) - Part 2, *Journal of Quality Technology*, 10(3): 99-103.

Stephens, K. S., and H. F. Dodge (1974), An Application of Markov Chains for the Evaluation of the Operating Characteristics of Chain Sampling Inspection Plans, *IAQR Journal*, 1(3): 131-138.

Stephens, K. S., and H. F. Dodge (1976), Comparison of Chain Sampling Plans with Single and Double Sampling Plans, *Journal of Quality Technology*, 8(1): 24-33.

Stephens, K. S., and H. F. Dodge (1976a), Two-Stage Chain Sampling Inspection Plans with Different Sample Sizes in the Two Stages, *Journal of Quality Technology*, 8(4): 207-224.

Troxell, J. R. (1972), An Investigation of Suspension Systems for Small Sample Inspections, Ph.D. Dissertation, Rutgers - The State University, New Brunswick, New Jersey.

COMPLIANCE SAMPLING

Consumer protection has always been a prime factor in the construction of industrial acceptance sampling plans. The methods and procedures presented in this text attest to that fact. A typical example is the set of Dodge-Romig LTPD plans developed as early as 1929. Increased use of acceptance sampling plans in connection with compliance testing to government standards, validation testing of supplier's inspection, and in the verification of extremely tight standards set by regulatory agencies, original equipment manufacturers, and consumers of all kinds suggests the need for sampling plans especially designed and adapted for this area of application.

It has been pointed out by M. G. Natrella in Muehlhause et al. (1975, pp. 42-43) that "there has been little experience to date to demonstrate the effectiveness of sampling schemes for compliance testing. Such experience, and related mathematical investigations, are needed for the formulation in general terms of the overall objectives of sampling schemes, so that the statistician and the regulator — given the standard — can make and explain an appropriate selection." In the area of compliance testing, and especially for safety-related items, the following features seem desirable in a sampling plan:

1. Rejection of the lot if any defective items are found in the sample
2. A well-defined relationship between the sampling plan and the size of the lot being inspected
3. A clear indication of the economic impact of the quality levels utilized in the plan
4. Simplicity and clarity in use

In safety and compliance testing an acceptance number of zero is particularly

desirable, since, to the uninformed, it would appear that the use of any
greater acceptance number implies passing lots which "have been shown to
have defectives in them." A. J. Duncan (1979) points out that in 1972,
the National Highway Traffic Safety Administration proposed to change the
rule regulating the performance of hazard warning flashers to eliminate an
acceptance number of c = 3. Quoting Duncan (1979, p. 21),

> ... It took the point of view that 'permissible failure
> rates raise difficult problems of interpretation and
> enforcement.' It was thus indicated that any sample
> the agency took would have to be 100% conformance
> for the lot to pass. In other words, the acceptance
> number would be zero.

The LSP and TNT plans presented here are illustrative of plans which
are particularly appropriate for use in compliance sampling as well as in
other areas of acceptance control. Plans for verification of quality
levels should be capable of demonstrating compliance to stated levels in
as economic a manner as possible. The simplified grand lot scheme is
particularly useful in this regard since it can be used to provide consumer
protection at very low quality levels while maintaining reasonable protection for the producer. The plans of H-109 can be used simply and effectively to verify inspections carried out previously by the supplier in a simple
straightforward manner.

LOT SENSITIVE SAMPLING PLAN (LSP)

The lot sensitive sampling plan developed by Schilling (1978) is applicable
in general acceptance sampling and is particularly useful in compliance and
safety-related testing. A consumer oriented LTPD plan, it is intended to
meet the objectives outlined for compliance testing. Based on the hypergeometric probability distribution, it gives the proportion of the lot that
must be sampled to guarantee that the fraction defective in the lot is less
than a prescribed limit with LTPD protection.

The LSP plan is easy to use and is based on the concept of acceptance
with zero defectives in the sample. It relates the sample size to lot size
in a straightforward way and provides, as a baseline, a minimum sample size
for sampling applications, since single-sampling plans allowing acceptance
with one or more defectives in the sample usually require larger sample
sizes. The economic impact of the plan vis-à-vis 100 percent inspection is

shown by the fraction of the lot to be inspected.

The disadvantages of plans allowing no defectives in the sample is, in an economic sense, in terms of good product rejected because of the severity of the acceptance criteria. Where possible, various acceptance sampling schemes and strategies should be considered as an alternative to plans of this type. However, if it is required that no defectives are to be allowed in the sample, the LSP plan has real advantages, particularly if the inspection is to be carried out on a unique lot.

Procedure

An LSP plan may be derived in the following manner:

1. Specify lot size N.
2. Specify the limiting quality level p_t that is to be protected against by the plan.
3. Compute the product $D = Np_t$.
4. Enter the body of Table 17-1 at the nearest value of D and read the corresponding value of f as the sum of the associated row and column headings.

    f = fraction of lot inspected.
5. The sampling plan is

    sample size = n = fN.

    acceptance number = c = 0.

    Sample size is always rounded up.

The plan is applied as follows:

1. *Randomly* sample n items from a lot of N items (i.e., sample a fraction f of the lot).
2. Reject if any defective units are found in the sample.

Protection

The use of the LSP plan as outlined provides LTPD protection to the consumer at the limiting fraction defective p_t specified. Specification of LTPD protection is equivalent to a reliability confidence coefficient of 90 percent. In other words, we can be 90 percent confident that a lot that has passed the plan has a fraction defective less than the value of p_t specified [or, equivalently, that it has a reliability of at least $(1 - p_t)$]. This statement is made in the sense that in repeated applications of the

TABLE 17-1 Values of $D = Np_t$ Corresponding to f

f	.00	.01	.02	.03	.04	.05	.06	.07	.08	.09
.9	1.0000	.9562	.9117	.8659	.8184	.7686	.7153	.6557	.5886	.5000
.8	1.4307	1.3865	1.3428	1.2995	1.2565	1.2137	1.1711	1.1286	1.0860	1.0432
.7	1.9125	1.8601	1.8088	1.7586	1.7093	1.6610	1.6135	1.5667	1.5207	1.4754
.6	2.5129	2.4454	2.3797	2.3159	2.2538	2.1933	2.1344	2.0769	2.0208	1.9660
.5	3.3219	3.2278	3.1372	3.0497	2.9652	2.8836	2.8047	2.7283	2.6543	2.5825
.4	4.5076	4.3640	4.2270	4.0963	3.9712	3.8515	3.7368	3.6268	3.5212	3.4196
.3	6.4557	6.2054	5.9705	5.7496	5.5415	5.3451	5.1594	4.9836	4.8168	4.6583
.2	10.3189	9.7682	9.2674	8.8099	8.3902	8.0039	7.6471	7.3165	7.0093	6.7231
.1	21.8543	19.7589	18.0124	16.5342	15.2668	14.1681	13.2064	12.3576	11.6028	10.9272
.0	*	229.1053	113.9741	75.5957	56.4055	44.8906	37.2133	31.7289	27.6150	24.4149

*For values of $f < .01$ use $f = 2.303/D$; for infinite lot size use sample size $n = 2.303/p_t$.

From Schilling (1978, p. 48). Reprinted by permission.

TABLE 17-2   Factors for Constructing the OC Curve

P_a	p	P_a	p	P_a	p	P_a	p
.999	.00043 p_t						
.995	.00218 p_t	.900	.046 p_t	.100	1.000 p_t	.005	2.300 p_t
.990	.0044 p_t	.750	.125 p_t	.050	1.301 p_t	.001	2.996 p_t
.975	.0110 p_t	.500	.301 p_t	.025	1.602 p_t		
.950	.0223 p_t	.250	.602 p_t	.010	2.000 p_t		

From Schilling (1978, p. 48). Reprinted by permission.

plan, lots that are composed of exactly p_t fraction defective would be rejected 90 percent of the time.

   To portray the probability of acceptance of the plan, it is possible
to approximate the Type B OC curve of the plan showing probability of acceptance P_a plotted against possible fractions defective p that could occur
in the manufacturing process from which the lot was taken. This may be
done using the factors given in Table 17-2 which, when multiplied by the
selected value of p_t, will give the approximate fractions defective associated with various probabilities of acceptance.

   Table 17-2 may also be used to approximate the well-known quantities
descriptive of the protection afforded by the plan such as indifference
quality, limiting quality, AQL (defined as having 95 percent probability of
acceptance), and so on, since these quantities are determined by probability
of acceptance. Furthermore, it provides the factors necessary to allow the
derivation of plans having probability of acceptance other than 10 percent
at the specified fraction defective.

   Suppose a plan was desired having approximately 5 percent probability
of acceptance for a specified fraction defective p*, that is, a plan that
would assure passing lots had at least 1 - p* reliability with 95 percent
confidence. Table 17-2 can be used to obtain such a plan as follows:

   1. Since Table 17-2 shows p = 1.301 p_t at P_a = .05, set p* = 1.301 p_t,
       and solve for p_t

       $$p_t = \frac{p*}{1.301}$$

   2. Use the value of p_t obtained to set up a sampling plan using the
       standard LSP procedure.

3. The resulting plan will have approximately P_a = .05 for fraction
   defective p*.

It should be noted that, for a stream of successive lots, the AOQL
can be approximated for LSP plans as

$$AOQL = \frac{.3679}{N}\left(\frac{1}{f} - 1\right).$$

Producer's Risk

Since acceptance is allowed only when no defectives are found in the sample
(c = 0), the producer must produce at a fraction defective that is less
than about 5 percent of the level, p_t, protected against by the plan in
order to assure a reasonably small probability (about 1 in 10 odds) of good
lots being rejected.

Clearly, a perfect lot has 100 percent probability of acceptance under
the LSP plan, since no defectives can be found in the sample. For such
lots the producer's risk of rejection is zero. A. J. Duncan (1977) has
shown that for lots containing only a single defective unit (i.e., lots of
fraction defective 1/N) the probability of acceptance is just

$$P_a = 1 - f$$

The corresponding producer's risk of such a lot being rejected is

$$1 - P_a = f$$

Thus, with a fraction of the lot inspected of f = .21 and lot size of 100,
as in Example 1, there is probability of acceptance of

$$P_a = 1 - .21 = .79$$

for lots containing a fraction defective

$$p = \frac{1}{100} = .01$$

and a corresponding producer's risk at that level of fraction defective of

$$1 - P_a = .21$$

This gives a minimum estimate of the producer's risk, since a lot
containing more than one defective unit would have a higher probability of
rejection. A. J. Duncan (1977, p. 6) has indicated that "computations of
producer's risks ... reveal that ... plans with zero acceptance numbers ...
can be hard on the producer unless most of his lots are perfect." It is
important to remember that single-sampling plans that require no defectives

in the sample for lot acceptance (such as LSP) should be used only when the state of the art permits near perfect quality levels to be economically produced.

Examples of LSP Applications

The following are examples of applications of the LSP plan.

EXAMPLE 1. A part is received at incoming inspection in lots of 100 items. Protection against a fraction defective of 10 percent is desired. The LSP plan is derived as follows:

1. $N = 100$
2. $p_t = .10$
3. $D = Np_t = 100(.10) = 10$
4. Table 17-1 gives $f = .21$ closest to $D = 10$
5. The sampling plan is

$$n = .21(100) = 21 \qquad c = 0$$

The plan is implemented in the following way:

1. Randomly sample 21 items from each lot of 100.
2. Reject the lot if any defectives are found; accept otherwise.

If rejected material is 100 percent inspected with rejected items replaced by good ones, the average outgoing quality limit is estimated as

$$AOQL = \frac{.3679}{100}\left(\frac{1}{.21} - 1\right) = .014$$

Also, from Table 17-2, it is possible to approximate other characteristics of the sampling plan, such as

1. Indifference quality (fraction defective having 50/50 chance of lot acceptance).

   $$IQ = .301p_t = .301(.10) = .0301$$

2. Limiting quality having probability of acceptance of 5 percent (fraction defective having 5 percent probability of acceptance).

   $$LQ = 1.301p_t = 1.301(.10) = .1301$$

3. Acceptable quality level (defined as fraction defective having 95 percent probability of acceptance).

   $$AQL = .0223p_t = .0223(.10) = .00223$$

This example illustrates the simplicity of calculation in deriving an LSP plan.

EXAMPLE 2. In bidding on a new contract, it is necessary to evaluate the consequences of quality requirements of 1 percent probability of acceptance at a fraction defective of 2 percent for products produced in lots of 100. Thus, $p^* = .02$, and

$$p^* = 2.0 p_t \quad \text{(from Table 17-2)}$$

$$p_t = \frac{p^*}{2.0} = \frac{.02}{2.0} = .01$$

The resulting sampling plan is derived as follows:

1. $N = 100$
2. $p_t = .01$
3. $D = N p_t = 100 (.01) = 1.0$
4. Table 17-1 gives $f = .90$ closest to $D = 1.0$
5. The sampling plan is

$$n = .90 (100) = 90 \qquad c = 0$$

This plan requires inspection of 90 percent of every lot, which may or may not be economically feasible. If this is the case, 100 percent inspection may be the only practical alternative. The LSP plan thus makes explicit the economic consequences of sampling in terms of the fraction of each lot to be inspected.

EXAMPLE 3. A lot of 10,000 items has been set aside for 100 percent inspection. It is uneconomical to inspect the lot if the fraction defective is 7 percent or more. Derive an LSP plan to test if 100 percent inspection is practical.

1. $N = 10,000$
2. $p_t = .07$
3. $D = N p_t = 10,000 (.07) = 700$
4. Since $f < .01$, use

$$f = \frac{2.303}{D} = \frac{2.303}{700} = .0033$$

5. The sampling plan is

$$n = fN = .0033 (10,000) = 33 \qquad c = 0$$

Further Considerations

As shown by Schilling (1978) derivation of the LSP plans is based on the
notion that, for the hypergeometric distribution, when c = 0 with k defectives pieces in the lot

$$P_a \leq \left(1 - \frac{n}{N}\right)^{Np}$$

Values of D for Table 17-1 are obtained as

$$D = \frac{\log P_a}{\log(1 - f)}$$

and, since

$$D = \frac{np}{f}$$

when the Poisson approximation applies

$$f = \frac{2.303}{D}$$

for a probability of acceptance of .10.

A useful set of curves for quick and easy assessment of LSP sample
sizes has been given by Hawkes (1979) which includes levels of consumer
protection of .01, .05, .10, .20, and .50. When a sample is required that
is less than 10 percent of the lot size (D > 22) and, of course, for con-
ceptually infinite populations, an excellent nomograph for use when c = 0,
based on the Poisson distribution, has been prepared by Nelson (1978).
This may be employed to obtain confidence limits on reliability from the
results of sampling inspection and to determine the sample size which will
give required protection.

TIGHTENED-NORMAL-TIGHTENED SCHEME (TNT)

While LSP plans are intended for application to unique lots, when product
is forthcoming in a stream of lots and a zero acceptance number is to be
maintained, the tightened-normal-tightened (TNT) scheme devised by Calvin
(1977) is particularly appropriate. This scheme utilizes two c = 0 sampling
plans of different sample size together with switching rules to build up the
shoulder of the OC curve after the manner of the switching rules of MIL-STD-
105D. This is done by a change in sample size rather than acceptance
number as in MIL-STD-105D. Calvin (1977) points out that, while increasing
producer protection, the switching rules have no real effect on LTPD which
remains essentially that of the tightened plan. Similar results were shown

by Schilling and Sheesley (1978) for MIL-STD-105D even when switching to reduced inspection was added. The procedure is as follows.

Procedure

A TNT scheme is specified by

n_1 = tightened (larger) sample size
n_2 = normal (smaller) sample size
t = criterion for switching to normal inspection
s = criterion for switching to tightened inspection

It is carried out as follows starting with tightened inspection

1. Inspect using tightened inspection with the larger sample size n_1, $c = 0$.
2. Switch to normal inspection when t lots in a row are accepted under tightened inspection.
3. Inspect using normal inspection with the smaller sample size n_2, $c = 0$.
4. Switch to tightened inspection after a rejection if an additional lot is rejected in the next s lots.

A diagrammatic representation of the switching rules for the TNT scheme is shown in Fig. 17-1.

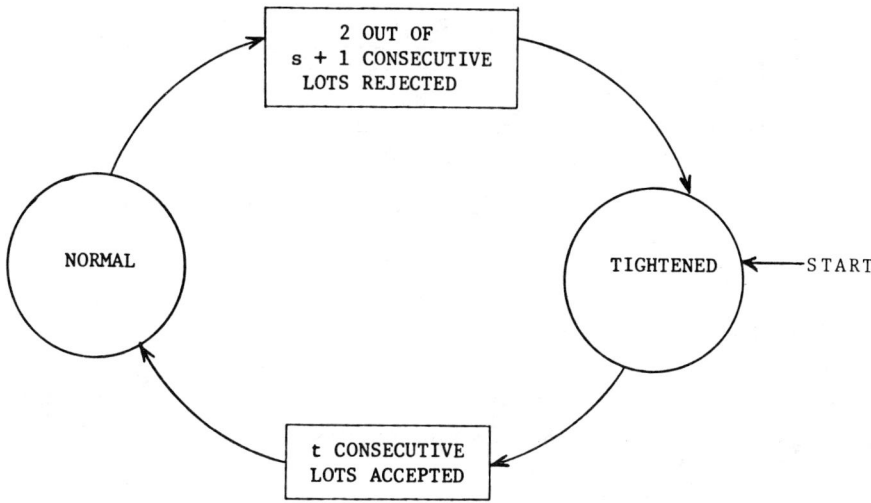

FIG. 17-1 Switching rules for TNT procedure.

Protection

The TNT plans correspond to the MIL-STD-105D normal-tightened plans when the switching criteria are set at $t = 5$, $s = 4$. In fact, TNT plans correspond directly to the MIL-STD-105D scheme (using normal-tightened switching only) when the normal plan has a zero acceptance number. For example, Code F, 0.65 percent AQL gives

$$\text{Normal:} \quad n = 20, \; c = 0$$
$$\text{Tightened:} \quad n = 32, \; c = 0$$

which correspond to the TNT plan with $t = 5$, $s = 4$, $n_1 = 32$, $n_2 = 20$. Calvin (1977) shows the scheme probability of acceptance of the TNT plan to be

$$P_a = \frac{P_1(1 - P_2^s)(1 - P_1^t)(1 - P_2) + P_2 P_1^t(1 - P_1)(2 - P_2^s)}{(1 - P_2^s)(1 - P_1^t)(1 - P_2) + P_1^t(1 - P_1)(2 - P_2^s)}$$

where

$$P_1 = (1 - p)^{n_1} = \text{probability of acceptance of tightened plan at}$$
$$\text{fraction defective } p$$

$$P_2 = (1 - p)^{n_2} = \text{probability of acceptance of normal plan at}$$
$$\text{fraction defective } p$$

The average sample number is

$$\text{ASN} = \bar{n} = \frac{n_1(1 - P_2^s)(1 - P_1^t)(1 - P_2) + n_2 P_1^t(1 - P_1)(2 - P_2^s)}{(1 - P_2^s)(1 - P_1^t)(1 - P_2) + P_1^t(1 - P_1)(2 - P_2^s)}$$

with average outgoing quality at fraction defective p:

$$\text{AOQ} = p P_a \left(\frac{N - \bar{n}}{N} \right) \quad \text{defectives replaced}$$

and

$$\text{AOQ} = \frac{p P_a (N - \bar{n})}{N - \bar{n}p - p(1 - P_a)(N - \bar{n})} \quad \text{defectives not replaced}$$

The improvement in the operating ratio of a TNT plan over that of its tightened component ($n_1 = n_2 = 20$) is shown in Fig. 17-2 taken from Calvin (1977).

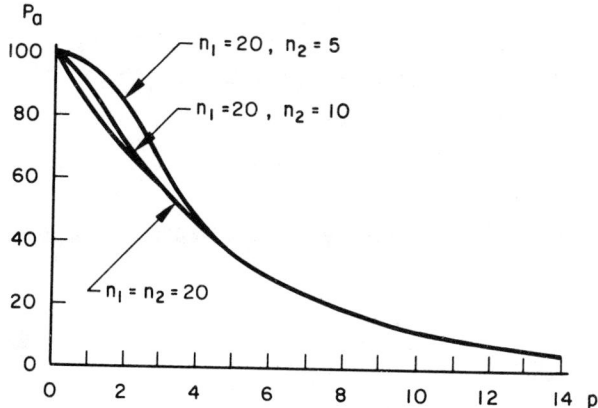

FIG. 17-2 Operating characteristic curves of TNT plans
(t = 5, s = 4). (Reproduced with permission from T. W.
Calvin, *TNT Zero Acceptance Number Sampling*, ASQC Thirty-
First Annual Technical Conference Transactions, p. 37.
Copyright(c)1977, American Society for Quality Control,
Inc.)

Selection

Schilling and Sheesley (1978, p. 78) have pointed out that "... the scheme
OC curve might be approximated by the normal OC curve for $P_a \geq 90$ percent
and the tightened OC curve for $P_a \leq 50$ percent with the intermediate
region appropriately interpolated." The suggestion is from their work on
MIL-STD-105D which includes switching to reduced inspection and so this
approximation should apply even better to the TNT plans. As an illustration, for the plan $n_1 = 20$, $n_2 = 5$, $t = 5$, $s = 4$, shown in Fig. 17-2, for
$p = .01$,

$$P_a = \frac{.8179(1 - .9510^4)(1 - .8179^5)(1 - .9510) + .9510(.8179^5)(1 - .8179)(2 - .9510^4)}{(1 - .9510^4)(1 - .8179^5)(1 - .9510) + (.8179^5)(1 - .8179)(2 - .9510^4)}$$

$$= .943$$

and for $p = .11$

$$P_a = \frac{.0972(1 - .5584^4)(1 - .0972^5)(1 - .5584) + .5584(.0972^5)(1 - .0972)(2 - .5584^4)}{(1 - .5584^4)(1 - .0972^5)(1 - .5584) + (.0972^5)(1 - .0972)(2 - .5584^4)}$$

$$= .097$$

For $n = 5$, $c = 0$ we have $p_{.95} = .01$ while for $n = 20$, $c = 0$, $p_{.10} = .11$. Thus, it would seem that, at least when using the analogous MIL-STD-105D switching criterion, a TNT plan can be derived using n_1 from the $c = 0$ plan having the desired LTPD and n_2 from a $c = 0$ plan having a producer's quality level at $P_a = .95$ equal to a specified value.

An easy way to find this value is to divide the $c = 0$ unity values for 10 and 95 percent probability of acceptance by the desired $p_{.10}$ and $p_{.95}$, respectively. Thus, the factors to obtain n_1 and n_2 are

$$n_1 p_{.10} = 2.303$$

$$n_2 p_{.95} = .0513$$

Further, the ratio of sample sizes will be

$$\frac{n_1}{n_2} = \frac{44.89}{R}$$

where R is the desired operating ratio. For example, to obtain a TNT plan having $p_{.10} = .11$ and $p_{.95} = .005$, we have

$$n_1 = \frac{2.303}{.11} = 20.9$$

$$n_2 = \frac{.0513}{.005} = 10.3$$

and, for the desired $R = 22$,

$$\frac{n_1}{n_2} = \frac{44.89}{22} = 2.04$$

Thus, the plans $n_1 = 20$, $c = 0$ having $p = .109$ at $P_a = .10$ and $n_2 = 10$, $c = 0$ having $p = .005$ at $P_a = .95$ would appear to suffice to give TNT: $t = 5$, $s = 4$, $n_1 = 20$, $n_2 = 10$. And they do as part of the Calvin (1977) tabulation. Thus, the approach, as suggested by Schilling and Sheesley can be used to quickly set up two-point TNT plans.

To find a TNT plan to match the plan $n = 20$, $c = 1$ which has $p_{.95} = .018$ and $p_{.10} = .18$ it is necessary to find $c = 0$ plans which have these probability points. Using the binomial distribution, they have sample sizes 3 and 12, respectively. Hence the plan is TNT: $t = 5$, $s = 4$, $n_1 = 12$, $n_2 = 3$. For this plan applied to lots of size $N = 100$ at $p = .018$:

$$P_a = \frac{\begin{pmatrix} .8042(1 - .9470^4)(1 - .8042^5)(1 - .9470) \\ + .9470(.8042^5)(1 - .8042)(2 - .9470^4) \end{pmatrix}}{\begin{pmatrix} (1 - .9470^4)(1 - .8042^5)(1 - .9470) \\ + (.8042^5)(1 - .8042)(2 - .9470^4) \end{pmatrix}}$$

$$= \frac{.8042(.006884) + .9470(.07875)}{(.006884) + (.07875)} = .936$$

$$\text{ASN} = \frac{12(.006884) + 3(.07875)}{.085634} = 3.72$$

$$\text{AOQ} = .018(.936)\left(\frac{100 - 3.72}{100}\right) = .016 \quad \text{with replacement}$$

$$\text{AOQ} = \frac{.018(.936)(100 - 3.72)}{100 - 3.72(.018) - .018(1 - .936)(100 - 3.72)} \quad \text{without replacement}$$

$$= .016$$

QUICK SWITCHING SYSTEM (QSS)

The TNT plans offer protection for the producer in situations in which a
zero acceptance number is required on a stream of lots. A similar procedure will be found in the quick switching system (QSS) proposed by H. F.
Dodge (1967) and studied extensively by L. D. Romboski (1969). The system
uses immediate switching to tightened inspection when a rejection occurs
under normal inspection. The QSS-1 plan involves an immediate switch back
to normal when a lot is accepted under tightened inspection. Other plans
allow a switch back to normal after two (QSS-2) or three (QSS-3) lots are
accepted under tightened inspection. Figure 17-3 shows how the quick

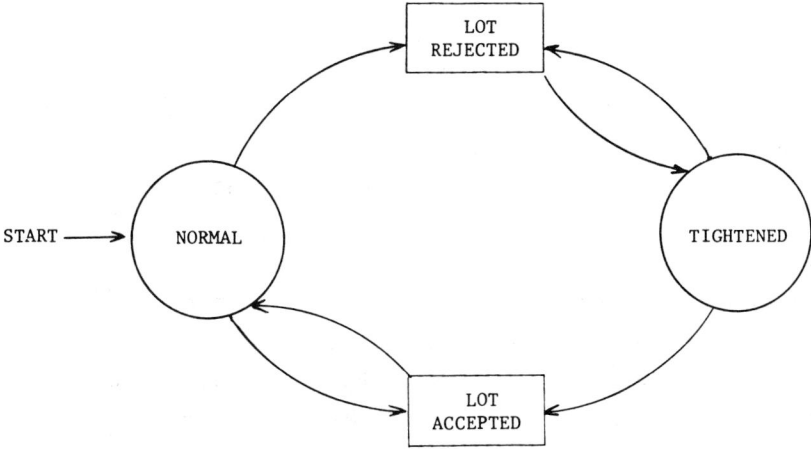

FIG. 17-3 Quick switching system.

switching system (QSS-1) is applied.

Romboski (1969) has tabulated unity values for a variety of QSS plans. A table for QSS-1 is given in Appendix Table T17-1. It shows acceptance numbers under normal, c_N, and tightened, c_T, inspection for a fixed sample size n on both tightened and normal. The unity values are used exactly as those for single, double, and multiple plans given earlier. For example, the plan n = 20, c_N = 1, c_T = 0 has an operating ratio R = 8.213 with

$$p_1 = \frac{np_{.95}}{20} = \frac{.308}{20} = .015$$

$$p_2 = \frac{np_{.10}}{20} = \frac{2.528}{20} = .126$$

and indifference quality

$$IQ = \frac{np_{.50}}{20} = \frac{1.146}{20} = .057$$

The value h_0 given in the table of unity values is the relative slope of the OC curve at the indifference quality level as defined by Hamaker (1950). Values of probability of acceptance for the individual normal P_N and the tightened P_T plans at the indifference quality level for the scheme are also given.

Note from the example that p_1 = .015 is approximately that of the plan n = 20, c = 1 which is .018. However p_2 = .126 approximates that of the plan n = 20, c = 0 which is .115. Thus, the QSS plan affects a favorable compromise in protection between its tightened and normal constituents.

Romboski has examined a variety of QSS plans including variations in normal and tightened sample sizes, acceptance numbers and switching rules. The TNT plans offer an ingenious application of a QSS procedure for the case when the acceptance number is restricted to zero. Thus, the QSS plans provide another vehicle for improvement of protection in situations in which small sample sizes are necessary but high levels of protection must be maintained.

SIMPLIFIED GRAND LOT PROCEDURE

Acceptance inspection and compliance testing often necessitate levels of protection for both the consumer and the producer that require large sample sizes relative to lot size. A given sample size can, however, be made to apply to several lots jointly if the lots can be shown to be homogeneous.

This reduces the economic impact of a necessarily large sample size. Grand lot schemes, as introduced by Simon (1941), can be used to affect such a reduction. The original grand lot scheme was later modified by Schilling (1979) to incorporate graphical analysis of means procedures in verifying the homogeneity of a grand lot. The resulting approach can be applied to attributes or variables data, is easy to use, provides high levels of protection economically, and can reduce sample size by as much as 80 percent. It may be applied to unique "one-off" lots, isolated lots from a continuing series, an isolated sequence of lots, or to a continuing series of lots.

MIL-STD-105D indicates that lots or batches should be formed in such a way that "each lot or batch shall, as far as is practicable, consist of units ... manufactured under essentially the same conditions ... ." This suggests that one way to reduce sample size relative to lot size is to increase the size of the lot as much as possible within limits of homogeneity of the product included therein. The grand lot scheme as originated by Simon, utilizes the power of the control chart to achieve a drastic reduction in sample size relative to lot size in the application of an acceptance sampling plan to homogeneous material.

As defined by Simon (1944), "A lot is an aggregation of articles which are essentially alike." The control chart is used to distinguish members of what may be considered a grand lot from a collection of sublots. Thus, the chart becomes the criterion for what is "alike" in Simon's definition. Use of a large sample size on the grand lot allows a precision in sampling which would often prove uneconomical if applied to each sublot separately. Thus, use of plans having low discrimination, such as c = 0 attributes plans, can be avoided.

For example, if the protection desired requires a sample size of 800 on each of 8 sublots, a total of 6400 units would be inspected. If the 8 sublots were sufficiently alike to allow aggregation into a single grand lot, the resulting sample of 800 on the grand lot would amount to an 87.5 percent reduction in sample size *with no decrease in protection*. In this way, very stringent levels of quality can be assured with great economy.

The concept of the grand lot has application in a number of areas of acceptance sampling:

1. Verification of the quality of large quantities of material on
    an acceptance sampling basis to very stringent quality levels
    (e.g., sampling for safety related defects when sample sizes

required are prohibitively large when applied to sublots).

2. Sampling quarantined or returned material to distinguish aberrant sublots and to determine disposition of the material (e.g., sampling many skids of material which have been rejected by a customer as unsatisfactory when it is believed that only a few skids may, in fact, be nonconforming).

3. Acceptance sampling of unique lots (sometimes called *one-off* lots) or of isolated lots from a continuing series, when the lot to be inspected may logically be divided into a set of sublots (e.g., acceptance of an order which is comprised of material produced by several different identifiable production units; the order may or may not be part of a continuing series).

4. Acceptance sampling of an isolated sequence of sublots which may logically be aggregated into a grand lot (e.g., acceptance of a week's production on the basis of results from individual days (or shifts)).

5. Acceptance sampling of a continuing series of sublots which can be shown to consistently comprise a grand lot (e.g., acceptance on an open order from a captive supplier).

Not only is the procedure useful in routine inspection of individual lots or series of lots of material, but it is also especially valuable in surveillance inspection and in compliance testing. The experience of the U.S. Army, Chemical Corps Material Command, in this regard has been set forth by Mandelson (1963).

The original grand lot scheme, as proposed by Simon, utilized a control chart approach to identifying the grand lot, but was complicated by the necessity of including extra procedures for assessing the compound probabilities associated with a simultaneous comparison of many points (representing the sublots) against the limits. Conventional control chart limits are set up to be used one point at a time. While Burr (1953) modified application of the approach somewhat by using approximations in place of Simon's original use of the incomplete beta function, the extra steps incurred by this part of the method remained. The procedure was simplified still further by Schilling (1979) by providing graphical control chart limits which automatically account for these compound probabilities and by incorporating an identical approach for both variables and attributes data. The simplified procedure is based on use of the analysis of means limits developed by Ott (1967), Ott and Lewis (1960), and Schilling (1973, 1973a).

Use of these limits retains the simplicity of the control chart without recourse to additional steps since they are designed to maintain specified probability levels when many points are compared to the limits.

Simon's Approach

The approach suggested by Simon (1941) is essentially as follows:

1. Determine an appropriate sublot sample size.
2. Sample the sublots.
3. Plot the sample results on a control chart of the form shown in Fig. 17-4.
4. Identify any points which plot beyond the extreme limits (E) as "extreme suspected mavericks." Eliminate these sublots, as outliers, from further consideration as part of the grand lot. Treat the eliminated lots separately, applying an appropriate sampling plan to each. Recompute the limits until there are no further extreme suspected mavericks outside the extreme limits (E).
5. Identify any points outside the maverick limits (M) as "suspected mavericks." Utilize the incomplete beta function to determine if the number of maverick points is significantly large on the basis of the compound probabilities inherent in conventional control limits. If it is, reject the grand lot hypothesis. (Note, this step is unnecessary when using analysis of means limits on the control chart, since a single point beyond the maverick limits is sufficient to reject the grand lot hypothesis when such limits are used.)

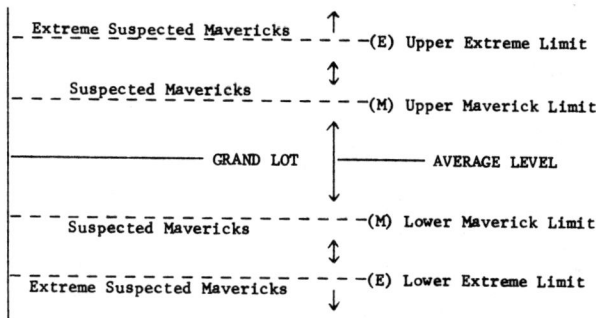

FIG. 17-4 Simon grand lot chart. (From Schilling, 1979, p. 117. Reprinted by permission.)

6. If the grand lot hypothesis is rejected, test each sublot sepa-
    rately using an appropriate sampling plan. This plan would normally
    be the same as that applied to the grand lot if the grand lot
    hypothesis had been accepted.

7. If the grand lot hypothesis is accepted, combine the sublots not
    determined to be "extreme suspected mavericks" into a grand lot
    and apply an acceptance sampling plan sufficient to give the con-
    sumer and the producer the protection desired, taking additional
    samples as necessary to complete the sample size required.

This procedure allows application of a very discriminating sampling
plan to lots made as large as possible. The sampling plan applied to the
grand lot should afford at least LTPD protection to the consumer with due
consideration for the producer's risk as evidenced by the OC curve. Application to large lots allows higher acceptance numbers to be used with
larger sample sizes which leads to better protection for both parties.

In discussing the grand lot approach, Simon (1941) points out that, if
the grand lot hypothesis is rejected, "... the grand lot judge is called
upon to revise his grand-lot grouping, if a logical basis for re-grouping
exists, or the grand lot must be abandoned and resort made to individual
sampling." This provides greater flexibility in application, however, re-
grouping should be allowed only on a documented rational basis and only
with the concurrence of both parties to the acceptance decision — the
producer and the consumer. Simon (1941) also states that "Very good grand
lot judgments are desirable but not essential to the operation of the
system, as very poor ones will almost inevitably be caught. Poor grand lot
judgments result in retesting ... and serve to decrease the efficiency of
the system."

Simplified Procedure: Attributes

Given a presumptive grand lot made up of k sublots, it is desired to obtain
LTPD protection against a process fraction defective p_t.

1. Determine the sublot sample size[*], n, as

[*]As suggested by Simon, this gives roughly 90 percent probability of obtaining at least one defective in the sublot sample if the process fraction
defective is, in fact, p_t. Alternatively, sublot sample size may be determined by sampling a fraction of the sublot, as obtained from a lot sensitive sampling plan. This relates sample size directly to sublot size and
maintains protection equivalent to the formula given above, with slightly
smaller samples. For large sublots, the results will be the same for both
approaches.

$$n = \frac{2.303}{p_t}$$

Round up.

2. Sample n items from each sublot and determine $\bar{p}$, the estimated fraction defective from the nk units sampled, as

$$\bar{p} = \frac{X}{N}$$

where

        X = total number nonconforming

        N = nk = total sample size

3. Construct an analysis of means chart in the form of Fig. 17-4 where

   a. Extreme limits are set at

$$E:\ \bar{p} \pm H_{.002} \sqrt{\frac{\bar{p}(1 - \bar{p})}{n}}$$

   b. Maverick limits are set at

$$M:\ \bar{p} \pm H_{.05} \sqrt{\frac{\bar{p}(1 - \bar{p})}{n}}$$

     using the H_{α} factors from Appendix Table T17-2.

4. Eliminate any sublot whose sample proportion plots beyond extreme limits (E) from further consideration as part of the grand lot. Dispose of such lots separately using an appropriate sampling plan. Sublots which plot below the lower extreme limit, however, may be accepted if the grand lot is accepted.

5. Recompute limits on the remaining points until all extreme suspected mavericks have been eliminated. Then, if any remaining points plot beyond the maverick limits (M), reject the grand lot hypothesis and test each sublot individually using an appropriate plan on each.

6. If all points plot within the maverick limits (M), accept the grand lot hypothesis and group the remaining sublots into a grand lot. Apply a standard sampling plan to the grand lot to obtain the required LTPD protection, using a sample size-acceptance number combination which will afford reasonable protection for the producer. Take additional samples as necessary to complete the required sample size.

Example: Attributes

Suppose a shipment consisting of 12 cartons, each containing 5000 parts, for
use in an assembly operation, is presented for incoming inspection. The
production process can tolerate 2.5 percent defective, but quality of 6 percent or more must be rejected. Inspection is on a go no-go basis. A grand
lot plan is to be used with p_t = .06.

1. The sublot sample size is determined to be

$$n = \frac{2.303}{.06} = 38.4 \sim 40$$

2. Sample results are as follows. (Data adapted from Simon (1941).)

Carton	Sample size	Defectives	Proportion defective
1	40	1	.025
2	40	2	.05
3	40	2	.05
4	40	5	.125
5	40	0	0
6	40	4	.10
7	40	3	.075
8	40	1	.025
9	40	7	.175
10	40	2	.05
11	40	1	.025
12	40	1	.025
Total	480	29	.06

3. Limits are set at

$$E: \ .06 \pm 3.60 \sqrt{\frac{.06(.94)}{40}}$$

$$.06 \pm .135$$

$$0 \text{ to } .195$$

$$M: \ .06 \pm 2.74 \sqrt{\frac{.06(.94)}{40}}$$

$$.06 \pm 103$$

$$0 \text{ to } .163$$

and the resulting analysis of means chart is shown in Fig. 17-5.

4. No sublots are identified as "extreme suspected mavericks" since

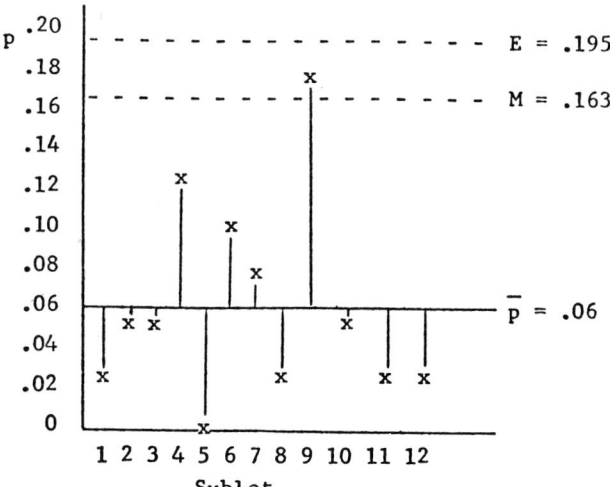

FIG. 17-5 Analysis of means chart - attributes data.
(From Schilling, 1979, p. 120. Reprinted by permission.)

      none plot beyond the extreme limit (E). It is, therefore, unnecessary to recompute the limits.

5.   Sublot 9 is identified as a "suspected maverick" since it plots beyond the maverick limit (M). The grand lot hypothesis is rejected and each sublot must be inspected separately. Using MIL-STD-105D, it is found from Table VI-A that for an AQL of 2.5 percent and an LTPD of 5.6 percent, the plan n = 500, c = 21 will give the desired protection on an isolated lot. Accordingly, an additional sample of 460 must be taken from each sublot and the plan applied to the samples of 500.

6.   If the grand lot hypothesis had been accepted, results from the individual lots could be aggregated. A further sample of 20 would be taken at random from the total shipment to reach the sample size of 500 necessary for application of the MIL-STD-105D plan.

It is interesting to note that rejection of the grand lot hypothesis resulted in inspection of 6000 units; whereas if the grand lot hypothesis had been accepted, inspection of only 500 units would have been required. This could have resulted in a 92 percent decrease in inspection effort; however, the procedure identified the lack of homogeneity of the cartons, making aggregation deceptive and unwarranted.

Simplified Procedure: Variables

Given a presumptive grand lot made up of k sublots, it is desired to apply a variables sampling plan for a measurement characteristic, X.

1. Determine the sublot sample size n as

$$n = \frac{120}{k} + 1$$

Round up. In no case sample less than five items from a sublot.

2. Sample n items from each sublot and compute $\overline{X}$ and s from each as follows:

$$\overline{X}_j = \frac{1}{n} \sum_{i=1}^{n} X_{ij}$$

$$s_j = \sqrt{\frac{\sum_{i=1}^{n} \left(X_{ij} - \overline{X}_j\right)^2}{n - 1}}$$

where X_{ij} is the ith observation of the measurement characteristic from the jth sublot, and $\overline{X}_j$ and s_j are the sample mean and standard deviation of the jth sublot. Also obtain

$$\overline{\overline{X}} = \frac{1}{k} \sum_{j=1}^{k} \overline{X}_j$$

$$\hat{s} = \sqrt{\frac{1}{k} \sum_{j=1}^{k} s_j^2}$$

3. Construct an analysis of means chart for s as in Fig. 17-4 where
   a. Extreme limits are set at

$$E: \hat{s} \pm H_{.002} \frac{\hat{s}}{\sqrt{2n}}$$

   b. Maverick limits are set at

$$M: \hat{s} \pm H_{.05} \frac{\hat{s}}{\sqrt{2n}}$$

   using the H_α factors of Appendix Table T17-2.

4. Eliminate any sublot whose standard deviation plots beyond the extreme limits (E) from further consideration as part of the grand lot. Dispose of such sublots separately using an appropriate sampling plan. Sublots which plot below the lower extreme limit, however, may be accepted if the grand lot is accepted, provided they are not disqualified on the basis of their mean.

5. Recompute limits on the remaining points until all extreme suspected mavericks have been eliminated. Then, if any remaining points plot beyond the maverick limits (M), reject the grand lot hypothesis and test each sublot individually using an appropriate sampling plan.

6. If all remaining points plot within the maverick limits (M), accept the grand lot hypothesis for the standard deviations and proceed to test the means against the grand lot hypothesis.

7. For the sublots not eliminated as extreme suspected mavericks in testing the standard deviations and using their estimated grand standard deviation, $\hat{s}$, and grand mean, $\bar{\bar{X}}$, plot an analysis of means chart with

   a. Extreme limits set at

   $$E: \bar{\bar{X}} \pm H_{.002} \frac{\hat{s}}{\sqrt{n}}$$

   b. Maverick limits set at

   $$M: \bar{\bar{X}} \pm H_{.05} \frac{\hat{s}}{\sqrt{n}}$$

8. Eliminate any sublot whose mean plots beyond extreme limits (E) from further consideration as part of the grand lot. Dispose of such lots separately. Recompute limits on the remaining points. Do not recompute limits for testing s against the grand lot hypothesis, however.

9. Recompute limits on the remaining points until all extreme suspected mavericks have been eliminated. Then, if any remaining points plot beyond maverick limits (M), reject the grand lot hypothesis and test each sublot individually using an appropriate sampling plan.

10. If all points plot within the maverick limits (M), accept the grand lot hypothesis and group the remaining sublots into a grand lot. Apply a standard sampling plan to the grand lot to obtain desired protection. Take additional samples as necessary to complete the required sample size.

Example: Variables

An arms wholesaler receives 30 consecutive lots of rounds of ammunition. These lots are to be tested for muzzle velocity. Specifications require

an individual round to be in the range 1670 to 1790 ft/sec. A grand lot
plan is to be employed.

1. The sublot sample size is

$$n = \frac{120}{30} + 1 = 5$$

2. Sample results are as follows (adapted and modified for illustra-
    tive purposes from Simon (1941, p. 367)).

Lot	$\overline{X}$	s	Lot	$\overline{X}$	s
1	1711	16.9	16	1783	20.6
2	1711	16.1	17	1777	3.6
3	1713	15.7	18	1794	6.0
4	1718	10.5	19	1773	14.9
5	1735	4.0	20	1789	21.8
6	1739	10.1	21	1798	6.0
7	1723	15.7	22	1789	11.7
8	1741	6.0	23	1788	15.7
9	1738	4.4	24	1799	12.1
10	1725	12.5	25	1807	17.7
11	1731	10.1	26	1784	4.3
12	1721	7.7	27	1775	15.7
13	1719	17.3	28	1787	12.8
14	1735	15.7	29	1770	6.1
15	1741	5.9	30	1796	19.7

$$\overline{\overline{X}} = \frac{52710}{30} = 1757$$

$$\hat{s} = \sqrt{\frac{5123.33}{30}}$$

$$= 13.1$$

3. Limits for s are set at

$$E: \quad 13.1 \pm 3.92 \sqrt{\frac{13.1}{10}}$$

$$13.1 \pm 16.2$$

$$0 \text{ to } 29.3$$

$$M: \quad 13.1 \pm 3.09 \sqrt{\frac{13.1}{10}}$$

$$13.1 \pm 12.8$$

$$0.3 \text{ to } 25.9$$

and the resulting analysis of means chart is shown in Fig. 17-6.

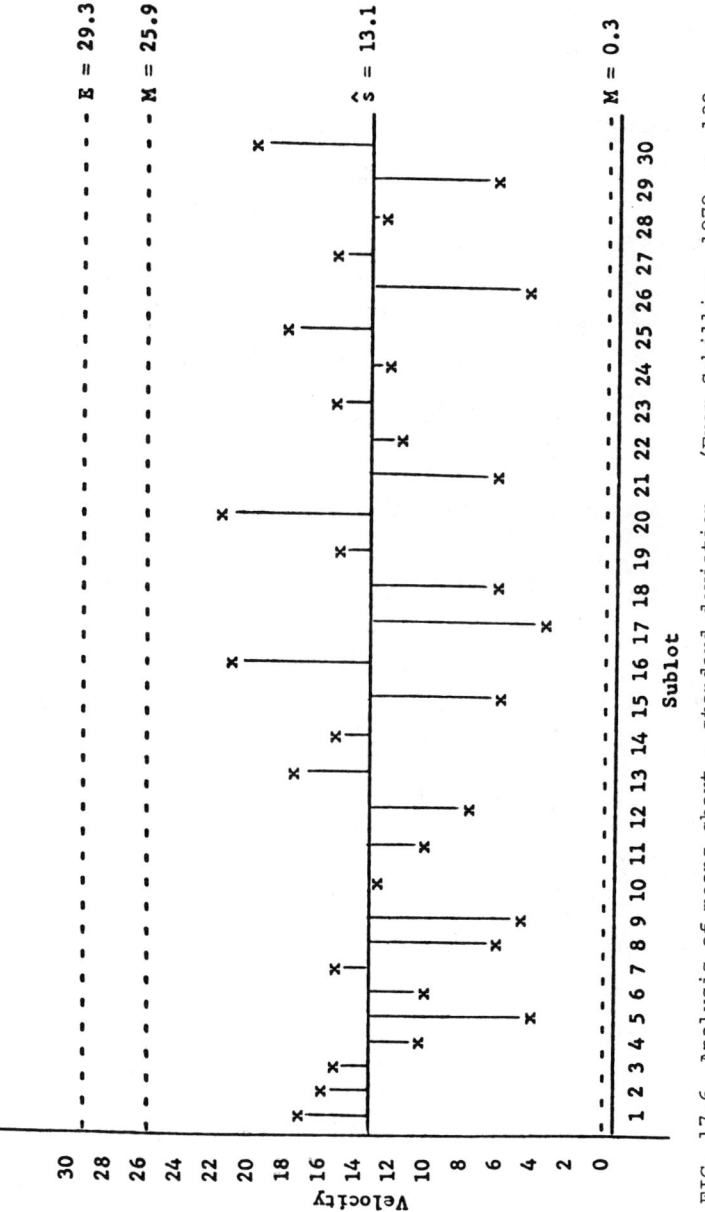

FIG. 17-6 Analysis of means chart - standard deviation. (From Schilling, 1979, p. 122. Reprinted by permission.)

4. The analysis of means plot for s shows no extreme suspected mavericks, so the limits need not be recomputed.

5. There are no suspected maverick lots on the basis of the analysis
   of means plot for s.

6. The grand lot hypothesis is accepted for standard deviations, and
   so the means are analyzed next.

7. Limits for $\overline{X}$ are set at

$$E: 1757 \pm 3.92 \frac{13.1}{\sqrt{5}}$$

$$1757 \pm 23.0$$

$$1734.0 \text{ to } 1780.0$$

$$M: 1757 \pm 3.09 \frac{13.1}{\sqrt{5}}$$

$$1757 \pm 18.1$$

$$1738.9 \text{ to } 1775.1$$

and the resulting analysis of means chart is shown in Fig. 17-7.

8. The analysis of means chart immediately shows a shift at sublot
   16. Twenty-one of the thirty points are extreme suspected mavericks.

9. Clearly, the grand lot hypothesis must be rejected. However, it is
   also evident that the shipment may be composed of two potential
   grand lots consisting of sublots 1-15 and 16-30, respectively.

10. The limits may be recalculated for these two groups as follows:
    Lots 1-15 (k = 15), $\overline{\overline{X}}$ = 1726.7

$$E: 1726.7 \pm 3.69 \frac{13.1}{\sqrt{5}}$$

$$1726.7 \pm 21.6$$

$$1705.1 \text{ to } 1748.3$$

$$M: 1726.7 \pm 2.84 \frac{13.1}{\sqrt{5}}$$

$$1726.7 \pm 16.6$$

$$1710.1 \text{ to } 1743.3$$

Lots 16-30 (k = 15), $\overline{\overline{X}}$ = 1787.3

$$E: 1787.3 \pm 3.69 \frac{13.1}{\sqrt{5}}$$

$$1787.3 \pm 21.6$$

$$1765.7 \text{ to } 1808.9$$

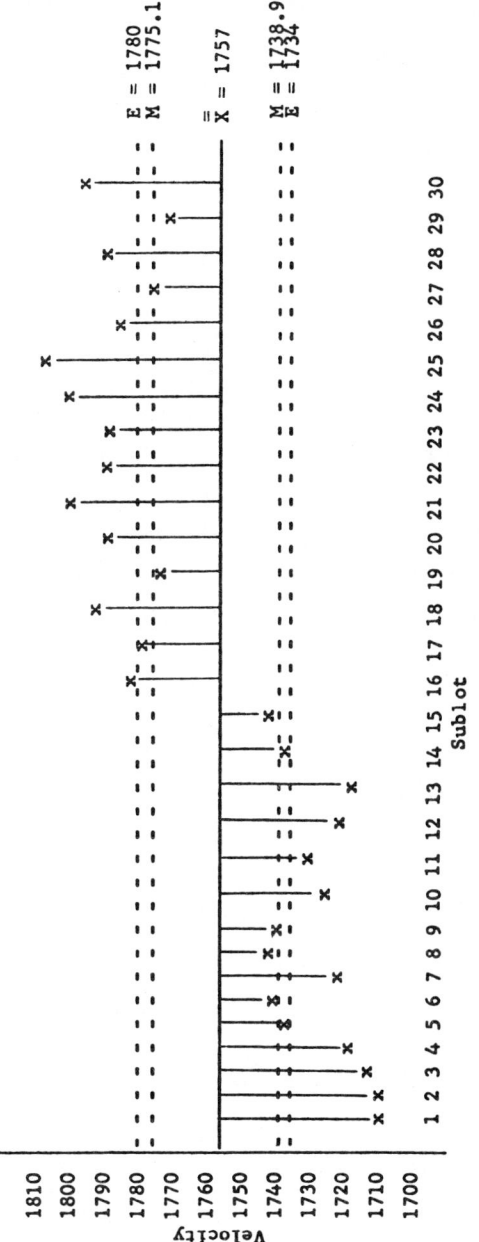

FIG. 17-7 First analysis of means chart - means. (From Schilling, 1979, p. 122. Reprinted by permission.)

$$M:\ 1787.3 \pm 2.84\ \sqrt{\frac{13.1}{5}}$$

$$1787.3 \pm 16.6$$

$$1770.7 \text{ to } 1803.9$$

The resulting analysis of means plot is shown in Fig. 17-8.

11. Figure 17-8 reveals that sublots 1-15 can be considered a grand lot, aggregated, and tested accordingly. Sublots 16-30, however, cannot be considered to form a grand lot since lots 25 and 29 are suspected mavericks.

12. Assuming normality of the underlying distribution of measurements, a variables plan from MIL-STD-414 may be selected from the OC curves to give an AQL of 0.1 percent and a consumer quality level of 1.0 percent with 10 percent probability of acceptance. Such a plan is Code N, 0.1 AQL, with standard deviation unknown. This requires a sample size of $n = 75$ with an acceptance constant $k = 2.66$, so no additional samples are needed. Standard variables acceptance procedures may then be applied separately to both specification limits, 1670 and 1790, respectively, since they are estimated to be more than 9 standard deviations apart, allowing a maximum standard deviation of 20.88. The acceptability criterion is

$$\frac{U - \overline{X}}{\hat{s}} > k$$

$$\frac{1790 - 1726.7}{13.1} > 2.66$$

$$4.83 > 2.66$$

and

$$\frac{\overline{X} - L}{\hat{s}} > k$$

$$\frac{1726.7 - 1670}{13.1} > 2.66$$

$$4.33 > 2.66$$

The acceptability criterion is met and so the grand lot consisting of sublots 1-15 is accepted. Note that the remaining lots 16-30 must be inspected separately, requiring an additional sample of 1050 if equivalent protection is to be maintained on each of them. This illustrates the leverage possible from the formation of a grand lot.

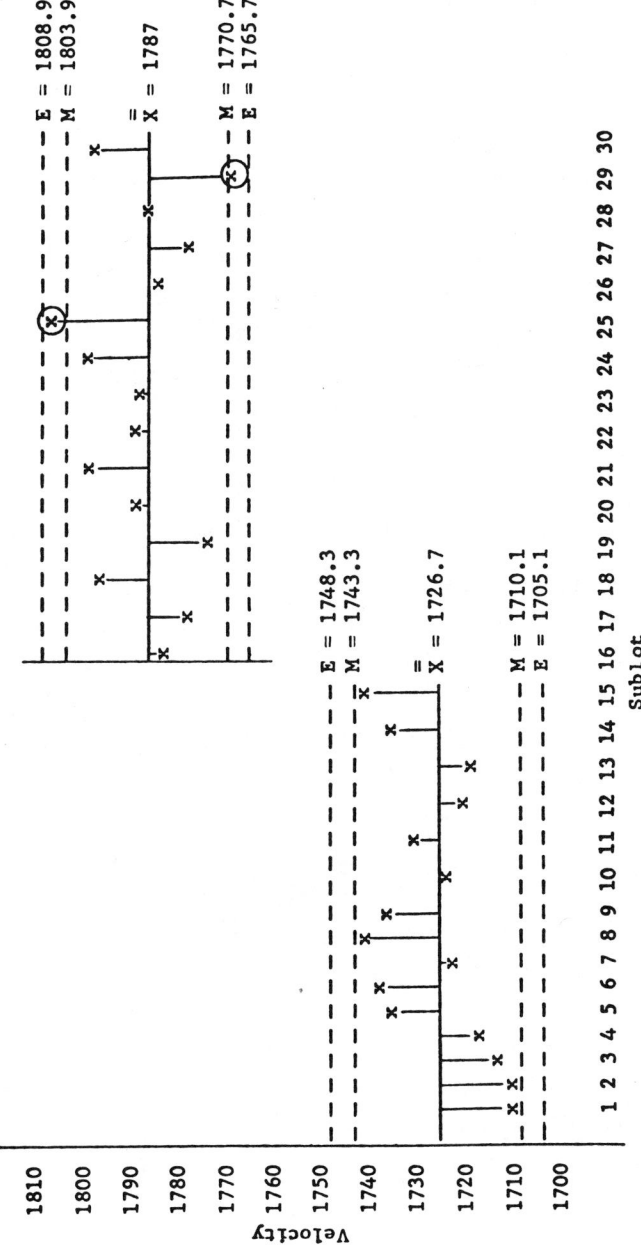

FIG. 17-8 Second analysis of means chart - means. (From Schilling, 1979, p. 123. Reprinted by permission.)

Continuing Series of Lots

In introducing the grand lot plan, Simon (1941, p. 33) pointed out that it can easily be applied to a continuing series of lots. He suggested the following approach:

> From the first few lots, or at least from the first lot, one must take a large sample in order to have a reliable estimate of the manufacturer's general level of quality as measured by the lot fraction defective ... From then on, one can treat his successive lots as additional members of the grand lot, testing each suspected maverick by a large sample to see if its quality is really satisfactory. However, the occurrence of an extreme suspected maverick or an excessive number of suspected mavericks should result in terminating the manufacturer's grand lot and in making him qualify all over again.

This can be carried out using the simplified procedure as follows:

1. Qualify the first 10 lots* using a standard sampling plan with a sample size-acceptance constant combination sufficient to protect both the consumer and the producer. Sample size for the qualification must equal or exceed that determined from the sublot sample size formulas given in the simplified method for attributes or for variables with k = 10.

2. Test the grand lot hypothesis on the first 10 lots using an analysis of means plot, as set forth in the simplified procedure.

3. If any of the 10 lots fail the standard sampling plan, or if the grand lot hypothesis is rejected, the producer must requalify subsequent lots.

4. If the grand lot hypothesis is accepted, construct a control chart as in Fig. 17-4 using probability limits to test subsequent lots. Use the overall values of $\bar{p}$ for attributes or $\bar{\bar{x}}$ and $\hat{s}$ for variables obtained from the ten qualification lots to set up the limits. Sample size from each subsequent lot is

$$\text{Attributes: } n = \frac{2.303}{p_t} \qquad \text{Variables: } n = 5$$

---

*This conforms to the criterion for switching from normal to reduced inspection under MIL-STD-105D and with the control chart approach of MIL-STD-105A.

Limits are set using the formulas given in the simplified procedure employing standard control limits, so that:

$$H_{.002} = 3.09$$
$$H_{.05} = 1.96$$

5. Lots which plot within the extreme limits are accepted. Lots which plot outside the extreme limits must be tested individually using an appropriate acceptance sampling plan.

6. The producer must requalify if any lot plots outside the extreme limits or if two out of any successive five points plot outside the maverick limits[*] in an undesirable direction.

Example: Continuing Series of Lots

Consider the attributes data given earlier. Suppose these constitute the next 12 from a continuing series of lots. The producer and the consumer agree to use an AQL of 2.5 percent and an LTPD of 6 percent. Table VI-A of MIL-STD-105D shows that for isolated lots, the plan n = 500, c = 21 is appropriate. Suppose sample results on the preceding 10 lots were

Lot	Sample size	Number defective	Proportion defective
-9	500	15	.03
-8	500	10	.02
-7	500	13	.026
-6	500	18	.036
-5	500	15	.03
-4	500	12	.024
-3	500	15	.03
-2	500	13	.026
-1	500	20	.04
0	500	19	.038
Total	5000	150	.03

1. All 10 qualification lots pass the standard plan.

2. The analysis of means plot to test the grand lot hypothesis has limits.

[*]This is essentially the same as the criterion for switching to tightened inspection under MIL-STD-105D.

$$E: \quad .03 \pm 3.53 \sqrt{\frac{.03(.97)}{500}}$$

$$.03 \pm .027$$

$$.003 \text{ to } .057$$

$$M: \quad .03 \pm 2.66 \sqrt{\frac{.03(.97)}{500}}$$

$$.03 \pm .020$$

$$.010 \text{ to } .050$$

and the resulting analysis of means chart is shown in Fig. 17-9.

3. The analysis of means plot shows the grand lot hypothesis is accepted, and since the 10 lots passed the standard plan, a control chart can be instituted.

4. Subsequent lots are sampled using a sample size of

$$n = \frac{2.303}{.06} = 38.4 \sim 40$$

and the resulting fractions defective plotted on a control chart with limits

$$E: \quad .03 \pm 3.09 \sqrt{\frac{.03(.97)}{40}}$$

$$.03 \pm .083$$

$$0 \text{ to } .113$$

$$M: \quad .03 \pm 1.96 \sqrt{\frac{.03(.97)}{40}}$$

$$.03 \pm .053$$

$$0 \text{ to } .083$$

The control chart for lots 1-4 is also shown in Fig. 17-9.

5. Lot 4 plots outside the extreme limit and so must be subjected to further testing.

6. The grand lot hypothesis is rejected at lot 4 and the producer would now have to requalify from the beginning of the procedure.

Further Considerations

Simon's (1941) original approach to testing the grand lot hypothesis after the extreme suspected mavericks were eliminated was as follows:

1. Prepare a control chart for the property being tested (p, s, or $\overline{X}$) with a specified probability, Q, of exceeding the maverick limits.

2. Count the actual number of points outside the maverick limits.

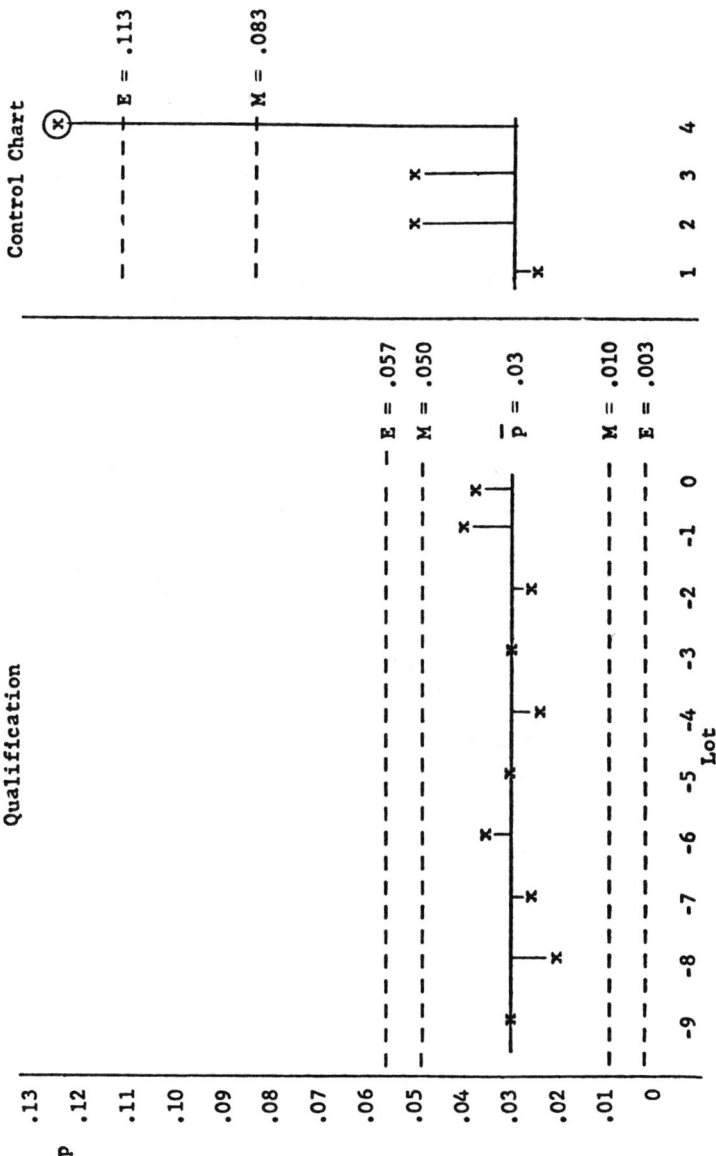

FIG. 17-9 Charts for continuing series of lots. (From Schilling, 1979, p. 125.
Reprinted by permission.)

3. Test the null hypothesis that the probability of exceeding the
    limits is equal to that specified, against an alternate hypothesis
    that it is greater. This is done by comparing the actual number
    of exceedances found against a critical value obtained from the
    incomplete beta function with parameter Q, sample size k, and
    Type I risk $\alpha = .10$.

4. Accept or reject the grand lot hypothesis as this null hypothesis
    is accepted or rejected.

This part of the procedure was intended to account for the degradation of
the Type I risk of a control chart as an increasing number of points is
compared to the limits. It is well known, for example, that if the probability that one point exceeding the limits is .05, the probability of an
exceedance in two points plotted is

$$1 - (.95)^2 = .0975$$

in three points is

$$1 - (.95)^3 = .1426$$

and so on.

Schilling (1979) pointed out that the advent of analysis of means procedures for constructing decision limits, which adjust the limits to take
account of the number of points involved in the comparison, made this part
of the original procedure unnecessary. The operation of the grand lot plan
was greatly simplified by straightforward comparison of the points against
the analysis of means limits. The H_α factors are upper bounds for the
studentized maximum absolute deviate as derived by Halperin, Greenhouse,
Cornfield and Zalokar (1955) which were incorporated into the analysis of
means procedure by Ott (1967), Ott and Lewis (1960), and Schilling (1973,
1973a). They were computed using a result attributed to Tukey (1953) in
the manner described by Schilling (1973). As such, the limits apply regardless of any correlation which might exist between the lots. The values
presented are for a Type I risk of .002 and .05, respectively. This corresponds to the British system of probability limits for control charts, as
presented, for example, by Pearson (1935), which employ these levels of
risk for action and warning limits. The risks are very close to those used
by Simon (1941) in his approach to variables data. They are also used in
the simplified procedure with attributes data for reasons of consistency

and uniformity.

The attributes sample size formula for sublots was chosen to conform with Simon's original recommendation to allow a 90 percent probability for at least one defect to occur when product quality is at the critical level. The formula is based on the Poisson distribution and so is conservative in cases where the binomial or hypergeometric distributions should apply. Chart limits for attributes employ the normal approximation to the binomial distribution but should be adequate in practice.

Sample size for variables data was chosen to give an estimate of the standard deviation with 120 degrees of freedom. A study of tables of H_α indicates that 120 degrees of freedom is sufficient to allow use of H_α factors for standard deviation known (i.e., using df = ∞ as an approximation). This corresponds to the practice of starting a control chart for subgroups of five after 30 points have been plotted (120 df) and using the estimate of the standard deviation as if it were known. In this way, one set of H_α factors could be presented for use with both variables and attributes data. Note that the values given in Appendix Table T17-2 are for infinite degrees of freedom and correspond to values of the extreme standardized deviate from the sample mean.

The method for dealing with a continuing series of lots was, of course, initially suggested by Simon (1941). A straightforward analysis of means is performed on the first ten lots since they are simultaneously compared to the decision limits. Thereafter, a control chart with British probability limits is utilized since subsequent lots will be compared to the limits individually, one at a time.

The simplified grand lot procedure suggested above can easily be modified to correspond to operating conditions. Sublots can be recombined to form new presumptive grand lots, where justified, when the grand lot hypothesis is rejected. Risks can be altered as appropriate and analysis of means limits computed using factors such as those given by Nelson (1974) for Schilling's h_α at other probability levels. These may be converted to values of H_α by the relation

$$H_\alpha = h_\alpha \sqrt{\frac{k-1}{k}}$$

While analysis of means is relatively insensitive to nonnormality, as shown by Schilling and Nelson (1976), special procedures are available for analysis of means when the variate departs substantially from the normal

form and have been given in Schilling (1973a). Also, the assumption of
shape in a variables sampling plan can be minimized by performing the
analysis of means procedure with variables data and using attributes to
inspect lots not belonging to the grand lot, or indeed, to inspect the
grand lot itself. Thus, the grand lot scheme can employ mixed variables-
attributes procedures as presented by Schilling and Dodge (1969).

The grand lot approach has great potential for increasing the efficiency and economy of acceptance sampling. The simplified graphical procedure facilitates its use in achieving the wide application Simon intended.

PROCEDURE FOR DETERMINING VALIDITY OF INSPECTION : H-109

Modern quality control practice recognizes the importance of process quality
in determining the level of quality sent to the consumer. Quality cannot be
"inspected into the product." This implies that acceptance sampling is best
carried on at the supplier's location so that feedback of the information
generated by acceptance inspection can lead to corrective action on the
process, if necessary. Further, the stringent quality levels imposed by
present day conditions of government regulations, consumerism, and competi-
tive pressures argue against repetition by the consumer of the large samples
necessary for verification testing. Accordingly, it has become more and
more common for acceptance sampling to be implemented by the producer, with
a consumer audit of the inspection results supplemented by a check inspection
as appropriate.

In April 1954, for example, the Department of Defense established a
policy promoting optimum use of the data generated by its suppliers in the
course of acceptance inspection. This resulted in military and federal procedures requiring the suppliers to perform the acceptance inspection while
maintaining records of their inspection results to be reviewed by government
inspectors. Inspection results have, of course, been subpoenaed in the course
of litigation with regard to safety and compliance regulations. This leads
to the problem of validation of inspection results by the consumer once the
quality system and associated acceptance sampling procedures of the producer
have been audited and accepted.

Quality and Reliability Handbook, H-109, entitled *Statistical Procedures
for Determining Validity of Suppliers' Attributes Inspection* was issued May
6, 1960 by the Department of Defense. In that document the military presented a system for validation of suppliers' inspections which can be used as

part of any quality assurance system. The procedures given validate the inspection on any given lot or lots, but not the acceptability of the material. That is left to the original supplier's inspection. If the inspections of a lot by the producer and the consumer are not found to correspond, action is taken to determine the cause of the discrepancy. The supplier's inspection is not necessarily impuned by a discrepant result, since such a result could occur from causes other than supplier negligence. Needless to say, the material coming from a questionable inspection should be reinspected with appropriate action taken to determine the cause of the discrepancy and correct it, even if this means going to another supplier.

The method of H-109 and its technical development has been thoroughly described by Ellner (1963). Its usefulness comes from the simplicity of the procedures employed, which are based on an "approximate" test for comparing two Poisson variates presented by Ellner.

Procedure

The H-109 procedure is as follows:

1. Supplier performs acceptance inspection and records results, say d_s defectives found in a sample of n (use first sample only as part of inspection validation in the case of double or multiple sampling).

2. Consumer performs check inspection using a sample size n/r and observes the number of defectives d_c found (r is the ratio of sample sizes and takes on the values r = 1, 2, 3, 5, 8).

3. The consumer uses H-109 Table I, shown here as Appendix Table T17-3, to obtain a lot action limit number $d_c(A)$ associated with the sample size ratio r and the number of defectives d_s found by the supplier.

4. If

   a. $d_c \leq d_c(A)$, no action is taken, there being no evidence of a discrepancy in the inspections.

   b. $d_c > d_c(A)$, action is taken by the consumer on the premise that a discrepancy exists between the inspections.

5. A cumulative results criterion is imposed by assigning a check rating to each sample result. The check ratings are found in H-109 Table II, partially shown here as Appendix Table T17-4. For a given r, by entering columns with d_s and rows with d_c, read the check rating c_i at the row-column intersection. Check ratings are

cumulated, so that over k inspections the cumulative check rating is

$$C_k = \sum_{i=1}^{k} c_i$$

6. After each inspection by the consumer the cumulative check rating is compared to the cumulative results warning $C_k(W)$ and action $C_k(A)$ criteria found in H-109, Table III, shown here in Appendix Table T17-5.

7. If

   a. $C_k < C_k(W)$, no action is taken, there being no evidence of a discrepancy in the inspections.

   b. $C_k(W) \leq C_k \leq C_k(A)$, the consumer's inspector is alerted to look for a possible discrepancy in the supplier's inspection system.

   c. $C_k > C_k(A)$, action is taken by the consumer on the premise that a discrepancy exists between the inspections.

When the action limit is exceeded on either an individual lot or for cumulative results, the consumer should take corrective action and not rely on the supplier's inspection results for determining the disposition of this or subsequent lots.

The operation of the H-109 procedure for validation of inspection results is depicted in Fig. 17-10, where

n = supplier sample size

r = ratio of supplier to consumer sample size

d_s = defectives from supplier inspection

d_c = defectives from consumer inspection

$d_c(A)$ = lot action limit number

C_k = cumulative check rating on kth lot

$C_k(W)$ = cumulative results warning criterion

$C_k(A)$ = cumulative results action criterion

As an example of application of the H-109 procedure, suppose MIL-STD-105D, Code F, 2.5 percent AQL was used on ten lots of size N = 125. The producer applied the plan n = 20, c = 1 and all lots were accepted. The consumer has duplicated the inspection (r = 1) and no discrepancy was found on individual lot inspections. A cumulative check rating of

$$C_{10} = 14.9$$

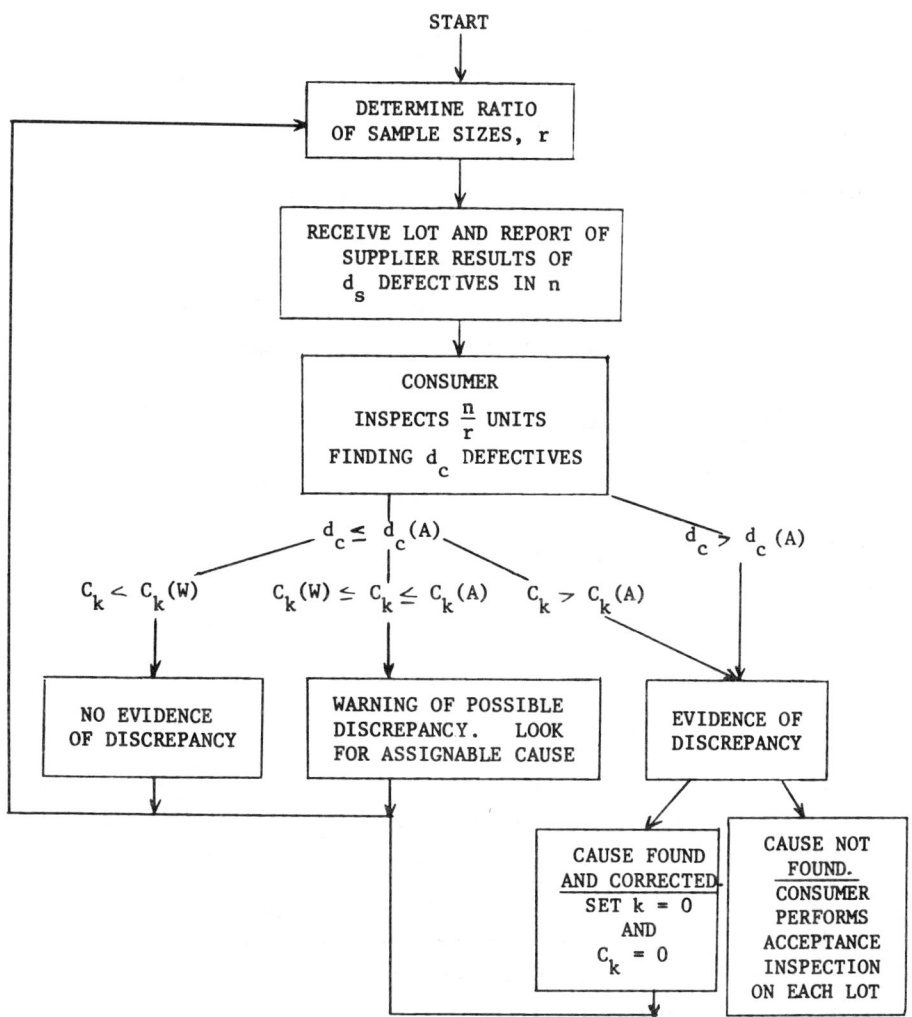

FIG. 17-10   Operation of H-109 procedure.

has been obtained. The consumer decides to inspect only half the sample
size taken by the producer on subsequent lots so that, $r = 2$. Lot size is
to be increased to $N = 300$ with double sampling to be employed.

A lot of 300 units is next received with documentation indicating that
Code H was used with 2 defectives found in a first sample of 32 and with
1 defective found in the second sample of 32 leading to lot acceptance.
Here $d_s = 2$ and the consumer takes a sample of $32/2 = 16$. Suppose 3 defectives are found. Reference to Appendix Table T17-3 shows $d_c(A) = 4$, hence

no action is taken at this point. However, referring to Appendix Table
T17-4, the check rating on this lot is c_{11} = 2.24, so that the cumulative
check rating is

$$C_{11} = 14.9 + 2.24 = 17.14$$

This exceeds the cumulative results warning criterion

$$C_{11}(W) = 16.96$$

shown in Appendix Table T17-5, but does not exceed the cumulative action
criterion of $C_{11}(A)$ = 20.14. As a result, the consumer's inspector begins
to look into the supplier's inspection system for the cause of a possible
discrepancy in the inspections.

Selection

The primary determinant in the operation of the procedure under control of
the consumer is r, the ratio of the supplier's to the consumer's sample
size. Of course, initial inspection should be carred out using r = 1 until
sufficient quality history has been accumulated to consider decreasing the
consumer's sample size with a resulting increase in r. Paralleling MIL-STD-
105D practice, probably at least 10 lots should be inspected before a change
in r is made.

   With small samples the supplier's inspection results would not be
expected to yield a large number of defectives upon which to base results.
This can be handled by pooling test results over a number of lots before a
test of inspection validity is made. Note that when the lot is substantially depleted by the supplier's inspection, the supplier's sample should
not be returned to the lot with defectives corrected until the consumer's
sample is taken. An alternate approach is to adjust the probability of
acceptance on a single lot to be such that a discrepancy will be detected
on a given number of lots with assigned (lower) probability. H-109 Table
IV shown here as Appendix Table T17-6 gives values of probability of acceptance in a single trial, P_{a_1}, which will give a compound probability of
acceptance, P_{a_k}, in k trials. It is based on the relation

$$P_{a_k} = P_{a_1}^{k}$$

assuming the inspection to be independent from lot to lot. For example,
if P_{a_1} is set at .90 on each inspection, the probability of no rejections

in 10 trials will be .35 since $.35 = .90^{10}$.

It is sometimes argued that the H-109 procedure places the producer in double jeopardy since two inspections are performed on the same lot, one by the producer and one by the consumer. However, the consumer's sampling is intended to check uniformity of inspection and not as a method of lot acceptance.

The value of r chosen depends upon a number of parameters in the sampling situation. Given

p'_s = quality level apparent from supplier's inspection

p'_c = quality level apparent from consumer's inspection

n_s = supplier's sample size

these parameters are

$p'_s n_s$ = number of defectives expected in supplier's inspection results

$\dfrac{p'_c}{p'_s}$ = expected apparent ratio of quality levels in inspection results

P_a = probability of acceptance of the null hypothesis of inspection uniformity.

Given these parameters, the value of r to be used may be determined from the operating characteristic curves of the H-109 procedure. These are given in the handbook. Figure 17-11 shows a typical curve for $r = 2$. The curves are labeled in terms of the expected number of defectives in the supplier's sample, $p'_s n_s$. They are plotted against probability of acceptance, P_a, and the expected apparent ratio of consumer's to the producer's quality level p'_c / p'_s.

To illustrate use of the OC curves in determining r, suppose MIL-STD-105D is to be used with double sampling on lots of 300 items. AQL is to be 2.5 percent. The resulting Code H plan has first sample size $n_1 = 32$. The supplier would be expected to provide AQL or better quality, so that $p'_s = .025$. Then the expected number of defectives in the supplier's (first) samples would be

$$n'_s p'_s = 32(.025) = .08$$

If the consumer wishes to detect a difference in inspections such that the consumer's apparent quality level would be two AQL classes higher when the supplier's inspection indicates 2.5 percent AQL, we have $p'_c = .065$ and

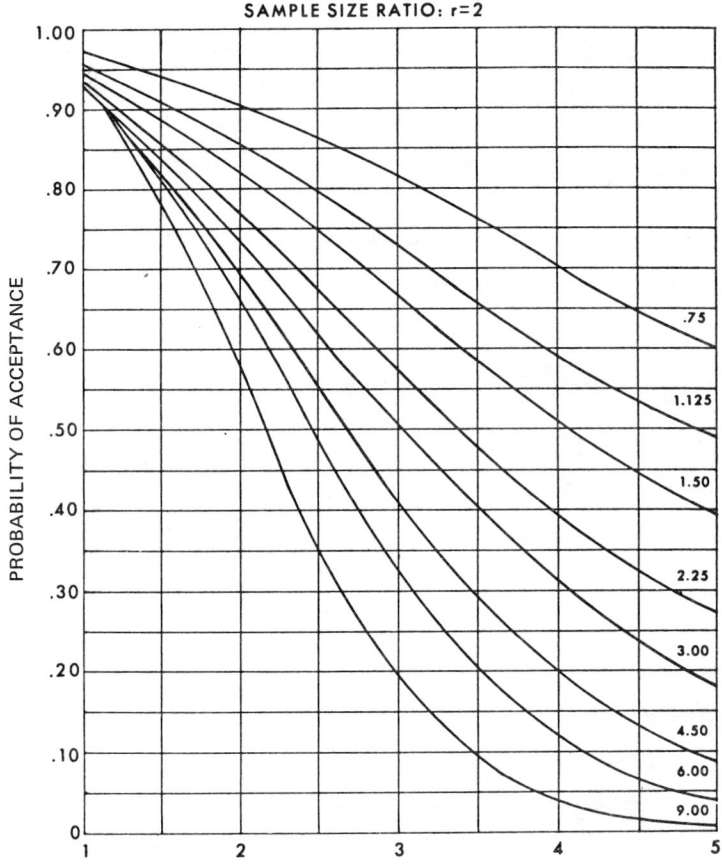

SAMPLE SIZE RATIO: r=2

RATIO OF FRACTIONS DEFECTIVE, p'_c/p'_s

(y-axis: PROBABILITY OF ACCEPTANCE)

FIG. 17-11 Operating characteristic curves of two-sample test for homogeneity. Figures on curves are the expected numbers of defectives (defects) in the supplier's sample. (From H-109, p. 33.)

$p'_s = .025$. Hence

$$\frac{p'_c}{p'_s} = \frac{.065}{.025} = 2.6$$

Assuming this is to be detected at least once in 10 lots, 4 out of 5 times, Appendix Table T17-6 shows

$$P_{a_1} = .85$$

should be used on an individual lot to achieve $P_{a_{10}} = .20$ in 10 lots. It can be seen from the curve labeled .75 in Fig. 17-11 that $r = 2$ will, in fact, meet these conditions.

H-109 Table V, reproduced here as Appendix Table T17-7, presents information on the OC curves in tabular form. It shows values of P_a indexed by r and p_c'/p_s' tabulated against $p_s'n_s$. For this example, interpolation between $p_c'/p_s' = 2$ and $p_c'/p_s' = 3$ when $r = 2$ and $p_s'n_s = .75$ shows $P_a \simeq .85$ as before. Similar interpolation for other values of r reveals $r = 2$ to be the best ratio shown.

Of course the OC curves and the table of values of probability of acceptance can be used in many other ways to evaluate any of the four values given the other three.

Further Considerations

The H-109 procedures are based on an "approximate" test for comparing two Poisson distributed observations. Ellner (1963) shows the ratio

$$\frac{2p_s'n_s\left(2d_c + 1\right)}{2p_c'n_c\left(2d_s + 1\right)}$$

to be distributed approximately F with $\nu_1 = 2d_s + 1$ and $\nu_2 = 2d_c + 1$ degrees of freedom. Since, under the null hypothesis of uniformity of inspection $p_s' = p_c'$ and since $r = n_s/n_c$, the null hypothesis may be tested by constructing the ratio

$$F = r\,\frac{d_c + 0.5}{d_s + 0.5}$$

and comparing the result to the F table with $\nu_1 = 2d_s + 1$ and $\nu_2 = 2d_c + 1$ degrees of freedom. Equivalently, the test may be expressed in terms of the incomplete β distribution (see Chap. 10) with critical region

$$I_x(a,b) \leq \alpha$$

where

$$x = \frac{1}{1 + r}$$
$$a = d_c + 0.5$$
$$b = d_s + 0.5$$

It is this test at the .05 level of significance that gives rise to Table I of Handbook H-109.

The check ratings of Table II of H-109 are determined as input to the

Fisher (1938, p. 104) method of combining probabilities of independent
tests which forms the basis of the cumulative results criterion of the H-109
procedure. Since r may vary from inspection to inspection and since the
"approximate" test treats the number of events d_s and d_c as continuous
variates, Fisher's method provides an ideal vehicle for the cumulation of
results. We then have

$$c_i = -\ln I_x(a,b)$$

and

$$2C_k(A) = 2 \sum_{i=1}^{k} c_i$$

distributed

$$\chi^2 \text{ with 2k degrees of freedom.}$$

The values given in H-109 Table III for warning and action limits are values
of $\chi^2/2$ with 2k degrees of freedom at the .05 and .01 significance levels.

As an illustration of the construction of these tables, take $d_s = 2$,
$d_c = 3$, and r = 2 as in the example above. Then

$$F = 2\frac{(3 + 0.5)}{(2 + 0.5)} = 2.8$$

with $v_1 = 2(2) + 1 = 5$ and $v_2 = 2(3) + 1 = 7$. When compared to the critical
value of F = 3.97 at the .05 level of significance the null hypothesis is
not rejected. Note that the critical value is 3.97 ~ 4 which is the value
listed in H-109 Table I (Appendix Table T17-3).

For these results

$$\begin{aligned}
c_i &= -\ln I_x(a,b) \\
&= -\ln I_{.33}(3.5,2.5) \\
&= -\ln(.106) \\
&= 2.24
\end{aligned}$$

which is the check rating for these values of d_s and d_c as it appears in
H-109 Table III (Appendix Table T17-4). Also H-109 Table III (Appendix
Table T17-5) shows

$$\begin{aligned}
C_{11}(W) &= 16.96 \\
C_{11}(A) &= 20.14
\end{aligned}$$

as critical values for the cumulative check ratings. Checking a χ^2 table
with 2k = 22 degrees of freedom we have

$$\alpha = .05: \quad \frac{\chi^2}{2} = \frac{33.9}{2} = 16.95$$

$$\alpha = .01: \quad \frac{\chi^2}{2} = \frac{40.3}{2} = 20.15$$

H-109 Table IV (Appendix Table T17-6) simply shows the relation

$$P_{a_k} = P_{a_1}^{\,k}$$

mentioned earlier.

Table V of H-109 (Appendix Table T17-7) presents values of

$$P_a = 1 - I_x(a,b)$$

evaluated for given values of the parameters of the plan. For example, when $r = 2$, $p_c'/p_s' = 2$, and $n_s p_s' = .75$, we have

$$n_s p_s' = .75$$
$$n_c p_c' = 2(.75) = 1.5$$
$$r = 2$$

so that

$$x = \frac{1}{1 + 2} = .33$$
$$a = 1.5 + .5 = 2.00$$
$$b = .75 + .5 = 1.25$$

and

$$P_a = 1 - I_{.33}(2.00,1.25) = 1 - .1445$$
$$P_a = .8555$$

as entered in Table V of H-109. The operating characteristic curves of H-109 are derived and plotted using the same procedure.

Some flaws have been noted in the application of H-109. For example, because of the approximate nature of the test, the level of significance is not always $\alpha = .05$ in Table I and may range as high as $\alpha = .12$. Furthermore, because of the nature of the check ratings employed, it is possible to exceed the action limit using the cumulative results criterion when neither the producer nor the consumer have found any defectives in their samples over the period cumulated. Nevertheless, the procedure remains a useful one and can be successfully employed even in view of these difficulties.

PROBLEMS

1. A lot 15,000 units is to be tested for a potential safety hazard. An LTPD of .065 percent is to be used. Construct a LSP. What is its AOQL?

2. Draw the OC curve of the LSP plan in Prob. 1.

3. If the level of .065 percent defective is to be guaranteed with 95 percent probability, what should be the sample size in Prob. 1?

4. A series of lots is to be inspected using a TNT plan and it is desired that $p_{.95} = .01$ and $p_{.10} = .08$. Construct the plan.

5. Compute the probability of acceptance, the ASN and the AOQ when defects are replaced for the TNT plan in Prob. 4 at $p = .05$ when lots are very large.

6. Derive a QSS-1 plan matching the criteria of Prob. 4. What is its indifference quality?

7. Product is sold in lots of 10,000. It is necessary to verify that quality is better than .005 proportion defective so that LTPD = 0.5 percent. However, when $p = 0.1$ percent defective, 95 percent of the lots should be accepted. These criteria would require use of the single-sampling plan $n = 1366$, $c = 3$. Construct a simplified grand lot plan to achieve these ends.

8. Samples of 9 units have been taken from each of 15 lots giving $\bar{\bar{X}} = 500$ and $\hat{s} = 45$. Construct extreme limits and maverick limits to test if the 15 lots constitute a grand lot. Assuming all sublots fall within the limits, test the grand lot against a lower specification limit of 400 using the plan $n = 135$, $k = 2.0$.

9. What should be the limits for a test of the grand lot hypothesis on a continuing source of supply from the vendor in Prob. 8?

10. Documents submitted with a lot indicate the supplier found 4 defective units in a sample of 100. The customer performs an H-109 inspection with $r = 5$ and finds 2 defectives. Should action be taken? The previous check rating after 20 lots was 27. What is the new check rating? Should action be taken?

REFERENCES

Burr, I. W. (1953), *Engineering Statistics and Quality Control*, McGraw-Hill, New York.

Calvin, T. W. (1977), *TNT Zero Acceptance Number Sampling*, American Society
    for Quality Control Technical Conference Transactions, Philadelphia,
    Pennsylvania.

Dodge, H. F. (1967), *A New Dual System of Acceptance Sampling*, Technical
    Report No. 16, The Statistics Center, Rutgers - The State University,
    New Brunswick, New Jersey.

Duncan, A. J. (1977), Addendum to Proposed Standard for Small Lot Sampling
    Plans Based on the Hypergeometric Probability Distribution, *The Relia-
    Com Review*, 2(2): 6.

Duncan, A. J. (1979), In the Federal Arena: E-11 and Regulatory Agencies,
    ASTM Standardization News, 7(1): 20-21, 39.

Ellner, H. (1963), Validating Results of Sampling Inspection by Attributes,
    Technometrics, 5(1): 23-46.

Fisher, R. A. (1938), *Statistical Methods for Research Workers*, Oliver and
    Boyd, London.

Halperin, M., S. W. Greenhouse, J. Cornfield, and J. Zalokar (1955), Tables
    of Percentage Points for the Studentized Maximum Absolute Deviate in
    Normal Samples, *Journal of the American Statistical Association*,
    50(269): 185-195.

Hamaker, H. C. (1950), Lot Inspection by Sampling, *Philips Technical Review*,
    11: 176-182.

Hawkes, C. J. (1979), Curves for Sample Size Determination in Lot Sensitive
    Sampling Plans, *Journal of Quality Technology*, 11(4): 205-210.

Mandelson, Joseph (1963), Use of the Grand Lot in Surveillance, *Industrial
    Quality Control*, 19(8): 10-12.

Muehlhause, C. O., V. L. Broussalian, A. J. Farrar, J. W. Lyons, M. G.
    Natrella, J. R. Rosenblatt, R. D. Stiehler, and J. H. Winger (1975),
    *Considerations in the Use of Sampling Plans for Effecting Compliance
    with Mandatory Safety Standards*, United States Department of Commerce,
    National Bureau of Standards, Report 75-697, pp. 42-43.

Nelson, L. S. (1974), Factors for the Analysis of Means, *Journal of Quality
    Technology*, 6(4): 175-181.

Nelson, L. S. (1978), Nomograph for Samples Having Zero Defectives, *Journal
    of Quality Technology*, 10(1): 42-43.

Ott, E. R. (1967), Analysis of Means — A Graphical Procedure, *Industrial
    Quality Control*, 24(2): 101-109.

Ott, E. R., and S. S. Lewis (1960), *Analysis of Means Applied to Percent Defective Data*, Technical Report No. 2, The Statistics Center, Rutgers - The State University, New Brunswick, New Jersey.

Pearson, E. S. (1935), *The Application of Statistical Methods to Industrial Standardization and Quality Control*, British Standard 600:1935, British Standards Institution, London.

Romboski, L. D. (1969), An Investigation of Quick Switching Acceptance Sampling Systems, Ph.D. Dissertation, Rutgers - The State University, New Brunswick, New Jersey.

Schilling, E. G. (1973), A Systematic Approach to the Analysis of Means, Part I: Analysis of Treatment Effects, *Journal of Quality Technology*, 5(3): 93-108.

Schilling, E. G. (1973a), A Systematic Approach to the Analysis of Means, Part II: Analysis of Contrasts; Part III: Analysis of Non-Normal Data, *Journal of Quality Technology*, 5(4): 147-159.

Schilling, E. G. (1978), A Lot Sensitive Sampling Plan for Compliance Testing and Acceptance Inspection, *Journal of Quality Technology*, 10(2): 47-51.

Schilling, E. G. (1979), A Simplified Graphical Grand Lot Acceptance Sampling Procedure, *Journal of Quality Technology*, 11(3): 116-127.

Schilling, E. G., and H. F. Dodge (1969), Procedures and Tables for Evaluating Dependent Mixed Acceptance Sampling Plans, *Technometrics*, 11(2): 341-372.

Schilling, E. G., and P. R. Nelson (1976), The Effect of Non-Normality on the Control Limits of $\bar{X}$ Charts, *Journal of Quality Technology*, 8(4): 183-188.

Schilling, E. G., and J. H. Sheesley (1978), The Performance of MIL-STD-105D Under the Switching Rules, *Journal of Quality Technology*, Part 1, 10(2): 76-83; Part 2, 10(3): 104-124.

Simon, L. E. (1941), *An Engineer's Manual of Statistical Methods*, John Wiley and Sons, New York.

Simon, L. E. (1944), The Industrial Lot and Its Sampling Implications, *Journal of the Franklin Institute*, 237: 359-370.

Tukey, J. W. (1953), *The Problem of Multiple Comparisons*, unpublished manuscript, Princeton University, Princeton, New Jersey.

United States Department of Defense (1960), *Statistical Procedures for Determining Validity of Suppliers' Attributes Inspection*, Supply and

Logistics (Interim) Handbook (H-109), Office of the Assistant
Secretary of Defense (Supply and Logistics), Washington, D.C.

United States Department of Defense (1963), *Military Standard, Sampling
Procedures and Tables for Inspection by Attributes* (MIL-STD-105D),
U.S. Government Printing Office, Washington, D.C.

United States Department of Defense (1957), *Military Standard, Sampling
Procedures and Tables for Inspection by Variables for Percent Defective*
(MIL-STD-414), U.S. Government Printing Office, Washington, D.C.

RELIABILITY SAMPLING

Historically, sampling plans have been used to assess the present quality of the material examined. They are employed to determine acceptability of the product against specifications at a given time. This has usually been the time of sale. Of course, the implication is that items presently acceptable will retain their utilitarian properties upon reaching the consumer. An important quality characteristic of some products, however, is degradation in use, that is, the useful life of the product with regard to some property.

The advent of considerations of reliability imposed by high technology programs such as space and atomic power, consumerism, and conformance testing to government mandatory standards, have placed a new dimension on the sampling problem, that of time. Reliability sampling plans are used to determine the acceptability of the product at some future point in its effective life. This usually involves some form of life testing.

An Advisory Group on Reliability of Electronic Equipment (AGREE) was formed in 1952 under the Assistant Secretary of Defense to "monitor and stimulate interest in reliability matters and recommend measures which would result in more reliable electronic equipment." AGREE (1957) defined reliability as follows: "Reliability is the probability of performing without failure a specified function under given conditions for a specified period of time." Reliability testing is to provide assurance of reliability. In this sense, it is not testing what the product is, but rather how it will operate, over time, in the hands of the consumer. The standard plans discussed so far determine whether the product is made to specifications. Reliability plans assess how it will perform.

The time dimension implicit in reliability testing is superimposed

on the sampling problem as an additional criterion. Samples must be tested
for a specified length of time. When all units are tested to failure, the
standard plans can be utilized to assess the results against specified requirements. If lifetimes are measured, these results can be used in a
variables sampling plan, such as MIL-STD-414, provided the distributional
assumption of the plan is satisfied. Also, the number failing before a
required time can be used with standard attributes plans in determining
disposition of the material, e.g., MIL-STD-105D.

In reliability and safety testing, extremely low levels of probability
of acceptance are often used. When a test based on a two-point plan (p_1,
p_2, α, β) has been passed, it is often said that a reliability of at least
$\pi = 1 - p_2$ has been demonstrated with $\gamma = 1 - \beta$ confidence. Specifications
are often written in this way. Clearly, a variety of plans could satisfy
such a requirement on what amounts to limiting quality. For example, it
follows from the Schilling-Johnson Appendix Table T5-2 that to demonstrate
.9995 confidence of .99 reliability, the plan n = 1000, c = 1 could be used
since, for c = 1, when P_a = .0005

$$\frac{np}{1 - \pi} = \frac{10.000}{.01} = 1000$$

For a discussion of this type of specification for sampling plans in reliability see Lloyd and Lipow (1962, p. 280).

When testing of the sample is terminated before the specified lifetime
with some units still unfailed, however, complications arise. A sample of
this sort is called *censored*, and implies that some units were tested without generating failures as such. In this case, the test termination time
should not be used as the failure time for the unfailed units since, clearly,
they would very probably have lasted longer, and consequently to do so would
bias the results. It is easily seen, from actuarial work, that mean lifetime would be grossly understated if, in a sample of 100 people, the first
death was used as average lifetime. Yet this is exactly the result if a
test is stopped at the first failure and the termination time used as the
failure time of the remaining elements in the sample.

Statistical methods have been developed for use with censored samples.
Life tests may be deliberately terminated after a given number of failures
or a specified period of time and analyzed using these procedures. This is
usually done to speed up the test or for economic considerations. The
methods for dealing with censored data can also be used in situations in

which some of the units have not been tested to failure because of difficulties with the test equipment, units failed for causes other than those
being tested, broken or stolen units, etc.

CENSORED SAMPLING

Analysis of censored data can be made with varying degrees of sophistication.
One of the most useful tools in this regard is a properly constructed probability plot. The actual failures observed are plotted against plotting
positions which have been adjusted for the amount and type of censoring in
the sample. The method is based on an empirical determination of the hazard
rate h(x), or instantaneous failure rate, associated with each of the observed failures. The cumulative hazard rate H(x) may then be transformed
into probability plotting positions. The development of the approach has
been aptly described by Nelson (1969) in an American Society for Quality
Control Brumbaugh Award winning paper. The method is as follows for a
sample of n:

1. Order the data, including failure and censoring times. Distinguish the actual failures by marking them with an asterisk.
2. Calculate the hazard value h(x) associated with each failure time
   as the reciprocal of the number k of units with failure and censoring times greater than or equal to the failure observed. That
   is

   $$h = \frac{1}{k}$$

3. Cumulate the hazard values to obtain the cumulative hazard value
   H for each observed failure. For the ith failure

   $$H_i = \sum_{j=1}^{i} h_j$$

4. Convert the cumulative hazard value to a plotting position using
   the relation

   $$P_i = (1 - e^{-H_i})100$$

   Appendix Table T18-1 is a tabulation of values of P_i for
   associated 100 H_i as calculated by Sheesley (1974).
5. Make a probability plot on appropriate paper to assess the shape

of the life distribution and, if the points fall on a straight line, to estimate its parameters.

Special hazard probability plot papers are available for this purpose upon which cumulative hazard may be plotted directly, avoiding steps 4 and 5. Plotting papers for the normal, lognormal, exponential, Weibull, and extreme value distributions are given by Nelson (1969).

The technique is illustrated by sample data taken from a life test of Class B insulation on small motors, or "motorettes" as given by Hahn and Nelson (1971). Ten motorettes were tested at 170°C to obtain information on the distribution of insulation life at elevated temperatures. The test was stopped at 5448 hr. with three motorettes still running. Results on the other seven were, in order of failure, 1764, 2772, 3444, 3542, 3780, 4860, 5196. Calculation of the probability positions is shown in Table 18-1. A normal probability plot of these data is shown in Fig. 18-1. Only the actual failures are plotted. The mean life of this sample of motorettes appears to be about 4400 hr.

Because of their frequently long duration, life tests are often concluded before all units have failed. Sometimes this is not done by design. In such situations, the hazard plotting procedure is an excellent technique for assessing failure distribution models preparatory to the initiation of a life test sampling plan. Plotting a substantial number of failures on a

TABLE 18-1 Probability Positions for Motorette Data

Ordered unit	k	Hours run x	Hazard $h = \frac{1}{k}$	Cumulative hazard $H = \Sigma h$	Percent cumulative probability $P = (1 - e^{-H})100$
1	10	1764	.10	.10	9.5
2	9	2772	.111	.211	19.0
3	8	3444	.125	.336	28.5
4	7	3542	.143	.479	38.1
5	6	3780	.167	.646	47.6
6	5	4860	.20	.846	57.1
7	4	5196	.25	1.096	66.6
8	3	5448*			
9	2	5448*			
10	1	5448*			

*Still running.

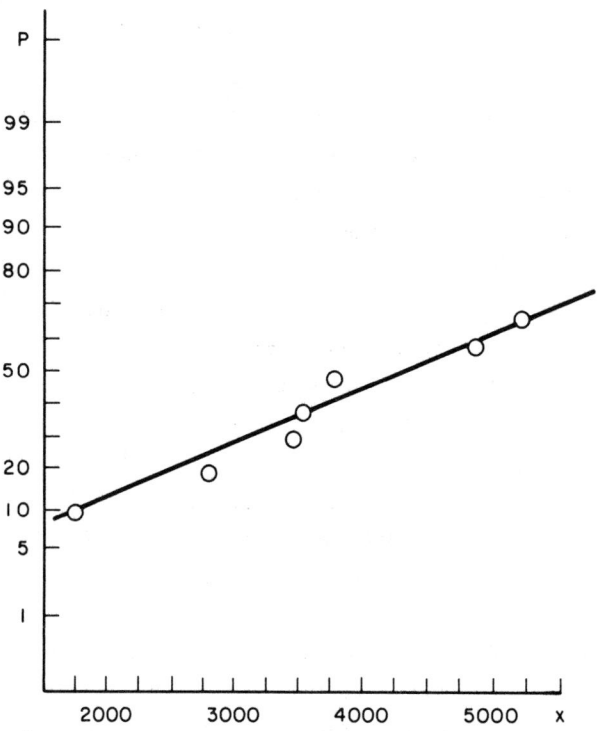

FIG. 18-1 Probability plot of motorette data.

variety of probability papers will give much insight into the probability
distribution involved. More sophisticated methods are of course available
and will be found in textbooks on reliability. See, for example, Locks
(1973), Mann, Schafer, and Singpurwalla (1974), or Nelson (1982).

The unique necessity for a time specification in reliability tests has
led to the development of a number of published sampling procedures. Two
will be discussed here and are based on the exponential and the Weibull
distributions, respectively.

VARIABLES PLANS FOR LIFE TESTING AND RELIABILITY[*]

Variables sampling plans for life and reliability testing are similar in
concept and operation to the plans previously described. They differ to

[*]This section is adapted and reprinted by permission from *Quality Control
Handbook*, edited by J. M. Juran, Frank M. Gryna, Jr. and Richard S. Bingham,
3rd ed., Section 25, Sampling by Variables by E. G. Schilling, McGraw-Hill,
New York, 1974, pp. 25-25 to 25-31, Copyright 1974, 1962 by McGraw-Hill, Inc.
All rights reserved.

TABLE 18-2 Life Characteristics for Two Failure Distributions

Exponential: $f(t) = \dfrac{1}{\mu} e^{-\frac{t}{\mu}}$

Weibull: $f(t) = \dfrac{\beta t^{\beta-1}}{\eta^\beta} e^{-\left(\frac{t}{\eta}\right)^\beta}$ where $\mu = \eta\Gamma\left(1 + \dfrac{1}{\beta}\right)$ and $g = \left[\Gamma\left(1 + \dfrac{1}{\beta}\right)\right]^\beta$

Life characteristic	Symbol	Exponential	Weibull
Proportion failing before time t	$F(t)$	$F(t) = 1 - e^{-t/\mu}$	$F(t) = 1 - e^{-g(t/\mu)^\beta}$
Proportion (r) of population surviving to time ρ	$\rho_r = 1 - F(\rho)$ $= r$	$\rho_r = e^{-\rho/\mu}$	$\rho_r = e^{-g(\rho/\mu)^\beta}$
Mean life or mean time between failures	μ	μ	μ
Hazard rate: instantaneous failure rate at time t	$z(t) = h(t)$	$z(t) = \dfrac{1}{\mu}$	$z(t) = \dfrac{\beta g t^{\beta-1}}{\mu^\beta}$
Cumulative hazard rate for period 0 to t	$H(t)$	$H(t) = t/\mu$	$H(t) = \dfrac{g t^\beta}{\mu^\beta}$

TABLE 18-3 Values of $g = \left[\Gamma \left(1 + \frac{1}{\beta} \right) \right]^{\beta}$ for Weibull Distribution

β	0.0	1.0	2.0	3.0	β	g
0.0		1.0000	0.7854	0.7121	0.33	1.8171
0.1	4.5287	0.9615	0.7750	0.7073	0.67	1.2090
0.2	2.6052	0.9292	0.7655	0.7028	1.33	0.8936
0.3	1.9498	0.9018	0.7568	0.6986	1.67	0.8289
0.4	1.6167	0.8782	0.7489	0.6947	3.33	0.6973
0.5	1.4142	0.8577	0.7415	0.6909	4.00	0.6750
0.6	1.2778	0.8397	0.7348	0.6874	5.00	0.6525
0.7	1.1794	0.8238	0.7285	0.6840		
0.8	1.1051	0.8096	0.7226	0.6809		
0.9	1.0468	0.7969	0.7172	0.6778		

the extent that, when units are not all run to failure, the length of the
test becomes an important parameter determining the characteristics of the
procedure. Further, time to failure tends to conform naturally to skewed
distributions such as the exponential or as approximated by the Weibull.
Accordingly, many life test plans are based on these distributions. When
time to failure is normally distributed and all units tested are run to
failure, the variables plans assuming normality, discussed above, apply;
attributes plans such as those in MIL-STD-105D may also be used.

Life tests, terminated before all the units have failed, may be

1. Failure terminated. A given sample size, n, is tested until the
   rth failure occurs. The test is then terminated.

2. Time terminated. A given sample size, n, is tested until a preassigned termination time, T, is reached. The test is then
   terminated.

Furthermore, these tests may be based upon specifications written in
terms of one of the following characteristics:

1. Mean life: the expected life of the product

2. Hazard rate: instantaneous failure rate at some specified time, t

3. Reliable life: the life beyond which some specified proportion
   of items in the lot or population will survive.

Several sets of plans are available for the testing of life and reliability.

Tables 18-2 and 18-3 will be found useful in converting life test characteristics. Formulas for various characteristics are shown in terms of mean life μ. Thus, using the tables, it will be found that a specification of mean life μ = 1000 hr. for a Rayleigh distribution (Weibull, β = 2) is equivalent to a hazard rate of .00157 at 1000 hr. or to a reliable life of 99 percent surviving at 113 hr.

HANDBOOK H-108

Quality Control and Reliability Handbook H-108 is intended to be used with quality characteristics that are exponentially distributed. Its title, Sampling Procedures and Tables for Life and Reliability Testing, suggests an emphasis on life testing and the standard deals primarily with a specification of mean life. While the procedures are quite general in application for the exponential distribution, they will be presented here in terms of the life testing problem.

H-108 is intended to test mean life, θ. Two values are specified:

θ_0 = acceptable mean life

θ_1 = unacceptable mean life

with risks

α = producer's risk

β = consumer's risk

respectively.

A given specific requirement on the mean of the exponential distribution can always be stated in terms of the proportion, p, of the population failing by a specified time, T. As shown in Table 18-2 the relation is

$$p = F(T) = 1 - e^{-T/\theta}$$

Analogous to the specifications for the mean, in the notation of H-108

p_0 = acceptable proportion of the lot failing before specified time, T

p_1 = unacceptable proportion of the lot failing before specified time, T

with associated risks α and β. Special tables are presented for use with this type of specification in time terminated tests.

A unique feature of the exponential distribution, constant failure rate, allows testing to be conducted either

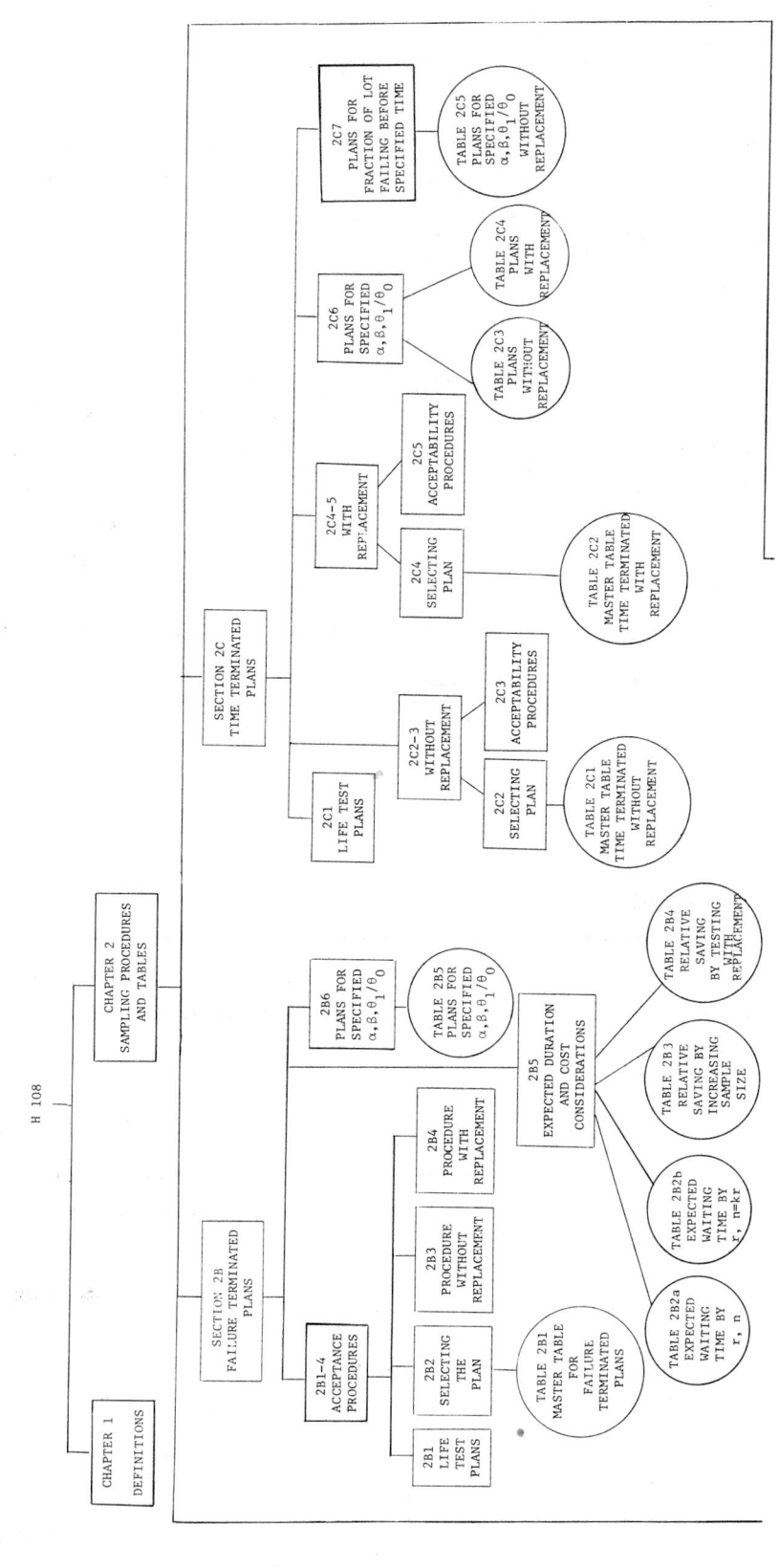

H 108

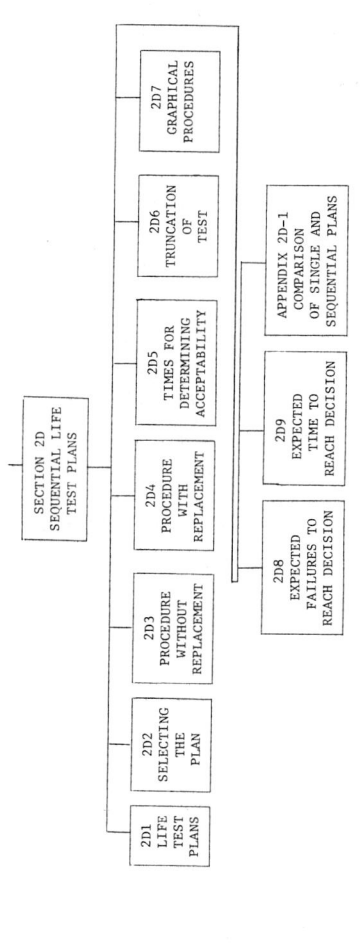

FIG. 18-2 Structure of H-108.

With replacement: units are replaced on test as they fail with the
replacements also contributing to total test time and the number
of failures.

Without replacement: units not replaced as they fail.

Beyond a necessary minimum number of units, n, the sample size is not
specified since, with constant failure rate, increasing the number of
units on test only decreases the length, or waiting time, of the test but
does not affect the number of failures per unit time. Thus, replacements
are possible without biasing the test. Provision is made in the handbook
for both failure terminated and time terminated tests. Sequential tests
are also provided.

The handbook contains a wealth of information on exponential life
testing. Its structure is shown in Fig. 18-2 which shows the location of
material on each of the three types of plans, including

Chapter 1. Definitions

Chapter 2. Section 2A, general description of life test sampling plans;
Section 2B, failure terminated plans for mean;
Section 2C, time terminated plans for mean and proportion
failing before specified time;
Section 2D, sequential plans for mean.

Operation

The three different types of tests provided in H-108 are conducted as
follows:

Failure Terminated

Place a sample of n units on test. Stop the test at the rth failure.
Record the successive failure times as $X_{i,n}$, i = 1,2 ... r. Compute the
estimated mean life $\hat{\theta}_{r,n}$ as

$$\hat{\theta}_{r,n} = \frac{1}{r}\left[\sum_{i=1}^{r} X_{i,n} + (n - r)X_{r,n}\right] \quad \text{(without replacement)}$$

or

$$\hat{\theta}_{r,n} = \frac{nX_{r,n}}{r} \quad \text{(with replacement)}$$

Compare $\hat{\theta}_{r,n}$ to the acceptance constant C.

If $\hat{\theta}_{r,n} \geq C$, accept

If $\hat{\theta}_{r,n} < C$, reject

Time Terminated

Place a sample of n units on test. Stop the test at time T. If r or fewer failures have occurred, accept. If more than r failures have occurred, reject.

Sequential

Place a sample of n units on test. Record the successive failure times as $X_{i,n}$, $i = 1,2 \ldots k$. Compute the total survival time $V(t)$ for k failures at time t as

$$V(t) = \sum_{i=1}^{k} X_{i,n} + (n - k)t \quad \text{(without replacement)}$$

or

$$V(t) = nt \quad \text{(with replacement)}$$

Compare to sequential limits. After the kth failure,

Accept if $V(t) \geq h_0 + ks$ or $V(t) \geq sr_0$.

Reject if $V(t) \leq h_1 + ks$ or $k = r_0$ and $V(t) < sr_0$.

Here r_0 is a truncation criterion for the sequential test. That is, the test is terminated at the r_0th failure and the truncation criterion applied.

Proportion Failing by Specified Time

Conduct a time terminated test (without replacement) as specified by the plan selected.

A summary of the operation of the H-108 test plans is shown in Table 18-4.

Selection

The selection of a plan begins with H-108 Table 2A-1 given here as Appendix Table T18-2. To find a plan the operating ratio

$$R = \frac{\theta_1}{\theta_0}$$

is formed for specified θ_0 and θ_1. The sampling plan code designation is then located under the risks α and β desired. For example, for $\alpha = .05$, $\beta = .10$ if

TABLE 18-4 Operation of H-108

STEP / CHARACTERISTIC	SECTION B	SECTION C - PARTS, I,II MEAN	SECTION D	SECTION C - PART III FRACTION FAILING BEFORE SPECIFIED TIME
TYPE TEST	FAILURE TERMINATED	TIME TERMINATED	SEQUENTIAL	TIME TERMINATED
SPECIFIED	θ_0 = ACCEPTABLE MEAN LIFE, α = PRODUCER'S RISK $\quad$ θ_1 = UNACCEPTABLE MEAN LIFE, β = CONSUMER'S RISK			T = time interval p_0 = satisfactory fraction failing in time T p_1 = unsatisfactory fraction failing in time T G = failure rate in time T where p = GT
CRITERIA	n = sample size r = termination number C = acceptability constant	n = sample size r = termination number T = termination time	n = sample size h_0 = acceptance intercept h_1 = rejection intercept s = slope r_0 = truncation criterion	n = sample size r = termination number T = termination time
OBSERVATION	$x_{1,n}$ = time of ith failure	$x_{r,n}$ = time of rth failure	$x_{1,n}$ = time of ith failure	$x_{r,n}$ = time of rth failure
WITHOUT REPLACEMENT	$\hat{\theta}_{r,n} = \dfrac{1}{r}\left[\sum_{i=1}^{r} x_{1,n} + (n-r)x_{r,n}\right]$ = estimate of mean life	$x_{r,n}$	$V(t) = \sum_{i=1}^{k} x_{1,n} + (n-k)t$ = total survival time at time t with k failures	$x_{r,n}$
WITH REPLACEMENT	$\hat{\theta}_{r,n} = \dfrac{nx_{r,n}}{r}$ = estimate of mean life	$x_{r,n}$	$V(t) = nt$ = total survival time at time t with k failures	$x_{r,n}$
ACCEPTABILITY CRITERION	$\hat{\theta}_{r,n} \geq C$ accept $\hat{\theta}_{r,n} < C$ reject	$x_{r,n} \geq T$ accept $x_{r,n} < T$ reject	Accept if $V(t) > h_0 + ks$ or $V(t) \geq sr_0$ Reject if $V(t) < h_1 + ks$ or $k = r_0$ and $V(t) < sr_0$	$x_{r,n} \geq T$ accept $x_{r,n} < T$ reject

CHARACTERISTIC

STATISTIC

θ_0 = 600 hr. θ_1 = 200 hr.

so that

$$R = \frac{\theta_1}{\theta_0} = \frac{200}{600} = \frac{1}{3}$$

Code B-8 would be selected.

Master tables for each type of test give the factors necessary to define the test. Indexed by sampling plan code, the factors are multiplied by θ_0 to give the test parameters. The master tables are

Table 2B1   Failure Terminated with or without replacement (gives r, C/θ_0)

Table 2C1   Time Terminated without Replacement (gives r, T/θ_0)

Table 2C2   Time Terminated with Replacement (gives r, T/θ_0)

Table 2D1   Sequential with or without replacement (gives r_0, h_0/θ_0, h_1/θ_0, s/θ_0)

Selected values from these tables for α = .05 are given here as Appendix Tables T18-3, T18-4, T18-5, and T18-6, respectively. The handbook also gives special tables in which additional plans are indexed by α, β, and $R = \theta_1/\theta_0$ for the failure and time terminated plans.

Time terminated plans for proportion failing before a specified time will be found in H-108 Table 2C5 given here as Appendix Table T18-7. It shows values of r and the factor D indexed by α, β, and $R = p_1/p_0$ where

$$n = \frac{D}{p_0}$$

The question of sample size is directly incorporated in time terminated tests which require n to be specified before a termination time can be determined. For failure terminated plans this is not the case. It would be theoretically possible to test, with replacement, one unit at a time until r failures were generated. This, of course, could take inordinately long. As a result a number of units, n, are tested simultaneously to speed up the test. Guidance in selecting n for failure terminated tests is given in H-108, Part II of Sec. B. Tables presented include

Table 2B-2(a)   Expected waiting time indexed by r and n showing ratio

$$\frac{\text{Expected waiting time for r failures in a sample of n}}{\text{Mean life of lot}}$$

Table 2B-2(b)   Expected waiting time indexed by r and n = kr, k = 1,2 ..., 10, and 20 showing ratio

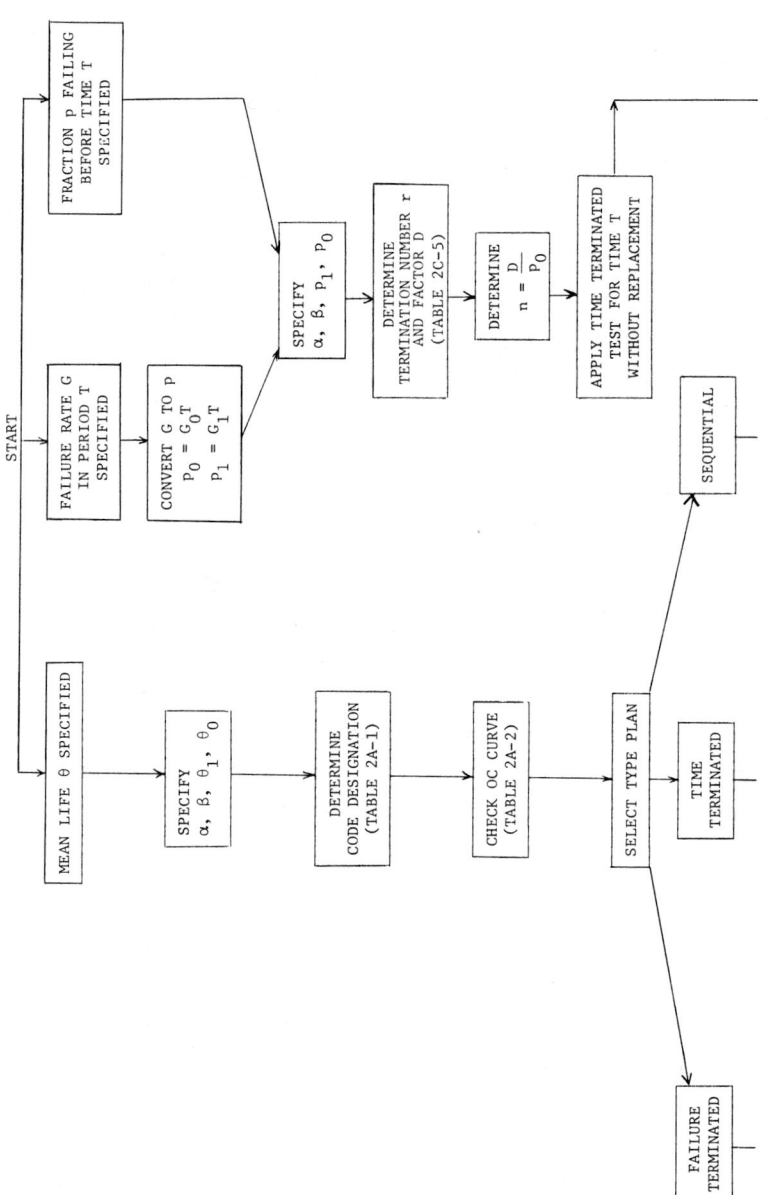

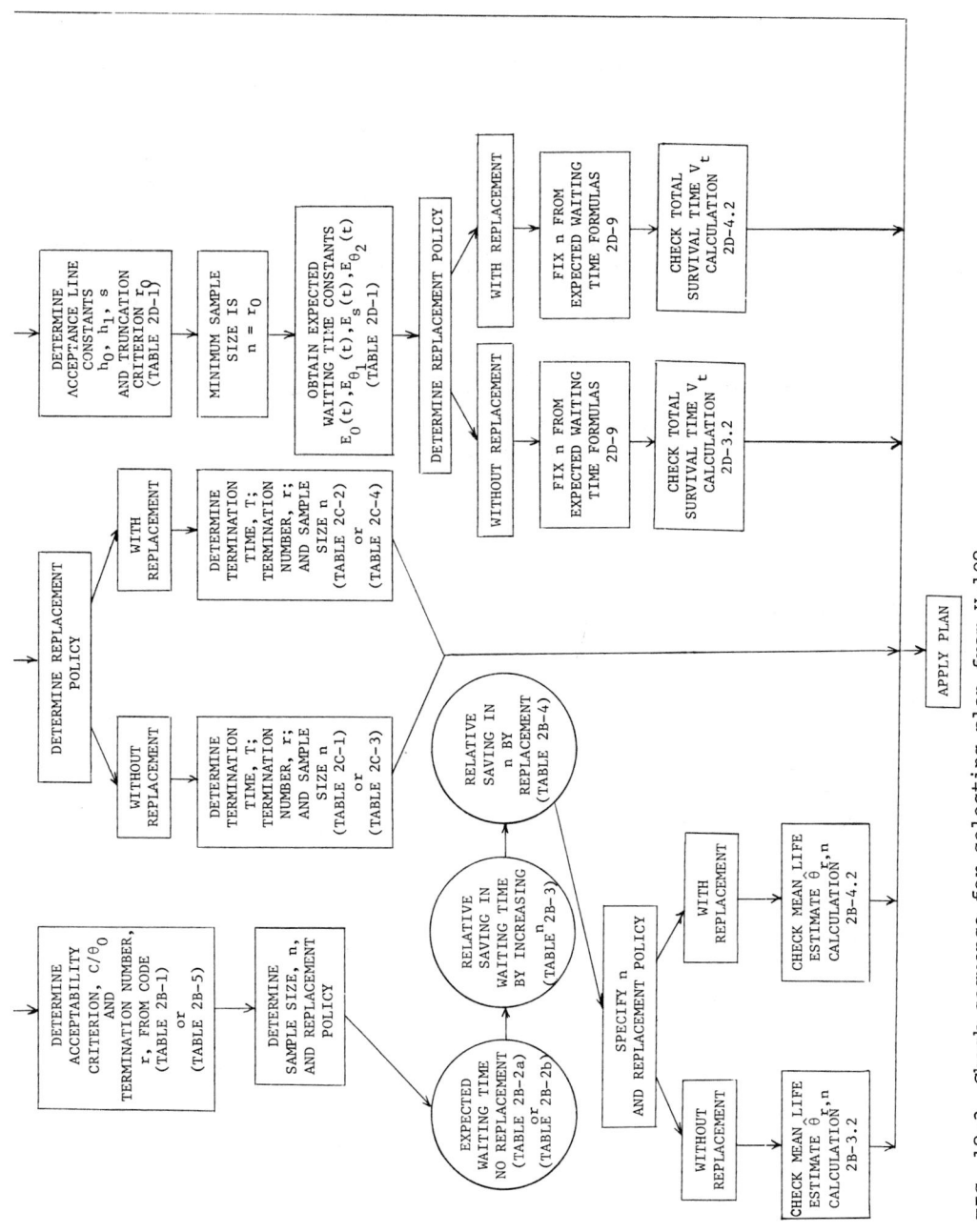

FIG. 18-3 Check sequence for selecting plan from H-108.

$$\left(\frac{\text{Expected waiting time for r failures in a sample of}}{\text{Mean life of lot}}\atop n = kr\right)$$

Table 2B-3   Expected relative saving in time by increasing sample

size indexed by r and n showing ratio

$$\frac{\text{Expected waiting time for r failures in a sample of n}}{\text{Expected waiting time for r failures in a sample of r}}$$

Table 2B-4   Expected relative saving in time by testing with

replacement indexed by n and r showing ratio

$$\left(\frac{\text{Expected waiting time for r failures in a sample of n}}{\text{when testing with replacement}}\middle/ {\text{Expected waiting time for r failures in a sample of n}\atop \text{when testing without replacement}}\right)$$

Sequential plans may also be conducted with a variety of sample sizes. Only minimum sample size is specified for tests without replacement. As a result, the sample size may also be chosen with regard to waiting time or economic considerations. Special factors are presented with the sequential plan to allow determination of the expected number of failures required for a decision and for calculation of the expected waiting time with any sample size.

A check sequence for the utilization of all these tables in determining a sampling plan is shown in Fig. 18-3.

Example of H-108 Application

After the manner of Hahn and Shapiro (1967, p. 107), consider a life test of a large lot of very expensive batteries. The test is to be conducted such that

$$\theta_0 = 70 \quad \text{with } \alpha = .05$$
$$\theta_1 = 7 \quad \text{with } \beta = .10$$

Then

$$R = \frac{\theta_1}{\theta_0} = \frac{7}{70} = .1$$

and H-108 Table 2A-1 shows Code B-2 to be relevant. Using the appropriate tables, possible test plans are as follows:

Failure Terminated

H-108 Table 2B-1 gives $r = 2$ and $C/\theta_0 = .178$. Hence

$$C = .178(70) = 12.5$$

Sample size would be determined from economic considerations. H-108 Table 2B-3 shows waiting time could be reduced by 61 percent by using a sample of 4 rather than 2. H-108 Table 2B-4 shows a further reduction in waiting time by 14 percent could be affected by testing with replacement. For a lot with mean life 70 hr., H-108 Table 2B-2(a) shows expected waiting time to be 40.8 hr. for a sample of 4 tested without replacement. Suppose a sample of 4 is tested and failures are observed at 10 and 40 hr. The test is stopped with the second failure. If the test was, in fact, conducted without replacement

$$\hat{\theta}_{r,4} = \frac{1}{2}[50 + 2(40)] = 65$$

Since 65 > C = 12.5, the lot is accepted.

Time Terminated

H-108 Table 2C-1(b) shows r = 2 and T/θ_0 = .104 for a test of n = 2r = 4 units without replacement. Here T = .104(70) = 7.28 hr. Suppose a sample of 4 is tested with no failures by 7 hr., 17 min. Since two failures would have been necessary to reject, the lot is accepted.

Sequential

H-108 Table 2D-1(b) shows for Code B-2,

$$r_0 = 6$$
$$h_0 = .2254(70) = 15.8$$
$$h_1 = -.2894(70) = -20.3$$
$$s = .2400(70) = 16.8$$

with expected number of failures to reach a decision

Mean Life	Expected failures to a decision
0	$E_0(r) = 1.2$
$\theta_1 = 7$	$E_{\theta_1}(r) = 1.6$
s = 16.8	$E_s(r) = 1.1$
$\theta_0 = 70$	$E_{\theta_0}(r) = 0.3$

The acceptance and rejection lines are

Acceptance: V(t) = 15.8 + 16.8k
Rejection : V(t) = -20.3 + 16.8k

The usual sequential diagram can be represented in tabular form by solving the equations as follows:

Failure	Reject	Accept
1	*	32.6
2	13.3	49.4

where * indicates no rejection can occur on the first failure. Suppose 4 units are placed on test with replacement. The first failure occurs after 10 hr., so that

$$V(t) = nt = 4(10) = 40$$

Since $V(t) = 40$ exceeds the acceptance line value of 32.6, the lot is accepted.

Proportion of Lot Failing by Specified Time

The specifications on mean life can be converted to proportion failing by a specified time by use of the relationship given in Table 18-2. In this case, for $T = 13$

$$P_0 = 1 - e^{-\frac{13}{70}} = .169$$

$$P_1 = 1 - e^{-\frac{13}{7}} = .844$$

Here

$$R = \frac{P_1}{P_0} = \frac{.844}{.169} = 4.99$$

H-108 Table 2C-5 gives $r = 4$ and $D = 1.37$ for $R = 5$ with $\alpha = .05$, $\beta = .10$ so

$$n = \frac{1.37}{.169} = 8.1 \sim 8$$

For this test 8 units are placed on test for a maximum of 13 hr. without replacement. If 4 or fewer units have failed at 13 hr., the lot is accepted.

Measures

OC curves are provided in Table 2A-2 of H-108 indexed by life test sampling plan code designation. The curve for Code B-2 is shown in Fig. 18-4. The OC curves are for failure terminated plans, the curves for sequential plans and time terminated tests are essentially equivalent.

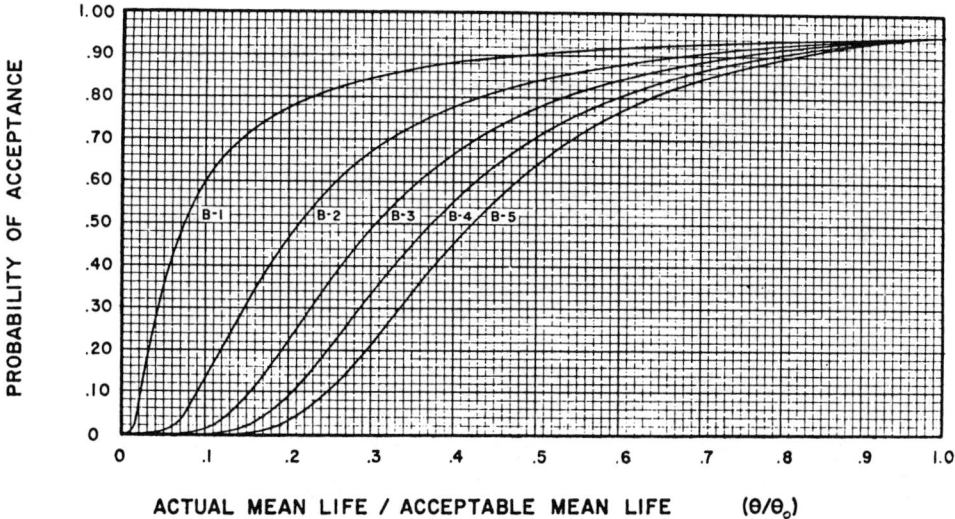

FIG. 18-4   H-108 Table 2A-2: Operating characteristic curves for life tests
terminated upon occurrence of preassigned number of failures. (From H-108,
p. 2.9.)

Further Considerations

The theory and development of the plans contained in H-108 will be found
in a comprehensive two-part paper published by B. Epstein (1960, 1960a) in
Technometrics.

TECHNICAL REPORT TR 7

Defense Department Quality Control and Reliability Technical Report TR 7
(1965) provides procedures and factors for adapting MIL-STD-105D plans to
life and reliability testing when a Weibull distribution of failure times
can be assumed. It allows appropriate test truncation times for specific
reliability criteria to be determined for use of the plans when all units
are not run to failure. The reliability criteria are

1.   Mean life μ: the expected life of the product
2.   Hazard rate $Z(t)$: the instantaneous failure rate at some
                specified time t
3.   Reliable life ρ_r: the life ρ beyond which some specified
                proportion r of the items in the population
                will survive.

Naturally, as with almost all variables criteria, these characteristics require Type B sampling.

All plans in TR 7 are based on an underlying Weibull distribution. Its cumulative probability distribution function may be written as

$$F(t_0) = p´ = P(t \leq t_0) = 1 - \exp\left[-\left(\frac{t_0 - \gamma}{\eta}\right)^\beta\right], \; t_0 \geq \gamma$$

with density function

$$f(t) = \frac{\beta(t - \gamma)^{\beta-1}}{\eta^\beta} \exp\left[-\left(\frac{t - \gamma}{\eta}\right)^\beta\right], \; t \geq \gamma$$

where γ = location (or threshold) parameter

  β = shape parameter

  η = scale parameter (characteristic life)

  μ = mean life

The mean μ of the Weibull distribution is at

$$\mu = \gamma + \eta\Gamma\left(1 + \frac{1}{\beta}\right)$$

with a hazard rate

$$Z(t) = \left(\frac{\beta}{\eta}\right)\left(\frac{t - \gamma}{\eta}\right)^{\beta-1}$$

and a reliable life

$$\rho_r = \gamma + \eta(-\ln r)^{1/\beta}$$

The location parameter γ is often taken to be zero. This is the case in TR 7. When it is not zero, e.g., when $\gamma = \gamma_0$, then the observations, t, are simply adjusted to $t´ = t - \gamma_0$ so that $\mu´ = \mu - \gamma_0$ and the analysis is performed in terms of $t´$ and $\mu´$. Naturally, final results are reported in terms of t and μ by reversing the process, to obtain

$$t = t´ + \gamma_0 \qquad \mu = \mu´ + \gamma_0$$

for final results $t´$ and $\mu´$. The procedures of TR 7 are independent of the scale parameter η, and so it need not be specified.

Probability plots and goodness of fit tests must be used to assure that individual measurements are distributed according to the Weibull model. When this distribution is found to be an appropriate approximation to the failure distribution, methods are available to characterize the product or a process by estimating the three parameters (γ, β, and η) of the Weibull distribution. These methods include estimates from the probability plots and also point and interval estimates.

The plans are given in TR 7 and are based on theoretical material and tables generated in three previous Defense Department quality control and reliability technical reports, each concerned with a specific reliability criterion used in TR 7. These are

Mean Life Criterion, TR 3 (1961)

Hazard Rate Criterion, TR 4 (1962)

Reliable Life Criterion, TR 6 (1963)

These three technical reports, written to be used with MIL-STD-105C, abound in excellent examples and detailed descriptions of the methods utilized in TR 7.

Once specified, the reliability criteria may be converted from one to the other using the relationships shown in Table 18-2. Mean life will be emphasized here because of its simplicity and the popularity of that criterion in nondefense life testing.

Mean Life Criterion

Technical Report TR 3 (1961) provided plans and procedures for developing and applying Weibull plans using mean life μ as the criterion for acceptance. The dimensionless ratio t/μ is related to the cumulative probability p'. Values of t or μ can easily be determined from the ratio t/μ once the other is specified. Since p' is the proportion of product failing before time t, it can be used in the role of "percent defective" in any attributes plan. The relationship of p' to t/μ, then, ties the "percent defective" to specified values of test time t and mean life μ.

The relationship is straightforward since

$$p' = F(t) = 1 - e^{-\left(\frac{t}{\eta}\right)^{\beta}}$$

and

$$\mu = \eta \Gamma\left(\frac{1}{\beta} + 1\right)$$

so that

$$p' = 1 - e^{-\left(\frac{t}{\mu}\Gamma\left(\frac{1}{\beta} + 1\right)\right)^{\beta}}$$

and solving for t/μ

$$\frac{t}{\mu} = \frac{(-\ln(1 - p'))^{1/\beta}}{\Gamma(\frac{1}{\beta} + 1)}$$

for proportion defective p' and any associated P_a.

Appendix Table T18-8 taken from TR 3 (1961) shows values of the ratio $100(t/\mu)$ corresponding to selected percents defective tabulated for a number of Weibull shape parameters. The table can be used to develop plans based on mean life or to convert measures of attributes sampling plans, notably their operating characteristics, in terms of mean life.

For example, suppose the plan $n = 20$, $c = 1$ is used in life testing of product with a Weibull shape parameter $\beta = 2$. A test termination time of 200 hr. is employed and the number of failures counted at that time. What are the operating characteristics of this plan in terms of mean life? Using the OC curve of the attributes plan and Appendix Table T18-8 we have

P_a	p	$100(t/\mu) = k$	$\mu = 100(200)\left(\frac{1}{k}\right)$
.983	.01	11.31	1768
.940	.02	16.03	1248
.810	.04	22.79	878
.517	.08	32.59	614
.289	.12	40.34	496
.176	.15	45.48	440
.069	.20	53.30	375
.024	.25	60.53	330

Further, suppose an attributes plan is to be derived having a PQL of 1250 hr. and a CQL of 400 hr. with risks $\alpha = .05$ and $\beta = .10$ respectively. Units are to be tested for 200 hr. and it is known that failures are distributed Weibull with $\beta = 2$. At

$$\text{PQL:} \quad 100\left(\frac{t}{\mu}\right) = 100\left(\frac{200}{1250}\right) = 16$$

and at

$$\text{CQL:} \quad 100\left(\frac{t}{\mu}\right) = 100\left(\frac{200}{400}\right) = 50$$

Using Appendix Table T18-8 we have

PQL: p = .02

CQL: p $\simeq$ .179

giving an operating ratio of

$$R = \frac{.179}{.02} = 8.95$$

and, using the table of unity values, the plan required is n = 20, c = 1. TR 3 contains other tables which facilitate the development of plans in this way, as do TR 4 and TR 6 for the other reliability criteria.

Hazard Rate Criterion

Technical Report TR 4 (1962) was patterned after TR 3, using the product $tZ(t) \times 100$ in place of the dimensionless ratio $(t/\mu) \times 100$. Note that the value of t given is the termination time of the test. For hazard rates specified for other times, tables were provided to convert the hazard rate specified into a corresponding hazard rate at termination time of the test. The cumulative probability p´ is related to $tZ(t) \times 100$ in a manner similar to TR 3. Resulting values and classifications useful in converting any attributes plan to a Weibull life test, where hazard rate is specified, are presented in TR 4.

Reliable Life Criterion

Technical Report TR 6 was also patterned somewhat after its predecessors, TR 3 and TR 4. It used the dimensionless quantity $(t/\rho) \times 100$ in the manner of $(t/\mu) \times 100$ and $tZ(t) \times 100$ in the previous reports. The cumulative probability p´ is related to $(t/\rho) \times 100$ and resulting values and classifications useful in converting any attributes plan to a Weibull life test where reliable life is specified, were tabulated.

TR 7 Tables

TR 7 combines the results of the preceding three technical reports in a document specifically intended to relate MIL-STD-105D to reliability testing where a Weibull distribution of failures can be assumed. Tables of the appropriate conversion factors are provided for the following criteria:

Table	Criterion	Conversion factor
1	Mean life	$(t/\mu) \times 100$
2	Hazard rate	$tZ(t) \times 100$
3	Reliable life $(r = .90)$	$(t/\rho) \times 100$
4	Reliable life $(r = .99)$	$(t/\rho) \times 100$

Each table is presented in three parts each of which is indexed by 10 values
of β (β = 1/3, 1/2, 2/3, 1, 1-1/3, 1-2/3, 2, 2-1/2, 3-1/3, 4). The three
parts are as follows for each criterion.

Part	Tabulation
A.	Values of the conversion factor corresponding to the AQL shown in MIL-STD-105D indexed by code letter and AQL
B.	Values of the conversion factor corresponding to a consumer's risk of P_a = .10 indexed by code letter and AQL
C.	Values of the conversion factor corresponding to a consumer's risk of P_a = .05 indexed by code letter and AQL

An additional table, TR 7 Table 2D, allows conversion of a specified hazard
rate to the corresponding hazard rate at test truncation time for use with
the tables which are in terms of hazard rate at test truncation time. TR 7
Table 2D presents values of $Z(t_2)/Z(t_1)$ indexed by the ratio of times
involved, t_2/t_1, and the various shape parameters.

The tables for the mean life criterion are reproduced here as follows.

Appendix Table T18-9	TR 7 Table 1A, 100 t/μ ratios at the AQL
Appendix Table T18-10	TR 7 Table 1B, 100 t/μ ratios at the limiting quality level, consumer's risk = .10
Appendix Table T18-11	TR 7 Table 1C, 100 t/μ ratios at the limiting quality level, consumer's risk = .05

The structure of TR 7 is shown in Fig. 18-5.

Operation

The conversion factors are employed in a manner identical to those presented
in TR 3, TR 4, and TR 6 and can be used to

1. Determine sample size necessary in testing for a fixed period of
   time to a specified value of the test criterion (mean life, hazard
   rate, reliable life) on the basis of desired AQL.

2. Determine the operating characteristics in terms of the test

criterion for given test times if a MIL-STD-105D plan has already been specified.

3. Determine a limiting quality level plan in terms of test time and MIL-STD-105D criteria for a specified value of the test criterion.

4. Determine a MIL-STD-105D plan most nearly matching the AQL, 10 percent limiting quality, or 5 percent limiting quality.

The report gives a detailed explanation with examples of how to reach these ends. A summary of the operation of TR 7 is given in Table 18-5.

A comparison of the criteria for the MIL-STD-105D normal plan, Code F, 2.5 percent AQL (n = 20, c = 1) for the case in which β = 2 may be instructive. TR 7 gives the following conversion factors.

Criterion	Factor	AQL	Limiting quality $P_a = .10$	$P_a = .05$
Percent defective (MIL-STD-105D)	$p' \times 100$	2.50	19.5	23.7
Mean life	$(t/\mu) \times 100$	18.0	50	56
Hazard rate	$tz(t) \times 100$	5.06	40	50
Reliable life (r = .90)	$(t/\rho) \times 100$	49	130	150
Reliable life (r = .99)	$(t/\rho) \times 100$	159	440	500

These criteria may be used to characterize a specific application. For example, if 125 units are to be tested for t = 10 hr., the converted values, in terms of the specified criteria become

Criterion	AQL	Limiting quality $P_a = .10$	$P_a = .05$
Percent defective (MIL-STD-105D)	2.50	19.5	23.7
Mean life	55.6	20	17.9
Hazard rate (t = 10 hr.)	.00506	.04	.05
Hazard rate at 20 hr.	.01012	.08	.10
Reliable life (r = .90)	20.4	7.7	6.7
Reliable life (r = .99)	6.3	2.3	2.0

The hazard conversion factors always give the hazard rate at the time of termination of the test. The hazard rate at 20 hr. was determined using Table 2D which shows

$$\frac{z(t_2)}{z(t_1)} = 2.00$$

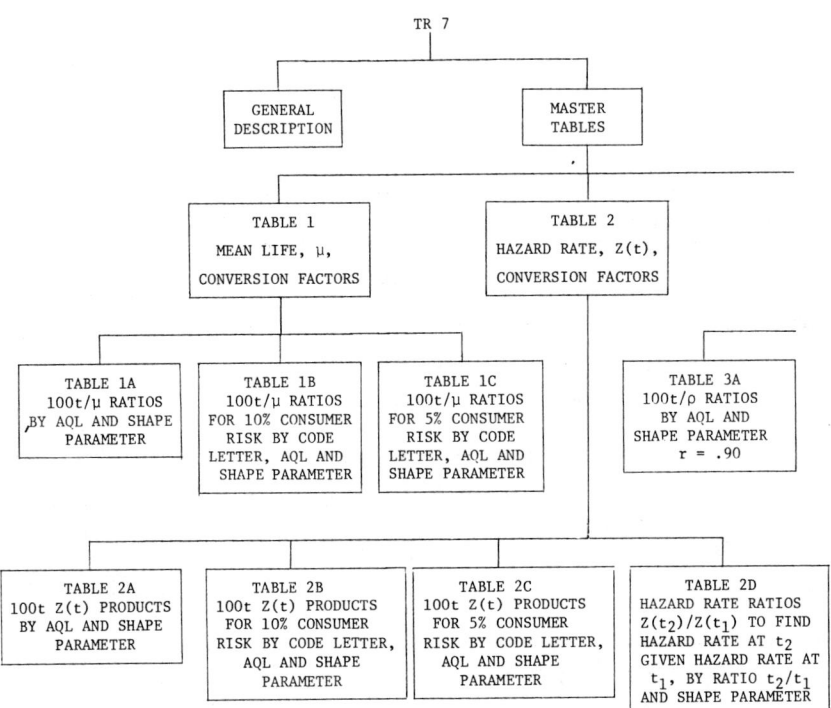

FIG. 18-5   Structure of TR 7.

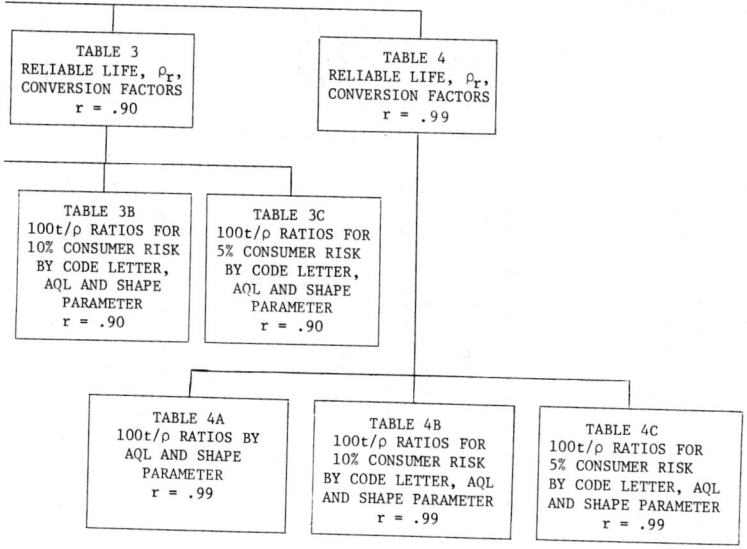

TABLE 18-5 Operation of TR 7

STEP\CRITERION	MEAN LIFE	HAZARD RATE	RELIABLE LIFE	MIL-STD-105D SCHEME
TYPE TEST	TIME TERMINATED			
SPECIFIED	μ_0 = acceptable mean life μ_1 = unacceptable mean life t = test time β = shape parameter	$z_0(t_0)$ = acceptable hazard rate at time t_0 $z_1(t_0)$ = unacceptable hazard rate at time t_0 t = test time β = shape parameter	ρ_0 = acceptable reliable life for proportion r surviving ρ_1 = unacceptable reliable life for proportion r surviving r = proportion surviving beyond life ρ t = test time β = shape parameter	o Code Letter from MIL-STD-105D Table 1 o AQL in terms of μ, $Z(t)$, or ρ_r o Shape parameter
CRITERIA	TR 7 Table 1A gives AQL corresponding to 100 t/μ_0 TR 7 Table 1B or 1C gives Code Letter corresponding to AQL and 100 t/μ_1 for consumer risk desired	If t $\neq$ t_0 convert hazard rates to time t using TR 7 Table 2D TR 7 Table 2A gives AQL corresponding to 100 t Z(t) TR 7 Table 2B or 2C gives Code Letter corresponding to AQL and 100 t Z(t) for consumer risk desired	TR 7 Table 3A or 4A gives AQL corresponding to 100 t/ρ for r specified TR 7 Table 3B,C or 4B,C gives Code Letter corresponding to AQL and 100 t/ρ for r specified and consumer risk desired	Obtain conversion factor from Table A of appropriate section given AQL and shape parameter Determine test time t algebraically from conversion factor and AQL Check limiting quality of normal plan through Tables B and C of appropriate section
PROCEDURE	SINGLE PLAN - Use criteria from MIL-STD-105D normal plan for Code Letter and AQL determined SCHEME - Use MIL-STD-105D system with AQL and Code Letter determined from normal plan above Test for time t and apply plan to failures observed			Use MIL-STD-105D scheme for Code Letter and AQL Test for time t and apply MIL-STD-105D to failures observed

when

$$\frac{t_2}{t_1} = 2$$

for the case of a Rayleigh distribution, $\beta = 2$. This is the effect of the linear increasing failure rate typical of a Rayleigh distribution.

The plans presented in TR 7 are, of course, time terminated. The conversion factors given in the technical report can be used to determine test termination time directly. Conversely, when a MIL-STD-105D test is conducted for a specified termination time, t, associated values of the desired reliability criterion can be found as

Mean life: $\qquad \mu = \dfrac{100}{\text{mean factor}} \times \text{termination time}$

Hazard: $\qquad Z(t) = \dfrac{\text{hazard factor}}{100 \times \text{termination time}}$

Reliable life: $\quad \rho_r = \dfrac{100}{\text{reliable life factor}} \times \text{termination time}$

Using the factors, test termination time can be determined from the specified reliability criterion as

Mean life: $\qquad t = \dfrac{\text{mean factor}}{100} \times \text{mean life}$

Reliable life: $\quad t = \dfrac{\text{reliable life factor}}{100} \times \text{reliable life}$

When hazard rate is specified at time t_0 the hazard rate must be transformed into the hazard rate at the specified test termination time t. This may be done through the use of Table 18-6, developed by the author, which shows values of

$$Q = \frac{t_2 \, Z(t_2)}{t_1 \, Z(t_1)}$$

To use Table 18-6, obtain the conversion factor from the appropriate table in TR 7. Calculate Q_0 as

$$Q_0 = \frac{100 \, t_0 \, Z(t_0)}{\text{hazard factor}}$$

Locate Q_0 in the column for the applicable shape parameter and read the corresponding value of t_2/t_1. The required test termination time is

TABLE 18-6 Values of $Q = t_2 Z(t_2)/t_1 Z(t_1)$ Corresponding to the Ratio t_2/t_1

t_2/t_1					Shape Parameter (β)					
	1/3	1/2	2/3	1	1-1/3	1-2/3	2	2-1/2	3-1/3	4
1.25	1.08	1.12	1.16	1.25	1.35	1.45	1.56	1.75	2.10	2.44
1.50	1.14	1.22	1.31	1.50	1.72	1.97	2.25	2.76	3.86	5.06
1.75	1.21	1.32	1.45	1.75	2.11	2.54	3.06	4.05	6.46	9.38
2.00	1.26	1.41	1.59	2.00	2.52	3.17	4.00	5.66	10.08	16.00
2.25	1.31	1.50	1.72	2.25	2.95	3.86	5.06	7.59	14.93	25.63
2.50	1.36	1.58	1.84	2.50	3.39	4.61	6.25	9.88	21.21	39.06
2.75	1.40	1.66	1.96	2.75	3.85	5.40	7.56	12.54	29.14	57.19
3.00	1.44	1.73	2.08	3.00	4.33	6.24	9.00	15.59	38.94	81.00
3.25	1.48	1.80	2.19	3.25	4.81	7.13	10.56	19.04	50.85	111.57
3.50	1.52	1.87	2.31	3.50	5.31	8.07	12.25	22.92	65.10	150.06
3.75	1.55	1.94	2.41	3.75	5.83	9.05	14.06	27.23	81.93	197.75
4.00	1.59	2.00	2.52	4.00	6.35	10.08	16.00	32.00	101.59	256.00
4.25	1.62	2.06	2.62	4.25	6.88	11.15	18.06	37.24	124.35	326.25
4.50	1.65	2.12	2.73	4.50	7.43	12.27	20.25	42.96	150.44	410.06
4.75	1.68	2.18	2.83	4.75	7.98	13.42	22.56	49.17	180.15	509.07
5.00	1.71	2.24	2.92	5.00	8.55	14.62	25.00	55.90	213.75	625.00

$$t_c\left(\frac{1}{t_2/t_1}\right)$$

Note that if the product $t_0 Z(t_0)$ is less than the factor it may be necessary to convert the hazard rate to a longer time period sufficient to make Q_0 larger than 1.

For example, suppose Code F, 2.5 percent AQL was used with a termination time 500 hr. The three reliability criteria corresponding to the AQL are to be estimated with a shape parameter 2. From Table A of each section we have

Reliability criterion	Factor
Mean life (100 t/μ)	18.0
Hazard rate (100 tZ(t))	5.06
Reliable life r = .90 (100 t/ρ)	49.0
Reliable life r = .99 (100 t/ρ)	159

For a 500 hr. test, these correspond to

Mean life	$\frac{100}{18} \times 500 = 2778$
Hazard rate (at 500 hr.)	$\frac{5.06}{100 \times 500} = .00010$
Reliable life (r = .90)	$\frac{100}{49} \times 500 = 1020$
Reliable life (r = .99)	$\frac{100}{159} \times 500 = 314$

Now, suppose Code F, 2.5 percent AQL is specified to have 10 percent limiting quality for the following reliability criteria. What would be the test time for each?

Reliability criterion	10% limiting quality
Mean life	1000 hr.
Reliable life r = .99	113 hr.
Hazard rate at 1000 hr.	.00157

The conversion factors from Table B of each section are

Mean life:	50
Reliable life:	440
Hazard rate:	40

so that the test termination times are

Mean life: $t = \dfrac{50}{100} \times 1000 = 500$ hr.

Reliable life: $t = \dfrac{440}{100} \times 113 \ = 497$ hr.

To determine the time necessary for testing the hazard rate, the ratio ϱ_0 is formed

$$\varrho_0 = \frac{100 \times 1000 \times .00157}{40} = 3.925$$

Table 18-6 indicates $t_2/t_1 = 2$. Hence, test time will be

$$t_0\left(\frac{1}{t_2/t_1}\right) = 1000\left(\frac{1}{2}\right) = 500 \text{ hr.}$$

Note that all three reliability criteria are equivalent as pointed out earlier in the chapter and they all lead to the same test termination time (500 hr.) for use with MIL-STD-105D, Code F, 2.5 percent AQL.

To select a normal plan from TR 7 for specified μ_1 and μ_2 given test time t and shape parameter β, proceed as follows:

1. Compute the ratios $100t/\mu_1$ and $100t/\mu_2$.
2. Enter TR 7 Table 1A (Appendix Table T18-9) to find the AQL corresponding to the value of $100t/\mu_1$ shown under the shape parameter.
3. Enter TR 7 Table 1B (Appendix Table T18-10) if P_a = .10 or Table 1C (Appendix Table T18-11) if P_a = .05 for the consumer's quality level. Moving down the column for the shape parameter, find the point where the AQL found in step 2 matches the ratio $100t/\mu_1$ listed in the column. This is the code letter for the plan.

Use the code letter and AQL to determine sample size and acceptance number from the normal plan given in MIL-STD-105D.

Thus, to determine a plan having AQL = 2800 hr. and 10 percent limiting quality of 1000 hr. for a test of 500 hr. for a life distribution with shape $\beta = 2$:

1. The ratios for μ_1 and μ_2 are 17.9 and 50, respectively.
2. AQL = 2.5.
3. Plan is Code F, 2.5 AQL.

The MIL-STD-105D normal plan n = 20, c = 1 testing for 500 hr. will give the protection desired.

TR 7 with the MIL-STD-105D System

The procedure of TR 7 and the preceding three technical reports was intended
to facilitate development of a single reliability or life test plan using
the criteria of a MIL-STD-105 normal plan. Utilization of TR 7 in conjunction with the MIL-STD-105D system and its switching rules is not described
in the technical reports. However, adaptation to use of the switching rules
is straightforward.

Once the sample size code letter and AQL have been determined together
with test time under the normal plan, the corresponding reduced and tightened acceptance criteria may be substituted for those of the normal plan as
appropriate to the switching rules. This leads to greater protection for
both the producer and the consumer. The OC curve of the resulting scheme
can be obtained by adapting the scheme OC curves given by Schilling and
Sheesley (1978) to the reliability criterion used through the conversion
factors given in TR 3, TR 4, and TR 6.

For example, in using the switching rules, the normal plan Code F,
2.5 AQL, t = 500 hr. would be incorporated into a scheme as follows:

Tightened: n = 32, Ac = 1, Re = 2, t = 500

Normal: n = 20, Ac = 1, Re = 2, t = 500

Reduced: n = 8, Ac = 0, Re = 2, t = 500

Here, the MIL-STD-105D limit numbers would not be used in switching to
reduced inspection. Using Appendix Table T18-8 for $\beta = 2$, the nominal AQL
would be

$$\frac{100t}{\mu} = 17.95 \qquad \mu = 2786 \text{ hr.}$$

From the Schilling-Sheesley tables (Appendix Table T11-21) the LTPD for
scheme performance would be about 12 percent, which converts to a mean
life of

$$\frac{100t}{\mu} = 40.34 \qquad \mu = 1239 \text{ hr.}$$

Note that use of the normal plan alone would result in a limiting quality
of 1000 hr., which shows the increased protection afforded by using the
switching rules.

The selection of a plan depends upon the use to which it will be put.
If a single plan is to be obtained, the procedure is simply that of determining a suitable match between the reliability criterion selected and the

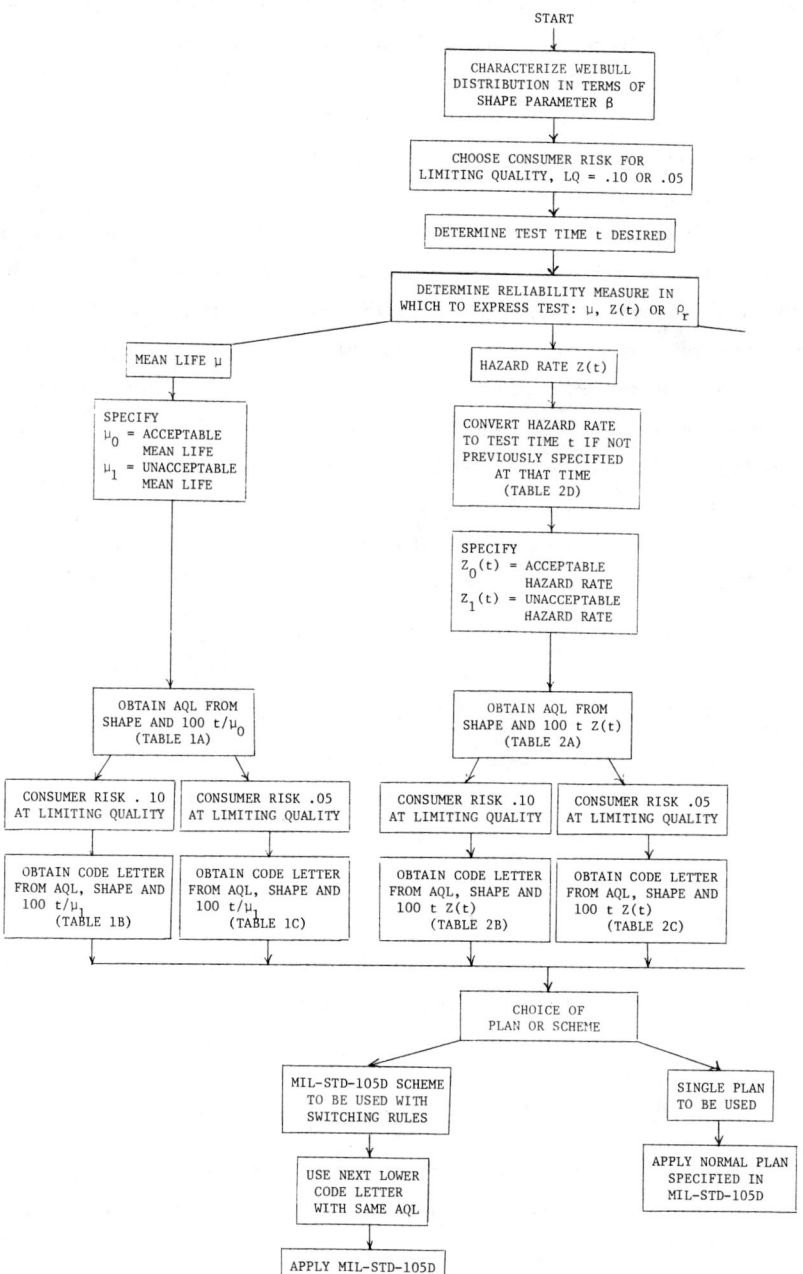

FIG. 18-6   Check sequence for determining procedure for TR 7.

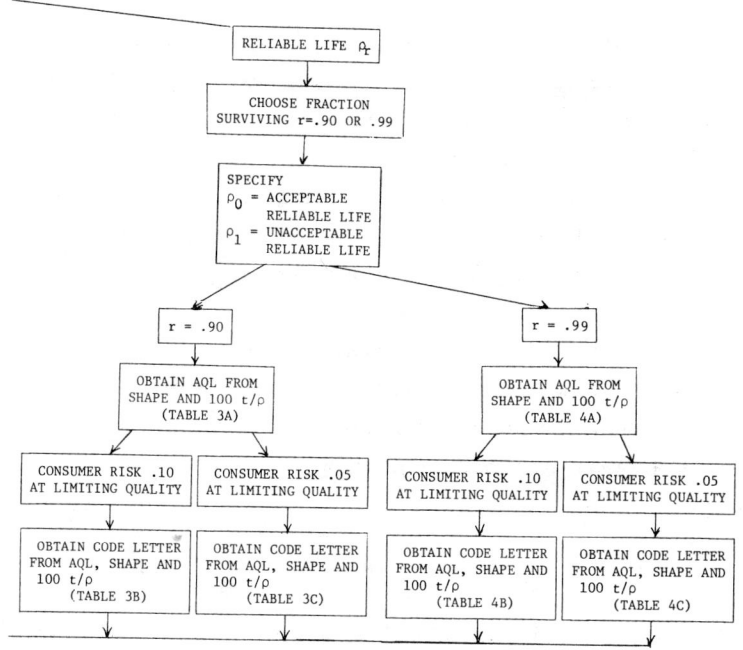

normal inspection attributes plans of MIL-STD-105D. If the MIL-STD-105D
system is to be used, with the switching rules, an appropriate AQL and code
letter combination must be found (see Table 18-5). Schilling and Sheesley
(1978) have shown that use of the system can, as a minimum, result in lowering the sample size code letter at least to the next lower category. This
has been incorporated in the check sequence for selecting a plan given in
Fig. 18-6. The procedure given is for matching producer's and consumer's
quality levels with the corresponding risks in a two-point procedure for
both a single plan and the MIL-STD-105D scheme.

Further Considerations

The development of TR 3, TR 4, and TR 6 which culminated in Technical
Report TR 7 was the outgrowth of the work of H. P. Goode and J. H. K. Kao
at Cornell University. Three papers were published in the *Proceedings of
the National Symposium on Reliability and Quality Control*. These classic
works were Sampling Plans Based on the Weibull Distribution (1961), Sampling
Plans and Tables for Life and Reliability Testing Based on the Weibull
Distribution (1962), and Weibull Tables for Bio-Assaying and Fatigue Testing
(1963).

   These papers led directly to straightforward application of the Weibull
distribution in acceptance sampling as typified by TR 7, which was prepared
by Professor Henry P. Goode. An analogous procedure for variables inspection based on MIL-STD-414 was subsequently developed by Kao (1964) while at
New York University.

PROBLEMS

1. If the eighth observation in the motorette data of Table 18-1 was 6000
   hr. with the other two units still running, how would the probability
   plotting positions be changed?

2. If mean life from a Weibull distribution with shape parameter $\beta = 3$
   is $\mu = 200$ hr., what is
   a. Proportion failing before 200 hr.
   b. Proportion surviving to 200 hr.
   c. Hazard rate at 200 hr.
   d. Cumulative hazard rate at 200 hr.

3. H-108 is to be used in a life test where $\theta_0 = 150$ hr. and $\theta_1 = 50$ hr.
   with $\alpha = .05$ and $\beta = .10$. Find the appropriate failure terminated

plan. Suppose 16 units are placed on test with replacement and the
eighth failure occurs at 40 hr., should the lot be accepted?

4. Find a time terminated test appropriate to the specifications of Prob.
    3. The test is made with replacement and a sample twice as big as r
    is used. Should the lot be accepted?

5. Suppose, by mistake, the test of Prob. 4 was performed without replace-
    ment. Should the lot be accepted?

6. The government inspector prefers that a sequential plan be used instead
    of the failure terminated plan of Prob. 3. The data is tested against
    a sequential plot. What are the parameters?

    a. r_0
    b. h_0
    c. h_1
    d. s

    Compute V(t) for the eighth failure. What would this value of V(t)
    indicate as to the disposition of the lot?

7. The specifications of Prob. 3 are identical to a proportion failing
    before 30 hr. of p_0 = .18 and p_1 = .45. Find a plan appropriate to
    these specifications with α = .05 and β = .10 if testing is performed
    without replacement.

8. Suppose the plan n = 32, c = 2 is used in life testing with a termina-
    tion time of 100 hr. Use $p_{.95}$ = .025, $p_{.10}$ = .15. For a Weibull dis-
    tribution with β = 1.67, what are the values of mean life having
    probability of acceptance .95 and .10?

9. MIL-STD-105D normal plan Code K, 6.5 AQL is being used in the life
    testing of a Weibull distribution having a shape parameter of β = 2.5
    using a termination time of 50 hr. Find, in terms of mean life, the

    a. AQL
    b. LQ at P_a = .10
    c. LQ at P_a = .05

10. Select a normal sampling plan from MIL-STD-105D having a high proba-
     bility (AQL) of passing units with mean life 1500 hr. and a low proba-
     bility (LQ at P_a = .10) of passing units with mean life 500 hr. The
     shape parameter of the applicable Weibull distribution is β = 1.
     Testing is to be for 15 hr.

REFERENCES

AGREE (1957), *Reliability of Military Electronic Equipment*, AGREE Task Group Report, U.S. Government Printing Office, Washington, D.C.

Epstein, B. (1960), Tests for the Validity of the Assumption that the Underlying Distribution of Life is Exponential, Part I, *Technometrics*, 2(1): 83-101.

Epstein, B. (1960a), Tests for the Validity of the Assumption that the Underlying Distribution of Life is Exponential, Part II, *Technometrics*, 2(2): 167-183.

Goode, H. P., and J. H. K. Kao (1961), *Sampling Plans Based on the Weibull Distribution*, Proceedings of the Seventh National Symposium on Reliability and Quality Control, Philadelphia, Pennsylvania, pp. 24-40.

Goode, H. P., and J. H. K. Kao (1962), *Sampling Procedures and Tables for Life and Reliability Testing Based on the Weibull Distribution (Hazard Rate Criterion)*, Proceedings of the Eighth National Symposium on Reliability and Quality Control, Washington, D.C., pp. 37-58.

Goode, H. P., and J. H. K. Kao (1963), *Weibull Tables for Bio-Assaying and Fatigue Testing*, Proceedings of the Ninth National Symposium on Reliability and Quality Control, San Francisco, California, pp. 270-286.

Hahn, G. J., and W. Nelson (1971), Graphical Analysis of Incomplete Accelerated Life Test Data, *Insulation/Circuits*, 17(10): 79-84.

Hahn, G. J., and S. S. Shapiro (1967), *Statistical Models in Engineering*, John Wiley and Sons, New York.

Juran, J. M. (1974), Ed., *Quality Control Handbook*, 3rd Ed., McGraw-Hill, New York.

Kao, J. H. K. (1964), *Sampling Procedures and Tables for Inspection by Variables for Percent Defectives (Based on the Weibull Distribution)*, Proceedings of the Tenth National Symposium on Reliability and Quality Control, Washington, D.C., pp. 41-56.

Lloyd, D. K., and M. Lipow (1962), *Reliability: Management, Methods, and Mathematics*, Prentice-Hall, Englewood Cliffs, New Jersey.

Locks, M. O. (1973), *Reliability, Maintainability, and Availability*, Hayden Book Company, Rochelle Park, New Jersey.

Mann, N. R., R. E. Schafer, and N. D. Singpurwalla (1974), *Methods for Statistical Analysis of Reliability and Life Data*, John Wiley and Sons, New York.

Nelson, W. (1969), Hazard Plotting for Incomplete Failure Data, *Journal of Quality Technology*, 1(1): 27-52.

Nelson, W. (1982), *Applied Life Data Analysis*, John Wiley and Sons, New York.

Schilling, E. G., and J. H. Sheesley (1978), The Performance of MIL-STD-105D Under the Switching Rules, *Journal of Quality Technology*, Part 1, 10(2): 76-83; Part 2, 10(3): 104-124.

Sheesley, J. H. (1974), *Tables to Convert Hazard Rates to Probabilities*, Report Number 1300-1119, General Electric Company, Cleveland, Ohio.

United States Department of Defense (1963), *Military Standard, Sampling Procedures and Tables for Inspection by Attributes* (MIL-STD-105D), U.S. Government Printing Office, Washington, D.C.

United States Department of Defense (1957), *Military Standard, Sampling Procedures and Tables for Inspection by Variables for Percent Defective* (MIL-STD-414), U.S. Government Printing Office, Washington, D.C.

United States Department of Defense (1960), Sampling Procedures and Tables for Life and Reliability Testing, *Quality Control and Reliability (Interim) Handbook* (H-108), Office of the Assistant Secretary of Defense (Supply and Logistics), Washington, D.C.

United States Department of Defense (1961), Sampling Procedures and Tables for Life and Reliability Testing Based on the Weibull Distribution (Mean Life Criterion), *Quality Control and Reliability Technical Report* (TR 3), Office of the Assistant Secretary of Defense (Installations and Logistics), U.S. Government Printing Office, Washington, D.C.

United States Department of Defense (1962), Sampling Procedures and Tables for Life and Reliability Testing Based on the Weibull Distribution (Hazard Rate Criterion), *Quality Control and Reliability Technical Report* (TR 4), Office of the Assistant Secretary of Defense (Installations and Logistics), U.S. Government Printing Office, Washington, D.C.

United States Department of Defense (1963), Sampling Procedures and Tables for Life and Reliability Testing Based on the Weibull Distribution (Reliable Life Criterion), *Quality Control and Reliability Technical Report* (TR 6), Office of the Assistant Secretary of Defense (Installations and Logistics), U.S. Government Printing Office, Washington, D.C.

United States Department of Defense (1965), Factors and Procedures for Applying MIL-STD-105D Sampling Plans to Life and Reliability Testing, *Quality Control and Reliability Assurance Technical Report* (TR 7), Office of the Assistant Secretary of Defense (Installations and Logistics), Washington, D.C.

ADMINISTRATION OF ACCEPTANCE SAMPLING

Effective acceptance sampling involves more than the selection and application of specific rules for lot inspection. As an integral part of the quality system, the acceptance sampling plan, applied on a lot by lot basis, becomes an element in the overall approach to maximizing quality at minimum cost. Acceptance sampling plans are, after all, action rules and as such must be adapted in a rational way to current results and the nature and history of the inspection performed. This is what we have called *acceptance control*, involving a continuing strategy of selection, application, and modification of acceptance sampling procedures to a changing inspection environment.

While acceptance sampling is sometimes regarded as a passive procedure for adjudication of quality, the active role of inspection was recognized early by H. F. Dodge. In accepting the Shewhart Award from the American Society for Quality Control, Dodge (1950, p. 6), pointed out that

> Using inspection results as a basis for action on the
> product at hand for deciding whether to accept or reject
> individual articles or lots of product as they come
> along is, of course, an immediate chore that we always
> have with us. However, inspection results also pro-
> vide a basis for action on the production process for
> the benefit of future product, for deciding whether the
> process should be left alone or action taken to find
> and eliminate disturbing causes.

As such, inspection should involve

1. Good data
2. Quick information

3. Incentives for the producer to provide quality at satisfactory levels

4. Quantity of inspection in keeping with quality history.

Indeed, according to Dodge (1950, p. 8),

> A product with a history of consistently good quality
> requires less inspection than one with no history or
> a history of erratic quality. Accordingly, it is
> good practice to include in inspection procedures
> provisions for reducing or increasing the amount of
> inspection, depending on the character and quantity
> of evidence at hand regarding the level of quality
> and the degree of control shown.

Figure 19-1 illustrates this principle in terms of the extent and nature of quality history. It shows roughly how representative sampling procedures could be changed as quality history is developed. It assumes quality levels have been appropriately set and that other suppliers are available. The overriding principle in acceptance control is to continually adapt the acceptance procedures to existing conditions. A control chart on inspection results is an excellent means to monitor the progress of the inspection or as a check inspection device if more formal procedures have been discontinued. It will indicate when results show a need for reassessment of inspection procedures. The stages in the application of a sampling procedure are shown in Table 19-1.

QUALITY HISTORY

PAST RESULTS	LITTLE	MODERATE	EXTENSIVE	CRITERION
EXCELLENT	AQL Plan	Cumulative Results Plan	Demerit Rating or Remove Inspection	Almost no (<1%) lots rejected
AVERAGE	Rectification or LTPD Plan	AQL Plan	Cumulative Results Plan	Few (<10%) lots rejected
POOR	100 Percent Inspection	Rectification or LTPD Plan with Cumulative Results Criterion	Discontinue Acceptance	Many ($\geq$10%) lots rejected
AMOUNT	Less than 10 lots	10 - 50 lots	More than 50 lots	

FIG. 19-1 Progression of sampling plans: extent of quality history.

TABLE 19-1 Life Cycle of Acceptance Control Application

STAGE	STEP	METHOD
PREPARATORY	Choose plan appropriate to purpose	Analysis of quality system to define the exact need for the procedure
	Determine producer capability	Process performance evaluation using control charts
	Determine consumer needs	Process capability study using control charts
	Set quality levels and risks	Economic analysis and negotiation
	Determine plan	Standard procedures if possible
INITIATION	Train inspector	Include plan, procedure, records, and action
	Apply plan properly	Insure random sampling
	Analyze results	Keep records and control charts
OPERATIONAL	Assess protection	Periodically check quality history and OC curves
	Adjust plan	When possible change severity to reflect quality history and cost
	Decrease sample size if warranted	Modify to use appropriate sampling plans taking advantage of credibility of supplier with cumulative results
PHASE OUT	Eliminate inspection effort where possible	Use demerit rating or check inspection procedures when quality is consistently good
		Keep control charts
ELIMINATION	Spot check only	Remove all inspection when warranted by extensive favorable history

The preparatory phase involves setting the specifications for acceptance sampling and selecting a plan. When the plan is initiated, care
should be taken to train the inspector and to analyze the results of initial
applications so that any discrepancies or problems can be worked out of the
procedure. Later, analysis of feedback information allows tightening up if
necessary, but should be geared toward a reduction of inspection effort if
justified by the history of the application. This may lead to the use of
skip-lotting, chain sampling, acceptance control charts or other special
procedures in the later stages of the application. Finally, the inspection
should be phased out altogether and replaced by such procedures as a check
inspection or a control chart. Sampling plans should be regarded as stop-
gap measures, instituted to correct an immediate problem or to give the
assurance desired on present product. The information the plans generate
should be used to lessen the need for future inspection as much as possible.
Sampling procedures should be designed to self destruct at the appropriate
time.

Too often a sampling plan is instituted, not to be changed for years.
Too often no one involved can tell when a plan was originated, why, or to
what criteria. "We've always used that plan." "It was written on the back
of an old envelope when I took over." Or, "Joe told us to sample in this
way before he retired — you remember Joe ..." These are clear indications
of lack of acceptance control. Acceptance sampling is not being controlled
in such cases — rigor mortis has set in.

It should be evident that the feedback of quality information is essential for a rational system of acceptance control. E. R. Ott (1975, pp.
181-2) has pointed out

> There are two standard procedures that, though often
> good in themselves, can serve to postpone careful
> analysis of the production process:
>   1. On-line inspection stations (100% screening).
>   These can become a way of life.
>   2. On-line acceptance sampling plans which pre-
>   vent excessively defective lots from pro-
>   ceeding on down the production line, but
>   have no feedback procedure included.
> These procedures become bad when they allow or encour-
> age carelessness in production. It gets easy for

production to shrug off responsibility for quality and
criticize inspection for letting bad quality proceed.

Sampling plans cost money to design and implement. They can be used
to perform more than a police function. The information generated is invaluable; it is regrettable that these results are often simply filed away or
never recorded. The institution of a sampling plan should have associated
with it effective procedures for the feedback and utilization of the data
resulting from the plan.

But above all, to be effective, a sampling procedure needs to be
enforced. There is no clearer signal to a supplier to relax quality standards than the consistent acceptance by the consumer of sub-standard material.
A sampling plan that cannot be enforced should be dropped, for such a plan
is nothing more than a costly exercise in futility.

SELECTION AND IMPLEMENTATION OF A SAMPLING PROCEDURE

Sampling plans are the basic tools of acceptance control. As in any field
much of the skill of the artisan is reflected in the ability to select the
tools appropriate for the job. The uses of some sampling plans are quite
broad, for example single sampling by attributes. Others are used to meet
a very specific need, such as H-108.

Table 19-2 presents a list of possible plans to meet varying needs.
The two-point plans shown could be single, double, multiple, or sequential
as determined by the requirements of the specific application. Note that
some plans, such as the two-point plans, would fit almost every category.
The table is simply suggestive of the type and variety of plans that could
be employed for the purposes shown.

The implementation of a specific application is shown by the check
sequence given in Fig. 19-2. This follows through the stages shown in
Table 19-1, but emphasizes the role of feedback in the continuing application of the plan. Plans may be installed in the areas of receiving inspection, process inspection, final inspection, or as a check inspection of a
small quantity of finished product. The approach remains much the same in
all areas. Of prime importance is the distinction between two prime purposes for sampling as spelled out in American War Standard Z1.3(1942).

a. To provide a basis for action on the *product* as it comes to the
    inspector; accept, reject (or rework),

b. To provide a basis for action on the *process* in the interests

TABLE 19-2 Selection of Plan

PURPOSE	SUPPLY	ATTRIBUTES	VARIABLES
Simple guarantee of producer's and consumer's quality levels at stated risks	Unique Lot	Two point plan (Type A)	Two point plan (Type B)
	Series of Lots	Dodge-Romig LTPD Two point plan (Type B)	Two point plan (Type B)
Maintain level of submitted quality at AQL or better	Series of Lots	MIL-STD-105D QSS Plan	MIL-STD-414 No Calc Plan
Rectification guaranteeing AOQL to consumer	Series of Lots	Dodge-Romig AOQL Anscombe Plan	Romig Variables Plans
	Flow of Individual Units	CSP-1, 2, 3 Multi-level Plan MIL-STD-1235A	Use measurements as Go-No go
	Flow of Segments of Production	Wald-Wolfowitz Girschik	Use measurements as Go-No go
Reduced inspection after good history	Series of Lots	Skip-Lot Chain Deferred Sentencing	Lot Plot Mixed Variables-Attributes Narrow Limit Gaging
Check Inspection	Series of Lots	Demerit Rating	Acceptance Control Chart
Compliance to Mandatory Standards	Unique Lot	Lot Sensitive Plan	Mixed Variables-Attributes with c = 0
	Series of Lots	TNT Plan	Simon Grand Lot Plan
Reliability Sampling	Unique Lot	Two point plan (Type B)	H-108 TR 7
	Series of Lots	LTPD Plan QSS System CRC Plan	TR 7 using MIL-STD-105D switching rules
Check Accuracy of Inspection	Series of Lots	H-109	Use measurements as Go-No go

of future product; leave the process alone or correct the process. This distinction will bear on the type of plan installed, how it is administered, and, of course, the type of OC curve calculated to assess its performance.

When a two-point plan is to be employed, a comparison of the administrative aspects of single, double, multiple, and sequential sampling is shown in Table 19-3, adapted from the Statistical Research Group (1948). Experience has shown single-sampling plans to be the most frequently employed while double sampling incorporates most of the advantages of repeated

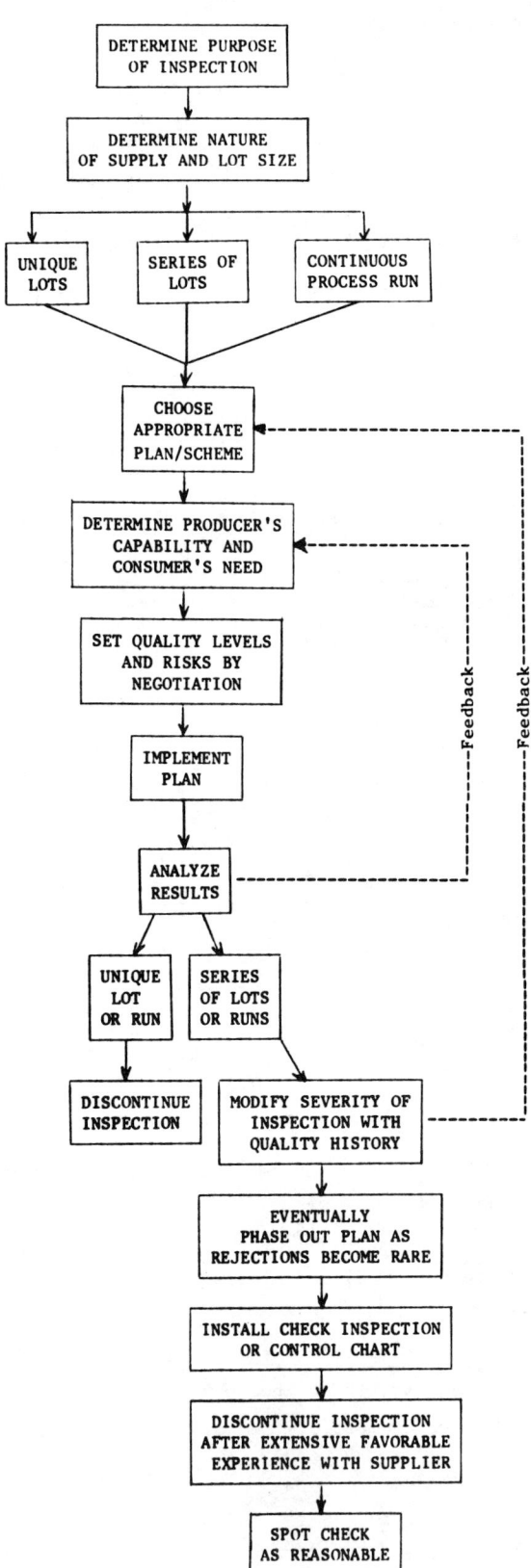

FIG. 19-2 Check sequence for implementation of sampling procedure.

TABLE 19-3 Comparison of Administrative Aspects of Single, Double, Multiple, and Sequential Sampling

	Single	Double	Multiple	Sequential
Average sample number	Most	Less	Still Less	Least
Number samples per lot	One	Two	Many	Most
Maximum number items inspected per lot	Least	More	Even more	Most
Variability in amount of inspection	None	More	Even more	Most

Adapted from Statistical Research Group (1948, p. 35).

samples without suffering many of the associated administrative burdens. Multiple and sequential plans seem less frequently employed because of variability in inspection load and complexity of administration even though they are, in terms of amount of inspection, the more efficient procedures.

DETERMINING QUALITY LEVELS

Before a sampling plan can be derived, one or more nominal quality levels must be set to define the protection to be afforded by the plan. These include

Acceptable Quality Level (AQL). Maximum fraction defective that, for purposes of acceptance sampling, can be considered satisfactory as a process average.

Average Outgoing Quality Limit (AOQL). Maximum average outgoing quality to the consumer under rectification.

Indifference Quality (IQ). Level of quality with equal chance of being accepted or rejected.

Lot Tolerance Percent Defective (LTPD). Objectionable level of quality that should be rejected at least 90 percent of the time. (Also 10 percent limiting quality.)

Producer's Quality Level (PQL). Level of quality which should be passed most of the time.

Consumer's Quality Level (CQL). Level of quality which should be rejected most of the time.

The AQL, AOQL, IQ, and LTPD are used to index many existing acceptance

sampling plans and schemes. Also, the PQL and CQL with associated risks
are used in the derivation of two-point plans. Risks must also be associated with the AQL, IQ, and LTPD; the latter two are fixed at .50 and .10 respectively, while the former sometimes varies over a range in the order of
0.5 to 13 percent as with MIL-STD-105D. These nominal quality levels are
necessarily fixed by the consumer to meet the consumer's needs with due
consideration to protecting reasonable producer quality levels from rejection. Sometimes this is done unilaterally, but more often by negotiation
between the consumer and the producer. The consumer should be as much
interested in good lots being accepted as the producer from the point of
view of scheduling and price.

In determining quality levels the consumer should attempt to minimize
the total cost in terms of purchase cost, inspection cost, assembly cost,
and eventual service costs. The first two costs increase as more and more
perfection is demanded, while the latter two decrease. The consumer should
not expect levels of quality better than those prevalent in industry without
special arrangements with the producer. Since it is usually not practical
to set quality levels for each customer, the producer must choose a level
acceptable to all the intended customers at prices they are willing to pay.
The sales, manufacturing, quality control, and engineering organizations of
both the producer and consumer should participate in setting quality levels
jointly weighing cost, feasibility, and customer acceptance. Quality levels
should be understood by both parties and form part of the purchasing specification either directly or by reference to recognized standards. It is
the producer's responsibility to perform sufficient inspection to assure
conformance. The consumer, however, should judge the producer's performance
on the basis of process average where possible and not on the results of a
single lot, since inspection of a single lot will seldom give a meaningful
estimate of longer run performance. These considerations have been amplified by the Electronic Industries Association (1956). Also, an excellent
discussion of some of the considerations important in setting quality levels
has been given by Hamaker (1949).

Setting AQL

The AQL is usually used as an index of sampling schemes and hence is used
with a series of lots. While the AQL is by nature associated with producer
risk, its magnitude must be established by the consumer. It represents the

consumer's estimate of the maximum fraction defective that can be tolerated for sampling purposes. Higher values are not acceptable. Lower values are desirable. Zero is seldom attainable at reasonable cost.

The state-of-the-art process average should be the starting point for determinining an AQL. This may be evaluated from past inspection results or by engineering estimate. In this regard, Bowker and Goode (1952, pp. 41-2) state

> The selection of an AQL range depends almost invariably
> upon a compromise between the quality that is likely to
> be submitted by the supplier and the quality that is
> ideal from a use standpoint (0 percent defective). The
> engineering and production staffs of the receiver can
> estimate the percent defective that can be tolerated
> from an economic or technical point of view. The qual-
> ity level one can reasonably expect from the supplier
> can best be determined from experience. In lieu of any
> experience with the item for which the plan is selected
> or with like items, some information might possibly be
> obtained from the supplier. An estimate of the quality
> currently obtainable should be made as close as possible
> in terms of percent defective, and if this estimate
> represents a satisfactory working quality.*

The state-of-the-art process average is *not* process capability. It should include allowances for differences between manufacturers and for variations in level of quality by a single manufacturer over time. It is the level at which quality can be expected to be maintained on a long-term basis, or at least for as long as the product in question is to be produced. If possible, in setting the AQL the consumer should perform a process per- formance evaluation on the data from previous inspection results on the same or similar material. Such procedures are outlined by Mentch (1980). In referring to the use of a process capability estimate from past data to set specifications, Mentch (1980, p. 121) points out

> This estimate of process capability is based on the
> assumption that it is feasible to bring the process

> into control in a technical and economic sense. Since
> this assumption is not always true, the use of this
> estimate of process capability to set specifications
> is not advisable. When it is necessary to set a
> specification to the process capability, this should
> be done on demonstrated performance in terms of con-
> sistently attainable levels ... and not on a collec-
> tion of historical data adjusted to be "in control."

The way to do this is to analyze process data over a sufficiently long
period to characterize the overall level of performance. This information
is used to set the AQL.

Bowker and Goode (1952, p. 42) suggest

> If the estimate of incoming quality is better (lower
> percent defective) than the quality one is willing to
> tolerate, and particularly if this estimate represents
> the best figure from a number of possible suppliers,
> it would be wise to make the AQL somewhat higher than
> this estimate, so that the acceptance criterion will
> be less exacting, fewer lots will be rejected, and
> costs will be reduced for all concerned. On the other
> hand, should the estimate for incoming quality be a
> higher percentage than the percent defective one can
> reasonably accept and use, the AQL class should be set
> at a lower percentage than the estimate, provided that
> the rejection of an excessive number of lots will not
> hamper the receiver's operations.[*]

This latter consequence would probably demand economic concessions to the
producer since an AQL lower than the state-of-the-art process average would
demand extensive screening of product. H-53 (1954, p. 13) points out

> Selecting extremely tight quality levels (low numerical
> values) might result in prohibitive inspection and end
> item costs, frequent rejection of products, or possible
> refusal by supplier to accept procurement orders or sign

---

[*]From *Sampling Inspection by Variables* by A. H. Bowker and H. P. Goode.
Copyright 1952, McGraw-Hill Book Company. Used with the permission of
McGraw-Hill Book Company.

contracts. On the other hand, selecting very liberal
quality levels (high numerical values) might result in
delivery of large quantities of unsatisfactory products
into the supply system.

Special considerations will, of course, motivate the consumer to move
the AQL away from the state-of-the-art process average. As listed by the
Statistical Research Group (1948, p. 84), some of these are

(a) Reduction in value of product occasioned by
    defectives. ... Sometimes the loss occasioned by
    a defective is so large that if there are more
    than a small percentage of defectives the product
    will be worth less than it costs. In such cases
    it may be desirable to fix the AQL at or below the
    breakeven percentage even if this should involve
    the rejection of a large proportion of submitted
    inspection lots.

(b) Class of defects. Major defects ordinarily reduce
    the value of product more than minor defects. Con-
    sequently the AQL should ordinarily be lower for
    major defectives than for minor defectives.

(c) Effect of defective product on later processing
    and assembling. If defective product results in
    marked waste of material and time during later pro-
    cessing and assembling, the AQL should be more
    exacting (lower). The number of items that are
    assembled may also play a part ...

(d) Suppliers' average quality and urgency of demand
    for product. If the quality that suppliers can
    furnish is poor and cannot readily be improved and
    if output is needed badly, the AQL may have to be
    higher than otherwise desired; if it is not higher
    excessive rejections may occur. If the suppliers'
    average quality can be expected to improve over a
    period of time, gradual lowering of the AQL may be
    desirable.

(e) Kind of defects included in the defects list. In
    order to permit consistent inspection and to keep

close control over the quality of product submitted,
it will sometimes be desirable to include in the
defects list defects whose effect on functioning is
questionable or to define defects more stringently
than is strictly necessary for the use to which the
product is to be put. When this is done, inspection
subjects the item to a severer test than the item
will receive when it is used and the AQL should
accordingly be more liberal than if each item were
subjected to a less severe test.

A further consideration is the number of different types of defects accumulated for test against a single AQL. H. M. Wadsworth (1970) has shown that
grouping of defects under a single inspection class (such as Majors) results
in an associated decrease in the effective AQL. For example, if N independent defect types are grouped together, each having fraction defective p, the
effective fraction defective P for the group would be

$$P = 1 - (1 - p)^N$$

As a consequence of this formula the AQL for the group, AQL_G, might be
increased to

$$\frac{AQL_G}{100} = 1 - \left(1 - \frac{AQL_I}{100}\right)^N$$

where AQL_I represents the AQL desired on each individual defect type.
This relationship is represented by Fig. 19-3 taken from Wadsworth (1970).
The AQL desired for each individual defect type is entered on the x-axis.
The AQL for the group (or class) is read from the y-axis. Thus, if two
defect types are classed as majors, each of which is to have a 4 percent
AQL, the AQL for the class should be 8 percent. This follows since

$$\frac{AQL}{100} = 1 - \left(1 - \frac{4}{100}\right)^2$$

AQL = 7.8 percent.

Setting AOQL

An AOQL should also reflect the consumer's need. Of course, this measure
of quality is meaningful only for a series of lots when rejected lots are

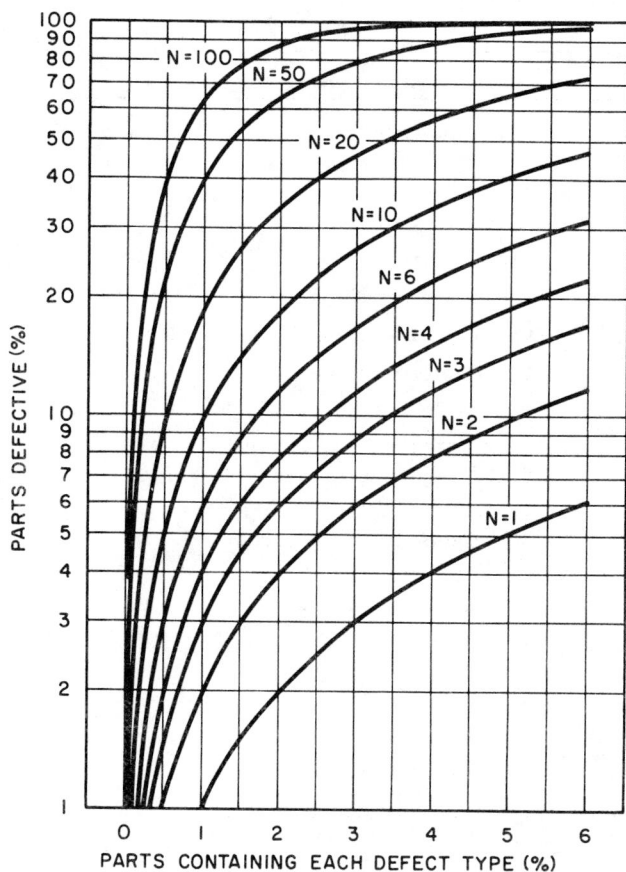

FIG. 19-3. Effect of grouping defects on percent
defective. (From Wadsworth, (1970, p. 182).
Reprinted by permission.)

100 percent inspected. Too high an AOQL will result in an uneconomic level
of defective material for the consumer. Too low an AOQL may cause excessive screening and higher costs particularly if set below the state-of-the-
art process average. Dodge (1948) suggests setting the AOQL about one and
one-half times higher than the state-of-the-art process average to avoid
excessive amounts of screening which results when the process average is
equal to the AOQL.

Dodge (1948) has described the administration of an AOQL plan in some
detail. As a guide to reasonable levels of AOQL, he tabulated a set of
values, "used in one shop under certain prescribed conditions." These are
shown here in Table 19-4.

TABLE 19-4   Examples of Choice of AOQL

Description	Inspection	AOQL
Metallic and Non-Metallic Rods, Tubes, Strips, and Sheets	Visual, Dimensional	3.0%
Milled Parts	Visual, Dimensional	0.75
Molded Plastic Parts	Visual, Dimensional	1.0
Die Cast Parts	Visual, Dimensional	1.0
Formed and Drawn Parts	Visual, Dimensional	2.0
Ferrous and Non-Ferrous Castings	Visual, Dimensional	3.0
Machine Screws	5 Dimensions	2.0
Hexagon Nuts	Visual Inspection after Zinc Plating	2.0
Twin Eyelets	6 Dimensions and 4 Visual Requirements	3.0
Ceramic Insulators	Visual, Dimensional	5.0
Misc. Inexpensive Electrical Apparatus	Breakdown	0.25
Misc. Completed Electrical Apparatus	Resistance	0.5
Relay Coils[*]	Inductance and Electrical Breakdown	1.0

[*]This is a process check for the specified requirements which is supplemented by another sampling inspection after assembly.

From Dodge (1948, p. 14). Reprinted by permission.

In general it is good practice to have the producer perform any screening of rejected lots — or at least pay for it. Thus in internal sampling, the receiving department may perform the sampling, but the producing department should be responsible for the 100 percent inspection.

Setting IQ

Indifference quality, or the "point of control," can be used as an element in the economic determination of quality levels as shown by Enell (1954). It is also important as the index of the Philips Standard Sampling System. It is, of course, the level of quality having fifty percent probability of acceptance, i.e., the 50:50 breakeven point between acceptance and rejection. Fortunately, the breakeven point between the producer and the consumer is

relatively easy to determine, one of the chief advantages of indifference
quality plans. As Hamaker, Taudin Chabot, and Willemeze (1950, p. 363)
have pointed out that

> The point of control may conveniently be interpreted as
> the point dividing "good" and "bad" lots. Experience
> has taught that producer and consumer readily agree as
> to a suitable choice of this parameter.

Thus, the indifference quality level can be a useful measure in
characterizing a plan.

Setting LTPD (or LQ)

An LTPD (or LQ) may be used as a quality level in inspecting a single lot
or it may be used with a series of lots. It should be borne in mind that
LTPD constitutes an extremely pessimistic view of the protection afforded
by a sampling plan. After all, nine to one odds are roughly akin to
obtaining three heads in three flips of a coin. It is not too likely, but
it can happen. However, no producer could stay in business if 90 percent
of the lots were rejected. Hence, the LTPD should be set well beyond the
AOQL or the AQL. In discussing an LTPD plan, Schilling (1978, p. 49)
points out that for c = 0 plans,

> The fraction defective to be protected against by the
> plan should be set at a level no less than 22 times
> the fraction defective that represents the state of
> the art. When the level to be protected against is
> necessarily closer to the state of the art fraction
> defective, sampling plans allowing one or more defects
> in the sample should be used. Thus, for example, by
> accepting if the sample contains three or fewer defec-
> tives, a single sampling plan can be derived for which
> the fraction defective protected against can reasonably
> be set at five times the state of the art fraction
> defective.

Setting the LTPD *at least* five times the state-of-the-art fraction defective
(usually reflected in AQL) is probably a good rule of thumb.

Dodge and Romig (1958, p. 6) recommend that LTPD be chosen at a level
that will almost surely be met by every lot. In fact, they suggest

In choosing a value of LTPD consider and compare the
cost of inspection with the economic loss that would
ensue if quality as bad as the LTPD were accepted often.
Even though the evaluation of economic loss may be
difficult, relative values for different levels of
percent defective may often be determined.

Thus the LTPD should be carefully chosen to be an extremely pessimistic
quality level which should be rejected most of the time. In a series of
independent lots, the probability that two successive lots would pass at
the LTPD level of quality is just one percent.

Relation of Levels

The relation of AQL (when defined as having probability of acceptance of
.95), AOQL, IQ, and LTPD varies among individual plans. Using the Poisson
approximation, it is possible to portray this relation by a modification of
the Thorndyke chart. This is shown in Fig. 19-4 and can be used in assessing the effect and interrelationships of these quantities. Here we have
drawn additional curves labeled L and M respectively for p_L = AOQL and p_M
(the fraction defective at which the AOQL occurs) on the chart. For example,
for higher values of acceptance number, the AOQL of a single sampling plan
approaches and exceeds the AQL, their being equal at about c = 17. Also
when c = 17

$$np_L = 11.6$$

so that, for a plan with sample size 500,

$$AOQL = \frac{11.6}{500} = .023$$

The AOQL occurs at

$$np_M = 13.5$$

so that

$$p_M = \frac{13.5}{500} = .027$$

Setting PQL and CQL

The producer quality level p_1 and the consumer quality level p_2 are used
in setting two-point plans. In general the PQL should be set much as the

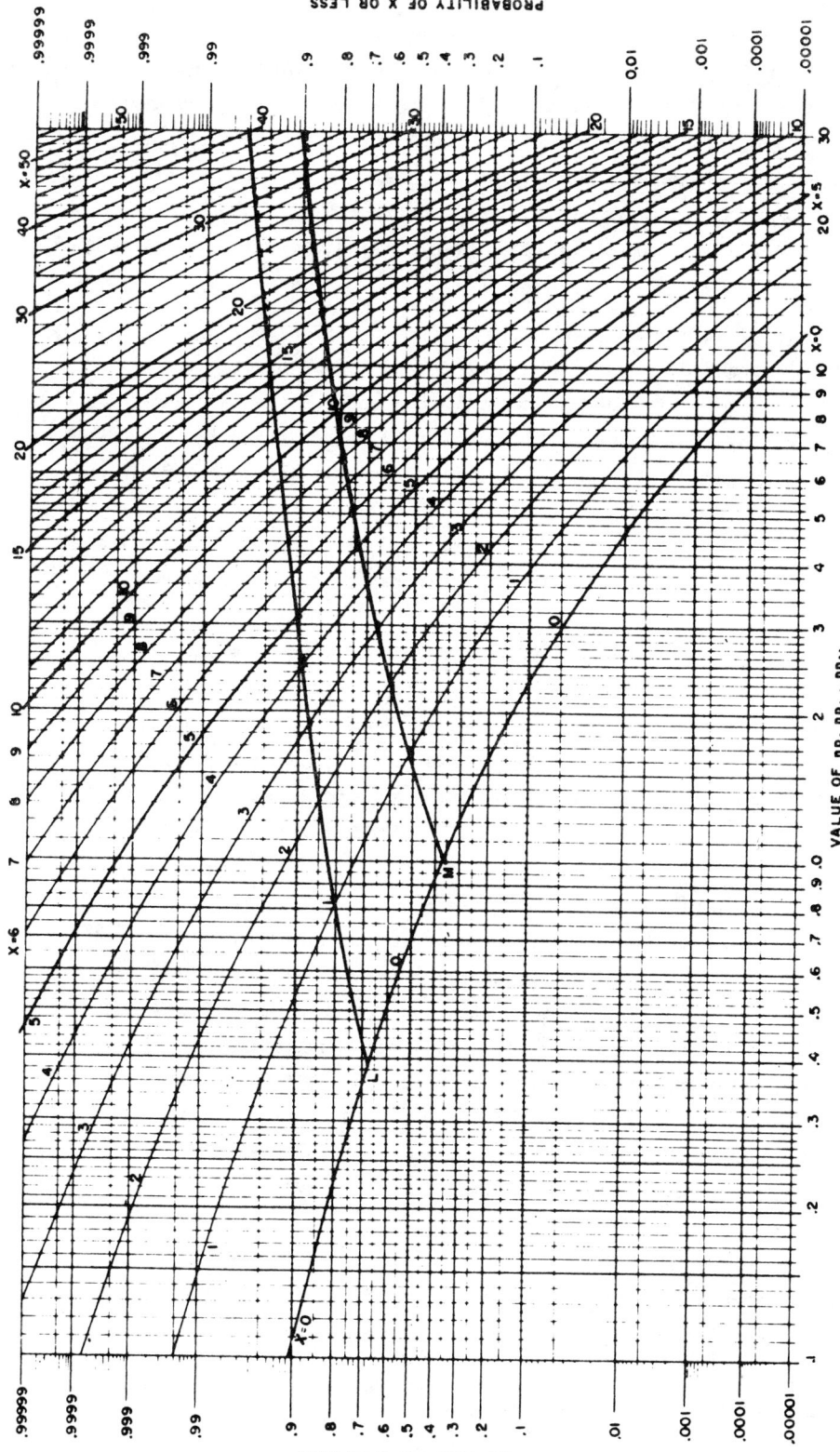

FIG. 19-4 Modified Thorndyke chart.

AQL is set for sampling schemes; that is, to reflect the state-of-the-art process average. The CQL, on the other hand should be a conservative level of unacceptable quality set much as the LTPD. The PQL and CQL are meaningless unless associated producer and consumer risks, α and β, are also quoted. It is common practice to set $\alpha = .05$ and $\beta = .10$.

In discussing two-point plans, Paul Peach (1947, pp. 27,33) suggests that

> ... Much as we would like to demand perfection in our purchased materials, such a demand is not practical. No method of inspection known can enforce such a standard. The common sense alternative is to decide in advance to tolerate some small proportion of defective material. This proportion ... is designated by the symbol p_1 or AQL.
>
> In general, p_1 should be known to the supplier, and should be part of the specification to be legally binding ... This does not apply to p_2; indeed, as a matter of policy, information about p_2 should never be given to an outsider. The vendor's contract sets a quality standard, which is, or should be p_1; his job is to meet that standard. The customer may, for the sake of economy in inspection, set some high value of p_2, thus taking a considerable risk of accepting sub-standard material; but it is not part of the vendor's business to inquire how great this risk is. Such information could only be used to cheat the customer, by enabling the vendor to manufacture, not to the standards of the contract, but to the loopholes in the customer's inspection. When a supplier asks a customer "How much inspection do you do?," the answer should be "That information is confidential" or "100 percent."
>
> In safeguarding his own rights, however, the buyer must not prejudice those of the supplier. If for his own convenience he changes from one inspection plan to another, he should be careful to keep p_1 and α the same; the supplier has a right to demand a fixed quality standard."[*]

*From P. Peach, *An Introduction to Industrial Statistics and Quality Control*, 2nd ed., 1947. Used by permission.

Military Standard 105D makes a similar point in Paragraph 4.4 when it states that, "The designation of an AQL shall not imply that the supplier has the right to supply knowingly any defective unit of product."

ECONOMIC CONSIDERATIONS

Ultimately, the selection of quality levels must be resolved by economic considerations. The consequences of various possible nominal levels must be weighed against costs, operating characteristics, and other factors. This may be done explicitly or implicitly.

Kavanagh (1946) has provided an in depth discussion of the procedure for determining unit cost of acceptance (in terms of saving by removing a defective) and unit cost of inspection. A simple model for balancing these costs has been presented by Enell (1954). Suppose costs are quoted on a per-unit basis and

A = unit cost of acceptance (i.e., the cost of one defective unit being accepted)

I = cost of inspection of one piece

C = cost of repairing or replacing one defective

p = fraction defective in the lot.

If R is the unit cost of rejection, then

$$R = \frac{I}{p} + C$$

which amounts to the cost of inspection to find one defective piece plus the cost of correcting it when found. At the breakeven point between the cost of acceptance and the cost of rejection

$$A = R \qquad A = \frac{I}{p} + C$$

with an associated fraction defective

$$p_B = \frac{I}{A - C}$$

at the breakeven point between the costs. But this is also the breakeven point between acceptance and rejection since when

$$p < p_B$$

the cost of rejection must exceed the cost of acceptance. Also, when

$$p > p_B$$

the cost of acceptance exceeds the cost of rejection. This can be seen in
Fig. 19-5 taken from Enell (1954).

Therefore, p_B may be regarded as the indifference quality since at p_B
the risk of acceptance and rejection would reasonably be the same (i.e.,
50:50). Single-sampling plans may be set up for this indifference quality
using the relation

$$p_B = \frac{c + 2/3}{n}$$

from the Poisson distribution as discovered by Campbell (1923). This gives

$$n = \frac{c + 2/3}{p_B}$$

where either n or c must be fixed before the formula can be used. Alternatively one of the Philips Standard Sampling System plans might be employed.

Enell, however, suggests that, using MIL-STD-105D, a plan be selected
having the indicated indifference quality for the sample size code letter
associated with the lot size in question. This might be done using the

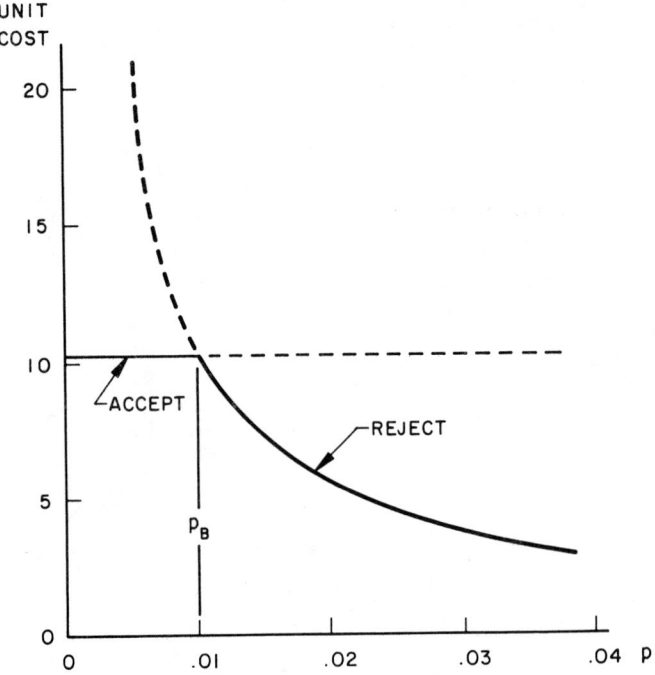

FIG. 19-5 Costs of acceptance and rejection. (From
Enell, (1954, p. 98). Reprinted by permission.)

Schilling and Sheesley (1978) tables for the MIL-STD-105D scheme.

For example, suppose

A = $1.00   I = $.026   C = $.50

then

$$p_B = \frac{.026}{1.00 - .50} = .052$$

Suppose lots of 100 are shipped so that Code F would be used. The Schilling-Sheesley tables show p_B = .055 is associated with AQL = 2.5 percent for Code F. Hence, the inspector would use Code F, 2.5 AQL for lots of 100.

Other, more sophisticated models for economic determination of quality levels and sampling plans have, of course, been presented. For example, see Smith (1965), Singh and Palanki (1976), or Liebesman (1979).

MANDATORY STANDARDS

Setting quality levels for sampling plans is sometimes regarded in terms of an adversary relationship between the producer and the consumer. Such a relationship is more apparent than real, however, and probably makes better semantics than sense.

The approach presented here implies consideration by the *consumer* of both the producer's and consumer's risks in setting up a sampling plan. The producer's risk must be given due consideration to protect the availability of supply to the consumer and to forestall price increases made necessary by a demand for unreasonable levels of quality.

The situation is analogous when sampling to mandatory standards for, in effect, the government represents the consumer. Where necessary, it is, after all, the state-of-the-art fraction defective which must be improved (not legislated). A cost-benefit analysis is clearly in order in setting quality levels for mandatory standards. Such an approach and its implications for sampling to mandatory standards have been discussed in an excellent seminal paper by Muehlhouse, Broussalian, Farrar, Lyons, Natrella, Rosenblatt, Stiehler, and Winger (1975).

Quality levels should be set in terms of the state-of-the-art fraction defective to be of greatest impact in the marketplace. To do otherwise would be to restrict supply and raise costs through rigid enforcement of unreasonable levels or to invite manufacturers to cheat on the standard through nonenforcement of unrealistic demands.

It should be remembered that the quality levels used in sampling are set for cost effective inspection and not as targets for performance. They should be changed as appropriate to reflect improvements in the state-of-the-art fraction defective.

COMPUTER PROGRAMS

The administration of acceptance sampling plans has been greatly simplified by the computer. Data bases can provide an excellent source for quality history, while individual computer programs can be used to set up and evaluate sampling plans and even to sentence individual lots.

A number of computer programs, useful in acceptance sampling, have been published in the literature. Edited through a 10-year period by K. E. Larson, R. P. Thayer, H. J. Mikulski, J. S. White, D. M. Olsson, and P. R. Nelson, the Computer Programs Department of the *Journal of Quality Technology* has been a continuing source of computer programs for use in quality control. An early representative set of these programs which illustrate acceptance sampling applications is given in Table 19-5. A variety is shown including programs for selecting plans, evaluating plans, and for arriving at acceptance decisions when complex procedures are involved.

The computer is a welcomed ally in making acceptance sampling applications fast and easy — a necessary requisite for application in industry.

A BASIC PRINCIPLE OF ADMINISTRATION

The need for simplicity and practicality in applications of acceptance sampling cannot be overstated. In no sense should integrity be sacrificed to expedience or theory to intuition. Nevertheless, it is often possible to devise simple, straightforward, theoretically correct methods which belie the complexity of more elegant procedures. A basic principle which applies to the administration of sampling plans is that the shop requires methods and procedures that are safe, sure, swift, and simple.

Unfortunately, straightforward mathematics seems often to lead to complicated procedures (e.g., MIL-STD-414) while simple methods (e.g., No-Calc) are often found almost intractable from the viewpoint of mathematical statistics. Nevertheless, the ultimate purpose of acceptance sampling is lot inspection on the factory floor. The methods must be understood and trusted by non-statisticians — inspectors, operators, supervisors, and the like before they are used. With Tukey (1959) we would agree with Churchill Eisenhart's

TABLE 19-5 Computer Programs for Acceptance Sampling (Optional Input/Output in Brackets)

PURPOSE	INPUT	OUTPUT	REFERENCE
SINGLE SAMPLING - DERIVATION	P_1, P_2, α, β, [lot size, p]	Plan [ATI at p, AOQ]	D. C. Snyder, R. F. Storer, JQT, July 1972
DOUBLE SAMPLING - DERIVATION	AQL, LTPD [lot size] (assumes α = .05, β = .10)	Plans for $n_2 = n_1$ and $n_2 = 2n_1$ with AOQL [P, P_a, AOQ, ATI, ASN]	B. Chow, P. C. Dickinson, H. Hughes, JQT, Oct. 1972, Oct. 1973
MULTIPLE SAMPLING - DERIVATION	AQL, LTPD [lot size] (assumes α = .05, β = .10)	Plans for equal n with AOQL [P, P_a, AOQ, ATI, ASN]	H. Hughes, P. C. Dickinson, B. Chow, JQT, January 1973
SINGLE, DOUBLE, MULTIPLE SAMPLING - EVALUATION P_a given p or p given P_a (Hypergeometric, Binomial Poisson, Normal)	Distribution, sample sizes, acceptance/rejection numbers, fractions defective to be evaluated, probabilities of acceptance to be determined [lot size]	Plan control table, P, P_a, ASN, AOQ, ATI	E. G. Schilling, J. H. Sheesley, P. R. Nelson, JQT, July 1978
DODGE CONTINUOUS SAMPLING - EVALUATION CSP-1,2,3	Fractions defective, index of plan (1,2,3), f, i, k	p, F, P_a, AOQ	J. H. Sheesley, JQT, January 1975
MIL-STD-414 - IMPLEMENTATION ACCEPTANCE/REJECTION Decision given data	Number lots, type inspection (T,N,R), measure of variability, lot size, specification limits, AQL, data	Plan, $\hat{p}$, M, accept or reject	P. R. Nelson, JQT, April 1977

definition of the practical power of a procedure as the product of its
mathematical power and the probability that the procedure will be used.

Examples of the importance of simplicity are legion. The popularity
of the range over the standard deviation in variables sampling is because
it is easy to understand and compute. So also with the $\overline{X}$, R chart. MIL-
STD-105 went from a complicated control chart approach for switching in
the A, B, and C versions to an easy counting rule in MIL-STD-105D because
of its simplicity. The intricacy of MIL-STD-414 has done much to forestall
wider application. Multiple sampling suffers, to a lesser extent, from the
same malady.

Acceptance sampling and the accompanying forms, methods, procedures,
and presentations must be made as uncomplicated as possible for successful
implementation in industry. Ease of application is not for the lazy but
for the industrious strapped by the tyranny of time. In the words of H. F.
Dodge (1973), "If you want a method or system used, keep it simple."

PROBLEMS

1. A variables sampling plan was instituted on the thickness of germanium
   pellets used in early transistors. Corrective action resulting from
   the feedback of information from this inspection, as plotted on control
   charts, led to extensive excellent quality history. What action should
   be taken?

2. About 100,000 components are manufactured each month for purchase by
   an original equipment manufacturer. A certain defect in a component
   could pose a potential safety problem. Accordingly, the customer has
   imposed a requirement that not more that 1 in 100,000 of these components may have the defect with 90 percent probability. What sampling
   procedure should be recommended?

3. Ten defect types are combined in the major defective category. Each
   has an AQL of 1 percent. What should be the combined AQL for the
   group?

4. If a 2 percent AOQL has been successfully used for machine screws, what
   might be a reasonable AQL = $p_{.95}$ if a sampling scheme is to be used?
   What might be a reasonable LTPD?

5. Experience has shown c = 2 to be a very desirable acceptance number
   for both producer and consumer. Using the indifference quality as a
   base, from the modified Thorndyke chart, what are the relative values
   of AOQL, IQ, LTPD?

6. A sampling plan is to be instituted on machine screws of a certain type. The unit cost of acceptance is $60 while the cost of inspecting a piece is $1. The cost to repair a defective unit is $6. What should be the indifference quality?

7. Using the results of Prob. 5, if c = 2 is to be used, what would be reasonable values of the AQL, AOQL, and LTPD in Prob. 6?

8. At present, a sampling inspection plan is applied manually at a cost of $.05 per piece using the plan n = 100, c = 2 on lots of 10,000 with a satisfactory level of outgoing quality. A computer is available which will perform the inspection on 100 percent of the product at a cost of $.001 per piece. Is it economical to purchase the computer? At what cost of inspection would the installation of the computer be worthwhile?

9. If replacement cost is negligible, we have $p_B = I/A$. Using the results of Prob. 5, convert this to a formula for AQL, for LTPD.

10. It has been traditional in some industries to "take a 10 percent sample of the lot," implying c = 0. This procedure has often been impuned since protection varies with lot size and to defeat the plan, it is necessary only to supply smaller lot sizes. If lots rejected under this procedure are 100 percent inspected, develop a formula for the AOQL from the formula

$$AOQL = \frac{Y}{n}\left(1 - \frac{n}{N}\right)$$

Does the result confirm or refute the criticism that protection varies with lot size?

REFERENCES

American Standards Association (1942), *Control Chart Method of Controlling Quality During Production, Z1.3*, American War Standard, New York, 1942.

Bowker, A. H., and H. P. Goode (1952), *Sampling Inspection by Variables*, McGraw-Hill, New York.

Campbell, G. A. (1923), Probability Curves Showing Poisson's Exponential Summation, *Bell System Technical Journal*, 2(1): 95-113.

Chow, B., P. C. Dickinson, and H. Hughes (1973), A Computer Program for the Solution of Double Sampling Plans, *Journal of Quality Technology*, 4(4): 205-209. Also, *Journal of Quality Technology*, 5(4): 166.

Dodge, H. F. (1948), Administration of a Sampling Inspection Plan, *Industrial Quality Control*, 5(3): 12-19.

Dodge, H. F. (1950), Inspection for Quality Assurance, *Industrial Quality Control*, 7(1): 6-10.

Dodge, H. F. (1973), Keep It Simple, *Quality Progress*, 6(8): 11-12.

Dodge, H. F., and H. G. Romig (1959), *Sampling Inspection Tables, Single Sampling and Double Sampling*, 2nd ed., John Wiley and Sons, New York.

Electronics Industries Association (1949), *Acceptable Quality Levels and How They Are Set Up*, Quality Acceptance Bulletin No. 2, Engineering Department, Electronics Industries Association, New York.

Enell, J. W. (1954), Which Sampling Plan Should I Choose, *Industrial Quality Control*, 10(6): 96-100.

Hamaker, H. C. (1949), Lot Inspection by Sampling, *Philips Technical Review*, 11(6): 176-182.

Hamaker, H. C., J. J. M. Taudin Chabot, and F. G. Willemze (1950), The Practical Application of Sampling Inspection Plans and Tables, *Philips Technical Review*, 11(12): 362-370.

Hughes, H., P. C. Dickinson, and B. Chow (1973), A Computer Program for the Solution of Multiple Sampling Plans, *Journal of Quality Technology*, 5(1): 39-42.

Kavanagh, A. J. (1946), On the Selection of an Inspection Plan, *Industrial Quality Control*, 2(5): 10-11.

Liebesman, B. S. (1979), The Use of MIL-STD-105D to Control Average Outgoing Quality, *Journal of Quality Technology*, 11(1): 36-43.

Mentch, C. C. (1980), Manufacturing Process Quality Optimization Studies, *Journal of Quality Technology*, 12(3): 119-129.

Muehlhause, C. O., V. L. Broussalian, A. J. Farrar, J. W. Lyons, M. G. Natrella, J. R. Rosenblatt, R. D. Stiehler, and J. H. Winger (1975), *Considerations in the Use of Sampling Plans for Effecting Compliance with Mandatory Safety Standards*, United States Department of Commerce, National Bureau of Standards, Report 75-697.

Nelson, P. R. (1977), A Computer Program for Military Standard 414: Sampling Procedures and Inspection by Variables for Percent Defective, *Journal of Quality Technology*, 9(2): 82-86.

Ott, E. R. (1975), *Process Quality Control*, McGraw-Hill, New York.

Peach, P. (1947), *An Introduction to Industrial Statistics and Quality Control*, 2nd ed., Edwards and Broughton Company, Raleigh, North Carolina.

Schilling, E. G. (1978), A Lot Sensitive Sampling Plan for Compliance Testing and Acceptance Sampling, *Journal of Quality Technology*, 10(2): 47-51.

Schilling, E. G., and J. H. Sheesley (1978), The Performance of MIL-STD-105D Under the Switching Rules, *Journal of Quality Technology*, Part 2, 10(3): 104-124.

Schilling, E. G., J. H. Sheesley, and P. R. Nelson (1978), GRASP: A General Routine for Attribute Sampling Plan Evaluation, *Journal of Quality Technology*, 10(3): 125-130.

Sheesley, J. H. (1975), A Computer Program to Evaluate Dodge's Continuous Sampling Plans, *Journal of Quality Technology*, 7(1): 43-45.

Singh, V. P., and H. R. Palanki (1976), Quality Levels in Acceptance Sampling, *Journal of Quality Technology*, 8(1): 37-47.

Smith, B. (1965), The Economics of Sampling Inspection, *Industrial Quality Control*, 21(9): 453-458.

Snyder, D. C., and R. F. Storer (1972), Single Sampling Plans Given an AQL, LTPD, Producer and Consumer Risks, *Journal of Quality Technology*, 4(3): 168-171.

Statistical Research Group, Columbia University (1948), *Sampling Inspection*, McGraw-Hill, New York.

Tukey, J. W. (1959), A Quick, Compact, Two Sample Test to Duckworth's Specifications, *Technometrics*, 1(1): 31-48.

United States Department of Defense (1954), *Guide for Sampling Inspection*, Quality and Reliability Assurance Handbook (H-53), Office of the Assistant Secretary of Defense (Installations and Logistics), Washington, D.C.

United States Department of Defense (1963), *Military Standard, Sampling Procedures and Tables for Inspection by Attributes* (MIL-STD-105D), U.S. Government Printing Office, Washington, D.C.

Wadsworth, H. M. (1970), The Effects of Class Inspection Under MIL-STD-105D, *Journal of Quality Technology*, 2(4): 181-185 (also April 1971, p. 106).

APPENDIX

TABLE T1-1 Control Chart Limits for Samples of n

PLOT	SAMPLE MEAN $\bar{X}$ AGAINST STANDARD μ WITH σ KNOWN	SAMPLE MEAN $\bar{X}$ AGAINST PAST DATA USING $\bar{X}$ AND $\bar{s}$ OR $\bar{R}$	SAMPLE STANDARD DEVIATION AGAINST STANDARD (KNOWN) σ	SAMPLE RANGE AGAINST STANDARD (KNOWN) σ	SAMPLE STANDARD DEVIATION s OR RANGE R AGAINST PAST DATA USING $\bar{s}$ OR $\bar{R}$	SAMPLE PROPORTIONS $\hat{p}$ OR DEFECTS PER UNIT $\hat{u}$ AGAINST STANDARD p OR u	SAMPLE* PROPORTIONS $\hat{p}$ OR DEFECTS PER UNIT $\hat{u}$ AGAINST PAST DATA USING $\bar{p}$ OR $\bar{u}$
UPPER CONTROL LIMIT	$\mu + 3\sigma/\sqrt{n}$ $= \mu + A\sigma$	$\bar{\bar{X}} + A_3\bar{s}$ $\bar{\bar{X}} + A_2\bar{R}$	$B_6\sigma$	$D_2\sigma$	$B_4\bar{s}$ $D_4\bar{R}$	$p + 3\sqrt{\dfrac{p(1-p)}{n}}$ $u + 3\sqrt{\dfrac{u}{n}}$	$\bar{p} \pm 3\sqrt{\dfrac{\bar{p}(1-\bar{p})}{n}}$ $\bar{u} \pm 3\sqrt{\dfrac{\bar{u}}{n}}$
CENTER LINE	μ	$\bar{\bar{X}}$	$c_4\sigma$	$d_2\sigma$	$\bar{s}$ $\bar{R}$	p u	$\bar{p}$ $\bar{u}$
LOWER CONTROL LIMIT	$\mu - 3\sigma/\sqrt{n}$ $\equiv \mu - A\sigma$	$\bar{\bar{X}} - A_3\bar{s}$ $\bar{\bar{X}} - A_2\bar{R}$	$B_5\sigma$	$D_1\sigma$	$B_3\bar{s}$ $D_3\bar{R}$	$p - 3\sqrt{\dfrac{p(1-p)}{n}}$ $u - 3\sqrt{\dfrac{u}{n}}$	$\bar{p} - 3\sqrt{\dfrac{\bar{p}(1-\bar{p})}{n}}$ $\bar{u} - 3\sqrt{\dfrac{\bar{u}}{n}}$

*For defects chart use u with n=1.

n	A	A_2	A_3	B_3	B_4	B_5	B_6	c_4	d_2	D_1	D_2	D_3	D_4
2	2.121	1.880	2.659	0	3.267	0	2.606	.7979	1.128	0	3.686	0	3.267
3	1.732	1.023	1.954	0	2.568	0	2.276	.8862	1.693	0	4.358	0	2.575
4	1.500	0.729	1.628	0	2.266	0	2.088	.9213	2.059	0	4.698	0	2.282
5	1.342	0.577	1.427	0	2.089	0	1.964	.9400	2.326	0	4.918	0	2.115
6	1.225	0.483	1.287	0.030	1.970	0.029	1.874	.9515	2.534	0	5.078	0	2.004
7	1.134	0.419	1.182	0.118	1.882	0.113	1.806	.9594	2.704	0.205	5.203	0.076	1.924
8	1.061	0.373	1.099	0.185	1.815	0.179	1.751	.9650	2.847	0.387	5.307	0.136	1.864
9	1.000	0.337	1.032	0.239	1.761	0.232	1.707	.9693	2.970	0.546	5.394	0.184	1.816
10	0.949	0.308	0.975	0.284	1.716	0.276	1.669	.9727	3.078	0.687	5.469	0.223	1.777

Adapted from ASQC Standard A1: Definitions, Symbols, Formulas, and Tables for Control Charts, 1970. Copyright 1970, American Society for Quality Control. Reprinted by permission.

TABLE T2-1   Random Numbers

1368	9621	9151	2066	1208	2664	9822	6599	6911	5112
5953	5936	2541	4011	0408	3593	3679	1378	5936	2651
7226	9466	9553	7671	8599	2119	5337	5953	6355	6889
8883	3454	6773	8207	5576	6386	7487	0190	0867	1298
7022	5281	1168	4099	8069	8721	8353	9952	8006	9045
4576	1853	7884	2451	3488	1286	4842	7719	5795	3953
8715	1416	7028	4616	3470	9938	5703	0196	3465	0034
4011	0408	2224	7626	0643	1149	8834	6429	8691	0143
1400	3694	4482	3608	1238	8221	5129	6105	5314	8385
6370	1884	0820	4854	9161	6509	7123	4070	6759	6113
4522	5749	8084	3932	7678	3549	0051	6761	6952	7041
7195	6234	6426	7148	9945	0358	3242	0519	6550	1327
0054	0810	2937	2040	2299	4198	0846	3937	3986	1019
5166	5433	0381	9686	5670	5129	2103	1125	3404	8785
1247	3793	7415	7819	1783	0506	4878	7673	9840	6629
8529	7842	7203	1844	8619	7404	4215	9969	6948	5643
8973	3440	4366	9242	2151	0244	0922	5887	4883	1177
9307	2959	5904	9012	4951	3695	4529	7197	7179	3239
2923	4276	9467	9868	2257	1925	3382	7244	1781	8037
6372	2808	1238	8098	5509	4617	4099	6705	2386	2830
6922	1807	4900	5306	0411	1828	8634	2331	7247	3230
9862	8336	6453	0545	6127	2741	5967	8447	3017	5709
3371	1530	5104	3076	5506	3101	4143	5845	2095	6127
6712	9402	9588	7019	9248	9192	4223	6555	7947	2474
3071	8782	7157	5941	8830	8563	2252	8109	5880	9912
4022	9734	7852	9096	0051	7387	7056	9331	1317	7833
9682	8892	3577	0326	5306	0050	8517	4376	0788	5443
6705	2175	9904	3743	1902	5393	3032	8432	0612	7972
1872	8292	2366	8603	4288	6809	4357	1072	6822	5611
2559	7534	2281	7351	2064	0611	9613	2000	0327	6145
4399	3751	9783	5399	5175	8894	0296	9483	0400	2272
6074	8827	2195	2532	7680	4288	6807	3101	6850	6410
5155	7186	4722	6721	0838	3632	5355	9369	2006	7681
3193	2800	6184	7891	9838	6123	9397	4019	8389	9508
8610	1880	7423	3384	4625	6653	2900	6290	9286	2396
4778	8818	2992	6300	4239	9595	4384	0611	7687	2088
3987	1619	4164	2542	4042	7799	9084	0278	8422	4330
2977	0248	2793	3351	4922	8878	5703	7421	2054	4391
1312	2919	8220	7285	5902	7882	1403	5354	9913	7109
3890	7193	7799	9190	3275	7840	1872	6232	5295	3148
0793	3468	8762	2492	5854	8430	8472	2264	9279	2128
2139	4552	3444	6462	2524	8601	3372	1948	1472	9667
8277	9153	2880	9053	6880	4284	5044	8931	0861	1517
2236	4778	6639	0862	9509	2141	0208	1450	1222	5281
8837	7686	1771	3374	2894	7314	6855	0440	3766	6047
6605	6380	4599	3333	0713	8401	7146	8940	2629	2006
8399	8175	3525	1646	4019	8390	4344	8975	4489	3423
8053	3046	9102	4515	2944	9763	3003	3408	1199	2791
9837	9378	3237	7016	7593	5958	0068	3114	0456	6840
2557	6395	9496	1884	0612	8102	4402	5498	0422	3335

TABLE 2-1 (continued)

2671	4690	1550	2262	2597	8034	0785	2978	4409	0237
9111	0250	3275	7519	9740	4577	2064	0286	3398	1348
0391	6035	9230	4999	3332	0608	6113	0391	5789	9926
2475	2144	1886	2079	3004	9686	5669	4367	9306	2595
5336	5845	2095	6446	5694	3641	1085	8705	5416	9066
6808	0423	0155	1652	7897	4335	3567	7109	9690	3739
8525	0577	8940	9451	6726	0876	3818	7607	8854	3566
0398	0741	8787	3043	5063	0617	1770	5048	7721	7032
3623	9636	3638	1406	5731	3978	8068	7238	9715	3363
0739	2644	4917	8866	3632	5399	5175	7422	2476	2607
6713	3041	8133	8749	8835	6745	3597	3476	3816	3455
7775	9315	0432	8327	0861	1515	2297	3375	3713	9174
8599	2122	6842	9202	0810	2936	1514	2090	3067	3574
7955	3759	5254	1126	5553	4713	9605	7909	1658	5490
4766	0070	7260	6033	7997	0109	5993	7592	5436	1727
5165	1670	2534	8811	8231	3721	7947	5719	2640	1394
9111	0513	2751	8256	2931	7783	1281	6531	7259	6993
1667	1084	7889	8963	7018	8617	6381	0723	4926	4551
2145	4587	8585	2412	5431	4667	1942	7238	9613	2212
2739	5528	1481	7528	9368	1823	6979	2547	7268	2467
8769	5480	9160	5354	9700	1362	2774	7980	9157	8788
6531	9435	3422	2474	1475	0159	3414	5224	8399	5820
2937	4134	7120	2206	5084	9473	3958	7320	9878	8609
1581	3285	3727	8924	6204	0797	0882	5945	9375	9153
6268	1045	7076	1436	4165	0143	0293	4190	7171	7932
4293	0523	8625	1961	1039	2856	4889	4358	1492	3804
6936	4213	3212	7229	1230	0019	5998	9206	6753	3762
5334	7641	3258	3769	1362	2771	6124	9813	7915	8960
9373	1158	4418	8826	5665	5896	0358	4717	8232	4859
6968	9428	8950	5346	1741	2348	8143	5377	7695	0685
4229	0587	8794	4009	9691	4579	3302	7673	9629	5246
3807	7785	7097	5701	6639	0723	4819	0900	2713	7650
4891	8829	1642	2155	0796	0466	2946	2970	9143	6590
1055	2968	7911	7479	8199	9735	8271	5339	7058	2964
2983	2345	0568	4125	0894	8302	0506	6761	7706	4310
4026	3129	2968	8053	2797	4022	9838	9611	0975	2437
4075	0260	4256	0337	2355	9371	2954	6021	5783	2827
8488	5450	1327	7358	2034	8060	1788	6913	6123	9405
1976	1749	5742	4098	5887	4567	6064	2777	7830	5668
2793	4701	9466	9554	8294	2160	7486	1557	4769	2781
0916	6272	6825	7188	9611	1181	2301	5516	5451	6832
5961	1149	7946	1950	2010	0600	5655	0796	0569	4365
3222	4189	1891	8172	8731	4769	2782	1325	4238	9270
1176	7834	4600	9992	9449	5824	5344	1008	6678	1921
2369	8971	2314	4806	5071	8908	8274	4936	3357	4441
0041	4329	9265	0352	4764	9070	7527	7791	1094	2008
0803	8302	6814	2422	6351	0637	0514	0246	1845	8594
9965	7804	3930	8803	0268	1426	3130	3613	3947	8086
0011	2387	3148	7559	4216	2946	2865	6333	1916	2259
1767	9871	3914	5790	5287	7915	8959	1346	5482	9251

TABLE T3-1 Values of e^{-x}

Tens/Hundredths Place	Units Place 0	1	2	3	4	5	6	7	8	9
0.00	1.0000	0.3679	0.1353	0.0498	0.0183	0.0067	0.0025	0.0009	0.0003	0.0001
0.05	0.9512	0.3499	0.1287	0.0474	0.0174	0.0064	0.0024	0.0009	0.0003	0.0001
0.10	0.9048	0.3329	0.1225	0.0450	0.0166	0.0061	0.0022	0.0008	0.0003	0.0001
0.15	0.8607	0.3166	0.1165	0.0429	0.0158	0.0058	0.0021	0.0008	0.0003	0.0001
0.20	0.8187	0.3012	0.1108	0.0408	0.0150	0.0055	0.0020	0.0007	0.0003	0.0001
0.25	0.7788	0.2865	0.1054	0.0388	0.0143	0.0052	0.0019	0.0007	0.0003	0.0001
0.30	0.7408	0.2725	0.1003	0.0369	0.0136	0.0050	0.0018	0.0007	0.0002	0.0001
0.35	0.7047	0.2592	0.0954	0.0351	0.0129	0.0047	0.0017	0.0006	0.0002	0.0001
0.40	0.6703	0.2466	0.0907	0.0334	0.0123	0.0045	0.0017	0.0006	0.0002	0.0001
0.45	0.6376	0.2346	0.0863	0.0317	0.0117	0.0043	0.0016	0.0006	0.0002	0.0001
0.50	0.6065	0.2231	0.0821	0.0302	0.0111	0.0041	0.0015	0.0006	0.0002	0.0001
0.55	0.5769	0.2122	0.0781	0.0287	0.0106	0.0039	0.0014	0.0005	0.0002	0.0001
0.60	0.5488	0.2019	0.0743	0.0273	0.0101	0.0037	0.0014	0.0005	0.0002	0.0001
0.65	0.5220	0.1920	0.0707	0.0260	0.0096	0.0035	0.0013	0.0005	0.0002	0.0001
0.70	0.4966	0.1827	0.0672	0.0247	0.0091	0.0033	0.0012	0.0005	0.0002	0.0001
0.75	0.4724	0.1738	0.0639	0.0235	0.0087	0.0032	0.0012	0.0004	0.0002	0.0001
0.80	0.4493	0.1653	0.0608	0.0224	0.0082	0.0030	0.0011	0.0004	0.0002	0.0001
0.85	0.4274	0.1572	0.0578	0.0213	0.0078	0.0029	0.0011	0.0004	0.0001	0.0001
0.90	0.4066	0.1496	0.0550	0.0202	0.0074	0.0027	0.0010	0.0004	0.0001	0.0001
0.95	0.3867	0.1423	0.0523	0.0193	0.0071	0.0026	0.0010	0.0004	0.0001	0.0000

TABLE T3-2   Cumulative Normal Probability, F(z)

z	.00	.01	.02	.03	.04	.05	.06	.07	.08	.09
−3.5	.0002	.0002	.0002	.0002	.0002	.0002	.0002	.0002	.0002	.0002
−3.4	.0003	.0003	.0003	.0003	.0003	.0003	.0003	.0003	.0003	.0002
−3.3	.0005	.0005	.0005	.0004	.0004	.0004	.0004	.0004	.0004	.0003
−3.2	.0007	.0007	.0006	.0006	.0006	.0006	.0006	.0005	.0005	.0005
−3.1	.0010	.0009	.0009	.0009	.0008	.0008	.0008	.0008	.0007	.0007
−3.0	.0013	.0013	.0013	.0012	.0012	.0011	.0011	.0011	.0010	.0010
−2.9	.0019	.0018	.0018	.0017	.0016	.0016	.0015	.0015	.0014	.0014
−2.8	.0026	.0025	.0024	.0023	.0023	.0022	.0021	.0021	.0020	.0019
−2.7	.0035	.0034	.0033	.0032	.0031	.0030	.0029	.0028	.0027	.0026
−2.6	.0047	.0045	.0044	.0043	.0041	.0040	.0039	.0038	.0037	.0036
−2.5	.0062	.0060	.0059	.0057	.0055	.0054	.0052	.0051	.0049	.0048
−2.4	.0082	.0080	.0078	.0075	.0073	.0071	.0069	.0068	.0066	.0064
−2.3	.0107	.0104	.0102	.0099	.0096	.0094	.0091	.0089	.0087	.0084
−2.2	.0139	.0136	.0132	.0129	.0125	.0122	.0119	.0116	.0113	.0110
−2.1	.0179	.0174	.0170	.0166	.0162	.0158	.0154	.0150	.0146	.0143
−2.0	.0228	.0222	.0217	.0212	.0207	.0202	.0197	.0192	.0188	.0183
−1.9	.0287	.0281	.0274	.0268	.0262	.0256	.0250	.0244	.0239	.0233
−1.8	.0359	.0351	.0344	.0336	.0329	.0322	.0314	.0307	.0301	.0294
−1.7	.0446	.0436	.0427	.0418	.0409	.0401	.0392	.0384	.0375	.0367
−1.6	.0548	.0537	.0526	.0516	.0505	.0495	.0485	.0475	.0465	.0455
−1.5	.0668	.0655	.0643	.0630	.0618	.0606	.0594	.0582	.0571	.0559
−1.4	.0808	.0793	.0778	.0764	.0749	.0735	.0721	.0708	.0694	.0681
−1.3	.0968	.0951	.0934	.0918	.0901	.0885	.0869	.0853	.0838	.0823
−1.2	.1151	.1131	.1112	.1093	.1075	.1056	.1038	.1020	.1003	.0985
−1.1	.1357	.1335	.1314	.1292	.1271	.1251	.1230	.1210	.1190	.1170
−1.0	.1587	.1562	.1539	.1515	.1492	.1469	.1446	.1423	.1401	.1379
−0.9	.1841	.1814	.1788	.1762	.1736	.1711	.1685	.1660	.1635	.1611
−0.8	.2119	.2090	.2061	.2033	.2005	.1977	.1949	.1922	.1894	.1867
−0.7	.2420	.2389	.2358	.2327	.2296	.2266	.2236	.2206	.2177	.2148
−0.6	.2743	.2709	.2676	.2643	.2611	.2578	.2546	.2514	.2483	.2451
−0.5	.3085	.3050	.3015	.2981	.2946	.2912	.2877	.2843	.2810	.2776
−0.4	.3446	.3409	.3372	.3336	.3300	.3264	.3228	.3192	.3156	.3121
−0.3	.3821	.3783	.3745	.3707	.3669	.3632	.3594	.3557	.3520	.3483
−0.2	.4207	.4168	.4129	.4090	.4052	.4013	.3974	.3936	.3897	.3859
−0.1	.4602	.4562	.4522	.4483	.4443	.4404	.4364	.4325	.4286	.4247
−0.0	.5000	.4960	.4920	.4880	.4840	.4801	.4761	.4721	.4681	.4641

TABLE T3-2   (continued)

z	.00	.01	.02	.03	.04	.05	.06	.07	.08	.09
+0.0	.5000	.5040	.5080	.5120	.5160	.5199	.5239	.5279	.5319	.5359
+0.1	.5398	.5438	.5478	.5517	.5557	.5596	.5636	.5675	.5714	.5753
+0.2	.5793	.5832	.5871	.5910	.5948	.5987	.6026	.6064	.6103	.6141
+0.3	.6179	.6217	.6255	.6293	.6331	.6368	.6406	.6443	.6480	.6517
+0.4	.6554	.6591	.6628	.6664	.6700	.6736	.6772	.6808	.6844	.6879
+0.5	.6915	.6950	.6985	.7019	.7054	.7088	.7123	.7157	.7190	.7224
+0.6	.7257	.7291	.7324	.7357	.7389	.7422	.7454	.7486	.7517	.7549
+0.7	.7580	.7611	.7642	.7673	.7704	.7734	.7764	.7794	.7823	.7852
+0.8	.7881	.7910	.7939	.7967	.7995	.8023	.8051	.8078	.8106	.8133
+0.9	.8159	.8186	.8212	.8238	.8264	.8289	.8315	.8340	.8365	.8389
+1.0	.8413	.8438	.8461	.8485	.8508	.8531	.8554	.8577	.8599	.8621
+1.1	.8643	.8665	.8686	.8708	.8729	.8749	.8770	.8790	.8810	.8830
+1.2	.8849	.8869	.8888	.8907	.8925	.8944	.8962	.8980	.8997	.9015
+1.3	.9032	.9049	.9066	.9082	.9099	.9115	.9131	.9147	.9162	.9177
+1.4	.9192	.9207	.9222	.9236	.9251	.9265	.9279	.9292	.9306	.9319
+1.5	.9332	.9345	.9357	.9370	.9382	.9394	.9406	.9418	.9429	.9441
+1.6	.9452	.9463	.9474	.9484	.9495	.9505	.9515	.9525	.9535	.9545
+1.7	.9554	.9564	.9573	.9582	.9591	.9599	.9608	.9616	.9625	.9633
+1.8	.9641	.9649	.9656	.9664	.9671	.9678	.9686	.9693	.9699	.9706
+1.9	.9713	.9719	.9726	.9732	.9738	.9744	.9750	.9756	.9761	.9767
+2.0	.9772	.9778	.9783	.9788	.9793	.9798	.9803	.9808	.9812	.9817
+2.1	.9821	.9826	.9830	.9834	.9838	.9842	.9846	.9850	.9854	.9857
+2.2	.9861	.9864	.9868	.9871	.9875	.9878	.9881	.9884	.9887	.9890
+2.3	.9893	.9896	.9898	.9901	.9904	.9906	.9909	.9911	.9913	.9916
+2.4	.9918	.9920	.9922	.9925	.9927	.9929	.9931	.9932	.9934	.9936
+2.5	.9938	.9940	.9941	.9943	.9945	.9946	.9948	.9949	.9951	.9952
+2.6	.9953	.9955	.9956	.9957	.9959	.9960	.9961	.9962	.9963	.9964
+2.7	.9965	.9966	.9967	.9968	.9969	.9970	.9971	.9972	.9973	.9974
+2.8	.9974	.9975	.9976	.9977	.9977	.9978	.9979	.9979	.9980	.9981
+2.9	.9981	.9982	.9982	.9983	.9984	.9984	.9985	.9985	.9986	.9986
+3.0	.9987	.9987	.9987	.9988	.9988	.9989	.9989	.9989	.9990	.9990
+3.1	.9990	.9991	.9991	.9991	.9992	.9992	.9992	.9992	.9993	.9993
+3.2	.9993	.9993	.9994	.9994	.9994	.9994	.9994	.9995	.9995	.9995
+3.3	.9995	.9995	.9995	.9996	.9996	.9996	.9996	.9996	.9996	.9997
+3.4	.9997	.9997	.9997	.9997	.9997	.9997	.9997	.9997	.9997	.9998
+3.5	.9998	.9998	.9998	.9998	.9998	.9998	.9998	.9998	.9998	.9998

From *Engineering Statistics and Quality Control* by I. W. Burr.
Copyright 1953 McGraw-Hill Book Company. Used with the permission
of McGraw-Hill Book Company.

TABLE T3-3 Lieberman-Owen Table of Hypergeometric Distribution, F(x), f(x) (N = 2, n = 1 through N = 100, n = 1.)

N	*n*	*k*	*x*	P(*x*)	p(*x*)
2	1	1	0	0.500000	0.500000
2	1	1	1	1.000000	0.500000
3	1	1	0	0.666667	0.666667
3	1	1	1	1.000000	0.333333
3	2	1	0	0.333333	0.333333
3	2	1	1	1.000000	0.666667
3	2	2	1	0.666667	0.666667
3	2	2	2	1.000000	0.333333
4	1	1	0	0.750000	0.750000
4	1	1	1	1.000000	0.250000
4	2	1	0	0.500000	0.500000
4	2	1	1	1.000000	0.500000
4	2	2	0	0.166667	0.166667
4	2	2	1	0.833333	0.666667
4	2	2	2	1.000000	0.166667
4	3	1	0	0.250000	0.250000
4	3	1	1	1.000000	0.750000
4	3	2	1	0.500000	0.500000
4	3	2	2	1.000000	0.500000
4	3	3	2	0.750000	0.750000
4	3	3	3	1.000000	0.250000
5	1	1	0	0.800000	0.800000
5	1	1	1	1.000000	0.200000
5	2	1	0	0.600000	0.600000
5	2	1	1	1.000000	0.400000
5	2	2	0	0.300000	0.300000
5	2	2	1	0.900000	0.600000
5	2	2	2	1.000000	0.100000
5	3	1	0	0.400000	0.400000
5	3	1	1	1.000000	0.600000
5	3	2	0	0.100000	0.100000
5	3	2	1	0.700000	0.600000
5	3	2	2	1.000000	0.300000
5	3	3	1	0.300000	0.300000
5	3	3	2	0.900000	0.600000
5	3	3	3	1.000000	0.100000
5	4	1	0	0.200000	0.200000
5	4	1	1	1.000000	0.800000
5	4	2	1	0.400000	0.400000
5	4	2	2	1.000000	0.600000
5	4	3	2	0.600000	0.600000
5	4	3	3	1.000000	0.400000
5	4	4	3	0.800000	0.800000
5	4	4	4	1.000000	0.200000
6	1	1	0	0.833333	0.833333
6	1	1	1	1.000000	0.166667
6	2	1	0	0.666667	0.666667
6	2	1	1	1.000000	0.333333
6	2	2	0	0.400000	0.400000
6	2	2	1	0.933333	0.533333

N	*n*	*k*	*x*	P(*x*)	p(*x*)
6	2	2	2	1.000000	0.066667
6	3	1	0	0.500000	0.500000
6	3	1	1	1.000000	0.500000
6	3	2	0	0.200000	0.200000
6	3	2	1	0.800000	0.600000
6	3	2	2	1.000000	0.200000
6	3	3	0	0.050000	0.050000
6	3	3	1	0.500000	0.450000
6	3	3	2	0.950000	0.450000
6	3	3	3	1.000000	0.050000
6	4	1	0	0.333333	0.333333
6	4	1	1	1.000000	0.666667
6	4	2	0	0.066667	0.066667
6	4	2	1	0.600000	0.533333
6	4	2	2	1.000000	0.400000
6	4	3	1	0.200000	0.200000
6	4	3	2	0.800000	0.600000
6	4	3	3	1.000000	0.200000
6	4	4	2	0.400000	0.400000
6	4	4	3	0.933333	0.533333
6	4	4	4	1.000000	0.066667
6	5	1	0	0.166667	0.166667
6	5	1	1	1.000000	0.833333
6	5	2	1	0.333333	0.333333
6	5	2	2	1.000000	0.666667
6	5	3	2	0.500000	0.500000
6	5	3	3	1.000000	0.500000
6	5	4	3	0.666667	0.666667
6	5	4	4	1.000000	0.333333
6	5	5	4	0.833333	0.833333
6	5	5	5	1.000000	0.166667
7	1	1	0	0.857143	0.857143
7	1	1	1	1.000000	0.142857
7	2	1	0	0.714286	0.714286
7	2	1	1	1.000000	0.285714
7	2	2	0	0.476190	0.476190
7	2	2	1	0.952381	0.476190
7	2	2	2	1.000000	0.047619
7	3	1	0	0.571429	0.571429
7	3	1	1	1.000000	0.428571
7	3	2	0	0.285714	0.285714
7	3	2	1	0.857143	0.571429
7	3	2	2	1.000000	0.142857
7	3	3	0	0.114286	0.114286
7	3	3	1	0.628571	0.514286
7	3	3	2	0.971428	0.342857
7	3	3	3	1.000000	0.028571
7	4	1	0	0.428571	0.428571
7	4	1	1	1.000000	0.571429
7	4	2	0	0.142857	0.142857

N	*n*	*k*	*x*	P(*x*)	p(*x*)
7	4	2	1	0.714286	0.571429
7	4	2	2	1.000000	0.285714
7	4	3	0	0.028571	0.028571
7	4	3	1	0.371429	0.342857
7	4	3	2	0.885714	0.514286
7	4	3	3	1.000000	0.114286
7	4	4	1	0.114286	0.114286
7	4	4	2	0.628571	0.514286
7	4	4	3	0.971428	0.342857
7	4	4	4	1.000000	0.028571
7	5	1	0	0.285714	0.285714
7	5	1	1	1.000000	0.714286
7	5	2	0	0.047619	0.047619
7	5	2	1	0.523809	0.476190
7	5	2	2	1.000000	0.476190
7	5	3	1	0.142857	0.142857
7	5	3	2	0.714286	0.571429
7	5	3	3	1.000000	0.285714
7	5	4	2	0.285714	0.285714
7	5	4	3	0.857143	0.571429
7	5	4	4	1.000000	0.142857
7	5	5	3	0.476190	0.476190
7	5	5	4	0.952381	0.476190
7	5	5	5	1.000000	0.047619
7	6	1	0	0.142857	0.142857
7	6	1	1	1.000000	0.857143
7	6	2	1	0.285714	0.285714
7	6	2	2	1.000000	0.714286
7	6	3	2	0.428571	0.428571
7	6	3	3	1.000000	0.571429
7	6	4	3	0.571429	0.571429
7	6	4	4	1.000000	0.428571
7	6	5	4	0.714286	0.714286
7	6	5	5	1.000000	0.285714
7	6	6	5	0.857143	0.857143
7	6	6	6	1.000000	0.142857
8	1	1	0	0.875000	0.875000
8	1	1	1	1.000000	0.125000
8	2	1	0	0.750000	0.750000
8	2	1	1	1.000000	0.250000
8	2	2	0	0.535714	0.535714
8	2	2	1	0.964286	0.428571
8	2	2	2	1.000000	0.035714
8	3	1	0	0.625000	0.625000
8	3	1	1	1.000000	0.375000
8	3	2	0	0.357143	0.357143
8	3	2	1	0.892857	0.535714
8	3	2	2	1.000000	0.107143
8	3	3	0	0.178571	0.178571
8	3	3	1	0.714286	0.535714

N	*n*	*k*	*x*	P(*x*)	p(*x*)
8	3	3	2	0.982143	0.267857
8	3	3	3	1.000000	0.017857
8	4	1	0	0.500000	0.500000
8	4	1	1	1.000000	0.500000
8	4	2	0	0.214286	0.214286
8	4	2	1	0.785714	0.571429
8	4	2	2	1.000000	0.214286
8	4	3	0	0.071429	0.071429
8	4	3	1	0.500000	0.428571
8	4	3	2	0.928571	0.428571
8	4	3	3	1.000000	0.071429
8	4	4	0	0.014286	0.014286
8	4	4	1	0.242857	0.228571
8	4	4	2	0.757143	0.514286
8	4	4	3	0.985714	0.228571
8	4	4	4	1.000000	0.014286
8	5	1	0	0.375000	0.375000
8	5	1	1	1.000000	0.625000
8	5	2	0	0.107143	0.107143
8	5	2	1	0.642857	0.535714
8	5	2	2	1.000000	0.357143
8	5	3	0	0.017857	0.017857
8	5	3	1	0.285714	0.267857
8	5	3	2	0.821429	0.535714
8	5	3	3	1.000000	0.178571
8	5	4	1	0.071429	0.071429
8	5	4	2	0.500000	0.428571
8	5	4	3	0.928571	0.428571
8	5	4	4	1.000000	0.071429
8	5	5	2	0.178571	0.178571
8	5	5	3	0.714286	0.535714
8	5	5	4	0.982143	0.267857
8	5	5	5	1.000000	0.017857
8	6	1	0	0.250000	0.250000
8	6	1	1	1.000000	0.750000
8	6	2	0	0.035714	0.035714
8	6	2	1	0.464286	0.428571
8	6	2	2	1.000000	0.535714
8	6	3	1	0.107143	0.107143
8	6	3	2	0.642857	0.535714
8	6	3	3	1.000000	0.357143
8	6	4	2	0.214286	0.214286
8	6	4	3	0.785714	0.571429
8	6	4	4	1.000000	0.214286
8	6	5	3	0.357143	0.357143
8	6	5	4	0.892857	0.535714
8	6	5	5	1.000000	0.107143
8	6	6	4	0.535714	0.535714
8	6	6	5	0.964286	0.428571
8	6	6	6	1.000000	0.035714

$N = 2\text{–}8$

TABLE T3-3 (continued)

Panel 1

N	n	k	x	P(x)	p(x)
8	7	1	0	0.125000	0.125000
8	7	1	1	1.000000	0.875000
8	7	2	1	0.250000	0.250000
8	7	2	2	1.000000	0.750000
8	7	3	2	0.375000	0.375000
8	7	3	3	1.000000	0.625000
8	7	4	3	0.500000	0.500000
8	7	4	4	1.000000	0.500000
8	7	5	4	0.625000	0.625000
8	7	5	5	1.000000	0.375000
8	8	6	5	0.750000	0.750000
8	8	6	6	1.000000	0.250000
8	8	7	6	0.875000	0.875000
8	8	7	7	1.000000	0.125000
8	8	1	0	0.888889	0.888889
9	1	1	1	1.000000	0.111111
9	1	1	0	0.888889	0.888889
9	1	2	1	1.000000	0.111111
9	2	2	0	0.777778	0.777778
9	2	2	1	1.000000	0.222222
9	2	2	0	0.583333	0.583333
9	2	1	1	0.972222	0.388889
9	2	2	2	1.000000	0.027778
9	3	1	0	0.666667	0.666667
9	3	1	1	1.000000	0.333333
9	3	2	0	0.416667	0.416667
9	3	2	1	0.916667	0.500000
9	3	2	2	1.000000	0.083333
9	3	3	0	0.238095	0.238095
9	3	3	1	0.773809	0.535714
9	3	3	2	0.988095	0.214286
9	3	3	3	1.000000	0.011905
9	4	1	0	0.555556	0.555556
9	4	1	1	1.000000	0.444444
9	4	2	0	0.277778	0.277778
9	4	2	1	0.833333	0.555556
9	4	2	2	1.000000	0.166667
9	4	3	0	0.119048	0.119048
9	4	3	1	0.595238	0.476190
9	4	3	2	0.952381	0.357143
9	4	3	3	1.000000	0.047619
9	4	4	0	0.357143	0.357143
9	5	4	1	0.833333	0.476190
9	5	4	2	0.992063	0.158730
9	5	4	3	1.000000	0.007936
9	5	1	0	0.444444	0.444444
9	5	1	1	1.000000	0.555556
9	5	2	0	0.166667	0.166667
9	5	2	1	0.722222	0.555556
9	5	2	2	1.000000	0.277778
9	5	3	0	0.047619	0.047619

Panel 2

N	n	k	x	P(x)	p(x)
9	5	3	1	0.404762	0.357143
9	5	3	2	0.880952	0.476190
9	5	3	3	1.000000	0.119048
9	5	4	0	0.007936	0.007936
9	5	4	1	0.166667	0.158730
9	5	4	2	0.642857	0.476190
9	5	4	3	0.960317	0.317460
9	5	4	4	1.000000	0.039683
9	5	5	1	0.039683	0.039683
9	5	5	2	0.357143	0.317460
9	6	5	3	0.476190	0.476190
9	6	5	4	0.992063	0.158730
9	6	5	0	0.047619	0.047619
9	6	1	1	0.333333	0.333333
9	6	1	0	0.666667	0.666667
9	6	2	0	0.083333	0.083333
9	6	2	1	0.583333	0.500000
9	6	2	2	1.000000	0.416667
9	7	3	0	0.011905	0.011905
9	7	3	1	0.226190	0.214286
9	7	1	0	0.761905	0.535714
9	7	1	1	1.000000	0.238095
9	7	2	0	0.047619	0.047619
9	7	2	1	0.404762	0.357143
9	7	2	2	0.880952	0.476190
9	7	3	3	1.000000	0.119048
9	7	3	0	0.119048	0.119048
9	7	3	1	0.595238	0.476190
9	7	3	2	0.952381	0.357143
9	7	3	3	1.000000	0.047619
9	7	1	0	0.238095	0.238095
9	7	1	1	0.773809	0.535714
9	7	2	0	0.980952	0.214286
9	7	2	1	1.000000	0.047619
9	7	2	2	0.222222	0.222222
9	7	2	3	1.000000	0.777778
9	8	3	0	0.416667	0.416667
9	8	3	1	1.000000	0.583333
9	8	3	2	0.083333	0.083333
9	9	1	0	0.238095	0.238095
9	9	1	1	0.773809	0.535714
9	9	2	0	0.988095	0.214286
9	9	2	1	1.000000	0.011905
9	9	2	2	0.083333	0.083333
9	9	3	0	0.583333	0.583333
9	9	3	1	1.000000	0.416667
9	9	3	2	0.722222	0.166667
9	9	3	3	1.000000	0.277778
9	9	4	0	0.277778	0.277778
9	9	4	1	0.833333	0.555556
9	9	4	2	1.000000	0.166667
9	9	4	3	0.416667	0.416667
9	9	4	4	0.916667	0.500000

Panel 3

N	n	k	x	P(x)	p(x)
9	7	6	6	1.000000	0.083333
9	7	6	5	0.583333	0.388889
9	7	7	6	0.972222	0.388889
9	7	7	7	1.000000	0.111111
9	8	1	0	0.111111	0.111111
9	8	1	1	1.000000	0.888889
9	8	2	1	0.222222	0.222222
9	8	2	2	1.000000	0.777778
9	8	3	2	0.333333	0.333333
9	8	3	3	1.000000	0.666667
9	8	4	4	0.444444	0.444444
9	8	4	3	1.000000	0.555556
9	8	5	4	0.555556	0.555556
9	8	5	5	1.000000	0.444444
9	8	6	5	0.666667	0.666667
9	8	6	6	1.000000	0.333333
9	8	7	6	0.777778	0.777778
9	8	7	7	1.000000	0.222222
9	8	8	7	0.888889	0.888889
10	1	8	8	1.000000	0.111111
10	1	1	0	0.900000	0.900000
10	1	1	1	1.000000	0.100000
10	2	1	0	0.800000	0.800000
10	2	1	1	1.000000	0.200000
10	2	2	0	0.622222	0.622222
10	2	2	1	0.977778	0.355556
10	2	2	2	1.000000	0.022222
10	2	1	0	0.700000	0.700000
10	3	1	1	1.000000	0.300000
10	3	2	0	0.466667	0.466667
10	3	2	1	0.933333	0.466667
10	3	3	2	0.166667	0.166667
10	3	3	3	0.666667	0.500000
10	3	3	1	0.291667	0.291667
10	3	3	2	0.816667	0.525000
10	4	4	0	0.991667	0.175000
10	4	4	1	1.000000	0.008333
10	4	1	0	0.600000	0.600000
10	4	1	1	1.000000	0.400000
10	4	2	0	0.333333	0.333333
10	4	2	1	0.866667	0.533333
10	4	2	2	0.133333	0.133333
10	4	3	0	0.166667	0.166667
10	4	3	1	0.666667	0.500000
10	4	3	2	1.000000	0.333333
10	4	4	0	0.071429	0.071429
10	4	4	1	0.452381	0.380952
10	4	4	2	0.880952	0.428571
10	4	4	3	0.995238	0.114286
10	4	4	4	1.000000	0.004762

Panel 4

N	n	k	x	P(x)	p(x)
10	5	1	0	0.500000	0.500000
10	5	1	1	1.000000	0.500000
10	5	2	0	0.222222	0.222222
10	5	2	1	0.777778	0.555556
10	5	2	2	1.000000	0.222222
10	5	3	0	0.083333	0.083333
10	5	3	1	0.500000	0.416667
10	5	3	2	0.916667	0.416667
10	5	3	3	1.000000	0.083333
10	5	4	0	0.023810	0.023810
10	5	4	1	0.261905	0.238095
10	5	4	2	0.738095	0.476190
10	5	4	3	0.976190	0.238095
10	5	4	0	0.003968	0.003968
10	5	5	1	0.103175	0.099206
10	5	5	2	0.500000	0.396825
10	5	5	3	0.896825	0.396825
10	5	5	4	0.996032	0.099206
10	5	5	5	1.000000	0.083333
10	5	5	0	0.023810	0.023810
10	6	1	1	0.400000	0.400000
10	6	1	2	1.000000	0.600000
10	6	2	3	0.133333	0.133333
10	6	2	4	0.666667	0.533333
10	6	2	0	1.000000	0.333333
10	6	3	1	0.033333	0.033333
10	6	3	2	0.333333	0.300000
10	6	3	3	0.833333	0.500000
10	6	3	0	1.000000	0.166667
10	6	4	1	0.004762	0.004762
10	6	4	2	0.119048	0.114286
10	6	4	3	0.547619	0.428571
10	6	4	4	0.928571	0.380952
10	6	4	5	1.000000	0.071429
10	6	5	2	0.023810	0.023810
10	7	5	3	0.261905	0.238095
10	7	5	4	0.738095	0.476190
10	7	5	5	0.976190	0.238095
10	7	5	2	1.000000	0.023810
10	7	6	3	0.071429	0.071429
10	7	6	4	0.452381	0.380952
10	7	6	5	0.880952	0.428571
10	7	6	0	0.995238	0.114286
10	7	1	1	1.000000	0.004762
10	7	1	2	0.300000	0.300000
10	7	2	0	1.000000	0.700000
10	7	2	1	0.066667	0.066667
10	7	2	2	0.533333	0.466667
10	7	3	0	1.000000	0.466667
10	7	3	1	0.008333	0.008333

TABLE T3-3 (continued)

$N = 10\text{-}11$

Reprinted from *Tables of the Hypergeometric Probability Distribution* by G. J. Lieberman and D. B. Owen, with the permission of the publishers, Stanford University Press. Copyright 1961 by the Board of Trustees of the Leland Stanford Junior University.

TABLE T3-4   Harvard Table of Binomial Distribution, 1 - F(r - 1), Pr(x ≥ r|n,p)

n	r	p=0.01	p=0.02	p=0.03	p=0.04	p=0.05	p=0.06	p=1/16	p=0.07	p=0.08	p=1/12
1	0	1.00000	1.00000	1.00000	1.00000	1.00000	1.00000	1.00000	1.00000	1.00000	1.00000
	1	0.01000	0.02000	0.03000	0.04000	0.05000	0.06000	0.06250	0.07000	0.08000	0.08333
2	0	1.00000	1.00000	1.00000	1.00000	1.00000	1.00000	1.00000	1.00000	1.00000	1.00000
	1	0.01990	0.03960	0.05910	0.07840	0.09750	0.11640	0.12109	0.13510	0.15360	0.15972
	2	0.00010	0.00040	0.00090	0.00160	0.00250	0.00360	0.00391	0.00490	0.00640	0.00694
3	0	1.00000	1.00000	1.00000	1.00000	1.00000	1.00000	1.00000	1.00000	1.00000	1.00000
	1	0.02970	0.05881	0.08733	0.11526	0.14263	0.16942	0.17603	0.19564	0.22131	0.22975
	2	0.00030	0.00118	0.00265	0.00467	0.00725	0.01037	0.01123	0.01401	0.01818	0.01968
	3	0.00000	0.00001	0.00003	0.00006	0.00013	0.00022	0.00024	0.00034	0.00051	0.00058
4	0	1.00000	1.00000	1.00000	1.00000	1.00000	1.00000	1.00000	1.00000	1.00000	1.00000
	1	0.03940	0.07763	0.11471	0.15065	0.18549	0.21925	0.22752	0.25195	0.28361	0.29393
	2	0.00059	0.00234	0.00519	0.00910	0.01402	0.01991	0.02153	0.02673	0.03443	0.03718
	3	0.00000	0.00003	0.00011	0.00025	0.00048	0.00083	0.00093	0.00130	0.00193	0.00217
	4		0.00000	0.00000	0.00000	0.00001	0.00001	0.00002	0.00002	0.00004	0.00005
5	0	1.00000	1.00000	1.00000	1.00000	1.00000	1.00000	1.00000	1.00000	1.00000	1.00000
	1	0.04901	0.09608	0.14127	0.18463	0.22622	0.26610	0.27580	0.30431	0.34092	0.35277
	2	0.00098	0.00384	0.00847	0.01476	0.02259	0.03187	0.03440	0.04249	0.05436	0.05858
	3	0.00001	0.00008	0.00026	0.00060	0.00116	0.00197	0.00222	0.00308	0.00453	0.00509
	4	0.00000	0.00000	0.00000	0.00001	0.00003	0.00006	0.00007	0.00011	0.00019	0.00023
	5				0.00000	0.00000	0.00000	0.00000	0.00000	0.00000	0.00000
6	0	1.00000	1.00000	1.00000	1.00000	1.00000	1.00000	1.00000	1.00000	1.00000	1.00000
	1	0.05852	0.11416	0.16703	0.21724	0.26491	0.31013	0.32107	0.35301	0.39364	0.40671
	2	0.00146	0.00569	0.01246	0.02155	0.03277	0.04592	0.04949	0.06082	0.07729	0.08309
	3	0.00002	0.00015	0.00050	0.00117	0.00223	0.00376	0.00423	0.00584	0.00851	0.00955
	4	0.00000	0.00000	0.00001	0.00004	0.00009	0.00018	0.00021	0.00032	0.00054	0.00063
	5			0.00000	0.00000	0.00000	0.00000	0.00001	0.00001	0.00002	0.00002
	6							0.00000	0.00000	0.00000	0.00000
7	0	1.00000	1.00000	1.00000	1.00000	1.00000	1.00000	1.00000	1.00000	1.00000	1.00000
	1	0.06793	0.13187	0.19202	0.24855	0.30166	0.35152	0.36350	0.39830	0.44215	0.45615
	2	0.00203	0.00786	0.01709	0.02938	0.04438	0.06178	0.06647	0.08127	0.10259	0.11006
	3	0.00003	0.00026	0.00086	0.00198	0.00376	0.00629	0.00706	0.00969	0.01401	0.01567
	4	0.00000	0.00001	0.00003	0.00008	0.00019	0.00039	0.00046	0.00071	0.00118	0.00137
	5		0.00000	0.00000	0.00000	0.00001	0.00001	0.00002	0.00003	0.00006	0.00007
	6							0.00000	0.00000	0.00000	0.00000
8	0	1.00000	1.00000	1.00000	1.00000	1.00000	1.00000	1.00000	1.00000	1.00000	1.00000
	1	0.07726	0.14924	0.21626	0.27861	0.33658	0.39043	0.40328	0.44042	0.48678	0.50147
	2	0.00269	0.01034	0.02234	0.03815	0.05724	0.07916	0.08503	0.10347	0.12976	0.13890
	3	0.00005	0.00042	0.00135	0.00308	0.00579	0.00962	0.01077	0.01470	0.02110	0.02354
	4	0.00000	0.00001	0.00005	0.00016	0.00037	0.00075	0.00087	0.00134	0.00220	0.00256
	5		0.00000	0.00000	0.00001	0.00002	0.00004	0.00005	0.00008	0.00015	0.00018
	6			0.00000	0.00000	0.00000	0.00000	0.00000	0.00000	0.00001	0.00001
	7									0.00000	0.00000
9	0	1.00000	1.00000	1.00000	1.00000	1.00000	1.00000	1.00000	1.00000	1.00000	1.00000
	1	0.08648	0.16625	0.23977	0.30747	0.36975	0.42701	0.44058	0.47959	0.52784	0.54301
	2	0.00344	0.01311	0.02816	0.04777	0.07121	0.09784	0.10492	0.12705	0.15832	0.16912
	3	0.00008	0.00061	0.00198	0.00448	0.00836	0.01380	0.01541	0.02091	0.02979	0.03315
	4	0.00000	0.00002	0.00009	0.00027	0.00064	0.00128	0.00149	0.00227	0.00372	0.00431
	5		0.00000	0.00000	0.00001	0.00003	0.00008	0.00010	0.00017	0.00031	0.00038
	6			0.00000	0.00000	0.00000	0.00000	0.00000	0.00001	0.00002	0.00002
	7								0.00000	0.00000	0.00000
10	0	1.00000	1.00000	1.00000	1.00000	1.00000	1.00000	1.00000	1.00000	1.00000	1.00000
	1	0.09562	0.18293	0.26258	0.33517	0.40126	0.46138	0.47554	0.51602	0.56561	0.58110
	2	0.00427	0.01618	0.03451	0.05815	0.08614	0.11759	0.12590	0.15173	0.18788	0.20027
	3	0.00011	0.00086	0.00276	0.00621	0.01150	0.01884	0.02101	0.02834	0.04008	0.04448
	4	0.00000	0.00003	0.00015	0.00044	0.00103	0.00203	0.00236	0.00358	0.00580	0.00672
	5		0.00000	0.00001	0.00002	0.00006	0.00015	0.00018	0.00031	0.00059	0.00071
	6			0.00000	0.00000	0.00000	0.00001	0.00001	0.00002	0.00004	0.00005
	7						0.00000	0.00000	0.00000	0.00000	0.00000

n	r	p=0.01	p=0.02	p=0.03	p=0.04	p=0.05	p=0.06	p=1/16	p=0.07	p=0.08	p=1/12
11	0	1.00000	1.00000	1.00000	1.00000	1.00000	1.00000	1.00000	1.00000	1.00000	1.00000
	1	0.10466	0.19927	0.28470	0.36176	0.43120	0.49370	0.50832	0.54990	0.60036	0.61600
	2	0.00518	0.01951	0.04135	0.06923	0.10189	0.13822	0.14775	0.17723	0.21810	0.23201
	3	0.00016	0.00117	0.00372	0.00829	0.01524	0.02476	0.02756	0.03698	0.05190	0.05747
	4	0.00000	0.00005	0.00023	0.00067	0.00155	0.00304	0.00353	0.00531	0.00854	0.00986
	5		0.00000	0.00001	0.00004	0.00011	0.00026	0.00032	0.00054	0.00100	0.00121
	6			0.00000	0.00000	0.00001	0.00002	0.00002	0.00004	0.00009	0.00011
	7					0.00000	0.00000	0.00000	0.00000	0.00001	0.00001
	8									0.00000	0.00000
12	0	1.00000	1.00000	1.00000	1.00000	1.00000	1.00000	1.00000	1.00000	1.00000	1.00000
	1	0.11362	0.21528	0.30616	0.38729	0.45964	0.52408	0.53905	0.58140	0.63233	0.64800
	2	0.00617	0.02311	0.04865	0.08094	0.11836	0.15954	0.17029	0.20332	0.24868	0.26401
	3	0.00021	0.00154	0.00485	0.01073	0.01957	0.03157	0.03507	0.04680	0.06520	0.07201
	4	0.00000	0.00007	0.00033	0.00098	0.00224	0.00434	0.00503	0.00753	0.01201	0.01383
	5		0.00000	0.00002	0.00006	0.00018	0.00043	0.00052	0.00088	0.00161	0.00193
	6				0.00000	0.00001	0.00003	0.00004	0.00008	0.00016	0.00020
	7					0.00000	0.00000	0.00000	0.00000	0.00001	0.00002
	8									0.00000	0.00000
13	0	1.00000	1.00000	1.00000	1.00000	1.00000	1.00000	1.00000	1.00000	1.00000	1.00000
	1	0.12248	0.23098	0.32697	0.41180	0.48666	0.55263	0.56786	0.61071	0.66175	0.67734
	2	0.00725	0.02695	0.05637	0.09319	0.13542	0.18142	0.19333	0.22978	0.27937	0.29601
	3	0.00027	0.00197	0.00616	0.01354	0.02451	0.03925	0.04353	0.05775	0.07987	0.08801
	4	0.00001	0.00010	0.00047	0.00137	0.00310	0.00598	0.00691	0.01028	0.01627	0.01868
	5	0.00000	0.00000	0.00003	0.00010	0.00029	0.00067	0.00080	0.00134	0.00244	0.00292
	6			0.00000	0.00001	0.00002	0.00006	0.00007	0.00013	0.00027	0.00034
	7					0.00000	0.00000	0.00000	0.00001	0.00002	0.00003
	8									0.00000	0.00000
14	0	1.00000	1.00000	1.00000	1.00000	1.00000	1.00000	1.00000	1.00000	1.00000	1.00000
	1	0.13125	0.24636	0.34716	0.43533	0.51233	0.57948	0.59487	0.63796	0.68881	0.70423
	2	0.00840	0.03103	0.06449	0.10593	0.15299	0.20369	0.21674	0.25645	0.30996	0.32779
	3	0.00034	0.00247	0.00767	0.01672	0.03005	0.04778	0.05289	0.06980	0.09583	0.10534
	4	0.00001	0.00014	0.00064	0.00185	0.00417	0.00797	0.00919	0.01360	0.02136	0.02446
	5	0.00000	0.00001	0.00004	0.00015	0.00043	0.00098	0.00118	0.00197	0.00354	0.00423
	6		0.00000	0.00000	0.00001	0.00003	0.00009	0.00012	0.00022	0.00045	0.00056
	7					0.00000	0.00001	0.00001	0.00002	0.00004	0.00006
	8							0.00000	0.00000	0.00000	0.00000
15	0	1.00000	1.00000	1.00000	1.00000	1.00000	1.00000	1.00000	1.00000	1.00000	1.00000
	1	0.13994	0.26143	0.36675	0.45791	0.53671	0.60471	0.62019	0.66330	0.71370	0.72887
	2	0.00963	0.03534	0.07297	0.11911	0.17095	0.22624	0.24038	0.28315	0.34027	0.35916
	3	0.00042	0.00304	0.00937	0.02029	0.03620	0.05713	0.06313	0.08286	0.11297	0.12388
	4	0.00001	0.00018	0.00085	0.00245	0.00547	0.01036	0.01193	0.01753	0.02731	0.03120
	5	0.00000	0.00001	0.00006	0.00022	0.00061	0.00140	0.00168	0.00278	0.00497	0.00592
	6		0.00000	0.00000	0.00001	0.00005	0.00015	0.00018	0.00034	0.00070	0.00086
	7				0.00000	0.00001	0.00001	0.00002	0.00003	0.00008	0.00010
	8						0.00000	0.00000	0.00000	0.00001	0.00001
	9									0.00000	0.00000
16	0	1.00000	1.00000	1.00000	1.00000	1.00000	1.00000	1.00000	1.00000	1.00000	1.00000
	1	0.14854	0.27620	0.38575	0.47960	0.55987	0.62843	0.64393	0.68687	0.73661	0.75147
	2	0.01093	0.03986	0.08179	0.13266	0.18924	0.24895	0.26411	0.30976	0.37015	0.38997
	3	0.00051	0.00369	0.01128	0.02424	0.04294	0.06728	0.07421	0.09688	0.13115	0.14349
	4	0.00002	0.00024	0.00110	0.00316	0.00700	0.01317	0.01513	0.02211	0.03417	0.03892
	5	0.00000	0.00001	0.00008	0.00031	0.00086	0.00194	0.00232	0.00381	0.00676	0.00803
	6		0.00000	0.00000	0.00002	0.00008	0.00022	0.00028	0.00051	0.00104	0.00129
	7				0.00000	0.00001	0.00002	0.00003	0.00005	0.00013	0.00016
	8					0.00000	0.00000	0.00000	0.00001	0.00001	0.00002
	9									0.00000	0.00000
17	0	1.00000	1.00000	1.00000	1.00000	1.00000	1.00000	1.00000	1.00000	1.00000	1.00000
	1	0.15706	0.29068	0.40417	0.50041	0.58188	0.65072	0.66618	0.70879	0.75768	0.77218
	2	0.01231	0.04459	0.09090	0.14654	0.20777	0.27171	0.28785	0.33616	0.39946	0.42009
	3	0.00061	0.00441	0.01339	0.02858	0.05025	0.07818	0.08608	0.11178	0.15027	0.16403
	4	0.00002	0.00031	0.00141	0.00401	0.00880	0.01641	0.01882	0.02734	0.04192	0.04763
	5	0.00000	0.00002	0.00011	0.00042	0.00116	0.00261	0.00312	0.00509	0.00895	0.01060
	6		0.00000	0.00001	0.00003	0.00012	0.00032	0.00040	0.00074	0.00149	0.00185
	7				0.00000	0.00001	0.00003	0.00004	0.00009	0.00020	0.00026
	8					0.00000	0.00000	0.00001	0.00001	0.00002	0.00003
	9								0.00000	0.00000	0.00000

n	r	p=0.01	p=0.02	p=0.03	p=0.04	p=0.05	p=0.06	p=1/16	p=0.07	p=0.08	p=1/12
18	0	1.00000	1.00000	1.00000	1.00000	1.00000	1.00000	1.00000	1.00000	1.00000	1.00000
	1	0.16549	0.30486	0.42205	0.52040	0.60279	0.67168	0.68704	0.72917	0.77706	0.79116
	2	0.01376	0.04951	0.10030	0.16069	0.22648	0.29445	0.31150	0.36224	0.42812	0.44943
	3	0.00073	0.00521	0.01572	0.03330	0.05813	0.08979	0.09869	0.12749	0.17020	0.18537
	4	0.00003	0.00039	0.00177	0.00499	0.01087	0.02012	0.02302	0.03325	0.05059	0.05733
	5	0.00000	0.00002	0.00015	0.00057	0.00155	0.00344	0.00411	0.00665	0.01159	0.01369
	6		0.00000	0.00001	0.00005	0.00017	0.00046	0.00057	0.00105	0.00209	0.00258
	7			0.00000	0.00000	0.00002	0.00005	0.00006	0.00013	0.00030	0.00039
	8					0.00000	0.00000	0.00001	0.00001	0.00004	0.00005
	9							0.00000	0.00000	0.00000	0.00000
19	0	1.00000	1.00000	1.00000	1.00000	1.00000	1.00000	1.00000	1.00000	1.00000	1.00000
	1	0.17383	0.31877	0.43939	0.53958	0.62265	0.69138	0.70660	0.74813	0.79490	0.80857
	2	0.01527	0.05462	0.10996	0.17508	0.24529	0.31709	0.33497	0.38793	0.45604	0.47791
	3	0.00086	0.00610	0.01826	0.03840	0.06655	0.10207	0.11199	0.14392	0.19084	0.20737
	4	0.00003	0.00049	0.00219	0.00612	0.01324	0.02430	0.02775	0.03985	0.06016	0.06800
	5	0.00000	0.00003	0.00020	0.00074	0.00201	0.00444	0.00529	0.00851	0.01471	0.01732
	6		0.00000	0.00001	0.00007	0.00024	0.00064	0.00079	0.00144	0.00285	0.00350
	7			0.00000	0.00001	0.00002	0.00007	0.00010	0.00020	0.00045	0.00057
	8				0.00000	0.00000	0.00001	0.00001	0.00002	0.00006	0.00008
	9						0.00000	0.00000	0.00000	0.00001	0.00001
	10									0.00000	0.00000
20	0	1.00000	1.00000	1.00000	1.00000	1.00000	1.00000	1.00000	1.00000	1.00000	1.00000
	1	0.18209	0.33239	0.45621	0.55800	0.64151	0.70989	0.72494	0.76576	0.81131	0.82452
	2	0.01686	0.05990	0.11984	0.18966	0.26416	0.33955	0.35820	0.41314	0.48314	0.50546
	3	0.00100	0.00707	0.02101	0.04386	0.07548	0.11497	0.12592	0.16100	0.21205	0.22992
	4	0.00004	0.00060	0.00267	0.00741	0.01590	0.02897	0.03302	0.04713	0.07062	0.07962
	5	0.00000	0.00004	0.00026	0.00096	0.00257	0.00563	0.00669	0.01071	0.01834	0.02155
	6		0.00000	0.00002	0.00010	0.00033	0.00087	0.00108	0.00193	0.00380	0.00465
	7			0.00000	0.00001	0.00003	0.00011	0.00014	0.00028	0.00064	0.00082
	8				0.00000	0.00000	0.00001	0.00001	0.00003	0.00009	0.00012
	9						0.00000	0.00000	0.00000	0.00001	0.00001
	10									0.00000	0.00000
21	0	1.00000	1.00000	1.00000	1.00000	1.00000	1.00000	1.00000	1.00000	1.00000	1.00000
	1	0.19027	0.34574	0.47252	0.57568	0.65944	0.72730	0.74213	0.78216	0.82640	0.83914
	2	0.01851	0.06535	0.12993	0.20440	0.28303	0.36177	0.38112	0.43783	0.50940	0.53205
	3	0.00116	0.00813	0.02397	0.04969	0.08492	0.12845	0.14044	0.17865	0.23374	0.25288
	4	0.00005	0.00073	0.00322	0.00887	0.01888	0.03413	0.03882	0.05510	0.08193	0.09214
	5	0.00000	0.00005	0.00033	0.00122	0.00324	0.00703	0.00834	0.01326	0.02253	0.02639
	6		0.00000	0.00003	0.00013	0.00044	0.00115	0.00143	0.00255	0.00496	0.00606
	7			0.00000	0.00001	0.00005	0.00015	0.00020	0.00040	0.00089	0.00113
	8				0.00000	0.00000	0.00002	0.00002	0.00005	0.00013	0.00018
	9						0.00000	0.00000	0.00001	0.00002	0.00002
	10									0.00000	0.00000
22	0	1.00000	1.00000	1.00000	1.00000	1.00000	1.00000	1.00000	1.00000	1.00000	1.00000
	1	0.19837	0.35883	0.48834	0.59265	0.67647	0.74366	0.75825	0.79741	0.84029	0.85255
	2	0.02023	0.07096	0.14021	0.21925	0.30185	0.38370	0.40368	0.46193	0.53476	0.55764
	3	0.00134	0.00927	0.02715	0.05588	0.09482	0.14245	0.15548	0.19679	0.25579	0.27614
	4	0.00006	0.00088	0.00384	0.01050	0.02218	0.03979	0.04517	0.06375	0.09408	0.10554
	5	0.00000	0.00006	0.00042	0.00152	0.00402	0.00866	0.01024	0.01619	0.02728	0.03187
	6		0.00000	0.00004	0.00018	0.00058	0.00151	0.00186	0.00330	0.00637	0.00776
	7			0.00000	0.00002	0.00007	0.00021	0.00027	0.00055	0.00122	0.00155
	8				0.00000	0.00001	0.00003	0.00003	0.00008	0.00019	0.00026
	9					0.00000	0.00000	0.00000	0.00001	0.00003	0.00004
	10								0.00000	0.00000	0.00000
23	0	1.00000	1.00000	1.00000	1.00000	1.00000	1.00000	1.00000	1.00000	1.00000	1.00000
	1	0.20639	0.37165	0.50369	0.60894	0.69264	0.75904	0.77336	0.81159	0.85307	0.86484
	2	0.02201	0.07671	0.15065	0.23418	0.32058	0.40530	0.42584	0.48541	0.55920	0.58222
	3	0.00152	0.01050	0.03054	0.06242	0.10517	0.15692	0.17100	0.21535	0.27811	0.29960
	4	0.00008	0.00104	0.00454	0.01232	0.02581	0.04595	0.05207	0.07307	0.10701	0.11975
	5	0.00000	0.00008	0.00052	0.00188	0.00493	0.01053	0.01243	0.01952	0.03262	0.03801
	6		0.00000	0.00005	0.00023	0.00075	0.00194	0.00238	0.00420	0.00804	0.00977
	7			0.00000	0.00002	0.00009	0.00029	0.00037	0.00074	0.00163	0.00206
	8				0.00000	0.00001	0.00004	0.00005	0.00011	0.00027	0.00036
	9					0.00000	0.00000	0.00001	0.00001	0.00004	0.00005
	10							0.00000	0.00000	0.00000	0.00001
	11										0.00000

TABLE T3-4   (continued)

n	r	p=0.01	p=0.02	p=0.03	p=0.04	p=0.05	p=0.06	p=1/16	p=0.07	p=0.08	p=1/12
24	0	1.00000	1.00000	1.00000	1.00000	1.00000	1.00000	1.00000	1.00000	1.00000	1.00000
	1	0.21432	0.38422	0.51858	0.62459	0.70801	0.77350	0.78752	0.82478	0.86482	0.87610
	2	0.02385	0.08261	0.16124	0.24917	0.33918	0.42652	0.44756	0.50825	0.58271	0.60577
	3	0.00173	0.01183	0.03415	0.06929	0.11594	0.17182	0.18692	0.23426	0.30060	0.32315
	4	0.00009	0.00123	0.00532	0.01432	0.02978	0.05260	0.05950	0.08303	0.12070	0.13474
	5	0.00000	0.00010	0.00064	0.00230	0.00597	0.01265	0.01490	0.02326	0.03857	0.04482
	6		0.00001	0.00006	0.00030	0.00096	0.00245	0.00301	0.00527	0.01001	0.01212
	7		0.00000	0.00000	0.00003	0.00013	0.00039	0.00050	0.00098	0.00214	0.00270
	8				0.00000	0.00001	0.00005	0.00007	0.00015	0.00038	0.00050
	9					0.00000	0.00001	0.00001	0.00002	0.00006	0.00008
	10						0.00000	0.00000	0.00000	0.00001	0.00001
	11									0.00000	0.00000
25	0	1.00000	1.00000	1.00000	1.00000	1.00000	1.00000	1.00000	1.00000	1.00000	1.00000
	1	0.22218	0.39654	0.53303	0.63960	0.72261	0.78709	0.80080	0.83704	0.87564	0.88642
	2	0.02576	0.08865	0.17196	0.26419	0.35762	0.44734	0.46881	0.53040	0.60528	0.62830
	3	0.00195	0.01324	0.03796	0.07648	0.12711	0.18711	0.20321	0.25344	0.32317	0.34670
	4	0.00011	0.00145	0.00619	0.01652	0.03409	0.05976	0.06746	0.09361	0.13509	0.15044
	5	0.00000	0.00012	0.00078	0.00278	0.00716	0.01505	0.01769	0.02745	0.04514	0.05231
	6		0.00001	0.00008	0.00038	0.00121	0.00306	0.00375	0.00653	0.01229	0.01484
	7		0.00000	0.00001	0.00004	0.00017	0.00051	0.00066	0.00128	0.00277	0.00349
	8			0.00000	0.00000	0.00002	0.00007	0.00010	0.00021	0.00052	0.00069
	9					0.00000	0.00001	0.00001	0.00003	0.00008	0.00011
	10						0.00000	0.00000	0.00000	0.00001	0.00002
	11									0.00000	0.00000
26	0	1.00000	1.00000	1.00000	1.00000	1.00000	1.00000	1.00000	1.00000	1.00000	1.00000
	1	0.22996	0.40860	0.54703	0.65402	0.73648	0.79986	0.81325	0.84845	0.88558	0.89589
	2	0.02772	0.09480	0.18279	0.27921	0.37587	0.46772	0.48956	0.55187	0.62691	0.64981
	3	0.00219	0.01475	0.04198	0.08399	0.13863	0.20272	0.21981	0.27283	0.34574	0.37017
	4	0.00013	0.00168	0.00714	0.01892	0.03874	0.06740	0.07595	0.10480	0.15014	0.16680
	5	0.00001	0.00015	0.00094	0.00333	0.00851	0.01773	0.02080	0.03208	0.05234	0.06049
	6	0.00000	0.00001	0.00010	0.00047	0.00151	0.00378	0.00462	0.00800	0.01492	0.01797
	7		0.00000	0.00001	0.00005	0.00022	0.00067	0.00085	0.00165	0.00353	0.00444
	8			0.00000	0.00001	0.00003	0.00010	0.00013	0.00029	0.00070	0.00092
	9				0.00000	0.00000	0.00001	0.00002	0.00004	0.00012	0.00016
	10						0.00000	0.00000	0.00001	0.00002	0.00002
	11								0.00000	0.00000	0.00000
27	0	1.00000	1.00000	1.00000	1.00000	1.00000	1.00000	1.00000	1.00000	1.00000	1.00000
	1	0.23766	0.42043	0.56062	0.66786	0.74966	0.81187	0.82492	0.85906	0.89474	0.90456
	2	0.02975	0.10108	0.19372	0.29420	0.39390	0.48765	0.50979	0.57263	0.64760	0.67031
	3	0.00244	0.01635	0.04620	0.09180	0.15049	0.21862	0.23667	0.29236	0.36823	0.39347
	4	0.00015	0.00194	0.00818	0.02152	0.04374	0.07552	0.08494	0.11656	0.16579	0.18375
	5	0.00001	0.00018	0.00113	0.00395	0.01002	0.02071	0.02425	0.03717	0.06016	0.06935
	6	0.00000	0.00001	0.00013	0.00059	0.00186	0.00462	0.00564	0.00968	0.01791	0.02151
	7		0.00000	0.00001	0.00007	0.00029	0.00085	0.00109	0.00209	0.00444	0.00556
	8			0.00000	0.00001	0.00004	0.00013	0.00018	0.00038	0.00093	0.00121
	9				0.00000	0.00000	0.00002	0.00002	0.00006	0.00017	0.00023
	10						0.00000	0.00000	0.00001	0.00003	0.00004
	11								0.00000	0.00000	0.00000
28	0	1.00000	1.00000	1.00000	1.00000	1.00000	1.00000	1.00000	1.00000	1.00000	1.00000
	1	0.24528	0.43202	0.57380	0.68114	0.76217	0.82316	0.83587	0.86892	0.90316	0.91252
	2	0.03182	0.10747	0.20473	0.30915	0.41169	0.50711	0.52949	0.59268	0.66737	0.68984
	3	0.00272	0.01805	0.05063	0.09990	0.16266	0.23476	0.25374	0.31198	0.39058	0.41654
	4	0.00017	0.00223	0.00932	0.02433	0.04907	0.08410	0.09442	0.12887	0.18198	0.20122
	5	0.00001	0.00021	0.00134	0.00466	0.01171	0.02400	0.02804	0.04273	0.06861	0.07888
	6	0.00000	0.00002	0.00016	0.00072	0.00227	0.00559	0.00680	0.01161	0.02129	0.02550
	7		0.00000	0.00001	0.00009	0.00036	0.00108	0.00137	0.00263	0.00552	0.00689
	8			0.00000	0.00001	0.00005	0.00018	0.00023	0.00050	0.00121	0.00158
	9				0.00000	0.00001	0.00003	0.00005	0.00008	0.00023	0.00031
	10						0.00000	0.00000	0.00001	0.00004	0.00005
	11								0.00000	0.00001	0.00001
	12									0.00000	0.00000

n	r	$p=0.01$	$p=0.02$	$p=0.03$	$p=0.04$	$p=0.05$	$p=0.06$	$p=1/16$	$p=0.07$	$p=0.08$	$p=1/12$
29	0	1.00000	1.00000	1.00000	1.00000	1.00000	1.00000	1.00000	1.00000	1.00000	1.00000
	1	0.25283	0.44338	0.58659	0.69390	0.77406	0.83377	0.84613	0.87810	0.91091	0.91981
	2	0.03396	0.11396	0.21580	0.32403	0.42922	0.52607	0.54863	0.61202	0.68624	0.70839
	3	0.00301	0.01984	0.05525	0.10827	0.17512	0.25110	0.27098	0.33163	0.41272	0.43932
	4	0.00019	0.00255	0.01056	0.02736	0.05475	0.09314	0.10438	0.14168	0.19867	0.21917
	5	0.00001	0.00025	0.00158	0.00544	0.01358	0.02761	0.03219	0.04876	0.07768	0.08908
	6	0.00000	0.00002	0.00019	0.00088	0.00274	0.00669	0.00813	0.01378	0.02508	0.02994
	7		0.00000	0.00002	0.00012	0.00046	0.00135	0.00171	0.00325	0.00678	0.00844
	8			0.00000	0.00001	0.00006	0.00023	0.00030	0.00065	0.00156	0.00202
	9				0.00000	0.00001	0.00003	0.00005	0.00011	0.00031	0.00041
	10					0.00000	0.00000	0.00001	0.00002	0.00005	0.00007
	11							0.00000	0.00000	0.00001	0.00001
	12									0.00000	0.00000
30	0	1.00000	1.00000	1.00000	1.00000	1.00000	1.00000	1.00000	1.00000	1.00000	1.00000
	1	0.26030	0.45452	0.59899	0.70614	0.78536	0.84374	0.85574	0.88663	0.91803	0.92649
	2	0.03615	0.12055	0.22692	0.33882	0.44646	0.54453	0.56723	0.63064	0.70421	0.72601
	3	0.00332	0.02172	0.06007	0.11690	0.18782	0.26760	0.28833	0.35125	0.43460	0.46174
	4	0.00022	0.00289	0.01190	0.03059	0.06077	0.10262	0.11479	0.15498	0.21579	0.23751
	5	0.00001	0.00030	0.00185	0.00632	0.01564	0.03154	0.03670	0.05526	0.08736	0.09992
	6	0.00000	0.00003	0.00023	0.00106	0.00328	0.00795	0.00963	0.01623	0.02929	0.03487
	7		0.00000	0.00002	0.00015	0.00057	0.00167	0.00211	0.00399	0.00825	0.01023
	8			0.00000	0.00002	0.00008	0.00030	0.00039	0.00083	0.00197	0.00255
	9				0.00000	0.00001	0.00005	0.00006	0.00015	0.00041	0.00055
	10					0.00000	0.00001	0.00001	0.00002	0.00007	0.00010
	11						0.00000	0.00000	0.00000	0.00001	0.00002
	12									0.00000	0.00000
31	0	1.00000	1.00000	1.00000	1.00000	1.00000	1.00000	1.00000	1.00000	1.00000	1.00000
	1	0.26770	0.46543	0.61102	0.71790	0.79609	0.85312	0.86476	0.89457	0.92459	0.93262
	2	0.03839	0.12723	0.23809	0.35351	0.46340	0.56248	0.58526	0.64856	0.73131	0.74272
	3	0.00365	0.02369	0.06507	0.12577	0.20075	0.28422	0.30576	0.37081	0.45617	0.48376
	4	0.00025	0.00327	0.01335	0.03405	0.06712	0.11252	0.12564	0.16872	0.23330	0.25620
	5	0.00001	0.00035	0.00215	0.00729	0.01789	0.03580	0.04158	0.06224	0.09764	0.11138
	6	0.00000	0.00003	0.00028	0.00127	0.00390	0.00936	0.01132	0.01896	0.03393	0.04029
	7		0.00000	0.00003	0.00018	0.00071	0.00205	0.00258	0.00485	0.00993	0.01229
	8			0.00000	0.00002	0.00011	0.00038	0.00050	0.00105	0.00248	0.00319
	9				0.00000	0.00001	0.00006	0.00008	0.00020	0.00053	0.00071
	10					0.00000	0.00001	0.00001	0.00003	0.00010	0.00014
	11						0.00000	0.00000	0.00000	0.00002	0.00002
	12									0.00000	0.00000
32	0	1.00000	1.00000	1.00000	1.00000	1.00000	1.00000	1.00000	1.00000	1.00000	1.00000
	1	0.27502	0.47612	0.62269	0.72918	0.80629	0.86193	0.87321	0.90195	0.93062	0.93823
	2	0.04068	0.13399	0.24927	0.36809	0.48004	0.57992	0.60273	0.66578	0.73758	0.75854
	3	0.00399	0.02577	0.07027	0.13488	0.21389	0.30091	0.32323	0.39025	0.47738	0.50534
	4	0.00029	0.00368	0.01490	0.03771	0.07381	0.12282	0.13690	0.18287	0.25113	0.27516
	5	0.00002	0.00041	0.00249	0.00836	0.02035	0.04041	0.04684	0.06970	0.10849	0.12345
	6	0.00000	0.00004	0.00034	0.00151	0.00460	0.01095	0.01321	0.02199	0.03903	0.04622
	7		0.00000	0.00004	0.00023	0.00087	0.00249	0.00313	0.00584	0.01185	0.01462
	8			0.00000	0.00004	0.00014	0.00048	0.00063	0.00132	0.00307	0.00395
	9				0.00000	0.00002	0.00008	0.00011	0.00026	0.00069	0.00092
	10					0.00000	0.00001	0.00002	0.00004	0.00013	0.00019
	11						0.00000	0.00000	0.00001	0.00002	0.00003
	12								0.00000	0.00000	0.00001
	13										0.00000
33	0	1.00000	1.00000	1.00000	1.00000	1.00000	1.00000	1.00000	1.00000	1.00000	1.00000
	1	0.28227	0.48659	0.63401	0.74001	0.81597	0.87022	0.88114	0.90881	0.93617	0.94338
	2	0.04303	0.14083	0.26048	0.38253	0.49635	0.59684	0.61963	0.68231	0.75302	0.77352
	3	0.00436	0.02793	0.07564	0.14421	0.22719	0.31765	0.34070	0.40954	0.49820	0.52644
	4	0.00032	0.00412	0.01656	0.04160	0.08081	0.13351	0.14854	0.19738	0.26923	0.29434
	5	0.00002	0.00048	0.00286	0.00954	0.02303	0.04535	0.05247	0.07762	0.11990	0.13609
	6	0.00000	0.00004	0.00040	0.00179	0.00539	0.01271	0.01532	0.02533	0.04459	0.05265
	7		0.00000	0.00005	0.00028	0.00106	0.00299	0.00376	0.00697	0.01402	0.01735
	8			0.00000	0.00004	0.00018	0.00060	0.00079	0.00164	0.00377	0.00484
	9				0.00000	0.00003	0.00010	0.00014	0.00033	0.00088	0.00117
	10					0.00000	0.00002	0.00002	0.00006	0.00018	0.00025
	11						0.00000	0.00000	0.00001	0.00003	0.00005
	12								0.00000	0.00000	0.00001
	13										0.00000

TABLE T3-5 Molina Table of Poisson Distribution,

$$1 - F(c - 1), \ Pr(x \geq c \,|\, a) = \sum_{x=c}^{\infty} \frac{a^x e^{-a}}{x!}$$

c	a = .001	a = .002	a = .003	a = .004
0	1.0000000	1.0000000	1.0000000	1.0000000
1	.0009995	.0019980	.0029955	.0039920
2	.0000005	.0000020	.0000045	.0000080

c	a = .005	a = .006	a = .007	a = .008
0	1.0000000	1.0000000	1.0000000	1.0000000
1	.0049875	.0059820	.0069756	.0079681
2	.0000125	.0000179	.0000244	.0000318
3			.0000001	.0000001

c	a = .009	a = .010	a = .02	a = .03
0	1.0000000	1.0000000	1.0000000	1.0000000
1	.0089596	.0099502	.0198013	.0295545
2	.0000403	.0000497	.0001973	.0004411
3	.0000001	.0000002	.0000013	.0000044

c	a = .04	a = .05	a = .06	a = .07
0	1.0000000	1.0000000	1.0000000	1.0000000
1	.0392106	.0487706	.0582355	.0676062
2	.0007790	.0012091	.0017296	.0023386
3	.0000104	.0000201	.0000344	.0000542
4	.0000001	.0000003	.0000005	.0000009

c	a = .08	a = .09	a = .10	a = .11
0	1.0000000	1.0000000	1.0000000	1.0000000
1	.0768837	.0860688	.0951626	.1041659
2	.0030343	.0038150	.0046788	.0056241
3	.0000804	.0001136	.0001547	.0002043
4	.0000016	.0000025	.0000038	.0000056
5				.0000001

c	a = .12	a = .13	a = .14	a = .15
0	1.0000000	1.0000000	1.0000000	1.0000000
1	.1130796	.1219046	.1306418	.1392920
2	.0066491	.0077522	.0089316	.0101858
3	.0002633	.0003323	.0004119	.0005029
4	.0000079	.0000107	.0000143	.0000187
5	.0000002	.0000003	.0000004	.0000006

TABLE T3-5   (continued)

c	a = .16	a = .17	a = .18	a = .19
0	1.0000000	1.0000000	1.0000000	1.0000000
1	.1478562	.1563352	.1647298	.1730409
2	.0115132	.0129122	.0143812	.0159187
3	.0006058	.0007212	.0008498	.0009920
4	.0000240	.0000304	.0000379	.0000467
5	.0000008	.0000010	.0000014	.0000018
6				.0000001

c	a = .20	a = .21	a = .22	a = .23
0	1.0000000	1.0000000	1.0000000	1.0000000
1	.1812692	.1894158	.1974812	.2054664
2	.0175231	.0191931	.0209271	.0227237
3	.0011485	.0013197	.0015060	.0017083
4	.0000568	.0000685	.0000819	.0000971
5	.0000023	.0000029	.0000036	.0000044
6	.0000001	.0000001	.0000001	.0000002

c	a = .24	a = .25	a = .26	a = .27
0	1.0000000	1.0000000	1.0000000	1.0000000
1	.2133721	.2211992	.2289484	.2366205
2	.0245815	.0264990	.0284750	.0305080
3	.0019266	.0021615	.0024135	.0026829
4	.0001142	.0001334	.0001548	.0001786
5	.0000054	.0000066	.0000080	.0000096
6	.0000002	.0000003	.0000003	.0000004

c	a = .28	a = .29	a = .30	a = .4
0	1.0000000	1.0000000	1.0000000	1.0000000
1	.2442163	.2517364	.2591818	.3296800
2	.0325968	.0347400	.0369363	.0615519
3	.0029701	.0032755	.0035995	.0079263
4	.0002049	.0002339	.0002658	.0007763
5	.0000113	.0000134	.0000158	.0000612
6	.0000005	.0000006	.0000008	.0000040
7				.0000002

c	a = .5	a = .6	a = .7	a = .8
0	1.000000	1.000000	1.000000	1.000000
1	.393469	.451188	.503415	.550671
2	.090204	.121901	.155805	.191208
3	.014388	.023115	.034142	.047423
4	.001752	.003358	.005753	.009080
5	.000172	.000394	.000786	.001411
6	.000014	.000039	.000090	.000184
7	.000001	.000003	.000009	.000021
8			.000001	.000002

TABLE T3-5   (continued)

c	a = .9	a = 1.0	a = 1.1	a = 1.2
0	1.000000	1.000000	1.000000	1.000000
1	.593430	.632121	.667129	.698806
2	.227518	.264241	.300971	.337373
3	.062857	.080301	.099584	.120513
4	.013459	.018988	.025742	.033769
5	.002344	.003660	.005435	.007746
6	.000343	.000594	.000968	.001500
7	.000043	.000083	.000149	.000251
8	.000005	.000010	.000020	.000037
9		.000001	.000002	.000005
10				.000001

c	a = 1.3	a = 1.4	a = 1.5	a = 1.6
0	1.000000	1.000000	1.000000	1.000000
1	.727468	.753403	.776870	.798103
2	.373177	.408167	.442175	.475069
3	.142888	.166502	.191153	.216642
4	.043095	.053725	.065642	.078813
5	.010663	.014253	.018576	.023682
6	.002231	.003201	.004456	.006040
7	.000404	.000622	.000926	.001336
8	.000064	.000107	.000170	.000260
9	.000009	.000016	.000028	.000045
10	.000001	.000002	.000004	.000007
11			.000001	.000001

c	a = 1.7	a = 1.8	a = 1.9	a = 2.0
0	1.000000	1.000000	1.000000	1.000000
1	.817316	.834701	.850431	.864665
2	.506754	.537163	.566251	.593994
3	.242777	.269379	.296280	.323324
4	.093189	.108708	.125298	.142877
5	.029615	.036407	.044081	.052653
6	.007999	.010378	.013219	.016564
7	.001875	.002569	.003446	.004534
8	.000388	.000562	.000793	.001097
9	.000072	.000110	.000163	.000237
10	.000012	.000019	.000030	.000046
11	.000002	.000003	.000005	.000008
12			.000001	.000001

TABLE T3-5   (continued)

c	a = 2.1	a = 2.2	a = 2.3	a = 2.4	a = 2.5
0	1.000000	1.000000	1.000000	1.000000	1.000000
1	.877544	.889197	.899741	.909282	.917915
2	.620385	.645430	.669146	.691559	.712703
3	.350369	.377286	.403961	.430291	.456187
4	.161357	.180648	.200653	.221277	.242424
5	.062126	.072496	.083751	.095869	.108822
6	.020449	.024910	.029976	.035673	.042021
7	.005862	.007461	.009362	.011594	.014187
8	.001486	.001978	.002589	.003339	.004247
9	.000337	.000470	.000642	.000862	.001140
10	.000069	.000101	.000144	.000202	.000277
11	.000013	.000020	.000029	.000043	.000062
12	.000002	.000004	.000006	.000008	.000013
13		.000001	.000001	.000002	.000002

c	a = 2.6	a = 2.7	a = 2.8	a = 2.9	a = 3.0
0	1.000000	1.000000	1.000000	1.000000	1.000000
1	.925726	.932794	.939190	.944977	.950213
2	.732615	.751340	.768922	.785409	.800852
3	.481570	.506375	.530546	.554037	.576810
4	.263998	.285908	.308063	.330377	.352768
5	.122577	.137092	.152324	.168223	.184737
6	.049037	.056732	.065110	.074174	.083918
7	.017170	.020569	.024411	.028717	.033509
8	.005334	.006621	.008131	.009885	.011905
9	.001487	.001914	.002433	.003058	.003803
10	.000376	.000501	.000660	.000858	.001102
11	.000087	.000120	.000164	.000220	.000292
12	.000018	.000026	.000037	.000052	.000071
13	.000004	.000005	.000008	.000011	.000016
14	.000001	.000001	.000002	.000002	.000003
15					.000001

c	a = 3.1	a = 3.2	a = 3.3	a = 3.4	a = 3.5
0	1.000000	1.000000	1.000000	1.000000	1.000000
1	.954951	.959238	.963117	.966627	.969803
2	.815298	.828799	.841402	.853158	.864112
3	.598837	.620096	.640574	.660260	.679153
4	.375160	.397480	.419662	.441643	.463367
5	.201811	.219387	.237410	.255818	.274555
6	.094334	.105408	.117123	.129458	.142386
7	.038804	.044619	.050966	.057853	.065288
8	.014213	.016830	.019777	.023074	.026739
9	.004683	.005714	.006912	.008293	.009874
10	.001401	.001762	.002195	.002709	.003315
11	.000383	.000497	.000638	.000810	.001019
12	.000097	.000129	.000171	.000223	.000289
13	.000023	.000031	.000042	.000057	.000076
14	.000005	.000007	.000010	.000014	.000019
15	.000001	.000001	.000002	.000003	.000004
16				.000001	.000001

TABLE T3-5   (continued)

c	a = 3.6	a = 3.7	a = 3.8	a = 3.9	a = 4.0
0	1.000000	1.000000	1.000000	1.000000	1.000000
1	.972676	.975276	.977629	.979758	.981684
2	.874311	.883799	.892620	.900815	.908422
3	.697253	.714567	.731103	.746875	.761897
4	.484784	.505847	.526515	.546753	.566530
5	.293562	.312781	.332156	.351635	.371163
6	.155881	.169912	.184444	.199442	.214870
7	.073273	.081809	.090892	.100517	.110674
8	.030789	.035241	.040107	.045402	.051134
9	.011671	.013703	.015984	.018533	.021363
10	.004024	.004848	.005799	.006890	.008132
11	.001271	.001572	.001929	.002349	.002840
12	.000370	.000470	.000592	.000739	.000915
13	.000100	.000130	.000168	.000216	.000274
14	.000025	.000034	.000045	.000059	.000076
15	.000006	.000008	.000011	.000015	.000020
16	.000001	.000002	.000003	.000004	.000005
17			.000001	.000001	.000001

c	a = 4.1	a = 4.2	a = 4.5	a = 4.4	a = 4.5
0	1.000000	1.000000	1.000000	1.000000	1.000000
1	.983427	.985004	.986431	.987723	.988891
2	.915479	.922023	.928087	.933702	.938901
3	.776186	.789762	.802645	.814858	.826422
4	.585818	.604597	.622846	.640552	.657704
5	.390692	.410173	.429562	.448816	.467896
6	.230688	.246857	.263338	.280088	.297070
7	.121352	.132536	.144210	.156355	.168949
8	.057312	.063943	.071032	.078579	.086586
9	.024492	.027932	.031698	.035803	.040257
10	.009540	.011127	.012906	.014890	.017093
11	.003410	.004069	.004825	.005688	.006669
12	.001125	.001374	.001666	.002008	.002404
13	.000345	.000431	.000534	.000658	.000805
14	.000098	.000126	.000160	.000201	.000252
15	.000026	.000034	.000045	.000058	.000074
16	.000007	.000009	.000012	.000016	.000020
17	.000002	.000002	.000003	.000004	.000005
18			.000001	.000001	.000001

c	2 = 4.6	a = 4.7	a = 4.8	a = 4.9	a = 5.0
0	1.000000	1.000000	1.000000	1.000000	1.000000
1	.989948	.990905	.991770	.992553	.993262
2	.943710	.948157	.952267	.956065	.959572
3	.837361	.847700	.857461	.866669	.875348
4	.674294	.690316	.705770	.720655	.734974

TABLE T3-5   (continued)

c	a = 4.6	a = 4.7	a = 4.8	a = 4.9	a = 5.0
5	.486766	.505391	.523741	.541788	.559507
6	.314240	.331562	.348994	.366499	.384039
7	.181971	.195395	.209195	.223345	.237817
8	.095051	.103969	.113334	.123138	.133372
9	.045072	.050256	.055817	.061761	.068094
10	.019527	.022206	.025141	.028345	.031828
11	.007777	.009022	.010417	.011971	.013695
12	.002863	.003389	.003992	.004677	.005453
13	.000979	.001183	.001422	.001699	.002019
14	.000312	.000385	.000473	.000576	.000698
15	.000093	.000118	.000147	.000183	.000226
16	.000026	.000034	.000043	.000055	.000069
17	.000007	.000009	.000012	.000015	.000020
18	.000002	.000002	.000003	.000004	.000005
19		.000001	.000001	.000001	.000001

c	a = 5.1	a = 5.2	a = 5.3	a = 5.4	a = 5.5
0	1.000000	1.000000	1.000000	1.000000	1.000000
1	.993903	.994483	.995008	.995483	.995913
2	.962810	.965797	.968553	.971094	.973436
3	.883522	.891213	.898446	.905242	.911624
4	.748732	.761935	.774590	.786709	.798301
5	.576875	.593872	.610482	.626689	.642482
6	.401580	.419087	.436527	.453868	.471081
7	.252580	.267607	.282866	.298329	.313964
8	.144023	.155078	.166523	.178341	.190515
9	.074818	.081935	.089446	.097350	.105643
10	.035601	.039674	.044056	.048755	.053777
11	.015601	.017699	.020000	.022514	.025251
12	.006328	.007310	.008409	.009632	.010988
13	.002387	.002809	.003289	.003835	.004451
14	.000841	.001008	.001202	.001427	.001685
15	.000278	.000339	.000412	.000498	.000599
16	.000086	.000108	.000133	.000164	.000200
17	.000025	.000032	.000041	.000051	.000063
18	.000007	.000009	.000012	.000015	.000019
19	.000002	.000002	.000003	.000004	.000005
20		.000001	.000001	.000001	.000001

c	a = 5.6	a = 5.7	a = 5.8	a = 5.9	a = 6.0
0	1.000000	1.000000	1.000000	1.000000	1.000000
1	.996302	.996654	.996972	.997261	.997521
2	.975594	.977582	.979413	.981098	.982649
3	.917612	.923227	.928489	.933418	.938031
4	.809378	.819952	.830037	.839647	.848796
5	.657850	.672785	.687282	.701335	.714943
6	.488139	.505015	.521685	.538127	.554320
7	.329742	.345634	.361609	.377639	.393697
8	.203025	.215851	.228974	.242371	.256020
9	.114322	.123382	.132814	.142611	.152763

TABLE T3-5 (continued)

c	a = 5.6	a = 5.7	a = 5.8	a = 5.9	a = 6.0
10	.059130	.064817	.070844	.077212	.083924
11	.028222	.031436	.034901	.038627	.042621
12	.012487	.014138	.015950	.017931	.020092
13	.005144	.005922	.006790	.007756	.008827
14	.001981	.002319	.002703	.003138	.003628
15	.000716	.000852	.001010	.001192	.001400
16	.000244	.000295	.000356	.000426	.000509
17	.000078	.000096	.000118	.000144	.000175
18	.000024	.000030	.000037	.000046	.000057
19	.000007	.000009	.000011	.000014	.000018
20	.000002	.000002	.000003	.000004	.000005
21		.000001	.000001	.000001	.000001

c	a = 6.1	a = 6.2	a = 6.3	a = 6.4	a = 6.5
0	1.000000	1.000000	1.000000	1.000000	1.000000
1	.997757	.997971	.998164	.998338	.998497
2	.984076	.985388	.986595	.987704	.988724
3	.942347	.946382	.950154	.953676	.956964
4	.857499	.865771	.873626	.881081	.888150
5	.728106	.740823	.753096	.764930	.776328
6	.570246	.585887	.601228	.616256	.630959
7	.409755	.425787	.441767	.457671	.473476
8	.269899	.283984	.298252	.312679	.327242
9	.163258	.174086	.185233	.196685	.208427
10	.090980	.098379	.106121	.114201	.122616
11	.046890	.051441	.056280	.061411	.066839
12	.022440	.024985	.027734	.030697	.033880
13	.010012	.011316	.012748	.014316	.016027
14	.004180	.004797	.005485	.006251	.007100
15	.001639	.001910	.002217	.002565	.002956
16	.000605	.000716	.000844	.000992	.001160
17	.000211	.000254	.000304	.000362	.000430
18	.000070	.000085	.000104	.000126	.000151
19	.000022	.000027	.000034	.000041	.000051
20	.000007	.000008	.000010	.000013	.000016
21	.000002	.000002	.000003	.000004	.000005
22	.000001	.000001	.000001	.000001	.000001

c	a = 6.6	a = 6.7	a = 6.8	a = 6.9	a = 7.0
0	1.000000	1.000000	1.000000	1.000000	1.000000
1	.998640	.998769	.998886	.998992	.999088
2	.989661	.990522	.991313	.992038	.992705
3	.960032	.962894	.965562	.968048	.970364
4	.894849	.901192	.907194	.912870	.918235
5	.787296	.797841	.807969	.817689	.827008
6	.645327	.659351	.673023	.686338	.699292
7	.489161	.504703	.520084	.535285	.550289
8	.341918	.356683	.371514	.386389	.401286
9	.220443	.232716	.245230	.257967	.270909

TABLE T3-5   (continued)

c	a = 6.6	a = 6.7	a = 6.8	a = 6.9	a = 7.0
10	.131361	.140430	.149816	.159510	.169504
11	.072567	.078598	.084934	.091575	.098521
12	.037291	.040937	.044825	.048961	.053350
13	.017889	.019910	.022097	.024458	.027000
14	.008038	.009072	.010208	.011452	.012811
15	.003395	.003886	.004434	.005042	.005717
16	.001352	.001569	.001816	.002094	.002407
17	.000509	.000599	.000703	.000822	.000958
18	.000182	.000217	.000258	.000306	.000362
19	.000062	.000075	.000090	.000108	.000130
20	.000020	.000024	.000030	.000037	.000044
21	.000006	.000008	.000010	.000012	.000014
22	.000002	.000002	.000003	.000004	.000005
23	.000001	.000001	.000001	.000001	.000001

c	a = 7.1	a = 7.2	a = 7.3	a = 7.4	a = 7.5
0	1.000000	1.000000	1.000000	1.000000	1.000000
1	.999175	.999253	.999324	.999389	.999447
2	.993317	.993878	.994393	.994865	.995299
3	.972520	.974526	.976393	.978129	.979743
4	.923301	.928083	.932594	.936847	.940855
5	.835937	.844484	.852660	.860475	.867938
6	.711881	.724103	.735957	.747443	.758564
7	.565080	.579644	.593968	.608038	.621845
8	.416183	.431059	.445893	.460667	.475361
9	.284036	.297332	.310776	.324349	.338033
10	.179788	.190350	.201180	.212265	.223592
11	.105771	.113323	.121175	.129323	.137762
12	.057997	.062906	.068081	.073526	.079241
13	.029730	.032655	.035782	.039117	.042666
14	.014292	.015901	.017645	.019531	.021565
15	.006463	.007285	.008188	.009178	.010260
16	.002757	.003149	.003586	.004071	.004608
17	.001113	.001288	.001486	.001709	.001959
18	.000426	.000500	.000584	.000680	.000790
19	.000155	.000184	.000218	.000258	.000303
20	.000054	.000065	.000078	.000093	.000111
21	.000018	.000022	.000026	.000032	.000039
22	.000006	.000007	.000009	.000011	.000013
23	.000002	.000002	.000003	.000003	.000004
24		.000001	.000001	.000001	.000001

c	a = 7.6	a = 7.7	a = 7.8	a = 7.9	a = 8.0
0	1.000000	1.000000	1.000000	1.000000	1.000000
1	.999500	.999547	.999590	.999629	.999665
2	.995696	.996060	.996394	.996700	.996981
3	.981243	.982636	.983930	.985131	.986246
4	.944629	.948181	.951523	.954666	.957620

TABLE T3-5   (continued)

c	a = 7.6	a = 7.7	a = 7.8	a = 7.9	a = 8.0
5	.875061	.881855	.888330	.894497	.900368
6	.769319	.779713	.789749	.799431	.808764
7	.635379	.648651	.661595	.674260	.686626
8	.489958	.504440	.518791	.532996	.547039
9	.351808	.365557	.379559	.393497	.407433
10	.235149	.246920	.258891	.271048	.283376
11	.146487	.155492	.164770	.174314	.184114
12	.085230	.091493	.098030	.104841	.111924
13	.046434	.050427	.054649	.059104	.063797
14	.023753	.026103	.028620	.031311	.034181
15	.011441	.012725	.014118	.015627	.017257
16	.005202	.005857	.006577	.007367	.008231
17	.002239	.002552	.002901	.003289	.003718
18	.000915	.001055	.001215	.001393	.001594
19	.000355	.000415	.000484	.000562	.000650
20	.000132	.000156	.000184	.000216	.000253
21	.000046	.000056	.000067	.000079	.000094
22	.000016	.000019	.000025	.000028	.000033
23	.000005	.000006	.000008	.000009	.000011
24	.000002	.000002	.000002	.000003	.000004
25		.000001	.000001	.000001	.000001

c	a = 8.1	a = 8.2	a = 8.3	a = 8.4	a = 8.5
0	1.000000	1.000000	1.000000	1.000000	1.000000
1	.999696	.999725	.999751	.999775	.999797
2	.997238	.997473	.997689	.997886	.998067
3	.987280	.988239	.989129	.989953	.990717
4	.960395	.963000	.965446	.967740	.969891
5	.905951	.911260	.916303	.921092	.925636
6	.817753	.826406	.834727	.842723	.850403
7	.698686	.710438	.721879	.733007	.743822
8	.560908	.574591	.588074	.601348	.614403
9	.421408	.435347	.449252	.463106	.476895
10	.295858	.308481	.321226	.334080	.347026
11	.194163	.204450	.214965	.225699	.236638
12	.119278	.126900	.134787	.142934	.151338
13	.068731	.073907	.079330	.084999	.090917
14	.037236	.040481	.043923	.047564	.051411
15	.019014	.020903	.022931	.025103	.027425
16	.009174	.010201	.011316	.012525	.013833
17	.004192	.004715	.005291	.005922	.006613
18	.001819	.002070	.002349	.002659	.003002
19	.000751	.000864	.000992	.001136	.001297
20	.000296	.000344	.000400	.000463	.000535
21	.000111	.000131	.000154	.000180	.000211
22	.000040	.000048	.000057	.000067	.000079
23	.000014	.000017	.000020	.000024	.000029
24	.000005	.000006	.000007	.000008	.000010
25	.000001	.000002	.000002	.000003	.000003
26		.000001	.000001	.000001	.000001

TABLE T3-5 (continued)

c	a = 8.6	a = 8.7	a = 8.8	a = 8.9	a = 9.0
0	1.000000	1.000000	1.000000	1.000000	1.000000
1	.999816	.999833	.999849	.999864	.999877
2	.998233	.998384	.998523	.998650	.998766
3	.991424	.992080	.992686	.993248	.993768
4	.971907	.973797	.975566	.977223	.978774
5	.929946	.934032	.937902	.941567	.945036
6	.857772	.864840	.871613	.878100	.884309
7	.754324	.764512	.774390	.783958	.793219
8	.627229	.639819	.652166	.664262	.676103
9	.490603	.504216	.517719	.531101	.544347
10	.360049	.373132	.386260	.399419	.412592
11	.247772	.259089	.270577	.282222	.294012
12	.159992	.168892	.178030	.187399	.196992
13	.097084	.103499	.110162	.117072	.124227
14	.055467	.059736	.064221	.068925	.073851
15	.029902	.032540	.035343	.038317	.041466
16	.015245	.016767	.018402	.020157	.022036
17	.007367	.008190	.009084	.010055	.011106
18	.003382	.003800	.004261	.004766	.005320
19	.001478	.001679	.001903	.002151	.002426
20	.000616	.000707	.000811	.000926	.001056
21	.000245	.000285	.000330	.000381	.000439
22	.000094	.000110	.000129	.000150	.000175
23	.000034	.000041	.000048	.000057	.000067
24	.000012	.000014	.000017	.000021	.000025
25	.000004	.000005	.000006	.000007	.000009
26	.000001	.000002	.000002	.000002	.000003
27		.000001	.000001	.000001	.000001

c	a = 9.1	a = 9.2	a = 9.3	a = 9.4	a = 9.5
0	1.000000	1.000000	1.000000	1.000000	1.000000
1	.999888	.999899	.999909	.999917	.999925
2	.998872	.998969	.999058	.999140	.999214
3	.994249	.994693	.995105	.995485	.995836
4	.980224	.981580	.982848	.984033	.985140
5	.948318	.951420	.954353	.957122	.959737
6	.890249	.895926	.901350	.906529	911472
7	.802177	.810835	.819197	.827267	.835051
8	.687684	.699000	.710050	.720829	.731337
9	.557448	.570391	.583166	.595765	.608177
10	.425765	.438924	.452054	.465142	.478174
11	.305933	.317974	.330119	.342356	.354672
12	.206800	.216815	.227029	.237430	.248010
13	.131624	.139261	.147133	.155238	.163570
14	.079001	.084376	.089978	.095807	.101864
15	.044795	.048309	.052010	.055903	.059992
16	.024044	.026188	.028470	.030897	.033473
17	.012242	.013468	.014788	.016206	.017727
18	.005924	.006584	.007302	.008083	.008928
19	.002731	.003066	.003435	.003840	.004284

TABLE T3-5   (continued)

c	a = 9.1	a = 9.2	a = 9.3	a = 9.4	a = 9.5
20	.001201	.001362	.001542	.001742	.001962
21	.000505	.000579	.000662	.000755	.000859
22	.000203	.000235	.000272	.000314	.000361
23	.000078	.000092	.000107	.000125	.000145
24	.000029	.000034	.000041	.000048	.000056
25	.000010	.000012	.000015	.000018	.000021
26	.000004	.000004	.000005	.000006	.000007
27	.000001	.000001	.000002	.000002	.000003
28			.000001	.000001	.000001

c	a = 9.6	a = 9.7	a = 9.8	a = 9.9	a = 10.0
0	1.000000	1.000000	1.000000	1.000000	1.000000
1	.999932	.999939	.999945	.999950	.999955
2	.999282	.999344	.999401	.999453	.999501
3	.996161	.996461	.996738	.996994	.997231
4	.986174	.987139	.988040	.988880	.989664
5	.962205	.964533	.966729	.968798	.970747
6	.916185	.920678	.924959	.929035	.932914
7	.842553	.849779	.856735	.863426	.869859
8	.741572	.751533	.761221	.770636	.779779
9	.620394	.632410	.644217	.655809	.667180
10	.491138	.504021	.516812	.529498	.542070
11	.367052	.379484	.391955	.404451	.416960
12	.258759	.269665	.280719	.291909	.303224
13	.172124	.180895	.189876	.199062	.208444
14	.108148	.114659	.121395	.128355	.135536
15	.064279	.068767	.073458	.078355	.083458
16	.036202	.039090	.042139	.045355	.048740
17	.019357	.021098	.022956	.024936	.027042
18	.009844	.010832	.011898	.013045	.014278
19	.004770	.005300	.005877	.006505	.007187
20	.002207	.002476	.002772	.003098	.003454
21	.000976	.001106	.001250	.001411	.001588
22	.000414	.000473	.000540	.000616	.000700
23	.000168	.000194	.000224	.000258	.000296
24	.000066	.000077	.000089	.000104	.000120
25	.000025	.000029	.000034	.000040	.000047
26	.000009	.000011	.000013	.000015	.000018
27	.000003	.000004	.000004	.000005	.000006
28	.000001	.000001	.000002	.000002	.000002
29			.000001	.000001	.000001

c	a = 10.1	a = 10.2	a = 10.3	a = 10.4	a = 10.5
0	1.000000	1.000000	1.000000	1.000000	1.000000
1	.999959	.999963	.999966	.999970	.999972
2	.999544	.999584	.999620	.999653	.999683
3	.997449	.997650	.997836	.998007	.998165
4	.990395	.991076	.991711	.992302	.992853

TABLE T5-1 Cameron Table of Unity Values for Constructing Single-Sampling Plans

c	Values of p_2/p_1 for: $\alpha = .05$ $\beta = .10$	Values of p_2/p_1 for: $\alpha = .05$ $\beta = .05$	Values of p_2/p_1 for: $\alpha = .05$ $\beta = .01$	np_1	c	Values of p_2/p_1 for: $\alpha = .01$ $\beta = -.10$	Values of p_2/p_1 for: $\alpha = .01$ $\beta = .05$	Values of p_2/p_1 for: $\alpha = .01$ $\beta = .01$	np_1
0	44.890	58.404	89.781	.052	0	229.105	298.073	458.210	.010
1	10.946	13.349	18.681	.355	1	26.184	31.933	44.686	.149
2	6.509	7.699	10.230	.818	2	12.206	14.439	19.278	.436
3	4.890	5.675	7.352	1.366	3	8.115	9.418	12.202	.823
4	4.057	4.646	5.890	1.970	4	6.249	7.156	9.072	1.279
5	3.549	4.023	5.017	2.613	5	5.195	5.889	7.343	1.785
6	3.206	3.604	4.435	3.286	6	4.520	5.082	6.253	2.330
7	2.957	3.303	4.019	3.981	7	4.050	4.524	5.506	2.906
8	2.768	3.074	3.707	4.695	8	3.705	4.115	4.962	3.507
9	2.618	2.895	3.462	5.426	9	3.440	3.803	4.548	4.130
10	2.497	2.750	3.265	6.169	10	3.229	3.555	4.222	4.771
11	2.397	2.630	3.104	6.924	11	3.058	3.354	3.959	5.428
12	2.312	2.528	2.968	7.690	12	2.915	3.188	3.742	6.099
13	2.240	2.442	2.852	8.464	13	2.795	3.047	3.559	6.782
14	2.177	2.367	2.752	9.246	14	2.692	2.927	3.403	7.477
15	2.122	2.302	2.665	10.035	15	2.603	2.823	3.269	8.181
16	2.073	2.244	2.588	10.831	16	2.524	2.732	3.151	8.895
17	2.029	2.192	2.520	11.633	17	2.455	2.652	3.048	9.616
18	1.990	2.145	2.458	12.442	18	2.393	2.580	2.956	10.346
19	1.954	2.103	2.403	13.254	19	2.337	2.516	2.874	11.082
20	1.922	2.065	2.352	14.072	20	2.287	2.458	2.799	11.825
21	1.892	2.030	2.307	14.894	21	2.241	2.405	2.733	12.574
22	1.865	1.999	2.265	15.719	22	2.200	2.357	2.671	13.329
23	1.840	1.969	2.226	16.548	23	2.162	2.313	2.615	14.088
24	1.817	1.942	2.191	17.382	24	2.126	2.272	2.564	14.853
25	1.795	1.917	2.158	18.218	25	2.094	2.235	2.516	15.623
26	1.775	1.893	2.127	19.058	26	2.064	2.200	2.472	16.397
27	1.757	1.871	2.098	19.900	27	2.035	2.168	2.431	17.175
28	1.739	1.850	2.071	20.746	28	2.009	2.138	2.393	17.957
29	1.723	1.831	2.046	21.594	29	1.985	2.110	2.358	18.742
30	1.707	1.813	2.023	22.444	30	1.962	2.083	2.324	19.532
31	1.692	1.796	2.001	23.298	31	1.940	2.059	2.293	20.324
32	1.679	1.780	1.980	24.152	32	1.920	2.035	2.264	21.120
33	1.665	1.764	1.960	25.010	33	1.900	2.013	2.236	21.919
34	1.653	1.750	1.941	25.870	34	1.882	1.992	2.210	22.721
35	1.641	1.736	1.923	26.731	35	1.865	1.973	2.185	23.525
36	1.630	1.723	1.906	27.594	36	1.848	1.954	2.162	24.333
37	1.619	1.710	1.890	28.460	37	1.833	1.936	2.139	25.143
38	1.609	1.698	1.875	29.327	38	1.818	1.920	2.118	25.955
39	1.599	1.687	1.860	30.196	39	1.804	1.903	2.098	26.770
40	1.590	1.676	1.846	31.066	40	1.790	1.887	2.079	27.587
41	1.581	1.666	1.833	31.938	41	1.777	1.873	2.060	28.406
42	1.572	1.656	1.820	32.812	42	1.765	1.859	2.043	29.228
43	1.564	1.646	1.807	33.686	43	1.753	1.845	2.026	30.051
44	1.556	1.637	1.796	34.563	44	1.742	1.832	2.010	30.877
45	1.548	1.628	1.784	35.441	45	1.731	1.820	1.994	31.704
46	1.541	1.619	1.773	36.320	46	1.720	1.808	1.980	32.534
47	1.534	1.611	1.763	37.200	47	1.710	1.796	1.965	33.365
48	1.527	1.603	1.752	38.082	48	1.701	1.785	1.952	34.198
49	1.521	1.596	1.743	38.965	49	1.691	1.775	1.938	35.032

From J. M. Cameron, *Industrial Quality Control*, 9(1): 1952, 38. Reprinted by permission.

TABLE T5-2 Cameron Table of Unity Values to Determine Probability of Acceptance

c	P(A)= .995	P(A)= .990	P(A)= .975	P(A)= .950	P(A)= .900	P(A)= .750	P(A)= .500	P(A)= .250	P(A)= .100	P(A)= .050	P(A)= .025	P(A)= .010	P(A)= .005
0	.00501	.0101	.0253	.0513	.105	.288	.693	1.386	2.303	2.996	3.689	4.605	5.298
1	.103	.149	.242	.355	.532	.961	1.678	2.693	3.890	4.744	5.572	6.638	7.430
2	.338	.436	.619	.818	1.102	1.727	2.674	3.920	5.322	6.296	7.224	8.406	9.274
3	.672	.823	1.090	1.366	1.745	2.535	3.672	5.109	6.681	7.754	8.768	10.045	10.978
4	1.078	1.279	1.623	1.970	2.433	3.369	4.671	6.274	7.994	9.154	10.242	11.605	12.594
5	1.537	1.785	2.202	2.613	3.152	4.219	5.670	7.423	9.275	10.513	11.668	13.108	14.150
6	2.037	2.330	2.814	3.286	3.895	5.083	6.670	8.558	10.532	11.842	13.060	14.571	15.660
7	2.571	2.906	3.454	3.981	4.656	5.956	7.669	9.684	11.771	13.148	14.422	16.000	17.134
8	3.132	3.507	4.115	4.695	5.432	6.838	8.669	10.802	12.995	14.434	15.763	17.403	18.578
9	3.717	4.130	4.795	5.426	6.221	7.726	9.669	11.914	14.206	15.705	17.085	18.783	19.998
10	4.321	4.771	5.491	6.169	7.021	8.620	10.668	13.020	15.407	16.962	18.390	20.145	21.398
11	4.943	5.428	6.201	6.924	7.829	9.519	11.668	14.121	16.598	18.208	19.682	21.490	22.779
12	5.580	6.099	6.922	7.690	8.646	10.422	12.668	15.217	17.782	19.442	20.962	22.821	24.145
13	6.231	6.782	7.654	8.464	9.470	11.329	13.668	16.310	18.958	20.668	22.230	24.139	25.496
14	6.893	7.477	8.396	9.246	10.300	12.239	14.668	17.400	20.128	21.886	23.490	25.446	26.836
15	7.566	8.181	9.144	10.035	11.135	13.152	15.668	18.486	21.292	23.098	24.741	26.743	28.166
16	8.249	8.895	9.902	10.831	11.976	14.068	16.668	19.570	22.452	24.302	25.984	28.031	29.484
17	8.942	9.616	10.666	11.633	12.822	14.986	17.668	20.652	23.606	25.500	27.220	29.310	30.792
18	9.644	10.346	11.438	12.442	13.672	15.907	18.668	21.731	24.756	26.692	28.448	30.581	32.092
19	10.353	11.082	12.216	13.254	14.525	16.830	19.668	22.808	25.902	27.879	29.671	31.845	33.383
20	11.069	11.825	12.999	14.072	15.383	17.755	20.668	23.883	27.045	29.062	30.888	33.103	34.668
21	11.791	12.574	13.787	14.894	16.244	18.682	21.668	24.956	28.184	30.241	32.102	34.355	35.947
22	12.520	13.329	14.580	15.719	17.108	19.610	22.668	26.028	29.320	31.416	33.309	35.601	37.219
23	13.255	14.088	15.377	16.548	17.975	20.540	23.668	27.098	30.453	32.586	34.512	36.841	38.485
24	13.995	14.853	16.178	17.382	18.844	21.471	24.668	28.167	31.584	33.752	35.710	38.077	39.745
25	14.740	15.623	16.984	18.218	19.717	22.404	25.667	29.234	32.711	34.916	36.905	39.308	41.000
26	15.490	16.397	17.793	19.058	20.592	23.338	26.667	30.300	33.836	36.077	38.096	40.535	42.252
27	16.245	17.175	18.606	19.900	21.469	24.273	27.667	31.365	34.959	37.234	39.284	41.757	43.497
28	17.004	17.957	19.422	20.746	22.348	25.209	28.667	32.428	36.080	38.389	40.468	42.975	44.738
29	17.767	18.742	20.241	21.594	23.229	26.147	29.667	33.491	37.198	39.541	41.649	44.190	45.976
30	18.534	19.532	21.063	22.444	24.113	27.086	30.667	34.552	38.315	40.690	42.827	45.401	47.210
31	19.305	20.324	21.888	23.298	24.998	28.025	31.667	35.613	39.430	41.838	44.002	46.609	48.440
32	20.079	21.120	22.716	24.152	25.885	28.966	32.667	36.672	40.543	42.982	45.174	47.813	49.666
33	20.856	21.919	23.546	25.010	26.774	29.907	33.667	37.731	41.654	44.125	46.344	49.015	50.888
34	21.638	22.721	24.379	25.870	27.664	30.849	34.667	38.788	42.764	45.266	47.512	50.213	52.108
35	22.422	23.525	25.214	26.731	28.556	31.792	35.667	39.845	43.872	46.404	48.676	51.409	53.324
36	23.208	24.333	26.052	27.594	29.450	32.736	36.667	40.901	44.978	47.540	49.840	52.601	54.538
37	23.998	25.143	26.891	28.460	30.345	33.681	37.667	41.957	46.033	48.676	51.000	53.791	55.748
38	24.791	25.955	27.733	29.327	31.241	34.626	38.667	43.011	47.187	49.808	52.158	54.979	56.956
39	25.586	26.770	28.576	30.196	32.139	35.572	39.667	44.065	48.289	50.940	53.314	56.164	58.160
40	26.384	27.587	29.422	31.066	33.038	36.519	40.667	45.118	49.390	52.069	54.469	57.347	59.363
41	27.184	28.406	30.270	31.938	33.938	37.466	41.667	46.171	50.490	53.197	55.622	58.528	60.563
42	27.986	29.228	31.120	32.812	34.839	38.414	42.667	47.223	51.589	54.324	56.772	59.717	61.761
43	28.791	30.051	31.970	33.686	35.742	39.363	43.667	48.274	52.686	55.449	57.921	60.884	62.956
44	29.598	30.877	32.824	34.563	36.646	40.312	44.667	49.325	53.782	56.572	59.068	62.059	64.150
45	30.408	31.704	33.678	35.441	37.550	41.262	45.667	50.375	54.878	57.695	60.214	63.231	65.340
46	31.219	32.534	34.534	36.320	38.456	42.212	46.667	51.425	55.972	58.816	61.358	64.402	66.529
47	32.032	33.365	35.392	37.200	39.363	43.163	47.667	52.474	57.065	59.936	62.500	65.571	67.716
48	32.848	34.198	36.250	38.082	40.270	44.115	48.667	53.522	58.158	61.054	63.641	66.738	68.901
49	33.664	35.032	37.111	38.965	41.179	45.067	49.667	54.571	59.249	62.171	64.780	67.903	70.084

From J. M. Cameron, *Industrial Quality Control*, 9(1): 1952, 39. Reprinted by permission.

TABLE T6-1 Unity Values for Construction and Evaluation of Single-, Double-, and Multiple-Sampling Plans. ($n_1 = n_2 \ldots n_k$, # indicates acceptance not allowed at a given stage)

PLAN	ACCEPTANCE NUMBERS	$R=p_2/p_1$	np_2		Probability of Acceptance												
					.99	.95	.90	.75	.50	.25	.10	.05	.01	.005	.001	.0005	.0001
OS	Ac = 0 Re = 1	44.893	2.303	np ASN n_1	.0101 1 1	.0513 1 1	.105 1 1	.288 1 1	.693 1 1	1.386 1 1	2.303 1 1	2.996 1 1	4.605 1 1	5.298 1 1	6.908 1 1	7.601 1 1	9.206 1 1
XD*	Ac = #1 Re = 1 2	32.655	1.636	np ASN n_1	.0100 1.990 1	.0501 1.951 1	.101 1.904 1	.259 1.772 1	.573 1.564 1	1.053 1.349 1	1.636 1.195 1	2.057 1.128 1	2.995 1.050 1	3.389 1.034 1	4.286 1.014 1	4.668 1.009 1	5.542 1.004 1
XM	Ac = ## 0 0 1 2 3 Re = 1 1 2 2 3 4 4	33.254	.838	np ASN n_1	.00501 2.995 1	.0252 2.973 1	.0508 2.941 1	.132 2.821 1	.294 2.538 1	.539 2.119 1	.838 1.732 1	1.057 1.536 1	1.566 1.271 1	1.788 1.205 1	2.312 1.111 1	2.541 1.086 1	3.071 1.049 1
XXD	Ac = 0 1 [FOR THIS PLAN ONLY USE $n_2=5n_1$] Re = 2 2	20.193	2.302	np ASN n_1	.0459 1.219 1	.114 1.507 1	.176 1.737 1	.347 2.226 1	.713 2.748 1	1.388 2.732 1	2.302 2.151 1	2.993 1.750 1	4.571 1.237 1	5.201 1.143 1	6.815 1.037 1	7.490 1.021 1	9.048 1.005 1
XXM	Ac = ## 0 0 1 2 3 Re = 1 2 2 2 3 4	20.204	.891	np ASN n_1	.00968 3.018 1	.0441 3.067 1	.0817 3.095 1	.183 3.072 1	.357 2.834 1	.602 2.383 1	.891 1.927 1	1.102 1.685 1	1.593 1.345 1	1.808 1.259 1	2.321 1.135 1	2.546 1.103 1	3.074 1.056 1
1S	Ac = 1 Re = 2	10.958	3.890	np ASN n_1	.149 1 1	.355 1 1	.532 1 1	.961 1 1	1.678 1 1	2.693 1 1	3.890 1 1	4.744 1 1	6.638 1 1	7.430 1 1	9.234 1 1	10.000 1 1	11.759 1 1
1D	Ac = 0 1 Re = 2 2	12.029	2.490	np ASN n_1	.0860 1.079 1	.207 1.168 1	.310 1.228 1	.566 1.321 1	1.006 1.368 1	1.661 1.316 1	2.490 1.206 1	3.124 1.137 1	4.649 1.045 1	5.324 1.026 1	6.914 1.007 1	7.604 1.004 1	9.209 1.001 1
1M	Ac = ## 0 0 1 1 2 Re = 2 2 2 3 3 3	8.903	.917	np ASN n_1	.0459 3.254 1	.103 3.501 1	.148 3.637 1	.252 3.774 1	.416 3.640 1	.643 3.169 1	.917 2.601 1	1.121 2.270 1	1.602 1.761 1	1.815 1.618 1	2.325 1.388 1	2.549 1.319 1	3.075 1.205 1
2S	Ac = 2 Re = 3	6.506	5.322	np ASN n_1	.436 1 1	.818 1 1	1.102 1 1	1.727 1 1	2.674 1 1	3.920 1 1	5.322 1 1	6.296 1 1	8.406 1 1	9.274 1 1	11.230 1 1	12.053 1 1	13.934 1 1
2D	Ac = 0 3 Re = 3 4	5.357	3.402	np ASN n_1	.363 1.298 1	.635 1.443 1	.827 1.511 1	1.231 1.581 1	1.816 1.564 1	2.566 1.450 1	3.402 1.306 1	3.986 1.222 1	5.290 1.097 1	5.852 1.066 1	7.201 1.025 1	7.810 1.016 1	9.295 1.005 1
2M	Ac = ## 0 0 1 2 3 4 Re = 2 3 3 4 4 5	6.244	1.355	np ASN n_1	.111 2.432 1	.217 2.789 1	.293 2.983 1	.451 3.207 1	.683 3.165 1	.988 2.776 1	1.355 2.261 1	1.635 1.950 1	2.343 1.470 1	2.671 1.344 1	3.458 1.167 1	3.803 1.122 1	4.602 1.060 1
3S	Ac = 3 Re = 4	4.891	6.681	np ASN n_1	.823 1 1	1.366 1 1	1.745 1 1	2.535 1 1	3.672 1 1	5.109 1 1	6.681 1 1	7.754 1 1	10.045 1 1	10.978 1 1	13.062 1 1	13.935 1 1	15.922 1 1
3D	Ac = 1 4 Re = 4 5	4.398	4.398	np ASN n_1	.635 1.130 1	1.000 1.245 1	1.246 1.316 1	1.750 1.421 1	2.465 1.470 1	3.373 1.414 1	4.398 1.293 1	5.130 1.211 1	6.808 1.084 1	7.542 1.053 1	9.270 1.017 1	10.019 1.010 1	11.757 1.003 1
3M	Ac = # 0 1 2 3 4 6 Re = 3 3 4 5 6 6 7	4.672	1.626	np ASN n_1	.200 2.461 1	.348 2.820 1	.446 3.026 1	.642 3.286 1	.910 3.288 1	1.246 2.935 1	1.626 2.450 1	1.901 2.156 1	2.553 1.693 1	2.848 1.559 1	3.566 1.340 1	3.887 1.274 1	4.650 1.163 1

TABLE T6-1 (continued)

PLAN	ACCEPTANCE NUMBERS	R=p2/p1	np2		99	Probability of Acceptance .95	.90	.75	.50	.25	.10	.05	.01	.005	.001	.0005	.0001
4S	Ac = 4	4.058	7.994	np	1.279	1.970	2.433	3.369	4.671	6.274	7.994	9.154	11.605	12.594	14.795	15.711	17.792
	Re = 5			ASN	1	1	1	1	1	1	1	1	1	1	1	1	1
				n1													
4D	Ac = 3 5	4.102	6.699	np	1.099	1.633	1.992	2.728	3.789	5.162	6.699	7.762	10.047	10.978	13.062	13.933	15.909
	Re = 6 6			ASN	1	1	1	1	1	1	1	1	1	1	1	1	1
				n1	1.025	1.077	1.125	1.233	1.341	1.345	1.242	1.164	1.055	1.033	1.009	1.005	1.001
4M	Ac = # 1 2 3 4 5 6	4.814	2.118	np	.266	.440	.558	.798	1.141	1.591	2.118	2.502	3.385	3.763	4.640	5.016	5.884
	Re = 3 4 4 6 6 7 7			ASN	2.128	2.300	2.417	2.590	2.618	2.384	2.021	1.792	1.427	1.326	1.174	1.132	1.070
				n1													
5S	Ac = 5	3.550	9.275	np	1.785	2.613	3.152	4.219	5.670	7.423	9.275	10.513	13.109	14.150	16.455	17.411	19.578
	Re = 6			ASN	1	1	1	1	1	1	1	1	1	1	1	1	1
				n1													
5D	Ac = 2 6	3.547	5.781	np	1.116	1.630	1.959	2.607	3.490	4.579	5.781	6.627	8.537	9.357	11.253	12.066	13.928
	Re = 5 7			ASN	1	1	1	1	1	1	1	1	1	1	1	1	1
				n1	1.097	1.199	1.263	1.360	1.405	1.352	1.243	1.171	1.064	1.039	1.012	1.007	1.002
5M	Ac = # 1 2 3 5 7 9	3.243	2.270	np	.490	.700	.830	1.079	1.410	1.814	2.270	2.604	3.411	3.776	4.642	5.017	5.884
	Re = 4 5 6 7 8 9 10			ASN	2.496	2.906	3.143	3.459	3.516	3.188	2.677	2.347	1.791	1.628	1.367	1.292	1.171
				n1													
6S	Ac = 6	3.206	10.532	np	2.330	3.285	3.895	5.083	6.670	8.558	10.532	11.842	14.571	15.660	18.062	19.056	21.302
	Re = 7			ASN	1	1	1	1	1	1	1	1	1	1	1	1	1
				n1													
6D	Ac = 3 7	3.217	6.914	np	1.559	2.149	2.525	3.262	4.268	5.519	6.914	7.898	10.087	11.000	13.068	13.936	15.903
	Re = 8 8			ASN	1	1	1	1	1	1	1	1	1	1	1	1	1
				n1	1.073	1.169	1.243	1.393	1.548	1.608	1.525	1.422	1.203	1.138	1.051	1.032	1.011
6M	Ac = 0 2 4 5 7 10 11	3.452	3.134	np	.604	.908	1.093	1.439	1.894	2.463	3.134	3.645	4.917	5.511	6.983	7.646	9.222
	Re = 4 5 8 9 10 12 12			ASN	1.584	1.928	2.134	2.425	2.519	2.288	1.902	1.663	1.304	1.211	1.083	1.054	1.018
				n1													
7S	Ac = 7	2.957	11.771	np	2.906	3.981	4.656	5.956	7.669	9.684	11.771	13.148	16.000	17.134	19.627	20.655	22.976
	Re = 8			ASN	1	1	1	1	1	1	1	1	1	1	1	1	1
				n1													
7D	Ac = 3 8	2.951	7.162	np	1.796	2.427	2.822	3.584	4.599	5.826	7.162	8.093	10.174	11.057	13.085	13.946	15.914
	Re = 7 9			ASN	1	1	1	1	1	1	1	1	1	1	1	1	1
				n1	1.106	1.215	1.288	1.409	1.492	1.467	1.352	1.262	1.110	1.072	1.024	1.014	1.004
7M	Ac = 0 1 3 5 7 10 13	2.892	2.959	np	.713	1.023	1.200	1.518	1.921	2.403	2.959	3.400	4.686	5.337	6.915	7.604	9.210
	Re = 4 6 8 10 11 12 14			ASN	2.022	2.586	2.882	3.255	3.325	2.966	2.397	2.019	1.406	1.261	1.091	1.057	1.018
				n1													

TABLE T6-1 (continued)

PLAN	ACCEPTANCE NUMBERS	$R=p_2/p_1$	np_2		99	.95	.90	.75	.50	.25	.10	.05	.01	.005	.001	.0005	.0001
8S	Ac = 8 Re = 9	2.768	12.995	np	3.507	4.695	5.432	6.838	8.669	10.802	12.995	14.435	17.403	18.578	21.157	22.218	24.600
8D	Ac = 3 11 Re = 7 12	2.668	8.248	np	2.268	3.092	3.583	4.489	5.628	6.925	8.248	9.121	10.964	11.722	13.470	14.232	16.046
				ASN/n_1	1.185	1.335	1.409	1.488	1.478	1.375	1.248	1.176	1.075	1.051	1.019	1.012	1.004
8M	Ac = 0 2 4 6 9 12 14 Re = 4 7 9 11 12 14 15	2.840	3.314	np	.787	1.167	1.375	1.739	2.190	2.720	3.314	3.761	4.936	5.517	6.983	7.646	9.219
				ASN/n_1	1.806	2.320	2.599	2.963	3.063	2.765	2.264	1.934	1.400	1.263	1.093	1.058	1.019
9S	Ac = 9 Re = 10	2.619	14.206	np	4.130	5.425	6.221	7.726	9.669	11.914	14.206	15.705	18.783	19.999	22.658	23.751	26.198
9D	Ac = 5 11 Re = 12 12	2.587	9.533	np	2.871	3.685	4.184	5.134	6.385	7.893	9.533	10.670	13.152	14.174	16.460	17.412	19.564
				ASN/n_1	1.071	1.167	1.243	1.401	1.584	1.694	1.662	1.573	1.328	1.241	1.105	1.071	1.026
9M	Ac = 1 3 5 8 11 13 15 Re = 5 8 10 12 14 16 16	2.813	4.219	np	1.117	1.500	1.719	2.123	2.659	3.349	4.219	4.924	6.682	7.454	9.239	10.000	11.754
				ASN/n_1	1.526	1.928	2.167	2.521	2.667	2.414	1.937	1.635	1.230	1.145	1.048	1.029	1.009
10S	Ac = 10 Re = 11	2.497	15.407	np	4.771	6.169	7.021	8.620	10.669	13.020	15.407	16.962	20.145	21.398	24.135	25.257	27.768
10D	Ac = 5 12 Re = 9 13	2.486	9.732	np	3.055	3.914	4.433	5.406	6.663	8.147	9.732	10.822	13.216	14.214	16.472	17.420	19.562
				ASN/n_1	1.085	1.183	1.248	1.357	1.426	1.394	1.286	1.206	1.081	1.051	1.016	1.009	1.003
10M	Ac = 0 3 6 8 11 14 18 Re = 5 8 10 13 15 17 19	2.516	3.927	np	1.144	1.561	1.792	2.199	2.701	3.286	3.927	4.391	5.498	6.003	7.266	7.851	9.303
				ASN/n_1	1.927	2.357	2.602	2.939	3.034	2.750	2.282	1.982	1.501	1.368	1.170	1.118	1.047
11S	Ac = 11 Re = 12	2.397	16.598	np	5.428	6.924	7.829	9.519	11.668	14.121	16.598	18.208	21.490	22.779	25.590	26.741	29.313
11D	Ac = 5 12 Re = 13 13	2.438	9.766	np	3.165	4.006	4.517	5.477	6.723	8.194	9.766	10.847	13.226	14.221	16.474	17.420	19.566
				ASN/n_1	1.101	1.216	1.299	1.463	1.642	1.752	1.737	1.664	1.429	1.332	1.163	1.115	1.047
11M	Ac = 1 4 7 10 13 16 20 Re = 6 8 11 14 17 21 21	2.567	4.657	np	1.362	1.814	2.074	2.539	3.125	3.834	4.657	5.285	6.847	7.561	9.273	10.022	11.758
				ASN/n_1	1.548	1.912	2.132	2.452	2.575	2.357	1.970	1.726	1.348	1.246	1.101	1.066	1.024
12S	Ac = 12 Re = 13	2.312	17.782	np	6.099	7.690	8.646	10.422	12.668	15.217	17.782	19.443	22.821	24.145	27.027	28.206	30.836
12D	Ac = 6 15 Re = 10 16	2.289	11.233	np	3.884	4.907	5.507	6.606	7.990	9.580	11.233	12.351	14.781	15.795	18.105	19.080	21.291
				ASN/n_1	1.092	1.196	1.261	1.358	1.403	1.353	1.246	1.175	1.068	1.043	1.014	1.008	1.002
12M	Ac = 0 3 7 10 14 18 21 Re = 6 9 12 15 17 20 22	2.249	4.400	np	1.510	1.956	2.203	2.640	3.172	3.774	4.400	4.828	5.802	6.242	7.372	7.916	9.316
				ASN/n_1	2.215	2.655	2.892	3.203	3.274	2.982	2.516	2.214	1.708	1.555	1.299	1.222	1.102

Probability of Acceptance

TABLE T6-1   (continued)

PLAN	ACCEPTANCE NUMBERS	$R=p_2/p_1$	np_2		Probability of Acceptance												
					.99	.95	.90	.75	.50	.25	.10	.05	.01	.005	.001	.0005	.0001
13S	Ac = 13	2.240	18.958	np	6.782	8.464	9.470	11.329	13.668	16.310	18.958	20.669	24.139	25.497	28.447	29.652	32.339
	Re = 14			ASN	1	1	1	1	1	1	1	1	1	1	1	1	1
13D	Ac = 5 14	2.227	10.474	np	3.797	4.704	5.246	6.249	7.520	8.976	10.474	11.474	13.624	14.523	16.608	17.509	19.589
	Re = 15 15			ASN	1	1	1	1	1	1	1	1	1	1	1	1	1
				n_1	1.184	1.332	1.427	1.591	1.750	1.842	1.838	1.789	1.603	1.511	1.312	1.242	1.122
13M	Ac = 1 3 5 8 11 14 19	2.474	4.298	np	1.244	1.737	1.987	2.408	2.920	3.537	4.298	4.947	6.682	7.454	9.239	10.001	11.752
	Re = 5 9 12 15 18 20 20			ASN	1	1	1	1	1	1	1	1	1	1	1	1	1
				n_1	1.669	2.326	2.691	3.188	3.368	3.009	2.315	1.852	1.266	1.161	1.050	1.030	1.009
14S	Ac = 14	2.177	20.128	np	7.477	9.246	10.300	12.239	14.668	17.400	20.128	21.886	25.446	26.836	29.853	31.084	33.824
	Re = 15			ASN	1	1	1	1	1	1	1	1	1	1	1	1	1
14D	Ac = 7 18	2.176	12.722	np	4.652	5.847	6.534	7.769	9.286	10.989	12.722	13.876	16.345	17.370	19.712	20.707	22.974
	Re = 11 19			ASN	1	1	1	1	1	1	1	1	1	1	1	1	1
				n_1	1.091	1.199	1.263	1.352	1.380	1.317	1.214	1.150	1.058	1.037	1.012	1.007	1.002
14M	Ac = 1 4 8 12 17 21 25	2.185	5.112	np	1.844	2.340	2.618	3.110	3.708	4.387	5.112	5.632	6.955	7.612	9.279	10.024	11.757
	Re = 7 10 13 17 20 23 26			ASN	1	1	1	1	1	1	1	1	1	1	1	1	1
				n_1	1.916	2.375	2.624	2.952	3.042	2.780	2.350	2.067	1.575	1.424	1.192	1.133	1.053
15S	Ac = 15	2.122	21.292	np	8.181	10.036	11.135	13.152	15.668	18.487	21.292	23.097	26.743	28.164	31.245	32.501	35.294
	Re = 16			ASN	1	1	1	1	1	1	1	1	1	1	1	1	1
15D	Ac = 5 16	2.091	11.405	np	4.476	5.455	6.033	7.094	8.419	9.908	11.405	12.381	14.405	15.224	17.088	17.890	19.782
	Re = 17 17			ASN	1	1	1	1	1	1	1	1	1	1	1	1	1
				n_1	1.293	1.463	1.559	1.710	1.838	1.904	1.898	1.861	1.716	1.640	1.459	1.384	1.235
15M	Ac = 2 7 13 18 23 28 30	2.142	6.795	np	2.553	3.173	3.529	4.167	4.953	5.850	6.795	7.453	8.971	9.658	11.362	12.132	13.948
	Re = 9 12 16 21 26 31 31			ASN	1	1	1	1	1	1	1	1	1	1	1	1	1
				n_1	1.606	1.914	2.089	2.339	2.443	2.301	2.028	1.842	1.507	1.397	1.205	1.148	1.064
18S	Ac = 18	1.990	24.756	np	10.346	12.442	13.672	15.907	18.668	21.731	24.756	26.692	30.581	32.091	35.353	36.679	39.622
	Re = 19			ASN	1	1	1	1	1	1	1	1	1	1	1	1	1
18D	Ac = 9 23	1.955	15.524	np	6.559	7.940	8.722	10.111	11.796	13.659	15.524	16.748	19.329	20.391	22.818	23.853	26.219
	Re = 14 24			ASN	1	1	1	1	1	1	1	1	1	1	1	1	1
				n_1	1.120	1.244	1.315	1.412	1.442	1.374	1.260	1.188	1.079	1.052	1.018	1.011	1.003
18M	Ac = 1 6 11 16 22 27 32	1.990	6.225	np	2.506	3.128	3.462	4.035	4.712	5.460	6.225	6.744	7.917	8.443	9.765	10.381	11.912
	Re = 8 12 17 22 25 29 33			ASN	1	1	1	1	1	1	1	1	1	1	1	1	1
				n_1	2.009	2.443	2.681	2.999	3.087	2.824	2.389	2.107	1.639	1.499	1.270	1.202	1.096

TABLE T6-1 (continued)

PLAN	ACCEPTANCE NUMBERS	R=p₂/p₁	np₂		.99	.95	.90	.75	.50	.25	.10	.05	.01	.005	.001	.0005	.0001
									Probability of Acceptance								
21S	Ac = 21	1.892	28.184	np	12.574	14.894	16.244	18.682	21.668	24.956	28.184	30.240	34.355	35.947	39.376	40.768	43.850
				ASN													
	Re = 22																
21D	Ac = 11 26	1.882	18.909	np	7.843	9.329	10.170	11.666	13.486	15.510	17.555	18.909	21.792	22.978	25.656	26.777	29.312
				ASN	1	1	1	1	1	1	1	1	1	1	1	1	1
	Re = 16 27			n_1	1.094	1.201	1.268	1.367	1.413	1.363	1.256	1.185	1.075	1.048	1.016	1.009	1.003
21M	Ac = 2 7 13 19 25 31 37	1.893	7.083	np	3.071	3.741	4.100	4.713	5.440	6.246	7.083	7.664	9.044	9.696	11.367	12.133	13.948
				ASN	1	1	1	1	1	1	1	1	1	1	1	1	1
	Re = 9 14 19 25 29 33 38			n_1	1.912	2.370	2.621	2.962	3.077	2.830	2.392	2.102	1.606	1.457	1.219	1.155	1.065
27S	Ac = 27	1.757	34.959	np	17.175	19.901	21.469	24.273	27.667	31.365	34.959	37.234	41.757	43.497	47.231	48.740	52.077
				ASN													
	Re = 28																
27D	Ac = 15 34	1.760	22.183	np	10.797	12.605	13.613	15.382	17.504	19.839	22.183	23.727	26.993	28.323	31.292	32.526	35.284
				ASN	1	1	1	1	1	1	1	1	1	1	1	1	1
	Re = 20 35			n_1	1.074	1.170	1.231	1.324	1.367	1.320	1.221	1.156	1.060	1.038	1.012	1.007	1.002
27M	Ac = 3 10 17 24 32 40 48	1.805	8.738	np	3.936	4.841	5.301	6.050	6.896	7.807	8.738	9.380	10.890	11.586	13.318	14.102	15.968
				ASN	1	1	1	1	1	1	1	1	1	1	1	1	1
	Re = 10 17 24 31 37 43 49			n_1	1.746	2.219	2.484	2.841	2.951	2.688	2.245	1.958	1.490	1.357	1.162	1.112	1.045
30S	Ac = 30	1.707	38.315	np	19.532	22.445	24.113	27.086	30.667	34.552	38.315	40.691	45.401	47.210	51.085	52.647	56.102
				ASN													
	Re = 31																
30D	Ac = 17 37	1.724	24.257	np	12.177	14.072	15.130	16.995	19.243	21.735	24.257	25.928	29.453	30.876	34.015	35.305	38.169
				ASN	1	1	1	1	1	1	1	1	1	1	1	1	1
	Re = 22 38			n_1	1.063	1.148	1.205	1.297	1.349	1.311	1.216	1.152	1.056	1.035	1.011	1.006	1.002
30M	Ac = 4 11 19 27 36 45 53	1.708	9.660	np	4.817	5.656	6.096	6.841	7.713	8.669	9.660	10.356	12.058	12.869	14.873	15.756	17.785
				ASN	1	1	1	1	1	1	1	1	1	1	1	1	1
	Re = 12 19 27 34 40 47 54			n_1	1.840	2.320	2.586	2.951	3.084	2.847	2.411	2.114	1.596	1.441	1.206	1.145	1.061
41S	Ac = 41	1.581	50.490	np	28.406	31.938	33.938	37.466	41.667	46.171	50.490	53.197	58.528	60.564	64.904	66.648	70.488
				ASN													
	Re = 42																
41D	Ac = 23 52	1.584	31.843	np	17.706	20.108	21.415	23.661	26.284	29.094	31.843	33.620	37.311	38.801	42.131	43.517	46.616
				ASN	1	1	1	1	1	1	1	1	1	1	1	1	1
	Re = 29 53			n_1	1.080	1.183	1.248	1.340	1.375	1.319	1.219	1.155	1.062	1.039	1.013	1.008	1.002
41M	Ac = 6 16 26 37 49 61 72	1.574	12.617	np	6.942	8.014	8.552	9.435	10.440	11.519	12.617	13.378	15.211	16.076	18.205	19.143	21.306
				ASN	1	1	1	1	1	1	1	1	1	1	1	1	1
	Re = 15 25 36 46 55 64 73			n_1	1.842	2.370	2.660	3.054	3.195	2.938	2.470	2.155	1.613	1.453	1.213	1.151	1.065
44S	Ac = 44	1.556	53.783	np	30.877	34.563	36.646	40.312	44.667	49.325	53.783	56.573	62.058	64.150	68.607	70.395	74.332
				ASN													
	Re = 45																
44D	Ac = 25 56	1.561	34.068	np	19.292	21.820	23.192	25.544	28.282	31.209	34.068	35.916	39.750	41.296	44.739	46.166	49.357
				ASN	1	1	1	1	1	1	1	1	1	1	1	1	1
	Re = 31 57			n_1	1.075	1.174	1.237	1.328	1.363	1.309	1.211	1.149	1.058	1.037	1.012	1.007	1.002
44M	Ac = 6 17 29 40 53 65 77	1.538	13.372	np	7.614	8.695	9.239	10.139	11.168	12.270	13.372	14.112	15.784	16.537	18.424	19.289	21.358
				ASN	1	1	1	1	1	1	1	1	1	1	1	1	1
	Re = 16 27 39 49 58 68 78			n_1	1.971	2.472	2.747	3.127	3.274	3.036	2.582	2.272	1.731	1.565	1.294	1.217	1.101

From E. G. Schilling and L. I. Johnson, *Journal of Quality Technology*, 12(4): 1980, 220-229. Reprinted by permission.

TABLE T7-1   Statistical Research Group: Table of Sequential Sampling Plans. Characteristic Quantities of Sequential Tests for the Binomial Distribution Computed for Various Combinations of p_1, p_2, α = .05, β = .10 and .50.

p_1	p_2	$\alpha=.05$ / β	h_2	h_1	s	$\bar{n}_0$	$\bar{n}_1$	$\bar{n}_{p_1}$	$\bar{n}_s$	$\bar{n}_{p_2}$
.0002	.0020	.10	1.2543	.9770	.000782	1250	2	1487	1568	847
.0003	.0025	.10	1.3618	1.0607	.001038	1022	2	1273	1393	766
"	.0040	.10	1.1143	.8679	.001429	608	2	681	677	356
"	.0055	.10	.9919	.7726	.001790	432	1	459	429	220
.0005	.002	.10	2.0827	1.6222	.001082	1499	3	2468	3125	1866
"	"	.50	1.6592	.4625	"	428	2	612	710	652
"	.003	.10	1.6109	1.2547	.001396	899	2	1241	1540	826
"	"	.50	1.2833	.3577	"	257	2	308	329	288
"	.004	.10	1.3876	1.0808	.001684	642	2	809	892	493
"	"	.50	1.1054	.3081	"	183	2	201	203	172
"	.005	.10	1.2528	.9758	.001956	499	2	594	626	338
"	"	.50	.9980	.2782	"	143	2	147	142	118
"	.006	.10	1.1606	.9040	.002216	408	2	467	475	252
"	"	.50	.9246	.2577	"	117	1	116	108	88
"	.007	.10	1.0925	.8510	.002466	346	2	383	378	198
"	"	.50	.8704	.2426	"	99	1	95	86	69
"	.008	.10	1.0937	.8098	.002709	299	2	325	312	162
"	"	.50	.8282	.2309	"	86	1	81	71	56
"	.009	.10	.9971	.7766	.002946	264	1	281	264	135
.00055	.0037	.10	1.5138	1.1791	.001653	714	2	947	1081	608
.001	.004	.10	2.0804	1.6204	.002165	749	3	1232	1561	932
"	.005	.10	1.7914	1.3953	.002487	562	2	831	1008	586
"	"	.50	1.4271	.3978	"	160	2	206	229	205
"	.0065	.10	1.5396	1.1992	.002941	408	2	547	630	356
"	.007	.10	1.4808	1.1534	.003086	374	2	490	555	311
"	"	.50	1.1796	.3288	"	107	2	121	126	109
"	.009	.10	1.3107	1.0209	.003646	281	2	342	368	201
"	.01	.10	1.2504	.9739	.003915	249	2	296	312	169
"	"	.50	.9961	.2777	"	71	2	73	71	59
"	.011	.10	1.2003	.9349	.004178	224	2	261	270	145
"	.013	.10	1.1216	.8736	.004689	187	2	210	210	111
"	.02	.10	.9587	.7467	.006369	118	1	123	113	58
"	"	.50	.7637	.2129	"	34	1	**	26	**
"	.03	.10	.8425	.6562	.008587	77	1	77	65	32
"	"	.50	.6712	.1871	"	22	1	**	15	**
"	.04	.10	.7752	.6038	.01068	51	1	55	44	22
"	"	.50	.6175	.1721	"	17	1	**	10	**
"	.05	.10	.7295	.5682	.01269	45	1	43	33	16
"	"	.50	.5811	.1620	"	13	1	**	8	**
"	.06	.10	.6956	.5418	.01465	37	1	35	26	13
"	"	.50	.5541	.1545	"	11	1	**	6	**
.0015	.0055	.10	2.2177	1.7274	.003080	561	3	968	1248	753
"	.0085	.10	1.6596	1.2926	.004039	321	2	451	533	306
"	.0110	.10	1.4437	1.1245	.004775	236	2	304	342	191
"	.0130	.10	1.3313	1.0370	.005336	195	2	239	260	143
"	.0150	.10	1.2479	.9720	.005877	166	2	197	208	112
"	.01875	.10	1.1365	.8852	.006852	130	2	147	148	79
.00175	.0125	.10	1.4617	1.1385	.005476	208	2	271	306	171

p_1	p_2	$\alpha=.05$ / β	h_2	h_1	s	$\bar{n}_0$	$\bar{n}_1$	$\bar{n}_{p_1}$	$\bar{n}_s$	$\bar{n}_{p_2}$
.002	.007	.10	2.2980	1.7899	.003993	449	3	795	1034	628
"	.010	.10	1.7870	1.3918	.004976	280	2	414	502	292
"	.013	.10	1.5351	1.1957	.005886	204	2	273	314	177
"	.016	.10	1.3806	1.0753	.006748	160	2	201	222	123
"	.019	.10	1.2741	.9924	.007574	132	2	158	168	92
"	.023	.10	1.1732	.9138	.008632	106	2	122	125	67
"	.025	.10	1.1339	.8832	.009147	97	2	109	110	59
.0023	.0188	.10	1.3649	1.0631	.007875	135	2	169	186	103
.0025	.0075	.10	2.6190	2.0399	.004553	449	3	880	1179	731
"	"	.50	2.0864	.5816	"	128	3	218	268	255
"	.01	.10	2.0737	1.6152	.005415	299	3	491	622	372
"	"	.50	1.6520	.4605	"	86	2	122	141	130
"	.015	.10	1.6019	1.2477	.006989	179	2	246	288	164
"	"	.50	1.2761	.3557	"	51	2	61	65	57
"	.02	.10	1.3782	1.0735	.008440	128	2	160	177	98
"	"	.50	1.0980	.3061	"	37	2	40	40	34
"	.03	.10	1.1502	.8959	.01113	81	2	92	94	50
"	"	.50	.9163	.2554	"	23	1	23	21	18
"	.04	.10	1.0283	.8009	.01363	59	2	64	61	32
"	"	.50	.8192	.2283	"	17	1	16	14	11
"	.05	.10	.9494	.7395	.01603	47	1	48	45	23
"	"	.50	.7563	.2108	"	14	1	**	10	**
"	.06	.10	.8928	.6954	.01834	38	1	39	34	18
"	"	.50	.7112	.1983	"	11	1	**	8	**
.0027	.025	.10	1.2856	1.0014	.01006	100	2	121	129	71
.003	.009	.10	2.6166	2.0380	.005464	373	3	733	981	608
"	.014	.10	1.8629	1.4510	.007151	203	2	310	381	224
"	.018	.10	1.5996	1.2459	.008390	149	2	205	240	137
"	.0186	.10	1.5706	1.2233	.008570	143	2	195	226	129
"	.022	.10	1.4368	1.1191	.009565	118	2	151	170	95
"	.026	.10	1.3241	1.0314	.01069	97	2	119	129	71
"	.030	.10	1.2405	.9662	.01178	83	2	97	103	56
"	.035	.10	1.1611	.9044	.01310	70	2	79	81	44
"	.036	.10	1.1476	.8939	.01336	67	2	76	78	42
.005	.01	.10	4.1398	3.2245	.007216	447	5	1289	1863	1222
"	"	.50	3.2980	.9193	"	128	4	320	423	427
"	.015	.10	2.6070	2.0305	.009111	223	3	438	586	364
"	"	.50	2.0768	.5789	"	64	3	109	133	127
"	.02	.10	2.0624	1.6064	.01084	149	3	244	309	185
"	"	.50	1.6430	.4580	"	43	2	60	70	65
"	.03	.10	1.5906	1.2389	.01400	89	2	122	143	82
"	"	.50	1.2671	.3532	"	26	2	30	32	29
"	.04	.10	1.3664	1.0643	.01693	63	2	79	87	49
"	"	.50	1.0886	.3034	"	18	2	20	20	17
"	.05	.10	1.2305	.9585	.01970	49	2	**	61	33
"	"	.50	.9803	.2733	"	14	2	14	14	12
"	.06	.10	1.1371	.8857	.02237	40	2	45	46	25
"	"	.50	.9059	.2525	"	12	1	11	10	9
"	.067	.10	1.0868	.8465	.02419	35	2	39	39	21

TABLE T7-1 (continued)

p_1	p_2	$a=.05$ β	h_2	h_1	*	$\bar{n}_0$	$\bar{n}_1$	$\bar{n}_{p_1}$	$\bar{n}_s$	$\bar{n}_{p_2}$
.005	.07	.10	1.0679	.8318	.02496	34	2	37	36	19
"	"	.50	.8507	.2371	*	10	1	9	8	7
.006	.012	.10	4.1338	3.2198	.008659	372	5	1073	1551	1017
"	.018	.10	2.6022	2.0268	.01093	186	3	364	488	303
"	.021	.10	2.2795	1.7755	.01199	149	3	262	342	208
"	.024	.10	2.0578	1.6028	.01301	124	3	203	257	194
"	.030	.10	1.7690	1.3779	.01496	93	2	136	165	97
"	.036	.10	1.5860	1.2353	.01682	74	2	101	119	68
"	.042	.10	1.4577	1.1354	.01860	62	2	80	91	51
"	.050	.10	1.3347	1.0396	.02091	50	2	62	68	38
"	.06	.10	1.2255	.9546	.02368	41	2	48	51	28
.0065	.027	.10	2.0004	1.5581	.01443	108	3	174	219	131
.0075	.015	.10	4.1248	3.2128	.01082	297	5	856	1238	812
"	"	.50	3.2860	.9160	*	85	4	212	281	284
"	.02	.10	2.9093	2.2660	.01276	178	3	382	523	330
"	"	.50	2.3176	.6461	*	51	3	95	119	115
"	.03	.10	2.0510	1.5975	.01627	99	3	161	205	123
"	"	.50	1.6339	.4555	*	28	2	40	46	43
"	.04	.10	1.6930	1.3186	.01950	68	2	97	117	68
"	"	.50	1.3487	.3760	*	20	2	24	27	24
"	.05	.10	1.4892	1.1599	.02255	52	2	68	78	45
"	"	.50	1.1864	.3307	*	15	2	17	18	16
"	.06	.10	1.3546	1.0551	.02547	42	2	52	58	32
"	"	.50	1.0791	.3008	*	12	2	13	13	11
"	.07	.10	1.2574	.9794	.02830	35	2	42	45	25
"	"	.50	1.0017	.2792	*	10	2	10	10	9
"	.08	.10	1.1831	.9215	.03105	30	2	35	36	20
"	"	.50	.9425	.2627	*	9	1	9	8	7
.01	.02	.10	4.1097	3.2010	.01444	222	5	639	925	607
"	"	.50	3.2740	.9126	*	64	4	159	210	212
"	.025	.10	3.1027	2.4167	.01639	148	4	335	465	296
"	"	.50	2.4718	.6890	*	43	3	83	106	104
"	.03	.10	2.5829	2.0118	.01824	111	3	216	290	181
"	"	.50	2.0577	.5736	*	32	3	54	66	63
"	.04	.10	2.0197	1.5887	.02172	74	3	120	153	92
"	"	.50	1.6249	.4529	*	21	2	30	35	32
"	.05	.10	1.7510	1.3639	.02499	55	2	81	98	58
"	"	.50	1.3949	.3888	*	16	2	20	22	20
"	.06	.10	1.5678	1.2211	.02811	44	2	60	70	40
"	"	.50	1.2490	.3482	*	13	2	15	16	14
"	.07	.10	1.4391	1.1209	.03113	37	2	47	53	30
"	"	.50	1.1465	.3196	*	11	2	12	12	11
"	.08	.10	1.3426	1.0458	.03406	31	2	38	43	24
"	"	.50	1.0696	.2982	*	9	2	10	10	8
.011	.020	.10	4.7619	3.7090	.01506	247	5	809	1191	793
"	.025	.10	3.4605	2.6954	.01707	158	4	393	556	359
"	.032	.10	2.6534	2.0667	.01970	105	3	210	284	177
"	.040	.10	2.1884	1.7046	.02253	76	3	131	169	103
"	.048	.10	1.9123	1.4895	.02523	60	2	93	116	69
"	.056	.10	1.7266	1.3448	.02782	49	2	71	86	50
"	.066	.10	1.5632	1.2176	.03095	40	2	54	63	37
"	.076	.10	1.4446	1.1252	.03398	34	2	43	50	28
"	.094	.10	1.2944	1.0082	.03925	26	2	32	35	19
.0115	.043	.10	2.1391	1.6661	.02397	70	3	118	152	92
.0118	.117	.10	1.1986	.9336	.04691	20	2	24	25	14
.012	.056	.10	1.8224	1.4195	.02872	50	2	75	93	55
.014	.102	.10	1.3900	1.0827	.04496	25	2	31	35	20
.015	.025	.10	5.5474	4.3209	.01958	221	6	835	1248	842
"	"	.50	4.4193	1.2319	*	63	4	207	284	294
"	.03	.10	4.0796	3.1776	.02166	147	5	423	612	402
"	"	.50	3.2500	.9059	*	42	4	105	139	141
"	.04	.10	2.8716	2.2367	.02554	88	3	188	258	163
"	"	.50	2.2876	.6377	*	25	3	47	59	57
"	.05	.10	2.3307	1.8153	.02917	63	3	113	149	92
"	"	.50	1.8567	.5176	*	18	2	28	34	32
"	.06	.10	2.0169	1.5710	.03263	49	3	79	100	61
"	"	.50	1.6068	.4479	*	14	2	20	23	21
"	.07	.10	1.8089	1.4089	.03596	40	2	60	74	44
"	"	.50	1.4410	.4017	*	12	2	15	17	15
.02	.03	.10	6.9527	5.4154	.02467	220	8	1027	1565	1073
"	"	.50	5.5388	1.5440	*	63	6	255	355	375
"	.035	.10	5.0264	3.9150	.02682	146	6	508	754	505
"	"	.50	4.0042	1.1162	*	42	5	126	171	177
"	.04	.10	4.0495	3.1541	.02889	110	5	314	455	300
"	"	.50	3.2260	.8992	*	32	4	78	103	105
"	.05	.10	3.0509	2.3763	.03282	73	4	164	228	146
"	"	.50	2.4305	.6775	*	21	3	41	52	51

p_1	p_2	$a=.05$ β	h_2	h_1	*	$\bar{n}_0$	$\bar{n}_1$	$\bar{n}_{p_1}$	$\bar{n}_s$	$\bar{n}_{p_2}$
.02	.06	.10	2.5348	1.9743	.03655	55	3	106	142	89
"	"	.50	2.0193	.5629	*	16	3	26	32	31
"	.07	.10	2.2146	1.7250	.04012	43	3	76	99	61
"	"	.50	1.7643	.4918	*	13	2	19	23	21
"	.08	.10	1.9941	1.5532	.04359	36	3	58	74	45
"	"	.50	1.5886	.4428	*	11	2	14	17	16
"	.0820	.10	1.9578	1.5249	.04427	35	3	56	71	43
"	.085	.10	1.9071	1.4855	.04528	33	2	52	66	39
"	.09	.10	1.8315	1.4265	.04696	31	2	47	58	35
"	"	.50	1.4590	.4067	*	9	2	12	13	12
"	.096	.10	1.7524	1.3650	.04894	28	2	42	51	31
"	.10	.10	1.7056	1.3285	.05025	27	2	39	47	28
"	"	.50	1.3588	.3788	*	8	2	10	11	10
"	.114	.10	1.5697	1.2227	.05476	23	2	31	37	22
"	.172	.10	1.2457	.9703	.07264	14	2	16	18	10
"	.178	.10	1.2238	.9532	.07444	13	2	16	17	10
.021	.037	.10	4.9588	3.8624	.02827	137	6	471	697	467
"	.043	.10	3.9090	3.0447	.03074	100	5	277	399	262
"	.052	.10	3.0785	2.3978	.03427	70	4	160	223	143
"	.062	.10	2.5683	2.0004	.03801	53	3	104	140	88
.022	.04	.10	4.6890	3.6522	.03014	122	5	398	586	391
"	.136	.10	1.4856	1.1571	.06370	19	2	25	29	17
.0255	.175	.10	1.3812	1.0758	.07058	14	2	18	20	12
.026	.107	.10	1.9249	1.4993	.05782	26	3	42	53	32
"	.115	.10	1.8263	1.4225	.06055	24	2	36	46	28
.027	.178	.10	1.4068	1.0957	.08208	14	2	18	20	12
.028	.225	.10	1.2510	.9744	.09803	10	2	12	14	8
.03	.04	.10	9.6978	7.5535	.03477	218	11	1403	2183	1524
"	"	.50	7.7256	2.1535	*	62	9	348	496	533
"	.045	.10	6.8647	5.3469	.03701	145	8	675	1030	707
"	"	.50	5.4687	1.5244	*	42	6	167	234	247
"	.05	.10	5.4365	4.2345	.03919	109	6	408	611	413
"	"	.50	4.3309	1.2073	*	31	5	101	139	144
"	.06	.10	3.9891	3.1071	.04336	72	5	206	299	197
"	"	.50	3.1779	.8856	*	21	4	51	68	69
"	.07	.10	3.2498	2.5312	.04735	54	4	129	182	118
"	"	.50	2.5889	.7217	*	16	3	32	41	41
"	.08	.10	2.7960	2.1778	.05119	43	3	91	125	80
"	"	.50	2.2274	.6209	*	13	3	23	28	28
"	.086	.10	2.5978	2.0234	.05345	38	3	76	104	66
"	.09	.10	2.4864	1.9367	.05493	36	3	69	93	58
"	"	.50	1.9808	.5522	*	11	3	17	21	20
"	.10	.10	2.2601	1.7604	.05857	31	3	55	72	45
"	"	.50	1.8005	.5019	*	9	2	14	16	16
"	.11	.10	2.0864	1.6251	.06213	27	3	45	58	36
"	"	.50	1.6621	.4633	*	8	2	11	13	13
"	.118	.10	1.9735	1.5371	.06494	24	3	39	50	31
"	.12	.10	1.9481	1.5174	.06563	24	3	38	48	29
"	"	.50	1.5520	.4326	*	7	2	9	11	10
"	.13	.10	1.8350	1.4293	.06908	21	2	32	41	25
"	"	.50	1.4618	.4075	*	6	2	8	9	9
"	.15	.10	1.6597	1.2927	.07583	18	2	25	31	18
"	"	.50	1.3222	.3686	*	5	2	6	7	6
"	.20	.10	1.3831	1.0773	.09220	12	2	15	18	11
"	"	.50	1.1018	.3071	*	4	2	4	4	4
.031	.076	.10	3.0609	2.3841	.05036	45	4	109	153	98
"	.155	.10	1.6551	1.2891	.07841	17	2	24	30	18
.032	.057	.10	4.7895	3.7305	.04336	87	6	291	441	294
"	.066	.10	3.8048	2.9635	.04707	63	4	174	25	24
.033	.145	.10	1.8027	1.4041	.07678	19	2	28	36	22
.035	.225	.10	1.3896	1.0823	.1054	11	2	14	16	10
.0375	.155	.10	1.8656	1.4531	.08404	18	3	28	35	22
.04	.06	.10	6.7767	5.2783	.04936	107	8	499	762	524
"	"	.50	5.3986	1.5049	*	31	6	124	173	183
"	.07	.10	4.8876	3.8069	.05369	71	6	246	366	246
"	"	.50	3.8937	1.0894	*	21	5	61	83	86
"	.08	.10	3.9287	3.0600	.05785	53	5	152	221	166
"	"	.50	3.1298	.8724	*	16	4	38	50	51
"	.09	.10	3.3437	2.6044	.06188	43	4	105	150	98
"	"	.50	2.6637	.7425	*	13	3	26	34	34
"	.10	.10	2.9469	2.2953	.06580	35	4	79	110	71
"	"	.50	2.3476	.6544	*	10	3	20	25	25
"	.11	.10	2.6583	2.0705	.06963	30	3	62	85	54
"	"	.50	2.1177	.5903	*	9	3	15	19	19
"	.118	.10	2.4777	1.9299	.07264	27	3	52	71	45

TABLE T7-1   (continued)

p_1	p_2	$\alpha=.05$ / β	h_2	h_1	*	$\bar{n}_0$	$\tilde{n}_1$	$\bar{n}_{p_1}$	$\bar{n}_0$	$\bar{n}_{p_2}$
.04	.12	.10	2.4178	1.8988	.07339	26	3	50	68	43
		.50	1.9421	.5414	*	8	3	12	15	15
	.13	.10	2.2632	1.7628	.07708	23	3	42	56	35
		.50	1.8030	.5026	*	7	2	10	13	12
	.138	.10	2.1473	1.6725	.08000	21	3	37	49	30
	.14	.10	2.1210	1.6520	.08072	21	3	36	47	29
		.50	1.6896	.4710	*	6	2	9	11	10
	.15	.10	2.0024	1.5597	.08431	19	3	31	40	25
		.50	1.5952	.4447	*	6	2	8	9	9
	.17	.10	1.8151	1.4137	.09137	16	2	24	31	19
		.50	1.4460	.4031	*	5	2	6	7	7
	.20	.10	1.6131	1.2565	.1018	13	2	18	22	13
		.50	1.2851	.3582	*	4	2	4	5	5
	.23	.10	1.4674	1.1429	.1120	11	2	14	17	10
	.25	.10	1.3900	1.0826	.1187	10	2	12	14	9
		.50	1.1073	.3073	*	3	2	3	3	3
	.317	.10	1.1991	.9340	.1412	7	2	8	9	6
.041	.092	.10	3.3497	2.6091	.06333	42	4	103	147	96
	.102	.10	2.9580	2.3040	.06726	35	4	78	109	70
.042	.072	.10	5.0636	3.9440	.05574	71	6	254	379	256
	.082	.10	4.0612	3.1633	.05993	53	5	156	228	151
	.195	.10	1.6909	1.3171	.1018	13	2	20	24	15
.0475	.1975	.10	1.8106	1.4103	.1073	14	3	21	27	17
.048	.192	.10	1.8644	1.4522	.1058	14	3	22	29	18
.049	.16	.10	2.2107	1.7219	.09493	19	3	33	44	28
.05	.07	.10	8.0793	6.2929	.05948	106	9	588	909	631
		.50	6.4363	1.7941	*	31	7	146	206	221
	.08	.10	5.7567	4.4838	.06301	71	7	286	431	294
		.50	4.5860	1.2784	*	21	5	71	98	103
	.09	.10	4.5820	3.5689	.06819	53	5	174	257	173
		.50	3.6502	1.0175	*	15	4	43	58	60
	.10	.10	3.8682	3.0129	.07236	42	5	119	174	115
		.50	3.0816	.8590	*	12	4	30	39	40
	.11	.10	3.3857	2.6371	.07642	35	4	88	126	83
		.50	2.6972	.7519	*	10	3	22	29	29
	.12	.10	3.0920	2.3648	.08040	30	4	69	97	63
		.50	2.4187	.6742	*	9	3	17	22	22
	.13	.10	2.7699	2.1575	.08430	26	4	56	77	50
		.50	2.2066	.6151	*	8	3	14	18	17
	.138	.10	2.5982	2.0237	.08738	24	3	48	66	42
	.14	.10	2.5598	1.9938	.08815	23	3	46	63	41
		.50	2.0392	.5684	*	7	3	11	14	14
	.15	.10	2.3891	1.8608	.09193	21	3	39	53	74
		.50	1.9032	.5305	*	6	3	10	12	12
	.16	.10	2.2472	1.7503	.09568	19	3	34	45	29
		.50	1.7902	.4990	*	6	2	8	10	10
	.17	.10	2.1271	1.6568	.09938	17	3	30	39	25
		.50	1.6946	.4724	*	5	2	7	9	9
	.20	.10	1.8550	1.4449	.1103	14	3	21	27	17
		.50	1.4778	.4119	*	4	2	5	6	6
	.23	.10	1.6648	1.2967	.1210	11	2	16	20	13
	.25	.10	1.5659	1.2197	.1281	10	2	14	17	11
		.50	1.2475	.3477	*	3	2	3	4	4
	.317	.10	1.3278	1.0342	.1516	7	2	9	11	7
	.325	.10	1.3058	1.0170	.1544	7	2	9	10	6
.051	.12	.10	3.1041	2.4177	.08107	30	4	71	101	66
.052	.10	.10	4.0947	3.1893	.07361	44	5	131	192	128
	.11	.10	3.5580	2.7713	.07771	36	4	95	138	91
.0575	.1875	.10	2.1725	1.6922	.1116	16	3	28	37	24
.058	.33	.10	1.3900	1.0827	.1639	7	2	9	11	7
.06	.08	.10	9.3483	7.2813	.06956	105	11	675	1052	736
		.50	7.4472	2.0759	*	30	9	167	239	257
	.09	.10	6.6005	5.1411	.07407	70	8	324	495	341
		.50	5.2582	1.4658	*	20	6	80	112	119
	.10	.10	5.2144	4.0614	.07845	52	6	195	293	199
		.50	4.1540	1.1579	*	15	5	48	67	70
	.11	.10	4.3741	3.4069	.08272	42	5	133	196	132
		.50	3.4846	.9713	*	12	4	33	45	46
	.12	.10	3.8076	2.9657	.08689	35	5	98	142	95
		.50	3.0333	.8455	*	10	4	24	32	33
	.13	.10	3.3981	2.6468	.09093	30	4	76	109	72
		.50	2.7071	.7546	*	9	3	19	25	25
	.136	.10	3.2022	2.4942	.09340	27	4	66	94	62
	.14	.10	3.0872	2.4046	.09500	26	4	61	86	56
		.50	2.4594	.6856	*	8	3	15	20	20
	.15	.10	2.8422	2.2138	.09897	23	4	50	71	46
		.50	2.2642	.6312	*	7	3	12	16	16

p_1	p_2	$\alpha=.05$ / β	h_2	h_1	*	$\bar{n}_0$	$\tilde{n}_1$	$\bar{n}_{p_1}$	$\bar{n}_0$	$\bar{n}_{p_2}$
.06	.16	.10	2.6437	2.0592	.1029	21	3	43	59	38
		.50	2.1061	.5871	*	6	3	11	13	13
	.17	.10	2.4791	1.9309	.1067	19	3	37	50	32
		.50	1.9749	.5505	*	6	3	9	11	11
	.18	.10	2.3400	1.8226	.1106	17	3	32	43	28
		.50	1.8642	.5196	*	5	3	8	10	10
	.20	.10	2.1171	1.6490	.1181	14	3	25	34	21
		.50	1.6866	.4701	*	4	2	6	8	7
	.22	.10	1.9452	1.5151	.1256	13	3	20	27	17
		.50	1.5497	.4320	*	4	2	5	6	6
	.25	.10	1.7486	1.3620	.1366	10	3	16	20	13
		.50	1.3930	.3883	*	3	2	4	5	4
	.30	.10	1.5179	1.1823	.1548	8	2	11	14	9
		.50	1.2092	.3371	*	3	2	3	3	3
.061	.14	.10	3.1463	2.4506	.09567	26	4	63	89	58
.062	.12	.10	3.9912	3.1087	.08814	36	5	105	154	103
	.13	.10	3.5436	2.7601	.09227	30	4	81	117	77
.063	.2125	.10	2.0800	1.6201	.1251	13	3	23	31	20
.065	.245	.10	1.8760	1.4612	.1388	11	3	18	23	15
	.560	.10	.9942	.7744	.2953	3	2	4	4	3
.07	.09	.10	10.5853	8.2448	.07962	104	12	759	1191	838
		.50	8.4327	2.3506	*	30	10	188	271	293
	.10	.10	7.4214	5.7805	.08419	69	9	361	556	386
		.50	5.9122	1.6480	*	20	7	89	126	135
	.11	.10	5.8280	4.5394	.08864	52	7	216	327	224
		.50	4.6428	1.2942	*	15	6	53	74	78
	.12	.10	4.8638	3.7884	.09299	41	6	148	218	148
		.50	3.8747	1.0801	*	12	5	36	50	52
	.13	.10	4.2150	3.2831	.09726	34	5	107	158	106
		.50	3.3579	.9360	*	10	4	26	36	37
	.14	.10	3.7469	2.9185	.1014	29	5	82	120	80
		.50	2.9849	.8321	*	9	4	20	27	28
	.15	.10	3.3921	2.6421	.1056	··	4	66	95	63
		.50	2.7023	.7533	*	··	4	16	22	22
	.16	.10	3.1131	2.4248	.1096	23	4	54	77	51
		.50	2.4800	.6913	*	7	3	13	18	18
	.17	.10	2.8873	2.2489	.1136	20	4	46	64	42
		.50	2.3001	.6412	*	6	3	11	15	15
	.18	.10	2.7004	2.1033	.1176	18	4	39	55	36
		.50	2.1513	.5997	*	6	3	10	12	12
	.20	.10	2.4079	1.8755	.1294	15	3	30	41	27
		.50	1.9182	.5347	*	5	3	7	9	9
	.22	.10	2.1880	1.7042	.1331	13	3	24	32	21
		.50	1.7430	.4859	*	4	3	6	7	7
	.25	.10	1.9424	1.5129	.1446	11	3	18	24	15
		.50	1.5474	.4313	*	3	2	4	5	5
	.29	.10	1.7090	1.3311	.1596	9	3	13	17	11
	.30	.10	1.6617	1.2943	.1633	8	2	12	16	10
		.50	1.3218	.3690	*	3	2	3	4	3
.075	.450	.10	1.2504	.9739	.2249	5	2	6	7	5
.08	.10	.10	11.7915	9.1844	.08966	103	13	842	1327	938
		.50	9.3936	2.6185	*	30	11	209	301	328
	.11	.10	8.2205	6.4029	.09429	68	10	397	616	430
		.50	6.5488	1.8255	*	20	8	98	140	150
	.12	.10	6.4242	5.0038	.09880	51	8	236	361	249
		.50	5.1178	1.4266	*	15	6	58	82	87
	.13	.10	5.3388	4.1584	.1032	41	6	159	240	164
		.50	4.2531	1.1856	*	12	5	39	54	57
	.14	.10	4.6094	3.5903	.1076	34	6	115	172	117
		.50	3.6721	1.0236	*	10	5	29	39	41
	.15	.10	4.0839	3.1809	.1118	29	5	89	131	88
		.50	3.2534	.9069	*	9	4	22	30	31
	.16	.10	3.6861	2.8711	.1160	25	5	71	103	69
		.50	2.9365	.8186	*	8	4	18	23	24
	.17	.10	3.3738	2.6278	.1202	22	4	58	84	56
		.50	2.6877	.7492	*	7	4	14	19	19
	.18	.10	3.1214	2.4312	.1243	20	4	49	70	46
		.50	2.4866	.6931	*	6	3	12	16	16
	.19	.10	2.9127	2.2687	.1283	18	4	42	59	39
		.50	2.3204	.6468	*	6	3	10	13	14
	.20	.10	2.7370	2.1318	.1323	17	4	36	51	33
		.50	2.1804	.6078	*	5	3	9	12	12
	.22	.10	2.4594	1.9133	.1403	14	3	28	39	25
		.50	1.9568	.5455	*	4	3	7	9	9
	.25	.10	2.1510	1.6754	.1520	12	3	21	28	18
		.50	1.7136	.4777	*	4	2	5	6	6
	.30	.10	1.8121	1.4114	.1713	9	3	14	18	12
		.50	1.4436	.4024	*	3	2	3	4	4
.09	.12	.10	8.9985	7.0089	.1044	68	11	432	675	473
		.50	7.1686	1.9983	*	20	9	107	153	165

TABLE T7-1   (continued)

p_1	p_2	$\alpha=.05$ / β	h_2	h_1	$*$	$\bar{n}_0$	$\bar{n}_1$	$\bar{n}_{p_1}$	$\bar{n}_s$	$\bar{n}_{p_2}$
.09	.13	.10	7.0040	5.4553	.1089	51	8	255	394	273
		.50	5.5796	1.5553		15	7	63	89	95
	.14	.10	5.7999	4.5175	.1134	40	7	171	261	179
		.50	4.6205	1.2880		12	6	42	59	63
	.15	.10	4.9917	3.8880	.1178	34	6	124	187	127
		.50	3.9766	1.1085		10	5	31	42	45
	.16	.10	4.4100	3.4350	.1221	29	6	95	141	96
		.50	3.5132	.9793		9	5	23	32	33
	.17	.10	3.9702	3.0924	.1264	25	5	75	111	75
		.50	3.1729	.8817		7	4	19	25	26
	.18	.10	3.6253	2.8237	.1306	22	5	62	90	60
		.50	2.8880	.8050		7	4	15	20	21
	.19	.10	3.3468	2.6068	.1348	20	4	52	75	50
		.50	2.6662	.7432		6	4	13	17	17
	.20	.10	3.1168	2.4277	.1389	18	4	44	63	42
		.50	2.4830	.6921		5	3	11	14	15
	.22	.10	2.7561	2.1482	.1471	15	4	33	47	31
		.50	2.1972	.6125		5	3	8	11	11
	.25	.10	2.3789	1.8529	.1592	12	3	24	33	22
		.50	1.8951	.5283		4	3	6	7	8
	.30	.10	1.9712	1.5353	.1789	9	3	15	21	13
		.50	1.5703	.4377		3	2	4	5	5
.10	.13	.10	9.7560	7.5989	.1144	67	12	466	732	515
		.50	7.7720	2.1665		19	9	116	166	180
	.14	.10	7.5677	5.8944	.1190	50	9	274	425	297
		.50	6.0287	1.6805		15	7	68	97	104
	.15	.10	6.2478	4.8664	.1236	40	8	183	281	194
		.50	4.9772	1.3874		12	6	45	64	68
	.16	.10	5.3625	4.1768	.1280	33	7	132	201	138
		.50	4.2720	1.1908		10	5	33	46	48
	.17	.10	4.7259	3.6810	.1324	28	6	101	151	103
		.50	3.7649	1.0495		8	5	25	34	36
	.18	.10	4.2451	3.3065	.1367	25	5	80	119	81
		.50	3.3818	.9427		7	4	20	27	28
	.19	.10	3.8682	3.0129	.1410	22	5	65	96	65
		.50	3.0816	.8590		7	4	16	22	23
	.20	.10	3.5643	2.7762	.1452	20	5	54	80	54
		.50	2.8394	.7915		6	4	13	18	19
	.22	.10	3.1027	2.4167	.1536	16	4	40	58	38
		.50	2.4718	.6890		5	3	10	13	13
	.25	.10	2.6309	2.0492	.1660	13	4	28	39	26
		.50	2.0959	.5842		4	3	7	9	9

p_1	p_2	$\alpha=.05$ / β	h_2	h_1	$*$	$\bar{n}_0$	$\bar{n}_1$	$\bar{n}_{p_1}$	$\bar{n}_s$	$\bar{n}_{p_2}$
.10	.27	.10	2.4034	1.8720	.1741	11	3	22	31	21
		.50	1.9147	.5337		4	3	6	7	7
	.30	.10	2.1411	1.6677	.1852	9	3	17	24	15
		.50	1.7057	.4755		3	3	4	5	5
.12	.15	.10	11.2104	8.7317	.1345	65	13	532	841	596
		.50	8.9307	2.4895		19	11	132	191	208
	.16	.10	8.6486	6.7363	.1392	49	11	311	486	342
		.50	6.8898	1.9206		14	9	77	110	119
	.17	.10	7.1051	5.5341	.1437	39	9	206	319	223
		.50	5.6602	1.5778		11	7	51	73	78
	.18	.10	6.0712	4.7288	.1483	32	8	148	227	158
		.50	4.8365	1.3482		10	6	37	52	55
	.20	.10	4.7685	3.7142	.1572	24	6	88	134	92
		.50	3.7988	1.0589		7	5	22	30	32
	.22	.10	3.9770	3.0977	.1660	19	5	60	89	61
		.50	3.1683	.8832		6	4	15	20	21
	.25	.10	3.2337	2.5187	.1788	15	4	38	55	37
		.50	2.5761	.7181		5	4	9	13	13
	.28	.10	2.7581	2.1482	.1915	12	4	27	38	26
		.50	2.1972	.6125		4	3	7	9	9
	.30	.10	2.5241	1.9660	.1998	10	4	22	31	21
		.50	2.0108	.5605		3	3	5	7	7
.15	.19	.10	10.1562	7.9106	.1694	47	13	362	571	405
		.50	8.0909	2.2554		14	10	90	130	141
	.20	.10	8.2984	6.4635	.1741	38	11	238	373	263
		.50	6.6108	1.8428		11	9	59	85	92
	.22	.10	6.1637	4.8009	.1833	27	8	128	198	138
		.50	4.9102	1.3687		8	7	32	45	48
	.25	.10	4.5447	3.5398	.1968	18	6	67	102	70
		.50	3.6205	1.0092		6	5	17	23	25
	.30	.10	3.2575	2.5372	.2188	12	5	33	48	33
		.50	2.5950	.7234		4	4	8	11	12
	.35	.10	2.5910	2.0181	.2405	9	4	20	29	19
		.50	2.0641	.5754		3	3	5	7	7
.20	.24	.10	12.3724	9.6368	.2196	44	16	436	696	498
		.50	9.8563	2.7475		13	13	108	158	174
	.25	.10	10.0471	7.8256	.2243	35	13	285	452	322
		.50	8.0039	2.2311		10	11	71	103	112
	.30	.10	5.3625	4.1768	.2477	17	8	77	120	84
		.50	4.2720	1.1908		5	6	19	27	29
	.35	.10	3.7672	2.9342	.2706	11	6	37	56	39
		.50	3.0011	.8366		4	5	9	13	14

From the Statistical Research Group, Sequential Analysis of Statistical Data: Applications, *AMP Report 30.2R*, Columbia University, 1945.

TABLE T7-2 Statistical Research Group: Table of Values of a and b for Sequential Sampling. (a and b in terms of α and β using common logarithms.)

$$a = \log \frac{1 - \beta}{\alpha} \qquad b = \log \frac{1 - \alpha}{\beta}$$

α for computing a, β for computing b

	.001	.01	.02	.03	.04	.05	.10	.15	.20	.30	.40
.001	3.000	2.000	1.699	1.522	1.398	1.301	1.000	.823	.699	.522	.398
.01	2.996	1.996	1.695	1.519	1.394	1.297	.996	.820	.695	.519	.394
.02	2.991	1.991	1.690	1.514	1.389	1.292	.991	.815	.690	.514	.389
.03	2.987	1.987	1.686	1.510	1.385	1.288	.987	.811	.686	.510	.385
.04	2.982	1.982	1.681	1.505	1.380	1.283	.982	.806	.681	.505	.380
.05	2.978	1.978	1.677	1.501	1.376	1.279	.978	.802	.677	.501	.376
.10	2.954	1.954	1.653	1.477	1.352	1.255	.954	.778	.653	.477	.352
.15	2.929	1.929	1.628	1.452	1.327	1.230	.929	.753	.628	.452	.327
.20	2.903	1.903	1.602	1.426	1.301	1.204	.903	.727	.602	.426	.301
.30	2.845	1.845	1.544	1.368	1.243	1.146	.845	.669	.544	.368	.243
.40	2.778	1.778	1.477	1.301	1.176	1.079	.778	.602	.477	.301	.176

β for computing a, α for computing a (row labels; left axis)

From the Statistical Research Group, Sequential Analysis of Statistical Data: Applications, *AMP Report 30.2R*, Columbia University, 1945.

TABLE T7-3 g_1 and g_2 in Terms of p_1 and p_2 Using Common Logarithms for Sequential Sampling

$$g_1 = \log\left(\frac{p_2}{p_1}\right) \qquad g_2 = \log\left(\frac{1-p_1}{1-p_2}\right)$$

Each cell shows g_1 (upper value) and g_2 (lower value).

p_1 \ p_2	.005	.01	.02	.03	.04	.05	.06	.07	.08	.09	.10	.15	.20
.001	.6990 / .0017	1.0000 / .0039	1.3010 / .0083	1.4771 / .0128	1.6021 / .0173	1.6990 / .0218	1.7782 / .0264	1.8451 / .0311	1.9031 / .0358	1.9542 / .0405	2.0000 / .0453	2.1761 / .0701	2.3010 / .0965
.005		.3010 / .0022	.6021 / .0066	.7782 / .0111	.9031 / .0156	1.0000 / .0201	1.0792 / .0247	1.1461 / .0293	1.2041 / .0340	1.2553 / .0388	1.3010 / .0436	1.4771 / .0684	1.6021 / .0947
.01			.3010 / .0044	.4771 / .0089	.6021 / .0134	.6990 / .0179	.7782 / .0225	.8451 / .0272	.9031 / .0318	.9542 / .0366	1.0000 / .0414	1.1761 / .0662	1.3010 / .0925
.02				.1761 / .0045	.3010 / .0090	.3979 / .0135	.4771 / .0181	.5441 / .0227	.6021 / .0274	.6532 / .0322	.6990 / .0370	.8751 / .0618	1.0000 / .0881
.03					.1249 / .0045	.2218 / .0090	.3010 / .0136	.3680 / .0183	.4260 / .0230	.4771 / .0277	.5229 / .0325	.6990 / .0574	.8239 / .0837
.04						.0969 / .0045	.1761 / .0091	.2430 / .0138	.3010 / .0185	.3522 / .0232	.3979 / .0280	.5740 / .0529	.6990 / .0792
.05							.0792 / .0046	.1461 / .0092	.2041 / .0139	.2553 / .0187	.3010 / .0235	.4771 / .0483	.6021 / .0746
.06								.0669 / .0046	.1249 / .0093	.1761 / .0141	.2218 / .0189	.3979 / .0437	.5229 / .0700
.07									.0580 / .0047	.1091 / .0094	.1549 / .0142	.3310 / .0391	.4559 / .0654
.08										.0512 / .0047	.0969 / .0095	.2730 / .0344	.3979 / .0607
.09											.0458 / .0048	.2218 / .0296	.3468 / .0560
.10												.1761 / .0248	.3010 / .0512
.15													.1249 / .0263

TABLE T8-1   Operating Characteristics of the One-Sided Normal Test for a
Level of Significance Equal to 0.05

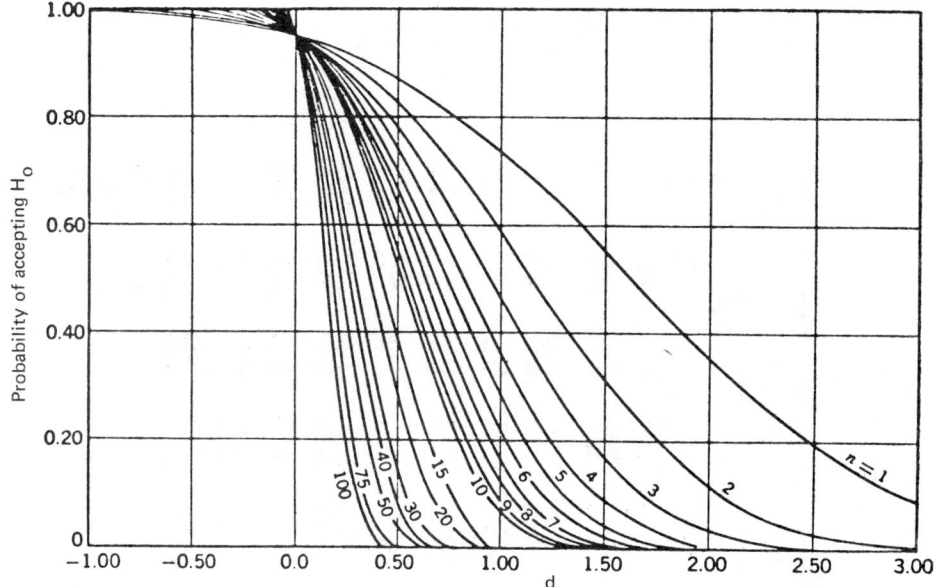

From A. H. Bowker and G. J. Lieberman, *Engineering Statistics*, 1959, p. 118.
Reprinted by permission of Prentice-Hall, Inc., Englewood Cliffs, New Jersey.

TABLE T8-2   Operating Characteristics of the Two-Sided Normal Test for a
Level of Significance Equal to 0.05

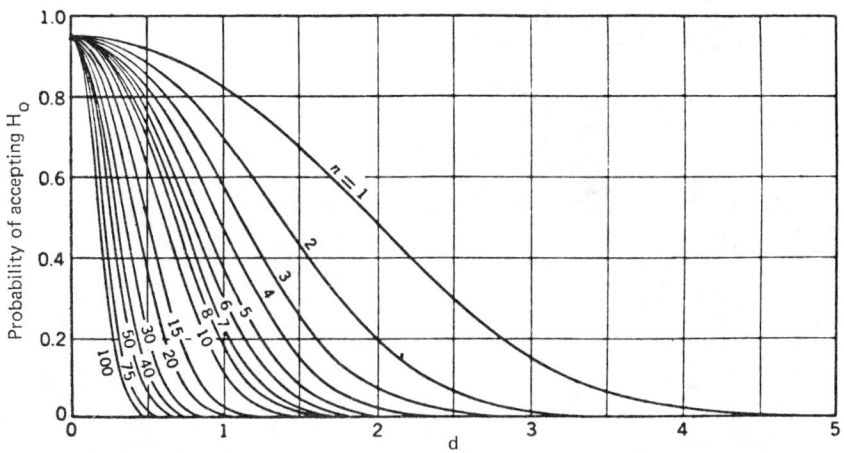

From C. L. Ferris, F. E. Grubbs, and C. L. Weaver, *Annals of Mathematical
Statistics*, 17: June 1946, p. 190.   Reprinted by permission.

TABLE T8-3 Operating Characteristics of the One-Sided t-Test for a Level
of Significance Equal to 0.05

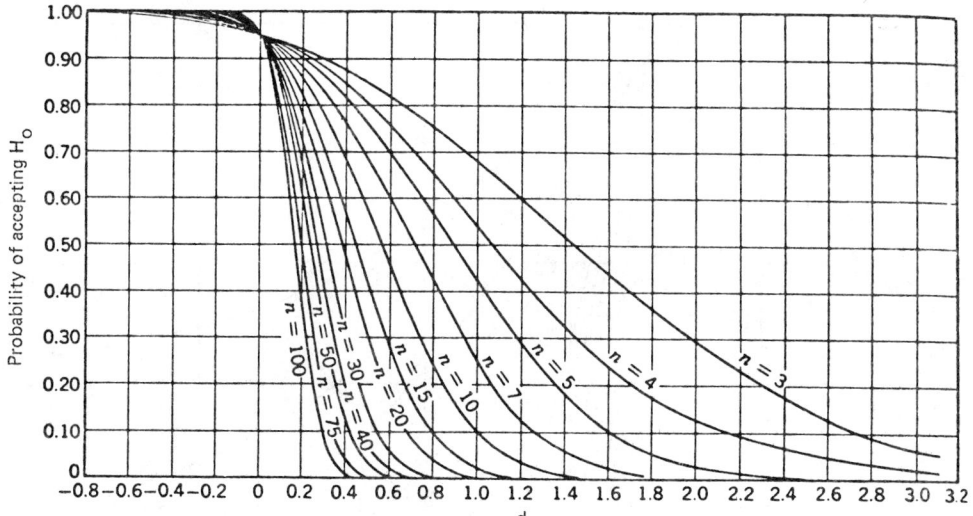

From A. H. Bowker and G. J. Lieberman, *Engineering Statistics*, 1959, p. 132.
Reprinted by permission of Prentice-Hall, Inc., Englewood Cliffs, New Jersey.

TABLE T8-4 Operating Characteristics of the Two-Sided
t-Test for a Level of Significance Equal to 0.05

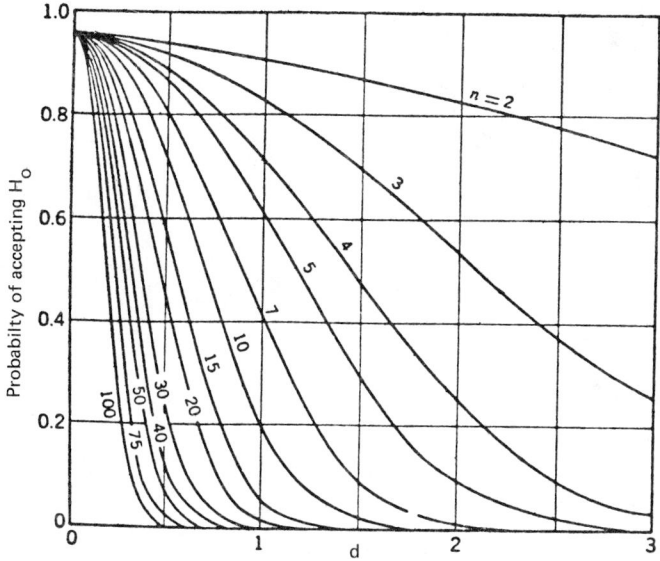

From C. L. Ferris, F. E. Grubbs, and C. L. Weaver,
Annals of Mathematical Statistics, 17: June 1946, p. 195.
Reprinted by permission.

TABLE T8-5 Operating Characteristics of the One-Sided (Upper Tail) Chi-Square
Test for a Level of Significance Equal to 0.05

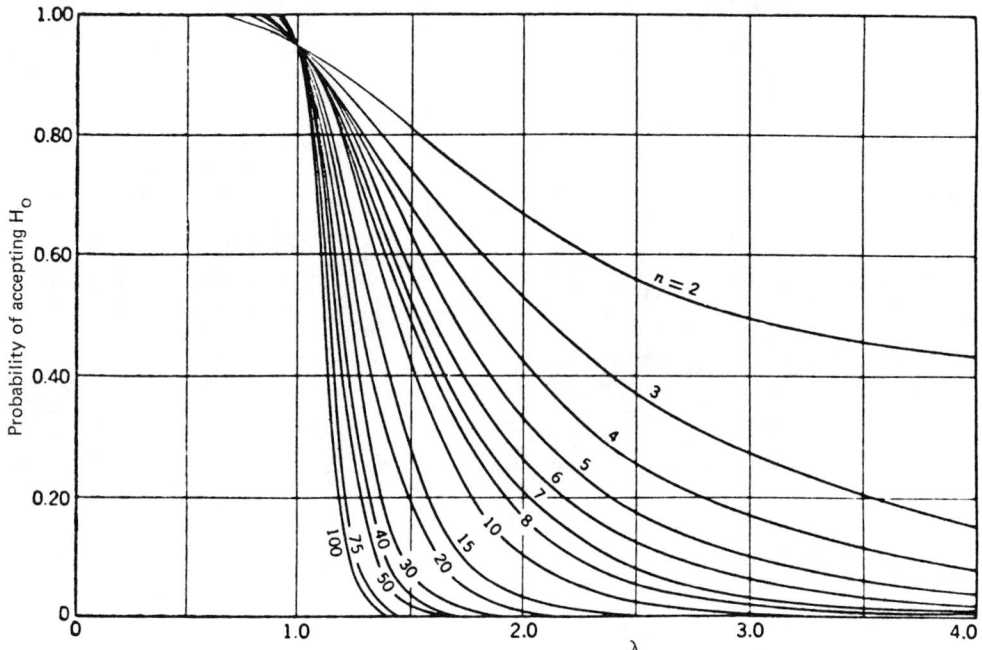

From C. L. Ferris, F. E. Grubbs, and C. L. Weaver, *Annals of Mathematical Statistics*, 17: June 1946, p. 181. Reprinted by permission.

TABLE T8-6   Factors for Acceptance Control Limits

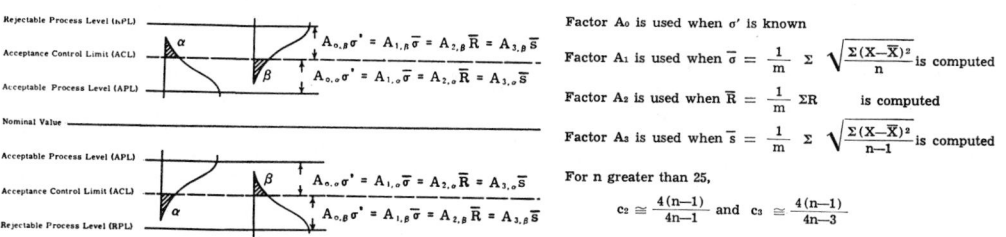

Factor A_0 is used when σ' is known

Factor A_1 is used when $\bar{\sigma} = \dfrac{1}{m} \Sigma \sqrt{\dfrac{\Sigma(X-\bar{X})^2}{n}}$ is computed

Factor A_2 is used when $\bar{R} = \dfrac{1}{m} \Sigma R$ is computed

Factor A_3 is used when $\bar{s} = \dfrac{1}{m} \Sigma \sqrt{\dfrac{\Sigma(X-\bar{X})^2}{n-1}}$ is computed

For n greater than 25,

$$c_2 \cong \frac{4(n-1)}{4n-1} \quad \text{and} \quad c_3 \cong \frac{4(n-1)}{4n-3}$$

Note: If the Acceptance Control Limits lie so close to the Nominal Value that two-tail probabilities must be used (within $\pm 2.5\ \sigma'/\sqrt{n}$ for $\alpha = 5\%$; $\pm 3.0\ \sigma'/\sqrt{n}$ for $\alpha = 1\%$; $\pm 3.2\ \sigma'/\sqrt{n}$ for $\alpha = 0.5\%$; $\pm 3.5\ \sigma'/\sqrt{n}$ for $\alpha = 0.1\%$), refer to Table III for correction terms to be applied to the factors in Table II.

n	$\alpha(\beta) = 5\%$				$\alpha(\beta) = 1\%$			
	$A_{0,05}$	$A_{1,05}$	$A_{2,05}$	$A_{3,05}$	$A_{0,01}$	$A_{1,01}$	$A_{2,01}$	$A_{3,01}$
2	1.163	2.062	1.031	1.458	1.644	2.915	1.458	2.062
3	0.950	1.313	0.561	1.071	1.343	1.856	0.793	1.515
4	0.822	1.031	0.400	0.893	1.163	1.458	0.565	1.262
5	0.736	0.875	0.316	0.782	1.040	1.237	0.447	1.106
6	0.672	0.773	0.265	0.706	0.950	1.093	0.374	0.998
7	0.622	0.700	0.230	0.648	0.879	0.990	0.325	0.916
8	0.582	0.644	0.205	0.603	0.823	0.911	0.289	0.852
9	0.548	0.600	0.185	0.566	0.775	0.848	0.261	0.800
10	0.520	0.564	0.169	0.535	0.736	0.797	0.239	0.756
11	0.496	0.533	0.156	0.508	0.702	0.754	0.221	0.719
12	0.475	0.507	0.146	0.486	0.671	0.717	0.206	0.687
13	0.456	0.485	0.137	0.466	0.645	0.685	0.193	0.659
14	0.440	0.465	0.129	0.448	0.622	0.657	0.182	0.633
15	0.425	0.447	0.122	0.433	0.601	0.633	0.173	0.612
16	0.411	0.432		0.418	0.581	0.611		0.592
17	0.399	0.418		0.405	0.564	0.591		0.573
18	0.388	0.405		0.394	0.548	0.572		0.557
19	0.377	0.393		0.383	0.533	0.556		0.541
20	0.368	0.382		0.373	0.520	0.540		0.527
21	0.359	0.372		0.364	0.508	0.526		0.514
22	0.351	0.363		0.355	0.496	0.513		0.502
23	0.343	0.355		0.347	0.485	0.502		0.491
24	0.336	0.347		0.339	0.474	0.490		0.480
25	0.329	0.339		0.332	0.465	0.480		0.470
over 25	$1.645/\sqrt{n}$	$1.645/c_2\sqrt{n}$		$1.645/c_3\sqrt{n}$	$2.326/\sqrt{n}$	$2.326/c_2\sqrt{n}$		$2.326/c_3\sqrt{n}$

n	$\alpha(\beta) = 0.5\%$				$\alpha(\beta) = 0.1\%$			
	$A_{0,005}$	$A_{1,005}$	$A_{2,005}$	$A_{3,005}$	$A_{0,001}$	$A_{1,001}$	$A_{2,001}$	$A_{3,001}$
2	1.821	3.229	1.614	2.283	2.185	3.874	1.937	2.740
3	1.487	2.056	0.878	1.678	1.784	2.467	1.054	2.013
4	1.288	1.614	0.626	1.398	1.545	1.937	0.751	1.677
5	1.152	1.370	0.495	1.225	1.382	1.644	0.594	1.470
6	1.052	1.211	0.415	1.105	1.262	1.453	0.498	1.326
7	0.974	1.097	0.360	1.015	1.168	1.316	0.432	1.218
8	0.911	1.009	0.320	0.944	1.093	1.211	0.384	1.132
9	0.859	0.939	0.289	0.886	1.030	1.127	0.347	1.063
10	0.815	0.883	0.265	0.837	0.977	1.059	0.316	1.005
11	0.777	0.836	0.245	0.796	0.932	1.002	0.294	0.955
12	0.744	0.794	0.228	0.761	0.892	0.953	0.274	0.913
13	0.714	0.759	0.214	0.730	0.857	0.911	0.257	0.876
14	0.688	0.728	0.202	0.702	0.826	0.874	0.242	0.842
15	0.665	0.701	0.192	0.678	0.798	0.841	0.230	0.813
16	0.644	0.677		0.655	0.773	0.812		0.786
17	0.625	0.654		0.635	0.750	0.785		0.761
18	0.607	0.634		0.617	0.728	0.760		0.740
19	0.591	0.616		0.599	0.709	0.739		0.719
20	0.576	0.599		0.584	0.691	0.718		0.701
21	0.562	0.583		0.569	0.675	0.700		0.683
22	0.550	0.568		0.556	0.659	0.682		0.667
23	0.538	0.556		0.544	0.645	0.667		0.652
24	0.526	0.543		0.532	0.631	0.651		0.638
25	0.515	0.532		0.520	0.618	0.638		0.624
over 25	$2.576/\sqrt{n}$	$2.576/c_2\sqrt{n}$		$2.576/c_3\sqrt{n}$	$3.090/\sqrt{n}$	$3.090/c_2\sqrt{n}$		$3.090/c_3\sqrt{n}$

From R. A. Freund, *Industrial Quality Control*, 14(4): 1957, p. 18.   Reprinted by permission.

TABLE T8-7 Correction Terms for Acceptance Control Factors

When the Acceptance Control Limits are too close to the Nominal Value (within $\pm 2.5\ \sigma'/\sqrt{n}$ for $\alpha = 5$ percent; $\pm 3.0\ \sigma'/\sqrt{n}$ for $\alpha = 1$ percent; $\pm 3.2\ \sigma'/\sqrt{n}$ for $\alpha = 0.5$ percent; $\pm 3.5\ \sigma'/\sqrt{n}$ for $\alpha = 0.1$ percent), corrections to the factors in Table II are required since two-tail probabilities must replace the one-tail probabilities otherwise applicable. The factors in Table II should be multiplied by the correction term (C.T.).*

$\Delta_1 =$ Deviations of the Acceptance Control Limit from the Nominal Value in terms of $\pm \sigma'/\sqrt{n}$. To be used when the APL values are to be determined from the Acceptance Control Limits.

$\Delta_2 =$ Deviations of the APL values from the Nominal Value in terms of $\pm \sigma'/\sqrt{n}$. To be used when the Acceptance Control Limits are to be determined from the APL values.

$\alpha = 5\%$			$\alpha = 1\%$			$\alpha = 0.5\%$			$\alpha = 0.1\%$		
Δ_1	Δ_2	C.T.	Δ_1	Δ_2	C.T.	Δ_1	Δ_2	C.F.	Δ_1	Δ_2	C.F.
1.960	0	1.1916	2.576	0	1.1072	2.807	0	1.0898	3.291	0	1.0648
1.970	0.100	1.1366	2.589	0.100	1.0700	2.821	0.100	1.0563	3.307	0.100	1.0378
1.999	0.200	1.0935	2.625	0.200	1.0426	2.862	0.200	1.0333	3.352	0.200	1.0202
2.045	0.300	1.0610	2.685	0.300	1.0252	2.922	0.300	1.0179	3.421	0.300	1.0100
2.107	0.400	1.0377	2.757	0.400	1.0134	3.000	0.400	1.0094	3.492	0.400	1.0005
2.182	0.500	1.0223	2.842	0.500	1.0068	3.088	0.500	1.0046	3.500	0.409	1.0004
2.267	0.600	1.0132	2.933	0.600	1.0032	3.181	0.600	1.0021			
2.356	0.700	1.0067	3.000	0.670	1.0018	3.200	0.619	1.0018			
2.451	0.800	1.0034									
2.500	0.851	1.0023									

*C.T. $= \dfrac{t_{\alpha_1}}{t}$ where:

α_1 is the risk of an average from a process centered at the APL value falling outside the nearer Acceptance Control Limit.

α_2 is the risk of that average falling outside the farther Acceptance Control Limit.

$\alpha = \alpha_1 + \alpha_2$

$t_{\alpha_2} = 2\Delta_1 + t_{\alpha_1}$

From R. A. Freund, *Industrial Quality Control*, 14(4): 1957, p. 19. Reprinted by permission.

TABLE T8-8 Boundary Values for Barnard's Sequential t-Test

$\alpha = .05 \quad \beta = .05$

k	D = .10 Y_1	D = .10 Y_2	D = .25 Y_1	D = .25 Y_2	D = .50 Y_1	D = .50 Y_2	D = .75 Y_1	D = .75 Y_2	D = 1.0 Y_1	D = 1.0 Y_2	D = 1.5 Y_1	D = 1.5 Y_2	D = 2.0 Y_1	D = 2.0 Y_2	D = 3.0 Y_1	D = 3.0 Y_2
2					[-6.96]	[3.01]	[-3.90]	[2.60]	[-2.14]	[2.13]	[-0.47]	[1.69]	0.37	[1.56]	0.95	[1.46]
4					[-3.13]	[2.73]	-1.49	[2.30]	-0.53	[2.03]	0.51	1.84	1.03	1.82	1.50	1.85
6					-2.07	2.56	-0.76	2.20	0.03	2.04	0.91	2.01	1.43	2.06	1.90	2.19
8			[-4.32]	[4.24]	-1.51	2.46	-0.35	2.16	0.37	2.09	1.23	2.18	1.74	2.29	2.22	2.47
10			[-3.67]	[3.91]	-1.15		-0.07	2.16	0.63	2.16	1.49	2.34	2.00	2.49	2.50	2.73
15	[-6.68]	[6.00]	-2.72	3.39	-0.57	2.34	0.44	2.23	1.11	2.34	2.01	2.70	2.54	2.94	3.10	3.29
20	[-5.87]	[5.55]	-2.17	3.10	-0.21	2.31	0.78	2.33	1.47	2.52	2.42	3.02	2.97	3.32		
25	-5.27	5.19	-1.77	2.90	0.07	2.30	1.05	2.44	1.76	2.70	2.78	3.32	3.36	3.67		
30	-4.81	4.91	-1.50	2.77	0.29	2.32	1.28	2.55	2.02	2.88	3.09	3.59	3.71	3.99		
35	-4.44	4.67	-1.28	2.67	0.48	2.36	1.49	2.66	2.24	3.05	3.38	3.84	4.03	4.29		
40	-4.14	4.47	-1.09	2.60	0.65	2.40	1.67	2.76	2.45	3.21	3.64	4.07	4.32	4.57		
45	-3.88	4.15	-0.93	2.55	0.79	2.44	1.84	2.87	2.64	3.36	3.89	4.29	4.60	4.83		
50	-3.47	3.90	-0.79	2.51	0.92	2.49	1.99	2.97	2.82	3.50	4.12	4.50	4.86	5.08		
60	-3.16	3.70	-0.56	2.44	1.16	2.58	2.27	3.17	3.16	3.77						
70	-2.88	3.55	-0.37	2.41	1.36	2.68	2.52	3.35	3.45	4.03						
80	-2.66	3.41	-0.20	2.39	1.54	2.78	2.76	3.53	3.73	4.27						
90	-2.47	2.99	-0.06	2.39	1.71	2.88	2.97	3.71	3.99	4.49						
100	-1.80	2.77	0.07	2.39	1.87	2.97	3.17	3.87	4.24	4.70						
150	-1.38		0.57	2.46	2.51	3.42	4.00	4.59	5.27	5.65						
200			0.93	2.57	3.03	3.83	4.75	5.23	6.15	6.48						
k_1	29		12		6		4		3		2		2		2	
k_2	31		13		7		5		5		4		3		3	
$\bar{k}_1$	600		100		30		20		10		<10		<10		<5	
$\bar{k}_2$	600		100		30		20		10		<10		<10		<5	

TABLE T10-1 d_2^* Factors and Degrees of Freedom ν for Estimating the Standard Deviation for the Average Range of k Samples of n

$$\bar{R}/d_2^* \rightarrow \sigma$$

No. of Samples, k		Size of Samples, n													
		2	3	4	5	6	7	8	9	10	11	12	13	14	15
1	d_2^*	1.41	1.91	2.24	2.48	2.67	2.83	2.96	3.08	3.18	3.27	3.35	3.42	3.49	3.55
	ν	1.00	1.98	2.93	3.83	4.68	5.48	6.25	6.98	7.68	8.35	8.99	9.61	10.2	10.8
2	d_2^*	1.28	1.81	2.15	2.40	2.60	2.77	2.91	3.02	3.13	3.22	3.30	3.38	3.45	3.51
	ν	1.92	3.83	5.69	7.47	9.16	10.8	12.3	13.8	15.1	16.5	17.8	19.0	20.2	21.3
3	d_2^*	1.23	1.77	2.12	2.38	2.58	2.75	2.89	3.01	3.11	3.21	3.29	3.37	3.44	3.50
	ν	2.82	5.66	8.44	11.1	13.6	16.0	18.3	20.5	22.6	24.6	26.5	28.4	30.1	31.9
4	d_2^*	1.21	1.75	2.11	2.37	2.57	2.74	2.88	3.00	3.10	3.20	3.28	3.36	3.43	3.49
	ν	3.71	7.49	11.2	14.7	18.1	21.3	24.4	27.3	30.1	32.7	35.3	37.7	40.1	42.4
5	d_2^*	1.19	1.74	2.10	2.36	2.56	2.73	2.87	2.99	3.10	3.19	3.28	3.35	3.42	3.49
	ν	4.59	9.31	13.9	18.4	22.6	26.6	30.4	34.0	37.5	40.8	44.0	47.1	50.1	52.9
6	d_2^*	1.18	1.73	2.09	2.35	2.56	2.73	2.87	2.99	3.09	3.19	3.27	3.35	3.42	3.49
	ν	5.47	11.1	16.7	22.0	27.0	31.8	36.4	40.8	45.0	49.0	52.8	56.5	60.1	63.5
7	d_2^*	1.17	1.73	2.09	2.35	2.55	2.72	2.86	2.99	3.09	3.19	3.27	3.35	3.42	3.48
	ν	6.35	12.9	19.4	25.6	31.5	37.1	42.5	47.6	52.4	57.1	61.6	65.9	70.0	74.0
8	d_2^*	1.17	1.72	2.08	2.35	2.55	2.72	2.86	2.98	3.09	3.19	3.27	3.35	3.42	3.48
	ν	7.23	14.8	22.1	29.2	36.0	42.4	48.5	54.3	59.9	65.2	70.3	75.2	80.0	84.6
9	d_2^*	1.16	1.72	2.08	2.34	2.55	2.72	2.86	2.98	3.09	3.18	3.27	3.35	3.42	3.48
	ν	8.11	16.6	24.9	32.9	40.4	47.7	54.5	61.1	67.3	73.3	79.1	84.6	90.0	95.1
10	d_2^*	1.16	1.72	2.08	2.34	2.55	2.72	2.86	2.98	3.09	3.18	3.27	3.34	3.42	3.48
	ν	8.99	18.4	27.6	36.5	44.9	52.9	60.6	67.8	74.8	81.5	87.8	94.0	99.9	106
11	d_2^*	1.16	1.71	2.08	2.34	2.55	2.72	2.86	2.98	3.09	3.18	3.27	3.34	3.41	3.48
	ν	9.87	20.2	30.4	40.1	49.4	58.2	66.6	74.6	82.2	89.6	96.6	103	110	116
12	d_2^*	1.15	1.71	2.07	2.34	2.55	2.72	2.86	2.98	3.09	3.18	3.27	3.34	3.41	3.48
	ν	10.7	22.0	33.1	43.7	53.8	63.5	72.6	81.3	89.7	97.7	105	113	120	127
13	d_2^*	1.15	1.71	2.07	2.34	2.55	2.71	2.86	2.98	3.09	3.18	3.27	3.34	3.41	3.48
	ν	11.6	23.8	35.8	47.3	58.3	68.7	78.6	88.1	97.2	106	114	122	130	137
14	d_2^*	1.15	1.71	2.07	2.34	2.54	2.71	2.86	2.98	3.08	3.18	3.27	3.34	3.41	3.48
	ν	12.5	25.7	38.6	51.0	62.8	74.0	84.7	94.9	105	114	123	131	140	148
15	d_2^*	1.15	1.71	2.07	2.34	2.54	2.71	2.86	2.98	3.08	3.18	3.26	3.34	3.41	3.48
	ν	13.4	27.5	41.3	54.6	67.2	79.3	90.7	102	112	122	132	141	150	158
20	d_2^*	1.14	1.70	2.07	2.33	2.54	2.71	2.85	2.98	3.08	3.18	3.26	3.34	3.41	3.48
	ν	17.8	36.5	55.0	72.7	89.6	106	121	135	149	163	175	188	200	211
30	d_2^*	1.14	1.70	2.07	2.33	2.54	2.71	2.85	2.97	3.08	3.18	3.26	3.34	3.41	3.47
	ν	26.5	54.7	82.4	109	134	158	181	203	224	244	263	281	299	316
50	d_2^*	1.13	1.70	2.06	2.33	2.54	2.71	2.85	2.97	3.08	3.17	3.26	3.34	3.41	3.47
	ν	44.0	91.0	137	181	224	264	302	338	373	406	438	469	499	527
	d_2	1.13	1.69	2.06	2.33	2.53	2.70	2.85	2.97	3.08	3.17	3.26	3.34	3.41	3.47
	c.d.	0.876	1.82	2.74	3.62	4.47	5.27	6.03	6.76	7.45	8.12	8.76	9.37	9.97	10.54

From L. S. Nelson, *Journal of Quality Technology*, 7(1): 1975, p. 48.   Reprinted by permission.

TABLE T10-2 Matched Single and Double, Known (σ) and Unknown (s) Standard Deviation, Variables Sampling Plans for Values of p_1 and p_2 with α = .05, β = .10 ($n_1 = n_2$, $k_t = k_r$)

p_1	p_2	SINGLE			DOUBLE					
		n_σ	n_s	k	n_σ	n_s	k_a	k_r	ASN_σ	ASN_s
.001	.0015	572	3180	3.02	422	2334	3.04	3.01	464.5	2568.5
	.002	191	1032	2.97	138	739	3.01	2.95	154.9	829.1
	.0025	107	567	2.93	75	391	3.00	2.90	87.5	455.6
	.003	74	381	2.90	51	260	2.98	2.86	59.4	302.4
	.004	45	226	2.84	33	163	2.92	2.80	36.8	181.2
	.005	33	160	2.80	24	115	2.89	2.75	26.8	128.0
	.006	26	124	2.77	18	84	2.90	2.70	20.9	97.8
	.007	22	102	2.73	15	69	2.90	2.66	17.7	81.8
	.008	19	87	2.71	14	64	2.79	2.65	15.1	69.4
	.009	17	76	2.68	12	53	2.81	2.61	13.4	59.5
	.01	15	67	2.66	10	44	2.88	2.57	12.0	53.1
	.012	13	55	2.62	9	39	2.76	2.54	10.0	43.5
	.015	11	44	2.57	7	29	2.84	2.46	8.4	35.1
	.02	8	34	2.51	6	24	2.65	2.41	6.6	26.4
	.025	7	27	2.46	5	19	2.62	2.35	5.5	21.0
	.03	6	23	2.41	4	15	2.80	2.26	4.9	18.5
	.035	6	20	2.37	4	15	2.53	2.26	4.3	16.3
	.04	5	18	2.34	4	15	2.42	2.26	4.2	15.7
	.05	5	15	2.28	3	10	2.51	2.14	3.3	11.2
	.06	4	13	2.23	3	10	2.32	2.14	3.1	10.5
.0025	.004	357	1678	2.72	286	1337	2.74	2.71	308.1	1439.7
	.005	161	736	2.68	111	501	2.74	2.65	132.1	596.0
	.006	99	443	2.64	71	313	2.70	2.61	80.6	354.9
	.0075	62	267	2.60	45	193	2.66	2.56	50.1	214.2
	.01	38	157	2.54	27	111	2.62	2.49	30.1	123.5
	.012	29	117	2.50	21	84	2.59	2.44	23.4	93.9
	.015	22	85	2.45	15	58	2.58	2.38	17.1	66.4
	.02	16	59	2.38	11	41	2.51	2.31	12.2	45.8
	.025	12	45	2.33	9	32	2.45	2.25	9.9	35.2
	.03	10	37	2.29	7	24	2.51	2.18	8.2	28.2
	.035	9	31	2.25	6	20	2.51	2.13	7.1	23.8
	.04	8	27	2.21	6	20	2.32	2.13	6.4	21.5
	.05	7	22	2.15	5	16	2.26	2.07	5.3	17.1
	.06	6	18	2.10	4	12	2.31	1.98	4.4	13.4
.005	.0075	417	1714	2.50	293	1195	2.53	2.48	341.2	1390.6
	.01	138	547	2.44	100	391	2.48	2.41	112.0	437.1
	.012	85	327	2.40	60	228	2.47	2.36	69.5	263.0
	.015	53	196	2.35	39	144	2.41	2.31	43.0	157.6
	.02	32	114	2.28	23	81	2.37	2.23	25.6	90.5
	.025	23	79	2.23	16	54	2.37	2.16	18.5	62.7
	.03	18	61	2.19	13	42	2.32	2.11	14.7	47.8
	.035	15	49	2.15	11	35	2.25	2.08	12.0	38.3
	.04	13	41	2.11	9	28	2.29	2.02	10.3	32.3
	.05	10	31	2.05	7	21	2.25	1.95	8.0	24.2
	.06	9	25	2.00	6	17	2.17	1.90	6.7	19.0
	.07	8	21	1.96	5	14	2.19	1.84	5.7	16.1

TABLE T10-2   (continued)

		SINGLE			DOUBLE					
p_1	p_2	n_σ	n_s	k	n_σ	n_s	k_a	k_r	ASN_σ	ASN_s
.0075	.01	763	2909	2.37	511	1935	2.41	2.36	640.4	2423.7
	.012	279	1040	2.33	213	787	2.36	2.32	233.1	860.2
	.015	125	450	2.29	91	324	2.33	2.26	101.4	360.2
	.02	60	208	2.22	43	146	2.30	2.18	49.2	166.0
	.025	39	129	2.17	28	91	2.25	2.12	31.3	102.1
	.03	29	92	2.12	20	63	2.25	2.06	23.2	73.3
	.035	23	71	2.08	16	49	2.20	2.02	18.0	55.4
	.04	19	58	2.05	14	42	2.14	1.99	15.2	45.9
	.05	14	42	1.99	10	29	2.13	1.91	11.2	32.7
	.06	12	33	1.94	8	22	2.10	1.85	9.0	24.9
	.07	10	27	1.90	7	19	2.03	1.81	7.7	20.9
	.08	9	23	1.86	6	16	2.01	1.76	6.6	17.7
.01	.015	351	1231	2.24	246	853	2.28	2.22	291.4	1009.5
	.02	116	388	2.17	87	289	2.21	2.15	94.7	313.6
	.025	64	208	2.12	45	143	2.21	2.08	52.6	166.2
	.03	44	137	2.08	31	95	2.16	2.03	35.0	107.5
	.035	33	100	2.04	23	69	2.15	1.98	26.5	79.6
	.04	26	78	2.00	19	55	2.11	1.94	21.4	62.2
	.045	22	64	1.97	15	43	2.13	1.90	17.6	50.7
	.05	19	54	1.94	13	36	2.09	1.87	14.9	41.5
	.06	15	41	1.89	11	30	1.98	1.83	11.8	32.4
	.07	12	33	1.85	9	24	1.96	1.77	9.8	26.3
	.08	11	27	1.81	7	18	2.05	1.70	8.3	21.4
	.09	9	23	1.77	6	15	2.09	1.65	7.4	18.5
	.10	8	20	1.74	6	15	1.87	1.65	6.5	16.3
.015	.02	633	2036	2.11	448	1427	2.14	2.09	544.3	1733.0
	.025	195	603	2.05	142	435	2.09	2.03	159.3	487.4
	.03	103	309	2.01	75	223	2.06	1.98	84.0	248.6
	.035	67	197	1.97	47	135	2.05	1.93	54.3	155.4
	.04	49	140	1.93	35	98	2.02	1.89	39.8	111.9
	.045	39	107	1.90	27	74	2.01	1.85	31.2	85.7
	.05	32	86	1.88	23	62	1.97	1.82	25.8	69.8
	.06	23	61	1.82	17	44	1.91	1.77	18.6	48.3
	.07	18	46	1.78	13	33	1.90	1.71	14.5	37.0
	.08	15	37	1.74	10	24	1.96	1.65	12.0	29.0
	.09	13	31	1.70	9	21	1.85	1.62	10.1	23.6
	.10	11	26	1.67	8	18	1.82	1.58	8.9	20.2
	.11	10	23	1.64	7	16	1.81	1.54	7.9	18.1
	.12	9	20	1.61	6	13	1.89	1.49	7.2	15.6
	.13	8	18	1.58	6	13	1.72	1.49	6.5	14.2
	.14	8	16	1.56	5	11	1.84	1.43	5.9	13.1
	.15	7	15	1.53	5	11	1.69	1.43	5.5	12.1
.02	.03	287	835	1.96	208	600	1.99	1.94	234.5	675.8
	.035	147	416	1.92	102	285	1.98	1.89	120.2	334.8
	.04	94	259	1.88	66	179	1.95	1.85	75.9	205.8
	.045	67	182	1.85	46	122	1.95	1.81	54.8	144.5
	.05	52	137	1.82	37	96	1.91	1.78	42.3	110.1
	.06	35	89	1.77	25	62	1.87	1.72	28.3	70.4
	.07	26	64	1.73	19	46	1.83	1.67	21.2	51.6
	.08	21	50	1.69	15	35	1.81	1.62	16.9	39.6
	.09	17	40	1.65	12	27	1.82	1.57	13.9	31.5
	.10	15	34	1.62	10	22	1.82	1.53	11.8	26.1
	.11	13	29	1.59	9	20	1.75	1.50	10.2	22.8
	.12	12	25	1.56	8	17	1.72	1.47	9.0	19.2
	.13	10	22	1.53	7	15	1.74	1.43	8.1	17.4
	.15	9	18	1.48	6	12	1.66	1.38	6.7	13.5
	.17	8	15	1.44	5	10	1.69	1.31	5.8	11.7
	.20	6	12	1.37	4	8	1.70	1.23	4.8	9.6

TABLE T10-2   (continued)

		SINGLE			DOUBLE					
p_1	p_2	n_σ	n_s	k	n_σ	n_s	k_a	k_r	ASN_σ	ASN_s
.03	.04	506	1333	1.81	411	1077	1.82	1.80	434.6	1138.7
	.045	250	643	1.78	185	472	1.81	1.76	206.6	526.6
	.05	154	389	1.75	105	261	1.82	1.72	127.7	316.5
	.06	81	197	1.70	56	134	1.78	1.66	65.8	156.5
	.07	53	124	1.65	37	85	1.74	1.61	42.4	97.7
	.08	38	88	1.61	28	63	1.69	1.57	31.0	69.9
	.09	30	66	1.58	22	48	1.65	1.53	24.0	52.5
	.10	24	53	1.54	17	36	1.67	1.48	19.4	41.3
	.11	20	43	1.51	14	29	1.67	1.44	16.3	34.0
	.12	18	37	1.48	13	27	1.58	1.42	14.2	29.6
	.13	16	32	1.46	11	22	1.59	1.38	12.3	24.8
	.15	13	24	1.41	9	17	1.52	1.33	9.8	18.7
	.20	8	15	1.30	6	11	1.44	1.20	6.6	12.1
	.25	6	11	1.20	4	7	1.61	1.05	5.0	8.7
	.30	5	8	1.12	4	7	1.18	1.05	4.1	7.2
.04	.06	224	524	1.64	159	368	1.68	1.62	180.6	417.6
	.07	114	258	1.60	83	186	1.64	1.57	91.8	205.0
	.08	72	159	1.56	51	110	1.63	1.52	58.4	125.8
	.09	51	110	1.52	37	78	1.59	1.48	41.2	87.2
	.10	39	82	1.49	28	58	1.57	1.44	31.3	65.1
	.11	32	65	1.46	22	44	1.57	1.40	25.3	50.7
	.12	26	53	1.43	19	37	1.52	1.37	21.1	41.2
	.13	22	44	1.40	17	33	1.47	1.35	18.3	35.6
	.14	20	38	1.37	14	27	1.48	1.31	15.5	30.0
	.15	17	33	1.35	12	22	1.50	1.27	13.7	25.3
	.17	14	25	1.30	10	18	1.42	1.23	11.0	19.9
	.20	11	19	1.24	8	14	1.35	1.16	8.7	15.2
	.25	8	13	1.15	5	8	1.52	1.01	6.3	10.0
	.30	6	9	1.06	4	6	1.35	0.92	4.7	7.0
	.35	5	7	.98	4	6	1.04	0.92	4.1	6.2
	.40	4	6	.91	3	4	1.05	0.80	3.2	4.3
.05	.07	300	660	1.55	204	443	1.60	1.53	246.7	535.4
	.08	149	319	1.51	113	239	1.54	1.49	122.3	258.0
	.09	93	194	1.47	66	135	1.54	1.44	75.8	154.4
	.10	65	133	1.44	46	92	1.52	1.40	52.9	106.1
	.11	49	98	1.41	36	70	1.48	1.37	40.0	78.1
	.12	39	76	1.38	28	53	1.48	1.33	32.0	60.9
	.13	32	62	1.35	23	43	1.45	1.30	25.9	48.6
	.14	27	51	1.33	20	37	1.42	1.27	22.2	41.3
	.15	24	43	1.30	17	31	1.41	1.24	19.0	34.8
	.16	21	37	1.28	15	27	1.38	1.22	16.5	29.9
	.17	18	33	1.26	13	23	1.40	1.18	14.8	26.4
	.20	14	23	1.19	10	17	1.30	1.12	10.9	18.6
	.25	10	15	1.10	7	11	1.21	1.02	7.5	11.9
	.30	7	11	1.01	5	8	1.21	0.90	5.6	9.0
	.35	6	8	.94	4	6	1.12	0.82	4.4	6.6
	.40	5	7	.86	3	4	1.31	0.69	3.7	4.9

From D. J. Sommers, *Journal of Quality Technology*, 13(1): 1981, pp. 26-29.
Reprinted by permission.

TABLE T10-3   Comparison of Approximate and Exact Values of N
and k for Variables Sampling Plans[*]

Exact			Approximate			
N	k	p_2	N	k	true p_1	true p_2
(1)	(2)	(3)	(4)	(5)	(6)	(7)
5	0.5445	0.5428	5	0.5150	0.0541	0.5519
10	0.8037	0.3774	10	0.7856	0.0524	0.3830
15	0.9292	0.3033	15	0.9163	0.0517	0.3071
20	1.0083	0.2604	20	0.9983	0.0513	0.2633
25	1.0643	0.2332	25	1.0560	0.0510	0.2344
30	1.1069	0.2119	30	1.1000	0.0509	0.2137
35	1.1409	0.1966	35	1.1348	0.0507	0.1981
40	1.1688	0.1846	40	1.1633	0.0507	0.1859
45	1.1922	0.1748	45	1.1874	0.0506	0.1759
50	1.2125	0.1666	50	1.2082	0.0505	0.1676
65	1.2592	0.1489	64	1.2556	0.0501	0.1500
75	1.2828	0.1404	74	1.2797	0.0501	0.1413
100	1.3264	0.1256	99	1.3241	0.0501	0.1262

[*]The exact values of k in column (2) have been computed from the non-
central t distribution, taking p_1 = 0.05, α = 0.01, and N as shown in
column (1). The exact values of p_2 for which β = 0.10 have then been
computed and entered in column (3). From these p_2, taking β = 0.10,
p_1 = 0.05, and α = 0.01, approximate values of N and k were computed
from the Wallis approximation. For the approximation the true values
of p_1 and p_2 for which α = 0.01 and β = 0.10 have been computed from
the non-central t distribution.

From Statistical Research Group, *Techniques of Statistical
Analysis*, McGraw-Hill, New York, 1947.

TABLE T10-4   Odeh-Owen Table 5: Two-Sided Sampling Plan Factors to Control
Equal Tails

GAMMA = 0.900

N \P →	0.20	0.10	0.05	0.025	0.02	0.01	0.005
2	6.987	10.253	13.090	15.586	16.331	18.500	20.486
3	3.039	4.258	5.311	6.244	6.523	7.340	8.092
4	2.295	3.188	3.957	4.637	4.841	5.438	5.988
5	1.976	2.742	3.400	3.981	4.156	4.666	5.136
6	1.806	2.494	3.092	3.620	3.779	4.243	4.669
7	1.721	2.334	2.894	3.389	3.538	3.972	4.372
8	1.666	2.227	2.755	3.227	3.369	3.783	4.164
9	1.626	2.158	2.652	3.106	3.242	3.641	4.009
10	1.595	2.112	2.576	3.012	3.144	3.532	3.888
11	1.570	2.075	2.520	2.938	3.066	3.444	3.792
12	1.550	2.045	2.479	2.879	3.004	3.371	3.712
13	1.533	2.020	2.446	2.833	2.953	3.312	3.646
14	1.519	1.999	2.419	2.796	2.912	3.261	3.589
15	1.506	1.981	2.395	2.767	2.880	3.219	3.541
16	1.496	1.965	2.374	2.742	2.853	3.184	3.499
17	1.486	1.950	2.356	2.720	2.830	3.155	3.463
18	1.478	1.938	2.340	2.701	2.810	3.130	3.433
19	1.470	1.927	2.325	2.683	2.791	3.109	3.406
20	1.463	1.916	2.312	2.667	2.775	3.090	3.383
21	1.457	1.907	2.300	2.653	2.760	3.073	3.364
22	1.451	1.899	2.290	2.640	2.746	3.057	3.346
23	1.446	1.891	2.280	2.628	2.733	3.043	3.330
24	1.441	1.884	2.270	2.617	2.721	3.029	3.315
25	1.437	1.877	2.262	2.606	2.711	3.017	3.301
30	1.419	1.851	2.227	2.565	2.667	2.967	3.245
35	1.406	1.831	2.202	2.534	2.634	2.929	3.203
40	1.396	1.816	2.182	2.510	2.609	2.901	3.171
45	1.387	1.804	2.166	2.491	2.589	2.878	3.146
50	1.381	1.794	2.154	2.476	2.573	2.859	3.124
60	1.370	1.778	2.133	2.451	2.547	2.829	3.091
70	1.362	1.766	2.118	2.433	2.528	2.807	3.066
80	1.356	1.757	2.106	2.418	2.513	2.790	3.047
90	1.351	1.750	2.097	2.407	2.500	2.776	3.031
100	1.347	1.744	2.089	2.397	2.490	2.764	3.018
120	1.341	1.734	2.076	2.382	2.474	2.746	2.997
150	1.334	1.723	2.062	2.365	2.457	2.726	2.975
300	1.317	1.699	2.030	2.326	2.416	2.678	2.922
500	1.309	1.686	2.013	2.306	2.394	2.654	2.895
600	1.306	1.682	2.008	2.300	2.388	2.647	2.887
700	1.304	1.679	2.005	2.295	2.383	2.641	2.880
800	1.303	1.677	2.002	2.292	2.379	2.637	2.875
900	1.301	1.675	1.999	2.289	2.376	2.633	2.871
1000	1.300	1.673	1.997	2.286	2.374	2.630	2.868
1500	1.297	1.668	1.990	2.278	2.365	2.620	2.856
2000	1.295	1.665	1.986	2.273	2.359	2.614	2.850
3000	1.292	1.661	1.981	2.267	2.353	2.607	2.842
5000	1.290	1.657	1.976	2.261	2.347	2.600	2.834
10000	1.287	1.654	1.971	2.255	2.341	2.593	2.826
∞	1.282	1.645	1.960	2.241	2.326	2.576	2.807

From R. E. Odeh and D. B. Owen, *Tables for Normal Tolerance Limits, Sample Plans, and Screening*, 1980, p. 146. Reprinted by courtesy of Marcel Dekker, Inc.

TABLE T10-5   Odeh-Owen Table 6:   Two-Sided Sampling Plan Factors to Control
Tails Separately

GAMMA = 0.900

N \P →	0.20	0.10	0.05	0.025	0.02	0.01	0.005
↓							
2	6.987	10.253	13.090	15.586	16.331	18.500	20.486
3	3.039	4.258	5.311	6.244	6.523	7.340	8.092
4	2.295	3.188	3.957	4.637	4.841	5.438	5.988
5	1.976	2.742	3.400	3.981	4.156	4.666	5.136
6	1.795	2.494	3.092	3.620	3.779	4.243	4.669
7	1.676	2.333	2.894	3.389	3.538	3.972	4.372
8	1.590	2.219	2.754	3.227	3.369	3.783	4.164
9	1.525	2.133	2.650	3.106	3.242	3.641	4.009
10	1.474	2.066	2.568	3.011	3.144	3.532	3.888
11	1.433	2.011	2.503	2.935	3.065	3.443	3.792
12	1.398	1.966	2.448	2.872	3.000	3.371	3.712
13	1.368	1.928	2.402	2.820	2.945	3.309	3.645
14	1.343	1.895	2.363	2.774	2.898	3.257	3.588
15	1.321	1.867	2.329	2.735	2.857	3.212	3.538
16	1.301	1.842	2.299	2.701	2.821	3.172	3.495
17	1.284	1.819	2.272	2.670	2.789	3.137	3.456
18	1.268	1.800	2.249	2.643	2.761	3.105	3.422
19	1.254	1.782	2.227	2.618	2.736	3.077	3.391
20	1.241	1.765	2.208	2.596	2.712	3.052	3.363
21	1.229	1.750	2.190	2.576	2.691	3.028	3.338
22	1.218	1.737	2.174	2.557	2.672	3.007	3.315
23	1.208	1.724	2.159	2.540	2.654	2.987	3.293
24	1.199	1.712	2.145	2.525	2.638	2.969	3.273
25	1.190	1.702	2.132	2.510	2.623	2.952	3.255
30	1.154	1.657	2.080	2.450	2.561	2.884	3.180
35	1.127	1.624	2.041	2.406	2.515	2.833	3.125
40	1.106	1.598	2.010	2.371	2.479	2.793	3.082
45	1.089	1.577	1.986	2.343	2.450	2.761	3.047
50	1.075	1.559	1.965	2.320	2.426	2.735	3.018
60	1.052	1.532	1.933	2.284	2.389	2.694	2.973
70	1.035	1.511	1.909	2.256	2.360	2.662	2.940
80	1.022	1.495	1.890	2.235	2.338	2.638	2.913
90	1.011	1.481	1.874	2.217	2.320	2.618	2.891
100	1.001	1.470	1.861	2.203	2.304	2.601	2.873
120	0.986	1.452	1.841	2.179	2.280	2.574	2.844
150	0.970	1.433	1.818	2.154	2.254	2.546	2.813
300	0.931	1.386	1.765	2.094	2.192	2.477	2.739
500	0.910	1.362	1.736	2.062	2.159	2.442	2.701
600	0.904	1.355	1.728	2.053	2.150	2.431	2.689
700	0.899	1.349	1.722	2.046	2.142	2.423	2.680
800	0.896	1.344	1.717	2.040	2.136	2.417	2.673
900	0.892	1.341	1.712	2.035	2.132	2.411	2.668
1000	0.890	1.338	1.709	2.031	2.127	2.407	2.663
1500	0.881	1.327	1.697	2.018	2.114	2.392	2.646
2000	0.875	1.321	1.690	2.010	2.105	2.383	2.637
3000	0.869	1.314	1.681	2.001	2.096	2.372	2.625
5000	0.863	1.306	1.673	1.991	2.086	2.362	2.614
10000	0.857	1.299	1.665	1.982	2.077	2.351	2.603
∞	0.842	1.282	1.645	1.960	2.054	2.326	2.576

From R. E. Odeh and D. B. Owen, *Tables for Normal Tolerance Limits, Sample Plans,
and Screening*, 1980, p. 147.   Reprinted by courtesy of Marcel Dekker, Inc.

TABLE T11-1 MIL-STD-105D Table VIII. Limit Numbers for Reduced Inspection

Acceptable Quality Level

Number of sample units from last 10 lots or batches	0.010	0.015	0.025	0.040	0.065	0.10	0.15	0.25	0.40	0.65	1.0	1.5	2.5	4.0	6.5	10	15	25	40	65	100	150	250	400	650	1000
20 - 29	•	•	•	•	•	•	•	•	•	•	•	•	•	•	•	0	0	2	4	8	14	22	40	68	115	181
30 - 49	•	•	•	•	•	•	•	•	•	•	•	•	•	•	0	0	1	3	7	13	22	36	63	105	178	277
50 - 79	•	•	•	•	•	•	•	•	•	•	•	•	•	0	0	2	3	7	14	25	40	63	110	181	301	
80 - 129	•	•	•	•	•	•	•	•	•	•	•	•	0	0	2	4	7	14	24	42	68	105	181	297		
130 - 199	•	•	•	•	•	•	•	•	•	•	•	0	0	2	4	7	13	25	42	72	115	177	301	490		
200 - 319	•	•	•	•	•	•	•	•	•	•	0	0	2	4	8	14	22	40	68	115	181	277	471			
320 - 499	•	•	•	•	•	•	•	•	•	0	0	1	4	8	14	24	39	68	113	189						
500 - 799	•	•	•	•	•	•	•	•	0	0	2	3	7	14	25	40	63	110	181							
800 - 1249	•	•	•	•	•	•	•	0	0	2	4	7	14	24	42	68	105	181								
1250 - 1999	•	•	•	•	•	•	0	0	2	4	7	13	24	40	69	110	169									
2000 - 3149	•	•	•	•	•	0	0	2	4	8	14	22	40	68	115	181										
3150 - 4999	•	•	•	•	0	0	1	4	8	14	24	38	67	111	186											
5000 - 7999	•	•	•	0	0	2	3	7	14	25	40	63	110	181												
8000 - 12499	•	•	0	0	2	4	7	14	24	42	68	105	181													
12500 - 19999	•	0	0	2	4	7	13	24	40	69	110	169														
20000 - 31499	0	0	2	4	8	14	22	40	68	115	181															
31500 - 49999	0	1	4	8	14	24	38	67	111	186																
50000 & Over	2	3	7	14	25	40	63	110	181	301																

From MIL-STD-105D, p. 28.

TABLE T11-2 MIL-STD-105D Table I. Sample Size Code Letters

Lot or batch size	Special inspection levels				General inspection levels		
	S-1	S-2	S-3	S-4	I	II	III
2 to 8	A	A	A	A	A	A	B
9 to 15	A	A	A	A	A	B	C
16 to 25	A	A	B	B	B	C	D
26 to 50	A	B	B	C	C	D	E
51 to 90	B	B	C	C	C	E	F
91 to 150	B	B	C	D	D	F	G
151 to 280	B	C	D	E	E	G	H
281 to 500	B	C	D	E	F	H	J
501 to 1200	C	C	E	F	G	J	K
1201 to 3200	C	D	E	G	H	K	L
3201 to 10000	C	D	F	G	J	L	M
10001 to 35000	C	D	F	H	K	M	N
35001 to 150000	D	E	G	J	L	N	P
150001 to 500000	D	E	G	J	M	P	Q
500001 and over	D	E	H	K	N	Q	R

From MIL-STD-105D, p. 9.

TABLE T11-3   MIL-STD-105D Table II-A.   Single-Sampling Plans for Normal Inspection (Master Table)

Acceptable Quality Levels (normal inspection)

Each cell below is shown as "Ac Re" (Acceptance number / Rejection number). ↓ = Use first sampling plan below arrow. ↑ = Use first sampling plan above arrow. If sample size equals, or exceeds, lot or batch size, do 100 percent inspection.

Code	Sample size	0.010	0.015	0.025	0.040	0.065	0.10	0.15	0.25	0.40	0.65	1.0	1.5	2.5	4.0	6.5	10	15	25	40	65	100	150	250	400	650	1000
A	2	↓	↓	↓	↓	↓	↓	↓	↓	↓	↓	↓	↓	↓	↓	↓	↓	0 1	1 2	2 3	3 4	5 6	7 8	10 11	14 15	21 22	30 31
B	3	↓	↓	↓	↓	↓	↓	↓	↓	↓	↓	↓	↓	↓	↓	↓	0 1	1 2	2 3	3 4	5 6	7 8	10 11	14 15	21 22	30 31	44 45
C	5	↓	↓	↓	↓	↓	↓	↓	↓	↓	↓	↓	↓	↓	↓	0 1	1 2	2 3	3 4	5 6	7 8	10 11	14 15	21 22	30 31	44 45	↑
D	8	↓	↓	↓	↓	↓	↓	↓	↓	↓	↓	↓	↓	↓	0 1	1 2	2 3	3 4	5 6	7 8	10 11	14 15	21 22	30 31	44 45	↑	↑
E	13	↓	↓	↓	↓	↓	↓	↓	↓	↓	↓	↓	↓	0 1	1 2	2 3	3 4	5 6	7 8	10 11	14 15	21 22	30 31	44 45	↑	↑	↑
F	20	↓	↓	↓	↓	↓	↓	↓	↓	↓	↓	↓	0 1	1 2	2 3	3 4	5 6	7 8	10 11	14 15	21 22	30 31	44 45	↑	↑	↑	↑
G	32	↓	↓	↓	↓	↓	↓	↓	↓	↓	↓	0 1	1 2	2 3	3 4	5 6	7 8	10 11	14 15	21 22	30 31	44 45	↑	↑	↑	↑	↑
H	50	↓	↓	↓	↓	↓	↓	↓	↓	↓	0 1	1 2	2 3	3 4	5 6	7 8	10 11	14 15	21 22	30 31	44 45	↑	↑	↑	↑	↑	↑
J	80	↓	↓	↓	↓	↓	↓	↓	↓	0 1	1 2	2 3	3 4	5 6	7 8	10 11	14 15	21 22	30 31	44 45	↑	↑	↑	↑	↑	↑	↑
K	125	↓	↓	↓	↓	↓	↓	↓	0 1	1 2	2 3	3 4	5 6	7 8	10 11	14 15	21 22	30 31	44 45	↑	↑	↑	↑	↑	↑	↑	↑
L	200	↓	↓	↓	↓	↓	↓	0 1	1 2	2 3	3 4	5 6	7 8	10 11	14 15	21 22	30 31	44 45	↑	↑	↑	↑	↑	↑	↑	↑	↑
M	315	↓	↓	↓	↓	↓	0 1	1 2	2 3	3 4	5 6	7 8	10 11	14 15	21 22	30 31	44 45	↑	↑	↑	↑	↑	↑	↑	↑	↑	↑
N	500	↓	↓	↓	↓	0 1	1 2	2 3	3 4	5 6	7 8	10 11	14 15	21 22	30 31	44 45	↑	↑	↑	↑	↑	↑	↑	↑	↑	↑	↑
P	800	↓	↓	↓	0 1	1 2	2 3	3 4	5 6	7 8	10 11	14 15	21 22	30 31	44 45	↑	↑	↑	↑	↑	↑	↑	↑	↑	↑	↑	↑
Q	1250	↓	↓	0 1	1 2	2 3	3 4	5 6	7 8	10 11	14 15	21 22	30 31	44 45	↑	↑	↑	↑	↑	↑	↑	↑	↑	↑	↑	↑	↑
R	2000	↓	0 1	1 2	2 3	3 4	5 6	7 8	10 11	14 15	21 22	30 31	44 45	↑	↑	↑	↑	↑	↑	↑	↑	↑	↑	↑	↑	↑	↑

↓ = Use first sampling plan below arrow. If sample size equals, or exceeds, lot or batch size, do 100 percent inspection.
↑ = Use first sampling plan above arrow.
Ac = Acceptance number.
Re = Rejection number.

From MIL-STD-105D, p. 10.

TABLE T11-4 MIL-STD-105D Table II-B. Single-Sampling Plans for Tightened Inspection (Master Table)

Acceptable Quality Levels (tightened inspection)

Sample size code letter	Sample size	0.010		0.015		0.025		0.040		0.065		0.10		0.15		0.25		0.40		0.65		1.0		1.5		2.5		4.0		6.5		10		15		25		40		65		100		150		250		400		650		1000	
		Ac	Re	Ac	Re	Ac	Re	Ac	Re	Ac	Re	Ac	Re	Ac	Re	Ac	Re	Ac	Re	Ac	Re	Ac	Re	Ac	Re	Ac	Re	Ac	Re	Ac	Re	Ac	Re	Ac	Re	Ac	Re	Ac	Re	Ac	Re	Ac	Re	Ac	Re	Ac	Re	Ac	Re	Ac	Re	Ac	Re
A	2	↓		↓		↓		↓		↓		↓		↓		↓		↓		↓		↓		↓		↓		↓		↓		↓		↓		0	1	1	2	2	3	3	4	5	6	8	9	12	13	18	19	27	28
B	3	↓		↓		↓		↓		↓		↓		↓		↓		↓		↓		↓		↓		↓		↓		↓		↓		0	1	1	2	2	3	3	4	5	6	8	9	12	13	18	19	27	28	41	42
C	5	↓		↓		↓		↓		↓		↓		↓		↓		↓		↓		↓		↓		↓		↓		↓		0	1	1	2	2	3	3	4	5	6	8	9	12	13	18	19	27	28	41	42	↑	
D	8	↓		↓		↓		↓		↓		↓		↓		↓		↓		↓		↓		↓		↓		↓		0	1	1	2	2	3	3	4	5	6	8	9	12	13	18	19	27	28	41	42	↑		↑	
E	13	↓		↓		↓		↓		↓		↓		↓		↓		↓		↓		↓		↓		↓		0	1	1	2	2	3	3	4	5	6	8	9	12	13	18	19	27	28	41	42	↑		↑		↑	
F	20	↓		↓		↓		↓		↓		↓		↓		↓		↓		↓		↓		↓		0	1	1	2	2	3	3	4	5	6	8	9	12	13	18	19	27	28	41	42	↑		↑		↑		↑	
G	32	↓		↓		↓		↓		↓		↓		↓		↓		↓		↓		↓		0	1	1	2	2	3	3	4	5	6	8	9	12	13	18	19	27	28	41	42	↑		↑		↑		↑		↑	
H	50	↓		↓		↓		↓		↓		↓		↓		↓		↓		↓		0	1	1	2	2	3	3	4	5	6	8	9	12	13	18	19	27	28	41	42	↑		↑		↑		↑		↑		↑	
J	80	↓		↓		↓		↓		↓		↓		↓		↓		↓		0	1	1	2	2	3	3	4	5	6	8	9	12	13	18	19	27	28	41	42	↑		↑		↑		↑		↑		↑		↑	
K	125	↓		↓		↓		↓		↓		↓		↓		↓		0	1	1	2	2	3	3	4	5	6	8	9	12	13	18	19	27	28	41	42	↑		↑		↑		↑		↑		↑		↑		↑	
L	200	↓		↓		↓		↓		↓		↓		↓		0	1	1	2	2	3	3	4	5	6	8	9	12	13	18	19	27	28	41	42	↑		↑		↑		↑		↑		↑		↑		↑		↑	
M	315	↓		↓		↓		↓		↓		↓		0	1	1	2	2	3	3	4	5	6	8	9	12	13	18	19	27	28	41	42	↑		↑		↑		↑		↑		↑		↑		↑		↑		↑	
N	500	↓		↓		↓		↓		↓		0	1	1	2	2	3	3	4	5	6	8	9	12	13	18	19	27	28	41	42	↑		↑		↑		↑		↑		↑		↑		↑		↑		↑		↑	
P	800	↓		↓		↓		↓		0	1	1	2	2	3	3	4	5	6	8	9	12	13	18	19	27	28	41	42	↑		↑		↑		↑		↑		↑		↑		↑		↑		↑		↑		↑	
Q	1250	↓		↓		↓		0	1	1	2	2	3	3	4	5	6	8	9	12	13	18	19	27	28	41	42	↑		↑		↑		↑		↑		↑		↑		↑		↑		↑		↑		↑		↑	
R	2000	↓		↓		0	1	1	2	2	3	3	4	5	6	8	9	12	13	18	19	27	28	41	42	↑		↑		↑		↑		↑		↑		↑		↑		↑		↑		↑		↑		↑		↑	
S	3150	↓		0	1	1	2	2	3	3	4	5	6	8	9	12	13	18	19	27	28	41	42	↑		↑		↑		↑		↑		↑		↑		↑		↑		↑		↑		↑		↑		↑		↑	

↓ = Use first sampling plan below arrow. If sample size equals or exceeds lot or batch size, do 100 percent inspection.

↑ = Use first sampling plan above arrow.

Ac = Acceptance number.

Re = Rejection number.

From MIL-STD-105D, p. 11.

TABLE T11-5   MIL-STD-105D Table II-C.   Single-Sampling Plans for Reduced Inspection (Master Table)

Acceptable Quality Levels (reduced inspection)†

(Each acceptable quality level column contains an acceptance number **Ac** and a rejection number **Re**, shown below as "Ac Re".)

Sample size code letter	Sample size	0.010	0.015	0.025	0.040	0.065	0.10	0.15	0.25	0.40	0.65	1.0	1.5	2.5	4.0	6.5	10	15	25	40	65	100	150	250	400	650	1000
A	2	↓	↓	↓	↓	↓	↓	↓	↓	↓	↓	↓	↓	↓	↓	↓	↓	0 1	1 2	2 3	3 4	5 6	7 8	10 11	14 15	21 22	30 31
B	2	↓	↓	↓	↓	↓	↓	↓	↓	↓	↓	↓	↓	↓	↓	↓	↓	0 2	1 3	2 4	3 5	5 6	7 8	10 11	14 15	21 22	30 31
C	2	↓	↓	↓	↓	↓	↓	↓	↓	↓	↓	↓	↓	↓	↓	0 1	0 2	1 3	1 4	2 5	3 6	5 8	7 10	10 13	14 17	21 24	↑
D	3	↓	↓	↓	↓	↓	↓	↓	↓	↓	↓	↓	↓	↓	0 1	0 2	1 3	1 4	2 5	3 6	5 8	7 10	10 13	14 17	21 24	↑	↑
E	5	↓	↓	↓	↓	↓	↓	↓	↓	↓	↓	↓	↓	0 1	0 2	1 3	1 4	2 5	3 6	5 8	7 10	10 13	14 17	21 24	↑	↑	↑
F	8	↓	↓	↓	↓	↓	↓	↓	↓	↓	↓	↓	0 1	0 2	1 3	1 4	2 5	3 6	5 8	7 10	10 13	↑	↑	↑	↑	↑	↑
G	13	↓	↓	↓	↓	↓	↓	↓	↓	↓	↓	0 1	0 2	1 3	1 4	2 5	3 6	5 8	7 10	10 13	↑	↑	↑	↑	↑	↑	↑
H	20	↓	↓	↓	↓	↓	↓	↓	↓	↓	0 1	0 2	1 3	1 4	2 5	3 6	5 8	7 10	10 13	↑	↑	↑	↑	↑	↑	↑	↑
J	32	↓	↓	↓	↓	↓	↓	↓	↓	0 1	0 2	1 3	1 4	2 5	3 6	5 8	7 10	10 13	↑	↑	↑	↑	↑	↑	↑	↑	↑
K	50	↓	↓	↓	↓	↓	↓	↓	0 1	0 2	1 3	1 4	2 5	3 6	5 8	7 10	10 13	↑	↑	↑	↑	↑	↑	↑	↑	↑	↑
L	80	↓	↓	↓	↓	↓	↓	0 1	0 2	1 3	1 4	2 5	3 6	5 8	7 10	10 13	↑	↑	↑	↑	↑	↑	↑	↑	↑	↑	↑
M	125	↓	↓	↓	↓	↓	0 1	0 2	1 3	1 4	2 5	3 6	5 8	7 10	10 13	↑	↑	↑	↑	↑	↑	↑	↑	↑	↑	↑	↑
N	200	↓	↓	↓	↓	0 1	0 2	1 3	1 4	2 5	3 6	5 8	7 10	10 13	↑	↑	↑	↑	↑	↑	↑	↑	↑	↑	↑	↑	↑
P	315	↓	↓	↓	0 1	0 2	1 3	1 4	2 5	3 6	5 8	7 10	10 13	↑	↑	↑	↑	↑	↑	↑	↑	↑	↑	↑	↑	↑	↑
Q	500	↓	↓	0 1	0 2	1 3	1 4	2 5	3 6	5 8	7 10	10 13	↑	↑	↑	↑	↑	↑	↑	↑	↑	↑	↑	↑	↑	↑	↑
R	800	↑	0 1	0 2	1 3	1 4	2 5	3 6	5 8	7 10	10 13	↑	↑	↑	↑	↑	↑	↑	↑	↑	↑	↑	↑	↑	↑	↑	↑

↓ = Use first sampling plan below arrow. If sample size equals or exceeds lot or batch size, do 100 percent inspection.

↑ = Use first sampling plan above arrow.

Ac = Acceptance number.

Re = Rejection number.

† = If the acceptance number has been exceeded, but the rejection number has not been reached, accept the lot, but reinstate normal inspection (see 10.1.4).

From MIL-STD-105D, p. 12.

TABLE T11-6 MIL-STD-105D Table III-A. Double-Sampling Plans for Normal Inspection (Master Table)

Acceptable Quality Levels (normal inspection)

Each AQL cell below shows the pair "Ac Re" (Ac = Acceptance number, Re = Rejection number). For each code letter the "First" row gives the first sampling plan and the "Second" row gives the second sampling plan. ↓ = use first sampling plan below arrow; ↑ = use first sampling plan above arrow; • = use corresponding single sampling plan.

Code	Sample	n	Cum n	0.010	0.015	0.025	0.040	0.065	0.10	0.15	0.25	0.40	0.65	1.0	1.5	2.5	4.0	6.5	10	15	25	40	65	100	150	250	400	650	1000
A				↓	↓	↓	↓	↓	↓	↓	↓	↓	↓	↓	↓	↓	↓	↓	↓	•	↓	↓	↓	↓	↓	↓	↓	↓	↓
B	First	2	2	↓	↓	↓	↓	↓	↓	↓	↓	↓	↓	↓	↓	↓	↓	↓	↓	•	0 2	0 3	1 4	2 5	3 7	5 9	7 11	11 16	17 22	25 31
B	Second	2	4																		1 2	3 4	4 5	6 7	8 9	12 13	18 19	26 27	37 38	56 57
C	First	3	3	↓	↓	↓	↓	↓	↓	↓	↓	↓	↓	↓	↓	↓	↓	•	0 2	0 3	1 4	2 5	3 7	5 9	7 11	11 16	17 22	25 31	↑
C	Second	3	6																1 2	3 4	4 5	6 7	8 9	12 13	18 19	26 27	37 38	56 57	
D	First	5	5	↓	↓	↓	↓	↓	↓	↓	↓	↓	↓	↓	↓	↓	•	0 2	0 3	1 4	2 5	3 7	5 9	7 11	11 16	17 22	25 31	↑	↑
D	Second	5	10															1 2	3 4	4 5	6 7	8 9	12 13	18 19	26 27	37 38	56 57		
E	First	8	8	↓	↓	↓	↓	↓	↓	↓	↓	↓	↓	↓	↓	•	0 2	0 3	1 4	2 5	3 7	5 9	7 11	11 16	17 22	25 31	↑	↑	↑
E	Second	8	16														1 2	3 4	4 5	6 7	8 9	12 13	18 19	26 27	37 38	56 57			
F	First	13	13	↓	↓	↓	↓	↓	↓	↓	↓	↓	↓	↓	•	0 2	0 3	1 4	2 5	3 7	5 9	7 11	11 16	17 22	25 31	↑	↑	↑	↑
F	Second	13	26													1 2	3 4	4 5	6 7	8 9	12 13	18 19	26 27	37 38	56 57				
G	First	20	20	↓	↓	↓	↓	↓	↓	↓	↓	↓	↓	•	0 2	0 3	1 4	2 5	3 7	5 9	7 11	11 16	17 22	25 31	↑	↑	↑	↑	↑
G	Second	20	40												1 2	3 4	4 5	6 7	8 9	12 13	18 19	26 27	37 38	56 57					
H	First	32	32	↓	↓	↓	↓	↓	↓	↓	↓	↓	•	0 2	0 3	1 4	2 5	3 7	5 9	7 11	11 16	17 22	25 31	↑	↑	↑	↑	↑	↑
H	Second	32	64											1 2	3 4	4 5	6 7	8 9	12 13	18 19	26 27	37 38	56 57						
J	First	50	50	↓	↓	↓	↓	↓	↓	↓	↓	•	0 2	0 3	1 4	2 5	3 7	5 9	7 11	11 16	17 22	25 31	↑	↑	↑	↑	↑	↑	↑
J	Second	50	100										1 2	3 4	4 5	6 7	8 9	12 13	18 19	26 27	37 38	56 57							
K	First	80	80	↓	↓	↓	↓	↓	↓	↓	•	0 2	0 3	1 4	2 5	3 7	5 9	7 11	11 16	17 22	25 31	↑	↑	↑	↑	↑	↑	↑	↑
K	Second	80	160									1 2	3 4	4 5	6 7	8 9	12 13	18 19	26 27	37 38	56 57								
L	First	125	125	↓	↓	↓	↓	↓	↓	•	0 2	0 3	1 4	2 5	3 7	5 9	7 11	11 16	17 22	25 31	↑	↑	↑	↑	↑	↑	↑	↑	↑
L	Second	125	250								1 2	3 4	4 5	6 7	8 9	12 13	18 19	26 27	37 38	56 57									
M	First	200	200	↓	↓	↓	↓	↓	•	0 2	0 3	1 4	2 5	3 7	5 9	7 11	11 16	17 22	25 31	↑	↑	↑	↑	↑	↑	↑	↑	↑	↑
M	Second	200	400							1 2	3 4	4 5	6 7	8 9	12 13	18 19	26 27	37 38	56 57										
N	First	315	315	↓	↓	↓	↓	•	0 2	0 3	1 4	2 5	3 7	5 9	7 11	11 16	17 22	25 31	↑	↑	↑	↑	↑	↑	↑	↑	↑	↑	↑
N	Second	315	630						1 2	3 4	4 5	6 7	8 9	12 13	18 19	26 27	37 38	56 57											
P	First	500	500	↓	↓	↓	•	0 2	0 3	1 4	2 5	3 7	5 9	7 11	11 16	17 22	25 31	↑	↑	↑	↑	↑	↑	↑	↑	↑	↑	↑	↑
P	Second	500	1000					1 2	3 4	4 5	6 7	8 9	12 13	18 19	26 27	37 38	56 57												
Q	First	800	800	↓	↓	•	0 2	0 3	1 4	2 5	3 7	5 9	7 11	11 16	17 22	25 31	↑	↑	↑	↑	↑	↑	↑	↑	↑	↑	↑	↑	↑
Q	Second	800	1600				1 2	3 4	4 5	6 7	8 9	12 13	18 19	26 27	37 38	56 57													
R	First	1250	1250	↓	•	0 2	0 3	1 4	2 5	3 7	5 9	7 11	11 16	17 22	25 31	↑	↑	↑	↑	↑	↑	↑	↑	↑	↑	↑	↑	↑	↑
R	Second	1250	2500			1 2	3 4	4 5	6 7	8 9	12 13	18 19	26 27	37 38	56 57														

⇧ = Use first sampling plan below arrow. If sample size equals or exceeds lot or batch size, do 100 percent inspection.

⇩ = Use first sampling plan above arrow.

Ac = Acceptance number

Re = Rejection number

• = Use corresponding single sampling plan (or alternatively, use double sampling plan below, where available).

From MIL-STD-105D, p. 13.

Acceptable Quality Levels (tightened inspection)

Sample size code letter	Sample	Sample size	Cumulative sample size
A			
B	First	2	2
B	Second	2	4
C	First	3	3
C	Second	3	6
D	First	5	5
D	Second	5	10
E	First	8	8
E	Second	8	16
F	First	13	13
F	Second	13	26
G	First	20	20
G	Second	20	40
H	First	32	32
H	Second	32	64
J	First	50	50
J	Second	50	100
K	First	80	80
K	Second	80	160
L	First	125	125
L	Second	125	250
M	First	200	200
M	Second	200	400
N	First	315	315
N	Second	315	630
P	First	500	500
P	Second	500	1000
Q	First	800	800
Q	Second	800	1600
R	First	1250	1250
R	Second	1250	2500
S	First	2000	2000
S	Second	2000	4000

AQL columns (each with Ac and Re): 0.010, 0.015, 0.025, 0.040, 0.065, 0.10, 0.15, 0.25, 0.40, 0.65, 1.0, 1.5, 2.5, 4.0, 6.5, 10, 15, 25, 40, 65, 100, 150, 250, 400, 650, 1000

⇩ = Use first sampling plan below arrow. If sample size equals or exceeds lot or batch size, do 100 percent inspection.

⇧ = Use first sampling plan above arrow.

Ac = Acceptance number

Re = Rejection number

• = Use corresponding single sampling plan (or, alternatively, use double sampling plan below, where available).

From MIL-STD-105D, p. 14.

TABLE T11-8 MIL-STD-105D Table III-C. Double-Sampling Plans for Reduced Inspection (Master Table)

Acceptable Quality Levels (reduced inspection)†

Sample size code letter	Sample	Sample size	Cumulative sample size	0.010 Ac Re	0.015 Ac Re	0.025 Ac Re	0.040 Ac Re	0.065 Ac Re	0.10 Ac Re	0.15 Ac Re	0.25 Ac Re	0.40 Ac Re	0.65 Ac Re	1.0 Ac Re	1.5 Ac Re	2.5 Ac Re	4.0 Ac Re	6.5 Ac Re	10 Ac Re	15 Ac Re	25 Ac Re	40 Ac Re	65 Ac Re	100 Ac Re	150 Ac Re	250 Ac Re	400 Ac Re	650 Ac Re	1000 Ac Re
A				↓	↓	↓	↓	↓	↓	↓	↓	↓	↓	↓	↓	↓	↓	•	↓	↓	•	•	•	•	•	•	•	•	•
B				↓	↓	↓	↓	↓	↓	↓	↓	↓	↓	↓	↓	↓	•	↑	↓	•	•	•	•	•	•	•	•	•	•
C				↓	↓	↓	↓	↓	↓	↓	↓	↓	↓	↓	↓	•	↑	↓	•	•	•	•	•	•	•	•	•	•	↑
D	First Second	2 2	2 4	↓	↓	↓	↓	↓	↓	↓	↓	↓	↓	↓	•	↑	↓	0 2 0 2	0 3 0 4	0 4 1 5	0 4 3 6	1 5 4 7	2 7 6 9	3 8 8 12	5 10 12 16	7 12 18 22	11 17 26 30	↑	↑
E	First Second	3 3	3 6	↓	↓	↓	↓	↓	↓	↓	↓	↓	↓	•	↑	↓	0 2 0 2	0 3 0 4	0 4 1 5	0 4 3 6	1 5 4 7	2 7 6 9	3 8 8 12	5 10 12 16	7 12 18 22	11 17 26 30	↑	↑	↑
F	First Second	5 5	5 10	↓	↓	↓	↓	↓	↓	↓	↓	↓	•	↑	↓	0 2 0 2	0 3 0 4	0 4 1 5	0 4 3 6	1 5 4 7	2 7 6 9	3 8 8 12	5 10 12 16	↑	↑	↑	↑	↑	↑
G	First Second	8 8	8 16	↓	↓	↓	↓	↓	↓	↓	↓	•	↑	↓	0 2 0 2	0 3 0 4	0 4 1 5	0 4 3 6	1 5 4 7	2 7 6 9	3 8 8 12	5 10 12 16	↑	↑	↑	↑	↑	↑	↑
H	First Second	13 13	13 26	↓	↓	↓	↓	↓	↓	↓	•	↑	↓	0 2 0 2	0 3 0 4	0 4 1 5	0 4 3 6	1 5 4 7	2 7 6 9	3 8 8 12	5 10 12 16	↑	↑	↑	↑	↑	↑	↑	↑
J	First Second	20 20	20 40	↓	↓	↓	↓	↓	↓	•	↑	↓	0 2 0 2	0 3 0 4	0 4 1 5	0 4 3 6	1 5 4 7	2 7 6 9	3 8 8 12	5 10 12 16	↑	↑	↑	↑	↑	↑	↑	↑	↑
K	First Second	32 32	32 64	↓	↓	↓	↓	↓	•	↑	↓	0 2 0 2	0 3 0 4	0 4 1 5	0 4 3 6	1 5 4 7	2 7 6 9	3 8 8 12	5 10 12 16	↑	↑	↑	↑	↑	↑	↑	↑	↑	↑
L	First Second	50 50	50 100	↓	↓	↓	↓	•	↑	↓	0 2 0 2	0 3 0 4	0 4 1 5	0 4 3 6	1 5 4 7	2 7 6 9	3 8 8 12	5 10 12 16	↑	↑	↑	↑	↑	↑	↑	↑	↑	↑	↑
M	First Second	80 80	80 160	↓	↓	↓	•	↑	↓	0 2 0 2	0 3 0 4	0 4 1 5	0 4 3 6	1 5 4 7	2 7 6 9	3 8 8 12	5 10 12 16	↑	↑	↑	↑	↑	↑	↑	↑	↑	↑	↑	↑
N	First Second	125 125	125 250	↓	↓	•	↑	↓	0 2 0 2	0 3 0 4	0 4 1 5	0 4 3 6	1 5 4 7	2 7 6 9	3 8 8 12	5 10 12 16	↑	↑	↑	↑	↑	↑	↑	↑	↑	↑	↑	↑	↑
P	First Second	200 200	200 400	↓	•	↑	↓	0 2 0 2	0 3 0 4	0 4 1 5	0 4 3 6	1 5 4 7	2 7 6 9	3 8 8 12	5 10 12 16	↑	↑	↑	↑	↑	↑	↑	↑	↑	↑	↑	↑	↑	↑
Q	First Second	315 315	315 630	•	↑	↓	0 2 0 2	0 3 0 4	0 4 1 5	0 4 3 6	1 5 4 7	2 7 6 9	3 8 8 12	5 10 12 16	↑	↑	↑	↑	↑	↑	↑	↑	↑	↑	↑	↑	↑	↑	↑
R	First Second	500 500	500 1000	↑	↑	0 2 0 2	0 3 0 4	0 4 1 5	0 4 3 6	1 5 4 7	2 7 6 9	3 8 8 12	5 10 12 16	↑	↑	↑	↑	↑	↑	↑	↑	↑	↑	↑	↑	↑	↑	↑	↑

↓ = Use first sampling plan below arrow. If sample size equals or exceeds lot or batch size, do 100 percent inspection.
↑ = Use first sampling plan above arrow.
Ac = Acceptance number.
Re = Rejection number.
• = Use corresponding single sampling plan (or alternatively, use double sampling plan below, when available.)
† = If, after the second sample, the acceptance number has been exceeded, but the rejection number has not been reached, accept the lot, but reinstate normal inspection (see 10.14).

From MIL-STD-105D, p. 15.

Acceptable Quality Levels (normal inspection)

Sample size code letter	Sample	Sample size	Cumulative sample size
A			
B			
C			
D	First	2	2
	Second	2	4
	Third	2	6
	Fourth	2	8
	Fifth	2	10
	Sixth	2	12
	Seventh	2	14
E	First	3	3
	Second	3	6
	Third	3	9
	Fourth	3	12
	Fifth	3	15
	Sixth	3	18
	Seventh	3	21
F	First	5	5
	Second	5	10
	Third	5	15
	Fourth	5	20
	Fifth	5	25
	Sixth	5	30
	Seventh	5	35
G	First	8	8
	Second	8	16
	Third	8	24
	Fourth	8	32
	Fifth	8	40
	Sixth	8	48
	Seventh	8	56
H	First	13	13
	Second	13	26
	Third	13	39
	Fourth	13	52
	Fifth	13	65
	Sixth	13	78
	Seventh	13	91
J	First	20	20
	Second	20	40
	Third	20	60
	Fourth	20	80
	Fifth	20	100
	Sixth	20	120
	Seventh	20	140

↓ = Use first sampling plan below arrow (refer to continuation of table on following page, when necessary). If sample size equals or exceeds lot or batch size, do 100 percent inspection.

↑ = Use first sampling plan above arrow.

Ac = Acceptance number.

Re = Rejection number.

• = Use corresponding single sampling plan (or alternatively, use multiple sampling plan below, where available).

++ = Use corresponding double sampling plan (or alternatively, use multiple sampling plan below, where available).

= Acceptance not permitted at this sample size.

TABLE T11-9 (continued)

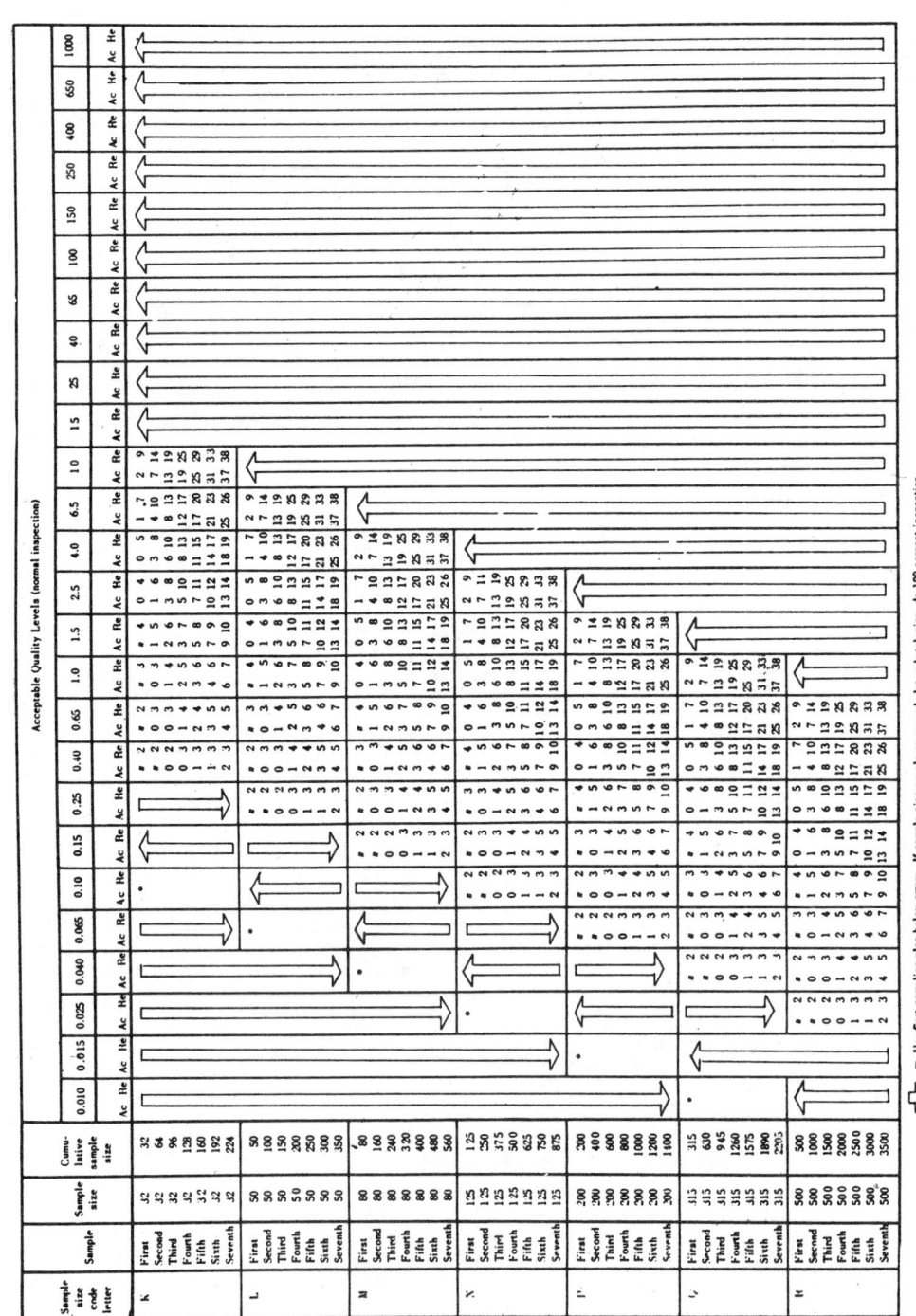

From MIL-STD-105D, pp. 16, 17.

TABLE T11-10 MIL-STD-105D Table IV-B. Multiple-Sampling Plans for Tightened Inspection (Master Table)

Acceptable Quality Levels (tightened inspection)

Notes:

$\Downarrow$ = Use first sampling plan below arrow (refer to continuation of table on following page, when necessary). If sample size equals or exceeds lot or batch size, do 100 percent inspection.

$\Uparrow$ = Use first sampling plan above arrow.

Ac = Acceptance number

Re = Rejection number

$\bullet\bullet\bullet\bullet\bullet\bullet\bullet$ = Use corresponding single sampling plan (or alternatively, use multiple sampling plan below, where available).

$\blacksquare\blacksquare\blacksquare\blacksquare\blacksquare\blacksquare\blacksquare$ = Use corresponding double sampling plan (or alternatively, use multiple sampling plan below, where available).

$\Diamond\!\Diamond$ Ac. $\ddagger$. = Acceptance not permitted at this sample size.

TABLE T11-10 (continued)

Acceptable Quality Levels (tightened inspection)

Sample size code letter	Sample	Sample size	Cumulative sample size
K	First Second Third Fourth Fifth Sixth Seventh	32 32 32 32 32 32 32	32 64 96 128 160 192 224
L	First Second Third Fourth Fifth Sixth Seventh	50 50 50 50 50 50 50	50 100 150 200 250 300 350
M	First Second Third Fourth Fifth Sixth Seventh	80 80 80 80 80 80 80	80 160 240 320 400 480 560
N	First Second Third Fourth Fifth Sixth Seventh	125 125 125 125 125 125 125	125 250 375 500 625 750 875
P	First Second Third Fourth Fifth Sixth Seventh	200 200 200 200 200 200 200	200 400 600 800 1000 1200 1400
Q	First Second Third Fourth Fifth Sixth Seventh	315 315 315 315 315 315 315	315 630 945 1260 1575 1890 2205
Q	First Second Third Fourth Fifth Sixth Seventh	500 500 500 500 500 500 500	500 1000 1500 2000 2500 3000 3500
S	First Second Third Fourth Fifth Sixth Seventh	800 800 800 800 800 800 800	800 1600 2400 3200 4000 4800 5600

⇩⇧ = Use first sampling plan below arrow. If sample size equals or exceeds lot or batch size, do 100 percent inspection.
= Use first sampling plan above arrow (refer to preceding page, when necessary).
Ac = Acceptance number
Re = Rejection number
• = Use corresponding single-sampling plan (or alternatively, use multiple sampling plan below, where available).
■ = Acceptance not permitted at this sample size.

From MIL-STD-105D, pp. 18, 19.

Multiple Sampling Plans for Reduced Inspection (Master Table)

Acceptable Quality Levels (reduced inspection) †

Sample size code letter	Sample	Sample size	Cumulative sample size
A			
B			
C			
D			
E			
F	First	2	2
	Second	2	4
	Third	2	6
	Fourth	2	8
	Fifth	2	10
	Sixth	2	12
	Seventh	2	14
G	First	3	3
	Second	3	6
	Third	3	9
	Fourth	3	12
	Fifth	3	15
	Sixth	3	18
	Seventh	3	21
H	First*	5	5
	Second	5	10
	Third	5	15
	Fourth	5	20
	Fifth	5	25
	Sixth	5	30
	Seventh	5	35
J	First	8	8
	Second	8	16
	Third	8	24
	Fourth	8	32
	Fifth	8	40
	Sixth	8	48
	Seventh	8	56
K	First	13	13
	Second	13	26
	Third	13	39
	Fourth	13	52
	Fifth	13	65
	Sixth	13	78
	Seventh	13	91

AQL column headings (left to right): 0.010, 0.015, 0.025, 0.040, 0.065, 0.10, 0.15, 0.25, 0.40, 0.65, 1.0, 1.5, 2.5, 4.0, 6.5, 10, 15, 25, 40, 65, 100, 150, 250, 400, 650, 1000 (each with Ac and Re sub-columns).

⇩⇧ = Use first sampling plan below arrow (refer to continuation of table on following page, when necessary). If sample size equals, or exceeds lot or batch size, do 100 percent inspection.

= Use first sampling plan above arrow.

Ac = Acceptance number

Re = Rejection number

• = Use corresponding single sampling plan (or alternatively, use multiple sampling plan below, where available).

‡ = Use corresponding double sampling plan (or alternatively, use multiple sampling plan below, where available).

* = Acceptance not permitted at this sample size.

† = If, after the final sample, the acceptance number has been exceeded, but the rejection number has not been reached, accept the lot but reinstate normal inspection (see 10.1.4).

TABLE T11-11 (continued)

Acceptable Quality Levels (reduced inspection)†

Legend:

◁▷ = Use first sampling plan below arrow. If sample size equals, or exceeds, lot or batch size, do 100 percent inspection.

◁▷ = Use first sampling plan above arrow (refer to preceding page when necessary).

Ac = Acceptance number

Re = Rejection number

* = Acceptance not permitted at this sample size.

† = If, after the final sample, the acceptance number has been exceeded, but the rejection number has not been reached, accept the lot, but reinstate normal inspection (see 10.1.4).

From MIL-STD-105D, pp. 20, 21.

TABLE T11-12 MIL-STD-105D Table V-A. Average Outgoing Quality Limit Factors for Normal Inspection (Single Sampling)

Code Letter	Sample Size	0.010	0.015	0.025	0.040	0.065	0.10	0.15	0.25	0.40	0.65	1.0	1.5	2.5	4.0	6.5	10	15	25	40	65	100	150	250	400	650	1000
																	Acceptable Quality Level										
A	2																17	28	42	69	97	160	220	330	470	730	1100
B	3															18	17	27	46	65	110	150	220	310	490	720	1100
C	5														12	11	15	24	39	63	90	130	190	290	430	660	
D	8													7.4	6.5	11	16	24	40	56	82	120	180	270	410		
E	13												4.6	4.2	6.9	9.7	14	22	34	50	72	110	170	250			
F	20											2.8	2.6	4.3	6.1	9.9	13	21	33	47	73						
G	32										1.8	1.7	2.7	3.9	6.3	9.0	12	19	29	46							
H	50									1.2	1.1	1.7	2.4	4.0	5.6	8.2	12	18	29								
J	80								0.74	0.67	1.1	1.6	2.5	3.6	5.2	7.5											
K	125							0.46	0.42	0.69	0.97	1.6	2.2	3.3	4.7	7.3											
L	200						0.29	0.27	0.44	0.62	1.00	1.4	2.1	3.0	4.7												
M	315					0.18	0.17	0.27	0.39	0.63	0.90	1.3	1.9	2.9													
N	500				0.12	0.11	0.17	0.24	0.40	0.56	0.82	1.2	1.8														
P	800			0.074	0.067	0.11	0.16	0.25	0.36	0.52	0.75	1.2															
Q	1250		0.046	0.042	0.069	0.097	0.16	0.22	0.33	0.47	0.73																
R	2000	0.029																									

Note: For the exact AOQL, the above values must be multiplied by $\left(1 - \dfrac{\text{Sample size}}{\text{Lot or Batch size}}\right)$

From MIL-STD-105D, p. 22.

TABLE T11-13 MIL-STD-105D Table V-B. Average Outgoing Quality Limit Factors for Tightened Inspection (Single Sampling)

Acceptable Quality Level

Code letter	Sample size	0.010	0.015	0.025	0.040	0.065	0.10	0.15	0.25	0.40	0.65	1.0	1.5	2.5	4.0	6.5	10	15	25	40	65	100	150	250	400	650	1000
A	2																			42	69	97	160	260	400	620	970
B	3																		28	46	65	110	170	270	410	650	1100
C	5																	17	27	39	63	100	160	250	390	610	
D	8															12	11	17	24	40	64	99	160	240	380		
E	13														7.4	6.5	11	15	24	40	61	95	150	240			
F	20													4.6	4.2	6.9	9.7	16	26	40	62						
G	32												2.8	2.6	4.3	6.1	9.9	16	25	39							
H	50											1.8	1.7	2.7	3.9	6.3	10	16	25								
J	80										1.2	1.1	1.7	2.4	4.0	6.4	9.9	16									
K	125									0.74	0.67	1.1	1.6	2.5	4.1	6.4	9.9										
L	200								0.46	0.42	0.69	0.97	1.6	2.6	4.0	6.2											
M	315							0.29	0.27	0.44	0.62	1.0	1.6	2.5	3.9												
N	500						0.18	0.17	0.27	0.39	0.63	1.0	1.6	2.5													
P	800					0.12	0.11	0.17	0.24	0.40	0.64	0.99	1.6														
Q	1250				0.074	0.067	0.11	0.16	0.25	0.41	0.64	0.99															
R	2000			0.046	0.042	0.069	0.097	0.16	0.26	0.40	0.62																
S	3150	0.018	0.029	0.027																							

Note: For the exact AOQL, the above values must be multiplied by $\left(1 - \dfrac{\text{Sample size}}{\text{Lot or Batch size}}\right)$

From MIL-STD-105D, p. 23.

TABLE T11-14 MIL-STD-105D Table VI-A. Limiting Quality (in percent defective) for which P_a = 10 percent (for Normal Inspection, Single Sampling)

Code letter	Sample size	Acceptable Quality Level															
		0.010	0.015	0.025	0.040	0.065	0.10	0.15	0.25	0.40	0.65	1.0	1.5	2.5	4.0	6.5	10
A	2															68	
B	3														54		
C	5													37			58
D	8												25			41	54
E	13											16			27	36	44
F	20										11			18	25	30	42
G	32									6.9			12	16	20	27	34
H	50								4.5			7.6	10	13	18	22	29
J	80							2.8			4.8	6.5	8.2	11	14	19	24
K	125						1.8			3.1	4.3	5.4	7.4	9.4	12	16	23
L	200					1.2			2.0	2.7	3.3	4.6	5.9	7.7	10	14	
M	315				0.73			1.2	1.7	2.1	2.9	3.7	4.9	6.4	9.0		
N	500			0.46			0.78	1.1	1.3	1.9	2.4	3.1	4.0	5.6			
P	800		0.29			0.49	0.67	0.84	1.2	1.5	1.9	2.5	3.5				
Q	1250	0.18			0.31	0.43	0.53	0.74	0.94	1.2	1.6	2.3					
R	2000			0.20	0.27	0.33	0.46	0.59	0.77	1.0	1.4						

From MIL-STD-105D, p. 24.

TABLE T11-15 MIL-STD-105D Table VI-B. Limiting Quality (in defects per hundred units) for Which P_a = 10 percent (for Normal Inspection, Single Sampling)

Acceptable Quality Level

Code letter	Sample size	0.010	0.015	0.025	0.040	0.065	0.10	0.15	0.25	0.40	0.65	1.0	1.5	2.5	4.0	6.5	10	15	25	40	65	100	150	250	400	650	1000
A	2															120			200	270	330	460	590	770	1000	1400	1900
B	3														77			130	180	220	310	390	510	670	940	1300	1800
C	5													46			78	110	130	190	240	310	400	560	770	1100	
D	8												29			49	67	84	120	150	190	250	350	480	670		
E	13											18			30	41	51	71	91	120	160	220	300	410			
F	20										12			20	27	33	46	59	77	100	140						
G	32									7.2			12	17	21	29	37	48	63	88							
H	50								4.6			7.8	11	13	19	24	31	40	56								
J	80							2.9			4.9	6.7	8.4	12	15	19	25	35									
K	125						1.8			3.1	4.3	5.4	7.4	9.4	12	16	23										
L	200					1.2			2.0	2.7	3.3	4.6	5.9	7.7	10	14											
M	315				0.73			1.2	1.7	2.1	2.9	3.7	4.9	6.4	9.0												
N	500			0.46			0.78	1.1	1.3	1.9	2.4	3.1	4.0	5.6													
P	800		0.29			0.49	0.67	0.84	1.2	1.5	1.9	2.5	3.5														
Q	1250	0.18			0.31	0.43	0.53	0.74	0.94	1.2	1.6	2.3															
R	2000			0.20	0.27	0.33	0.46	0.59	0.77	1.0	1.4																

From MIL-STD-105D, p. 25.

TABLE T11-16 MIL-STD-105D Table VII-A. Limiting Quality (in percent defective) for Which P_a = 5 percent (for Normal Inspection, Single sampling)

Code letter	Sample size	Acceptable Quality Level															
		0.010	0.015	0.025	0.040	0.065	0.10	0.15	0.25	0.40	0.65	1.0	1.5	2.5	4.0	6.5	10
A	2																
B	3															78	
C	5											21	31	45	63		66
D	8										14					47	60
E	13														32	41	50
F	20													22	28	34	46
G	32						2.4		5.8	8.9			14	18	23	30	37
H	50					1.5		3.7				9.1	12	15	20	25	32
J	80										5.8	7.7	9.4	13	16	20	26
K	125		0.38							3.8	5.0	6.2	8.4	11	14	18	24
L	200			0.60					2.4	3.2	3.9	5.3	6.6	8.5	11	15	
M	315				0.95			1.5	2.0	2.5	3.3	4.2	5.4	7.0	9.6		
N	500						0.95	1.3	1.6	2.1	2.6	3.4	4.4	6.1			
P	800					0.59	0.79	0.97	1.3	1.6	2.1	2.7	3.8				
Q	1250	0.24			0.38	0.50	0.62	0.84	1.1	1.4	1.8	2.4					
R	2000			0.24	0.32	0.39	0.53	0.66	0.85	1.1	1.5						

From MIL-STD-105D, p. 26.

TABLE T11-17 MIL-STD-105D Table VII-B. Limiting Quality (in defects per hundred units) for which P_a = 5 percent (for Normal Inspection, Single Sampling)

Acceptable Quality Level

Code letter	Sample size	0.010	0.015	0.025	0.040	0.065	0.10	0.15	0.25	0.40	0.65	1.0	1.5	2.5	4.0	6.5	10	15	25	40	65	100	150	250	400	650	1000
A	2															150			240	320	390	530	660	850	1100	1500	2000
B	3														100			160	210	260	350	440	570	730	1000	1400	1900
C	5													60			95	130	160	210	260	340	440	610	810	1100	
D	8												38			59	79	97	130	160	210	270	380	510	710		
E	13											23			37	48	60	81	100	130	170	230	310	440			
F	20										15			24	32	39	53	66	85	110	150						
G	32									9.4			15	20	24	33	41	53	68	95							
H	50								6.0			9.5	13	16	21	26	34	44	61								
J	80							3.8			5.9	7.9	9.7	13	16	21	27	38									
K	125						2.4			3.8	5.0	6.2	8.4	11	14	18	24										
L	200					1.5			2.4	3.2	3.9	5.3	6.6	8.5	11	15											
M	315				0.95			1.5	2.0	2.5	3.3	4.2	5.4	7.0	9.6												
N	500			0.60			0.95	1.3	1.6	2.1	2.6	3.4	4.4	6.1													
P	800		0.38			0.59	0.79	0.97	1.3	1.6	2.1	2.7	3.8														
Q	1250	0.24			0.38	0.50	0.62	0.84	1.1	1.4	1.8	2.4															
R	2000			0.24	0.32	0.39	0.53	0.66	0.85	1.1	1.5																

TABLE T11-18 MIL-STD-105D Table IX. Average Sample Size Curves for Double and Multiple Sampling (Normal and Tightened Inspection)

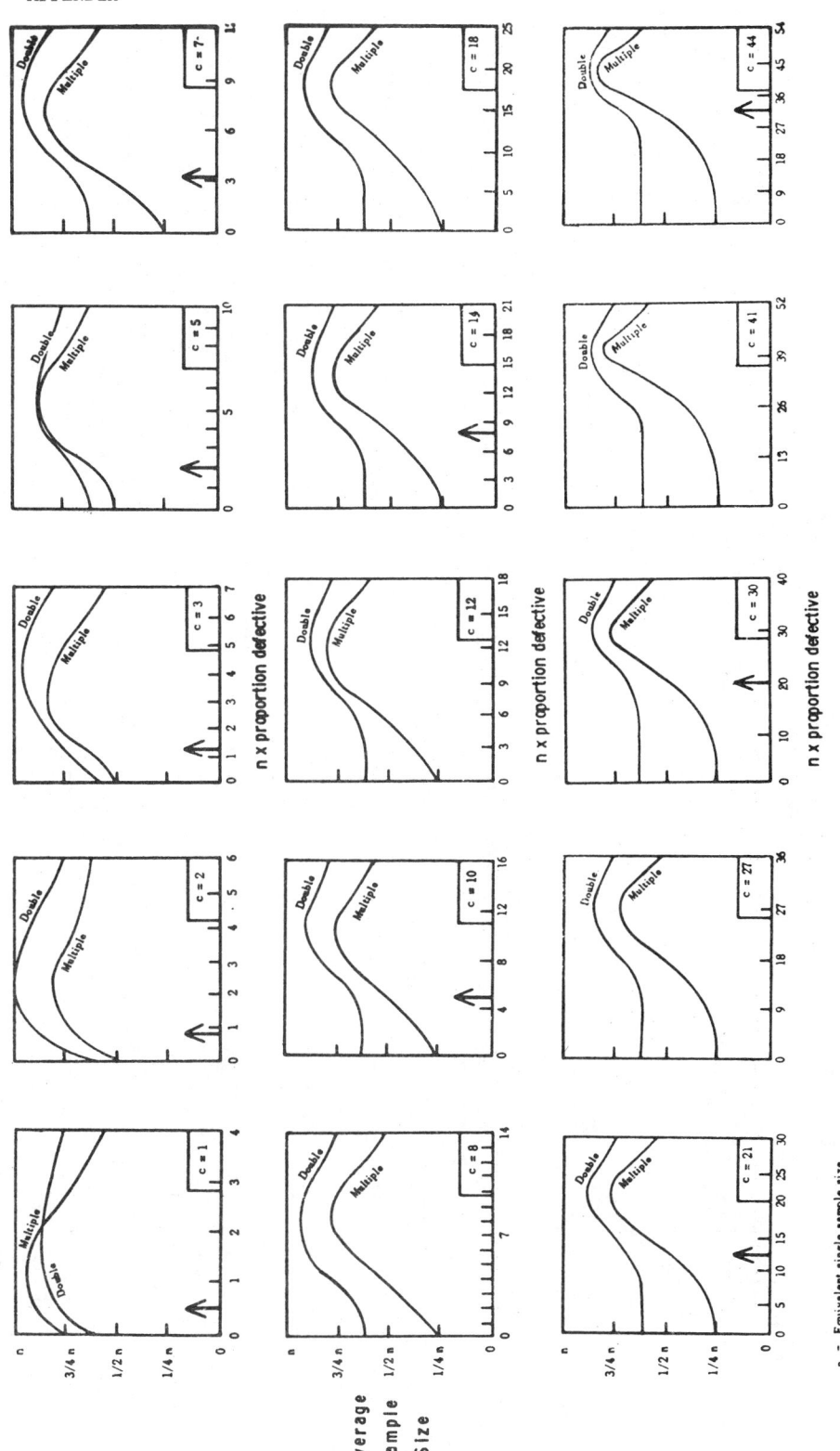

From MIL-STD-105D, p. 29.

TABLE T11-19 MIL-STD-105D Table X-F. Tables for Sample Size Code Letter: F

CHART F - OPERATING CHARACTERISTIC CURVES FOR SINGLE SAMPLING PLANS

(Curves for double and multiple sampling are matched as closely as practicable)

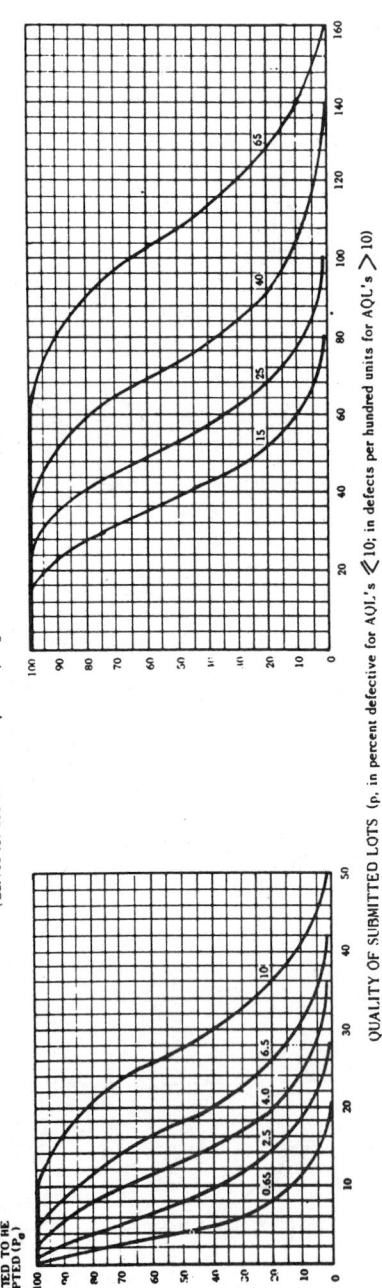

PERCENT OF LOTS
EXPECTED TO BE
ACCEPTED (P_a)

QUALITY OF SUBMITTED LOTS (p, in percent defective for AQL's $\ll$ 10; in defects per hundred units for AQL's $>$ 10)

Note: Figures on curves are Acceptable Quality Levels (AQL's) for normal inspection.

TABLE X-F-1 - TABULATED VALUES FOR OPERATING CHARACTERISTIC CURVES FOR SINGLE SAMPLING PLANS

P_a	Acceptable Quality Levels (normal inspection)												
	0.65	2.5	4.0	6.5	10	0.65	2.5	4.0	6.5	10	15	25	40	65
	p (in percent defective)					p (in defects per hundred units)								
99.0	0.050	0.75	2.25	4.31	9.75	0.051	0.75	2.18	4.12	8.92	14.5	17.5	23.9	30.5	37.4	51.7	62.9
95.0	0.256	1.80	4.22	7.13	14.0	0.257	1.78	4.09	6.83	13.1	19.9	23.5	30.8	38.5	46.2	62.2	74.5
90.0	0.525	2.69	5.64	9.03	16.6	0.527	2.66	5.51	8.73	15.8	23.3	27.2	35.1	43.2	51.5	68.4	81.2
75.0	1.43	4.81	8.70	12.8	21.6	1.44	4.81	8.68	12.7	21.1	29.8	34.2	43.1	52.1	61.2	79.5	93.4
50.0	3.41	8.25	13.1	18.1	27.9	3.47	8.39	13.4	18.4	28.4	38.3	43.3	53.3	63.3	73.3	93.3	108
25.0	6.70	12.9	18.7	24.2	34.8	6.93	13.5	19.6	25.5	37.1	48.4	54.0	65.1	76.1	87.0	109	125
10.0	10.9	18.1	24.5	30.4	41.5	11.5	19.5	26.6	33.4	46.4	58.9	65.0	77.0	88.9	101	124	141
5.0	13.9	21.6	28.3	34.4	45.6	15.0	23.7	31.5	38.8	52.6	65.7	72.2	84.8	97.2	109	133	151
1.0	20.6	28.9	35.6	42.0	53.4	23.0	33.2	42.0	50.2	65.5	80.0	87.0	101	114	127	153	172
	1.0	4.0	6.5	10	✕	1.0	4.0	6.5	10	15	✕	25	40	✕	65
	Acceptable Quality Levels (tightened inspection)												

Note: Binomial distribution used for percent defective computations; Poisson for defects per hundred units.

TABLE T11-19 (continued)

Acceptable Quality Levels (normal inspection) — Ac = Acceptance number, Re = Rejection number (each AQL cell shows "Ac Re")

Type of sampling plan	Cumulative sample size	Less than 0.65	0.65	1.0	1.5	2.5	4.0	6.5	10	15	25	40	65	Higher than 65
Single	20	▽	Use Letter	Use Letter E	0 1	1 2	2 3	3 4	5 6	7 8	10 11	14 15	21 22	△
Double	13	▽	•	Use Letter H	Use Letter H	0 2	0 3	1 4	2 5	3 7	5 9	7 11	11 16	△
	26					1 2	3 4	4 5	6 7	8 9	12 13	18 19	26 27	
Multiple	5	▽	•	Use Letter G	Use Letter G	∗ 2	∗ 2	∗ 3	∗ 4	0 4	0 5	1 6	2 9	△
	10					∗ 2	0 3	0 3	1 5	1 6	3 8	4 10	7 14	
	15					0 2	0 3	1 4	2 6	3 8	6 10	8 13	13 19	
	20					0 3	1 4	2 5	3 7	5 10	8 13	12 17	19 25	
	25					1 3	2 4	3 6	5 8	7 11	11 15	17 20	25 29	
	30					1 3	3 5	4 6	7 9	10 12	14 17	21 23	31 33	
	35					2 3	4 5	6 7	9 10	13 14	18 19	25 26	37 38	
Acceptable Quality Levels (tightened inspection)		Less than 1.0	✕	1.0	1.5	2.5	4.0	6.5	10	15	25	40	✕	Higher than 65

△ = Use next preceding sample size code letter for which acceptance and rejection numbers are available.

▽ = Use next subsequent sample size code letter for which acceptance and rejection numbers are available.

Ac = Acceptance number

Re = Rejection number

• = Use single sampling plan above (or alternatively use letter J).

∗ = Acceptance not permitted at this sample size.

From MIL-STD-105D, pp. 40, 41.

TABLE T11-20   MIL-STD-105D Scheme Average Outgoing Quality Limit Factors (in defects per hundred units).   (Also applicable to percent defective for AQL less than 15 with specific values for percent defective shown in parenthesis.)

Acceptable Quality Level

Code Letter	0.010	0.015	0.025	0.040	0.065	0.10	0.15	0.25	0.40	0.65	1.0	1.5	2.5	4.0	6.5	10	15	25	40	65	100	150	250	400	650	1000
A															(11)13			30	48	78	130	200	310	450	710	1100
B														(6.8)7.5			19	32	52	84	130	210	300	480	710	1100
C													(4.4)4.7			(12)12	20	31	51	78	130	180	290	430	660	
D												(2.8)2.9			(7.0)7.0	(12)12	20	32	49	76	120	180	270	410		
E											(1.9)1.9			(4.5)4.5	(7.5)7.4	(13)12	20	30	47	69	110	170	260			
F										(1.2)1.2			(2.9)2.9	(4.9)4.8	(7.9)7.8	(14)13	20	31	45	71						
G									(.74).75			(1.8)1.8	(3.0)3.0	(4.9)4.9	(8.1)7.9	(13)13	19	28	45							
H								(.47).47				(2.0)2.0	(3.2)3.1	(5.1)5.1	(8.0)7.8	(13)13	18	29								
J							(.30).30			(.72).72	(1.2)1.2	(2.0)2.0	(3.2)3.2	(5.0)4.9	(7.7)7.6	(12)12	18									
K						.19			.46	.77	1.3	2.1	3.2	4.9	7.2	12										
L					.12			.29	.48	.78	1.3	2.0	3.1	4.5	7.1											
M				.075			.18	.31	.50	.80	1.3	2.0	2.9	4.5												
N			.047			.12	.20	.31	.51	.78	1.3	1.8	2.9													
P		.030			.072	.12	.20	.32	.49	.76	1.2	1.8														
Q	.019			.046	.077	.13	.21	.32	.49	.72	1.2															
R			.029	.048	.078	.13	.20	.31	.45	.71																

NOTE: For a better approximation to the AOQL, the values must be multiplied by (1 – normal plan sample size/lot or batch size).

From E. G. Schilling and J. H. Sheesley, Journal of Quality Technology, 10(3): 1978, p. 106.   Reprinted by permission.

TABLE T11-21 MIL-STD-105D Scheme Limiting Quality (in defects per hundred units) for Which P_a = 10 Percent. (Also applicable to percent defective for AQL less than 15 with specific values for percent defective shown in parenthesis.)

Code Letter	Acceptable Quality Level																									
	0.010	0.015	0.025	0.040	0.065	0.10	0.15	0.25	0.40	0.65	1.0	1.5	2.5	4.0	6.5	10	15	25	40	65	100	150	250	400	650	1000
A															(53.6) 76.7		77.8	130	194	266	334	464	650	889	1240	1750
B														36.9 46.0			77.8	130	177	223	309	433	593	825	1170	1680
C													(25.0) 28.8			(40.6) 48.6	66.5	106	134	185	260	356	495	699	1010	
D												(16.2) 17.7			(26.8) 29.9	(40.6) 48.6	51.4	83.5	116	162	222	309	437	631		
E											(10.9) 11.5		(11.6) 12.2	(18.1) 19.4	(26.8) 29.9	(36.0) 40.9	46.4	71.3	100	137	190	269	388			
F										(6.94) 7.19		(7.56) 7.78	(11.6) 12.2	(18.1) 19.4	(24.5) 26.6	(30.4) 33.4	40.6	65.0	88.9	124						
G									(4.50) 4.60			(7.56) 7.78		(15.8) 16.6	(19.7) 20.9	(27.1) 29.0	35.6	55.6	77.4							
H								(2.84) 2.88			(4.77) 4.86		(10.3) 10.6	(12.9) 13.4	(17.8) 18.5	(24.7) 26.0	30.9	49.5								
J							(1.83) 1.84			(3.08) 3.11	(4.77) 4.86	(6.52) 6.65	(8.16) 8.35	(11.3) 11.6	(15.7) 16.2	(21.4) 22.2										
K						1.15			1.94	3.11	4.26	5.34	7.42	10.4	14.2	19.8										
L					.731			1.23	1.94	2.66	3.34	4.64	6.50	8.89	12.4											
M				.460				1.23	1.69	2.12	2.94	4.13	5.64	7.86												
N						.486	.778	1.06	1.34	1.85	2.60	3.56	4.95													
P			.288		.311	.486	.665	.835	1.16	1.62	2.22	3.09														
Q		.184		.194	.311	.426	.534	.742	1.04	1.42	1.98															
R	.115		.123	.194	.266	.334	.464	.650	.889	1.24																

From E. G. Schilling and J. H. Sheesley, *Journal of Quality Technology*, 10(3): 1978, p. 107. Reprinted by permission.

TABLE T11-22 MIL-STD-105D Scheme Limiting Quality (in defects per hundred units) for Which P_a = 5 Percent. (Also applicable to percent defective for AQL less than 15 with specific values for percent defective shown in parenthesis.)

Acceptable Quality Level

Code Letter	0.010	0.015	0.025	0.040	0.065	0.10	0.15	0.25	0.40	0.65	1.0	1.5	2.5	4.0	6.5	10	15	25	40	65	100	150	250	400	650	1000
A															(63.2) 99.8			158	237	315	388	526	722	972	1340	1860
B														(45.1) 59.9			94.9	158	210	258	350	481	648	890	1240	1770
C													(31.2) 37.4			(47.1) 59.3	94.9	126	155	210	289	389	534	745	1060	
D												(20.6) 23.0			(31.6) 36.5	(47.1) 59.3	78.7	96.9	131	180	243	334	465	665		
E											(13.9) 15.0			(21.6) 23.7	(31.6) 36.5	(41.0) 48.4	59.6	80.9	111	150	205	286	409			
F										(8.94) 9.36			(14.0) 14.8	(21.6) 23.7	(22.5) 24.2	(34.4) 38.8	52.6	72.2	97.2	133						
G									(5.81) 5.99			(9.14) 9.49	(14.0) 14.8	(19.9) 21.0	(22.5) 24.2	(30.1) 32.9	45.1	60.8	83.4							
H								(3.68) 3.74			(5.79) 5.93	(9.14) 9.49	(12.1) 12.6	(19.9) 21.0	(27.0) 28.9	(28.3) 31.5	38.9	53.4								
J							(2.37) 2.40			(3.74) 3.79	(5.79) 5.93	(7.66) 7.87	(9.41) 9.69	(12.7) 13.1	(17.3) 18.0	(23.2) 24.3	33.4									
K						1.50			2.37	3.79	5.04	6.20	8.41	11.5	15.6	21.4										
L					.951			1.51	2.37	3.15	3.88	5.26	7.22	9.72	13.3											
M				.599			.949	1.51	2.00	2.46	3.34	4.58	6.17	8.47												
N			.374			.593	.949	1.26	1.55	2.10	2.89	3.89	5.34													
P		.240			.379	.593	.787	.969	1.31	1.80	2.43	3.34														
Q	.150			.237	.379	.504	.620	.841	1.15	1.56	2.14															
R		.151	.237	.315	.388	.526	.722	.972	1.33																	

From E. G. Schilling and J. H. Sheesley, *Journal of Quality Technology*, 10(3): 1978, p. 108. Reprinted by permission.

TABLE A4-F-1. Tabulated Values for Operating Characteristic Curves for Scheme

P_a	Acceptable Quality Levels (normal inspection)														
	.65	2.5	4.0	6.5	10	.65	2.5	4.0	6.5	10	15	25	40	65
	p (in percent defective)					p (in defects per hundred units)								
99.0	0.104	.978	2.94	4.93	10.1	0.104	.958	2.84	4.72	9.41	15.0	25.0	39.5	64.9
95.0	0.357	1.85	4.11	6.94	13.0	0.358	1.82	4.02	6.69	12.3	19.2	30.2	45.7	73.4
90.0	0.571	2.47	4.91	8.24	14.4	0.572	2.45	4.82	8.00	13.8	21.4	33.3	49.7	78.3
75.0	1.11	3.66	6.40	10.4	16.5	1.11	3.66	6.37	10.3	16.2	24.8	38.0	56.0	85.5
50.0	2.22	5.40	8.71	13.6	19.2	2.24	5.46	8.85	13.8	19.5	29.4	44.3	64.2	94.8
25.0	4.24	8.21	12.9	18.7	24.3	4.34	8.43	13.5	19.6	25.6	37.2	54.0	76.1	109
10.0	6.94	11.6	18.1	24.5	30.4	7.19	12.2	19.4	26.6	33.4	46.4	65.0	88.9	124
5.0	8.94	14.0	21.6	28.3	34.4	9.36	14.8	23.7	31.5	38.8	52.6	72.2	97.2	133
1.0	13.4	19.0	28.9	35.8	42.1	14.4	20.7	33.2	42.0	50.2	65.5	87.1	114	153

TABLE A4-F-2. Tabulated Values for Average Sample Number Curves for Scheme

P_a	Acceptable Quality Levels (normal inspection)														
	.65	2.5	4.0	6.5	10	.65	2.5	4.0	6.5	10	15	25	40	65
	p (in percent defective)					p (in defects per hundred units)								
99.0	9.5	14.6	13.4	15.7	17.9	9.5	14.5	13.2	15.3	16.8	17.8	16.2	15.1	15.7
95.0	14.4	19.1	18.5	19.5	19.9	14.4	19.0	18.3	19.3	19.8	20.0	19.9	19.8	19.9
90.0	18.6	21.5	19.7	19.9	20.0	18.6	21.5	19.6	19.9	20.0	20.0	20.0	20.0	20.0
75.0	26.1	26.2	20.0	20.0	20.0	26.0	26.2	20.0	20.0	20.0	20.0	20.0	20.0	20.0
50.0	31.0	30.9	20.0	20.0	20.0	31.0	30.9	20.0	20.0	20.0	20.0	20.0	20.0	20.0
25.0	32.0	32.0	20.0	20.0	20.0	32.0	32.0	20.0	20.0	20.0	20.0	20.0	20.0	20.0
10.0	32.0	32.0	20.0	20.0	20.0	32.0	32.0	20.0	20.0	20.0	20.0	20.0	20.0	20.0
5.0	32.0	32.0	20.0	20.0	20.0	32.0	32.0	20.0	20.0	20.0	20.0	20.0	20.0	20.0
1.0	32.0	32.0	20.0	20.0	20.0	32.0	32.0	20.0	20.0	20.0	20.0	20.0	20.0	20.0

TABLE A4-F-3. Tabulated Values for Average Outgoing Quality Curves for Scheme
(Lot Size = 120)

P_a	Acceptable Quality Levels (normal inspection)														
	.65	2.5	4.0	6.5	10	.65	2.5	4.0	6.5	10	15	25	40	65
	p (in percent defective)					p (in defects per hundred units)								
99.0	0.088	0.85	2.6	4.2	8.5	0.088	0.83	2.5	4.1	8.0	13	21	34	56
95.0	0.30	1.5	3.3	5.5	10	0.30	1.5	3.2	5.3	9.8	15	24	36	58
90.0	0.44	1.8	3.7	6.2	11	0.44	1.8	3.6	6.0	10	16	25	37	59
75.0	0.65	2.2	4.0	6.5	10	0.66	2.2	4.0	6.5	10	15	24	35	53
50.0	0.83	2.0	3.6	5.7	8.0	0.83	2.0	3.7	5.8	8.1	12	18	27	39
25.0	0.78	1.5	2.7	3.9	5.1	0.80	1.5	2.8	4.1	5.3	7.7	11	16	23
10.0	0.51	0.85	1.5	2.0	2.5	0.53	0.89	1.6	2.2	2.8	3.9	5.4	7.4	10
5.0	0.33	0.51	0.90	1.2	1.4	0.34	0.54	0.99	1.3	1.6	2.2	3.0	4.1	5.6
1.0	0.099	0.14	0.24	0.30	0.35	0.11	0.15	0.28	0.35	0.42	0.55	0.72	0.94	1.3
AOQL	0.85	2.2	4.1	6.6	11	0.86	2.2	4.0	6.5	11	17	26	38	59

TABLE A4-F-4. Tabulated Values for Average Total Inspection Curves for Scheme
(Lot Size = 120)

P_a	Acceptable Quality Levels (normal inspection)														
	.65	2.5	4.0	6.5	10	.65	2.5	4.0	6.5	10	15	25	40	65
	p (in percent defective)					p (in defects per hundred units)								
99.0	10.5	15.6	14.4	16.7	18.9	10.5	15.5	14.2	16.3	17.8	18.8	17.2	16.1	16.8
95.0	19.5	24.1	23.5	24.5	25.0	19.5	24.0	23.3	24.3	24.8	25.0	24.9	24.8	24.9
90.0	28.5	31.2	29.7	29.9	30.0	28.5	31.1	29.6	29.9	30.0	30.0	30.0	30.0	30.0
75.0	49.2	49.2	45.0	45.0	45.0	49.2	49.2	45.0	45.0	45.0	45.0	45.0	45.0	45.0
50.0	75.4	75.2	70.0	70.0	70.0	75.4	75.2	70.0	70.0	70.0	70.0	70.0	70.0	70.0
25.0	98.0	98.0	95.0	95.0	95.0	98.0	98.0	95.0	95.0	95.0	95.0	95.0	95.0	95.0
10.0	111	111	110	110	110	111	111	110	110	110	110	110	110	110
5.0	116	116	115	115	115	116	116	115	115	115	115	115	115	115
1.0	119	119	119	119	119	119	119	119	119	119	119	119	119	119

From E. G. Schilling and J. H. Sheesley, *Journal of Quality Technology*, 10(3): 1978, p. 114. Reprinted by permission.

TABLE T11-24 Operating Ratios for the MIL-STD-105D Scheme ($R = p_{.10}/p_{.95}$, calculated using Poisson distribution)

Code Letter	0.010	0.015	0.025	0.040	0.065	0.10	0.15	0.25	0.40	0.65	1.0	1.5	2.5	4.0	6.5	10	15	25	40	65	100	150	250	400	650	1000
A															30.32			7.43	5.01	4.02	2.72	2.42	2.15	1.95	1.69	1.58
B														23.23			6.71	5.04	3.99	2.72	2.41	2.15	1.95	1.69	1.58	1.46
C													20.14			6.67	4.96	3.98	2.72	2.41	2.15	1.95	1.68	1.57	1.47	
D												19.34			6.54	4.81	4.01	2.71	2.42	2.14	1.95	1.68	1.57	1.46		
E											20.54			6.88	4.83	4.01	2.72	2.41	2.15	1.94	1.68	1.57	1.46			
F										20.08			6.70	4.83	3.98	2.72	2.42	2.15	1.95	1.69	1.57	1.46				
G									20.63			7.01	4.86	3.97	2.72	2.42	2.15	1.94	1.69							
H								20.14			6.67	4.96	3.98	2.72	2.41	2.15	1.95	1.68								
J							20.51			6.81	4.86	4.01	2.71	2.42	2.14	1.95	1.68									
K						20.07			6.64	4.84	4.02	2.71	2.42	2.15	1.94	1.69										
L					20.42			6.72	4.83	3.98	2.72	2.42	2.15	1.95	1.69											
M				20.18			6.88	4.82	3.98	2.71	2.41	2.15	1.94	1.69												
N			20.14			6.67	4.96	3.98	2.72	2.41	2.15	1.95	1.68													
P		20.42			6.81	4.81	4.01	2.71	2.42	2.14	1.95	1.68														
Q	20.07			6.64	4.84	4.02	2.71	2.42	2.15	1.94	1.69															
R			6.72	4.83	3.98	2.72	2.42	2.15	1.95	1.69																

From E. G. Schilling and L. I. Johnson, *Journal of Quality Technology*, 12(4): 1980, p. 226. Reprinted by permission.

TABLE T12-1 MIL-STD-414 Table B-6. Values of T for Tightened Inspection: Standard Deviation Method

Sample size code letter	Acceptable Quality Levels (in percent defective)														Number of Lots
	.04	.065	.10	.15	.25	.40	.65	1.0	1.5	2.5	4.0	6.5	10.0	15.0	
B	*	*	*	*	*	*	*	*	*	2 4 5	3 5 6	4 6 8	4 7 9	4 8 11	5 10 15
C	*	*	*	*	*	*	*	2 3 5	2 4 6	3 5 7	3 6 8	4 7 9	4 7 10	4 8 11	5 10 15
D	*	*	*	*	*	*	2 4 5	3 4 6	3 5 7	3 6 8	4 6 9	4 7 10	4 7 10	4 8 11	5 10 15
E	*	*	*	*	2 4 5	3 4 6	3 5 6	3 5 7	4 6 8	4 6 9	4 7 9	4 7 10	4 8 11	4 8 11	5 10 15
F	*	*	*	3 4 6	3 5 6	3 5 7	3 6 8	4 6 8	4 6 9	4 7 9	4 7 10	4 8 11	4 8 11	4 8 11	5 10 15
G	3 4 6	3 5 6	3 5 6	3 5 7	3 6 7	4 6 8	4 6 9	4 7 9	4 7 9	4 7 10	4 7 10	4 8 11	4 8 11	4 8 11	5 10 15
H	3 5 6	3 5 7	3 5 7	3 6 8	4 6 8	4 6 9	4 7 9	4 7 9	4 7 10	4 7 10	4 8 11	4 8 11	4 8 11	4 8 11	5 10 15
I	3 5 7	3 6 7	4 6 8	4 6 8	4 6 9	4 7 9	4 7 9	4 7 10	4 7 10	4 7 10	4 8 11	4 8 11	4 8 11	4 8 11	5 10 15
J	3 6 8	4 6 8	4 6 8	4 6 9	4 7 9	4 7 9	4 7 10	4 7 10	4 7 10	4 8 11	4 8 11	4 8 11	4 8 11	4 8 11	5 10 15
K	4 6 8	4 6 8	4 6 9	4 6 9	4 7 9	4 7 9	4 7 10	4 7 10	4 8 10	4 8 11	4 8 11	4 8 11	4 8 11	4 8 11	5 10 15
L	4 6 8	4 6 9	4 6 9	4 7 9	4 7 9	4 7 10	4 7 10	4 7 10	4 8 10	4 8 11	4 8 11	4 8 11	4 8 11	4 8 11	5 10 15
M	4 6 9	4 7 9	4 7 9	4 7 9	4 7 10	4 7 10	4 7 10	4 7 10	4 8 11	4 8 11	4 8 11	4 8 11	4 8 11	4 8 11	5 10 15
N	4 7 9	4 7 9	4 7 10	4 7 10	4 7 10	4 7 10	4 8 11	4 8 11	4 8 11	4 8 11	4 8 11	4 8 11	4 8 11	4 8 11	5 10 15
O	4 7 10	4 7 10	4 7 10	4 7 10	4 7 10	4 8 11	4 8 11	4 8 11	4 8 11	4 8 11	4 8 11	4 8 11	4 8 11	4 8 11	5 10 15

*There are no sampling plans provided in this Standard for these code letters and AQL values.

TABLE T12-1   (continued)

Sample size code letter	Acceptable Quality Levels (in percent defective)														Number of Lots
	.04	.065	.10	.15	.25	.40	.65	1.0	1.5	2.5	4.0	6.5	10.0	15.0	
P	4	4	4	4	4	4	4	4	4	4	4	4	4	4	5
	7	7	7	8	8	8	8	8	8	8	8	8	8	8	10
	10	10	10	10	11	11	11	11	11	11	1î	11	11	12	15
Q	4	4	4	4	4	4	4	4	4	4	4	4	4	4	5
	7	8	8	8	8	8	8	8	8	8	8	8	8	8	10
	10	11	11	11	11	11	11	11	11	11	11	11	11	12	15

The top figure in each block refers to the preceding 5 lots, the middle figure to the preceding 10 lots and the bottom figure to the preceding 15 lots.

Tightened inspection is required when the number of lots with estimates of percent defective above the AQL from the preceding 5, 10, or 15 lots is greater than the given value of T in the table, and the process average from these lots exceeds the AQL.

All estimates of the lot percent defective are obtained from Table B-5.

From MIL-STD-414, pp. 54, 55.

TABLE T12-2   MIL-STD-414 Table B-7.   Limits of Estimated Lot Percent Defective for Reduced Inspection: Standard Deviation Method

Sample size code letter	.04	.065	.10	.15	.25	.40	.65	1.0	1.5	2.5	4.0	6.5	10.0	15.0	Number of Lots
										Acceptable Quality Levels					
B	*	*	*	*	*	*	*	*	*	[42]**	[28]**	[18]**	[12]**	[9]**	5 10 15
C	*	*	*	*	*	*	*	[45]**	[31]**	[22]**	[15]**	[10]**	[7]**	.77 15.00 ▲	5 10 15
D	*	*	*	*	*	*	[33]**	[25]**	[18]**	[13]**	[9]**	0.00 4.40 6.50	.74 9.96 10.00	6.06 15.00 ▲	5 10 15
E	*	*	*	*	[25]**	[18]**	[14]**	[11]**	.00 .10 .88	.00 .88 2.49	.13 2.65 4.00	1.38 5.96 6.50	4.24 10.00 ▲	9.09 15.00 ▲	5 10 15
F	*	*	*	▶ .000 .002	.000 .001 .029	.000 .016 .123	.000 .101 .369	.003 .317 .81	.044 .74 1.50	.306 1.80 2.50	1.05 3.56 4.00	2.81 6.50 ▲	5.79 10.00 ▲	10.47 15.00 ▲	5 10 15
G	▶ .000 .003	.000 .002 .010	.000 .006 .028	.000 .018 .062	.002 .057 .151	.011 .143 .315	.047 .330 .626	.136 .643 1.00	.323 1.14 1.50	.84 2.23 2.50	1.84 3.94 4.00	3.80 6.50 ▲	6.86 10.00 ▲	11.52 15.00 ▲	5 10 15
H	.000 .004 .013	.000 .010 .029	.002 .023 .058	.005 .048 .105	.017 .111 .215	.048 .225 .396	.123 .445 .65	.266 .785 1.00	.521 1.31 1.50	1.14 2.40 2.50	2.24 4.00 ▲	4.29 6.50 ▲	7.40 10.00 ▲	12.07 15.00 ▲	5 10 15
I	.001 .009 .021	.002 .020 .043	.006 .039 .077	.014 .071 .133	.037 .146 .248	.083 .274 .40	.185 .509 .65	.360 .863 1.00	.653 1.39 1.50	1.33 2.48 2.50	2.49 4.00 ▲	4.59 6.50 ▲	7.74 10.00 ▲	12.43 15.00 ▲	5 10 15
J	.002 .013 .027	.005 .027 .052	.012 .050 .089	.023 .087 .146	.054 .169 .25	.113 .306 .40	.233 .550 .65	.431 .909 1.00	.750 1.44 1.50	1.47 2.50	2.66 4.00 ▲	4.81 6.50 ▲	7.98 10.00 ▲	12.69 15.00 ▲	5 10 15

*There are no sampling plans provided in this Standard for these code letters and AQL values.

TABLE T12-2 (continued)

Sample size code letter	Acceptable Quality Levels														Number of Lots
	.04	.065	.10	.15	.25	.40	.65	1.0	1.5	2.5	4.0	6.5	10.0	15.0	
K	.004	.008	.017	.032	.069	.137	.270	.483	.821	1.57	2.79	4.96	8.15	12.88	5
	.017	.033	.059	.099	.186	.328	.577	.940	1.47	2.50	4.00	6.50	10.00	15.00	10
	.032	.058	.097	.15	.25	.40	.65	1.00	1.50	▲	▲	▲	▲	▲	15
L	.005	.011	.022	.040	.082	.157	.300	.525	.876	1.64	2.88	5.08	8.29	13.03	5
	.020	.038	.065	.108	.199	.343	.596	.961	1.49	2.50	4.00	6.50	10.00	15.00	10
	.035	.063	.10	.15	.25	.40	.65	1.00	1.50	▲	▲	▲	▲	▲	15
M	.008	.016	.030	.052	.102	.187	.345	.587	.959	1.76	3.03	5.27	8.50	13.25	5
	.025	.045	.075	.120	.215	.364	.621	.989	1.50	2.50	4.00	6.50	10.00	15.00	10
	.04	.065	.10	.15	.25	.40	.65	1.00	▲	▲	▲	▲	▲	▲	15
N	.014	.026	.044	.072	.134	.235	.414	.681	1.082	1.92	3.24	5.52	8.81	13.60	5
	.031	.054	.087	.136	.236	.389	.65	1.00	1.50	2.50	4.00	6.50	10.00	15.00	10
	.04	.065	.10	.15	.25	.40	▲	▲	▲	▲	▲	▲	▲	▲	15
O	.018	.032	.053	.085	.153	.261	.453	.733	1.149	2.01	3.36	5.67	8.98	13.80	5
	.034	.058	.093	.143	.245	.40	.65	1.00	1.50	2.50	4.00	6.50	10.00	15.00	10
	.04	.065	.10	.15	.25	▲	▲	▲	▲	▲	▲	▲	▲	▲	15
P	.023	.039	.064	.101	.177	.296	.501	.799	1.237	2.13	3.52	5.87	9.22	14.07	5
	.038	.064	.10	.15	.25	.40	.65	1.00	1.50	2.50	4.00	6.50	10.00	15.00	10
	.04	.065	▲	▲	▲	▲	▲	▲	▲	▲	▲	▲	▲	▲	15
Q	.025	.044	.069	.108	.188	.312	.525	.830	1.276	2.19	3.59	5.96	9.32	14.19	5
	.04	.065	.10	.15	.25	.40	.65	1.00	1.50	2.50	4.00	6.50	10.00	15.00	10
	▲	▲	▲	▲	▲	▲	▲	▲	▲	▲	▲	▲	▲	▲	15

All AQL and table values, except those in the brackets, are in percent defective.

▼▲Use the first figure in direction of arrow and corresponding number of lots. In each block the top figure refers to the preceding 5 lots, the middle figure to the preceding 10 lots, and the bottom figure to the preceding 15 lots.

Reduced inspection may be instituted when every estimated lot percent defective from the preceding 5, 10, or 15 lots is below the figure given in the table; reduced inspection for sampling plans marked (**) in the table requires that the estimated lot percent defective is equal to zero for the number of consecutive lots indicated in brackets. In addition, all other conditions for reduced inspection, in Part III of Section B, must be satisfied.

All estimates of the lot percent defective are obtained from Table B-5.

From MIL-STD-414, pp. 56, 57.

TABLE T12-3   MIL-STD-414 Table A-1.
AQL Conversion Table

For specified AQL values falling within these ranges	Use this AQL value
——— to 0.049	0.04
0.050 to 0.069	0.065
0.070 to 0.109	0.10
0.110 to 0.164	0.15
0.165 to 0.279	0.25
0.280 to 0.439	0.40
0.440 to 0.699	0.65
0.700 to 1.09	1.0
1.10 to 1.64	1.5
1.65 to 2.79	2.5
2.80 to 4.39	4.0
4.40 to 6.99	6.5
7.00 to 10.9	10.0
11.00 to 16.4	15.0

From MIL-STD-414, p. 4.

TABLE T12-4   MIL-STD-414 Table A-2.
Sample Size Code Letters[1]

Lot Size	Inspection Levels				
	I	II	III	IV	V
3 to 8	B	B	B	B	C
9 to 15	B	B	B	B	D
16 to 25	B	B	B	C	E
26 to 40	B	B	B	D	F
41 to 65	B	B	C	E	G
66 to 110	B	B	D	F	H
111 to 180	B	C	E	G	I
181 to 300	B	D	F	H	J
301 to 500	C	E	G	I	K
501 to 800	D	F	H	J	L
801 to 1,300	E	G	I	K	L
1,301 to 3,200	F	H	J	L	M
3,201 to 8,000	G	I	L	M	N
8,001 to 22,000	H	J	M	N	O
22,001 to 110,000	I	K	N	O	P
110,001 to 550,000	I	K	O	P	Q
550,001 and over	I	K	P	Q	Q

[1]Sample size code letters given in body of table are applicable when the indicated inspection levels are to be used.

From MIL-STD-414, p. 4.

TABLE T12-5 MIL-STD-414 Table B-3. Master Table for Normal and Tightened Inspection for Plans Based on Variability Unknown: Standard Deviation Method (Double Specification Limit and Form 2, Single Specification Limit)

Sample size code letter	Sample size	Acceptable Quality Levels (normal inspection)													
		.04	.065	.10	.15	.25	.40	.65	1.00	1.50	2.50	4.00	6.50	10.00	15.00
		M	M	M	M	M	M	M	M	M	M	M	M	M	M
B	3	↓	↓	↓	↓	↓	↓	↓	↓	↓	7.59	18.86	26.94	33.69	40.47
C	4	↓	↓	↓	↓	↓	↓	↓	1.53	5.50	10.92	16.45	22.86	29.45	36.90
D	5	↓	↓	↓	↓	↓	↓	1.33	3.32	5.83	9.80	14.39	20.19	26.56	33.99
E	7	↓	↓	↓	↓	0.422	1.06	2.14	3.55	5.35	8.40	12.20	17.35	23.29	30.50
F	10	↓	↓	↓	0.349	0.716	1.30	2.17	3.26	4.77	7.29	10.54	15.17	20.74	27.57
G	15	0.099	0.186	0.312	0.503	0.818	1.31	2.11	3.05	4.31	6.56	9.46	13.71	18.94	25.61
H	20	0.135	0.228	0.365	0.544	0.846	1.29	2.05	2.95	4.09	6.17	8.92	12.99	18.03	24.53
I	25	0.155	0.250	0.380	0.551	0.877	1.29	2.00	2.86	3.97	5.97	8.63	12.57	17.51	23.97
J	30	0.179	0.280	0.413	0.581	0.879	1.29	1.98	2.83	3.91	5.86	8.47	12.36	17.24	23.58
K	35	0.170	0.264	0.388	0.535	0.847	1.23	1.87	2.68	3.70	5.57	8.10	11.87	16.65	22.91
L	40	0.179	0.275	0.401	0.566	0.873	1.26	1.88	2.71	3.72	5.58	8.09	11.85	16.61	22.86
M	50	0.163	0.250	0.363	0.503	0.789	1.17	1.71	2.49	3.45	5.20	7.61	11.23	15.87	22.00
N	75	0.147	0.228	0.330	0.467	0.720	1.07	1.60	2.29	3.20	4.87	7.15	10.63	15.13	21.11
O	100	0.145	0.220	0.317	0.447	0.689	1.02	1.53	2.20	3.07	4.69	6.91	10.32	14.75	20.66
P	150	0.134	0.203	0.293	0.413	0.638	0.949	1.43	2.05	2.89	4.43	6.57	9.88	14.20	20.02
Q	200	0.135	0.204	0.294	0.414	0.637	0.945	1.42	2.04	2.87	4.40	6.53	9.81	14.12	19.92
		.065	.10	.15	.25	.40	.65	1.00	1.50	2.50	4.00	6.50	10.00	15.00	
		Acceptability Quality Levels (tightened inspection)													

All AQL and table values are in percent defective.

↓ Use first sampling plan below arrow, that is, both sample size as well as M value. When sample size equals or exceeds lot size, every item in the lot must be inspected.

From MIL-STD-414, p. 45.

TABLE T12-6 MIL-STD-414 Table B-4. Master Table for Reduced Inspection for Plans Based on Variability Unknown: Standard Deviation Method (Double Specification Limit and Form 2, Single Specification Limit)

Sample size code letter	Sample size	Acceptable Quality Levels												
		.04	.065	.10	.15	.25	.40	.65	1.00	1.50	2.50	4.00	6.50	10.00
		M	M	M	M	M	M	M	M	M	M	M	M	M
B	3	↓	↓	↓	↓	↓	↓	↓	↓	7.59	18.86	26.94	33.69	40.47
C	3	↓	↓	↓	↓	↓	↓	↓	↓	7.59	18.86	26.94	33.69	40.47
D	3	↓	↓	↓	↓	↓	↓	↓	↓	7.59	18.86	26.94	33.69	40.47
E	3	↓	↓	↓	↓	↓	↓	↓	↓	7.59	18.86	26.94	33.69	40.47
F	4	↓	↓	↓	↓	↓	↓	1.53	5.50	10.92	16.45	22.86	29.45	36.90
G	5	↓	↓	↓	↓	↓	1.33	3.32	5.83	9.80	14.39	20.19	26.56	33.99
H	7	↓	↓	↓	0.422	1.06	2.14	3.55	5.35	8.40	12.20	17.35	23.29	30.50
I	10	↓	↓	0.349	0.716	1.30	2.17	3.26	4.77	7.29	10.54	15.17	20.74	27.57
J	10	↓	↓	0.349	0.716	1.30	2.17	3.26	4.77	7.29	10.54	15.17	20.74	27.57
K	15	0.186	0.312	0.503	0.818	1.31	2.11	3.05	4.31	6.56	9.46	13.71	18.94	25.61
L	20	0.228	0.365	0.544	0.846	1.29	2.05	2.95	4.09	6.17	8.92	12.99	18.03	24.53
M	20	0.228	0.365	0.544	0.846	1.29	2.05	2.95	4.09	6.17	8.92	12.99	18.03	24.53
N	25	0.250	0.380	0.551	0.877	1.29	2.00	2.86	3.97	5.97	8.63	12.57	17.51	23.97
O	30	0.280	0.413	0.581	0.879	1.29	1.98	2.83	3.91	5.86	8.47	12.36	17.24	23.58
P	50	0.250	0.363	0.503	0.789	1.17	1.71	2.49	3.45	5.20	7.61	11.23	15.87	22.00
Q	75	0.228	0.330	0.467	0.720	1.07	1.60	2.29	3.20	4.87	7.15	10.63	15.13	21.11

All AQL and table values are in percent defective.

↓ Use first sampling plan below arrow, that is, both sample size as well as M value. When sample size equals or exceeds lot size, every item in the lot must be inspected.

From MIL-STD-414, p. 46.

TABLE T12-7 MIL-STD-414 Table B-5. Table for Estimating the Lot Percent Defective Using Standard Deviation Method[1]

Q_U or Q_L	3	4	5	7	10	15	20	25	30	35	40	50	75	100	150	200
0	50.00	50.00	50.00	50.00	50.00	50.00	50.00	50.00	50.00	50.00	50.00	50.00	50.00	50.00	50.00	50.00
.1	47.24	46.67	46.44	46.26	46.16	46.10	46.08	46.06	46.05	46.05	46.04	46.04	46.03	46.03	46.02	46.02
.2	44.46	43.33	42.90	42.54	42.35	42.24	42.19	42.16	42.15	42.13	42.13	42.11	42.10	42.09	42.08	42.08
.3	41.63	40.00	39.37	38.87	38.60	38.44	38.37	38.33	38.31	38.29	38.28	38.27	38.25	38.24	38.22	38.22
.31	41.35	39.67	39.02	38.50	38.23	38.06	37.99	37.95	37.93	37.91	37.90	37.89	37.87	37.86	37.84	37.84
.32	41.06	39.33	38.67	38.14	37.86	37.69	37.62	37.58	37.55	37.54	37.52	37.51	37.49	37.48	37.46	37.46
.33	40.77	39.00	38.32	37.78	37.49	37.31	37.24	37.20	37.18	37.16	37.15	37.13	37.11	37.10	37.09	37.08
.34	40.49	38.67	37.97	37.42	37.12	36.94	36.87	36.83	36.80	36.78	36.77	36.75	36.73	36.72	36.71	36.71
.35	40.20	38.33	37.62	37.06	36.75	36.57	36.49	36.45	36.43	36.41	36.40	36.38	36.36	36.35	36.33	36.33
.36	39.91	38.00	37.28	36.69	36.38	36.20	36.12	36.08	36.05	36.04	36.02	36.01	35.98	35.97	35.96	35.96
.37	39.62	37.67	36.93	36.33	36.02	35.83	35.75	35.71	35.68	35.66	35.65	35.63	35.61	35.60	35.59	35.58
.38	39.33	37.33	36.58	35.98	35.65	35.46	35.38	35.34	35.31	35.29	35.28	35.26	35.24	35.23	35.22	35.21
.39	39.03	37.00	36.23	35.62	35.29	35.10	35.01	34.97	34.94	34.93	34.91	34.89	34.87	34.86	34.85	34.84
.40	38.74	36.67	35.88	35.26	34.93	34.73	34.65	34.60	34.58	34.56	34.54	34.53	34.50	34.49	34.48	34.47
.41	38.45	36.33	35.54	34.90	34.57	34.37	34.28	34.24	34.21	34.19	34.18	34.16	34.13	34.12	34.11	34.10
.42	38.15	36.00	35.19	34.55	34.21	34.00	33.92	33.87	33.85	33.83	33.81	33.79	33.77	33.76	33.74	33.74
.43	37.85	35.67	34.85	34.19	33.85	33.64	33.56	33.51	33.48	33.46	33.45	33.43	33.40	33.39	33.38	33.37
.44	37.56	35.33	34.50	33.84	33.49	33.28	33.20	33.15	33.12	33.10	33.09	33.07	33.04	33.03	33.02	33.01
.45	37.26	35.00	34.16	33.49	33.13	32.92	32.84	32.79	32.76	32.74	32.73	32.71	32.68	32.67	32.66	32.65
.46	36.96	34.67	33.81	33.13	32.78	32.57	32.48	32.43	32.40	32.38	32.37	32.35	32.32	32.31	32.30	32.29
.47	36.66	34.33	33.47	32.78	32.42	32.21	32.12	32.07	32.04	32.02	32.01	31.99	31.96	31.95	31.94	31.93
.48	36.35	34.00	33.12	32.43	32.07	31.85	31.77	31.72	31.69	31.67	31.65	31.63	31.61	31.60	31.58	31.58
.49	36.05	33.67	32.78	32.08	31.72	31.50	31.41	31.36	31.33	31.31	31.30	31.28	31.25	31.24	31.23	31.22
.50	35.75	33.33	32.44	31.74	31.37	31.15	31.06	31.01	30.98	30.96	30.95	30.93	30.90	30.89	30.87	30.87
.51	35.44	33.00	32.10	31.39	31.02	30.80	30.71	30.66	30.63	30.61	30.60	30.57	30.55	30.54	30.52	30.52
.52	35.13	32.67	31.76	31.04	30.67	30.45	30.36	30.31	30.28	30.26	30.25	30.23	30.20	30.19	30.17	30.17
.53	34.82	32.33	31.42	30.70	30.32	30.10	30.01	29.96	29.93	29.91	29.90	29.86	29.85	29.84	29.83	29.82
.54	34.51	32.00	31.08	30.36	29.98	29.76	29.67	29.62	29.59	29.57	29.55	29.53	29.51	29.49	29.48	29.48
.55	34.20	31.67	30.74	30.01	29.64	29.41	29.32	29.27	29.24	29.22	29.21	29.19	29.16	29.15	29.14	29.13
.56	33.88	31.33	30.40	29.67	29.29	29.07	28.98	28.93	28.90	28.88	28.87	28.85	28.82	28.81	28.79	28.79
.57	33.57	31.00	30.06	29.33	28.95	28.73	28.64	28.59	28.56	28.54	28.53	28.51	28.48	28.47	28.45	28.45
.58	33.25	30.67	29.73	28.99	28.61	28.39	28.30	28.25	28.22	28.20	28.19	28.17	28.14	28.13	28.12	28.11
.59	32.93	30.33	29.39	28.66	28.28	28.05	27.96	27.92	27.89	27.87	27.85	27.83	27.81	27.79	27.78	27.77
.60	32.61	30.00	29.05	28.32	27.94	27.72	27.63	27.58	27.55	27.53	27.52	27.50	27.47	27.46	27.45	27.44
.61	32.28	29.67	28.72	27.98	27.60	27.39	27.30	27.25	27.22	27.20	27.18	27.16	27.14	27.13	27.11	27.11
.62	31.96	29.33	28.39	27.65	27.27	27.05	26.96	26.92	26.89	26.87	26.85	26.83	26.81	26.80	26.78	26.78
.63	31.63	29.00	28.05	27.32	26.94	26.72	26.63	26.59	26.56	26.54	26.52	26.50	26.48	26.47	26.45	26.45
.64	31.30	28.67	27.72	26.99	26.61	26.39	26.31	26.26	26.23	26.21	26.20	26.18	26.15	26.14	26.13	26.12
.65	30.97	28.33	27.39	26.66	26.28	26.07	25.98	25.93	25.90	25.88	25.87	25.85	25.83	25.82	25.80	25.80
.66	30.63	28.00	27.06	26.33	25.96	25.74	25.66	25.61	25.58	25.56	25.55	25.53	25.51	25.49	25.48	25.48
.67	30.30	27.67	26.73	26.00	25.63	25.42	25.33	25.29	25.26	25.24	25.23	25.21	25.19	25.17	25.16	25.16
.68	29.96	27.33	26.40	25.68	25.31	25.10	25.01	24.97	24.94	24.92	24.91	24.89	24.87	24.86	24.84	24.84
.69	29.61	27.00	26.07	25.35	24.99	24.78	24.70	24.65	24.62	24.60	24.59	24.57	24.55	24.54	24.53	24.52

[1] Values tabulated are read in percent.

TABLE T12-7   (continued)

Q_u or Q_l	Sample Size															
	3	4	5	7	10	15	20	25	30	35	40	50	75	100	150	200
.70	29.27	26.67	25.74	25.03	24.67	24.46	24.38	24.33	24.31	24.29	24.28	24.26	24.24	24.23	24.21	24.21
.71	28.92	26.33	25.41	24.71	24.35	24.15	24.06	24.02	23.99	23.98	23.96	23.95	23.92	23.91	23.90	23.90
.72	28.57	26.00	25.09	24.39	24.03	23.83	23.75	23.71	23.68	23.67	23.65	23.64	23.61	23.60	23.59	23.59
.73	28.22	25.67	24.76	24.07	23.72	23.52	23.44	23.40	23.37	23.36	23.34	23.33	23.31	23.30	23.29	23.28
.74	27.86	25.33	24.44	23.75	23.41	23.21	23.13	23.09	23.07	23.05	23.04	23.02	23.00	22.99	22.98	22.98
.75	27.50	25.00	24.11	23.44	23.10	22.90	22.83	22.79	22.76	22.75	22.73	22.72	22.70	22.69	22.68	22.67
.76	27.13	24.67	23.79	23.12	22.79	22.60	22.52	22.48	22.46	22.44	22.43	22.42	22.40	22.39	22.38	22.37
.77	26.77	24.33	23.47	22.81	22.48	22.30	22.22	22.18	22.16	22.14	22.13	22.12	22.10	22.09	22.08	22.08
.78	26.39	24.00	23.15	22.50	22.18	21.99	21.92	21.89	21.86	21.85	21.84	21.82	21.80	21.79	21.78	21.78
.79	26.02	23.67	22.83	22.19	21.87	21.70	21.63	21.59	21.57	21.55	21.54	21.53	21.51	21.50	21.49	21.49
.80	25.64	23.33	22.51	21.88	21.57	21.40	21.33	21.29	21.27	21.26	21.25	21.23	21.22	21.21	21.20	21.20
.81	25.25	23.00	22.19	21.58	21.27	21.10	21.04	21.00	20.98	20.97	20.96	20.94	20.93	20.92	20.91	20.91
.82	24.86	22.67	21.87	21.27	20.98	20.81	20.75	20.71	20.69	20.68	20.67	20.65	20.64	20.63	20.62	20.62
.83	24.47	22.33	21.56	20.97	20.68	20.52	20.46	20.42	20.40	20.39	20.38	20.37	20.35	20.35	20.34	20.34
.84	24.07	22.00	21.24	20.67	20.39	20.23	20.17	20.14	20.12	20.11	20.10	20.09	20.07	20.06	20.06	20.05
.85	23.67	21.67	20.93	20.37	20.10	19.94	19.89	19.86	19.84	19.82	19.82	19.80	19.79	19.78	19.78	19.77
.86	23.26	21.33	20.62	20.07	19.81	19.66	19.60	19.57	19.56	19.54	19.54	19.53	19.51	19.51	19.50	19.50
.87	22.84	21.00	20.31	19.78	19.52	19.38	19.32	19.30	19.28	19.27	19.26	19.25	19.24	19.23	19.22	19.22
.88	22.42	20.67	20.00	19.48	19.23	19.10	19.04	19.02	19.00	18.99	18.98	18.98	18.96	18.96	18.95	18.95
.89	21.99	20.33	19.69	19.19	18.95	18.82	18.77	18.74	18.73	18.72	18.71	18.70	18.69	18.69	18.68	18.68
.90	21.55	20.00	19.38	18.90	18.67	18.54	18.50	18.47	18.46	18.45	18.44	18.43	18.42	18.42	18.41	18.41
.91	21.11	19.67	19.07	18.61	18.39	18.27	18.22	18.20	18.19	18.18	18.17	18.17	18.16	18.15	18.15	18.15
.92	20.66	19.33	18.77	18.33	18.11	18.00	17.96	17.94	17.92	17.92	17.91	17.90	17.89	17.89	17.88	17.88
.93	20.20	19.00	18.46	18.04	17.84	17.73	17.69	17.67	17.66	17.65	17.65	17.64	17.63	17.63	17.62	17.62
.94	19.74	18.67	18.16	17.76	17.57	17.46	17.43	17.41	17.40	17.39	17.39	17.38	17.37	17.37	17.36	17.36
.95	19.25	18.33	17.86	17.48	17.29	17.20	17.17	17.15	17.14	17.13	17.13	17.12	17.12	17.11	17.11	17.11
.96	18.76	18.00	17.56	17.20	17.03	16.94	16.91	16.89	16.88	16.88	16.87	16.87	16.86	16.86	16.86	16.85
.97	18.25	17.67	17.25	16.92	16.76	16.68	16.65	16.63	16.63	16.62	16.62	16.61	16.61	16.61	16.60	16.60
.98	17.74	17.33	16.96	16.65	16.49	16.42	16.39	16.38	16.37	16.37	16.37	16.36	16.36	16.36	16.36	16.36
.99	17.21	17.00	16.66	16.37	16.23	16.16	16.14	16.13	16.12	16.12	16.12	16.12	16.11	16.11	16.11	16.11
1.00	16.67	16.67	16.36	16.10	15.97	15.91	15.89	15.88	15.88	15.87	15.87	15.87	15.87	15.87	15.87	15.87
1.01	16.11	16.33	16.07	15.83	15.72	15.66	15.64	15.63	15.63	15.63	15.63	15.63	15.62	15.62	15.62	15.62
1.02	15.53	16.00	15.78	15.56	15.46	15.41	15.40	15.39	15.39	15.39	15.39	15.38	15.38	15.38	15.38	15.38
1.03	14.93	15.67	15.48	15.30	15.21	15.17	15.15	15.15	15.15	15.15	15.15	15.15	15.15	15.15	15.15	15.15
1.04	14.31	15.33	15.19	15.03	14.96	14.92	14.91	14.91	14.91	14.91	14.91	14.91	14.91	14.91	14.91	14.91
1.05	13.66	15.00	14.91	14.77	14.71	14.68	14.67	14.67	14.67	14.67	14.67	14.68	14.68	14.68	14.68	14.68
1.06	12.98	14.67	14.62	14.51	14.46	14.44	14.44	14.44	14.44	14.44	14.44	14.45	14.45	14.45	14.45	14.45
1.07	12.27	14.33	14.33	14.26	14.22	14.20	14.20	14.21	14.21	14.21	14.21	14.22	14.22	14.22	14.22	14.23
1.08	11.51	14.00	14.05	14.00	13.97	13.97	13.97	13.98	13.98	13.98	13.99	13.99	13.99	14.00	14.00	14.00
1.09	10.71	13.67	13.76	13.75	13.73	13.74	13.74	13.75	13.75	13.76	13.76	13.77	13.77	13.77	13.78	13.78

TABLE T12-7 (continued)

Q_U or Q_L	Sample Size															
	3	4	5	7	10	15	20	25	30	35	40	50	75	100	150	200
1.10	9.84	13.33	13.48	13.49	13.50	13.51	13.52	13.52	13.53	13.54	13.54	13.54	13.55	13.55	13.56	13.56
1.11	8.89	13.00	13.20	13.25	13.26	13.28	13.29	13.30	13.31	13.31	13.32	13.32	13.33	13.34	13.34	13.34
1.12	7.82	12.67	12.93	13.00	13.03	13.05	13.07	13.08	13.09	13.10	13.10	13.11	13.12	13.12	13.12	13.13
1.13	6.60	12.33	12.65	12.75	12.80	12.83	12.85	12.86	12.87	12.88	12.89	12.89	12.90	12.91	12.91	12.92
1.14	5.08	12.00	12.37	12.51	12.57	12.61	12.63	12.65	12.66	12.67	12.67	12.68	12.69	12.70	12.70	12.70
1.15	0.29	11.67	12.10	12.27	12.34	12.39	12.42	12.44	12.45	12.46	12.46	12.47	12.48	12.49	12.49	12.50
1.16	0.00	11.33	11.83	12.03	12.12	12.18	12.21	12.22	12.24	12.25	12.25	12.26	12.28	12.28	12.29	12.29
1.17	0.00	11.00	11.56	11.79	11.90	11.96	12.00	12.02	12.03	12.04	12.05	12.06	12.07	12.08	12.08	12.09
1.18	0.00	10.67	11.29	11.56	11.68	11.75	11.79	11.81	11.82	11.84	11.84	11.85	11.87	11.88	11.88	11.89
1.19	0.00	10.33	11.02	11.33	11.46	11.54	11.58	11.61	11.62	11.63	11.64	11.65	11.67	11.68	11.69	11.69
1.20	0.00	10.00	10.76	11.10	11.24	11.34	11.38	11.41	11.42	11.43	11.44	11.46	11.47	11.48	11.49	11.49
1.21	0.00	9.67	10.50	10.87	11.03	11.13	11.18	11.21	11.22	11.24	11.25	11.26	11.28	11.29	11.30	11.30
1.22	0.00	9.33	10.23	10.65	10.82	10.93	10.98	11.01	11.03	11.04	11.05	11.07	11.09	11.09	11.10	11.11
1.23	0.00	9.00	9.97	10.42	10.61	10.73	10.78	10.81	10.84	10.85	10.86	10.88	10.90	10.91	10.91	10.92
1.24	0.00	8.67	9.72	10.20	10.41	10.53	10.59	10.62	10.64	10.66	10.67	10.69	10.71	10.72	10.73	10.73
1.25	0.00	8.33	9.46	9.98	10.21	10.34	10.40	10.43	10.46	10.47	10.48	10.50	10.52	10.53	10.54	10.55
1.26	0.00	8.00	9.21	9.77	10.00	10.15	10.21	10.25	10.27	10.29	10.30	10.32	10.34	10.35	10.36	10.37
1.27	0.00	7.67	8.96	9.55	9.81	9.96	10.02	10.06	10.09	10.10	10.12	10.13	10.16	10.17	10.18	10.19
1.28	0.00	7.33	8.71	9.34	9.61	9.77	9.84	9.88	9.90	9.92	9.94	9.95	9.98	9.99	10.00	10.01
1.29	0.00	7.00	8.46	9.13	9.42	9.58	9.65	9.70	9.72	9.74	9.76	9.78	9.80	9.82	9.83	9.83
1.30	0.00	6.67	8.21	8.93	9.22	9.40	9.48	9.52	9.55	9.57	9.58	9.60	9.63	9.64	9.65	9.66
1.31	0.00	6.33	7.97	8.72	9.03	9.22	9.30	9.34	9.37	9.39	9.41	9.43	9.46	9.47	9.48	9.49
1.32	0.00	6.00	7.73	8.52	8.85	9.04	9.12	9.17	9.20	9.22	9.24	9.26	9.29	9.30	9.31	9.32
1.33	0.00	5.67	7.49	8.32	8.66	8.86	8.95	9.00	9.03	9.05	9.07	9.09	9.12	9.13	9.15	9.15
1.34	0.00	5.33	7.25	8.12	8.48	8.69	8.78	8.83	8.86	8.88	8.90	8.92	8.95	8.97	8.98	8.99
1.35	0.00	5.00	7.02	7.92	8.30	8.52	8.61	8.66	8.69	8.72	8.74	8.76	8.79	8.81	8.82	8.83
1.36	0.00	4.67	6.79	7.73	8.12	8.35	8.44	8.50	8.53	8.55	8.57	8.60	8.63	8.65	8.66	8.67
1.37	0.00	4.33	6.56	7.54	7.95	8.18	8.28	8.33	8.37	8.39	8.41	8.44	8.47	8.49	8.50	8.51
1.38	0.00	4.00	6.33	7.35	7.77	8.01	8.12	8.17	8.21	8.24	8.25	8.28	8.31	8.33	8.35	8.35
1.39	0.00	3.67	6.10	7.17	7.60	7.85	7.96	8.01	8.05	8.08	8.10	8.12	8.16	8.18	8.19	8.20
1.40	0.00	3.33	5.88	6.98	7.44	7.69	7.80	7.86	7.90	7.92	7.94	7.97	8.01	8.02	8.04	8.05
1.41	0.00	3.00	5.66	6.80	7.27	7.53	7.64	7.70	7.74	7.77	7.79	7.82	7.86	7.87	7.89	7.90
1.42	0.00	2.67	5.44	6.62	7.10	7.37	7.49	7.55	7.59	7.62	7.64	7.67	7.71	7.73	7.74	7.75
1.43	0.00	2.33	5.23	6.45	6.94	7.22	7.34	7.40	7.44	7.47	7.50	7.52	7.56	7.58	7.60	7.61
1.44	0.00	2.00	5.01	6.27	6.78	7.07	7.19	7.26	7.30	7.33	7.35	7.38	7.42	7.44	7.46	7.47
1.45	0.00	1.67	4.81	6.10	6.63	6.92	7.04	7.11	7.15	7.18	7.21	7.24	7.28	7.30	7.31	7.33
1.46	0.00	1.33	4.60	5.93	6.47	6.77	6.90	6.97	7.01	7.04	7.07	7.10	7.14	7.16	7.18	7.19
1.47	0.00	1.00	4.39	5.77	6.32	6.63	6.75	6.83	6.87	6.90	6.93	6.96	7.00	7.02	7.04	7.05
1.48	0.00	.67	4.19	5.60	6.17	6.48	6.61	6.69	6.73	6.77	6.79	6.82	6.86	6.88	6.90	6.91
1.49	0.00	.33	3.99	5.44	6.02	6.34	6.48	6.55	6.60	6.63	6.65	6.69	6.73	6.75	6.77	6.78

TABLE T12-7   (continued)

Q0 of QL	Sample Size															
	3	4	5	7	10	15	20	25	30	35	40	50	75	100	150	200
1.50	0.00	0.00	3.80	5.28	5.87	6.20	6.34	6.41	6.46	6.50	6.52	6.55	6.60	6.62	6.64	6.65
1.51	0.00	0.00	3.61	5.13	5.73	6.06	6.20	6.28	6.33	6.36	6.39	6.42	6.47	6.49	6.51	6.52
1.52	0.00	0.00	3.42	4.97	5.59	5.93	6.07	6.15	6.20	6.23	6.26	6.29	6.34	6.36	6.38	6.39
1.53	0.00	0.00	3.23	4.82	5.45	5.80	5.94	6.02	6.07	6.11	6.13	6.17	6.21	6.24	6.26	6.27
1.54	0.00	0.00	3.05	4.67	5.31	5.67	5.81	5.89	5.95	5.98	6.01	6.04	6.09	6.11	6.13	6.15
1.55	0.00	0.00	2.87	4.52	5.18	5.54	5.69	5.77	5.82	5.86	5.88	5.92	5.97	5.99	6.01	6.02
1.56	0.00	0.00	2.69	4.38	5.05	5.41	5.56	5.65	5.70	5.74	5.76	5.80	5.85	5.87	5.89	5.90
1.57	0.00	0.00	2.52	4.24	4.92	5.29	5.44	5.53	5.58	5.62	5.64	5.68	5.73	5.75	5.78	5.79
1.58	0.00	0.00	2.35	4.10	4.79	5.16	5.32	5.41	5.46	5.50	5.53	5.56	5.61	5.64	5.66	5.67
1.59	0.00	0.00	2.19	3.96	4.66	5.04	5.20	5.29	5.34	5.38	5.41	5.45	5.50	5.52	5.54	5.56
1.60	0.00	0.00	2.03	3.83	4.54	4.92	5.09	5.17	5.23	5.27	5.30	5.33	5.38	5.41	5.43	5.44
1.61	0.00	0.00	1.87	3.69	4.41	4.81	4.97	5.06	5.12	5.16	5.18	5.22	5.27	5.30	5.32	5.33
1.62	0.00	0.00	1.72	3.57	4.30	4.69	4.86	4.95	5.01	5.04	5.07	5.11	5.16	5.19	5.21	5.22
1.63	0.00	0.00	1.57	3.44	4.18	4.58	4.75	4.84	4.90	4.94	4.97	5.01	5.06	5.08	5.11	5.12
1.64	0.00	0.00	1.42	3.31	4.06	4.47	4.64	4.73	4.79	4.83	4.86	4.90	4.95	4.98	5.00	5.01
1.65	0.00	0.00	1.28	3.19	3.95	4.36	4.53	4.62	4.68	4.72	4.75	4.79	4.85	4.87	4.90	4.91
1.66	0.00	0.00	1.15	3.07	3.84	4.25	4.43	4.52	4.58	4.62	4.65	4.69	4.74	4.77	4.80	4.81
1.67	0.00	0.00	1.02	2.95	3.73	4.15	4.32	4.42	4.48	4.52	4.55	4.59	4.64	4.67	4.70	4.71
1.68	0.00	0.00	0.89	2.84	3.62	4.05	4.22	4.32	4.38	4.42	4.45	4.49	4.55	4.57	4.60	4.61
1.69	0.00	0.00	0.77	2.73	3.52	3.94	4.12	4.22	4.28	4.32	4.35	4.39	4.45	4.47	4.50	4.51
1.70	0.00	0.00	0.66	2.62	3.41	3.84	4.02	4.12	4.18	4.22	4.25	4.30	4.35	4.38	4.41	4.42
1.71	0.00	0.00	0.55	2.51	3.31	3.75	3.93	4.02	4.09	4.13	4.16	4.20	4.26	4.29	4.31	4.32
1.72	0.00	0.00	0.45	2.41	3.21	3.65	3.83	3.93	3.99	4.04	4.07	4.11	4.17	4.19	4.22	4.23
1.73	0.00	0.00	0.36	2.30	3.11	3.56	3.74	3.84	3.90	3.94	3.98	4.02	4.08	4.10	4.13	4.14
1.74	0.00	0.00	0.27	2.20	3.02	3.46	3.65	3.75	3.81	3.85	3.89	3.93	3.99	4.01	4.04	4.05
1.75	0.00	0.00	0.19	2.11	2.93	3.37	3.56	3.66	3.72	3.77	3.80	3.84	3.90	3.93	3.95	3.97
1.76	0.00	0.00	0.12	2.01	2.83	3.28	3.47	3.57	3.63	3.68	3.71	3.76	3.81	3.84	3.87	3.88
1.77	0.00	0.00	0.06	1.92	2.74	3.20	3.38	3.48	3.55	3.59	3.63	3.67	3.73	3.76	3.78	3.80
1.78	0.00	0.00	0.02	1.83	2.66	3.11	3.30	3.40	3.47	3.51	3.54	3.59	3.64	3.67	3.70	3.71
1.79	0.00	0.00	0.00	1.74	2.57	3.03	3.21	3.32	3.38	3.43	3.46	3.51	3.56	3.59	3.63	3.63
1.80	0.00	0.00	0.00	1.65	2.49	2.94	3.13	3.24	3.30	3.35	3.38	3.43	3.48	3.51	3.54	3.55
1.81	0.00	0.00	0.00	1.57	2.40	2.86	3.05	3.16	3.22	3.27	3.30	3.35	3.40	3.43	3.46	3.47
1.82	0.00	0.00	0.00	1.49	2.32	2.79	2.98	3.08	3.15	3.19	3.22	3.27	3.33	3.36	3.38	3.40
1.83	0.00	0.00	0.00	1.41	2.25	2.71	2.90	3.00	3.07	3.11	3.15	3.19	3.25	3.28	3.31	3.32
1.84	0.00	0.00	0.00	1.34	2.17	2.63	2.82	2.93	2.99	3.04	3.07	3.12	3.18	3.21	3.23	3.25
1.85	0.00	0.00	0.00	1.26	2.09	2.56	2.75	2.85	2.92	2.97	3.00	3.05	3.10	3.13	3.16	3.17
1.86	0.00	0.00	0.00	1.19	2.02	2.48	2.68	2.78	2.85	2.89	2.93	2.97	3.03	3.06	3.09	3.10
1.87	0.00	0.00	0.00	1.12	1.95	2.41	2.61	2.71	2.78	2.82	2.86	2.90	2.96	2.99	3.02	3.03
1.88	0.00	0.00	0.00	1.06	1.88	2.34	2.54	2.64	2.71	2.75	2.79	2.83	2.89	2.92	2.95	2.96
1.89	0.00	0.00	0.00	0.99	1.81	2.28	2.47	2.57	2.64	2.69	2.72	2.77	2.83	2.85	2.88	2.90

TABLE T12-7 (continued)

Q_U or Q_L	Sample Size															
	3	4	5	7	10	15	20	25	30	35	40	50	75	100	150	200
1.90	0.00	0.00	0.00	0.93	1.75	2.21	2.40	2.51	2.57	2.62	2.65	2.70	2.76	2.79	2.82	2.83
1.91	0.00	0.00	0.00	0.87	1.68	2.14	2.34	2.44	2.51	2.56	2.59	2.63	2.69	2.72	2.75	2.77
1.92	0.00	0.00	0.00	0.81	1.62	2.08	2.27	2.38	2.45	2.49	2.52	2.57	2.63	2.66	2.69	2.70
1.93	0.00	0.00	0.00	0.76	1.56	2.02	2.21	2.32	2.38	2.43	2.46	2.51	2.57	2.60	2.62	2.64
1.94	0.00	0.00	0.00	0.70	1.50	1.96	2.15	2.25	2.32	2.37	2.40	2.45	2.51	2.54	2.56	2.58
1.95	0.00	0.00	0.00	0.65	1.44	1.90	2.09	2.19	2.26	2.31	2.34	2.39	2.45	2.48	2.50	2.52
1.96	0.00	0.00	0.00	0.60	1.38	1.84	2.03	2.14	2.20	2.25	2.28	2.33	2.39	2.42	2.44	2.46
1.97	0.00	0.00	0.00	0.56	1.33	1.78	1.97	2.08	2.14	2.19	2.22	2.27	2.33	2.36	2.39	2.40
1.98	0.00	0.00	0.00	0.51	1.27	1.73	1.92	2.02	2.09	2.13	2.17	2.21	2.27	2.30	2.33	2.34
1.99	0.00	0.00	0.00	0.47	1.22	1.67	1.86	1.97	2.03	2.08	2.11	2.16	2.22	2.25	2.27	2.29
2.00	0.00	0.00	0.00	0.43	1.17	1.62	1.81	1.91	1.98	2.03	2.06	2.10	2.16	2.19	2.22	2.23
2.01	0.00	0.00	0.00	0.39	1.12	1.57	1.76	1.86	1.93	1.97	2.01	2.05	2.11	2.14	2.17	2.18
2.02	0.00	0:00	0.00	0.36	1.07	1.52	1.71	1.81	1.87	1.92	1.95	2.00	2.06	2.09	2.11	2.13
2.03	0.00	0.00	0.00	0.32	1.03	1.47	1.66	1.76	1.82	1.87	1.90	1.95	2.01	2.04	2.06	2.08
2.04	0.00	0.00	0.00	0.29	0.98	1.42	1.61	1.71	1.77	1.82	1.85	1.90	1.96	1.99	2.01	2.03
2.05	0.00	0.00	0.00	0.26	0.94	1.37	1.56	1.66	1.73	1.77	1.80	1.85	1.91	1.94	1.96	1.98
2.06	0.00	0.00	0.00	0.23	0.90	1.33	1.51	1.61	1.68	1.72	1.76	1.80	1.86	1.89	1.92	1.93
2.07	0.00	0.00	0.00	0.21	0.86	1.28	1.47	1.57	1.63	1.68	1.71	1.76	1.81	1.84	1.87	1.88
2.08	0.00	0.00	0.00	0.18	0.82	1.24	1.42	1.52	1.59	1.63	1.66	1.71	1.77	1.79	1.82	1.84
2.09	0.00	0.00	0.00	0.16	0.78	1.20	1.38	1.48	1.54	1.59	1.62	1.66	1.72	1.75	1.78	1.79
2.10	0.00	0.00	0.00	0.14	0.74	1.16	1.34	1.44	1.50	1.54	1.58	1.62	1.68	1.71	1.73	1.75
2.11	0.00	0.00	0.00	0.12	0.71	1.12	1.30	1.39	1.46	1.50	1.53	1.58	1.63	1.66	1.69	1.70
2.12	0.00	0.00	0.00	0.10	0.67	1.08	1.26	1.35	1.42	1.46	1.49	1.54	1.59	1.62	1.65	1.66
2.13	0.00	0.00	0.00	0.08	0.64	1.04	1.22	1.31	1.38	1.42	1.45	1.50	1.55	1.58	1.61	1.62
2.14	0.00	0.00	0.00	0.07	0.61	1.00	1.18	1.28	1.34	1.38	1.41	1.46	1.51	1.54	1.57	1.58
2.15	0.00	0.00	0.00	0.06	0.58	0.97	1.14	1.24	1.30	1.34	1.37	1.42	1.47	1.50	1.53	1.54
2.16	0.00	0.00	0.00	0.05	0.55	0.93	1.10	1.20	1.26	1.30	1.34	1.38	1.43	1.46	1.49	1.50
2.17	0.00	0.00	0.00	0.04	0.52	0.90	1.07	1.16	1.22	1.27	1.30	1.34	1.40	1.42	1.45	1.46
2.18	0.00	0.00	0.00	0.03	0.49	0.87	1.03	1.13	1.19	1.23	1.26	1.30	1.36	1.39	1.41	1.42
2.19	0.00	0.00	0.00	0.02	0.46	0.83	1.00	1.09	1.15	1.20	1.23	1.27	1.32	1.35	1.38	1.39
2.20	0.000	0.000	0.000	0.015	0.437	0.803	0.968	1.061	1.120	1.161	1.192	1.233	1.287	1.314	1.340	1.352
2.21	0.000	0.000	0.000	0.010	0.413	0.772	0.936	1.028	1.087	1.128	1.158	1.199	1.253	1.279	1.305	1.318
2.22	0.000	0.000	0.000	0.006	0.389	0.743	0.905	0.996	1.054	1.095	1.125	1.166	1.219	1.245	1.271	1.283
2.23	0.000	0.000	0.000	0.003	0.366	0.715	0.875	0.965	1.023	1.063	1.093	1.134	1.186	1.212	1.238	1.250
2.24	0.000	0.000	0.000	0.002	0.345	0.687	0.845	0.935	0.992	1.032	1.061	1.102	1.154	1.180	1.205	1.218
2.25	0.000	0.000	0.000	0.001	0.324	0.660	0.816	0.905	0.962	1.002	1.031	1.071	1.123	1.148	1.173	1.186
2.26	0.000	0.000	0.000	0.000	0.304	0.634	0.789	0.876	0.933	0.972	1.001	1.041	1.092	1.117	1.142	1.155
2.27	0.000	0.000	0.000	0.000	0.285	0.609	0.762	0.848	0.904	0.943	0.972	1.011	1.062	1.087	1.112	1.124
2.28	0.000	0.000	0.000	0.000	0.267	0.585	0.735	0.821	0.876	0.915	0.943	0.982	1.033	1.058	1.082	1.094
2.29	0.000	0.000	0.000	0.000	0.250	0.561	0.710	0.794	0.849	0.887	0.915	0.954	1.004	1.029	1.053	1.065

TABLE T12-7　(continued)

Q_U or Q_L							Sample Size									
	3	4	5	7	10	15	20	25	30	35	40	50	75	100	150	200
2.30	0.000	0.000	0.000	0.000	0.233	0.538	0.685	0.769	0.823	0.861	0.888	0.927	0.977	1.001	1.025	1.037
2.31	0.000	0.000	0.000	0.000	0.218	0.516	0.661	0.743	0.797	0.834	0.862	0.900	0.949	0.974	0.997	1.009
2.32	0.000	0.000	0.000	0.000	0.203	0.495	0.637	0.719	0.772	0.809	0.836	0.874	0.923	0.947	0.971	0.982
2.33	0.000	0.000	0.000	0.000	0.189	0.474	0.614	0.695	0.748	0.784	0.811	0.848	0.897	0.921	0.944	0.956
2.34	0.000	0.000	0.000	0.000	0.175	0.454	0.592	0.672	0.724	0.760	0.787	0.824	0.872	0.895	0.915	0.930
2.35	0.000	0.000	0.000	0.000	0.163	0.435	0.571	0.650	0.701	0.736	0.763	0.799	0.847	0.870	0.893	0.905
2.36	0.000	0.000	0.000	0.000	0.151	0.416	0.550	0.628	0.678	0.714	0.740	0.776	0.823	0.846	0.869	0.880
2.37	0.000	0.000	0.000	0.000	0.139	0.398	0.530	0.606	0.656	0.691	0.717	0.753	0.799	0.822	0.845	0.856
2.38	0.000	0.000	0.000	0.000	0.128	0.381	0.510	0.586	0.635	0.670	0.695	0.730	0.777	0.799	0.822	0.833
2.39	0.000	0.000	0.000	0.000	0.118	0.364	0.491	0.566	0.614	0.648	0.674	0.709	0.754	0.777	0.799	0.810
2.40	0.000	0.000	0.000	0.000	0.109	0.348	0.473	0.546	0.594	0.628	0.653	0.687	0.732	0.755	0.777	0.787
2.41	0.000	0.000	0.000	0.000	0.100	0.332	0.455	0.527	0.575	0.608	0.633	0.667	0.711	0.733	0.755	0.766
2.42	0.000	0.000	0.000	0.000	0.091	0.317	0.437	0.509	0.555	0.588	0.613	0.646	0.691	0.712	0.734	0.744
2.43	0.000	0.000	0.000	0.000	0.083	0.302	0.421	0.491	0.537	0.569	0.593	0.627	0.670	0.692	0.713	0.724
2.44	0.000	0.000	0.000	0.000	0.076	0.288	0.404	0.474	0.519	0.551	0.575	0.608	0.651	0.672	0.693	0.703
2.45	0.000	0.000	0.000	0.000	0.069	0.275	0.389	0.457	0.501	0.533	0.556	0.589	0.632	0.653	0.673	0.684
2.46	0.000	0.000	0.000	0.000	0.063	0.262	0.373	0.440	0.484	0.516	0.539	0.571	0.613	0.634	0.654	0.664
2.47	0.000	0.000	0.000	0.000	0.057	0.249	0.359	0.425	0.468	0.499	0.521	0.553	0.595	0.615	0.635	0.646
2.48	0.000	0.000	0.000	0.000	0.051	0.237	0.344	0.409	0.452	0.482	0.505	0.536	0.577	0.597	0.617	0.627
2.49	0.000	0.000	0.000	0.000	0.046	0.226	0.331	0.394	0.436	0.466	0.488	0.519	0.560	0.580	0.600	0.609
2.50	0.000	0.000	0.000	0.000	0.041	0.214	0.317	0.380	0.421	0.451	0.473	0.503	0.543	0.563	0.582	0.592
2.51	0.000	0.000	0.000	0.000	0.037	0.204	0.304	0.366	0.407	0.436	0.457	0.487	0.527	0.546	0.565	0.575
2.52	0.000	0.000	0.000	0.000	0.033	0.193	0.292	0.352	0.392	0.421	0.442	0.472	0.511	0.530	0.549	0.558
2.53	0.000	0.000	0.000	0.000	0.029	0.184	0.280	0.339	0.379	0.407	0.428	0.457	0.495	0.514	0.533	0.542
2.54	0.000	0.000	0.000	0.000	0.026	0.174	0.268	0.326	0.365	0.393	0.413	0.442	0.480	0.499	0.517	0.527
2.55	0.000	0.000	0.000	0.000	0.023	0.165	0.257	0.314	0.352	0.379	0.400	0.428	0.465	0.484	0.502	0.511
2.56	0.000	0.000	0.000	0.000	0.020	0.156	0.246	0.302	0.340	0.366	0.386	0.414	0.451	0.469	0.487	0.496
2.57	0.000	0.000	0.000	0.000	0.017	0.148	0.236	0.291	0.327	0.354	0.373	0.401	0.437	0.455	0.473	0.482
2.58	0.000	0.000	0.000	0.000	0.015	0.140	0.226	0.279	0.316	0.341	0.361	0.388	0.424	0.441	0.459	0.468
2.59	0.000	0.000	0.000	0.000	0.013	0.133	0.216	0.269	0.304	0.330	0.349	0.375	0.410	0.428	0.445	0.454
2.60	0.000	0.000	0.000	0.000	0.011	0.125	0.207	0.258	0.293	0.318	0.337	0.363	0.398	0.415	0.432	0.441
2.61	0.000	0.000	0.000	0.000	0.009	0.118	0.198	0.248	0.282	0.307	0.325	0.351	0.385	0.402	0.419	0.428
2.62	0.000	0.000	0.000	0.000	0.008	0.112	0.189	0.238	0.272	0.296	0.314	0.339	0.373	0.390	0.406	0.415
2.63	0.000	0.000	0.000	0.000	0.007	0.105	0.181	0.229	0.262	0.285	0.303	0.328	0.361	0.378	0.394	0.402
2.64	0.000	0.000	0.000	0.000	0.005	0.099	0.172	0.220	0.252	0.275	0.293	0.317	0.350	0.366	0.382	0.390
2.65	0.000	0.008	0.000	0.000	0.005	0.094	0.165	0.211	0.243	0.265	0.282	0.307	0.339	0.355	0.371	0.379
2.66	0.000	0.000	0.000	0.000	0.004	0.088	0.157	0.202	0.233	0.256	0.273	0.296	0.328	0.344	0.359	0.367
2.67	0.000	0.000	0.000	0.000	0.003	0.083	0.150	0.194	0.224	0.246	0.263	0.286	0.317	0.333	0.348	0.356
2.68	0.000	0.000	0.000	0.000	0.002	0.078	0.143	0.186	0.216	0.237	0.254	0.277	0.307	0.322	0.338	0.345
2.69	0.000	0.000	0.000	0.000	0.002	0.073	0.136	0.179	0.208	0.229	0.245	0.267	0.297	0.312	0.327	0.335

TABLE T12-7   (continued)

								Sample Size								
	3	4	5	7	10	15	20	25	30	35	40	50	75	100	150	200
2.70	0.000	0.000	0.000	0.000	0.001	0.069	0.130	0.171	0.200	0.220	0.236	0.258	0.288	0.302	0.317	0.325
2.71	0.000	0.000	0.000	0.000	0.001	0.064	0.124	0.164	0.192	0.212	0.227	0.249	0.278	0.293	0.307	0.315
2.72	0.000	0.000	0.000	0.000	0.000	0.060	0.118	0.157	0.184	0.204	0.219	0.241	0.269	0.283	0.298	0.305
2.73	0.000	0.000	0.000	0.000	0.000	0.057	0.112	0.151	0.177	0.197	0.211	0.232	0.260	0.274	0.288	0.296
2.74	0.000	0.000	0.000	0.000	0.000	0.053	0.107	0.144	0.170	0.189	0.204	0.224	0.252	0.266	0.279	0.286
2.75	0.000	0.000	0.000	0.000	0.000	0.049	0.102	0.138	0.163	0.182	0.196	0.216	0.243	0.257	0.271	0.277
2.76	0.000	0.000	0.000	0.000	0.000	0.046	0.097	0.132	0.157	0.175	0.189	0.209	0.235	0.249	0.262	0.269
2.77	0.000	0.000	0.000	0.000	0.000	0.043	0.092	0.126	0.151	0.168	0.182	0.201	0.227	0.241	0.254	0.260
2.78	0.000	0.000	0.000	0.000	0.000	0.040	0.087	0.121	0.145	0.162	0.175	0.194	0.220	0.233	0.246	0.252
2.79	0.000	0.000	0.000	0.000	0.000	0.037	0.083	0.115	0.139	0.156	0.169	0.187	0.212	0.225	0.238	0.244
2.80	0.000	0.000	0.000	0.000	0.000	0.035	0.079	0.110	0.133	0.150	0.162	0.181	0.205	0.218	0.230	0.237
2.81	0.000	0.000	0.000	0.000	0.000	0.032	0.075	0.105	0.128	0.144	0.156	0.174	0.198	0.211	0.223	0.229
2.82	0.000	0.000	0.000	0.000	0.000	0.030	0.071	0.101	0.122	0.138	0.150	0.168	0.192	0.204	0.216	0.222
2.83	0.000	0.000	0.000	0.000	0.000	0.028	0.067	0.096	0.117	0.133	0.145	0.162	0.185	0.197	0.209	0.215
2.84	0.000	0.000	0.000	0.000	0.000	0.026	0.064	0.092	0.112	0.128	0.139	0.156	0.179	0.190	0.202	0.208
2.85	0.000	0.000	0.000	0.000	0.000	0.024	0.060	0.088	0.108	0.122	0.134	0.150	0.173	0.184	0.195	0.201
2.86	0.000	0.000	0.000	0.000	0.000	0.022	0.057	0.084	0.103	0.118	0.129	0.145	0.167	0.178	0.189	0.195
2.87	0.000	0.000	0.000	0.000	0.000	0.020	0.054	0.080	0.099	0.113	0.124	0.139	0.161	0.172	0.183	0.188
2.88	0.000	0.000	0.000	0.000	0.000	0.019	0.051	0.076	0.094	0.108	0.119	0.134	0.155	0.166	0.177	0.182
2.89	0.000	0.000	0.000	0.000	0.000	0.017	0.048	0.073	0.090	0.104	0.114	0.129	0.150	0.160	0.171	0.176
2.90	0.000	0.000	0.000	0.000	0.000	0.016	0.046	0.069	0.087	0.100	0.110	0.125	0.145	0.155	0.165	0.171
2.91	0.000	0.000	0.000	0.000	0.000	0.015	0.043	0.066	0.083	0.096	0.106	0.120	0.140	0.150	0.160	0.165
2.92	0.000	0.000	0.000	0.000	0.000	0.013	0.041	0.063	0.079	0.092	0.101	0.115	0.135	0.145	0.155	0.160
2.93	0.000	0.000	0.000	0.000	0.000	0.012	0.038	0.060	0.076	0.088	0.097	0.111	0.130	0.140	0.149	0.154
2.94	0.000	0.000	0.000	0.000	0.000	0.011	0.036	0.057	0.072	0.084	0.093	0.107	0.125	0.135	0.144	0.149
2.95	0.000	0.000	0.000	0.000	0.000	0.010	0.034	0.054	0.069	0.081	0.090	0.103	0.121	0.130	0.140	0.144
2.96	0.000	0.000	0.000	0.000	0.000	0.009	0.032	0.051	0.066	0.077	0.086	0.099	0.117	0.126	0.135	0.140
2.97	0.000	0.000	0.000	0.000	0.000	0.009	0.030	0.049	0.063	0.074	0.083	0.095	0.112	0.121	0.130	0.135
2.98	0.000	0.000	0.000	0.000	0.000	0.008	0.028	0.046	0.060	0.071	0.079	0.091	0.108	0.117	0.126	0.130
2.99	0.000	0.000	0.000	0.000	0.000	0.007	0.027	0.044	0.057	0.068	0.076	0.088	0.104	0.113	0.122	0.126
3.00	0.000	0.000	0.000	0.000	0.000	0.006	0.025	0.042	0.055	0.065	0.073	0.084	0.101	0.109	0.118	0.122
3.01	0.000	0.000	0.000	0.000	0.000	0.006	0.024	0.040	0.052	0.062	0.070	0.081	0.097	0.105	0.114	0.118
3.02	0.000	0.000	0.000	0.000	0.000	0.005	0.022	0.038	0.050	0.059	0.067	0.078	0.093	0.101	0.110	0.114
3.03	0.000	0.000	0.000	0.000	0.000	0.005	0.021	0.036	0.048	0.057	0.064	0.075	0.090	0.098	0.106	0.110
3.04	0.000	0.000	0.000	0.000	0.000	0.004	0.019	0.034	0.045	0.054	0.061	0.072	0.087	0.094	0.102	0.106
3.05	0.000	0.000	0.000	0.000	0.000	0.004	0.018	0.032	0.043	0.052	0.059	0.069	0.083	0.091	0.099	0.103
3.06	0.000	0.000	0.000	0.000	0.000	0.003	0.017	0.030	0.041	0.050	0.056	0.066	0.080	0.088	0.095	0.099
3.07	0.000	0.000	0.000	0.000	0.000	0.003	0.016	0.029	0.039	0.047	0.054	0.064	0.077	0.085	0.092	0.096
3.08	0.000	0.000	0.000	0.000	0.000	0.003	0.015	0.027	0.037	0.045	0.052	0.061	0.074	0.081	0.089	0.092
3.09	0.000	0.000	0.000	0.000	0.000	0.002	0.014	0.026	0.036	0.043	0.049	0.059	0.072	0.079	0.086	0.089

TABLE T12-7   (continued)

C_p or Q_L	Sample Size															
	3	4	5	7	10	15	20	25	30	35	40	50	75	100	150	200
3.10	0.000	0.000	0.000	0.000	0.000	0.002	0.013	0.024	0.034	0.041	0.047	0.056	0.069	0.076	0.083	0.086
3.11	0.000	0.000	0.000	0.000	0.000	0.002	0.012	0.023	0.032	0.039	0.045	0.054	0.066	0.073	0.080	0.083
3.12	0.000	0.000	0.000	0.000	0.000	0.002	0.011	0.022	0.031	0.038	0.043	0.052	0.064	0.070	0.077	0.080
3.13	0.000	0.000	0.000	0.000	0.000	0.002	0.011	0.021	0.029	0.036	0.041	0.050	0.061	0.068	0.074	0.077
3.14	0.000	0.000	0.000	0.000	0.000	0.001	0.010	0.019	0.028	0.034	0.040	0.048	0.059	0.065	0.071	0.075
3.15	0.000	0.000	0.000	0.000	0.000	0.001	0.009	0.018	0.026	0.033	0.038	0.046	0.057	0.063	0.069	0.072
3.16	0.000	0.000	0.000	0.000	0.000	0.001	0.009	0.017	0.025	0.031	0.036	0.044	0.055	0.060	0.066	0.069
3.17	0.000	0.000	0.000	0.000	0.000	0.001	0.008	0.016	0.024	0.030	0.035	0.042	0.053	0.058	0.064	0.067
3.18	0.000	0.000	0.000	0.000	0.000	0.001	0.007	0.015	0.022	0.028	0.033	0.040	0.050	0.056	0.062	0.065
3.19	0.000	0.000	0.000	0.000	0.000	0.001	0.007	0.015	0.021	0.027	0.032	0.038	0.049	0.054	0.059	0.062
3.20	0.000	0.000	0.000	0.000	0.000	0.001	0.006	0.014	0.020	0.026	0.030	0.037	0.047	0.052	0.057	0.060
3.21	0.000	0.000	0.000	0.000	0.000	0.000	0.006	0.013	0.019	0.024	0.029	0.035	0.045	0.050	0.055	0.058
3.22	0.000	0.000	0.000	0.000	0.000	0.000	0.005	0.012	0.018	0.023	0.027	0.034	0.043	0.048	0.053	0.056
3.23	0.000	0.000	0.000	0.000	0.000	0.000	0.005	0.011	0.017	0.022	0.026	0.032	0.041	0.046	0.051	0.054
3.24	0.000	0.000	0.000	0.000	0.000	0.000	0.005	0.011	0.016	0.021	0.025	0.031	0.040	0.044	0.049	0.052
3.25	0.000	0.000	0.000	0.000	0.000	0.000	0.004	0.010	0.015	0.020	0.024	0.030	0.038	0.043	0.048	0.050
3.26	0.000	0.000	0.000	0.000	0.000	0.000	0.004	0.009	0.015	0.019	0.023	0.028	0.037	0.041	0.046	0.048
3.27	0.000	0.000	0.000	0.000	0.000	0.000	0.004	0.009	0.014	0.019	0.022	0.027	0.035	0.040	0.044	0.046
3.28	0.000	0.000	0.000	0.000	0.000	0.000	0.003	0.008	0.013	0.017	0.021	0.026	0.034	0.038	0.042	0.045
3.29	0.000	0.000	0.000	0.000	0.000	0.000	0.003	0.008	0.012	0.016	0.020	0.025	0.032	0.037	0.041	0.043
3.30	0.000	0.000	0.000	0.000	0.000	0.000	0.003	0.007	0.012	0.015	0.019	0.024	0.031	0.035	0.039	0.042
3.31	0.000	0.000	0.000	0.000	0.000	0.000	0.003	0.007	0.011	0.015	0.018	0.023	0.030	0.034	0.038	0.040
3.32	0.000	0.000	0.000	0.000	0.000	0.000	0.002	0.006	0.010	0.014	0.017	0.022	0.029	0.032	0.036	0.039
3.33	0.000	0.000	0.000	0.000	0.000	0.000	0.002	0.006	0.010	0.013	0.016	0.021	0.027	0.031	0.035	0.037
3.34	0.000	0.000	0.000	0.000	0.000	0.000	0.002	0.006	0.009	0.013	0.015	0.020	0.026	0.030	0.034	0.036
3.35	0.000	0.000	0.000	0.000	0.000	0.000	0.002	0.005	0.009	0.012	0.015	0.019	0.025	0.029	0.032	0.034
3.36	0.000	0.000	0.000	0.000	0.000	0.000	0.002	0.005	0.008	0.011	0.014	0.018	0.024	0.028	0.031	0.033
3.37	0.000	0.000	0.000	0.000	0.000	0.000	0.002	0.005	0.008	0.011	0.013	0.017	0.023	0.026	0.030	0.032
3.38	0.000	0.000	0.000	0.000	0.000	0.000	0.001	0.004	0.007	0.010	0.013	0.016	0.022	0.025	0.029	0.031
3.39	0.000	0.000	0.000	0.000	0.000	0.000	0.001	0.004	0.007	0.010	0.012	0.016	0.021	0.024	0.028	0.029
3.40	0.000	0.000	0.000	0.000	0.000	0.000	0.001	0.004	0.007	0.009	0.011	0.015	0.020	0.023	0.027	0.028
3.41	0.000	0.000	0.000	0.000	0.000	0.000	0.001	0.003	0.006	0.009	0.011	0.014	0.020	0.022	0.026	0.027
3.42	0.000	0.000	0.000	0.000	0.000	0.000	0.001	0.003	0.006	0.008	0.010	0.014	0.019	0.022	0.025	0.026
3.43	0.000	0.000	0.000	0.000	0.000	0.000	0.001	0.003	0.005	0.008	0.010	0.013	0.018	0.021	0.024	0.025
3.44	0.000	0.000	0.000	0.000	0.000	0.000	0.001	0.003	0.005	0.007	0.009	0.012	0.017	0.020	0.023	0.024
3.45	0.000	0.000	0.000	0.000	0.000	0.000	0.001	0.003	0.005	0.007	0.009	0.012	0.016	0.019	0.022	0.023
3.46	0.000	0.000	0.000	0.000	0.000	0.000	0.001	0.002	0.005	0.007	0.008	0.011	0.016	0.018	0.021	0.022
3.47	0.000	0.000	0.000	0.000	0.000	0.000	0.001	0.002	0.004	0.006	0.008	0.011	0.015	0.017	0.020	0.022
3.48	0.000	0.000	0.000	0.000	0.000	0.000	0.001	0.002	0.004	0.006	0.007	0.010	0.014	0.017	0.019	0.021
3.49	0.000	0.000	0.000	0.000	0.000	0.000	0.000	0.002	0.004	0.005	0.007	0.010	0.014	0.016	0.019	0.020

TABLE T12-7 (continued)

Sample Size (rows indexed by $\frac{Q_U}{\sigma}$ or $\frac{Q_L}{\sigma}$)

	3	4	5	7	10	15	20	25	30	35	40	50	75	100	150	200
3.50	0.000	0.000	0.000	0.000	0.000	0.000	0.000	0.002	0.003	0.005	0.007	0.009	0.013	0.015	0.018	0.019
3.51	0.000	0.000	0.000	0.000	0.000	0.000	0.000	0.002	0.003	0.005	0.006	0.009	0.013	0.015	0.017	0.018
3.52	0.000	0.000	0.000	0.000	0.000	0.000	0.000	0.002	0.003	0.005	0.006	0.008	0.012	0.014	0.017	0.018
3.53	0.000	0.000	0.000	0.000	0.000	0.000	0.000	0.001	0.003	0.004	0.006	0.008	0.012	0.014	0.016	0.017
3.54	0.000	0.000	0.000	0.000	0.000	0.000	0.000	0.001	0.003	0.004	0.006	0.008	0.011	0.013	0.016	0.016
3.55	0.000	0.000	0.000	0.000	0.000	0.000	0.000	0.001	0.003	0.004	0.005	0.007	0.011	0.013	0.015	0.016
3.56	0.000	0.000	0.000	0.000	0.000	0.000	0.000	0.001	0.002	0.004	0.005	0.007	0.010	0.012	0.014	0.015
3.57	0.000	0.000	0.000	0.000	0.000	0.000	0.000	0.001	0.002	0.003	0.005	0.006	0.010	0.012	0.013	0.014
3.58	0.000	0.000	0.000	0.000	0.000	0.000	0.000	0.001	0.002	0.003	0.004	0.006	0.009	0.011	0.013	0.014
3.59	0.000	0.000	0.000	0.000	0.000	0.000	0.000	0.001	0.002	0.003	0.004	0.006	0.009	0.010	0.012	0.013
3.60	0.000	0.000	0.000	0.000	0.000	0.000	0.000	0.001	0.002	0.003	0.004	0.006	0.008	0.010	0.012	0.013
3.61	0.000	0.000	0.000	0.000	0.000	0.000	0.000	0.001	0.002	0.003	0.004	0.005	0.008	0.010	0.011	0.012
3.62	0.000	0.000	0.000	0.000	0.000	0.000	0.000	0.001	0.002	0.003	0.003	0.005	0.008	0.009	0.011	0.012
3.63	0.000	0.000	0.000	0.000	0.000	0.000	0.000	0.001	0.002	0.002	0.003	0.005	0.007	0.009	0.010	0.011
3.64	0.000	0.000	0.000	0.000	0.000	0.000	0.000	0.001	0.001	0.002	0.003	0.004	0.007	0.008	0.010	0.011
3.65	0.000	0.000	0.000	0.000	0.000	0.000	0.000	0.000	0.001	0.002	0.003	0.004	0.007	0.008	0.010	0.010
3.66	0.000	0.000	0.000	0.000	0.000	0.000	0.000	0.000	0.001	0.002	0.002	0.004	0.006	0.008	0.009	0.010
3.67	0.000	0.000	0.000	0.000	0.000	0.000	0.000	0.000	0.001	0.002	0.002	0.004	0.006	0.007	0.009	0.010
3.68	0.000	0.000	0.000	0.000	0.000	0.000	0.000	0.000	0.001	0.002	0.002	0.004	0.006	0.007	0.008	0.009
3.69	0.000	0.000	0.000	0.000	0.000	0.000	0.000	0.000	0.001	0.002	0.002	0.003	0.005	0.007	0.008	0.009

	3	4	5	7	10	15	20	25	30	35	40	50	75	100	150	200
3.70	0.000	0.000	0.000	0.000	0.000	0.000	0.000	0.000	0.001	0.002	0.002	0.003	0.005	0.006	0.008	0.008
3.71	0.000	0.000	0.000	0.000	0.000	0.000	0.000	0.000	0.001	0.001	0.002	0.003	0.005	0.006	0.007	0.008
3.72	0.000	0.000	0.000	0.000	0.000	0.000	0.000	0.000	0.001	0.001	0.002	0.003	0.005	0.006	0.007	0.008
3.73	0.000	0.000	0.000	0.000	0.000	0.000	0.000	0.000	0.001	0.001	0.002	0.003	0.005	0.006	0.007	0.007
3.74	0.000	0.000	0.000	0.000	0.000	0.000	0.000	0.000	0.001	0.001	0.002	0.002	0.004	0.005	0.007	0.007
3.75	0.000	0.000	0.000	0.000	0.000	0.000	0.000	0.000	0.001	0.001	0.001	0.002	0.004	0.005	0.006	0.007
3.76	0.000	0.000	0.000	0.000	0.000	0.000	0.000	0.000	0.001	0.001	0.001	0.002	0.004	0.005	0.006	0.007
3.77	0.000	0.000	0.000	0.000	0.000	0.000	0.000	0.000	0.001	0.001	0.001	0.002	0.004	0.005	0.006	0.006
3.78	0.000	0.000	0.000	0.000	0.000	0.000	0.000	0.000	0.001	0.001	0.001	0.002	0.004	0.004	0.006	0.006
3.79	0.000	0.000	0.000	0.000	0.000	0.000	0.000	0.000	0.001	0.001	0.001	0.002	0.003	0.004	0.005	0.006
3.80	0.000	0.000	0.000	0.000	0.000	0.000	0.000	0.000	0.000	0.001	0.001	0.002	0.003	0.004	0.005	0.006
3.81	0.000	0.000	0.000	0.000	0.000	0.000	0.000	0.000	0.000	0.001	0.001	0.002	0.003	0.004	0.005	0.005
3.82	0.000	0.000	0.000	0.000	0.000	0.000	0.000	0.000	0.000	0.001	0.001	0.002	0.003	0.004	0.005	0.005
3.83	0.000	0.000	0.000	0.000	0.000	0.000	0.000	0.000	0.000	0.001	0.001	0.002	0.003	0.004	0.005	0.005
3.84	0.000	0.000	0.000	0.000	0.000	0.000	0.000	0.000	0.000	0.001	0.001	0.001	0.002	0.003	0.004	0.005
3.85	0.000	0.000	0.000	0.000	0.000	0.000	0.000	0.000	0.000	0.001	0.001	0.001	0.002	0.003	0.004	0.004
3.86	0.000	0.000	0.000	0.000	0.000	0.000	0.000	0.000	0.000	0.001	0.001	0.001	0.002	0.003	0.004	0.004
3.87	0.000	0.000	0.000	0.000	0.000	0.000	0.000	0.000	0.000	0.001	0.001	0.001	0.002	0.003	0.004	0.004
3.88	0.000	0.000	0.000	0.000	0.000	0.000	0.000	0.000	0.000	0.000	0.001	0.001	0.002	0.003	0.004	0.004
3.89	0.000	0.000	0.000	0.000	0.000	0.000	0.000	0.000	0.000	0.000	0.001	0.001	0.002	0.003	0.003	0.004
3.90	0.000	0.000	0.000	0.000	0.000	0.000	0.000	0.000	0.000	0.000	0.001	0.001	0.002	0.003	0.003	0.004

From MIL-STD-414, pp. 47-51.

TABLE T12–8   MIL–STD–414 Table B–1.   Master Table for Normal and Tightened Inspection for Plans Based on Variability Unknown: Standard Deviation Method (Single Specification Limit, Form 1)

Sample size code letter	Sample size	\<-- Acceptable Quality Levels (normal inspection) -->													
		.04	.065	.10	.15	.25	.40	.65	1.00	1.50	2.50	4.00	6.50	10.00	15.00
		k	k	k	k	k	k	k	k	k	k	k	k	k	k
B	3	↓	↓	↓	↓	↓	↓	↓	▼	▼	1.12	.958	.765	.566	.341
C	4	↓	↓	↓	↓	↓	↓	↓	1.45	1.34	1.17	1.01	.814	.617	.393
D	5	↓	↓	↓	↓	↓	↓	1.65	1.53	1.40	1.24	1.07	.874	.675	.455
E	7	↓	↓	↓	↓	2.00	1.88	1.75	1.62	1.50	1.33	1.15	.955	.755	.536
F	10	↓	↓	↓	2.24	2.11	1.98	1.84	1.72	1.58	1.41	1.23	1.03	.828	.611
G	15	2.64	2.53	2.42	2.32	2.20	2.06	1.91	1.79	1.65	1.47	1.30	1.09	.886	.664
H	20	2.69	2.58	2.47	2.36	2.24	2.11	1.96	1.82	1.69	1.51	1.33	1.12	.917	.695
I	25	2.72	2.61	2.50	2.40	2.26	2.14	1.98	1.85	1.72	1.53	1.35	1.14	.936	.712
J	30	2.73	2.61	2.51	2.41	2.28	2.15	2.00	1.86	1.73	1.55	1.36	1.15	.946	.723
K	35	2.77	2.65	2.54	2.45	2.31	2.18	2.03	1.89	1.76	1.57	1.39	1.18	.969	.745
L	40	2.77	2.66	2.55	2.44	2.31	2.18	2.03	1.89	1.76	1.58	1.39	1.18	.971	.746
M	50	2.83	2.71	2.60	2.50	2.35	2.22	2.08	1.93	1.80	1.61	1.42	1.21	1.00	.774
N	75	2.90	2.77	2.66	2.55	2.41	2.27	2.12	1.98	1.84	1.65	1.46	1.24	1.03	.804
O	100	2.92	2.80	2.69	2.58	2.43	2.29	2.14	2.00	1.86	1.67	1.48	1.26	1.05	.819
P	150	2.96	2.84	2.73	2.61	2.47	2.33	2.18	2.03	1.89	1.70	1.51	1.29	1.07	.841
Q	200	2.97	2.85	2.73	2.62	2.47	2.33	2.18	2.04	1.89	1.70	1.51	1.29	1.07	.845
		.065	.10	.15	.25	.40	.65	1.00	1.50	2.50	4.00	6.50	10.00	15.00	
		\<-- Acceptable Quality Levels (tightened inspection) -->													

All AQL values are in percent defective.

↓ Use first sampling plan below arrow, that is, both sample size as well as k value. When sample size equals or exceeds lot size, every item in the lot must be inspected.

From MIL–STD–414, p. 39.

TABLE T12-9 MIL-STD-414 Table B-2. Master Table for Reduced Inspection for Plans Based on Variability Unknown: Standard Deviation Method (Single Specification Limit, Form 1)

Sample size code letter	Sample size	Acceptable Quality Levels												
		.04	.065	.10	.15	.25	.40	.65	1.00	1.50	2.50	4.00	6.50	10.00
		k	k	k	k	k	k	k	k	k	k	k	k	k
B	3	↓	↓	↓	↓	↓	↓	↓	↓	1.12	.958	.765	.566	.341
C	3	↓	↓	↓	↓	↓	↓	↓	↓	1.12	.958	.765	.566	.341
D	3	↓	↓	↓	↓	↓	↓	↓	↓	1.12	.958	.765	.566	.341
E	3	↓	↓	↓	↓	↓	↓	↓	↓	1.12	.958	.765	.566	.341
F	4	↓	↓	↓	↓	↓	↓	1.45	1.34	1.17	1.01	.814	.617	.393
G	5	↓	↓	↓	↓	↓	1.65	1.53	1.40	1.24	1.07	.874	.675	.455
H	7	↓	↓	↓	2.00	1.88	1.75	1.62	1.50	1.33	1.15	.955	.755	.536
I	10	↓	↓	2.24	2.11	1.98	1.84	1.72	1.58	1.41	1.23	1.03	.828	.611
J	10	↓	↓	2.24	2.11	1.98	1.84	1.72	1.58	1.41	1.23	1.03	.828	.611
K	15	2.53	2.42	2.32	2.20	2.06	1.91	1.79	1.65	1.47	1.30	1.09	.886	.664
L	20	2.58	2.47	2.36	2.24	2.11	1.96	1.82	1.69	1.51	1.33	1.12	.917	.695
M	20	2.58	2.47	2.36	2.24	2.11	1.96	1.82	1.69	1.51	1.33	1.12	.917	.695
N	25	2.61	2.50	2.40	2.26	2.14	1.98	1.85	1.72	1.53	1.35	1.14	.936	.712
O	30	2.61	2.51	2.41	2.28	2.15	2.00	1.86	1.73	1.55	1.36	1.15	.946	.723
P	50	2.71	2.60	2.50	2.35	2.22	2.08	1.93	1.80	1.61	1.42	1.21	1.00	.774
Q	75	2.77	2.66	2.55	2.41	2.27	2.12	1.98	1.84	1.65	1.46	1.24	1.03	.804

All AQL values are in percent defective.

↓ Use first sampling plan below arrow, that is, both sample size as well as k value. When sample size equals or exceeds lot size, every item in the lot must be inspected.

From MIL-STD-414, p. 40.

TABLE T12-10   MIL-STD-414   Table B-8.   Values of F for Maximum Standard Deviation (MSD)

Sample size code letter	Sample size	Acceptable Quality Levels (in percent defective)													
		.04	.065	.10	.15	.25	.40	.65	1.00	1.50	2.50	4.00	6.50	10.00	15.00
B	3										.436	.453	.475	.502	.538
C	4								.339	.353	.374	.399	.432	.472	.528
D	5							.294	.308	.323	.346	.372	.408	.452	.511
E	7					.242	.253	.266	.280	.295	.318	.345	.381	.425	.485
F	10				.214	.224	.235	.248	.261	.276	.298	.324	.359	.403	.460
G	15	.182	.188	.195	.202	.211	.222	.235	.248	.262	.284	.309	.344	.386	.442
H	20	.177	.183	.190	.197	.206	.216	.229	.242	.255	.277	.302	.336	.377	.432
I	25	.174	.180	.187	.193	.203	.212	.225	.238	.251	.273	.297	.331	.372	.426
J	30	.173	.179	.185	.192	.201	.210	.223	.236	.249	.270	.295	.328	.369	.423
K	35	.170	.176	.183	.189	.198	.208	.220	.232	.245	.266	.291	.323	.364	.416
L	40	.169	.176	.182	.188	.198	.207	.219	.232	.245	.266	.290	.323	.363	.416
M	50	.166	.172	.178	.184	.194	.203	.214	.227	.241	.261	.284	.317	.356	.408
N	75	.162	.168	.174	.181	.189	.199	.211	.223	.235	.255	.279	.310	.348	.399
O	100	.160	.166	.172	.179	.187	.197	.208	.220	.233	.253	.276	.307	.345	.395
P	150	.158	.163	.170	.175	.185	.193	.206	.216	.230	.249	.271	.302	.341	.388
Q	200	.157	.163	.168	.175	.183	.193	.203	.215	.228	.248	.269	.302	.338	.386

The MSD may be obtained by multiplying the factor F by the difference between the upper specification limit U and lower specification limit L. The formula is MSD = F(U-L). The MSD serves as a guide for the magnitude of the estimate of lot standard deviation when using plans for the double specification limit case, based on the estimate of lot standard deviation of unknown variability. The estimate of lot standard deviation, if it is less than the MSD, helps to insure, but does not guarantee, lot acceptability.

NOTE: There is a corresponding acceptability constant in Table B-1 for each value of F. For reduced inspection, find the acceptability constant of Table B-2 in Table B-1 and use the corresponding value of F.

From MIL-STD-414, p. 58.

TABLE T13-1   Values of Plotting Positions (p_i) to be Used in Plotting on Normal Probability Paper for the No-Calc Procedure

n	p_1	p_2	p_3	p_4	p_5	p_6	p_7	p_8	p_9	p_{10}
2	18.775									
3	14.020	50.000								
4	10.982	38.288								
5	8.940	31.271	50.000							
6	7.490	26.485	42.231							
7	6.416	22.979	36.620	50.000						
8	5.592	20.290	32.350	44.140						
9	4.942	18.159	28.979	39.537	50.000					
10	4.419	16.426	26.245	35.816	45.282					
11	3.988	14.990	23.980	32.740	41.392	50.000				
12	3.629	13.779	22.073	30.151	38.125	46.047				
13	3.326	12.746	20.444	27.941	35.339	42.682	50.000			
14	3.066	11.853	19.036	26.032	32.933	39.779	46.596			
15	2.841	11.075	17.807	24.365	30.834	37.248	43.631	50.000		
16	2.645	10.390	16.724	22.897	28.985	35.021	41.024	47.010		
17	2.473	9.783	15.764	21.595	27.345	33.045	38.712	44.361	50.000	
18	2.321	9.241	14.906	20.431	25.879	31.279	36.648	41.996	47.333	
19	2.185	8.7545	14.136	19.384	24.561	29.692	34.793	39.872	44.939	50.000
20	2.063	8.3158	13.439	18.438	23.370	28.258	33.110	37.962	42.779	47.589

Note: When $i > n/2$ use $p_i = 100 - p_{n-i+1}$

For $n > 20$ use $p_i = \dfrac{2i-1}{2n}$

From H. Chernoff and G. J. Lieberman, *Industrial Quality Control*, 13(7): 1957, p. 5.   Reprinted by permission.

TABLE T13-2   Values of Maximum Estimated Percentage Defective Allowing Acceptance of the Lot (p^*)

Code Letter	Sample Size	AQL								
		0.40	0.65	1.00	1.50	2.50	4.00	6.50	10.00	15.00
B	3			5.70	7.33	10.24	13.95	19.35	26.12	35.02
C	4			5.21	6.76	10.12	13.69	18.81	25.13	33.46
D	5		3.97	5.21	6.76	9.38	12.76	17.60	23.60	31.41
E	7	2.49	3.42	4.54	5.93	8.33	11.45	15.96	21.56	28.81
F	10	2.11	2.92	3.88	5.19	7.36	10.23	14.43	19.71	26.48
G	15	1.78	2.56	3.44	4.60	6.66	9.34	13.33	18.36	24.94
H	20	1.62	2.37	3.23	4.31	6.26	8.86	12.74	17.63	24.06
I	25	1.56	2.27	3.11	4.18	6.09	8.63	12.44	17.26	23.65
J	30	1.50	2.19	3.01	4.07	5.94	8.46	12.23	17.00	23.29
K	35	1.40	2.03	2.83	3.82	5.63	8.08	11.74	16.44	22.65
L	40	1.40	2.02	2.83	3.81	5.62	8.05	11.72	16.40	22.61
M	50	1.27	1.81	2.58	3.51	5.22	7.57	11.12	15.69	21.78
N	75	1.13	1.66	2.33	3.23	4.86	7.11	10.54	14.99	20.95
O	100	1.06	1.57	2.22	3.09	4.68	6.87	10.24	14.63	20.52
P	150	0.97	1.45	2.07	2.89	4.42	6.54	9.82	14.11	19.92
Q	200	0.96	1.43	2.05	2.87	4.38	6.49	9.76	14.04	19.84

From H. Chernoff and G. J. Lieberman, *Industrial Quality Control*, 13(7): 1957, p. 5.   Reprinted by permission.

TABLE T13-3   Matched Attributes, Narrow Limit, Known (σ) and Unknown (s) Standard Deviation Variables Plans for Values of p_1 and p_2 with α = .05, β = .10

p_1	p_2	Attributes		NL-Gauge			Variables		
		n	c	n	c	t	n_σ	n_s	k
.001	.0015	40071	50	855	440	3.06	572	3180	3.02
	.002	11729	17	285	135	2.91	191	1032	2.97
	.0025	6114	10	160	80	2.94	107	567	2.93
	.003	3888	7	110	55	2.91	74	381	2.90
	.004	1976	4	67	35	2.92	45	226	2.84
	.005	1319	3	49	23	2.75	33	160	2.80
	.006	1099	3	38	19	2.80	26	124	2.77
	.007	749	2	32	14	2.62	22	102	2.73
	.008	655	2	28	14	2.75	19	87	2.71
	.009	582	2	25	12	2.68	17	76	2.68
	.01	524	2	22	11	2.72	15	67	2.66
	.012	318	1	19	8	2.49	13	55	2.62
	.015	254	1	15	7	2.58	11	44	2.57
	.02	190	1	12	6	2.61	8	34	2.51
	.025	152	1	10	5	2.58	7	27	2.46
	.03	127	1	9	3	2.13	6	23	2.41
	.035	108	1	8	4	2.52	6	20	2.37
	.04	56	0	7	3	2.34	5	18	2.34
	.05	44	0	6	3	2.48	5	15	2.28
	.06	37	0	5	2	2.23	4	13	2.23
.0025	.004	11467	37	533	268	2.73	357	1678	2.72
	.005	4689	17	240	113	2.61	161	736	2.68
	.006	2743	11	148	71	2.60	99	443	2.64
	.0075	1554	7	91	44	2.57	62	267	2.60
	.01	789	4	56	28	2.56	38	157	2.54
	.012	549	3	42	20	2.47	29	117	2.50
	.015	439	3	32	16	2.49	22	85	2.45
	.02	261	2	23	12	2.49	16	59	2.38
	.025	209	2	18	9	2.40	12	45	2.33
	.03	127	1	15	7	2.29	10	37	2.29
	.035	108	1	13	6	2.25	9	31	2.25
	.04	95	1	12	7	2.53	8	27	2.21
	.05	76	1	10	3	1.77	7	22	2.15
	.06	63	1	8	4	2.26	6	18	2.10
.005	.0075	8011	50	622	314	2.51	417	1714	2.50
	.01	2343	17	206	111	2.54	138	547	2.44
	.012	1370	11	128	62	2.37	85	327	2.40
	.015	776	7	78	41	2.43	53	196	2.35
	.02	394	4	47	22	2.23	32	114	2.28
	.025	263	3	34	17	2.27	23	79	2.23
	.03	219	3	27	14	2.28	18	61	2.19
	.035	149	2	22	10	2.09	15	49	2.15
	.04	130	2	19	9	2.11	13	41	2.11
	.05	104	2	15	8	2.22	10	31	2.05
	.06	63	1	12	6	2.11	9	25	2.00
	.07	54	1	11	4	1.73	8	21	1.96

TABLE T13-3   (continued)

p_1	p_2	Attributes		NL-Gauge			Variables		
		n	c	n	c	t	n_σ	n_s	k
.0075	.01	11158	98	1137	571	2.38	763	2909	2.37
	.012	3820	37	420	212	2.35	279	1040	2.33
	.015	1561	17	186	98	2.36	125	450	2.29
	.02	703	9	90	43	2.18	60	208	2.22
	.025	416	6	58	29	2.19	39	129	2.17
	.03	262	4	43	23	2.24	29	92	2.12
	.035	187	3	33	15	2.01	23	71	2.08
	.04	164	3	28	12	1.92	19	58	2.05
	.05	104	2	21	10	1.99	14	42	1.99
	.06	86	2	17	8	1.94	12	33	1.94
	.07	74	2	14	6	1.81	10	27	1.90
	.08	47	1	12	6	1.96	9	23	1.86
.01	.015	4003	50	525	258	2.22	351	1231	2.24
	.02	1170	17	173	85	2.16	116	388	2.17
	.025	609	10	96	52	2.24	64	208	2.12
	.03	387	7	65	31	2.04	44	137	2.08
	.035	261	5	49	24	2.04	33	100	2.04
	.04	196	4	39	20	2.07	26	78	2.00
	.045	174	4	32	17	2.09	22	64	1.97
	.05	131	3	28	14	1.99	19	54	1.94
	.06	109	3	22	11	1.95	15	41	1.89
	.07	74	2	18	8	1.78	12	33	1.85
	.08	64	2	15	7	1.81	11	27	1.81
	.09	57	2	13	6	1.78	9	23	1.77
	.10	37	1	12	6	1.84	8	20	1.74
.015	.02	5576	98	945	470	2.10	633	2036	2.11
	.025	1566	31	290	143	2.04	195	603	2.05
	.03	779	17	154	79	2.05	103	309	2.01
	.035	468	11	101	52	2.02	67	197	1.97
	.04	320	8	73	37	1.97	49	140	1.93
	.045	257	7	57	29	1.95	39	107	1.90
	.05	207	6	47	24	1.93	32	86	1.88
	.06	130	4	34	16	1.79	23	61	1.82
	.07	93	3	27	13	1.78	18	46	1.78
	.08	81	3	22	11	1.80	15	37	1.74
	.09	57	2	19	8	1.57	13	31	1.70
	.10	51	2	16	8	1.75	11	26	1.67
	.11	47	2	14	7	1.73	10	23	1.64
	.12	43	2	13	6	1.61	9	20	1.61
	.13	39	2	12	5	1.48	8	18	1.58
	.14	36	2	11	4	1.33	8	16	1.56
	.15	25	1	10	4	1.41	7	15	1.53

TABLE T13-3   (continued)

p_1	p_2	Attributes n	c	NL-Gauge n	c	t	Variables n_σ	n_s	k
.02	.03	1963	49	429	223	2.01	287	835	1.96
	.035	958	26	219	110	1.93	147	416	1.92
	.04	584	17	140	67	1.84	94	259	1.88
	.045	390	12	100	49	1.84	67	182	1.85
	.05	304	10	77	40	1.89	52	137	1.82
	.06	193	7	52	26	1.80	35	89	1.77
	.07	130	5	39	19	1.73	26	64	1.73
	.08	97	4	31	15	1.69	21	50	1.69
	.09	86	4	26	13	1.70	17	40	1.65
	.10	65	3	22	12	1.79	15	34	1.62
	.11	59	3	19	9	1.59	13	29	1.59
	.12	43	2	17	8	1.56	12	25	1.56
	.13	39	2	15	7	1.53	10	22	1.53
	.15	34	2	12	5	1.38	9	18	1.48
	.17	30	2	11	4	1.21	8	15	1.44
	.20	18	1	9	3	1.09	6	12	1.37
.03	.04	2732	96	756	378	1.81	506	1333	1.81
	.045	1283	48	372	177	1.72	250	643	1.78
	.05	781	31	230	110	1.70	154	389	1.75
	.06	388	17	121	61	1.72	81	197	1.70
	.07	233	11	78	39	1.67	53	124	1.65
	.08	159	8	57	29	1.66	38	88	1.61
	.09	128	7	44	21	1.55	30	66	1.58
	.10	90	5	36	18	1.58	24	53	1.54
	.11	82	5	30	16	1.64	20	43	1.51
	.12	64	4	26	12	1.44	18	37	1.48
	.13	59	4	23	10	1.35	16	32	1.46
	.15	43	3	18	9	1.48	13	24	1.41
	.20	25	2	12	6	1.40	8	15	1.30
	.25	20	2	9	5	1.48	6	11	1.20
	.30	12	1	7	4	1.48	5	8	1.12
.04	.06	961	48	334	165	1.63	224	524	1.64
	.07	462	25	170	84	1.59	114	258	1.60
	.08	276	16	108	54	1.57	72	159	1.56
	.09	194	12	77	37	1.49	51	110	1.52
	.10	139	9	58	29	1.51	39	82	1.49
	.11	115	8	47	23	1.46	32	65	1.46
	.12	85	6	39	19	1.43	26	53	1.43
	.13	78	6	33	17	1.48	22	44	1.40
	.14	64	5	28	14	1.42	20	38	1.37
	.15	51	4	25	12	1.35	17	33	1.35
	.17	45	4	20	10	1.37	14	25	1.30
	.20	32	3	16	7	1.16	11	19	1.24
	.25	20	2	11	5	1.15	8	13	1.15
	.30	16	2	9	3	.78	6	9	1.06
	.35	14	2	7	4	1.34	5	7	.98
	.40	9	1	6	2	.69	4	6	.91

TABLE T13-3   (continued)

p_1	p_2	Attributes		NL-Gauge			Variables		
		n	c	n	c	t	n_σ	n_s	k
.05	.07	1131	68	448	227	1.57	300	660	1.55
	.08	542	35	222	106	1.46	149	319	1.51
	.09	333	23	138	71	1.52	93	194	1.47
	.10	220	16	97	46	1.39	65	133	1.44
	.11	158	12	73	36	1.41	49	98	1.41
	.12	125	10	58	28	1.36	39	76	1.38
	.13	97	8	48	23	1.33	32	62	1.35
	.14	81	7	40	20	1.36	27	51	1.33
	.15	68	6	35	19	1.45	24	43	1.30
	.16	63	6	30	14	1.24	21	37	1.28
	.17	52	5	27	13	1.26	18	33	1.26
	.20	38	4	20	10	1.26	14	23	1.19
	.25	25	3	14	8	1.37	10	15	1.10
	.30	16	2	10	5	1.14	7	11	1.01
	.35	14	2	8	4	1.09	6	8	.94
	.40	12	2	7	2	.51	5	7	.86

From E. G. Schilling and D. J. Sommers, *Journal of Quality Technology*, 13(2): 1981, Table 1, pp. 84-85.   Reprinted by permission.

TABLE T13-4   Tightened Inspection Optimal Narrow Limit Plans for MIL-STD-105D

Sample size code letter		Acceptable Quality Levels (tightened inspection)															
		0.010	0.015	0.025	0.040	0.065	0.10	0.15	0.25	0.40	0.65	1.0	1.5	2.5	4.0	6.5	10
A	n																
	t																
	Ac																
	Re																
B	n															3	
	t															0.00	
	Ac															0	
	Re															1	
C	n														3		
	t														1.19		
	Ac														1		
	Re														2		
D	n													4			6
	t													1.14			1.07
	Ac													1			3
	Re													2			4
E	n												5			8	10
	t												1.20			1.67	0.98
	Ac												1			5	5
	Re												2			6	6
F	n											5			9	12	14
	t											1.92			1.43	1.04	1.02
	Ac											2			4	5	7
	Re											3			5	6	8
G	n										6			11	15	18	22
	t										1.92			1.67	1.25	1.02	0.89
	Ac										2			5	6	7	10
	Re										3			6	7	8	11
H	n									7			13	17	21	27	34
	t									1.94			2.27	1.65	1.36	1.23	0.99
	Ac									2			8	8	9	13	17
	Re									3			9	9	10	14	18
J	n								7			14	20	25	33	43	53
	t								2.48			2.17	1.93	1.71	1.41	1.25	1.01
	Ac								3			7	10	12	15	21	26
	Re								4			8	11	13	16	22	27
K	n							8			16	22	28	38	50	64	79
	t							2.79			2.34	2.12	1.96	1.81	1.52	1.34	1.14
	Ac							4			8	11	14	20	25	33	42
	Re							5			9	12	15	21	26	34	43
L	n						9			18	26	32	45	59	76	96	
	t						3.08			2.37	2.39	2.15	1.98	1.81	1.49	1.26	
	Ac						5			8	14	16	23	31	36	45	
	Re						6			9	15	17	24	32	37	46	
M	n					10		20	29	37	51	69	89	114			
	t					3.33		2.54	2.33	2.36	2.16	1.93	1.73	1.58			
	Ac					6		9	13	19	26	34	43	57			
	Re					7		10	14	20	27	35	44	58			
N	n				11		22	32	41	57	77	102	134				
	t				3.56		2.70	2.47	2.40	2.34	2.09	1.90	1.78				
	Ac				7		10	14	19	29	37	48	66				
	Re				8		11	15	20	30	38	49	67				
P	n			11		24	35	45	64	89	118	156					
	t			3.23		2.96	2.89	2.63	2.45	2.28	2.21	1.97					
	Ac			5		12	19	22	31	43	61	76					
	Re			6		13	20	23	32	44	62	77					
Q	n		12			26	37	49	70	99	132	175					
	t		3.46			3.00	2.89	2.83	2.75	2.42	2.32	2.22					
	Ac		6			12	18	25	38	47	65	90					
	Re		7			13	19	26	39	48	66	91					
R	n	14			27	41	54	78	109	147	198						
	t	3.20			3.28	2.91	2.86	2.83	2.75	2.50	2.30						
	Ac	5			14	18	25	40	59	73	94						
	Re	6			15	19	26	41	60	74	95						
S	n				30												
	t				3.36												
	Ac				15												
	Re				16												

From E. G. Schilling and D. J. Sommers, *Journal of Quality Technology*, 13(2): 1981, Table 2, p. 86. Reprinted by permission.

TABLE T13-5   Normal Inspection Optimal Narrow Limit Plans for MIL-STD-105D

Sample size code letter		Acceptable Quality Levels (normal inspection)															
		0.010	0.015	0.025	0.040	0.065	0.10	0.15	0.25	0.40	0.65	1.0	1.5	2.5	4.0	6.5	10
A	n															2	
	t															0.00	
	Ac															0	
	Re															1	
B	n														3		
	t														0.00		
	Ac														0		
	Re														1		
C	n													3			5
	t													1.19			0.00
	Ac													1			1
	Re													2			2
D	n												4			6	7
	t												1.14			1.07	0.48
	Ac												1			3	3
	Re												2			4	4
E	n											5			8	10	11
	t											1.20			1.67	0.98	0.61
	Ac											1			5	5	5
	Re											2			6	6	6
F	n										5			9	12	14	17
	t										1.92			1.43	1.04	1.02	0.60
	Ac										2			4	5	7	8
	Re										3			5	6	8	9
G	n									6			11	15	18	22	26
	t									1.92			1.67	1.25	1.02	0.89	0.77
	Ac									2			5	6	7	10	13
	Re									3			6	7	8	11	14
H	n								7			13	17	21	27	32	38
	t								1.94			2.27	1.65	1.36	1.23	1.00	0.84
	Ac								2			8	8	9	13	15	19
	Re								3			9	9	10	14	16	20
J	n							7			14	20	25	33	40	48	56
	t							2.48			2.17	1.93	1.71	1.41	1.29	1.20	0.98
	Ac							3			7	10	12	15	19	25	29
	Re							4			8	11	13	16	20	26	30
K	n							8		16	22	28	38	46	56	69	86
	t							2.79		2.34	2.12	1.96	1.81	1.64	1.45	1.16	1.05
	Ac							4		8	11	14	20	24	29	33	46
	Re							5		9	12	15	21	25	30	34	47
L	n					9			18	26	32	45	54	68	82	108	
	t					3.08			2.37	2.39	2.15	1.98	1.90	1.57	1.63	1.30	
	Ac					5			8	14	16	23	29	32	46	56	
	Re					6			9	15	17	24	30	33	47	57	
M	n				10			20	29	37	51	62	79	98	127		
	t				3.33			2.54	2.33	2.36	2.16	1.84	1.90	1.70	1.55		
	Ac				6			9	13	19	26	27	41	49	66		
	Re				7			10	14	20	27	28	42	50	67		
N	n				11			22	32	41	57	72	90	115	148		
	t				3.56			2.70	2.47	2.40	2.34	2.19	2.05	1.92	1.66		
	Ac				7			10	14	19	29	36	45	58	70		
	Re				8			11	15	20	30	37	46	59	71		
P	n		11			24	35	45	64	81	103	131	171				
	t		3.23			2.96	2.89	2.63	2.45	2.23	2.25	2.21	1.96				
	Ac		5			12	19	22	31	36	52	71	87				
	Re		6			13	20	23	32	37	53	72	88				
Q	n	12				26	37	49	70	89	116	148	196				
	t	3.46				3.00	2.89	2.83	2.75	2.46	2.47	2.23	2.06				
	Ac	6				12	18	25	38	42	61	71	93				
	Re	7				13	19	26	39	43	62	72	94				
R	n			27	41	54	78	99	129	163	222						
	t			3.28	2.91	2.86	2.83	2.78	2.62	2.40	2.25						
	Ac			14	18	25	40	53	67	78	106						
	Re			15	19	26	41	54	68	79	107						
	n																
	t																
	Ac																
	Re																

From E. G. Schilling and D. J. Sommers, *Journal of Quality Technology*, 13(2): 1981, Table 3, p. 87. Reprinted by permission.

TABLE T13-6 Reduced Inspection Optimal Narrow Limit Plans for MIL-STD-105D

Sample size code letter		Acceptable Quality Levels (reduced inspection)														
		0.010	0.015	0.025	0.040	0.065	0.10	0.15	0.25	0.40	0.65	1.0	1.5	2.5	4.0	6.5	10
A	n															2	
	t															0.00	
	Ac															0	
	Re															1	
B	n														2		
	t														0.00		
	Ac														0		
	Re														1		
C	n													2			2
	t													0.00			0.00
	Ac													0			0
	Re													1			2
D	n												3			3	3
	t												0.00			0.00	0.00
	Ac												0			0	1
	Re												1			2	3
E	n											3			5	5	5
	t											1.19			0.00	0.00	0.00
	Ac											1			0	1	1
	Re											2			2	3	4
F	n										4			6	7	8	8
	t										1.14			1.07	0.48	0.00	0.00
	Ac										1			1	2	1	2
	Re										2			4	4	4	5
G	n									5			8	10	11	12	13
	t									1.20			1.67	.98	0.61	0.50	0.00
	Ac									1			3	3	2	4	3
	Re									2			6	6	6	6	6
H	n								5			9	12	14	16	17	19
	t								1.92			1.43	1.25	1.02	0.83	0.60	0.45
	Ac								2			2	4	4	5	6	8
	Re								3			5	7	8	9	9	11
J	n							6			11	15	18	20	22	26	28
	t							1.91			1.90	1.58	1.02	1.14	0.89	0.77	0.49
	Ac							2			4	6	4	7	7	10	11
	Re							3			7	9	8	11	11	14	14
K	n							7		13	17	21	24	27	32	36	41
	t							1.94		1.69	1.65	1.48	1.29	1.32	1.00	0.91	0.86
	Ac							2		3	6	7	8	11	12	15	21
	Re							3		6	9	11	12	15	16	19	24
L	n					7		14	20	25	29	33	40	45	52		
	t					2.48		2.17	1.93	1.71	1.59	1.49	1.29	1.07	0.89		
	Ac					3		5	8	9	11	13	16	17	20		
	Re					4		8	11	13	15	17	20	21	24		
M	n				8		16	23	28	34	39	47	54	64			
	t				2.80		2.34	2.06	1.87	1.84	1.71	1.61	1.41	1.34			
	Ac				4		6	9	10	14	16	21	23	30			
	Re				5		9	12	14	18	20	25	27	34			
N	n			9		18	26	33	39	45	55	64	77				
	t			2.80		2.51	2.20	2.19	2.01	1.98	1.92	1.73	1.60				
	Ac			4		7	10	14	16	20	27	30	37				
	Re			5		10	13	18	20	24	31	34	41				
P	n		10		20	29	37	44	51	64	75	90					
	t		2.57		2.54	2.42	2.36	2.28	2.21	2.12	1.88	1.77					
	Ac		3		7	12	15	20	24	32	34	42					
	Re		4		10	15	20	24	28	36	38	46					
Q	n	10		22	32	41	49	58	73	86	104						
	t	3.22		2.70	2.47	2.34	2.32	2.36	2.21	2.18	1.95						
	Ac	5		8	12	14	19	27	34	43	48						
	Re	6		11	15	19	24	31	38	47	52						
R	n			24	35	45	55	64	82	97	119						
	t			2.86	2.89	2.63	2.54	2.45	2.31	2.29	2.22						
	Ac			9	17	18	23	27	36	46	59						
	Re			12	20	23	28	32	40	50	63						
	n																
	t																
	Ac																
	Re																

From E. G. Schilling and D. J. Sommers, *Journal of Quality Technology*, 13(2): 1981, Table 4, p. 88. Reprinted by permission.

TABLE T13-7 MIL-STD-105D Scheme Probability of Acceptance (P_a) and Average Sample Number (ASN) at AQL Using Narrow Limit Plans. (Limit Numbers for Switching to Reduced Inspection Not Used.)

Sample Size Code Letter		Acceptable Quality Levels (normal inspection)															
		0.010	0.015	0.025	0.040	0.065	0.10	0.15	0.25	0.40	0.65	1.0	1.5	2.5	4.0	6.5	10
A	P_a															.863	
	ASN															2.21	
B	P_a														.885		
	ASN														2.68		
C	P_a													.901			.921
	ASN													2.73			4.48
D	P_a												.901			.908	.987
	ASN												3.67			5.69	4.06
E	P_a										.899				.908	.978	.981
	ASN										4.18				7.55	6.68	8.02
F	P_a										.896			.913	.980	.978	.996
	ASN										4.75			8.64	8.66	11.20	10.60
G	P_a									.899			.922	.978	.977	.994	.996
	ASN									5.74			10.56	11.71	14.76	15.45	16.48
H	P_a							.903				.912	.983	.979	.994	.995	.999
	ASN							6.15				12.31	13.54	17.63	19.55	21.34	21.02
J	P_a							.909			.903	.977	.982	.994	.995	.997	.998
	ASN							6.68			13.74	16.68	21.48	24.29	27.24	29.22	31.12
K	P_a					.903				.911	.975	.976	.996	.995	.998	.998	.999
	ASN					7.71				15.66	18.72	24.63	28.13	32.35	35.21	40.59	44.98
L	P_a				.897			.912		.977	.973	.993	.996	.997	.997	.997	
	ASN				8.35			17.44		22.01	28.78	34.39	38.42	43.94	50.14	59.43	
M	P_a				.902		.924	.978	.976	.992	.994	.999	.997	.998			
	ASN				9.29		19.29	24.93	32.69	39.94	45.67	50.52	59.67	71.01			
N	P_a			.903			.911	.982	.977	.993	.993	.997	.998	.998			
	ASN			10.15			21.45	27.81	37.13	45.12	53.65	60.16	69.42	85.28			
P	P_a		.910			.903	.977	.981	.993	.994	.996	.996	.999				
	ASN		10.67			23.54	30.91	40.90	50.64	59.98	70.47	82.94	96.77				
Q	P_a	.903			.911	.975	.977	.995	.994	.997	.995	.997					
	ASN	11.43			25.31	33.60	45.07	55.33	67.07	79.46	96.36	115.26					
R	P_a			.912	.976	.972	.993	.996	.996	.997	.996						
	ASN			26.79	36.88	49.74	62.84	73.26	89.09	107.25	135.31						

From E. G. Schilling and D. J. Sommers, *Journal of Quality Technology*, 13(2): 1981, Table 5, p. 89. Reprinted by permission.

TABLE T13-8   Joint Probabilities for Mixed Plans

n = 5
i = 0

z_A	FRACTION DEFECTIVE, p						
	.005	.01	.02	.05	.10	.15	.20
-2.50	.9752	.9510	.9039	.7738	.5905	.4437	.3277
-2.45	.9752	.9510	.9039	.7738	.5905	.4437	.3277
-2.40	.9752	.9510	.9039	.7738	.5905	.4437	.3277
-2.35	.9752	.9510	.9039	.7738	.5905	.4437	.3277
-2.30	.9752	.9510	.9039	.7738	.5905	.4437	.3277
-2.25	.9752	.9510	.9039	.7738	.5905	.4437	.3277
-2.20	.9752	.9510	.9039	.7738	.5905	.4437	.3277
-2.15	.9752	.9510	.9039	.7738	.5905	.4437	.3277
-2.10	.9752	.9510	.9039	.7738	.5905	.4437	.3277
-2.05	.9752	.9510	.9039	.7738	.5905	.4437	.3277
-2.00	.9752	.9510	.9039	.7738	.5905	.4437	.3277
-1.95	.9752	.9510	.9039	.7738	.5905	.4437	.3277
-1.90	.9752	.9510	.9039	.7738	.5905	.4437	.3277
-1.85	.9752	.9510	.9039	.7738	.5905	.4437	.3277
-1.80	.9752	.9510	.9039	.7738	.5905	.4437	.3277
-1.75	.9752	.9509	.9039	.7737	.5904	.4437	.3276
-1.70	.9752	.9509	.9038	.7737	.5904	.4436	.3276
-1.65	.9751	.9509	.9038	.7737	.5904	.4436	.3276
-1.60	.9751	.9508	.9037	.7737	.5903	.4435	.3275
-1.55	.9750	.9507	.9037	.7735	.5902	.4434	.3274
-1.50	.9749	.9506	.9035	.7734	.5901	.4433	.3273
-1.45	.9747	.9504	.9033	.7732	.5899	.4431	.3271
-1.40	.9744	.9501	.9030	.7729	.5896	.4428	.3268
-1.35	.9740	.9497	.9027	.7725	.5892	.4425	.3264
-1.30	.9734	.9492	.9021	.7720	.5887	.4419	.3259
-1.25	.9727	.9484	.9013	.7712	.5879	.4412	.3252
-1.20	.9716	.9473	.9003	.7701	.5869	.4401	.3242
-1.15	.9702	.9459	.8989	.7687	.5855	.4388	.3228
-1.10	.9683	.9440	.8970	.7669	.5836	.4370	.3211
-1.05	.9658	.9416	.8945	.7644	.5812	.4346	.3188
-1.00	.9626	.9383	.8913	.7612	.5780	.4315	.3159
-0.95	.9584	.9342	.8871	.7571	.5740	.4276	.3121
-0.90	.9532	.9289	.8819	.7518	.5689	.4227	.3075
-0.85	.9466	.9223	.8753	.7453	.5626	.4167	.3018
-0.80	.9384	.9142	.8672	.7373	.5548	.4093	.2949
-0.75	.9285	.9043	.8573	.7275	.5454	.4004	.2867
-0.70	.9165	.8923	.8453	.7158	.5342	.3899	.2771
-0.65	.9022	.8780	.8311	.7018	.5209	.3776	.2660
-0.60	.8854	.8613	.8144	.6855	.5055	.3634	.2533
-0.55	.8659	.8418	.7951	.6666	.4878	.3473	.2391
-0.50	.8436	.8195	.7729	.6451	.4678	.3294	.2235

TABLE T13-8   (continued)

n = 5
i = 0

z_A	FRACTION DEFECTIVE, p						
	.005	.01	.02	.05	.10	.15	.20
-0.45	.8182	.7942	.7478	.6208	.4456	.3096	.2067
-0.40	.7899	.7660	.7198	.5939	.4211	.2883	.1889
-0.35	.7586	.7348	.6890	.5644	.3947	.2656	.1703
-0.30	.7245	.7008	.6554	.5325	.3666	.2419	.1513
-0.25	.6877	.6642	.6193	.4985	.3371	.2176	.1323
-0.20	.6486	.6254	.5811	.4628	.3067	.1930	.1137
-0.15	.6075	.5846	.5411	.4258	.2758	.1687	.0959
-0.10	.5649	.5424	.4998	.3880	.2449	.1452	.0792
-0.05	.5213	.4992	.4578	.3500	.2146	.1227	.0639
0.00	.4771	.4557	.4155	.3123	.1854	.1018	.0503
0.05	.4331	.4123	.3736	.2755	.1577	.0828	.0386
0.10	.3897	.3696	.3326	.2401	.1320	.0659	.0287
0.15	.3475	.3282	.2930	.2066	.1086	.0511	.0206
0.20	.3070	.2886	.2554	.1754	.0876	.0387	.0143
0.25	.2685	.2512	.2201	.1468	.0694	.0285	.0095
0.30	.2326	.2163	.1875	.1210	.0537	.0203	.0060
0.35	.1993	.1842	.1577	.0982	.0407	.0140	.0036
0.40	.1690	.1551	.1311	.0784	.0300	.0093	.0020
0.45	.1418	.1291	.1075	.0615	.0216	.0059	.0010
0.50	.1176	.1061	.0869	.0474	.0151	.0036	.0005
0.55	.0964	.0861	.0693	.0358	.0102	.0020	.0002
0.60	.0781	.0691	.0545	.0265	.0066	.0011	.0001
0.65	.0625	.0546	.0421	.0191	.0042	.0005	.0000
0.70	.0494	.0426	.0321	.0135	.0025	.0002	.0000
0.75	.0386	.0328	.0240	.0093	.0014	.0001	.0000
0.80	.0297	.0249	.0177	.0063	.0008	.0000	.0000
0.85	.0226	.0186	.0128	.0041	.0004	.0000	.0000
0.90	.0169	.0136	.0091	.0026	.0002	.0000	.0000
0.95	.0125	.0099	.0063	.0016	.0001	.0000	.0000
1.00	.0091	.0070	.0043	.0009	.0000	.0000	.0000
1.05	.0066	.0049	.0028	.0005	.0000	.0000	.0000
1.10	.0046	.0034	.0018	.0003	.0000	.0000	.0000
1.15	.0032	.0023	.0012	.0001	.0000	.0000	.0000
1.20	.0022	.0015	.0007	.0001	.0000	.0000	.0000
1.25	.0015	.0010	.0004	.0000	.0000	.0000	.0000
1.30	.0010	.0006	.0003	.0000	.0000	.0000	.0000
1.35	.0006	.0004	.0001	.0000	.0000	.0000	.0000
1.40	.0004	.0002	.0001	.0000	.0000	.0000	.0000
1.45	.0003	.0001	.0000	.0000	.0000	.0000	.0000
1.50	.0002	.0001	.0000	.0000	.0000	.0000	.0000
1.55	.0001	.0000	.0000	.0000	.0000	.0000	.0000
1.60	.0001	.0000	.0000	.0000	.0000	.0000	.0000
1.65	.0000	.0000	.0000	.0000	.0000	.0000	.0000
1.70	.0000	.0000	.0000	.0000	.0000	.0000	.0000
1.75	.0000	.0000	.0000	.0000	.0000	.0000	.0000
1.80	.0000	.0000	.0000	.0000	.0000	.0000	.0000

TABLE T13-8   (continued)

n = 5
i = 1

z_A	FRACTION DEFECTIVE, p						
	.005	.01	.02	.05	.10	.15	.20
-2.50	.024	.048	.092	.204	.328	.391	.410
-2.45	.024	.048	.092	.204	.328	.391	.410
-2.40	.024	.048	.092	.204	.328	.391	.410
-2.35	.024	.048	.092	.204	.328	.391	.410
-2.30	.024	.048	.092	.204	.328	.391	.410
-2.25	.024	.048	.092	.204	.328	.391	.410
-2.20	.024	.048	.092	.204	.328	.391	.410
-2.15	.024	.048	.092	.204	.328	.391	.410
-2.10	.024	.048	.092	.204	.328	.391	.410
-2.05	.024	.048	.092	.204	.328	.391	.410
-2.00	.024	.048	.092	.204	.328	.391	.410
-1.95	.024	.048	.092	.204	.328	.391	.410
-1.90	.024	.048	.092	.204	.328	.391	.410
-1.85	.024	.048	.092	.204	.328	.391	.410
-1.80	.024	.048	.092	.204	.328	.391	.410
-1.75	.024	.048	.092	.204	.328	.391	.410
-1.70	.024	.048	.092	.204	.328	.391	.410
-1.65	.024	.048	.092	.204	.328	.391	.410
-1.60	.024	.048	.092	.204	.328	.391	.410
-1.55	.024	.048	.092	.204	.328	.391	.410
-1.50	.024	.048	.092	.204	.328	.391	.410
-1.45	.024	.048	.092	.204	.328	.391	.410
-1.40	.024	.048	.092	.204	.328	.391	.410
-1.35	.024	.048	.092	.204	.328	.391	.410
-1.30	.024	.048	.092	.204	.328	.391	.410
-1.25	.024	.048	.092	.204	.328	.391	.409
-1.20	.024	.048	.092	.204	.328	.391	.409
-1.15	.024	.048	.092	.204	.328	.391	.409
-1.10	.024	.048	.092	.204	.328	.391	.409
-1.05	.024	.048	.092	.204	.328	.391	.409
-1.00	.024	.048	.092	.204	.328	.391	.409
-0.95	.024	.048	.092	.203	.328	.391	.408
-0.90	.024	.048	.092	.203	.327	.390	.408
-0.85	.024	.048	.092	.203	.327	.390	.407
-0.80	.024	.048	.092	.203	.327	.389	.406
-0.75	.024	.048	.092	.203	.326	.388	.404
-0.70	.024	.048	.092	.203	.326	.387	.402
-0.65	.024	.048	.092	.202	.325	.385	.398
-0.60	.024	.048	.092	.202	.323	.382	.394
-0.55	.024	.048	.092	.201	.321	.379	.389
-0.50	.024	.048	.091	.201	.319	.374	.383

TABLE T13-8   (continued)

n = 5
i = 1

z_A	.005	.01	.02	.05	.10	.15	.20
-0.45	.024	.048	.091	.199	.316	.369	.375
-0.40	.024	.047	.091	.198	.312	.362	.365
-0.35	.024	.047	.090	.196	.307	.353	.353
-0.30	.024	.047	.090	.194	.301	.343	.339
-0.25	.024	.047	.089	.191	.294	.331	.322
-0.20	.024	.046	.088	.187	.285	.317	.304
-0.15	.024	.046	.086	.183	.275	.302	.283
-0.10	.023	.045	.085	.178	.263	.284	.261
-0.05	.023	.044	.083	.172	.250	.264	.237
0.00	.023	.043	.081	.165	.236	.243	.212
0.05	.022	.042	.078	.158	.220	.221	.187
0.10	.022	.041	.075	.149	.203	.198	.161
0.15	.021	.039	.072	.140	.185	.175	.136
0.20	.020	.038	.068	.130	.167	.151	.113
0.25	.019	.036	.064	.120	.148	.129	.091
0.30	.018	.034	.060	.109	.129	.108	.072
0.35	.017	.032	.055	.098	.111	.088	.055
0.40	.016	.029	.051	.087	.094	.070	.041
0.45	.015	.027	.046	.077	.078	.055	.029
0.50	.014	.025	.041	.066	.064	.041	.020
0.55	.013	.022	.037	.057	.051	.030	.014
0.60	.011	.020	.032	.047	.039	.022	.009
0.65	.010	.018	.028	.039	.030	.015	.005
0.70	.009	.015	.024	.032	.022	.010	.003
0.75	.008	.013	.020	.025	.016	.006	.002
0.80	.007	.011	.017	.020	.011	.004	.001
0.85	.006	.009	.014	.015	.007	.002	.000
0.90	.005	.008	.011	.011	.005	.001	.000
0.95	.004	.006	.009	.008	.003	.001	.000
1.00	.003	.005	.007	.006	.002	.000	.000
1.05	.003	.004	.005	.004	.001	.000	.000
1.10	.002	.003	.004	.003	.001	.000	.000
1.15	.002	.002	.003	.002	.000	.000	.000
1.20	.001	.002	.002	.001	.000	.000	.000
1.25	.001	.001	.001	.001	.000	.000	.000
1.30	.001	.001	.001	.000	.000	.000	.000
1.35	.001	.001	.001	.000	.000	.000	.000
1.40	.000	.000	.000	.000	.000	.000	.000
1.45	.000	.000	.000	.000	.000	.000	.000
1.50	.000	.000	.000	.000	.000	.000	.000
1.55	.000	.000	.000	.000	.000	.000	.000
1.60	.000	.000	.000	.000	.000	.000	.000
1.65	.000	.000	.000	.000	.000	.000	.000
1.70	.000	.000	.000	.000	.000	.000	.000
1.75	.000	.000	.000	.000	.000	.000	.000
1.80	.000	.000	.000	.000	.000	.000	.000

TABLE T13-8   (continued)

n = 5
i = 2

z_A	.005	.01	.02	.05	.10	.15	.20
−2.50	.000	.001	.004	.021	.073	.138	.205
−2.45	.000	.001	.004	.021	.073	.138	.205
−2.40	.000	.001	.004	.021	.073	.138	.205
−2.35	.000	.001	.004	.021	.073	.138	.205
−2.30	.000	.001	.004	.021	.073	.138	.205
−2.25	.000	.001	.004	.021	.073	.138	.205
−2.20	.000	.001	.004	.021	.073	.138	.205
−2.15	.000	.001	.004	.021	.073	.138	.205
−2.10	.000	.001	.004	.021	.073	.138	.205
−2.05	.000	.001	.004	.021	.073	.138	.205
−2.00	.000	.001	.004	.021	.073	.138	.205
−1.95	.000	.001	.004	.021	.073	.138	.205
−1.90	.000	.001	.004	.021	.073	.138	.205
−1.85	.000	.001	.004	.021	.073	.138	.205
−1.80	.000	.001	.004	.021	.073	.138	.205
−1.75	.000	.001	.004	.021	.073	.138	.205
−1.70	.000	.001	.004	.021	.073	.138	.205
−1.65	.000	.001	.004	.021	.073	.138	.205
−1.60	.000	.001	.004	.021	.073	.138	.205
−1.55	.000	.001	.004	.021	.073	.138	.205
−1.50	.000	.001	.004	.021	.073	.138	.205
−1.45	.000	.001	.004	.021	.073	.138	.205
−1.40	.000	.001	.004	.021	.073	.138	.205
−1.35	.000	.001	.004	.021	.073	.138	.205
−1.30	.000	.001	.004	.021	.073	.138	.205
−1.25	.000	.001	.004	.021	.073	.138	.205
−1.20	.000	.001	.004	.021	.073	.138	.205
−1.15	.000	.001	.004	.021	.073	.138	.205
−1.10	.000	.001	.004	.021	.073	.138	.205
−1.05	.000	.001	.004	.021	.073	.138	.205
−1.00	.000	.001	.004	.021	.073	.138	.205
−0.95	.000	.001	.004	.021	.073	.138	.205
−0.90	.000	.001	.004	.021	.073	.138	.205
−0.85	.000	.001	.004	.021	.073	.138	.205
−0.80	.000	.001	.004	.021	.073	.138	.205
−0.75	.000	.001	.004	.021	.073	.138	.205
−0.70	.000	.001	.004	.021	.073	.138	.205
−0.65	.000	.001	.004	.021	.073	.138	.205
−0.60	.000	.001	.004	.021	.073	.138	.205
−0.55	.000	.001	.004	.021	.073	.138	.204
−0.50	.000	.001	.004	.021	.073	.138	.204

TABLE T13-8 (continued)

n = 5
i = 2

z_A	.005	.01	.02	.05	.10	.15	.20
-0.45	.000	.001	.004	.021	.073	.138	.204
-0.40	.000	.001	.004	.021	.073	.138	.203
-0.35	.000	.001	.004	.021	.073	.137	.202
-0.30	.000	.001	.004	.021	.073	.137	.201
-0.25	.000	.001	.004	.021	.072	.136	.199
-0.20	.000	.001	.004	.021	.072	.135	.197
-0.15	.000	.001	.004	.021	.072	.134	.194
-0.10	.000	.001	.004	.021	.072	.133	.191
-0.05	.000	.001	.004	.021	.071	.131	.186
0.00	.000	.001	.004	.021	.070	.128	.180
0.05	.000	.001	.004	.021	.069	.125	.173
0.10	.000	.001	.004	.021	.068	.121	.165
0.15	.000	.001	.004	.021	.066	.116	.155
0.20	.000	.001	.004	.020	.065	.111	.144
0.25	.000	.001	.004	.020	.062	.105	.132
0.30	.000	.001	.004	.020	.060	.097	.119
0.35	.000	.001	.004	.019	.056	.090	.105
0.40	.000	.001	.004	.018	.053	.081	.091
0.45	.000	.001	.003	.018	.049	.072	.078
0.50	.000	.001	.003	.017	.045	.063	.065
0.55	.000	.001	.003	.016	.041	.054	.052
0.60	.000	.001	.003	.015	.036	.046	.041
0.65	.000	.001	.003	.014	.031	.037	.031
0.70	.000	.001	.003	.012	.027	.030	.023
0.75	.000	.001	.003	.011	.023	.023	.017
0.80	.000	.001	.002	.010	.018	.018	.012
0.85	.000	.001	.002	.009	.015	.013	.008
0.90	.000	.001	.002	.007	.012	.009	.005
0.95	.000	.001	.002	.006	.009	.006	.003
1.00	.000	.001	.002	.005	.007	.004	.002
1.05	.000	.000	.001	.004	.005	.003	.001
1.10	.000	.000	.001	.003	.003	.002	.001
1.15	.000	.000	.001	.003	.002	.001	.000
1.20	.000	.000	.001	.002	.001	.001	.000
1.25	.000	.000	.001	.001	.001	.000	.000
1.30	.000	.000	.001	.001	.001	.000	.000
1.35	.000	.000	.000	.001	.000	.000	.000
1.40	.000	.000	.000	.000	.000	.000	.000
1.45	.000	.000	.000	.000	.000	.000	.000
1.50	.000	.000	.000	.000	.000	.000	.000
1.55	.000	.000	.000	.000	.000	.000	.000
1.60	.000	.000	.000	.000	.000	.000	.000
1.65	.000	.000	.000	.000	.000	.000	.000
1.70	.000	.000	.000	.000	.000	.000	.000
1.75	.000	.000	.000	.000	.000	.000	.000
1.80	.000	.000	.000	.000	.000	.000	.000

From E. G. Schilling and H. F. Dodge, *Technometrics*, 11(2): 1969, pp. 362-365. Reprinted by permission.

TABLE T14-1   Values of x and y for Determining AOQL

Given c	x	y	Given c	x	y	Given c	x	y	Given c	x	y
0	1.00	0.3679	10	8.05	6.528	20	15.92	13.89	30	24.11	21.70
1	1.62	0.8400	11	8.82	7.233	21	16.73	14.66	31	24.95	22.50
2	2.27	1.371	12	9.59	7.948	22	17.54	15.43	32	25.78	23.30
3	2.95	1.942	13	10.37	8.670	23	18.35	16.20	33	26.62	24.10
4	3.64	2.544	14	11.15	9.398	24	19.17	16.98	34	27.45	24.90
5	4.35	3.168	15	11.93	10.13	25	19.99	17.76	35	28.29	25.71
6	5.07	3.812	16	12.72	10.88	26	20.81	18.54	36	29.13	26.52
7	5.80	4.472	17	13.52	11.62	27	21.63	19.33	37	29.97	27.33
8	6.55	5.146	18	14.31	12.37	28	22.46	20.12	38	30.82	28.14
9	7.30	5.831	19	15.12	13.13	29	23.29	20.91	39	31.66	28.96
10	8.05	6.528	20	15.92	13.89	30	24.11	21.70	40	32.51	29.77

From *Sampling Inspection Tables*, 2nd ed., by H. F. Dodge and H. G. Romig.   Copyright 1944, Bell Telephone Laboratories.   Reprinted by permission of John Wiley and Sons, Inc. and Bell Telephone Laboratories.

TABLE T16-1 Values of Y for Determining AOQL, for SkSP-2 Plans

c	f	n/N=0 i			
		4	6	8	10
1	2/3	0.8682	0.8479	0.8421	0.8405
	1/2	0.8954	0.8564	0.8443	0.8411
	1/3	0.9443	0.8784	0.8493	0.8423
	1/4	0.9861	0.8939	0.8549	0.8436
	1/5	1.0219	0.9125	0.8613	0.8450
2	2/3	1.4281	1.3935	1.3794	1.3741
	1/2	1.4785	1.4163	1.3884	1.3773
	1/3	1.5619	1.4604	1.4081	1.3844
	1/4	1.6284	1.5000	1.4291	1.3927
	1/5	1.6835	1.5349	1.4501	1.4021
3	2/3	2.0294	1.9835	1.9610	1.9505
	1/2	2.1023	2.0229	1.9806	1.9593
	1/3	2.2177	2.0927	2.0205	1.9789
	1/4	2.3067	2.1511	2.0582	2.0004
	1/5	2.3971	2.2006	2.0925	2.0223
4	2/3	2.6604	2.6054	2.5754	2.5594
	1/2	2.7547	2.6615	2.6076	2.5764
	1/3	2.8998	2.7561	2.6683	2.6124
	1/4	3.0097	2.8320	2.7217	2.6482
	1/5	3.0980	2.8948	2.7681	2.6817
5	2/3	3.3140	3.2516	3.2151	3.1939
	1/2	3.4286	3.3242	3.2605	3.2207
	1/3	3.6018	3.4423	3.3417	3.2742
	1/4	3.7312	3.5346	3.4098	3.3236
	1/5	3.8344	3.6100	3.4674	3.3677
6	2/3	3.9857	3.9171	3.8751	3.8491
	1/2	4.1197	4.0058	3.9338	3.8866
	1/3	4.3195	4.1463	4.0347	3.9575
	1/4	4.4673	4.2543	4.1168	4.0198
	1/5	4.5846	4.3416	4.1850	4.0738
7	2/3	4.6726	4.5986	4.5518	4.5215
	1/2	4.8250	4.7029	4.6237	4.5700
	1/3	5.0502	4.8650	4.7437	4.6580
	1/4	5.2156	4.9878	4.8390	4.7326
	1/5	5.3462	5.0865	4.9174	4.7959
8	2/3	5.3722	5.2937	5.2425	5.2083
	1/2	5.5424	5.4130	5.3275	5.2680
	1/3	5.7919	5.5958	5.4658	5.3727
	1/4	5.9740	5.7330	5.5739	5.4590
	1/5	6.1174	5.8425	5.6620	5.5313
9	2/3	6.0829	6.0003	5.9454	5.9077
	1/2	6.2702	6.1343	6.0431	5.9785
	1/3	6.5430	6.3372	6.1993	6.0993
	1/4	6.7413	6.4881	6.3197	6.1970
	1/5	6.8969	6.6080	6.4171	6.2779
10	2/3	6.8033	6.7172	6.6588	6.6180
	1/2	7.0070	6.8654	6.7690	6.7000
	1/3	7.3024	7.0876	6.9425	6.8363
	1/4	7.5162	7.2518	7.0748	6.9450
	1/5	7.6836	7.3819	7.1812	7.0341

From R. L. Perry, Ph.D. Dissertation, Rutgers University, 1970. Reprinted by permission.

TABLE T16-2 Unity Values for SkSP-2 and Matched Single Sampling Plans

Matched Single Sampling Plan			Skip-Lot Plan SkSP-2				Ratio of SkSP-2 Sample Size to Matched Single Sampling Plan Sample Size
c^*	OR	$np^*_{.95}$	f,i	c	OR	$np_{.95}$	
2	6.500	0.818	(1/5,8)	1	6.505	0.598	.731
3	4.890	1.366	(1/5,14)	2	4.883	1.090	.731
4	4.057	1.970	(1/2,4)	3	4.063	1.645	.830
5	3.549	2.613	(1/2,6)	4	3.522	2.270	.868
			(1/2,8)	4	3.574	2.237	.856
			(1/5,8)	3	3.561	1.876	.718
6	3.206	3.285	(1/2,10)	5	3.207	2.892	.880
			(1/4,8)	4	3.191	2.505	.762
7	2.957	3.981	(1/2,12)	6	2.951	3.569	.894
			(1/4,10)	5	2.930	3.166	.795
			(1/5,14)	5	2.963	3.130	.789
			(1/5,6)	4	2.982	2.681	.673
8	2.768	4.695	(2/3,4)	7	2.757	4.270	.909
			(2/3,6)	7	2.777	5.238	.902
			(1/2,14)	7	2.759	4.266	.908
			(1/4,14)	6	2.778	3.791	.807
			(1/5,8)	5	2.782	3.334	.709
9	2.618	5.425	(2/3,6)	8	2.611	4.977	.917
			(2/3,8)	8	2.627	4.947	.912
			(1/2,4)	7	2.627	4.482	.626
			(1/3,10)	7	2.597	4.533	.835
			(1/5,10)	6	2.629	4.007	.738
			(1/5,4)	5	2.594	3.578	.659
10	2.497	6.169	(2/3,8)	9	2.493	5.698	.924
			(2/3,10)	9	2.505	5.670	.919
			(1/3,14)	8	2.507	5.184	.840
			(1/5,12)	7	2.506	4.698	.761
			(1/5,6)	6	2.499	4.215	.683

From H. F. Dodge and R. L. Perry, in *ASQC Technical Conference Transactions*, 1971. Copyright 1971 American Society for Quality Control, Inc. Reprinted by permission.

TABLE T16-3 Poisson Unity Values for Constructing ChSP-1 Plans

i	np_1 for $L(p_1)=0.95$	np_2 for $L(p_2)=0.10$	p_2/p_1	nAOQL	$\dfrac{AOQL}{p_1}$	np_M
1	0.207	2.490	12.029	0.5033	2.431	1.000
2	0.162	2.325	14.352	0.4190	2.586	0.897
3	0.139	2.303	16.568	0.3889	2.798	0.902
4	0.124	2.303	18.573	0.3764	3.036	0.943
5	0.114	2.303	20.202	0.3717	3.261	0.972
6	0.106	2.303	21.726	0.3689	3.500	0.990
7	0.100	2.303	23.030	0.3683	3.483	0.994
8	0.094	2.303	24.500	0.3680	3.915	0.998
9	0.090	2.303	25.589	0.3679	4.088	0.999
10	0.087	2.303	26.471	0.3679	4.229	0.999
∞	0.051	2.303	44.890	0.3680	7.214	1.000

From V. Soundararajan, *Journal of Quality Technology*, 10(3): 1978, p. 58. Reprinted by permission.

TABLE T16-4   ChSP-1 Plans Indexed by AQL($p_{.95}$) and LTPD($p_{.10}$)

LTPD in percent	Sample size	AQL in percent									
		0.10	0.15	0.25	0.40	0.65	1.00	1.50	2.50	4.00	6.50
1.0	228	2									
1.5	152	4	1								
2.0	114	7	2								
2.5	91		3	1							
3.0	76		4	2							
3.5	65			2							
4.0	57			3	1						
4.5	50			4	2						
5.0	45			5	2						
5.5	41			7	3						
6.0	38			9	3						
6.5	35				4	1					
7.0	32				5	1					
7.5	30				5	1					
8.0	28				6	2					
8.5	26				7	2					
9.0	25					2					
9.5	23					3					
10.0	22					3	1				
11.0	20					4	2	1			
12.0	18					5	2	1			
13.0	17					5	2	1	1		
14.0	16						2	1	1		
15.0	15						3	1	1		
16.0	14						3	2	1	1	
17.0	13						4	2	1	1	
18.0	12						5	2	1	1	
19.0	11						6	3	1	1	
20.0	11						6	3	1	1	
21.0	10						7	3	1	1	
22.0	10						7	3	1	1	
23.0	9							4	1	1	
24.0	9							4	1	1	
25.0	8							5	2	1	
30.0	7							7	2	1	
35.0	6								2	1	
40.0	5								4	2	1
50.0	4								7	3	1
60.0	3								6	2	
70.0	2								8	4	

From V. Soundararajan, *Journal of Quality Technology*, 10(3): 1978, p. 101.   Reprinted by permission.

TABLE T16-5 ChSP-1 Plans Indexed by AQL($p_{.95}$) and AOQL

AQL in percent	AOQL in percent																							
	0.10		0.25		0.50		0.75		1.0		1.5		2.0		2.5		3.0		3.5		4.0		4.5	
	n	i	n	i	n	i	n	i	n	i	n	i	n	i	n	i	n	i	n	i	n	i	n	i
0.05	504	1	147																					
0.075			149	5																				
0.10			168	2	73																			
0.15					74	3	49																	
0.20					89	2	50	7	36															
0.25					101	1	51	4	37	9														
0.30							56	2	38	5	24													
0.35							68	1	39	3	25	10												
0.40									42	2	25	7												
0.45									51	1	25	5	18											
0.50									51	1	26	4	19	9										
0.55											26	3	19	6	14									
0.60											28	2	19	5	15	9								
0.65											34	1	19	4	15	8	12							
0.70											34	1	20	3	15	6	13	10						
0.75													20	3	15	5	13	9	10					
0.80													22	2	16	4	13	7	11	10				
0.85													26	1	16	4	13	6	11	9				
0.90													26	1	16	3	13	5	11	8	9			
0.95													26	1	17	2	13	4	11	7	10	10		
1.0													26	1	17	2	13	4	11	6	10	9	8	
1.5																	17	1	15	1	11	2	9	4
2.0																					13	1	12	1

AQL in percent	AOQL in percent																					
	5.0		5.5		6.0		6.5		7.0		7.5		8.0		8.5		9.0		9.5		10.0	
	n	i	n	i	n	i	n	i	n	i	n	i	n	i	n	i	n	i	n	i	n	i
1.0	7		7		6		6															
1.5	8	5	7	7	7	9	6	10	5		5		5		5							
2.0	9	2	7	3	7	4	6	5	6	6	5	7	5	9	5	10	5		4		4	
2.5	11	1	10	1	9	1	7	2	6	3	5	4	5	5	5	6	5	7	4	8	4	9
3.0					9	1	8	1	8	1	7	1	6	2	5	3	5	4	4	4	4	5
3.5									8	1	7	1	7	1	6	1	5	2	5	3	4	3
4.0													7	1	6	1	6	1	6	1	5	2
4.5																	6	1	6	1	5	1
5.0																					5	1

AQL in percent	AOQL in percent																					
	10.5		11.0		11.5		12.0		12.5		13.0		13.5		14.0		14.5		15.0		15.5	
	n	i	n	i	n	i	n	i	n	i	n	i	n	i	n	i	n	i	n	i	n	i
2.00	4																					
2.50	4	10	3		3		3		3		3											
3.00	4	6	4	7	4	8	3	9	3	9	3	10	3		3		3		3			
3.50	4	4	4	4	4	5	3	6	3	6	3	7	3	8	3	9	3	10	3	10	3	
4.00	4	2	4	3	4	3	4	4	3	4	3	5	3	5	3	6	3	7	3	8	3	8
4.50	5	1	5	1	4	2	4	2	4	3	3	3	3	4	3	4	3	5	3	5	3	6
5.00	5	1	5	1	5	1	5	1	4	2	4	2	3	3	3	3	3	3	3	4	3	4
5.50			5	1	5	1	5	1	4	1	4	1	3	2	3	2	3	3	3	3	3	3
6.00							5	1	4	1	4	1	4	1	4	1	4	1	3	2	3	2
6.50											4	1	4	1	4	1	4	1	4	1	4	1
7.00															4	1	4	1	4	1	4	1
7.50																			4	1	4	1

TABLE T16-5   (continued)

AQL in percent	AOQL in percent 16.00		16.50		17.00		17.50		18.00		18.50		19.00		19.50		20.00		30.00	
	n	i	n	i	n	i	n	i	n	i	n	i	n	i	n	i	n	i	n	i
3.50	3,		3,		3,		2,													
4.00	3,	9	3,	9	3,	10	3,	10	2,		2,		2,		2,					
4.50	3,	6	3,	7	3,	8	3,	8	2,	9	2,	9	2,	10	2,	10	2,			
5.00	3,	5	3,	5	3,	6	3,	6	2,	7	2,	7	2,	8	2,	8	2,	9		
5.50	3,	3	3,	4	3,	4	3,	5	2,	5	2,	6	2,	6	2,	6	2,	7		
6.00	3,	3	3,	3	3,	3	3,	4	2,	4	2,	5	2,	5	2,	5	2,	6		
6.50	4,	1	3,	2	3,	2	3,	3	3,	3	3,	3	2,	3	2,	4	2,	4	2,	
7.00	4,	1	4,	1	3,	1	3,	2	3,	2	3,	3	2,	3	2,	3	2,	3	2,	10
7.50	4,	1	4,	1	3,	1	3,	1	3,	1	3,	2	3,	2	3,	2	2,	2	2,	9
8.00			4,	1	3,	1	3,	1	3,	1	3,	1	3,	1	3,	1	2,	2	2,	7
8.50					3,	1	3,	1	3,	1	3,	1	3,	1	3,	1	3,	1	2,	6
9.00									3,	1	3,	1	2,	1	3,	1	3,	1	2,	5
9.50													2,	1	3,	1	3,	1	2,	4
10.00																	3,	1	2,	4

From V. Soundararajan, *Journal of Quality Technology*, 10(3): 1978, pp. 100-101.   Reprinted by permission.

TABLE T17-1   Unity Values for the QSS System

c_N	c_T	$np_{.95}$	$np_{.10}$	OR	$np_{.50}$	h_o	P_N at $np_{.50}$	P_T at $np_{.50}$
*0	0	0.051	2.303	44.891	0.693	0.693	.5000	.5000
1	0	0.308	2.528	8.213	1.146	1.230	.6822	.3178
*1	1	0.355	3.890	10.946	1.678	1.052	.5000	.5000
2	0	0.644	2.821	4.383	1.568	1.748	.7916	.2084
2	1	0.770	4.080	5.301	2.156	1.530	.6346	.3654
*2	2	0.818	5.322	6.509	2.674	1.319	.5000	.5000
3	0	1.005	3.149	3.134	1.976	2.259	.8614	.1386
3	1	1.210	4.335	3.581	2.608	2.012	.7342	.2658
3	2	1.318	5.496	4.170	3.159	1.769	.6116	.3884
*3	3	1.366	6.681	4.890	3.672	1.541	.5000	.5000
4	0	1.375	3.494	2.540	2.376	2.766	.9071	.0929
4	1	1.653	4.633	2.803	3.046	2.496	.8076	.1924
4	2	1.823	5.729	3.142	3.625	2.229	.7017	.2983
4	3	1.921	6.844	3.562	4.161	1.974	.5974	.4026
*4	4	1.970	7.994	4.057	4.671	1.735	.5000	.5000
5	0	1.750	3.849	2.199	2.771	3.271	.9374	.0626
5	1	2.091	4.958	2.371	3.473	2.982	.8612	.1388
5	2	2.320	6.006	2.588	4.077	2.695	.7730	.2270
5	3	2.472	7.062	2.857	4.634	2.418	.6798	.3202
5	4	2.564	8.150	3.179	5.162	2.156	.5875	.4125
*5	5	2.613	9.275	3.549	5.670	1.909	.5000	.5000
6	0	2.127	4.210	1.979	3.162	3.775	.9577	.0423
6	1	2.524	5.300	2.100	3.893	3.470	.9002	.0998
6	2	2.806	6.314	2.251	4.519	3.166	.8286	.1714
6	3	3.009	7.324	2.434	5.095	2.872	.7481	.2519
6	4	3.149	8.357	2.654	5.640	2.589	.6639	.3361
6	5	3.236	9.426	2.913	6.163	2.321	.5801	.4199
*6	6	3.285	10.532	3.206	6.670	2.069	.5000	.5000
7	0	2.505	4.574	1.826	3.550	4.278	.9713	.0287
7	1	2.951	5.654	1.916	4.307	3.960	.9285	.0715
7	2	3.279	6.643	2.026	4.954	3.642	.8714	.1286
7	3	3.530	7.618	2.158	5.547	3.332	.8036	.1964
7	4	3.716	8.606	2.316	6.107	3.033	.7291	.2709
7	5	3.847	9.625	2.502	6.644	2.746	.6516	.3484
7	6	3.932	10.680	2.716	7.164	2.474	.5744	.4256
*7	7	3.981	11.771	2.957	7.669	2.218	.5000	.5000
8	0	2.883	4.941	1.714	3.936	4.781	.9805	.0195
8	1	3.373	6.014	1.783	4.716	4.451	.9489	.0511
8	2	3.743	6.987	1.867	5.382	4.121	.9041	.0959

*Indicates values for single sampling plans.

TABLE T17-1 (continued)

c_N	c_T	$np_{.95}$	$np_{.10}$	OR	$np_{.50}$	h_0	P_N at $np_{.50}$	P_T at $np_{.50}$
8	3	4.036	7.935	1.966	5.992	3.798	.8481	.1519
8	4	4.266	8.889	2.084	6.566	3.484	.7836	.2164
8	5	4.440	9.864	2.222	7.116	3.182	.7139	.2861
8	6	4.564	10.872	2.382	7.647	2.892	.6417	.3583
8	7	4.646	11.915	2.565	8.164	2.617	.5697	.4303
*8	8	4.695	12.995	2.768	8.669	2.357	.5000	.5000
9	0	3.261	5.310	1.628	4.320	5.283	.9867	.0133
9	1	3.790	6.380	1.683	5.123	4.943	.9635	.0365
9	2	4.197	7.342	1.749	5.806	4.602	.9288	.0712
9	3	4.529	8.270	1.826	6.430	4.268	.8833	.1167
9	4	4.798	9.196	1.917	7.018	3.941	.8287	.1713
9	5	5.013	10.135	2.022	7.580	3.625	.7672	.2328
9	6	5.177	11.102	2.144	8.122	3.321	.7013	.2987
9	7	5.296	12.102	2.285	8.649	3.030	.6335	.3665
9	8	5.376	13.137	2.443	9.164	2.752	.5658	.4342
*9	9	5.425	14.206	2.618	9.669	2.488	.5000	.5000
10	0	3.639	5.679	1.561	4.703	5.784	.9909	.0091
10	1	4.203	6.750	1.606	5.526	5.436	.9740	.0260
10	2	4.645	7.704	1.659	6.225	5.086	.9474	.0526
10	3	5.010	8.618	1.720	6.864	4.741	.9109	.0891
10	4	5.315	9.522	1.791	7.464	4.404	.8653	.1347
10	5	5.568	10.434	1.874	8.037	4.076	.8122	.1878
10	6	5.770	11.366	1.970	8.590	3.758	.7533	.2467
10	7	5.927	12.326	2.080	9.127	3.453	.6908	.3092
10	8	6.042	13.318	2.204	9.651	3.159	.6266	.3734
10	9	6.120	14.346	2.344	10.164	2.879	.5625	.4375
*10	10	6.169	15.407	2.497	10.669	2.613	.5000	.5000
11	0	4.017	6.050	1.506	5.085	6.285	.9938	.0062
11	1	4.614	7.121	1.544	5.926	5.929	.9815	.0185
11	2	5.085	8.071	1.587	6.641	5.571	.9612	.0388
11	3	5.481	8.974	1.637	7.293	5.218	.9323	.0677
11	4	5.819	9.862	1.695	7.905	4.871	.8948	.1052
11	5	6.105	10.752	1.761	8.489	4.532	.8496	.1504
11	6	6.344	11.656	1.837	9.052	4.203	.7979	.2021
11	7	6.537	12.581	1.925	9.598	3.884	.7414	.2586
11	8	6.687	13.536	2.024	10.131	3.577	.6818	.3182
11	9	6.798	14.523	2.136	10.653	3.282	.6206	.3794
11	10	6.875	15.545	2.261	11.165	3.001	.5596	.4404
*11	11	6.924	16.598	2.397	11.668	2.732	.5000	.5000
12	0	4.394	6.420	1.461	5.466	6.786	.9958	.0042
12	1	5.021	7.495	1.493	6.325	6.423	.9869	.0131
12	2	5.520	8.442	1.529	7.054	6.058	.9715	.0285
12	3	5.945	9.338	1.571	7.718	5.697	.9488	.0512

*Indicates values for single sampling plans.

TABLE T17-1   (continued)

c_N	c_T	$np_{.95}$	$np_{.10}$	OR	$np_{.50}$	h_o	P_N at $np_{.50}$	P_T at $np_{.50}$
12	4	6.312	10.214	1.618	8.342	5.341	.9183	.0817
12	5	6.629	11.086	1.672	8.936	4.993	.8804	.1196
12	6	6.900	11.966	1.734	9.509	4.653	.8357	.1643
12	7	7.127	12.864	1.805	10.064	4.324	.7855	.2145
12	8	7.312	13.785	1.885	10.605	4.004	.7310	.2690
12	9	7.457	14.736	1.976	11.134	3.696	.6739	.3261
12	10	7.565	15.719	2.078	11.654	3.400	.6155	.3845
12	11	7.640	16.734	2.190	12.165	3.117	.5571	.4429
*12	12	7.690	17.782	2.312	12.668	2.846	.5000	.5000
13	0	4.771	6.792	1.423	5.845	7.287	.9971	.0029
13	1	5.425	7.870	1.451	6.721	6.918	.9907	.0093
13	2	5.951	8.816	1.482	7.464	6.546	.9792	.0208
13	3	6.401	9.707	1.517	8.141	6.178	.9615	.0385
13	4	6.794	10.573	1.556	8.775	5.814	.9369	.0631
13	5	7.139	11.432	1.601	9.379	5.458	.9054	.0946
13	6	7.440	12.294	1.652	9.960	5.109	.8673	.1327
13	7	7.699	13.168	1.710	10.524	4.769	.8233	.1767
13	8	7.917	14.061	1.776	11.073	4.439	.7745	.2255
13	9	8.095	14.979	1.850	11.610	4.119	.7219	.2781
13	10	8.235	15.926	1.934	12.137	3.810	.6670	.3330
13	11	8.340	16.905	2.027	12.655	3.513	.6109	.3891
13	12	8.415	17.916	2.129	13.165	3.228	.5549	.4451
*13	13	8.464	18.958	2.240	13.668	2.956	.5000	.5000
14	0	5.148	7.163	1.391	6.224	7.788	.9980	.0020
14	1	5.828	8.246	1.415	7.116	7.413	.9934	.0066
14	2	6.377	9.193	1.442	7.872	7.036	.9848	.0152
14	3	6.850	10.080	1.472	8.560	6.661	.9711	.0289
14	4	7.268	10.940	1.505	9.204	6.290	.9515	.0485
14	5	7.639	11.787	1.543	9.818	5.926	.9257	.0743
14	6	7.967	12.634	1.586	10.408	5.569	.8936	.1064
14	7	8.255	13.489	1.634	10.980	5.220	.8555	.1445
14	8	8.503	14.359	1.689	11.537	4.880	.8122	.1878
14	9	8.713	15.249	1.750	12.081	4.550	.7646	.2354
14	10	8.885	16.165	1.819	12.615	4.229	.7138	.2862
14	11	9.021	17.109	1.897	13.139	3.920	.6608	.3392
14	12	9.124	18.085	1.982	13.656	3.622	.6069	.3931
14	13	9.197	19.091	2.076	14.165	3.335	.5529	.4471
*14	14	9.246	20.128	2.177	14.668	3.062	.5000	.5000
15	0	5.524	7.535	1.364	6.603	8.289	.9986	.0014
15	1	6.228	8.623	1.385	7.509	7.909	.9953	.0047
15	2	6.799	9.571	1.408	8.278	7.526	.9889	.0111
15	3	7.295	10.456	1.433	8.976	7.145	.9784	.0216
15	4	7.735	11.311	1.462	9.631	6.769	.9629	.0371

*Indicates values for single sampling plans.

TABLE T17-1 (continued)

c_N	c_T	$np_{.95}$	$np_{.10}$	OR	$np_{.50}$	h_o	P_N at $np_{.50}$	P_T at $np_{.50}$
15	5	8.129	12.150	1.495	10.253	6.398	.9419	.0581
15	6	8.482	12.985	1.531	10.852	6.033	.9151	.0849
15	7	8.796	13.824	1.572	11.432	5.676	.8826	.1174
15	8	9.073	14.674	1.617	11.996	5.327	.8447	.1553
15	9	9.312	15.541	1.669	12.547	4.987	.8022	.1978
15	10	9.515	16.429	1.727	13.088	4.656	.7558	.2442
15	11	9.682	17.342	1.791	13.619	4.336	.7065	.2935
15	12	9.814	18.285	1.863	14.141	4.026	.6553	.3447
15	13	9.914	19.257	1.942	14.657	3.727	.6032	.3968
15	14	9.987	20.260	2.029	15.165	3.439	.5512	.4488
*15	15	10.036	21.292	2.122	15.668	3.164	.5000	.5000
16	0	5.900	7.906	1.340	6.980	8.789	.9991	.0009
16	1	6.627	9.000	1.358	7.902	8.405	.9967	.0033
16	2	7.218	9.950	1.378	8.682	8.017	.9920	.0080
16	3	7.734	10.835	1.401	9.391	7.631	.9839	.0161
16	4	8.195	11.686	1.426	10.054	7.249	.9718	.0282
16	5	8.610	12.518	1.454	10.686	6.872	.9548	.0452
16	6	8.986	13.344	1.485	11.292	6.501	.9326	.0674
16	7	9.324	14.170	1.520	11.880	6.136	.9051	.0949
16	8	9.627	15.004	1.559	12.451	5.779	.8724	.1276
16	9	9.894	15.851	1.602	13.009	5.430	.8348	.1652
16	10	10.125	16.715	1.651	13.556	5.090	.7930	.2070
16	11	10.322	17.602	1.705	14.094	4.759	.7478	.2522
16	12	10.484	18.514	1.766	14.622	4.438	.6999	.3001
16	13	10.613	19.454	1.833	15.143	4.128	.6503	.3497
16	14	10.711	20.424	1.907	15.657	3.828	.5999	.4001
16	15	10.783	21.424	1.987	16.165	3.540	.5496	.4504
*16	16	10.832	22.452	2.073	16.668	3.263	.5000	.5000
17	0	6.276	8.278	1.319	7.358	9.290	.9994	.0006
17	1	7.024	9.378	1.335	8.292	8.901	.9977	.0023
17	2	7.635	10.330	1.353	9.084	8.509	.9942	.0058
17	3	8.169	11.215	1.373	9.803	8.118	.9881	.0119
17	4	8.649	12.064	1.395	10.476	7.731	.9786	.0214
17	5	9.084	12.891	1.419	11.115	7.348	.9650	.0350
17	6	9.481	13.710	1.446	11.730	6.971	.9468	.0532
17	7	9.841	14.526	1.476	12.325	6.600	.9237	.0763
17	8	10.167	15.346	1.509	12.903	6.235	.8957	.1043
17	9	10.460	16.176	1.547	13.468	5.879	.8629	.1371
17	10	10.718	17.020	1.588	14.021	5.530	.8257	.1743
17	11	10.943	17.883	1.634	14.564	5.190	.7847	.2153
17	12	11.133	18.768	1.686	15.098	4.859	.7405	.2595
17	13	11.291	19.679	1.743	15.625	4.538	.6939	.3061
17	14	11.417	20.618	1.806	16.145	4.227	.6458	.3542
17	15	11.514	21.585	1.875	16.658	3.926	.5969	.4031

*Indicates values for single sampling plans.

TABLE T17-1   (continued)

c_N	c_T	$np_{.95}$	$np_{.10}$	QR	$np_{.50}$	h_o	P_N at $np_{.50}$	P_T at $np_{.50}$
17	16	11.585	22.582	1.949	17.165	3.637	.5481	.4519
*17	17	11.634	23.606	2.029	17.668	3.359	.5000	.5000
18	0	6.651	8.650	1.300	7.734	9.790	.9996	.0004
18	1	7.419	9.755	1.315	8.682	9.397	.9984	.0016
18	2	8.048	10.711	1.331	9.485	9.001	.9958	.0042
18	3	8.600	11.597	1.348	10.214	8.606	.9912	.0088
18	4	9.098	12.444	1.368	10.895	8.214	.9838	.0162
18	5	9.552	13.268	1.389	11.543	7.826	.9730	.0270
18	6	9.967	14.081	1.413	12.165	7.443	.9582	.0418
18	7	10.348	14.888	1.439	12.766	7.066	.9390	.0610
18	8	10.696	15.698	1.468	13.351	6.695	.9153	.0847
18	9	11.012	16.514	1.500	13.923	6.331	.8869	.1131
18	10	11.295	17.341	1.535	14.482	5.975	.8541	.1459
18	11	11.546	18.183	1.575	15.031	5.626	.8173	.1827
18	12	11.764	19.045	1.619	15.571	5.286	.7770	.2230
18	13	11.950	19.929	1.668	16.103	4.956	.7338	.2662
18	14	12.104	20.839	1.722	16.627	4.634	.6884	.3116
18	15	12.227	21.776	1.781	17.146	4.323	.6416	.3584
18	16	12.322	22.742	1.846	17.658	4.022	.5942	.4058
18	17	12.393	23.736	1.915	18.165	3.731	.5468	.4532
*18	18	12.442	24.756	1.990	18.668	3.453	.5000	.5000
19	0	7.027	9.022	1.284	8.111	10.291	.9997	.0003
19	1	7.813	10.133	1.297	9.071	9.894	.9988	.0012
19	2	8.459	11.092	1.311	9.884	9.494	.9970	.0030
19	3	9.028	11.980	1.327	10.622	9.095	.9935	.0065
19	4	9.542	12.826	1.344	11.312	8.699	.9878	.0122
19	5	10.013	13.648	1.363	11.967	8.306	.9792	.0208
19	6	10.447	14.456	1.384	12.597	7.918	.9673	.0327
19	7	10.847	15.257	1.407	13.205	7.535	.9515	.0485
19	8	11.215	16.057	1.432	13.797	7.159	.9315	.0685
19	9	11.551	16.862	1.460	14.374	6.788	.9072	.0928
19	10	11.857	17.674	1.491	14.939	6.425	.8786	.1214
19	11	12.132	18.500	1.525	15.494	6.068	.8459	.1541
19	12	12.376	19.341	1.563	16.039	5.720	.8095	.1905
19	13	12.589	20.202	1.605	16.577	5.380	.7699	.2301
19	14	12.770	21.085	1.651	17.107	5.049	.7276	.2724
19	15	12.921	21.994	1.702	17.630	4.728	.6834	.3166
19	16	13.042	22.930	1.758	18.147	4.416	.6378	.3622
19	17	13.136	23.894	1.819	18.659	4.114	.5917	.4083
19	18	13.205	24.885	1.884	19.166	3.823	.5455	.4545
*19	19	13.255	25.903	1.954	19.668	3.544	.5000	.5000

*Indicates values for single sampling plans.

TABLE T17-1   (continued)

c_N	c_T	$np_{.95}$	$np_{.10}$	OR	$np_{.50}$	h_o	P_N at $np_{.50}$	P_T at $np_{.50}$
20	0	7.402	9.394	1.269	8.486	10.791	.9998	.0002
20	1	8.206	10.511	1.281	9.459	10.391	.9992	.0008
20	2	8.869	11.474	1.294	10.282	9.988	.9978	.0022
20	3	9.453	12.363	1.308	11.029	9.585	.9952	.0048
20	4	9.983	13.210	1.323	11.727	9.184	.9908	.0092
20	5	10.470	14.030	1.340	12.390	8.788	.9841	.0159
20	6	10.920	14.835	1.358	13.027	8.395	.9745	.0255
20	7	11.337	15.631	1.379	13.642	8.007	.9616	.0384
20	8	11.724	16.423	1.401	14.240	7.625	.9449	.0551
20	9	12.080	17.218	1.425	14.823	7.248	.9242	.0758
20	10	12.407	18.019	1.452	15.394	6.878	.8995	.1005
20	11	12.705	18.829	1.482	15.954	6.515	.8707	.1293
20	12	12.973	19.653	1.515	16.505	6.159	.8382	.1618
20	13	13.210	20.494	1.551	17.047	5.811	.8022	.1978
20	14	13.418	21.354	1.591	17.582	5.472	.7633	.2367
20	15	13.595	22.237	1.636	18.110	5.141	.7219	.2781
20	16	13.742	23.145	1.684	18.632	4.819	.6787	.3213
20	17	13.861	24.080	1.737	19.148	4.507	.6343	.3657
20	18	13.953	25.041	1.795	19.659	4.205	.5893	.4107
20	19	14.023	26.030	1.856	20.166	3.913	.5444	.4556
*20	20	14.072	27.045	1.922	20.668	3.632	.5000	.5000

*Indicates values for single sampling plans.

From L. D. Romboski, Ph.D. Dissertation, Rutgers University, 1969.   Reprinted by permission.

TABLE T17-2   H$_\alpha$ Values for Simplified Grand Lot Sampling

α = .002

k	0	1	2	3	4	5	6	7	8	9
0	0.	0.	2.19	2.78	3.01	3.17	3.28	3.36	3.48	3.48
10	3.53	3.57	3.60	3.64	3.66	3.69	3.71	3.74	3.76	3.77
20	3.79	3.81	3.82	3.84	3.85	3.86	3.88	3.89	3.90	3.91
30	3.92	3.93	3.94	3.95	3.96	3.97	3.97	3.98	3.99	4.00
40	4.00	4.01	4.02	4.02	4.03	4.04	4.04	4.05	4.06	4.06
50	4.07	4.07	4.08	4.08	4.09	4.09	4.10	4.10	4.11	4.11
60	4.11	4.12	4.12	4.13	4.13	4.14	4.14	4.14	4.15	4.15
70	4.15	4.16	4.16	4.17	4.17	4.17	4.18	4.18	4.18	4.19
80	4.19	4.19	4.19	4.20	4.20	4.20	4.21	4.21	4.21	4.21
90	4.22	4.22	4.22	4.23	4.23	4.23	4.23	4.24	4.24	4.24
100	4.24	4.25	4.25	4.25	4.25	4.26	4.26	4.26	4.26	4.26
110	4.27	4.27	4.27	4.27	4.28	4.28	4.28	4.28	4.28	4.29
120	4.29	4.29	4.29	4.29	4.30	4.30	4.30	4.30	4.30	4.30
130	4.31	4.31	4.31	4.31	4.31	4.32	4.32	4.32	4.32	4.32
140	4.32	4.33	4.33	4.33	4.33	4.33	4.33	4.34	4.34	4.34
150	4.34	4.34	4.34	4.34	4.35	4.35	4.35	4.35	4.35	4.35
160	4.35	4.36	4.36	4.36	4.36	4.36	4.36	4.36	4.37	4.37
170	4.37	4.37	4.37	4.37	4.37	4.38	4.38	4.38	4.38	4.38
180	4.38	4.38	4.38	4.39	4.39	4.39	4.39	4.39	4.39	4.39
190	4.39	4.40	4.40	4.40	4.40	4.40	4.40	4.40	4.40	4.40
200	4.41	4.41	4.41	4.41	4.41	4.41	4.41	4.41	4.41	4.42

α = .05

k	0	1	2	3	4	5	6	7	8	9
0	0.	0.	1.39	1.96	2.16	2.30	2.41	2.49	2.56	2.61
10	2.66	2.71	2.74	2.78	2.81	2.84	2.86	2.89	2.91	2.93
20	2.95	2.97	2.98	3.00	3.01	3.03	3.04	3.06	3.07	3.08
30	3.09	3.10	3.11	3.12	3.13	3.14	3.15	3.16	3.17	3.18
40	3.19	3.19	3.20	3.21	3.22	3.22	3.23	3.24	3.24	3.25
50	3.26	3.26	3.27	3.28	3.28	3.29	3.29	3.30	3.30	3.31
60	3.31	3.32	3.32	3.33	3.33	3.34	3.34	3.35	3.35	3.36
70	3.36	3.36	3.37	3.37	3.38	3.38	3.38	3.39	3.39	3.40
80	3.40	3.40	3.41	3.41	3.41	3.42	3.42	3.42	3.43	3.43
90	3.43	3.44	3.44	3.44	3.45	3.45	3.45	3.45	3.46	3.46
100	3.46	3.47	3.47	3.47	3.47	3.48	3.48	3.48	3.49	3.49
110	3.49	3.49	3.50	3.50	3.50	3.50	3.51	3.51	3.51	3.51
120	3.51	3.52	3.52	3.52	3.52	3.53	3.53	3.53	3.53	3.53
130	3.54	3.54	3.54	3.54	3.55	3.55	3.55	3.55	3.55	3.56
140	3.56	3.56	3.56	3.56	3.56	3.57	3.57	3.57	3.57	3.57
150	3.58	3.58	3.58	3.58	3.58	3.58	3.59	3.59	3.59	3.59
160	3.59	3.60	3.60	3.60	3.60	3.60	3.60	3.61	3.61	3.61
170	3.61	3.61	3.61	3.61	3.62	3.62	3.62	3.62	3.62	3.62
180	3.63	3.63	3.63	3.63	3.63	3.63	3.63	3.64	3.64	3.64
190	3.64	3.64	3.64	3.64	3.65	3.65	3.65	3.65	3.65	3.65
200	3.65	3.65	3.66	3.66	3.66	3.66	3.66	3.66	3.66	3.66

*Computed as in Schilling (1973a).

From E. G. Schilling, *Journal of Quality Technology*, 11(3): 1979, p. 119.   Reprinted by permission.

TABLE T17-3   H109 Table 1.   Lot Action Limit Numbers, $d_c(A)$.   Limits for
Determining Discrepancies Between Supplier's and Consumer's Paired
Attributes Sampling Inspections

d_s	r = 1 $d_c(A)$	r = 2 $d_c(A)$	r = 3 $d_c(A)$	r = 5 $d_c(A)$	r = 8 $d_c(A)$
0	3	2	2	1	1
1	5	3	3	2	2
2	7	4	3	3	2
3	9	5	4	3	2
4	11	6	5	3	3
5	12	7	5	4	3
6	14	8	6	4	3
7	15	9	6	5	3
8	17	9	7	5	4
9	18	10	7	5	4
10	19	11	8	6	4
11	21	12	8	6	4
12	22	12	9	6	5
13	23	13	9	7	5
14	25	14	10	7	5
15	26	14	10	7	5
16	27	15	11	7	5
17	28	16	11	8	6
18	30	16	12	8	6
19	31	17	12	8	6
20	32	18	13	9	6
21	34	18	13	9	6
22	35	19	14	9	7
23	36	20	14	9	7
24	37	20	15	10	7
25	39	21	15	10	7
26	40	22	15	10	7
27	41	22	16	11	8
28	42	23	16	11	8
29	43	24	17	11	8
30	45	24	17	11	8
31	46	25	18	12	8
32	47	25	18	12	8
33	48	26	18	12	9
34	49	27	19	13	9
35	51	27	19	13	9

From H109, pp. 21, 22.

TABLE T17-4   H109 Table II.   Check Ratings for Paired Attributes Sampling Inspections

d_c \ d_s	0	1	2	3	4	5	6	7	8	9	10	11	12	13	14	15
0	0.69	0.20	0.08	0.03	0.02	0.01	0.00	0.00	0.00	0.00	0.00	0.00	0.00	0.00	0.00	0.00
1	1.70	0.69	0.34	0.17	0.09	0.05	0.03	0.01	0.01	0.00	0.00	0.00	0.00	0.00	0.00	0.00
2	2.58	1.24	0.69	0.40	0.23	0.14	0.08	0.05	0.03	0.02	0.01	0.00	0.00	0.00	0.00	0.00
3	3.41	1.83	1.11	0.69	0.44	0.28	0.17	0.11	0.07	0.04	0.02	0.01	0.01	0.00	0.00	0.00
4	4.20	2.43	1.57	1.04	0.69	0.46	0.31	0.20	0.13	0.09	0.05	0.03	0.02	0.01	0.01	0.00
5	4.98	3.05	2.06	1.42	0.99	0.69	0.48	0.33	0.23	0.15	0.10	0.07	0.04	0.03	0.02	0.01
6	5.74	3.68	2.57	1.84	1.33	0.95	0.69	0.50	0.35	0.25	0.17	0.12	0.08	0.05	0.04	0.02
7	6.50	4.31	3.10	2.28	1.70	1.26	0.94	0.69	0.51	0.37	0.27	0.19	0.13	0.09	0.06	0.04
8	7.25	4.96	3.65	2.75	2.09	1.59	1.21	0.92	0.69	0.52	0.38	0.28	0.21	0.15	0.11	0.07
9	7.99	5.60	4.21	3.23	2.51	1.95	1.51	1.17	0.90	0.69	0.53	0.40	0.30	0.22	0.16	0.12
10		6.25	4.78	3.73	2.94	2.33	1.84	1.45	1.14	0.89	0.69	0.53	0.41	0.31	0.23	0.17
11		6.81	5.34	4.24	3.40	2.73	2.19	1.75	1.40	1.11	0.88	0.69	0.54	0.42	0.32	0.24
12		7.46	5.91	4.76	3.86	3.14	2.56	2.08	1.68	1.36	1.09	0.87	0.69	0.55	0.43	0.33
13		8.10	6.50	5.29	4.34	3.57	2.94	2.42	1.99	1.62	1.32	1.07	0.87	0.69	0.55	0.44
14			7.09	5.83	4.83	4.01	3.34	2.78	2.30	1.91	1.58	1.29	1.06	0.86	0.69	0.56
15			7.69	6.38	5.33	4.47	3.75	3.15	2.64	2.21	1.84	1.53	1.27	1.04	0.85	0.69
16				6.94	5.84	4.94	4.18	3.54	2.99	2.52	2.13	1.78	1.49	1.24	1.03	0.85
17				7.50	6.36	5.41	4.62	3.94	3.36	2.86	2.43	2.06	1.73	1.46	1.22	1.01
18				8.07	6.89	5.90	5.07	4.35	3.73	3.20	2.74	2.34	2.00	1.69	1.43	1.20
19					7.42	6.39	5.52	4.77	4.12	3.56	3.07	2.64	2.26	1.94	1.65	1.40
20						7.96	6.90	5.99	5.21	4.53	3.93	3.41	2.95	2.55	2.20	1.88	1.61
21						7.41	6.47	5.65	4.94	4.31	3.76	3.28	2.85	2.47	2.14	1.84
22						7.93	6.95	6.10	5.36	4.70	4.13	3.61	3.16	2.76	2.40	2.08
23							7.44	6.56	5.79	5.11	4.50	3.96	3.48	3.06	2.67	2.34
24							7.94	7.03	6.23	5.52	4.89	4.32	3.82	3.36	2.96	2.60
25								7.51	6.68	5.94	5.28	4.69	4.16	3.69	3.26	2.88
26								7.99	7.13	6.37	5.68	5.07	4.51	4.02	3.57	3.17
27									7.59	6.80	6.09	5.45	4.88	4.36	3.89	3.46
28									8.06	7.25	6.51	5.85	5.25	4.71	4.21	3.77
29										7.70	6.94	6.25	5.63	5.07	4.55	4.09
30											7.37	6.66	6.02	5.43	4.90	4.41
31											7.81	7.08	6.41	5.81	5.25	4.75
32												7.51	6.82	6.19	5.62	5.09
33												7.93	7.23	6.58	5.99	5.44
34													7.65	6.98	6.37	5.80
35														7.38	6.75	6.17
36														7.79	7.14	6.54
37															7.54	6.92
38															7.94	7.30
39																7.70

r=1, d_s = 0 TO 15 (ENLARGED SECTOR)

TABLE T17-4   (continued)

d_c \ d_s	0	1	2	3	4	5	6	7	8	9	10	11	12	13	14	15
0	0.94	0.37	0.19	0.11	0.07	0.04	0.02	0.02	0.01	0.01	0.00	0.00	0.00	0.00	0.00	0.00
1	2.39	1.23	0.76	0.49	0.33	0.22	0.15	0.10	0.07	0.05	0.03	0.03	0.02	0.01	0.01	0.01
2	3.68	2.17	1.46	1.02	0.74	0.53	0.39	0.28	0.21	0.15	0.11	0.08	0.06	0.04	0.03	0.02
3	4.93	3.15	2.24	1.66	1.25	0.95	0.73	0.56	0.43	0.33	0.25	0.19	0.15	0.11	0.08	0.06
4	6.13	4.15	3.08	2.36	1.85	1.45	1.15	0.91	0.72	0.57	0.45	0.36	0.28	0.22	0.17	0.14
5	7.32	5.17	3.96	3.12	2.50	2.02	1.63	1.33	1.08	0.88	0.72	0.58	0.47	0.38	0.31	0.25
6		6.19	4.86	3.92	3.20	2.64	2.18	1.81	1.51	1.26	1.04	0.86	0.71	0.59	0.49	0.40
7		7.23	5.79	4.75	3.95	3.30	2.78	2.35	1.98	1.68	1.42	1.19	1.01	0.85	0.71	0.60
8		8.28	6.73	5.60	4.72	4.01	3.42	2.92	2.50	2.15	1.84	1.58	1.35	1.15	0.98	0.84
9			7.69	6.48	5.52	4.74	4.09	3.54	3.06	2.66	2.31	2.00	1.73	1.50	1.29	1.12
10				7.37	6.35	5.50	4.79	4.19	3.66	3.20	2.81	2.46	2.15	1.88	1.65	1.44
11				8.29	7.20	6.29	5.52	4.86	4.28	3.78	3.34	2.95	2.61	2.31	2.03	1.80
12					8.06	7.09	6.27	5.56	4.94	4.39	3.91	3.48	3.10	2.76	2.46	2.18
13						7.91	7.04	6.28	5.62	5.03	4.51	4.03	3.62	3.25	2.91	2.60
14							7.83	7.02	6.31	5.69	5.12	4.62	4.17	3.76	3.39	3.06
15								7.78	7.03	6.37	5.77	5.22	4.74	4.30	3.90	3.53
16									7.77	7.06	6.43	5.85	5.33	4.85	4.42	4.03
17										7.78	7.11	6.50	5.94	5.44	4.98	4.56
18											7.78	7.16	6.58	6.04	5.55	5.10
19												7.84	7.23	6.66	6.14	5.66
20													7.90	7.30	6.75	6.25
21														7.95	7.38	6.85
22														8.60	8.02	7.46
23																8.10

$r=2$, $d_s = 0$ TO 15 (ENLARGED SECTOR)

TABLE T17-4   (continued)

d_c \ d_s	0	1	2	3	4	5	6	7	8	9	10	11	12	13	14	15
0	1.10	0.50	0.29	0.19	0.12	0.09	0.06	0.04	0.03	0.02	0.02	0.01	0.01	0.01	0.00	0.00
1	2.85	1.63	1.10	0.78	0.57	0.43	0.32	0.24	0.19	0.14	0.11	0.09	0.07	0.05	0.04	0.03
2	4.45	2.85	2.06	1.57	1.21	0.95	0.75	0.60	0.48	0.38	0.31	0.25	0.20	0.16	0.13	0.10
3	5.98	4.12	3.13	2.46	1.98	1.61	1.32	1.08	0.89	0.74	0.61	0.51	0.42	0.35	0.29	0.24
4	7.48	5.40	4.24	3.44	2.84	2.36	1.98	1.67	1.41	1.20	1.02	0.86	0.73	0.62	0.53	0.45
5	8.96	6.71	5.40	4.47	3.76	3.20	2.73	2.35	2.02	1.74	1.51	1.29	1.12	0.97	0.84	0.73
6		8.02	6.59	5.55	4.74	4.09	3.55	3.09	2.70	2.36	2.07	1.81	1.58	1.39	1.22	1.07
7			7.80	6.66	5.76	5.03	4.41	3.89	3.44	3.04	2.70	2.38	2.12	1.88	1.67	1.49
8				7.80	6.82	6.01	5.32	4.74	4.23	3.78	3.38	3.02	2.71	2.43	2.18	1.96
9					7.90	7.02	6.28	5.62	5.06	4.56	4.11	3.70	3.35	3.03	2.74	2.48
10						8.07	7.26	6.55	5.93	5.38	4.87	4.42	4.03	3.67	3.35	3.05
11						9.13	8.26	7.50	6.82	6.22	5.68	5.20	4.76	4.36	4.00	3.67
12									7.76	7.11	6.53	6.00	5.52	5.08	4.68	4.31
13										8.03	7.40	6.83	6.31	5.84	5.40	5.00
14												7.69	7.14	6.63	6.16	5.72
15													7.99	7.46	6.94	6.48
16															7.75	7.25

r=3, d_s= 0 15 (ENLARGED SECTOR)

d_c \ d_s	0	1	2	3	4	5	6	7	8	9	10	11	12	13	14	15
0	1.32	0.68	0.45	0.32	0.24	0.18	0.14	0.11	0.09	0.07	0.05	0.05	0.04	0.03	0.02	0.02
1	3.49	2.21	1.62	1.26	1.00	0.81	0.66	0.54	0.45	0.38	0.32	0.27	0.23	0.19	0.17	0.14
2	5.49	3.83	2.98	2.41	1.99	1.67	1.41	1.21	1.04	0.89	0.77	0.66	0.57	0.50	0.43	0.38
3	7.43	5.49	4.43	3.69	3.13	2.69	2.33	2.03	1.78	1.57	1.38	1.20	1.06	0.94	0.84	0.74
4		7.18	5.94	5.06	4.37	3.83	3.37	2.99	2.66	2.37	2.12	1.87	1.68	1.51	1.37	1.23
5		8.88	7.50	6.48	5.69	5.04	4.50	4.03	3.63	3.28	2.97	2.66	2.41	2.20	2.00	1.83
6				7.96	7.06	6.33	5.70	5.16	4.69	4.28	3.90	3.53	3.23	2.97	2.73	2.51
7					8.48	7.66	6.95	6.34	5.81	5.33	4.91	4.47	4.13	3.82	3.53	3.27
8							8.25	7.58	6.98	6.45	5.97	5.49	5.09	4.73	4.41	4.10
9									8.20	7.62	7.08	6.55	6.11	5.71	5.34	5.00
10											8.24	7.66	7.18	6.73	6.32	5.94
11											9.39	8.81	8.29	7.80	7.35	6.93
12															8.42	7.97

r=5, d_s= 0 TO 15 (ENLARGED SECTOR)

d_c \ d_s	0	1	2	3	4	5	6	7	8	9	10	11	12	13	14	15
0	1.53	0.88	0.63	0.48	0.38	0.31	0.25	0.21	0.18	0.15	0.13	0.11	0.09	0.08	0.07	0.06
1	4.12	2.80	2.17	1.77	1.48	1.26	1.08	0.94	0.82	0.72	0.63	0.55	0.49	0.44	0.39	0.35
2	6.53	4.82	3.92	3.31	2.85	2.49	2.20	1.95	1.74	1.56	1.41	1.22	1.11	1.01	0.92	0.83
3	8.88	6.89	5.77	4.99	4.39	3.90	3.49	3.15	2.86	2.60	2.37	2.10	1.92	1.77	1.63	1.51
4		8.98	7.69	6.75	6.02	5.42	4.92	4.48	4.11	3.78	3.48	3.13	2.90	2.69	2.50	2.33
5				8.58	7.73	7.03	6.44	5.92	5.47	5.07	4.71	4.28	4.00	3.74	3.50	3.28
6						8.71	8.03	7.44	6.92	6.45	6.03	5.54	5.20	4.88	4.60	4.33
7								9.02	8.43	7.90	7.42	6.88	6.48	6.11	5.78	5.47
8											8.88	8.27	7.83	7.42	7.04	6.68
9															8.36	7.96

r=8, d_s = 0 TO 15 (ENLARGED SECTOR)

From H109, pp. 24, 26, 28.

TABLE T17-5 H109 Table III. Cumulative Check Rating Criteria

No. Lots Verified	Median Value	Critical Values	
		Warning	Action
3	2.67	6.30	8.41
4	3.67	7.75	10.05
5	4.67	9.15	11.60
6	5.67	10.51	13.11
7	6.67	11.84	14.57
8	7.67	13.15	16.00
9	8.67	14.43	17.40
10	9.67	15.70	18.78
11	10.67	16.96	20.14
12	11.67	18.21	21.49
13	12.67	19.44	22.82
14	13.67	20.67	24.14
15	14.67	21.89	25.45
16	15.67	23.10	26.74
17	16.67	24.30	28.03
18	17.67	25.50	29.31
19	18.67	29.69	30.58
20	19.67	27.88	31.84
21	20.67	29.06	33.10
22	21.67	30.24	34.36
23	22.67	31.42	35.60
24	23.67	32.59	36.84
25	24.67	33.75	38.08
26	25.67	34.92	39.31
27	26.67	36.08	40.54
28	27.67	37.23	41.76
29	28.67	38.39	42.98
30	29.67	39.54	44.19

From H109, p. 29.

TABLE T17-6   H109 Table IV.   Probability of Acceptance in k Trials

Probability of Acceptance in a Single Trial	Probability of Acceptance of Hypothesis of Homogeneity in All K Trials		
	K = 3	K = 5	K = 10
0.95	0.86	0.77	0.60
0.90	0.73	0.59	0.35
0.85	0.61	0.44	0.20
0.80	0.51	0.33	0.11
0.75	0.42	0.24	0.06
0.70	0.34	0.17	0.03
0.65	0.28	0.12	0.01
0.60	0.22	0.08	0.01
0.55	0.17	0.05	0.00
0.50	0.12	0.03	0.00
0.45	0.09	0.02	0.00
0.40	0.06	0.01	0.00
0.35	0.04	0.00	0.00

From H-109, p. 30.

TABLE T17-7   H109 Table V.   Probability of Acceptance of Inspection Verification Plans

$p'_s n_s$	r = 1 p'_c/p'_s				r = 2 p'_c/p'_s				r = 3 p'_c/p'_s				r = 5 p'_c/p'_s				r = 8 p'_c/p'_s			
	1	2	3	4.5	1	2	3	4.5	1	2	3	4.5	1	2	3	4.5	1	2	3	4.5
0.75	.98	.90	.79	.59	.97	.90	.81	.65	.99	.95	.90	.80	.93	.86	.80	.72	.96	.91	.86	.80
1.125	.96	.84	.68	.45	.96	.86	.73	.54	.98	.92	.84	.69	.93	.85	.78	.67	.95	.90	.84	.76
1.50	.95	.80	.61	.35	.95	.82	.66	.45	.97	.89	.78	.59	.93	.85	.76	.64	.95	.89	.82	.72
2.25	.94	.75	.51	.23	.93	.77	.57	.32	.96	.84	.68	.45	.94	.85	.74	.57	.95	.88	.79	.65
3.00	.94	.72	.44	.14	.93	.74	.50	.24	.95	.80	.61	.35	.94	.84	.68	.48	.95	.86	.74	.58
4.50	.94	.66	.30	.04	.93	.69	.41	.13	.94	.75	.50	.21	.95	.81	.61	.35	.95	.83	.68	.46
6.00	.95	.59	.19	.01	.94	.66	.32	.06	.94	.71	.40	.12	.95	.77	.54	.26	.95	.80	.63	.38
9.00	.95	.46	.07	.00	.95	.58	.19	.01	.94	.62	.26	.04	.95	.72	.41	.14	.95	.77	.53	.25

NOTE: (a) p'_c/p'_s is the ratio of the expected fractions defective in the consumer's and supplier's samples of size n_c and n_s, respectively.

(b) $p'_s n_s$ is the expected number of defectives in the supplier's sample size n_s, where $n_s = r n_c$.

From H109, p. 31.

TABLE T18-1 Hazard Values Corresponding to Probability Plotting Positions for Censored Data.

Probability Tenth of Percent

	0.0	0.1	0.2	0.3	0.4	0.5	0.6	0.7	0.8	0.9
0.	0.	0.10	0.20	0.30	0.40	0.50	0.60	0.70	0.80	0.90
1.	1.01	1.11	1.21	1.31	1.41	1.51	1.61	1.71	1.82	1.92
2.	2.02	2.12	2.22	2.33	2.43	2.53	2.63	2.74	2.84	2.94
3.	3.05	3.15	3.25	3.36	3.46	3.56	3.67	3.77	3.87	3.98
4.	4.08	4.19	4.29	4.40	4.50	4.60	4.71	4.81	4.92	5.02
5.	5.13	5.23	5.34	5.45	5.55	5.66	5.76	5.87	5.97	6.08
6.	6.19	6.29	6.40	6.51	6.61	6.72	6.83	6.94	7.04	7.15
7.	7.26	7.36	7.47	7.58	7.69	7.80	7.90	8.01	8.12	8.23
8.	8.34	8.45	8.56	8.66	8.77	8.88	8.99	9.10	9.21	9.32
9.	9.43	9.54	9.65	9.76	9.87	9.98	10.09	10.20	10.31	10.43
10.	10.54	10.65	10.76	10.87	10.98	11.09	11.20	11.32	11.43	11.54
11.	11.65	11.77	11.88	11.99	12.10	12.22	12.33	12.44	12.56	12.67
12.	12.78	12.90	13.01	13.12	13.24	13.35	13.47	13.58	13.70	13.81
13.	13.93	14.04	14.16	14.27	14.39	14.50	14.62	14.73	14.85	14.97
14.	15.08	15.20	15.32	15.43	15.55	15.67	15.78	15.90	16.02	16.13
15.	16.25	16.37	16.49	16.61	16.72	16.84	16.96	17.08	17.20	17.32
16.	17.44	17.55	17.67	17.79	17.91	18.03	18.15	18.27	18.39	18.51
17.	18.63	18.75	18.87	19.00	19.12	19.24	19.36	19.48	19.60	19.72
18.	19.85	19.97	20.09	20.21	20.33	20.46	20.58	20.70	20.83	20.95
19.	21.07	21.20	21.32	21.44	21.57	21.69	21.82	21.94	22.06	22.19
20.	22.31	22.44	22.56	22.69	22.82	22.94	23.07	23.19	23.32	23.45
21.	23.57	23.70	23.83	23.95	24.08	24.21	24.33	24.46	24.59	24.72
22.	24.85	24.97	25.10	25.23	25.36	25.49	25.62	25.75	25.88	26.01
23.	26.14	26.27	26.40	26.53	26.66	26.79	26.92	27.05	27.18	27.31
24.	27.44	27.58	27.71	27.84	27.97	28.10	28.24	28.37	28.50	28.63
25.	28.77	28.90	29.04	29.17	29.30	29.44	29.57	29.71	29.84	29.98
26.	30.11	30.25	30.38	30.52	30.65	30.79	30.92	31.06	31.20	31.33
27.	31.47	31.61	31.75	31.88	32.02	32.16	32.30	32.43	32.57	32.71
28.	32.85	32.99	33.13	33.27	33.41	33.55	33.69	33.83	33.97	34.11
29.	34.25	34.39	34.53	34.67	34.81	34.96	35.10	35.24	35.38	35.52
30.	35.67	35.81	35.95	36.10	36.24	36.38	36.53	36.67	36.82	36.96
31.	37.11	37.25	37.40	37.54	37.69	37.83	37.98	38.13	38.27	38.42
32.	38.57	38.71	38.86	39.01	39.16	39.30	39.45	39.60	39.75	39.90
33.	40.05	40.20	40.35	40.50	40.65	40.80	40.95	41.10	41.25	41.40
34.	41.55	41.70	41.86	42.01	42.16	42.31	42.46	42.62	42.77	42.92
35.	43.08	43.23	43.39	43.54	43.70	43.85	44.01	44.16	44.32	44.47
36.	44.63	44.79	44.94	45.10	45.26	45.41	45.57	45.73	45.89	46.04
37.	46.20	46.36	46.52	46.68	46.84	47.00	47.16	47.32	47.48	47.64
38.	47.80	47.96	48.13	48.29	48.45	48.61	48.78	48.94	49.10	49.27
39.	49.43	49.59	49.76	49.92	50.09	50.25	50.42	50.58	50.75	50.92
40.	51.08	51.25	51.42	51.58	51.75	51.92	52.09	52.26	52.42	52.59
41.	52.76	52.93	53.10	53.27	53.44	53.61	53.79	53.96	54.13	54.30
42.	54.47	54.65	54.82	54.99	55.16	55.34	55.51	55.69	55.86	56.04
43.	56.21	56.39	56.56	56.74	56.92	57.09	57.27	57.45	57.63	57.80
44.	57.98	58.16	58.34	58.52	58.70	58.88	59.06	59.24	59.42	59.60
45.	59.78	59.97	60.15	60.33	60.51	60.70	60.88	61.06	61.25	61.43
46.	61.62	61.80	61.99	62.18	62.36	62.55	62.74	62.92	63.11	63.30
47.	63.49	63.68	63.87	64.06	64.25	64.44	64.63	64.82	65.01	65.20
48.	65.39	65.59	65.78	65.97	66.16	66.36	66.55	66.75	66.94	67.14
49.	67.33	67.53	67.73	67.92	68.12	68.32	68.52	68.72	68.92	69.11
50.	69.31	69.51	69.72	69.92	70.12	70.32	70.52	70.72	70.93	71.13

Probability Percent

TABLE T18-1 (continued)

Probability Percent	Probability Tenth of Percent									
	0.0	0.1	0.2	0.3	0.4	0.5	0.6	0.7	0.8	0.9
51.	71.33	71.54	71.74	71.95	72.15	72.36	72.57	72.77	72.98	73.19
52.	73.40	73.61	73.81	74.02	74.23	74.44	74.65	74.87	75.08	75.29
53.	75.50	75.72	75.93	76.14	76.36	76.57	76.79	77.00	77.22	77.44
54.	77.65	77.87	78.09	78.31	78.53	78.75	78.97	79.19	79.41	79.63
55.	79.85	80.07	80.30	80.52	80.74	80.97	81.19	81.42	81.64	81.87
56.	82.10	82.33	82.55	82.78	83.01	83.24	83.47	83.70	83.93	84.16
57.	84.40	84.63	84.86	85.10	85.33	85.57	85.80	86.04	86.27	86.51
58.	86.75	86.99	87.23	87.47	87.71	87.95	88.19	88.43	88.67	88.92
59.	89.16	89.40	89.65	89.89	90.14	90.39	90.63	90.88	91.13	91.38
60.	91.63	91.88	92.13	92.38	92.63	92.89	93.14	93.39	93.65	93.90
61.	94.16	94.42	94.67	94.93	95.19	95.45	95.71	95.97	96.23	96.50
62.	96.76	97.02	97.29	97.55	97.82	98.08	98.35	98.62	98.89	99.16
63.	99.43	99.70	99.97	100.24	100.51	100.79	101.06	101.34	101.61	101.89
64.	102.17	102.44	102.72	103.00	103.28	103.56	103.85	104.13	104.41	104.70
65.	104.98	105.27	105.56	105.84	106.13	106.42	106.71	107.00	107.29	107.59
66.	107.88	108.18	108.47	108.77	109.06	109.36	109.66	109.96	110.26	110.56
67.	110.87	111.17	111.47	111.78	112.09	112.39	112.70	113.01	113.32	113.63
68.	113.94	114.26	114.57	114.89	115.20	115.52	115.84	116.16	116.48	116.80
69.	117.12	117.44	117.77	118.09	118.42	118.74	119.07	119.40	119.73	120.06
70.	120.40	120.73	121.07	121.40	121.74	122.08	122.42	122.76	123.10	123.44
71.	123.79	124.13	124.48	124.83	125.18	125.53	125.88	126.23	126.58	126.94
72.	127.30	127.65	128.01	128.37	128.74	129.10	129.46	129.83	130.20	130.56
73.	130.93	131.30	131.68	132.05	132.43	132.80	133.18	133.56	133.94	134.32
74.	134.71	135.09	135.48	135.87	136.26	136.65	137.04	137.44	137.83	138.23
75.	138.63	139.03	139.43	139.84	140.24	140.65	141.06	141.47	141.88	142.30
76.	142.71	143.13	143.55	143.97	144.39	144.82	145.24	145.67	146.10	146.53
77.	146.97	147.40	147.84	148.28	148.72	149.17	149.61	150.06	150.51	150.96
78.	151.41	151.87	152.33	152.79	153.25	153.71	154.18	154.65	155.12	155.59
79.	156.06	156.54	157.02	157.50	157.99	158.47	158.96	159.45	159.95	160.45
80.	160.94	161.45	161.95	162.46	162.96	163.48	163.99	164.51	165.03	165.55
81.	166.07	166.60	167.13	167.66	168.20	168.74	169.28	169.83	170.37	170.93
82.	171.48	172.04	172.60	173.16	173.73	174.30	174.87	175.45	176.03	176.61
83.	177.20	177.79	178.38	178.98	179.58	180.18	180.79	181.40	182.02	182.64
84.	183.26	183.89	184.52	185.15	185.79	186.43	187.08	187.73	188.39	189.05
85.	189.71	190.38	191.05	191.73	192.41	193.10	193.79	194.49	195.19	195.90
86.	196.61	197.33	198.05	198.78	199.51	200.25	200.99	201.74	202.50	203.26
87.	204.02	204.79	205.57	206.36	207.15	207.94	208.75	209.56	210.37	211.20
88.	212.03	212.86	213.71	214.56	215.42	216.28	217.16	218.04	218.93	219.82
89.	220.73	221.64	222.56	223.49	224.43	225.38	226.34	227.30	228.28	229.26
90.	230.26	231.26	232.28	233.30	234.34	235.39	236.45	237.52	238.60	239.69
91.	240.79	241.91	243.04	244.18	245.34	246.51	247.69	248.89	250.10	251.33
92.	252.57	253.83	255.10	256.39	257.70	259.03	260.37	261.73	263.11	264.51
93.	265.93	267.36	268.82	270.31	271.81	273.34	274.89	276.46	278.06	279.69
94.	281.34	283.02	284.73	286.47	288.24	290.04	291.88	293.75	295.65	297.59
95.	299.57	301.59	303.66	305.76	307.91	310.11	312.36	314.66	317.01	319.42
96.	321.89	324.42	327.02	329.68	332.42	335.24	338.14	341.12	344.20	347.38
97.	350.66	354.05	357.56	361.19	364.97	368.89	372.97	377.23	381.67	386.32
98.	391.20	396.33	401.74	407.45	413.52	419.97	426.87	434.28	442.28	450.99
99.	460.52	471.05	482.83	496.18	511.60	529.83	552.15	580.91	621.46	690.77

From J. H. Sheesley, Report Number 1300-1119, General Electric Company, Cleveland, Ohio, 1974. Reprinted by permission.

TABLE T18-2 H108 Table 2A-1. Life Test Sampling Plan Code Designation

$\alpha=0.01$ $\beta=0.10$		$\alpha=0.05$ $\beta=0.10$		$\alpha=0.10$ $\beta=0.10$		$\alpha=0.25$ $\beta=0.10$		$\alpha=0.50$ $\beta=0.10$	
Code	θ_1/θ_0	Code	θ_1/θ_0	Code	θ_1/θ_0	Code	θ_1/θ_0	Code	θ_1/θ_0
A-1	0.004	B-1	0.022	C-1	0.046	D-1	0.125	E-1	0.301
A-2	.038	B-2	.091	C-2	.137	D-2	.247	E-2	.432
A-3	.082	B-3	.154	C-3	.207	D-3	.325	E-3	.502
A-4	.123	B-4	.205	C-4	.261	D-4	.379	E-4	.550
A-5	.160	B-5	.246	C-5	.304	D-5	.421	E-5	.584
A-6	.193	B-6	.282	C-6	.340	D-6	.455	E-6	.611
A-7	.221	B-7	.312	C-7	.370	D-7	.483	E-7	.633
A-8	.247	B-8	.338	C-8	.396	D-8	.506	E-8	.652
A-9	.270	B-9	.361	C-9	.418	D-9	.526	E-9	.667
A-10	.291	B-10	.382	C-10	.438	D-10	.544	E-10	.681
A-11	.371	B-11	.459	C-11	.512	D-11	.608	E-11	.729
A-12	.428	B-12	.512	C-12	.561	D-12	.650	E-12	.759
A-13	.470	B-13	.550	C-13	.597	D-13	.680	E-13	.781
A-14	.504	B-14	.581	C-14	.624	D-14	.703	E-14	.798
A-15	.554	B-15	.625	C-15	.666	D-15	.737	E-15	.821
A-16	.591	B-16	.658	C-16	.695	D-16	.761	E-16	.838
A-17	.653	B-17	.711	C-17	.743	D-17	.800	E-17	.865
A-18	.692	B-18	.745	C-18	.774	D-18	.824	E-18	.882

Producer's risk α is the probability of rejecting lots with mean life θ_0.
Consumer's risk β is the probability of accepting lots with mean life θ_1.

From H108, p. 2.2.

TABLE T18-3 H108 Table 2B-1. Master Table for Life Tests Terminated upon Occurrence of Preassigned Number of Failures

r	Producer's risk (α)									
	0.01 Code	0.01 C/θ₀	0.05 Code	0.05 C/θ₀	0.10 Code	0.10 C/θ₀	0.25 Code	0.25 C/θ₀	0.50 Code	0.50 C/θ₀
1	A-1	0.010	B-1	0.052	C-1	0.106	D-1	0.288	E-1	0.693
2	A-2	.074	B-2	.178	C-2	.266	D-2	.481	E-2	.839
3	A-3	.145	B-3	.272	C-3	.367	D-3	.576	E-3	.891
4	A-4	.206	B-4	.342	C-4	.436	D-4	.634	E-4	.918
5	A-5	.256	B-5	.394	C-5	.487	D-5	.674	E-5	.934
6	A-6	.298	B-6	.436	C-6	.525	D-6	.703	E-6	.945
7	A-7	.333	B-7	.469	C-7	.556	D-7	.726	E-7	.953
8	A-8	.363	B-8	.498	C-8	.582	D-8	.744	E-8	.959
9	A-9	.390	B-9	.522	C-9	.604	D-9	.760	E-9	.963
10	A-10	.413	B-10	.543	C-10	.622	D-10	.773	E-10	.967
15	A-11	.498	B-11	.616	C-11	.687	D-11	.816	E-11	.978
20	A-12	.554	B-12	.663	C-12	.726	D-12	.842	E-12	.983
25	A-13	.594	B-13	.695	C-13	.754	D-13	.859	E-13	.987
30	A-14	.625	B-14	.720	C-14	.774	D-14	.872	E-14	.989
40	A-15	.669	B-15	.755	C-15	.803	D-15	.889	E-15	.992
50	A-16	.701	B-16	.779	C-16	.824	D-16	.901	E-16	.993
75	A-17	.751	B-17	.818	C-17	.855	D-17	.920	E-17	.996
100	A-18	.782	B-18	.841	C-18	.874	D-18	.931	E-18	.997

Producer's risk α is the probability of rejecting lots with mean life θ_0.
Acceptance criterion: Accept lot if $\hat{\theta} \geq \theta_0(C/\theta_0)$. [a]
For explanation of the code, see par. 2A3.2 and table 2A-1.

From H108, p. 2.28.

TABLE T18-4 H108 Table 2C-1(b). Master Table for Life Tests Terminated at Preassigned Time: Testing Without Replacement (Values of T/θ_0 for $\alpha = 0.05$)

Code	r	Sample size									
		2r	3r	4r	5r	6r	7r	8r	9r	10r	20r
B-1	1	0.026	0.017	0.013	0.010	0.009	0.007	0.006	0.006	0.005	0.003
B-2	2	.104	.065	.048	.038	.031	.026	.023	.020	.018	.009
B-3	3	.168	.103	.075	.058	.048	.041	.036	.031	.028	.014
B-4	4	.217	.132	.095	.074	.061	.052	.045	.040	.036	.017
B-5	5	.254	.153	.110	.086	.071	.060	.052	.046	.041	.020
B-6	6	.284	.170	.122	.095	.078	.066	.057	.051	.045	.022
B-7	7	.309	.185	.132	.103	.084	.072	.062	.055	.049	.024
B-8	8	.330	.197	.141	.110	.090	.076	.066	.058	.052	.025
B-9	9	.348	.207	.148	.115	.094	.080	.069	.061	.055	.027
B-10	10	.363	.216	.154	.120	.098	.083	.072	.064	.057	.028
B-11	15	.417	.246	.175	.136	.112	.094	.082	.072	.065	.032
B-12	20	.451	.266	.189	.147	.120	.102	.088	.078	.070	.034
B-13	25	.475	.280	.199	.154	.126	.107	.093	.082	.073	.036
B-14	30	.493	.290	.206	.160	.131	.111	.096	.085	.076	.037
B-15	40	.519	.305	.216	.168	.137	.116	.101	.089	.079	.039
B-16	50	.536	.315	.223	.173	.142	.120	.104	.092	.082	.040
B-17	75	.564	.331	.235	.182	.149	.126	.109	.096	.086	.042
B-18	100	.581	.340	.242	.187	.153	.130	.112	.099	.089	.043

For explanation of the code, see par. 2A3.2 and table 2A-1.

From H108, p. 2.45.

TABLE T18-5 H108 Table 2C-2(b). Master Table for Life Tests Terminated at Preassigned Time: Testing With Replacement (Values of T/θ_0 for $\alpha = 0.05$)

Code	r	Sample size									
		2r	3r	4r	5r	6r	7r	8r	9r	10r	20r
B-1	1	0.026	0.017	0.013	0.010	0.009	0.007	0.006	0.006	0.005	0.003
B-2	2	.089	.059	.044	.036	.030	.025	.022	.020	.018	.009
B-3	3	.136	.091	.068	.055	.045	.039	.034	.030	.027	.014
B-4	4	.171	.114	.085	.068	.057	.049	.043	.038	.034	.017
B-5	5	.197	.131	.099	.079	.066	.056	.049	.044	.039	.020
B-6	6	.218	.145	.109	.087	.073	.062	.054	.048	.044	.022
B-7	7	.235	.156	.117	.094	.078	.067	.059	.052	.047	.023
B-8	8	.249	.166	.124	.100	.083	.071	.062	.055	.050	.025
B-9	9	.261	.174	.130	.104	.087	.075	.065	.058	.052	.026
B-10	10	.271	.181	.136	.109	.090	.078	.068	.060	.054	.027
B-11	15	.308	.205	.154	.123	.103	.088	.077	.068	.062	.031
B-12	20	.331	.221	.166	.133	.110	.095	.083	.074	.066	.033
B-13	25	.348	.232	.174	.139	.116	.099	.087	.077	.070	.035
B-14	30	.360	.240	.180	.144	.120	.103	.090	.080	.072	.036
B-15	40	.377	.252	.189	.151	.126	.108	.094	.084	.075	.038
B-16	50	.390	.260	.195	.156	.130	.111	.097	.087	.078	.039
B-17	75	.409	.273	.204	.164	.136	.117	.102	.091	.082	.041
B-18	100	.421	.280	.210	.168	.140	.120	.105	.093	.084	.042

For explanation of the code, see par. 2A3.2 and table 2A-1.

From H108, p. 2.47.

TABLE T18-6 H108 Table 2D-1(b). Master Table for Sequential Life Tests ($\alpha = 0.05$)

Code	r_0	h_0/θ_0	h_1/θ_0	s/θ_0	$F_0(r)$	$E\theta_1(r)$	$E_\infty(r)$	$E\theta_0(r)$
B-1	3	0.0506	−0.0650	0.0859	0.8	0.8	0.4	0.0
B-2	6	.2254	−.2894	.2400	1.2	1.6	1.1	.3
B-3	9	.4098	−.5261	.3405	1.5	2.3	1.9	.6
B-4	12	.5805	−.7453	.4086	1.8	3.0	2.6	.9
B-5	15	.7345	−.9430	.4576	2.1	3.7	3.3	1.2
B-6	18	.8842	−1.1352	.4972	2.3	4.3	4.1	1.6
B-7	21	1.0209	−1.3107	.5282	2.5	5.0	4.8	1.9
B-8	24	1.1495	−1.4757	.5538	2.7	5.6	5.5	2.3
B-9	27	1.2719	−1.6329	.5756	2.8	6.3	6.3	2.7
B-10	30	1.3916	−1.7866	.5948	3.0	6.9	7.0	3.0
B-11	45	1.9101	−2.4523	.6607	3.7	10.0	10.7	5.0
B-12	60	2.3620	−3.0325	.7024	4.3	13.1	14.5	7.0
B-13	75	2.7516	−3.5327	.7307	4.8	16.1	18.2	9.1
B-14	90	3.1217	−4.0079	.7530	5.8	19.2	22.1	11.2
B-15	120	3.7522	−4.8173	.7833	6.2	25.0	29.5	15.3
B-16	150	4.3314	−5.5610	.8053	6.9	31.0	37.1	19.7
B-17	225	5.5386	−7.1109.	.8391	8.5	45.6	55.9	30.5
B-18	300	6.5773	−8.4444	.8600	9.8	60.4	75.1	41.6

For explanation of the code, see par. 2A3.2 and table 2A-1.

From H108, p. 2.63.

TABLE T18-7 H108 Table 2C-5. Master Table for Proportion Failing Before Specified Time. Life Test Sampling Plans for Specified α, β, and p_1/p_0

Values of r (upper numbers) and of D (lower numbers)*

p_1/p_0	$\alpha=0.01$			$\alpha=0.05$			$\alpha=0.10$		
	$\beta=0.01$	0.05	0.10	0.01	0.0b	0.10	0.01	0.05	0.10
3/2	136	101	83	95	67	55	77	52	41
	110.4	79.1	63.3	79.6	54.1	43.4	66.0	43.0	33.0
2	46	35	30	33	23	19	26	18	15
	31.7	22.7	18.7	24.2	15.7	12.4	19.7	12.8	10.3
5/2	27	21	18	19	14	11	15	11	9
	16.4	11.8	9.62	12.4	8.46	6.17	10.3	7.02	5.43
3	19	15	13	13	10	8	11	8	6
	10.3	7.48	6.10	7.69	5.43	3.98	7.02	4.66	3.15
4	12	10	9	9	7	6	7	5	4
	5.43	4.13	3.51	4.70	3.29	2.61	3.90	2.43	1.75
5	9	8	7	7	5	4	5	4	3
	3.51	2.91	2.33	3.29	1.97	1.37	2.43	1.75	1.10
10	5	4	4	4	3	3	3	2	2
	1.28	.823	.823	1.37	.818	.818	1.10	.532	.532

*The sample size n is obtained by taking the largest integer less than or equal to the tabled value divided by p_0, i.e., $n=[D/p_0]$. Producer's risk α is the probability of rejecting lots with acceptable proportion of lot failing before specified time, p_0. Consumer's risk β is the probability of accepting lots with unacceptable proportion of lot failing before specified time, p_1.

From H108, p. 2.55.

TABLE T18-8 TR3 Table 1. Table of Values for Percent Truncation, $(t/\mu) \times 100$

p' (in %)	Shape Parameter - β								
	1/3	1/2	1	1 2/3	2	2 1/2	3 1/3	4	5
.010			.010	.45	1.13	2.83	7.03	11.03	17.26
.012			.012	.49	1.24	3.04	7.42	11.55	17.91
.015			.015	.57	1.38	3.32	7.94	12.21	18.72
.020			.020	.67	1.59	3.73	8.66	13.12	19.83
.025			.025	.77	1.78	4.08	9.26	13.87	20.74
.030			.030	.86	1.95	4.40	9.77	14.52	21.50
.040			.040	1.02	2.26	4.93	10.65	15.60	22.77
.050			.050	1.18	2.53	5.39	11.40	16.49	23.82
.065			.065	1.37	2.88	5.98	12.32	17.62	25.10
.080			.080	1.56	3.19	6.50	13.13	18.56	26.16
.100			.10	1.78	3.57	7.11	14.03	19.62	27.36
.12			.12	1.98	3.92	7.65	14.82	20.53	28.37
.15			.15	2.26	4.37	8.36	15.84	21.71	29.67
.20			.20	2.69	5.07	9.39	17.27	23.33	31.43
.25			.25	3.08	5.64	10.27	18.47	24.68	32.87
.30			.30	3.44	6.18	11.05	19.51	25.83	34.09
.40			.40	4.07	7.14	12.39	21.27	27.76	36.12
.50		.001	.50	4.67	7.99	13.55	22.75	29.36	37.76
.65		.002	.65	5.46	9.12	15.06	24.62	31.35	39.81
.80		.003	.80	6.19	10.11	16.36	26.21	33.03	41.50
1.00		.005	1.01	7.08	11.31	17.90	28.03	34.93	43.40
1.2		.007	1.21	7.90	12.40	19.26	29.62	36.57	45.02
1.5		.011	1.51	9.07	13.87	21.08	31.68	38.68	47.09
2.0		.020	2.02	10.77	16.03	23.67	34.56	41.59	49.90
2.5		.032	2.53	12.33	17.95	25.90	36.98	44.01	52.21
3.0		.047	3.05	13.78	19.69	27.89	39.09	46.09	54.17
4.0	.001	.083	4.08	16.42	22.79	31.35	42.69	49.59	57.45
5.0	.002	.13	5.13	18.84	25.58	34.35	45.71	52.50	60.13
6.5	.005	.23	6.72	22.15	29.25	38.28	49.57	56.18	63.46
8.0	.010	.35	8.34	25.20	32.59	41.72	52.88	59.29	66.26
10.0	.020	.56	10.54	29.01	36.63	45.82	56.73	62.85	69.44
12	.034	.82	12.78	32.58	40.34	49.50	60.11	65.96	72.18
15	.070	1.32	16.25	37.63	45.48	54.49	64.60	70.05	75.73
20	.18	2.49	22.31	45.51	53.30	61.85	71.04	75.83	80.68
25	.40	4.14	28.77	52.99	60.53	68.47	76.67	80.80	84.89
30	.76	6.36	35.37	60.29	67.39	74.62	81.79	85.26	88.62
40	2.22	13.04	51.08	74.79	80.64	86.15	91.09	93.27	95.22
50	5.55	24.02	69.31	89.82	93.95	97.33	99.82	100.67	101.21
65	19.28	55.10	104.98	115.23	115.61	114.92	113.06	111.68	109.98
80	69.48	129.52	160.94	148.91	143.14	136.34	128.53	124.27	119.79

From TR3, p. 26.

TABLE T18-9   TR7 Table 1A.   100t/μ Ratios at the Acceptable Quality
Level (normal inspection) for the MIL-STD-105D Plans

Acceptable Quality Level p'(%)	Shape Parameter, β									
	$\frac{1}{3}$	$\frac{1}{2}$	$\frac{2}{3}$	1	$1\frac{1}{3}$	$1\frac{2}{3}$	2	$2\frac{1}{2}$	$3\frac{1}{3}$	4
0.010	17-12	50-8	75-6	.010	.11	.45	1.13	2.83	7.03	11.0
0.015	56-12	11-7	14-5	.015	.15	.57	1.38	3.32	7.94	12.2
0.025	26-11	31-7	30-5	.025	.22	.77	1.78	4.08	9.26	13.9
0.040	11-10	80-7	60-5	.040	.31	1.02	2.26	4.93	10.7	15.6
0.065	46-10	21-6	13-4	.065	.44	1.37	2.88	5.98	12.3	17.6
0.10	17-9	50-6	25-4	.10	.61	1.78	3.57	7.11	14.0	19.6
0.15	56-9	11-5	44-4	.15	.83	2.26	4.37	8.36	15.8	21.7
0.25	26-8	31-5	94-4	.25	1.22	3.08	5.64	10.3	18.5	24.7
0.40	11-7	80-5	.019	.40	1.73	4.07	7.14	12.4	21.3	27.8
0.65	46-7	21-4	.040	.65	2.50	5.46	9.12	15.1	24.6	31.4
1.0	17-6	51-4	.076	1.01	3.45	7.08	11.3	17.9	28.0	34.9
1.5	59-6	.011	.14	1.51	4.69	9.07	13.9	21.1	31.7	38.7
2.5	27-5	.032	.30	2.53	6.91	12.3	18.0	25.9	37.0	44.0
4.0	11-4	.083	.62	4.08	9.88	16.4	22.8	31.4	42.7	49.6
6.5	51-4	.23	1.31	6.72	14.4	22.2	29.3	38.3	49.6	56.2
10	.019	.56	2.57	10.5	20.1	29.0	36.6	45.8	56.7	62.9

The negative figure after a ratio shows the number of decimal points to
provide.   Thus 13-4 = .0013.

From TR7, p. 14.

TABLE T18-10 TR7 Table 1B. 100t/μ Ratios at the Limiting Quality
Level for the MIL-STD-105D Plans: Consumer's Risk = 0.10

Code Letter	AQL	Shape Parameter, β									
		$\frac{1}{3}$	$\frac{1}{2}$	$\frac{2}{3}$	1	$1\frac{1}{3}$	$1\frac{2}{3}$	2	$2\frac{1}{2}$	$3\frac{1}{3}$	4
A	6.5	25	68	92	120	120	118	118	116	115	115
B	4.0	7.2	29	50	77	89	95	98	100	102	103
C	2.5	1.6	10	23	46	61	70	77	82	88	91
C	10	11	40	62	89	100	102	103	105	106	106
D	1.5	.38	4.1	11.6	28	43	53	60	68	76	80
D	6.5	2.4	13	28	53	67	76	81	86	91	94
D	10	7.2	29	50	77	89	95	98	100	102	103
E	1.0	.094	1.5	5.6	17	30	39	47	56	66	71
E	4.0	.49	4.8	13	31	45	55	63	70	78	82
E	6.5	1.5	10	22	45	59	68	76	80	86	90
E	10	3.5	17	37	60	73	82	87	90	95	97
F	0.65	.026	.66	2.9	11	22	30	38	47	58	64
F	2.5	.14	2.0	6.7	20	33	42	50	58	68	72
F	4.0	.36	4.0	11	28	42	52	59	67	75	80
F	6.5	.80	6.5	16	36	51	61	68	73	81	85
F	10	2.6	14	29	54	68	77	82	87	92	95
G	0.40	62-4	.26	1.4	7.2	15	23	30	39	50	57
G	1.5	.032	.76	3.2	12	22	31	39	48	59	65
G	2.5	.086	1.4	5.3	17	29	38	47	55	65	70
G	4.0	.18	2.4	7.7	22	35	45	53	60	70	74
G	6.5	.52	5.0	13	31	46	56	63	70	78	82
G	10	1.2	8.8	20	42	57	66	73	78	85	89
H	0.25	16-4	.11	.74	4.6	11	17	24	33	44	51
H	1.0	84-4	.31	1.6	7.8	16	24	31	40	51	58
H	1.5	.021	.59	2.6	11	20	29	37	46	57	63
H	2.5	.046	.97	3.9	14	25	34	42	51	61	67
H	4.0	.12	1.8	6.5	19	32	42	49	58	67	72
H	6.5	.27	3.2	9.7	25	39	49	57	65	73	78
H	10	.68	6.0	15	34	49	58	67	73	80	85
J	0.15	40-5	.042	.37	2.9	7.5	13	19	27	38	45
J	0.65	20-4	.12	.80	4.9	11	18	24	33	45	52
J	1.0	54-4	.23	1.3	6.7	14	22	29	38	49	57
J	1.5	.010	.36	1.8	8.3	17	25	32	42	53	59
J	2.5	.030	.72	3.1	12	22	31	39	48	58	64
J	4.0	.063	1.2	4.5	15	27	36	44	53	63	68
J	6.5	.16	2.3	7.5	21	34	44	52	60	69	74
J	10	.34	3.8	11	27	41	51	59	67	75	80

A negative figure after a ratio shows the number of decimal points
to provide. Thus 62-4 = .0062.

TABLE T18-10 (continued)

Code Letter	AQL	Shape Parameter, β									
		$\frac{1}{3}$	$\frac{1}{2}$	$\frac{2}{3}$	1	$1\frac{1}{3}$	$1\frac{2}{3}$	2	$2\frac{1}{2}$	$3\frac{1}{3}$	4
K	0.10	10-5	.017	.19	1.8	5.5	10	15	23	33	40
K	.40	50-5	.049	.41	3.1	8.0	14	20	28	39	46
K	.65	13-4	.093	.67	4.3	10	17	23	32	43	50
K	1.0	27-4	.15	.94	5.4	12	19	26	35	46	53
K	1.5	76-4	.29	1.5	7.6	15	23	31	40	51	58
K	2.5	.015	.47	2.2	9.8	19	27	35	44	55	61
K	4.0	.039	.85	3.5	13	23	33	41	50	60	66
K	6.5	.092	1.5	5.5	17	29	39	47	56	65	70
K	10	.27	3.2	9.7	25	39	49	57	65	73	78
L	0.065	25-6	67-4	.093	1.1	3.8	7.7	12	19	29	36
L	0.25	12-5	.019	.20	1.9	5.7	10	15	23	34	41
L	0.40	33-5	.036	.33	2.7	7.2	12	18	26	37	45
L	0.65	66-5	.058	.47	3.4	8.5	14	20	29	40	47
L	1.0	18-4	.11	.79	4.8	11	18	24	33	44	52
L	1.5	40-4	.18	1.1	6.0	13	20	27	36	48	55
L	2.5	91-4	.32	1.7	8.0	16	24	32	40	52	59
L	4.0	.020	.56	2.6	10	20	29	36	45	56	62
L	6.5	.060	1.1	4.4	15	26	36	44	52	62	68
M	0.040	60-7	26-4	.047	.73	2.7	5.8	9.6	16	25	32
M	0.15	30-6	78-4	.10	1.2	4.0	8.0	12	19	30	37
M	0.25	80-6	.015	.17	1.7	5.1	9.7	14	22	33	40
M	0.40	16-5	.023	.23	2.1	6.0	11	16	24	35	42
M	0.65	45-5	.045	.39	3.0	7.8	13	19	27	38	46
M	1.0	95-5	.074	.56	3.8	9.3	15	22	30	42	49
M	1.5	22-4	.12	.85	5.0	11	18	25	34	45	52
M	2.5	51-4	.22	1.3	6.6	14	22	29	38	49	56
M	4.0	.013	.45	2.1	9.4	18	27	34	43	54	61
N	0.025	14-7	10-4	.024	.46	1.9	4.4	7.6	13	22	28
N	0.10	72-7	31-4	.052	.79	2.8	6.1	10	16	26	32
N	0.15	19-6	56-4	.082	1.0	3.5	7.3	11	18	28	35
N	0.25	40-6	92-4	.11	1.3	4.3	8.4	13	20	30	37
N	0.40	11-5	.017	.19	1.8	5.5	10	15	23	33	40
N	0.65	22-5	.028	.27	2.4	6.6	12	17	25	36	43
N	1.0	50-5	.049	.41	3.1	8.0	14	20	28	39	46
N	1.5	12-4	.083	.62	4.0	9.8	16	22	31	42	49
N	2.5	35-4	.17	1.0	5.9	13	20	27	36	47	54
P	0.015	35-8	40-5	.012	.29	1.3	3.3	6.0	11	19	25
P	0.065	17-7	12-4	.026	.49	2.0	4.6	7.8	13	22	29
P	0.10	44-7	22-4	.041	.67	2.5	5.5	9.2	15	25	31
P	0.15	92-7	34-4	.057	.84	3.0	6.3	10	17	26	33
P	0.25	25-6	68-4	.094	1.1	3.8	7.7	12	19	29	36
P	0.40	51-6	.011	.13	1.4	4.6	8.9	13	20	31	38
P	0.65	12-5	.019	.20	1.9	5.7	10	15	23	34	41
P	1.0	28-5	.033	.30	2.5	6.9	12	18	26	37	44
P	1.5	77-5	.063	.50	3.5	8.8	15	21	29	41	48

TABLE T18-10   (continued)

Code Letter	AQL	Shape Parameter, β									
		$\frac{1}{3}$	$\frac{1}{2}$	$\frac{2}{3}$	1	$1\frac{1}{3}$	$1\frac{2}{3}$	2	$2\frac{1}{2}$	$3\frac{1}{3}$	4
Q	0.010	90-9	16-5	63-4	.18	.96	2.5	4.9	9.0	16	23
Q	0.040	44-8	48-5	.013	.31	1.4	3.5	6.2	11	19	26
Q	0.065	11-7	90-5	.021	.43	1.8	4.2	7.4	12	21	28
Q	0.10	22-7	14-4	.029	.53	2.1	4.8	8.2	14	23	30
Q	0.15	62-7	28-4	.048	.75	2.7	5.9	9.7	16	25	32
Q	0.25	13-6	45-4	.069	.95	3.3	6.8	11	17	27	34
Q	0.40	30-6	78-4	.10	1.2	4.1	8.0	12	19	30	37
Q	0.65	70-6	.013	.15	1.6	4.9	9.4	14	22	32	39
Q	1.0	19-5	.026	.26	2.3	6.4	11	17	25	35	43
R	0.025	10-8	18-5	68-4	.19	1.0	2.6	5.0	9.3	17	23
R	0.040	26-8	35-5	.010	.26	1.2	3.2	5.8	10	18	25
R	0.065	54-8	55-5	.015	.33	1.5	3.6	6.5	11	20	26
R	0.10	15-7	11-4	.024	.47	1.9	4.5	7.7	13	22	29
R	0.15	30-7	17-4	.034	.59	2.3	5.1	8.7	14	24	30
R	0.25	70-7	30-4	.051	.78	2.8	6.0	10	16	26	33
R	0.40	17-6	52-4	.075	1.0	3.4	7.1	11	18	28	35
R	0.65	46-6	.010	.12	1.4	4.5	8.7	13	20	31	38

From TR7, pp. 15-17.

TABLE T18-11   TR7 Table 1C. $100t/\mu$ Ratios at the Limiting Quality Level for the MIL-STD-105D Plans: Consumer's Risk = 0.05

Code Letter	AQL	Shape Parameter, β									
		$\frac{1}{3}$	$\frac{1}{2}$	$\frac{2}{3}$	1	$1\frac{1}{3}$	$1\frac{2}{3}$	2	$2\frac{1}{2}$	$3\frac{1}{3}$	4
A	6.5	55	120	130	140	140	130	130	120	120	120
B	4.0	16	50	73	100	110	110	110	110	110	110
C	2.5	3.5	18	35	60	74	82	87	90	96	97
C	10	20	59	84	110	120	110	110	110	110	110
D	1.5	.84	6.9	17	36	52	61	69	76	82	86
D	6.5	4.3	20	37	64	77	85	90	93	97	99
D	10	13	43	65	93	100	100	100	110	100	100
E	1.0	.22	2.8	8.6	23	37	47	55	63	72	76
E	4.0	.95	7.4	18	39	53	63	70	76	83	87
E	6.5	2.5	14	28	53	67	76	82	86	92	95
E	10	5.5	24	43	69	82	89	94	97	99	100
F	0.65	.059	1.1	4.4	15	26	35	43	52	62	68
F	2.5	.25	3.1	9.3	25	38	48	56	64	73	77
F	4.0	.60	5.4	14	33	48	57	65	72	79	83
F	6.5	1.2	8.6	20	42	57	66	73	78	85	89
F	10	3.8	19	36	62	75	83	88	92	96	98
G	0.40	.013	.43	2.1	9.3	18	27	34	43	54	61
G	1.5	.059	1.1	4.4	15	26	35	43	52	62	68
G	2.5	.13	1.9	6.7	20	32	42	50	58	67	72
G	4.0	.29	3.4	10	26	30	50	57	65	74	78
G	6.5	.76	6.3	16	35	50	60	67	74	81	85
G	10	1.6	10	23	47	61	70	77	82	88	91
H	0.25	37-4	.18	1.1	5.9	13	20	27	36	47	54
H	1.0	.014	.46	2.2	9.6	18	27	34	42	55	61
H	1.5	.034	.82	3.4	12	23	32	40	49	60	66
H	2.5	.070	1.3	4.9	16	27	37	45	54	64	70
H	4.0	.18	2.5	7.9	22	35	45	53	61	71	75
H	6.5	.40	4.1	11	28	43	53	60	68	76	80
H	10	.93	7.4	18	39	54	63	70	76	83	87
J	0.15	90-5	.072	.55	3.7	9.3	15	22	30	41	49
J	0.65	37-4	.18	1.1	5.9	13	20	27	36	47	54
J	1.0	92-4	.32	1.7	8.0	16	24	32	40	52	58
J	1.5	.016	.48	2.3	9.9	19	28	35	44	57	61
J	2.5	.046	.95	3.9	14	25	34	42	51	61	67
J	4.0	.089	1.5	5.5	17	29	39	47	55	65	70
J	6.5	.18	2.5	7.9	22	35	45	53	61	71	75
J	10	.45	4.6	12	30	44	54	62	69	77	81

A negative figure after a ratio shows the number of decimal points to provide. Thus 92-4 = .0092

CHAPTER 1

Answers to the problems of Chap. 1 are given directly in the text.

CHAPTER 2

1. Probability of acceptance is $98/100 = .98$.

2. Yes, each tablet has an equal chance to be selected.

3. $P_2^6 = \dfrac{6!}{4!} = 30$.

4. $C_2^6 = \dfrac{6!}{2!4!} = 15$; $C_2^2 = 1$. Probability $= 1/15$.

5.

6. The two draws are not independent. Should be $(2/6)(1/5) = 1/15$.

    Probability both good is $(4/6)(3/5) = 12/30 = 6/15$.

    Probability both the same is $1/15 + 6/15 = 7/15$.

    Can be added since they are mutually exclusive.

7. $.95 + .95 - (.95)(.95) = .9975$.

8. $AQL = .017$, $IQ = .206$, $LTPD = .536$ (Binomial)

9. $.10 + .90(.10) = .19$.

10. Probability all fail $= .5^5 = .031$.

    Probability all pass $= .5^5 = .031$.

    Probability at least one failure $= 1 - .031 = .969$.

CHAPTER 3

1. $f(0) = \dfrac{C_5^{98} \, C_0^{2}}{C_5^{100}} = \dfrac{98!}{5!93!} \cdot \dfrac{2!}{0!2!} \cdot \dfrac{95!5!}{100!} = \dfrac{98 \cdot 97 \cdot 96 \cdot 95 \cdot 94}{100 \cdot 99 \cdot 98 \cdot 97 \cdot 96} = .9020$

$f(1) = \dfrac{C_4^{98} \, C_1^{2}}{C_5^{100}} = \dfrac{98!}{4!94!} \cdot \dfrac{2!}{1!1!} \cdot \dfrac{5!95!}{100!} = 10 \, \dfrac{98!95!}{94!100!}$

$\qquad = 10 \cdot \dfrac{98 \cdot 97 \cdot 96 \cdot 95}{100 \cdot 99 \cdot 98 \cdot 97 \cdot 96} = .0960$

$\qquad F(\leq 1) = .9980$

$\qquad \mu = np = 5\left(\dfrac{2}{100}\right) = .1$

$\qquad \sigma = \sqrt{npq} \; \sqrt{\dfrac{N - n}{N - 1}} = \sqrt{5(.02)(.98)} \sqrt{\dfrac{100 - 5}{100 - 1}} = .3067$

2. $f(0) = C_0^5 (.02)^0 (.98)^5 = .9039$

$f(1) = C_1^5 (.02)^1 (.98)^4 = .0922$

$F(\leq 1) = .9961$

$\mu = np = 5(.02) = .1$
$\sigma = \sqrt{npq} = \sqrt{5(.02)(.98)} = .3130$

3. $f(0) = \dfrac{2^0 e^{-2}}{0!} = .1353$

$f(1) = \dfrac{2^1 e^{-2}}{1!} = .2707$

$f(2) = \dfrac{2^2 e^{-2}}{2!} = .2707$

$F(\leq 2) = .6767, \qquad$ Poisson distribution not symmetric
$\sigma = \sqrt{2} = 1.4142$

4. $b^{-1}(5|1,.05) = C_0^4 .05^1 .95^4 = .0407$

$\mu = \dfrac{2(.95)}{.05} = 38$

5. $F(20000) = 1 - e^{-20000/10000} = 1 - .1353 = .8647$

$\sigma = \mu = 10000$ hr.

6. $\gamma = 0 \qquad \beta = 1 \qquad \eta = 10000$

$F(10000) = .6321$

7. $z = -2.5$

$P(\leq 1.005) = .0062$

$C_1^3 \; .0062^1 \; .9938^3 = .0184$

8. $\sigma_{\overline{X}} = \dfrac{.0003}{\sqrt{9}} = .0001$

$z = \dfrac{3.001 - 3}{.0001} = 10$

   Very unlikely

9. "f-binomial"

$f(0) = C_0^2 \left(\dfrac{5}{25}\right)^0 \left(1 - \dfrac{5}{25}\right)^{2-0} = .64$

$f(1) = C_1^2 \left(\dfrac{5}{25}\right)^1 \left(1 - \dfrac{5}{25}\right)^{2-1} = .32$

$P(\leq 1) = .96$

10. $\hat{\sigma} \simeq \dfrac{48 - 24}{2} = 12$

$\hat{\mu} \simeq \dfrac{48 + 24}{2} = 36$

CHAPTER 4

1. Type A, hypergeometric, Poisson

2.

p	P_a Type A	P_a Type B
.125	.50	.5863
.25	.2143	.3164
.375	.0714	.1526
.50	.0143	.0625

3. .5500, .2720, .1154, .0385. For $n > 16$, they would be even closer
    to the Type B probabilities of Example 2.

4. $\mu = 2/100$ units

   .1353, .2707, .2707, .1804

   $P_a = .6767$

5. Inspector sees fraction defective .10, .20, .30, .40. Effective OC curve using binomial distribution is:

   Actual p: .125, .25, .375, .50

   Apparent p: .10, .20, .30, .40

   P_a: .9185, .7373, .5282, .3370

6. Type B

   p: .125   .25   .375   .50

   AOQ: .0366, .0396, .0286, .0156

7.   p: .125   .25   .375   .50

   ATI: 5.65, 6.73, 7.39, 7.75

8. Using the Poisson approximation

   $$ASN_c = 15\ F(2|15) + \frac{3}{.1}(1 - F(3|16))$$

   $$= 15(.8088) + 30(.0788) = 14.5$$

9. $\hat{p} = \dfrac{2}{11 - 1} = .20$

10. PQL = .0166   CQL = .122

CHAPTER 5

1. Binomial: Poisson:

   a. n = 30, c = 3   n = 35, c = 3

   b. n = 59, c = 4   n = 66, c = 4

   c. n = 193, c = 7   n = 200, c = 7

2. a.

P_a	.95	.75	.50	.25	.10
p(binomial)	.028	.074	.126	.194	.268
p(Poisson)	.027	.074	.129	.207	.299

   b.

P_a	.95	.75	.50	.25	.10
p(binomial)	.002	.009	.021	.042	.069
p(Poisson)	.002	.009	.022	.043	.071

   c.

P_a	.95	.75	.50	.25	.10
p(binomial or Poisson)	.006	.014	.021	.031	.043

3. a. $n = .2(200) = 40$, $c = 2$

    b. P_a .75   .50   .25

         p   .045   .065   .09

4. $n = 32$, $c = 1$, $\mu_{.50} = 5.2$

5. a. $n = 13$, $c = 1$

         P_a .95   .50   .10

         AOQ   .027   .062   .027

         ATI   62.35   506.5   901.3

                 AOQL = .064

    b. $n = 32$, $c = 0$

         P_a .95   .50   .10

         AOQ   .002   .010   .007

         ATI   80.4   516   903.2

                 AOQL = .011

    c. $n = 125$, $c = 2$

         P_a .95   .50   .10

         AOQ   .005   .009   .004

         ATI   168.8   562.5   912.5

                 AOQL = .0096

6. $n = 5$, $c = 0$ $n = 5$, $c = 1$

P_a	.95	.50	.10
p	.010	.129	.369
AOQ	.010	.062	.036
ATI	14.75	102.5	180.5

         AOQL = .072

P_a	.95	.50	.10
p	.076	.314	.584
AOQ	.070	.153	.057
ATI	14.75	102.5	180.5

         AOQL = .164

7. $n = 5$, $c = 1$ $n = 10$, $c = 1$

P_a	.95	.50	.10
p	.071	.336	.778
AOQ	.066	.164	.076
ATI	14.75	102.5	180.5

         AOQL = .164

P_a	.95	.50	.10
p	.036	.168	.389
AOQ	.032	.080	.037
ATI	19.5	105.0	181.0

         AOQL = .080

8. Binomial: $n = 128$, $c = 7$

    Hypergeometric: $n = .2(500) = 100$, $c = 5$

9. $n = 131$, $c = 7$. The Poisson is a conservative approximation of the binomial.

10. $P(x < np) = 1 - P\left(\dfrac{\chi^2}{2} < \chi_\nu^2 = 2c + 2\right)$

    Hence

    $$np = \chi_\nu^2 = 2c + 2 \text{ for given probability of acceptance}$$

    Since

    $$F(\chi_8^2 < 2.73) = .05 \qquad F(\chi_8^2 < 15.5) = .95$$

    we have

    $$np_{.95} = \frac{2.73}{2} = 1.36 \qquad np_{.05} = \frac{15.5}{2} = 7.75$$

    and

    $$R = \frac{np_{.05}}{np_{.95}} = \frac{15.5}{2.73} = 5.68$$

CHAPTER 6

1. a. $n_i = 17$ Ac = 0, 3
                          Re = 3, 4

    b. $n_i = 34$ Ac = 1, 4
                          Re = 4, 5

    c. $n_i = 120$ Ac = 3, 8
                          Re = 7, 9

2. a. $n_i = 7$ Ac = #, 0, 0, 1, 2, 3, 4
                          Re = 2, 3, 3, 4, 4, 5, 5

    b. $n_i = 18$ Ac = #, 1, 2, 3, 5, 7, 9
                          Re = 4, 5, 6, 7, 8, 9, 10

    c. $n_i = 50$ Ac = 0, 1, 3, 5, 7, 10, 13
                          Re = 4, 6, 8, 10, 11, 12, 14

3. | P_a | p |
    |------|------|
    | .95 | .026 |
    | .50 | .126 |
    | .10 | .311 |

4.

P_a	p
.95	.034
.50	.139
.10	.306

5.

P_a	ASN	AOQ	ATI
.95	9.34	.025	58.87
.50	10.94	.063	505.47
.10	9.65	.031	900.96

6.

P_a	ASN	AOQ	ATI
.95	10.50	.032	59.32
.50	10.92	.070	508.04
.10	7.80	.031	901.34

7. $n = 35$, $c = 3$ has $p_2 = .19$. Corresponding matched plans are:

Double: $n_i = 24$,   Ac = 1, 4   Re = 4, 5; ASN at p_1 is 29.9

Multiple: $n_i = 9$,   Ac = #,0,1,2,3,4,6   Re = 3,3,4,5,6,6,7;

ASN at p_1 is 25.4

Savings of 15 and 27 percent are possible.

8. Double: $n = 12$, $c = 1.25 \sim 1$

Multiple: $n = 12$, $c = 1.43 \sim 1$

9. 9.21, hardly worthwhile.

10.

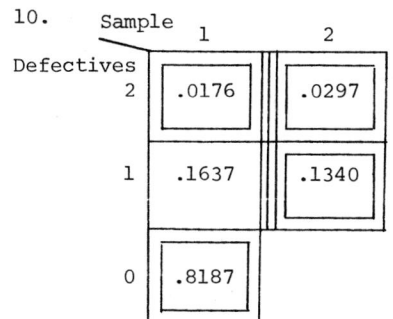

Sample	1	2	Total
Accept A_j	.8187	.1340	.9527
Reject R_j	.0176	.0297	.0473
Terminate T_j	.8363	.1637	1.0
Indecision I_j	.1637	0	X

$P_a = .9527$ ASN = 9.31

CHAPTER 7

1. $Y_2 = 1.6131 + .1018k$

$Y_1 = -1.2565 + .1018k$

2. $Y_2 = 1.6617 + .1633k$

$Y_1 = -1.2943 + .1633k$

3. $Y_2 = 2.5348 + .0365k$

$Y_1 = -1.9743 + .0365k$

4. Increasing slope raises probability of acceptance. Decreasing slope increases probability of rejection. Increasing h_2 decreases probability of rejection. Increasing h_1 decreases probability of acceptance.

5.

p	P_a	ASN
.05	.95	119
.0723	.56	174
.10	.10	115

6.

p	ASN	AOQ
.03	38	.028
.12	29	.012

7. $Y_2 = 1.6284 + 2.7606k$

$Y_1 = -1.2684 + 2.7606k$

8.

Defects/100	Defects/Unit	ASN/100	ASN
1	.01	.683	68.3
5.9	.059	.426	42.6

9. .459/100 Units or 45.9 Units

10.

Sample	n	Ac	Re
1	2	4.25 ~ 5	7.14 ~ 8
2	2	9.77 ~ 10	12.67 ~ 13
3	2	15.29 ~ 16	18.19 ~ 19

One possible solution is to change Re = 19 to 17 at the last stage.

CHAPTER 8

1. d = 2, t-test with n = 5.

2. Chi-square. $\chi^2 = 14\left(\frac{7}{6}\right)^2 = 19.05 < 23.7$, accept.

3. $n = \left(\dfrac{(1.64 + 1.28)(1.5)}{90 - 87}\right)^2 = 2.1 \sim 3$

   $d = \dfrac{1.64}{1.64 + 1.28}|90 - 87| = 1.68$

   Lower ACL at $90 - 1.68 = 88.32$

   Reject lot means 88 and 87

4. NCL = 89.5, $\quad \Delta_2 = \dfrac{\sqrt{n}(\text{APL} - \text{NCL})}{\sigma} = \dfrac{\sqrt{3}(2.5)}{1.5} = 2.887$ not $< .619$ so $n = 3$.

5. No necessary meaning since acceptance control charts may be set up to allow a drift in the mean.

6. $\mu_1 = 400 \qquad \alpha = .025$

   $\mu_2 = 408 \qquad \beta = .10$

   $Y_2 = 28.67 + 404k$

   $Y_1 = -18.22 + 404k$

   Accept on third sample.

7. See Prob. 6

   $Y_2' = -28.67 - 4k$ $\qquad\qquad$ $Y_2 = 28.67 + 4k$

   $Y_1' = 18.22 - 4k$ $\qquad\qquad$ $Y_1 = -18.22 + 4k$

8. $Y_2 = 1027.7 + 79.3k$

   $Y_1 = -800.5 + 79.3k$

9. $D = 1.0$, $\Sigma(x - \mu) = -1, -1, -1.17, -1.55, -1.31, -0.72, -0.20, -0.39,$
   $$-0.55, -0.55$$

   Accept on ninth sample.

10. $\tan \theta = 404$ so $\theta = 89^\circ\, 51'$, clearly a rescaling is needed.

   $d = h_2/s = 28.67/404 = .071$

CHAPTER 9

1. One sample from each of the 10 compartments.

2. $d = (7 - 5)/3 = .67$, $n = 20$, need 10 more samples, 1 per compartment.

3. $t = \dfrac{5.5 - 5}{2/\sqrt{20}} = 1.12 < 1.73$, accept the shipment.

4. Testing: $s_3^2 = .7$; Reduction: $s_4^2 = .45 - \dfrac{.7}{2} = .1$

5. $s_2^2 = 2.2 - .7 = 1.5$; $\quad s_1^2 = 4.75 - \dfrac{2.2}{2} = 3.65$

6.

Source	SS	df	MS
Between segments	114	24	4.75
Increments within segments	55	25	2.2

7. $n_2 = \dfrac{1.5}{16\left(\dfrac{(5-7)^2}{8.567} - \dfrac{3.65}{16} - \dfrac{.7}{4} - \dfrac{.1}{2}\right)} = 6.8 \sim 8$ to be even.

8. $\sigma_{\overline{X}} = \sqrt{\dfrac{3.65}{16} + \dfrac{1.5}{128} + \dfrac{.7}{4} + \dfrac{.1}{2}} = \sqrt{.4648} = .68$

$z = \dfrac{5.9 - 5.0}{.68} = \dfrac{.9}{.68} = 1.32 < 1.645$ accept the shipment

9. $\sigma^2 = 3.65 + 1.5 + .7 + .1$ $5.9 \pm 1.96(.68)$

$\sigma^2 = 5.95$ 5.9 ± 1.33

$\sigma = 2.44$ 4.57 to 7.23

10. $n_2 = \sqrt{\dfrac{1.5}{3.65}} = .64 \sim 1$

$n_1 = \dfrac{16(1.5 + 1(3.65))}{16(1)\left(\dfrac{1}{1.96}\right)^2 + (1)(3.65)} = 10.55 \sim 11$

CHAPTER 10

1. a. $n = 10, k = 2$

b. $n = 30, k = 2$

2. $\overline{X} = 6.83, \sigma = .08$

Accept since $6.83 < 6.84$

3. $n = 10, \quad M = .017, \quad \hat{p} = .017, \quad$ accept.

4. $\dfrac{M}{100} = I_{.31}(14,14) = .019; \quad \hat{p}_U = I_{.22}(14,14) = .0006; \quad$ accept.

5. Take 8 subgroups of 5, MAR $= .92(U - L$, use plan with $\hat{s} = \overline{R}/2.35$

hence $\overline{X} + 2s \le U$ becomes $\overline{X} + .85\,\overline{R} \le U$

6. $n = 13, k = 1.63,$ MSD $= (U - L)/3.9$

7. $\overline{X} = 65, s = 3.37 > 2.30 =$ MSD, reject.

8. For $n = 13, k = 1.75$ by interpolation

$65 + 1.75(3.37) = 70.9 > 69,$ reject

$65 - 1.75(3.37) = 59.1 < 60,$ reject

9. $T_{U_1} = \dfrac{U - \overline{X}}{s_1} = \dfrac{7.0 - 6.83}{.09} = 1.89.$ Yes, resample.

10. $p_{.50} = .0228$ from $k = 2.$

CHAPTER 11

1. Code H, 1.0 AQL:

    a. Normal: $n = 50$; $Ac = 1$; $Re = 2$

        Tightened: $n = 80$; $Ac = 1$; $Re = 2$

        Reduced: $n = 20$; $Ac = 0$; $Re = 2$

    b. Normal: $n_i = 32$; $Ac = 0,1$; $Re = 2,2$

        Tightened: $n_i = 50$; $Ac = 0,1$; $Re = 2,2$

        Reduced: $n_i = 13$; $Ac = 0,0$; $Re = 2,2$

    c. Normal: $n_i = 13$; $Ac = \#,\#,0,0,1,1,2$; $Re = 2,2,2,3,3,3,3$

        Tightened: $n_i = 20$; $Ac = \#,\#,0,0,1,1,2$; $Re = 2,2,2,3,3,3,3$

        Reduced: $n_i = 5$; $Ac = \#,\#,0,0,0,0,1$; $Re = 2,2,2,3,3,3,3$

2. a. $1.7\left(1 - \dfrac{50}{390}\right) = 1.48$ percent

    b. $1.1\left(1 - \dfrac{80}{390}\right) = 0.87$ percent

        AOQL of tightened is about AQL.

        AOQL of scheme is about AOQL tightened.

3. a. 7.6 percent

    b. 7.8 defects per hundred units

4. Double

5. a. No action.

    b. No action, already back to normal.

    c. Switch to normal.

6. a. 95 percent.

    b. 18.3 percent.

    c. 4.0 percent.

    d. 23.3 percent.

7. a. $.5^{10} = .001$

    b. $.9^{10} = .349$

8. $C_1^2 \, .1^1 \, .9^1 = .18$

9. 5000, 1, .015 percent.

10. $n = 35$, $c = 3$; Code D, 10.0 AQL.

CHAPTER 12

1. Code I, 1.0 AQL:

    a. Normal: $n = 25$, $k = 1.85$

    b. Tightened: $n = 25$, $k = 1.98$

    c. Reduced: $n = 10$, $k = 1.58$

2. a. $2 > 1.85$, accept

    b. $2 > 1.98$, accept

    c. $2 > 1.58$, accept

3. Code I, 1.0 AQL

    a. Normal: $n = 25$, $M = 2.86$

    b. Tightened: $n = 25$, $M = 2.00$

    c. Reduced: $n = 10$, $M = 4.77$

4. a. Normal: $3.82 > 2.86$, reject

    b. Tightened: $3.82 > 2.00$, reject

    c. Reduced: $2.34 < 4.77$. accept

5. MSD $= 9.52$, $s = 10$, does not pass.

6. LTPD $= 11.2$ percent, IQ $= 2.8$ percent.

7. No action. Must have 8 lots of 10 to switch to tightened.

    Minimum process average is $\dfrac{7(4.0) + 3(0)}{10} = 2.8$ percent.

    Cannot switch to reduced, all lots must have $\hat{p} < 3.94$ percent.
    But, 7 lots have $\hat{p} > 4.0$ percent.

8. Code M, 0.65 AQL.

9. Reject since $1.88 + 1.88 = 3.76 > M = 2.81$

10. Reject since $Q_U = Q_L = 2$ giving $p_U = p_L = 2.275$ percent and
    $\hat{p} = 4.55 > 2.59$.

CHAPTER 13

1. $P(i) = 97.9$, $2.063 < 3.23$, accept

2. $z = 2$, $\hat{p} = .023$, reject

3. NLG $= 110 - 1.5(6) = 101$, reject

4.

p	z_p	z_g	p_g	np_g	P_a (Poisson)	P_a (binomial)
.0025	2.81	.89	.19	.95	.93	.95
.034	1.82	-.10	.54	2.7	.49	.43
.109	1.23	-.69	.75	3.75	.28	.10

    Poisson only an approximation and at $p = .109$ n not large, p not
    small.

5. $n \approx 25$, $c \approx 9$, $t \approx 1.15$

6. Tightened: $n = 14$, $Ac = 7$, $Re = 8$, $t = 2.17$

    Normal: $n = 13$, $Ac = 8$, $Re = 9$, $t = 2.27$

    Reduced: $n = 9$, $Ac = 2$, $Re = 5$, $t = 1.43$

7. a. .98

    b. 12

    c. .039

8. a. .46

    b. 22.8

    c. .023

9. $n = 55$, $c = 1$

10. $p_0 = \dfrac{c + 2/3}{n}$

c	Formula	Table
0	.0067	.0069
1	.0167	.0168
2	.0267	.0267
3	.0367	.0367
4	.0467	.0467
5	.0567	.0567

CHAPTER 14

1. a. 1.7 percent

    b. 1.1 percent

    c. 1.8 percent

2. $n = 28$, $c = 2$

3. a. $n = 19$, $c = 1$; $n_1 = 15$, $n_2 = 17$, $c_1 = 0$, $c_2 = 2$

    b. $n = 34$, $c = 2$; $n_1 = 21$, $n_2 = 44$, $c_1 = 0$, $c_2 = 4$

    c. $n = 210$, $c = 13$; $n_1 = 22$, $n_2 = 58$, $c_1 = 0$, $c_2 = 5$

4. a. $n = 50$, $c = 0$; $n_1 = 55$, $n_2 = 30$, $c_1 = 0$, $c_2 = 1$

    b. $n = 195$, $c = 4$; $n_1 = 110$, $n_2 = 195$, $c_1 = 1$, $c_2 = 7$

    c. $n = 575$, $c = 17$; $n_1 = 320$, $n_2 = 585$, $c_1 = 7$, $c_2 = 27$

5. $n = 195$, $c = 4$, ATI $= 237.5$

6. $n = 34$, $c = 2$, $p_M = .067$

7. AOQL $\simeq 1.5 \, p_0$

8. $f_1 = .3690$, $f_2 = .1900$, ASN $= 54.9$, AOQL $= 1.1$ percent, $p_M = 2.4$ percent

9. $n = 50$, $c = 0$, AOQL $= 0.55$ percent, $I_{min} = 70$

10. ASN/$N = f_1$, AOQ $= 0$, NAOQL $= 4.69$, $Np_M = 10.3$

CHAPTER 15

1. $f = .10$, $i = 27$, UAOQL = 24.3 percent
2. $f = .10$, $i = 36$, UAOQL = 33.3 percent
3. a. .51
   b. .52
4. $r = 2.9 \sim 3$
5. $i = 12$, $f = .23$
6. $i = 58$, $f = .1$, AOQL = .039
7. $N_0 = 576$, $k = 24$, $f = .042$, $M^* = 2$
8. $N = 24$, $m = 2$
9. $f = 1/7$, $i = 14$, $S = 59$
10. No.

CHAPTER 16

1. $n = 40$, $c = 2$, $i = 14$, $f = .20$
2. $P_a = .972$, $F = .55$, ASN = 90.8, $AOQL_1 = .013$, $AOQL_2 = .06$
3. $n = 160$, $c = 7$; $n_1 = n_2 = 98$, Ac = 3,7, Re = 8,9;
   $n_i = 41$, Ac = 0,1,3,5,7,10,13, Re = 4,6,8,10,11,12,14
   ASN(single) = 160, ASN(double) = 119, ASN(multiple) = 106,
   $ASN_{sk} = 92.8$
4. $P_a = .99$, .95, .50, .10, .05, .01 have $p = .003$, .007, .038, .115,
   .150, .230, AOQL = .02, P_a(chain) = .122
5. $n = 7$, $i = 2$
6. $n = 8$, $i = 5$, $p_{.95} = .014$, AOQL = .046, $p_M = .122$
7. $n = 5$, $D = 47$, IQ = .07
8. $n = 142$
9. $U_S = .74$, $C_S = 39.54$; LIMITS $.74 \pm .597$, $D = 114$, $D/n = .114$
   Yes, out of control low.
10. Discontinue criterion since sample result meets CRC_2.

CHAPTER 17

1. $n = 3150$, $c = 0$, AOQL = .01 percent
2. P_a: .95, .75, .50, .25, .10; $p = .0015$ percent, .008 percent,
   .02 percent, .04 percent, .065 percent
3. $n = 3900$, $c = 0$
4. $t = 5$, $s = 4$, $n_1 = 29$, $n_2 = 6$

5. P_a = .228, ASN = 28.9, AOQ = .011

6. C_N = 1, C_T = 0, n = 32, IQ = .036

7. Sample 460 from each lot. If three lots form grand lot use plan
    n = 1380, c = 3, on grand lots to demonstrate compliance.

8. s: E 5.9, 84.1, M 14.9, 75.1; $\bar{X}$: E 444.6, 555.4, M 457.4, 542.6;
    Accept.

9. n = 5, s: E 1.0, 90.0, M 17.1, 72.9; $\bar{X}$: E 437.8, 562.2, M 460.6, 539.4

10. No action, C_k = 28.99. No action.

CHAPTER 18

1. The eighth ordered unit would show h = .33, H = 1.426, P = 76.0

2. a. .509

    b. .491

    c. .01

    d. .7121

3. Code B-8, r = 8, c = 74.7; $\hat{\theta}$ = 80 ≥ 74.7, accept.

4. Code B-8, r = 8, T = 37; < 8 failures at 37 hr., accept.

5. Code B-8, r = 8, T = 50; > 8 failures at 50 hr., reject.

6. Code B-8, r_0 = 24, h_0 = 172, h_1 = -221, s = 83, V(t) = 640,
    continue testing.

7. p_1/p_0 = 2.5, r = 11, n = 35, T = 30 hr.

8. $\mu_{.95}$ = 811, $\mu_{.10}$ = 266

9. AQL = 130.6, LQ(10) = 89.3, LQ(5) = 86.2

10. Code N, 1.0 AQL.

CHAPTER 19

1. Institute demerit rating and consider discontinuing the inspection.

2. LSP indicates 90 percent of a lot of 100,000 must be sampled. To
    reduce the sampling frequency to .2, D = Np_t = 10.3 which implies
    N = 1,030,000. Use grand lot scheme to combine 10 months production
    with acceptance number of zero.

3. 9.6 ~ 10 percent.

4. AQL = 2/1.5 = 1.3 percent, LTPD = 5(1.3) = 7.5 percent.

5. AQL: AOQL: IQ: LTPD = .3: .5: 1: 2

6. p_B = .015

7. AQL = .45 percent, AOQL = .75 percent, LTPD = 3.0 percent.

8. No. Manual costs $5.00 per lot. Computer costs $10.00 per lot.
    Breakeven at $.10 per piece.

9. AQL = .3 I/A, LTPD = 2 I/A.

10. Use $n = .1 N$ to obtain AOQL = 3.311/N.

AUTHOR INDEX

SUBJECT INDEX

Appendix Tables Underlined

Acceptable process level, 166,178,
  180,183-184,186-187,189,215
Acceptable quality level, 10,28,
  210,261,277-280,283,286,288,
  291-295,297,299-301,303,307,309,
  311,313,317,319-320,326-327,
  344-345,361,373,426,432,435,455,
  464,478,480,495,502,505,512,515,
  548-549,555-556,560,571-575,
  579-580,582-583,585,677,712-713,
  738
Acceptance constant, 116,173,175,
  229,237,251,308-309,344,502,
  504,534
Acceptance control chart, 169-170,
  172-173,210,567
Acceptance control limit, 169,171,
  173,635
Acceptance criteria, 232,237,
  258-260,265,280,290,301,311,315,
  317,321,335,346,378,442,476,557
Acceptance criterion, 246,311,319,
  329,349,361,502,574
Acceptance limit, 173,222,225,229,
  242,352,356
Acceptance number, 82-83,107,111,
  113-114,117,119,120-121,127-128,
  131-133,145,156,162,222,260,283,
  290-291,294,302,335,340,346,352,
  363,373,378,387,392,394,397,427,
  450-451,475-476,482,488,492-493,
  556,580
Acceptance polygon, 233-234,254-255
Acceptance quality control, 1,2,4-5,
  8,98,189,343,475,564,567-568
Acceptance sampling, 222,260,
  264-265,326
Acceptance sampling plan, 1,7,100,
  278,442

Action limit, 511-512,514,518-519,
  722
Administration, 128,154,564,569,
  577,586
Advisory Group on Reliability of
  Electronic Equipment (AGREE), 524
Altman diagram, 373
American Institute of Electrical
  Engineers, 9
American Mathematical Society, 9
American National Standards
  Institute, 323
American Society for Quality Control,
  170,400,526,564
American Society for Testing and
  Materials, 9
American Society of Mechanical
  Engineers, 9
American Statistical Association,
  9,323
Analysis of means, 489-491,493-494,
  496-498,500,502,504-506,508-509
Analysis of variance, nested, 201
Anscombe rectifying inspection
  scheme, 373,393
ANSI Committee Z-1, 322
ANSI Z1.4, 322
ANSI Z1.9, 322
AOQL curves, 416
AOQL factors, 291,344
AOQL plan, 277,386-387,392,577
AOQL plan, double sampling, 375-376,
  378,385
AOQL plan, single sampling, 375-376,
  378,385
AOQL scheme, 372,668
Apparent level of quality, 85-86
Approximations, 64
AQL group, 576